FIREFIGHTER'S HAZARDOUS MATERIALS REFERENCE BOOK

FIREFIGHTER'S HAZARDOUS MATERIALS REFERENCE BOOK

By
Daniel J. Davis
and
Grant T. Christianson

VNR VAN NOSTRAND REINHOLD
New York

Library of Congress Catalog Card Number 90-22498
ISBN 0-442-00377-3

Manufactured in the United States of America

Published by Van Nostrand Reinhold
115 Fifth Avenue
New York, New York 10003

Chapman and Hall
2-6 Boundary Row
London, SE1 8HN

Thomas Nelson Australia
102 Dodds Street
South Melbourne 3205
Victoria, Australia

Nelson Canada
1120 Birchmount Road
Scarborough, Ontario M1K 5G4, Canada

16 15 14 13 12 11 10 9 8 7 6 5 4 3 2 1

Library of Congress Cataloging-in-Publication Data

Davis, Daniel J.
Firefighter's hazardous materials reference book/by Daniel J. Davis & Grant T. Christianson.
p. cm.
Includes bibliographical references and index.
ISBN 0-442-00377-3
1. Hazardous substances—Fires and fire prevention—Handbooks, manuals, etc. I. Christianson, Grant T. II. Title.
TH9446.H38DC38 1991
628.9'2—dc20 90-22498
CIP

DEDICATION

This book is dedicated to Lawrence W. Adams, Base Fire Chief at Minot Air Force Base, North Dakota, for his courageous support in allowing us to attend the many hazardous materials seminars the last few years. If we had not been able to attend, we would not have known a need existed for this material.

To our wives, Julie A. Davis, and Patricia Ann Christianson, without whose patience and support over the last two years this book would not have been possible.

DISCLAIMER

Daniel J. Davis and Grant T. Christianson have prepared the *Firefighter's Hazardous Materials Reference Book* with what they believe is the best currently available information. The *Firefighter's Hazardous Materials Reference Book* is subject to revision as additional knowledge and experience is gained. Daniel J. Davis and Grant T. Christianson cannot guarantee the accuracy of information used to develop the *Firefighter's Hazardous Materials Reference Book* contained herein, and the *Firefighter's Hazardous Materials Reference Book* does not constitute endorsement by Daniel J. Davis and Grant T. Christianson. Daniel J. Davis and Grant T. Christianson accept no responsibility for damages or liabilities of any kind which may be claimed to result from the use of the *Firefighter's Hazardous Materials Reference Book*.

CONTENTS

Preface	ix
Acknowledgments	xi
List of Terms	xiii
Reference Sheets	1–878
Chemical Name Index	879
DOT Number Index	886
Synonym Index	893

PREFACE

Approximately three years ago, attending numerous hazardous materials seminars, we heard the same question repeated again and again: Why hasn't anyone tried to compile all of the necessary data for first responders into one simple, easy-to-read, and understandable book instead of a dozen books with conflicting information? First responders' needs differ with each chemical emergency, and basic information is always needed to expeditiously, efficiently, and safely handle any hazardous material.

With this in mind, we decided to compile as much needed information as possible on specific chemicals. Realizing that different tests and procedures resulted in variable data, the *safest* information was gathered. For example, published sources list lower explosive limits varying by as much as 50 degrees for the same chemical. In our compilation, we have used the lowest explosive limit to ensure users were well within the lower explosive range.

We as authors are by no mean experts in the field of hazardous materials. Therefore, two years ago, we started compiling information on specific chemicals that we as first responders needed initially on the scene of a hazardous material incident.

After we had put all the information together, we saw that *all* first responders could benefit from the information compiled. This book in no way implies that it is the only source of information to be used at a hazardous materials incident, but we feel that it is a vital step in the right direction for initial assessment, containment, neutralization, and safety in dealing with these chemicals.

There are many other books that could give you the same information as you will find here. However, in dealing with chemicals as potent or hazardous as the ones of today, *time* is of the essence. Researching 12 or more books to get the same information as we have in one book loses time in starting the operation. Also, many reference books dealing with hazardous materials refer to equipment that first responders do not understand or have access to initially. Our book gives the first responder basic information and basic understanding when confronted with a hazardous materials situation.

After many attempts to design an information sheet, we decided to go with the format used here because information is easy to find and read. It is appealing to the eye and each page represents one chemical.

The *Firefighter's Hazardous Materials Reference Book* is designed to be used in conjunction with the *D.O.T. Emergency Response Guidebook;* D.O.T. I.D. Numbers and Guide Numbers are given.

This book is no substitute for your own knowledge and judgment. The information listed within has been assembled from many governmental agencies and private organizations. The information is correct to the absolute best of our knowledge.

Any comments or questions concerning the *Firefighter's Hazardous Materials Reference Book* should be directed to the authors through Van Nostrand Reinhold Co., 115 Fifth Ave, New York, New York 10003. Attention: Mr. Esposito.

The *Firefighter's Hazardous Materials Reference Book* sheets are very easy to understand and follow. They are broken down into nine sections:

1. Hazardous Material
   a. Chemical Name
   b. Synonyms
   c. D.O.T. Hazard Class
   d. I.D. Number
   e. D.O.T. Guide Number
   f. S.T.C.C. Number
   g. Reportable Quantity
   h. Manufacturer's Name
   i. Manufacturer's Phone Number
2. Physical Description
   a. Normal Physical Form
   b. Color
   c. Odor
   d. Other Information
3. Chemical Properties
   a. Specific Gravity
   b. Boiling Point
   c. Melting Point

- d. Vapor Density
- e. Melting Point
- f. Vapor Pressure
- g. Solubility in Water
- h. Degree of Solubility
- i. Other Information

4. Health Hazards
   - a. Inhalation
   - b. Ingestion
   - c. Absorption
   - d. Hazards to Wildlife
   - e. Decontamination Procedures
   - f. First Aid Procedures
5. Fire Hazards
   - a. Flashpoint
   - b. Ignition Temperature
   - c. Flammable Explosive High Range
   - d. Flammable Explosive Low Range
   - e. Toxic Products of Combustion
   - f. Other Hazards
   - g. Possible Extinguishing Agents
6. Reactivity Hazards
   - a. Reactive with What
   - b. Other Reactions
7. Corrosivity Hazards
   - a. Corrosive with What
   - b. Neutralizing Agents
8. Radioactivity Hazards
   - a. Type of Radiation Emitted
   - b. Other Hazards
9. Recommended Protection of Response Personnel

The List of Terms incorporated with the sheets is a guide explaining each category in simple terms. It should be noted that the decontamination procedures and first aid procedures have been standardized to basic aid for the first responder level. There are medical books published for hazardous materials that should be consulted by more qualified medical personnel when they arrive at a scene of a hazardous materials incident.

There are some instances where a category has an n/a. This means that there was no information *available* for that type of chemical.

Under Physical Description, the category of Other Information is what the chemical is used for in industry. There are some instances where this information was not available. Also, it should be noted under Odor that this is not a very safe way to identify the material involved. In most cases, if you can smell the chemical, it is too late for your safety!

Under Chemical Properties, the category of Other Information is used to let the reader know if the chemical is soluble in water, if the vapors are heavier than air, if the chemical is lighter or heavier than water, etc.

Under Health Hazards, Decontaminated Procedures, it is important that decontamination solutions *not be scented with lemon* (e.g., *Lemon Fresh Joy*) as the lemon soap vapors may interfere with readings used when ensuring that proper decontamination has been accomplished. Hazards to Wildlife pertain to wildlife that live in the water; this does not include wildlife that might drink from the water.

Under Radioactivity Hazard, there are three types of radiation that would be harmful to the first responder upon contact: alpha, beta, and gamma radiation.

Alpha radiation involves the alpha particle, a positively charged particle that is emitted by some radioactive materials. It is less penetrating than beta and gamma radiation and is not considered to be dangerous unless ingested. If ingested, it will attack internal organs.

Beta radiation involves the beta particle, which is much smaller but more penetrating than the alpha particle. Beta particles can damage skin tissue, and they can damage internal organs if they enter the body. Full protective clothing, including self-contained breathing apparatus, will protect against most beta radiation.

Gamma radiation is especially harmful since it has great penetrating power. Gamma rays are a form of ionizing radiation with high energy that travels at the speed of light. It can cause skin burns and can severely injure internal body organs. Protective clothing is inadequate in preventing penetrating of gamma radiation.

Under Recommended Protection for Response Personnel, there is guidance for the proper type of suit to wear for the type of chemical involved and any other precautions the first responder must take when the chemical involved is in a hazardous materials situation.

ACKNOWLEDGMENTS

The authors of the *Firefighter's Hazardous Materials Reference Book,* Daniel J. Davis and Grant T. Christianson, wish to thank all of the personnel who gave permission to use their material for reference in the summation of our book.

National Fire Protection Association for the use of *Fire Hazard Properties of Flammable Liquids, Gases and Volatile Solids*. NFPA 325M-1984. Copyright 1984 National Fire Protection Association, Quincy Massachusetts, 02269-9110. This reprinted material is not the official position of the National Fire Protection Association, which is represented only by the standard of its entirety.

Association of American Railroads for the use of *Emergency Handling of Hazardous Materials in Surface Transportation*. Copyright 1987 Association of American Railroads, Hazardous Materials Systems (BOE), 50 F Street NW, Washington DC 20001.

Association of American Railroads for the use of *Emergency Action Guides*. Copyright 1984 Association of American Railroads, Hazardous Materials Systems (BOE), 50 F Street NW, Washington, DC 20001.

Department of Transportation for the use of the *Emergency Response Guide Book*. Copyright 1987 Office of Hazardous Materials Transportation (DHM-51) Research and Special Programs Administration, U.S. Dept of Transportation, Washington, DC 20590.

NIOSH for the use of *The Pocket Guide to Chemical Hazards*. Copyright September, 1985. Division of Standards Development and Technology Transfer, NIOSH, 4676 Columbia Parkway, Cincinnati, OH 45226-1998.

United States Department of Transportation—U.S. Coast Guard for the use of the *CHRIS—Chemical Hazard Response Information System*. Copyright November 14, 1984, Commandant (G-WER-2). U.S. Coast Guard, Washington, DC 20593.

LIST OF TERMS

HAZARDOUS MATERIAL

CHEMICAL NAME—The common name of the chemical used in identification.

D.O.T. Hazard Class—One of the nine classes:

1. EXPLOSIVE
2. GASES
3. FLAMMABLE LIQUIDS
4. FLAMMABLE SOLIDS
5. OXIDIZERS AND OR ORGANIC PEROXIDES
6. POISONOUS AND ETIOLOGIC MATERIALS
7. RADIOACTIVE MATERIALS
8. CORROSIVES
9. OTHER REGULATED MATERIALS

I.D. Number—A number given to a chemical by the Department of Transportation. Example: Acetone—1090.

NA North America

UN United Nations

D.O.T. Guide Number—A guide for initial action to be taken to protect yourself and the general public when you are called upon to handle incidents involving hazardous materials.

S.T.C.C. Number—Standard Transportation Commodity Code. Example: Acetone—4908105

Reportable Quantity—The amount of material spilled that has to be reported to the Environmental Protection Agency. (EPA) Example: Acetone—5000/2270. The 5000 represents pounds, the 2270 represents kilos.

Mfg Name—The name of a company that manufactures the chemical, and a point of contact in case of an emergency.

Telephone No.—The telephone number where a manufacturer of the chemical can be reached.

PHYSICAL DESCRIPTION:

Normal Physical Form—Solid, liquid, or gas. Example: Acetone—Liquid.

Color—What color is the material? Example: Acetone—colorless.

Odor—What does the material smell like? Example: Acetone—pleasant odor. *Caution:* Smell is not an advisable means of identification.

Other Information—What the material is used for. Example: Acetone—used for making other chemicals.

CHEMICAL PROPERTIES:

Specific Gravity—The ratio of the density of a substance to the density of a reference substance. The usual ratio is the weight of a solid or liquid as compared with the weight of an equal volume of water.

Vapor Density—The ratio of the relative density of a gas or vapor (with no air present) as compared with clean, dry air.

Boiling Point—The temperature at which the vapor pressure of a liquid just equals atmospheric pressure. Example: Acetone—133°F.

Melting Point—The temperature at which a solid changes to a liquid. Example: Acetone— −169°F.

Vapor Pressure—That amount of pressure the chemical will produce when it is exposed to the air at a certain temperature.

Solubility in Water—Will the material mix with water? Example: Acetone—soluble.

Degree of Solubility—The amount from .0001% to 100% by weight that the material will mix with water.

OTHER INFORMATION—Any pertinent information that falls into this category. Example: Acetone—soluble in water, but quite violent.

HEALTH HAZARDS:

Inhalation Hazard—What hazard to expect from this chemical if you would happen to breathe in the vapors or dust. Example: Acetone—irritating to the nose and throat.

Ingestion Hazard—What hazard to expect if you would accidentally swallow or ingest in your body. Example: Acetone—will cause severe headaches and dizziness.

Absorption Hazard—What hazard to expect if the material would absorb through your skin, eyes, open sores or wounds. Example: Acetone—irritating to the skin and eyes.

Hazards to Wildlife—What harm would come to aquatic life or waterfowl if the chemical would enter a water environment. Example: Acetone—harmful to aquatic life in high concentrations.

Decontamination Procedures—The basic procedure for removing the material from your body, clothing, and equipment before you contaminate anything else. *Note:* All water runoff from decontamination procedures *must* be collected for proper disposal. Example: Acetone—Wash away any material with copious amount of soap and water.

FIRST AID PROCEDURES—The basic procedure to care for an overcome person in case of exposure to this material until proper medical personnel arrive. Example: Acetone—remove the victim to fresh air, call emergency medical care. If not breathing, give CPR. If breathing is difficult, administer oxygen. Treat for shock.

FIRE HAZARDS:

Flashpoint—The minimum temperature of the liquid at which it will produce vapors sufficient to form an ignitable mixture with air near the surface of the liquid or the container. Example: Acetone—0°F (-18°C).

Ignition Temperature—The temperature at which the material will ignite and burn. Example: Acetone—869°F (465°C).

Flammable Explosive High Range—Also known as the Upper Flammable Limit and as the Upper Explosive Limit. This is the maximum concentration of gas, vapor, fumes, or dust in air above which ignition will not occur. On a scale of 1% to 100%, if the percentage is above the upper limit, the air mixture is considered too rich to burn or explode. Example: Acetone—13%. Any mixture above 13% would be too rich to burn.

Flammable Explosive Low Range—Also known as the Lower Flammable Limit and as the Lower Explosive Limit. This is the minimum concentration of gas, vapor, fumes or dust in air below which ignition will not occur. On a scale from 1% to 100%, if the percentage is below the lower limit, the air mixture is considered too lean to burn or explode. Example: Acetone—2.5%. Any mixture below 2.5% is considered too lean to burn.

Toxic Products of Combustion—What toxic/hazardous byproducts the material would form if involved in a fire. Examples of byproducts include phosgene gases, oxides of nitrogen, irritating gases, etc. Carbon monoxide is found in the burning of organic materials. However, it is not included in this category.

Other Hazards—What other hazards could you expect in case of fire involvement and non fire involvement. Example: Do not extinguish the fire unless the flow can be stopped. Cool affected containers with flooding quantities of water, etc.

Possible Extinguishing Agents—Preferred agents used to extinguish the fire if possible. Example: Acetone—Use large quantities of flooding water from as far a distance as possible. Use dry chemical, alcohol foam, or carbon dioxide.

REACTIVITY HAZARDS:

Reactive with What—What does the chemical react with and how, and what to expect from the reaction. Example: Strong oxidizers, bleach, gold, silver, etc. *Note:* This category pertains to water and air; however, some other common compounds are listed for your information.

Other Reactions—Anything else the chemical would react to. Example: Water to form an acidic solution.

CORROSIVITY HAZARDS:

Corrosive with What—What the material is corrosive with. Example—Metals and tissue.

Neutralizing Agent—The type of material to use to neutralize the spill. Examples: Crushed limestone, soda ash, or lime.

RADIOACTIVITY HAZARDS:

Type Radiation Emitted—Alpha, beta, or gamma rays

Other Hazards—What other hazards would you expect with this material in radiation exposure.

Recommended Protection for Response Personnel:

The suggested guidance for response personnel to follow in the event of a hazardous materials incident.

FIREFIGHTER'S HAZARDOUS MATERIALS REFERENCE BOOK

ACETAL

DOT Number: UN 1088 *DOT Hazard Class:* Flammable liquid *DOT Guide Number:* 26
Synonyms: none given
STCC Number: 4908103 *Reportable Qty:* None *CHEMTREC Phone No:* 1-800-424-9300

Physical Description:

Physical Form: Liquid *Color:* Colorless *Odor:* Pleasant
Other Information: n/a

Chemical Properties:

Specific Gravity: .8 *Vapor Density:* 4.1 *Boiling Point: 215° F(101.6° C)*
Melting Point: n/a *Vapor Pressure:* n/a *Solubility in water:* Slight
Other Information: Lighter than and slightly soluble in water. Vapors are heavier than air.

Health Hazards:

Inhalation Hazard: Harmful if inhaled.
Ingestion Hazard: n/a
Absorption Hazard: n/a
Hazards to Wildlife: n/a
Decontamination Procedures: Wash away any material with copious amounts of soap and water.
First Aid Procedures: Remove victim to fresh air, call emergency medical care. If not breathing give CPR. If breathing is difficult administer oxygen. Treat for shock.

Fire Hazards:

Flashpoint: –5° F(–20.5° C) *Ignition temperature:* 446° F(230° C)
Flammable Explosive High Range: 10.4 *Low Range:* 1.6
Toxic Products of Combustion: n/a
Other Hazards: Do not extinguish the fire unless the flow can be stopped. Cool all affected containers with flooding quantities of water.
Possible extinguishing agents: Use alcohol foam, dry chemical, carbon dioxide.

Reactivity Hazards:

Reactive With: n/a *Other Reactions:* n/a

Corrosivity Hazards:

Corrosive With: n/a *Neutralizing Agent:* n/a

Radioactivity Hazards:

Radiation Emitted: n/a *Other Hazards:* n/a

Recommended Protection for Response Personnel:

Avoid breathing vapors, keep upwind. Structural protective clothing provides limited protection. wash away any material which may have come into contact with the body with copious amounts of water. if the fire becomes uncontrollable, consider appropriate evacuation.

ACETALDEHYDE

DOT Number: UN 1089 *DOT Hazard Class:* Flammable liquid *DOT Guide Number:* 26
Synonyms: acetic aldehyde, ethanal, ethyl aldehyde
STCC Number: 4907210 *Reportable Qty:* 1000/454
MFG Name: Plastic Manufacturers *Phone No:* 1-215-438-1082

Physical Description:

Physical Form: Liquid *Color:* Colorless *Odor:* Pungent, choking odor
Other Information: A colorless liquid at temperatures below 69° F(20° C) but rapidly volatilizes at this temperature. It has a suffocating and fruity odor.

Chemical Properties:

Specific Gravity: .78 *Vapor Density:* 1.5 *Boiling Point: 70° F (20° C)*
Melting Point: −193° F(−125° C) *Vapor Pressure:* 755 psig at 68° F (20° C) *Solubility in water:* Yes
Other Information: It is easily oxidized by air to form unstable peroxides which may explode. (polymerizable) Lighter than and soluble in water. Weighs 6.5 lbs/2.9 kg per gallon/3.8 l.

Health Hazards:

Inhalation Hazard: Headache, difficulty in breathing, loss of consciousness
Ingestion Hazard: Harmful if swallowed. Absorption Hazard: Will burn the skin and eyes.
Hazards to Wildlife: Dangerous to aquatic life.
Decontamination Procedures: Wash away any material with copious amounts of soap and water.
First Aid Procedures: Remove victim to fresh air, call emergency medical care. If not breathing give CPR. If breathing is difficult administer oxygen. Treat for shock.

Fire Hazards:

Flashpoint: -38° F(-38.8° C) *Ignition temperature:* 347° F(175° C)
Flammable Explosive High Range: 60 *Low Range:* 4.0
Toxic Products of Combustion: n/a
Other Hazards: Do not extinguish the fire unless the flow can be stopped. Cool all affected containers with flooding quantities of water.
Possible extinguishing agents: Use alcohol foam, dry chemical, carbon dioxide.

Reactivity Hazards:

Reactive With: n/a *Other Reactions:* n/a

Corrosivity Hazards:

Corrosive With: n/a *Neutralizing Agent:* n/a

Radioactivity Hazards:

Radiation Emitted: n/a *Other Hazards:* n/a

Recommended Protection for Response Personnel:

Avoid breathing vapors, keep upwind. Wear a sealed chemical suit, (butyl rubber, chlorinated polyethylene). Wash away any material which may have come into contact with the body with copious amounts of soap and water. Consider appropriate evacuation.

ACETIC ACID (aqueous)

DOT Number: UN 2790 *DOT Hazard Class:* Corrosive solution *DOT Guide Number:* 60
Synonyms: vinegar acid
STCC Number: 4931401 *Reportable Qty:* 5000/2270
MFG Name: Allied Corp. *Phone No:* 1-201-455-2000

Physical Description:

Physical Form: Aqueous Solution *Color:* Colorless *Odor:* Vinegar
Other Information: Weighs 8.8 lbs/3.8 kg per gallon/3.8 l.

Chemical Properties:

Specific Gravity: .9 *Vapor Density:* 3.5 *Boiling Point: 215° F(101.6° C)*
Melting Point: 62° F(16.6° C) *Vapor Pressure:* n/a *Solubility in water:* Slight
Other Information: n/a

Health Hazards:

Inhalation Hazard: Coughing, nausea, vomiting, difficulty in breathing.
Ingestion Hazard: Harmful if swallowed.
Absorption Hazard: Will burn the skin and eyes.
Hazards to Wildlife: Dangerous to aquatic life.
Decontamination Procedures: Wash away any material with copious amounts of soap and water.
First Aid Procedures: Remove victim to fresh air, call emergency medical care. If not breathing give CPR. If breathing is difficult administer oxygen. Treat for shock.

Fire Hazards:

Flashpoint: 55° F(12.7° C) *Ignition temperature:* 842° F(450° C)
Flammable Explosive High Range: 8 *Low Range:* 1.7
Toxic Products of Combustion: n/a
Other Hazards: Use flooding quantities of water as a fog to cool affected containers.
Possible extinguishing agents: Use alcohol foam, dry chemical, carbon dioxide.

Reactivity Hazards:

Reactive With: Attacks most common metals including most stainless steels. *Other Reactions:* n/a

Corrosivity Hazards:

Corrosive With: Steel and tissue *Neutralizing Agent:* Crushed limestone, soda ash, or lime

Radioactivity Hazards:

Radiation Emitted: n/a *Other Hazards:* n/a

Recommended Protection for Response Personnel:

Avoid breathing vapors, keep upwind. Wear a sealed chemical suit, (polycarbonate, butyl rubber, chlorinated polyethylene). Wash away any material which may have come into contact with the body with copious amounts of soap and water. Consider appropriate evacuation.

ACETIC ACID (glacial)

DOT Number: UN 2789 *DOT Hazard Class:* Corrosive *DOT Guide Number:* 29
Synonyms: glacial acetic acid, vinegar acid
STCC Number: 4913303 *Reportable Qty:* 5000/2270
MFG Name: Allied Corp. *Phone No:* 1-201-455-2000

Physical Description:

Physical Form: Liquid *Color:* Colorless *Odor:* Vinegar
Other Information: It is used to make other chemicals, as a food additive, and in petroleum production.

Chemical Properties:

Specific Gravity: 1 *Vapor Density:* 2.1 *Boiling Point: 245° F(118.3° C)*
Melting Point: 62° F(16.6° C) *Vapor Pressure:* 11 mm Hg at 68° F(20° C) *Solubility in water:* Yes
Other Information: Soluble in water with releases of heat. Weighs 8.8 lbs/3.8 kg per gallon/3.8 l.

Health Hazards:

Inhalation Hazard: Coughing, nausea, vomiting, difficulty in breathing.
Ingestion Hazard: Harmful if swallowed.
Absorption Hazard: Will burn the skin and eyes.
Hazards to Wildlife: Dangerous to aquatic life.
Decontamination Procedures: Wash away any material with copious amounts of soap and water.
First Aid Procedures: Remove victim to fresh air, call emergency medical care. If not breathing give CPR. If breathing is difficult administer oxygen. Treat for shock.

Fire Hazards:

Flashpoint: 103° F(39.4° C) *Ignition temperature:* 867° F(463.8° C)
Flammable Explosive High Range: 19.9 *Low Range:* 4
Toxic Products of Combustion: n/a
Other Hazards: Vapors may explode if ignited in an enclosed area.
Possible extinguishing agents: Apply water from as far a distance as possible. Use alcohol foam, dry chemical, or carbon dioxide.

Reactivity Hazards:

Reactive With: n/a *Other Reactions:* n/a

Corrosivity Hazards:

Corrosive With: Metals and tissue. *Neutralizing Agent:* Crushed limestone, soda ash, or lime.

Radioactivity Hazards:

Radiation Emitted: n/a *Other Hazards:* n/a

Recommended Protection for Response Personnel:

Avoid breathing vapors, keep upwind. Wear a sealed chemical suit, (polycarbonate, butyl rubber, chlorinated polyethylene). Wash away any material which may come into contact with the body with copious amounts of soap and water. Consider appropriate evacuation.

ACETIC ANHYDRIDE

DOT Number: UN 1715 *DOT Hazard Class:* Corrosive *DOT Guide Number:* 39
Synonyms: ethanoic anhydride
STCC Number: 4931304 *Reportable Qty:* 5000/2270
MFG Name: Union Carbide *Phone No:* 1-203-794-2000

Physical Description:

Physical Form: Liquid *Color:* Colorless *Odor:* Vinegar
Other Information: Used to make fibers, plastics, pharmaceuticals, dyes, and explosives.

Chemical Properties:

Specific Gravity: 1.1 *Vapor Density:* 3.5 *Boiling Point: 284° F(140° C)*
Melting Point: -99° F(-72.7° C) *Vapor Pressure:* 4 mm Hg at 68° F(20° C)
Solubility in water: Yes *Degree of Solubility:* 12%
Other Information: It is soluble in water, alcohol, benzene, chloroform, and ether. Readily decomposed by water to acetic acid with the release of Heat. Weighs 9 lbs/3.9 kg to the gallon/3.8 l.

Health Hazards:

Inhalation Hazard: Nausea, vomiting, difficulty in breathing. *Ingestion Hazard: Harmful if swallowed.*
Absorption Hazard: Will burn the skin and eyes.Hazards to Wildlife: Dangerous to aquatic life.
Decontamination Procedures: Wash away any material with copious amounts of soap and water.
First Aid Procedures: Remove victim to fresh air, call emergency medical care. If not breathing give CPR. If breathing is difficult administer oxygen. Treat for shock.

Fire Hazards:

Flashpoint: 120° F(48.8° C) *Ignition temperature:* 600° F(315.5° C)
Flammable Explosive High Range: 10.3 *Low Range:* 2.7
Toxic Products of Combustion: Toxic and irritating products in a fire.
Other Hazards: Vapor forms explosive mixture with air. Reacts with water to form acetic acid.
Possible extinguishing agents: Use alcohol foam, dry chemical, carbon dioxide.

Reactivity Hazards:

Reactive With: Attacks most common metals including most stainless steels.
Other Reactions: Reacts slowly with water to evolve heat and acetic acid.

Corrosivity Hazards:

Corrosive With: Steel and tissue. Attacks some forms of plastics, rubber, and coatings.
Neutralizing Agent: Crushed limestone, soda ash, or lime.

Radioactivity Hazards:

Radiation Emitted: n/a *Other Hazards:* n/a

Recommended Protection for Response Personnel:

Avoid breathing vapors, keep upwind. Wear a sealed chemical suit, (polycarbonate, butyl rubber, neoprene, chlorinated polyethylene). Wash away any material which may have come into contact with the body with copious amounts of soap and water. Consider appropriate evacuation.

ACETONE

DOT Number: UN 1090 *DOT Hazard Class:* Flammable liquid *DOT Guide Number:* 26
Synonyms: dimethyl ketone, β-ketopropane, 2-propanone, pyroacetic ether
STCC Number: 4908105 *Reportable Qty:* 5000/2270
MFG Name: Union Carbide *Phone No:* 1-203-794-2000

Physical Description:

Physical Form: Liquid *Color:* Colorless *Odor:* Pleasant
Other Information: Used to make other chemicals, paint, nail polish removers, solvent, and for many other uses.

Chemical Properties:

Specific Gravity: .08 *Vapor Density:* 2.0 *Boiling Point: 133° F(56.1° C)*
Melting Point: -138° F(-94.4° C) *Vapor Pressure:* 181 mm Hg at 68° F(20° C) *Solubility in water:* Yes
Other Information: Lighter than and soluble in water but quite volatile. Vapors are heavier than air.

Health Hazards:

Inhalation Hazard: Irritation to the nose and throat.
Ingestion Hazard: Severe headaches, dizziness.
Absorption Hazard: Irritating to the skin and eyes.
Hazards to Wildlife: Dangerous to aquatic life in high concentrations.
Decontamination Procedures: Wash away any material with copious amounts of soap and water.
First Aid Procedures: Remove victim to fresh air, call emergency medical care. If not breathing give CPR. If breathing is difficult administer oxygen. Treat for shock.

Fire Hazards:

Flashpoint: 0° F(-17.7° C) *Ignition temperature:* 869° F(465° C)
Flammable Explosive High Range: 13 *Low Range:* 2.5
Toxic Products of Combustion: n/a
Other Hazards: Containers may rupture violently in fire. Flashback along vapor trail may occur.
Possible extinguishing agents: Large quantities of flooding water from as far a distance as possible. Use dry chemical, alcohol foam, carbon dioxide.

Reactivity Hazards:

Reactive With: n/a *Other Reactions:* n/a

Corrosivity Hazards:

Corrosive With: Attacks many forms of plastics and rubber to include rayon. *Neutralizing Agent:* n/a

Radioactivity Hazards:

Radiation Emitted: n/a *Other Hazards:* n/a

Recommended Protection for Response Personnel:

Avoid breathing vapors, keep upwind. Wash away any material which may come into contact with the body with copious amounts of soap and water. Wear a sealed chemical suit, (neoprene, butyl rubber, chlorinated polyethylene). If the fire becomes uncontrollable, or if the container is exposed to direct flame, consider appropriate evacuation.

ACETONE CYANOHYDRIN

DOT Number: UN 1541 *DOT Hazard Class:* Poison B *DOT Guide Number:* 55
Synonyms: 2-hydroxy-2-pethyl propionitrile
STCC Number: 4921401 *Reportable Qty:* 10/4.54
MFG Name: Aldrich Chemical *Phone No:* 1-414-273-3850

Physical Description:

Physical Form: Liquid *Color:* Colorless *Odor:* Mild almond odor
Other Information: Dissociates to acetone, a flammable liquid, and hydrogen cyanide, a flammable poisonous gas under normal storage and transportation conditions.

Chemical Properties:

Specific Gravity: .09 *Vapor Density:* 2.9 *Boiling Point: 248° F(120° C)*
Melting Point: -2° F(-18.8° C) *Vapor Pressure:* 20 mm Hg at 194° F(90° C) *Solubility in water:* Yes
Other Information: It is lethal by Inhalation, and less readily by skin absorption. Weighs 7.8 lbs/3.3 kg per gallon/3.8 l.

Health Hazards:

Inhalation Hazard: Poisonous if inhaled.
Ingestion Hazard: Poisonous if swallowed.
Absorption Hazard: Will burn the skin and eyes.
Hazards to Wildlife: n/a
Decontamination Procedures: Wash away any material with copious amounts of soap and water.
First Aid Procedures: Remove victim to fresh air, call emergency medical care. If not breathing give CPR. If breathing is difficult administer oxygen. Treat for shock.

Fire Hazards:

Flashpoint: 165° F(73.8° C) *Ignition temperature:* 1270° F(687.7° C)
Flammable Explosive High Range: 12.0 *Low Range:* 2.2
Toxic Products of Combustion: Oxides of nitrogen are produced.
Other Hazards: Rate of dissociation is increased by contact with alkalis and/or heat. Containers may rupture violently in a fire.
Possible extinguishing agents: Use alcohol foam, dry chemical, carbon dioxide.

Reactivity Hazards:

Reactive With: n/a *Other Reactions:* n/a

Corrosivity Hazards:

Corrosive With: n/a *Neutralizing Agent:* n/a

Radioactivity Hazards:

Radiation Emitted: n/a *Other Hazards:* n/a

Recommended Protection for Response Personnel:

Avoid breathing vapors, keep upwind. Wash away any material which may have come into contact with the body with copious amounts of soap and water. Wear a sealed chemical suit, (polycarbonate, butyl rubber). If the fire becomes uncontrollable, or if the container is exposed to direct flame, consider appropriate evacuation.

ACETONITRILE

DOT Number: NA 1648 *DOT Hazard Class:* Flammable liquid *DOT Guide Number:* 28
Synonyms: cyano-methane, ethyl nitrile, ethanenitrile, methyl cyanide
STCC Number: 4907405 *Reportable Qty:* 5000/2270
MFG Name: E.I. Du Pont *Phone No:* 1-800-441-3637

Physical Description:

Physical Form: Liquid *Color:* Colorless *Odor:* Ethereal
Other Information: n/a

Chemical Properties:

Specific Gravity: .8 *Vapor Density:* 1.4 *Boiling Point: 179° F(81.6° C)*
Melting Point: 42° F(5.5° C) *Vapor Pressure:* 73 mm Hg at 68° F(20° C) *Solubility in water:* Yes
Other Information: Lighter than and soluble in water. Vapors are heavier than air.

Health Hazards:

Inhalation Hazard: Difficulty in breathing.
Ingestion Hazard: Harmful if swallowed.
Absorption Hazard: Irritating to the skin and eyes.
Hazards to Wildlife: Dangerous to aquatic life.
Decontamination Procedures: Wash away any material with copious amounts of soap and water.
First Aid Procedures: Remove victim to fresh air, call emergency medical care. If not breathing give CPR. If breathing is difficult administer oxygen. Treat for shock.

Fire Hazards:

Flashpoint: 42° F(5.5° C) *Ignition temperature:* 975° F(523.8° C)
Flammable Explosive High Range: 16.0 *Low Range:* 3.0
Toxic Products of Combustion: Toxic oxides of nitrogen are produced during combustion of this material.
Other Hazards: n/a
Possible extinguishing agents: Use water from as far a distance as possible. Use alcohol foam, dry chemical, or carbon dioxide. Do not extinguish the fire unless the flow can be stopped. Cool all affected containers with water.

Reactivity Hazards:

Reactive With: n/a *Other Reactions:* n/a

Corrosivity Hazards:

Corrosive With: n/a *Neutralizing Agent:* n/a

Radioactivity Hazards:

Radiation Emitted: n/a *Other Hazards:* n/a

Recommended Protection for Response Personnel:

Avoid breathing vapors, keep upwind. Wash away any material which may come into contact with the body with copious amounts of soap and water. Wear a sealed chemical suit, (polycarbonate, butyl rubber). If the fire becomes uncontrollable, or if the containers are exposed to direct flame, consider appropriate evacuation.

ACETOPHENONE

DOT Number: NA 9207 *DOT Hazard Class:* Combustible liquid *DOT Guide Number:* 27
Synonyms: acetylbenzene, hypone, phenyl methyl ketone
STCC Number: n/a *Reportable Qty:* n/a
MFG Name: Union Carbide *Phone No:* 1-203-794-2000

Physical Description:

Physical Form: Liquid *Color:* Colorless *Odor:* Flowery sweet
Other Information: Slowly sinks in water. Freezing point is 68° F(20° C)

Chemical Properties:

Specific Gravity: 1.0 + *Vapor Density:* 4.1 *Boiling Point: 396° F(202.2° C)*
Melting Point: n/a *Vapor Pressure:* 1 mm Hg at 15° F(-9.4° C) *Solubility in water:* No
Other Information: n/a

Health Hazards:

Inhalation Hazard: n/a
Ingestion Hazard: Harmful if swallowed.
Absorption Hazard: Irritating to the skin and eyes.
Hazards to Wildlife: n/a
Decontamination Procedures: Wash away any material with copious amounts of soap and water.
First Aid Procedures: Remove victim to fresh air, call emergency medical care. If not breathing give CPR. If breathing is difficult administer oxygen. Treat for shock.

Fire Hazards:

Flashpoint: 170° F(76.6° C) *Ignition temperature:* 1058° F(570° C)
Flammable Explosive High Range: n/a *Low Range:* n/a
Toxic Products of Combustion: n/a
Other Hazards: n/a
Possible extinguishing agents: Use alcohol foam, dry chemical, carbon dioxide.

Reactivity Hazards:

Reactive With: n/a *Other Reactions:* n/a

Corrosivity Hazards:

Corrosive With: n/a *Neutralizing Agent:* n/a

Radioactivity Hazards:

Radiation Emitted: n/a *Other Hazards:* n/a

Recommended Protection for Response Personnel:

Avoid breathing vapors, keep upwind. Structural protective clothing provides limited protection. Wash away any material which may come into contact with the body with copious amounts of soap and water. If the fire becomes uncontrollable, or if the container is exposed to direct flame, consider appropriate evacuation.

ACETYL BROMIDE

DOT Number: UN 1716 *DOT Hazard Class:* Corrosive *DOT Guide Number:* 60
Synonyms: none given
STCC Number: 4931705 *Reportable Qty:* 5000/2270
MFG Name: Aldrich Chemical *Phone No:* 1-414-273-3850

Physical Description:

Physical Form: Liquid *Color:* Colorless *Odor:* Sharp, unpleasant
Other Information: n/a

Chemical Properties:

Specific Gravity: 1.33 *Vapor Density:* 4.2 *Boiling Point: 170° F(76.6° C)*
Melting Point: 142° F(61.1° C) *Vapor Pressure:* n/a *Solubility in water:* n/a
Other Information: Decomposed by water to acetic and hydrobromic acids with release of heat. Weighs 13.9 lbs/6.0 kg per gallon/3.8 l.

Health Hazards:

Inhalation Hazard: Difficulty in breathing.
Ingestion Hazard: Harmful if swallowed.
Absorption Hazard: Will burn the skin and eyes.
Hazards to Wildlife: n/a
Decontamination Procedures: Wash away any material with copious amounts of soap and water.
First Aid Procedures: Remove victim to fresh air, call emergency medical care. If not breathing give CPR. If breathing is difficult administer oxygen. Treat for shock.

Fire Hazards:

Flashpoint: n/a *Ignition temperature:* n/a
Flammable Explosive High Range: n/a *Low Range:* n/a
Toxic Products of Combustion: Toxic, irritating hydrogen bromide fumes may be formed in fires.
Other Hazards: n/a
Possible extinguishing agents: Do not use water on the material itself. Use dry chemical or carbon dioxide.

Reactivity Hazards:

Reactive With: Water, wood, and most metals. *Other Reactions:* n/a

Corrosivity Hazards:

Corrosive With: Metals and tissue *Neutralizing Agent:* Crushed limestone, soda ash, or lime.

Radioactivity Hazards:

Radiation Emitted: n/a *Other Hazards:* n/a

Recommended Protection for Response Personnel:

Avoid breathing vapors, keep upwind. Wear a sealed chemical suit, (polycarbonate, butyl rubber). Wash away any material which may have come into contact with the body with copious amounts of soap and water. Consider appropriate evacuation.

ACETYL CHLORIDE

DOT Number: UN 1717
DOT Hazard Class: Flammable liquid
DOT Guide Number: 29
Synonyms: ethanoyl chloride
STCC Number: 4907601
Reportable Qty: 5000/2270
MFG Name: Mallinckrodt Inc.
Phone No: 1-314-895-2000

Physical Description:

Physical Form: Liquid
Color: Colorless
Odor: Pungent
Other Information: n/a

Chemical Properties:

Specific Gravity: 1.1+
Vapor Density: 2.7
Boiling Point: 124° F(51.1° C)
Melting Point: n/a
Vapor Pressure: n/a
Solubility in water: Yes
Other Information: Violently decomposes with water to acetic and hydrochloric acids with release of heat. Vapors are heavier than air. Weighs 9.2 lbs/3.9 kg per gallon/3.8 l.

Health Hazards:

Inhalation Hazard: Difficulty in breathing.
Ingestion Hazard: Harmful if swallowed.
Absorption Hazard: Will Burn the skin and eyes.
Hazards to Wildlife: n/a
Decontamination Procedures: Wash away any material with copious amounts of soap and water.
First Aid Procedures: Remove victim to fresh air, call emergency medical care. If not breathing give CPR. If breathing is difficult administer oxygen. Treat for shock.

Fire Hazards:

Flashpoint: 40° F(4.4° C)
Ignition temperature: 734° F(390° C)
Flammable Explosive High Range: n/a
Low Range: n/a
Toxic Products of Combustion: n/a
Other Hazards: Apply no water due to violent decomposition of material.
Possible extinguishing agents: Do not use water or foam. Use dry chemical, carbon dioxide.

Reactivity Hazards:

Reactive With: Water to form acetic and hydrochloric acid.
Other Reaction: n/a

Corrosivity Hazards:

Corrosive With: Most metals and tissue
Neutralizing Agent: Crushed limestone, soda ash, or lime.

Radioactivity Hazards:

Radiation Emitted: n/a
Other Hazards: n/a

Recommended Protection for Response Personnel:

Avoid breathing vapors, keep upwind. Wear a sealed chemical suit, (butyl rubber, viton, chlorinated polyethylene). Wash away any material which may come in contact with the body with copious amounts of soap and water. If the fire becomes uncontrollable, or if the container is exposed to direct flame, consider appropriate evacuation.

ACETYL PEROXIDE

DOT Number: UN 2084 *DOT Hazard Class:* Organic peroxide *DOT Guide Number:* 49
Synonyms: diacetyl acetic acid
STCC Number: 4919510 *Reportable Qty:* n/a
MFG Name: Aztec Chemical *Phone No:* 1-713-682-5300

Physical Description:

Physical Form: Liquid *Color:* Colorless *Odor:* Sharp
Other Information: The solution cannot contain more than 25% of the chemical in a nonviolent solvent.

Chemical Properties:

Specific Gravity: 1.2 *Vapor Density:* 4.1 *Boiling Point: Decomposes*
Melting Point: n/a *Vapor Pressure:* n/a *Solubility in water:* n/a
Other Information: The dry material can form explosive mixtures.

Health Hazards:

Inhalation Hazard: n/a
Ingestion Hazard: Harmful if swallowed.
Absorption Hazard: Irritating to the skin and eyes.
Hazards to Wildlife: n/a
Decontamination Procedures: Wash away any material with copious amounts of soap and water.
First Aid Procedures: Remove victim to fresh air, call emergency medical care. If not breathing give CPR. If breathing is difficult administer oxygen. Treat for shock.

Fire Hazards:

Flashpoint: 113° F(45° C) *Ignition temperature:* n/a
Flammable Explosive High Range: n/a *Low Range:* n/a
Toxic Products of Combustion: n/a
Other Hazards: May explode on contact with combustibles. Containers may explode in a fire.
Possible extinguishing agents: Apply water from as far a distance as possible. Use alcohol foam, dry chemical, or carbon dioxide.

Reactivity Hazards:

Reactive With: Combustible Materials. *Other Reactions:* n/a

Corrosivity Hazards:

Corrosive With: n/a *Neutralizing Agent:* n/a

Radioactivity Hazards:

Radiation Emitted: n/a *Other Hazards:* n/a

Recommended Protection for Response Personnel:

Avoid breathing vapors, keep upwind. Wear a sealed chemical suit, (polycarbonate, butyl rubber, chlorinated polyethylene). Wash away any material which may come into contact with the body with copious amounts of soap and water. If the fire becomes uncontrollable, consider appropriate evacuation.

ACETYL CHLORIDE

DOT Number: UN 1717 *DOT Hazard Class:* Flammable liquid *DOT Guide Number:* 29
Synonyms: ethanoyl chloride
STCC Number: 4907601 *Reportable Qty:* 5000/2270
MFG Name: Mallinckrodt Inc. *Phone No:* 1-314-895-2000

Physical Description:

Physical Form: Liquid *Color:* Colorless *Odor:* Pungent
Other Information: n/a

Chemical Properties:

Specific Gravity: 1.1+ *Vapor Density:* 2.7 *Boiling Point: 124° F(51.1° C)*
Melting Point: n/a *Vapor Pressure:* n/a *Solubility in water:* Yes
Other Information: Violently decomposes with water to acetic and hydrochloric acids with release of heat. Vapors are heavier than air. Weighs 9.2 lbs/3.9 kg per gallon/3.8 l.

Health Hazards:

Inhalation Hazard: Difficulty in breathing.
Ingestion Hazard: Harmful if swallowed.
Absorption Hazard: Will Burn the skin and eyes.
Hazards to Wildlife: n/a
Decontamination Procedures: Wash away any material with copious amounts of soap and water.
First Aid Procedures: Remove victim to fresh air, call emergency medical care. If not breathing give CPR. If breathing is difficult administer oxygen. Treat for shock.

Fire Hazards:

Flashpoint: 40° F(4.4° C) *Ignition temperature:* 734° F(390° C)
Flammable Explosive High Range: n/a *Low Range:* n/a
Toxic Products of Combustion: n/a
Other Hazards: Apply no water due to violent decomposition of material.
Possible extinguishing agents: Do not use water or foam. Use dry chemical, carbon dioxide.

Reactivity Hazards:

Reactive With: Water to form acetic and hydrochloric acid.
Other Reaction: n/a

Corrosivity Hazards:

Corrosive With: Most metals and tissue
Neutralizing Agent: Crushed limestone, soda ash, or lime.

Radioactivity Hazards:

Radiation Emitted: n/a *Other Hazards:* n/a

Recommended Protection for Response Personnel:

Avoid breathing vapors, keep upwind. Wear a sealed chemical suit, (butyl rubber, viton, chlorinated polyethylene). Wash away any material which may come in contact with the body with copious amounts of soap and water. If the fire becomes uncontrollable, or if the container is exposed to direct flame, consider appropriate evacuation.

ACETYL PEROXIDE

DOT Number: UN 2084 *DOT Hazard Class:* Organic peroxide *DOT Guide Number:* 49
Synonyms: diacetyl acetic acid
STCC Number: 4919510 *Reportable Qty:* n/a
MFG Name: Aztec Chemical *Phone No:* 1-713-682-5300

Physical Description:

Physical Form: Liquid *Color:* Colorless *Odor:* Sharp
Other Information: The solution cannot contain more than 25% of the chemical in a nonviolent solvent.

Chemical Properties:

Specific Gravity: 1.2 *Vapor Density:* 4.1 *Boiling Point: Decomposes*
Melting Point: n/a *Vapor Pressure:* n/a *Solubility in water:* n/a
Other Information: The dry material can form explosive mixtures.

Health Hazards:

Inhalation Hazard: n/a
Ingestion Hazard: Harmful if swallowed.
Absorption Hazard: Irritating to the skin and eyes.
Hazards to Wildlife: n/a
Decontamination Procedures: Wash away any material with copious amounts of soap and water.
First Aid Procedures: Remove victim to fresh air, call emergency medical care. If not breathing give CPR. If breathing is difficult administer oxygen. Treat for shock.

Fire Hazards:

Flashpoint: 113° F(45° C) *Ignition temperature:* n/a
Flammable Explosive High Range: n/a *Low Range:* n/a
Toxic Products of Combustion: n/a
Other Hazards: May explode on contact with combustibles. Containers may explode in a fire.
Possible extinguishing agents: Apply water from as far a distance as possible. Use alcohol foam, dry chemical, or carbon dioxide.

Reactivity Hazards:

Reactive With: Combustible Materials. *Other Reactions:* n/a

Corrosivity Hazards:

Corrosive With: n/a *Neutralizing Agent:* n/a

Radioactivity Hazards:

Radiation Emitted: n/a *Other Hazards:* n/a

Recommended Protection for Response Personnel:

Avoid breathing vapors, keep upwind. Wear a sealed chemical suit, (polycarbonate, butyl rubber, chlorinated polyethylene). Wash away any material which may come into contact with the body with copious amounts of soap and water. If the fire becomes uncontrollable, consider appropriate evacuation.

ACETYL PEROXIDE solution

DOT Number: UN 2084 *DOT Hazard Class:* Organic peroxide (not over 25% peroxide)
Synonyms: diacetyl peroxide solution *DOT Guide Number:* 49
STCC Number: 4919510 *Reportable Qty:* n/a
MFG Name: Aztec Chemical *Phone No:* 1-713-682-5300

Physical Description:

Physical Form: Liquid *Color:* Colorless *Odor:* Sharp
Other Information: n/a

Chemical Properties:

Specific Gravity: 1.2 *Vapor Density:* 4 *Boiling Point: Decomposes*
Melting Point: n/a *Vapor Pressure:* n/a *Solubility in water:* n/a
Other Information: The solution cannot contain more than 25% of the chemical name acetyl peroxide in a nonviolent solvent.

Health Hazards:

Inhalation Hazard: n/a
Ingestion Hazard: Harmful if swallowed.
Absorption Hazard: Irritating to the skin and eyes.
Hazards to Wildlife: n/a
Decontamination Procedures: Wash away any material with copious amounts of soap and water.
First Aid Procedures: Remove victim to fresh air, call emergency medical care. If not breathing give CPR. If breathing is difficult administer oxygen. Treat for shock.

Fire Hazards:

Flashpoint: 113° F(45° C) *Ignition temperature:* n/a
Flammable Explosive High Range: n/a *Low Range:* n/a
Toxic Products of Combustion: n/a
Other Hazards: May explode upon contact with combustible containers, may explode in a fire.
Possible extinguishing agents: Use alcohol foam, dry chemical, or carbon dioxide. Use water in flooding quantities as a fog.

Reactivity Hazards:

Reactive With: Combustible materials *Other Reactions:* n/a

Corrosivity Hazards:

Corrosive With: n/a *Neutralizing Agent:* n/a

Radioactivity Hazards:

Radiation Emitted: n/a *Other Hazards:* n/a

Recommended Protection for Response Personnel:

Avoid breathing vapors, keep upwind. Structural protective clothing provides limited protection. Wash away any material which may have come into contact with the body with copious amounts of soap and water. Under prolonged exposure to heat or fire, containers of the material may explode. Consider appropriate evacuation.

ACETYLACETONE

DOT Number: UN 2080 *DOT Hazard Class:* Flammable liquid *DOT Guide Number:* 48
Synonyms: diacetylmethane, 2,4-pentanedione
STCC Number: n/a *Reportable Qty:* n/a
MFG Name: Mackenzie Chemical Works *Phone No:* 1-516-234-8600

Physical Description:

Physical Form: Liquid *Color:* Colorless *Odor:* Unpleasant
Other Information: n/a

Chemical Properties:

Specific Gravity: .97 *Vapor Density:* 3.4 *Boiling Point: 284° F(140° C)*
Melting Point: -10° F(-23.3° C) *Vapor Pressure:* n/a *Solubility in water:* Yes
Other Information: Floats and mixes slowly with water. Flammable, irritating vapor is produced.

Health Hazards:

Inhalation Hazard: Dizziness, headache, loss of consciousness.
Ingestion Hazard: Harmful if swallowed.
Absorption Hazard: Irritating to the skin and eyes.
Hazards to Wildlife: n/a
Decontamination Procedures: Wash away any material with copious amounts of soap and water.
First Aid Procedures: Remove victim to fresh air, call emergency medical care. If not breathing give CPR. If breathing is difficult administer oxygen. Treat for shock.

Fire Hazards:

Flashpoint: 93° F(33.8° C) *Ignition temperature:* 644° F(340° C)
Flammable Explosive High Range: 11.6 *Low Range:* 2.4
Toxic Products of Combustion: n/a
Other Hazards: Flashback along vapor trail may occur. Vapors may explode if ignited in an enclosed area.
Possible extinguishing agents: Alcohol foam, dry chemical or carbon dioxide. WATER MAY BE INEFFECTIVE ON FIRES!

Reactivity Hazards:

Reactive With: May dissolve plastics. *Other Reactions:* n/a

Corrosivity Hazards:

Corrosive With: n/a *Neutralizing Agent:* n/a

Radioactivity Hazards:

Radiation Emitted: n/a *Other Hazards:* n/a

Recommended Protection for Response Personnel:

Avoid breathing vapors, keep upwind. Structural protective clothing provides limited protection. Wash away any material which may have come into contact with the body with copious amounts of soap and water. Consider appropriate evacuation.

ACETYLENE

DOT Number: UN 1001 *DOT Hazard Class:* Flammable gas *DOT Guide Number:* 17
Synonyms: ethine, ethyne
STCC Number: 4905701 *Reportable Qty:* n/a
MFG Name: Rexarc Inc. *Phone No:* 1-513-839-4604

Physical Description:

Physical Form: Gas *Color:* Colorless *Odor:* Faint garlic
Other Information: Easily ignites and burns with a sooty flame. Only shipped in cylinders that cannot be used for any other commodity.

Chemical Properties:

Specific Gravity: .9 *Vapor Density: .91* *Boiling Point: -118° F(-83.3° C)*
Melting Point: n/a *Vapor Pressure:* 635 psig at 68° F(20° C) *Solubility in water:* No
Other Information: Under certain conditions, acetylene forms explosive compounds with copper, silver, and mercury. Also forms spontaneously explosive acetylene chloride with chlorine.

Health Hazards:

Inhalation Hazard: Nontoxic but causes asphyxiation.
Ingestion Hazard: n/a
Absorption Hazard: n/a
Hazards to Wildlife: Not harmful to wildlife.
Decontamination Procedures: n/a
First Aid Procedures: Remove victim to fresh air, call emergency medical care. If not breathing give CPR. If breathing is difficult administer oxygen. Treat for shock.

Fire Hazards:

Flashpoint: Gas *Ignition temperature:* 571° F(299.4° C)
Flammable Explosive High Range: 100 *Low Range:* 2.5
Toxic Products of Combustion: n/a
Other Hazards: Under fire conditions, cylinders may violently rupture and rocket. (BLEVE). Easily ignites and burns with a sooty flame.
Possible extinguishing agents: Large quantities of flooding water from as far distance as possible. Stop the flow of gas.

Reactivity Hazards:

Reactive With: Chlorine and oxidizing gases. *Other Reactions:* n/a

Corrosivity Hazards:

Corrosive With: n/a *Neutralizing Agent:* n/a

Radioactivity Hazards:

Radiation Emitted: n/a *Other Hazards:* n/a

Recommended Protection for Response Personnel:

Avoid breathing vapors, keep upwind. Approach fire with caution. If the fire becomes uncontrollable, or if the container is exposed to direct flame, consider appropriate evacuation.

ACRIDINE

DOT Number: UN 2713 *DOT Hazard Class:* n/a *DOT Guide Number:* 32
Synonyms: 10-azaanthracene, dibenzo(b,e)pyridine, benzo(b)quinoline
STCC Number: n/a *Reportable Qty:* n/a
MFG Name: Aldrich Chemical *Phone No:* 1-414-273-3850

Physical Description:

Physical Form: Solid *Color:* Yellow *Odor:* Weak, irritating
Other Information: n/a

Chemical Properties:

Specific Gravity: 1.2 *Vapor Density:* n/a *Boiling Point: 655° F(346.1° C)*
Melting Point: 230° F(110° C) *Vapor Pressure:* 1 mm Hg at 129° F(53.8° C) *Solubility in water:* n/a
Other Information: Sinks in water

Health Hazards:

Inhalation Hazard: Coughing, difficulty in breathing.
Ingestion Hazard: Harmful if swallowed.
Absorption Hazard: Irritating to the skin and eyes.
Hazards to Wildlife: Dangerous to aquatic Life.
Decontamination Procedures: Wash away any material with copious amounts of soap and water.
First Aid Procedures: Remove victim to fresh air, call emergency medical care. If not breathing give CPR. If breathing is difficult administer oxygen. Treat for shock.

Fire Hazards:

Flashpoint: n/a (combustible solid) *Ignition temperature:* n/a
Flammable Explosive High Range: n/a *Low Range:* n/a
Toxic Products of Combustion: Toxic oxides of nitrogen may be formed in fire.
Other Hazards: n/a
Possible extinguishing agents: Water, foam, monoammonium phosphate, and dry chemical.

Reactivity Hazards:

Reactive With: n/a *Other Reactions:* n/a

Corrosivity Hazards:

Corrosive With: n/a *Neutralizing Agent:* n/a

Radioactivity Hazards:

Radiation Emitted: n/a *Other Hazards:* n/a

Recommended Protection for Response Personnel:

Avoid breathing vapors, keep upwind. Structural protective clothing provides limited protection. Wash away any material which may have come into contact with the body with copious amounts of soap and water. Consider appropriate evacuation.

ACROLEIN

DOT Number: UN 1092 *DOT Hazard Class:* Flammable liquid *DOT Guide Number:* 30
Synonyms: acraldehyde, acryaldehyde, acrylic aldehyde, allyl aldehyde, ethylene aldehyde, 2-propenal
STCC Number: 4906410 *Reportable Qty:* 1/.454
MFG Name: Union Carbide *Phone No:* 1-203-794-2000

Physical Description:

Physical Form: Liquid *Color:* Colorless to yellow *Odor:* Pungent
Other Information: Poisonous, polymerizable, used to make other chemicals, plastics, herbicide, pharmaceuticals, and antibacterial agents.

Chemical Properties:

Specific Gravity: .8 *Vapor Density:* 1.9 *Boiling Point: 125° F(51.6° C)*
Melting Point: -125° F(-87.2° C) *Vapor Pressure:* 215 mm Hg at 68° F(20° C)
Solubility in water: Yes *Degree of Solubility:* .07
Other Information: Soluble in water, alcohol, ether and acetone. Vapors are heavier than air. Weighs 7 lbs/3 kg per gallon/3.8 l.

Health Hazards:

Inhalation Hazard: Poisonous if inhaled. *Ingestion Hazard:* Poisonous if swallowed.
Absorption Hazard: Will burn the eyes. Irritating to the skin. *Hazards to Wildlife:* Dangerous to aquatic life.
Decontamination Procedures: Wash away any material with copious amounts of soap and water.
First Aid Procedures: Remove victim to fresh air, call emergency medical care. If not breathing give CPR. If breathing is difficult administer oxygen. Treat for shock.

Fire Hazards:

Flashpoint: -15° F(-26.1° C) *Ignition temperature:* 428° F(220° C)
Flammable Explosive High Range: 31 *Low Range:* 2.8
Toxic Products of Combustion: Unburned acrolein, carbon monoxide, and peroxides.
Other Hazards: Vapors are heavier than air, and may travel a considerable distance to a source of ignition and flash-back.
Possible extinguishing agents: Large quantities of flooding water from as far a distance as far a distance as possible. Use dry chemical, foam, or carbon dioxide.

Reactivity Hazards:

Reactive With: Heat (polymerization can occur) *Other Reactions:* Air, alkaline materials, amines, thiourea, oxidants, light, sulfur dioxide, metal salts, acids, 2-aminoethanol, ethylene diamine, and ethylenimine.

Corrosivity Hazards:

Corrosive With: n/a *Neutralizing Agent:* n/a

Radioactivity Hazards:

Radiation Emitted: n/a *Other Hazards:* n/a

Recommended Protection for Response Personnel:

Avoid breathing vapors, keep upwind. Approach fire with caution. If the fire becomes uncontrollable, or if the container is exposed to direct flame, consider appropriate evacuation. Wear a sealed chemical suit, (polycarbonate, butyl rubber)

ACRYLAMIDE

DOT Number: UN 2074 *DOT Hazard Class:* Flammable liquid *DOT Guide Number:* 55
Synonyms: acrylic acid amine(50%), acrylic amide(50%), propenamide(50%)
STCC Number: 4909183 *Reportable Qty:* 5000/2270
MFG Name: Dow Chemical *Phone No:* 1-415-432-5555

Physical Description:

Physical Form: Liquid *Color:* Colorless *Odor:* Odorless
Other Information: It is used in sewage and waste treatment, to make dyes, adhesives, and for many other uses.

Chemical Properties:

Specific Gravity: 1.05 *Vapor Density:* 2.45 *Boiling Point: Decomposes*
Melting Point: 183° F(83.8° C) *Vapor Pressure:* .007 mm Hg at 68° F(20° C)
Solubility in water: Yes *Degree of Solubility:* 216%
Other Information: Lighter than and insoluble in water. Vapors are heavier than air.

Health Hazards:

Inhalation Hazard: Harmful if inhaled.
Ingestion Hazard: Harmful if swallowed.
Absorption Hazard: Will burn the skin and eyes.
Hazards to Wildlife: Dangerous to aquatic Life.
Decontamination Procedures: Wash away any material with copious amounts of soap and water.
First Aid Procedures: Remove victim to fresh air, call emergency medical care. If not breathing give CPR. If breathing is difficult administer oxygen. Treat for shock.

Fire Hazards:

Flashpoint: n/a *Ignition temperature:* n/a
Flammable Explosive High Range: n/a *Low Range:* n/a
Toxic Products of Combustion: Toxic oxides of nitrogen are produced.
Other Hazards: N/A
Possible extinguishing agents: Apply water from as far a distance as possible. Use alcohol foam, dry chemical, or carbon dioxide.

Reactivity Hazards:

Reactive With: Polymerization may occur at temperatures above 125° F(51.6° C) *Other Reactions:* None

Corrosivity Hazards:

Corrosive With: n/a *Neutralizing Agent:* n/a

Radioactivity Hazards:

Radidtion Emitted: n/a *Other Hazards:* n/a

Recommended Protection for Response Personnel:

Avoid breathing vapors, keep upwind. Wear a sealed chemical suit, (nitrile, chlorinated polyethylene, butyl rubber, neoprene). Wash away any material which may have come into contact with the body with copious amounts of soap and water. Consider appropriate evacuation.

ACRYLIC ACID

DOT Number: UN 2218 *DOT Hazard Class:* Corrosive *DOT Guide Number:* 29
Synonyms: none given
STCC Number: 4931405 *Reportable Qty:* 5000/2270
MFG Name: Rohm and Haas Co. *Phone No:* 1-215-592-3000

Physical Description:

Physical Form: Liquid *Color:* Colorless *Odor:* Irritating
Other Information: Prolonged exposure to heat can cause this material to polymerize.

Chemical Properties:

Specific Gravity: 1.1 *Vapor Density:* 2.5 *Boiling Point: 287° F(141.6° C)*
Melting Point: n/a *Vapor Pressure:* 10 mm Hg at 104° F(40° C) *Solubility in water:* Yes
Other Information: Vapors form explosive mixtures with air.

Health Hazards:

Inhalation Hazard: n/a
Ingestion Hazard: Harmful if swallowed.
Absorption Hazard: Will burn the skin and eyes.
Hazards to Wildlife: n/a
Decontamination Procedures: Wash away any material with copious amounts of soap and water.
First Aid Procedures: Remove victim to fresh air, call emergency medical care. If not breathing give CPR. If breathing is difficult administer oxygen. Do not induce vomiting! Treat for shock.

Fire Hazards:

Flashpoint: 122° F(50° C) *Ignition temperature:* 820° F(437.7° C)
Flammable Explosive High Range: 8.0 *Low Range:* 2.4
Toxic Products of Combustion: Vapor forms explosive mixtures with air.
Other Hazards: If polymerization takes place in a container, violent rupture may take place.
Possible extinguishing agents: Large quantities of flooding water from as far a distance as possible. Use dry chemical, alcohol foam, or carbon dioxide.

Reactivity Hazards:

Reactive With: Air *Other Reactions:* n/a

Corrosivity Hazards:

Corrosive With: Metals or tissue *Neutralizing Agent:* Crushed limestone, soda ash, or lime

Radioactivity Hazards:

Radiation Emitted: n/a *Other Hazards:* n/a

Recommended Protection for Response Personnel:

Avoid breathing vapors, keep upwind. Wear a sealed chemical suit, (polycarbonate, butyl rubber, chlorinated polyethylene). Approach the fire with caution. If the fire becomes uncontrollable, or if the containers are exposed to direct flame, consider appropriate evacuation.

ACRYLONITRILE

DOT Number: UN 1093 *DOT Hazard Class:* Flammable liquid *DOT Guide Number:* 30
Synonyms: vinyl cyanide
STCC Number: 4906420 *Reportable Qty:* 100/45.4
MFG Name: Monsanto Chemical *Phone No:* 1-314-694-1000

Physical Description:

Physical Form: Liquid *Color:* Colorless to pale yellow *Odor:* Strong, pungent
Other Information: Used to make insecticides, plastics, fibers, and other chemicals. Weighs 6.7 lbs/2.9 kg per gallon/3.8 l.

Chemical Properties:

Specific Gravity: .08 *Vapor Density:* 1.8 *Boiling Point: 171° F(77.2° C)*
Melting Point: -117° F(-82.7° C) *Vapor Pressure:* 102 mm Hg at 73.4° F(23° C)
Solubility in water: Yes *Degree of Solubility:* 7.1%
Other Information: May polymerize when contaminated to strong bases or if container is subject to heat.

Health Hazards:

Inhalation Hazard: Poisonous if inhaled. *Ingestion Hazard:* Poisonous if swallowed.
Absorption Hazard: Irritating to the skin and eyes. *Hazards to Wildlife:* Dangerous to aquatic life.
Decontamination Procedures: Wash away any material with copious amounts of soap and water.
First Aid Procedures: Remove victim to fresh air, call emergency medical care. If not breathing give CPR. If breathing is difficult administer oxygen. Treat for shock.

Fire Hazards:

Flashpoint: 32° F(0° C) *Ignition temperature:* 898° F(481.1° C)
Flammable Explosive High Range: 17 *Low Range:* 3.0
Toxic Products of Combustion: Toxic oxides of nitrogen are produced.
Other Hazards: Vapors form explosive mixtures with air. Possible rupture of container if involved in fire.
Possible extinguishing agents: Large quantities of flooding water from as far a distance as possible. Use dry chemical, alcohol foam, or carbon dioxide.

Reactivity Hazards:

Reactive With: Copper and copper alloys
Other Reactions: Penetrates leather, attacks high concentrations of aluminum.

Corrosivity Hazards:

Corrosive With: n/a *Neutralizing Agent:* n/a

Radioactivity Hazards:

Radiation Emitted: n/a *Other Hazards:* n/a

Recommended Protection for Response Personnel:

Avoid breathing vapors, keep upwind. Wear a sealed chemical suit (polycarbonate, butyl rubber, chlorinated polyethylene). Approach the fire with caution. if the fire becomes uncontrollable, or if the containers are exposed to direct flame, consider appropriate evacuation.

ADIPIC ACID

DOT Number: NA 9077 *DOT Hazard Class:* Combustible Solid *DOT Guide Number:* 31
Synonyms: adipinic acid
STCC Number: 4966110 *Reportable Qty:* 5000/2270
Mfg Name: Monsanto Chemical *Phone No:* 1-314-694-1000

Physical Description:

Physical Form: Solid *Color:* White *Odor:* n/a
Other Information: Used to make plastics, foams, and for other uses.

Chemical Properties:

Specific Gravity: 1.37 *Vapor Density:* 5.04 *Boiling Point: 509° F(265° C)*
Melting Point: n/a *Vapor Pressure:* n/a *Solubility in water:* No
Other Information: Adipic acid is a white crystalline solid.

Health Hazards:

Inhalation Hazard: Coughing, difficulty in breathing.
Ingestion Hazard: Harmful if swallowed.
Absorption Hazard: Irritating to the skin and eyes.
Hazards to Wildlife: n/a
Decontamination Procedures: Wash away any material with copious amounts of soap and water.
First Aid Procedures: Remove victim to fresh air, call emergency medical care. If not breathing give CPR. If breathing is difficult administer oxygen. Treat for shock.

Fire Hazards:

Flashpoint: 385° F196.1° C) *Ignition temperature:* 788° F(420° C)
Flammable Explosive High Range: n/a *Low Range:* n/a
Toxic Products of Combustion: n/a
Other Hazards: n/a
Possible extinguishing agents: Large quantities of flooding water from as far a distance as possible. Use dry chemical, alcohol foam, or carbon dioxide.

Reactivity Hazards:

Reactive With: n/a *Other Reactions:* n/a

Corrosivity Hazards:

Corrosive With: n/a *Neutralizing Agent:* n/a

Radioactivity Hazards:

Radiation Emitted: n/a *Other Hazards:* n/a

Recommended Protection for Response Personnel:

Avoid breathing vapors or dust, keep upwind. Wear a sealed chemical suit (viton, butyl rubber, chlorinated polyethylene, pvc, nitrile, neoprene). Wash away any material which may have come into contact with the body with copious amounts of soap and water. Consider appropriate evacuation.

ADIPONITRILE

DOT Number: UN 2205 *DOT Hazard Class:* Flammable liquid *DOT Guide Number:* 55
Synonyms: 1,4-dicyanobutane, tetramethylenecyanide
STCC Number: n/a *Reportable Qty:* None
Mfg Name: Monsanto Chemical *Phone No:* 1-314-694-1000

Physical Description:

Physical Form: Liquid *Color:* Colorless to light yellow *Odor:* Odorless
Other Information: n/a

Chemical Properties:

Specific Gravity: .96 *Vapor Density:* 3.7 *Boiling Point: 554° F(290° C)*
Melting Point: 36° F(2.2° C) *Vapor Pressure:* n/a *Solubility in water:* No
Other Information: Floats on water

Health Hazards:

Inhalation Hazard: n/a
Ingestion Hazard: Nausea and vomiting.
Absorption Hazard: Irritating to the skin and eyes.
Hazards to Wildlife: Dangerous to aquatic life.
Decontamination Procedures: Wash away any material with copious amounts of soap and water.
First Aid Procedures: Remove victim to fresh air, call emergency medical care. If not breathing give CPR. If breathing is difficult administer oxygen. Treat for shock.

Fire Hazards:

Flashpoint: 199° F(92.7° C) *Ignition temperature:* n/a
Flammable Explosive High Range: n/a *Low Range:* 1
Toxic Products of Combustion: Toxic gases are generated in a fire.
Other Hazards: Vapors may explode if ignited in an enclosed area.
Possible extinguishing agents: Use foam, dry chemical, carbon dioxide and water.

Reactivity Hazards:

Reactive With: n/a *Other Reactions:* n/a

Corrosivity Hazards:

Corrosive With: n/a *Neutralizing Agent:* n/a

Radioactivity Hazards:

Radiation Emitted: n/a *Other Hazards:* n/a

Recommended Protection for Response Personnel:

Avoid breathing vapors, keep upwind. Wear a sealed chemical suit, (polycarbonate, butyl rubber). Wash away any material which may have come into contact with the body with copious amounts of water. If the fire becomes uncontrollable, consider appropriate evacuation.

ALDRIN

DOT Number: NA 2761 *DOT Hazard Class:* Poison B *DOT Guide Number:* 55
Synonyms: HHDN, octalene
STCC Number: 4921403 *Reportable Qty:* 1/.0454
Mfg Name: Triangle Chemical Co. *Phone No:* 1-912-743-1548

Physical Description:

Physical Form: Solid *Color:* Brown to white *Odor:* Mild chemical
Other Information: Noncombustible

Chemical Properties:

Specific Gravity: 1.6 *Vapor Density:* 12 *Boiling Point: n/a*
Melting Point: 220° F(104.4° C) *Vapor Pressure:* n/a *Solubility in water:* No
Other Information: In case of damage or leaking from these containers, contact the Pesticide Safety Team Network at 1-800-424-9300.

Health Hazards:

Inhalation Hazard: n/a
Ingestion Hazard: Poisonous if swallowed.
Absorption Hazard: Poisonous if absorbed.
Hazards to Wildlife: n/a
Decontamination Procedures: Wash away any material with copious amounts of soap and water.
First Aid Procedures: Remove victim to fresh air, call emergency medical care. If not breathing give CPR. If breathing is difficult administer oxygen. Treat for shock.

Fire Hazards:

Flashpoint: n/a *Ignition temperature:* n/a
Flammable Explosive High Range: n/a *Low Range:* n/a
Toxic Products of Combustion: n/a
Other Hazards: n/a
Possible extinguishing agents: Use suitable agent for the type of surrounding fire.

Reactivity Hazards:

Reactive With: n/a *Other Reactions:* n/a

Corrosivity Hazards:

Corrosive With: n/a *Neutralizing Agent:* n/a

Radioactivity Hazards:

Radiation Emitted: n/a *Other Hazards:* n/a

Recommended Protection for Response Personnel:

Avoid breathing vapors or dust, keep upwind. Avoid body contact with this material. Wear a sealed chemical suit (polycarbonate, butyl rubber). Wash away any material which may have come into contact with the body with copious amounts of soap and water. Consider appropriate evacuation.

ALLYL ALCOHOL

DOT Number: UN 1098 *DOT Hazard Class:* Flammable liquid *DOT Guide Number:* 57
Synonyms: vinyl carbinol
STCC Number: 4907425 *Reportable Qty:* 100/45.4
Mfg Name: Shell Chemical Corp. *Phone No:* 1-713-241-6161

Physical Description:

Physical Form: Liquid *Color:* Colorless *Odor:* Mustard
Other Information: Vapors are heavier than air, may travel a considerable distance back and flash back.

Chemical Properties:

Specific Gravity: .9 *Vapor Density:* 2.0 *Boiling Point: 206° F(96.6° C)*
Melting Point: -200° F(-128.8° C) *Vapor Pressure:* 840 psia at 68° F(20° C) *Solubility in water:* Yes
Other Information: Lighter, soluble in water.

Health Hazards:

Inhalation Hazard: Poisonous if inhaled.
Ingestion Hazard: Poisonous if swallowed.
Absorption Hazard: Poisonous if absorbed.
Hazards to Wildlife: n/a
Decontamination Procedures: Wash away any material with copious amounts of soap and water.
First Aid Procedures: Remove victim to fresh air, call emergency medical care. If not breathing give CPR. If breathing is difficult administer oxygen. Treat for shock.

Fire Hazards:

Flashpoint: 70° F(21.1° C) *Ignition temperature:* 713° F(378.3° C)
Flammable Explosive High Range: 18 *Low Range:* 2.5
Toxic Products of Combustion: n/a
Other Hazards: n/a
Possible extinguishing agents: Do not extinguish the fire unless the flow can be stopped. Use alcohol foam, dry chemical, carbon dioxide, or flooding quantities of water.

Reactivity Hazards:

Reactive With: Strong Oxidizers *Other Reactions:* n/a

Corrosivity Hazards:

Corrosive With: n/a *Neutralizing Agent:* n/a

Radioactivity Hazards:

Radiation Emitted: n/a *Other Hazards:* n/a

Recommended Protection for Response Personnel:

Avoid breathing vapors, keep upwind. Avoid bodily contact with this material. Wear a sealed chemical suit, (pvc, viton, nitrile, butyl rubber, polycarbonate, chlorinated polyethylene, neoprene). Wash away any material which may have come into contact with the body with copious amounts of soap and water. Consider appropriate evacuation.

ALLYL BROMIDE

DOT Number: UN 1099 *DOT Hazard Class:* Flammable liquid *DOT Guide Number:* 57
Synonyms: broallylene, 2-bromopropane
STCC Number: 4907410 *Reportable Qty:* n/a
Mfg Name: White Chemical Corp. *Phone No:* 1-201-437-0050

Physical Description:

Physical Form: Liquid *Color:* Colorless to light yellow *Odor:* Irritating
Other Information: n/a

Chemical Properties:

Specific Gravity: 1.04 *Vapor Density:* 4.2 *Boiling Point: 160° F(71.1° C)*
Melting Point: n/a *Vapor Pressure:* n/a *Solubility in water:* Slight
Other Information: Heavier, slightly soluble in water.

Health Hazards:

Inhalation Hazard: Headache, dizziness, coughing, loss of consciousness
Ingestion Hazard: n/a
Absorption Hazard: Irritating to the skin and eyes.
Hazards to Wildlife: n/a
Decontamination Procedures: Wash away any material with copious amounts of soap and water.
First Aid Procedures: Remove victim to fresh air, call emergency medical care. If not breathing give CPR. If breathing is difficult administer oxygen. Treat for shock.

Fire Hazards:

Flashpoint: 30° F(-1.1° C) *Ignition temperature:* 563° F(295° C)
Flammable Explosive High Range: 7.3 *Low Range:* 4.4
Toxic Products of Combustion: Poisonous gases may be produced in a fire.
Other Hazards: Flashback along vapor trail may occur, Vapors may explode if ignited in an enclosed area.
Possible extinguishing agents: Use water from as far a distance as possible. Use alcohol foam, dry chemical, or carbon dioxide.

Reactivity Hazards:

Reactive With: n/a *Other Reactions:* n/a

Corrosivity Hazards:

Corrosive With: n/a *Neutralizing Agent:* n/a

Radioactivity Hazards:

Radiation Emitted: n/a *Other Hazards:* n/a

Recommended Protection for Response Personnel:

Avoid breathing vapors, keep upwind. Structural protective clothing provides limited protection. Wash away any material which may have come into contact with the body with copious amounts of soap and water. If the fire becomes uncontrollable, consider appropriate evacuation.

ALLYL CHLORIDE

DOT Number: UN 1100 *DOT Hazard Class:* Flammable liquid *DOT Guide Number:* 57
Synonyms: chloro-2-propene
STCC Number: 4907412 *Reportable Qty:* 1000/454
Mfg Name: Shell Chemical Corp. *Phone No:* 1-713-241-6161

Physical Description:

Physical Form: Liquid *Color:* Colorless to yellow to purple *Odor:* Unpleasant, pungent
Other Information: Weighs 7.8 lbs/3.3 kg per gallon/3.8 l.

Chemical Properties:

Specific Gravity: .9 *Vapor Density:* 2.6 *Boiling Point: 113° F(45° C)*
Melting Point: -209° F(-133.8° C) *Vapor Pressure:* 295 mm Hg at 68° F(20° C)
Solubility in water: No *Degree of Solubility:* .4%
Other Information: n/a

Health Hazards:

Inhalation Hazard: Poisonous if inhaled.
Ingestion Hazard: Poisonous if swallowed.
Absorption Hazard: Poisonous to the skin and eyes.
Hazards to Wildlife: n/a
Decontamination Procedures: Wash away any material with copious amounts of soap and water.
First Aid Procedures: Remove victim to fresh air, call emergency medical care. If not breathing give CPR. If breathing is difficult administer oxygen. Treat for shock.

Fire Hazards:

Flashpoint: -25° F(-31.6° C) *Ignition temperature:* 737° F(391.6° C)
Flammable Explosive High Range: 11.1 *Low Range:* 2.9
Toxic Products of Combustion: n/a
Other Hazards: n/a
Possible extinguishing agents: Do not extinguish the fire unless the flow can be stopped. Use alcohol foam, dry chemical, carbon dioxide, or flooding quantities of water.

Reactivity Hazards:

Reactive With: Strong oxidizers, peroxides, and aluminum *Other Reactions:* n/a

Corrosivity Hazards:

Corrosive With: n/a *Neutralizing Agent:* n/a

Radioactivity Hazards:

Radiation Emitted: n/a *Other Hazards:* n/a

Recommended Protection for Response Personnel:

Avoid breathing vapors, keep upwind. Avoid bodily contact with this material. Wear a sealed chemical suit, (butyl rubber). Wash away any material which may have come into contact with the body with copious amounts of soap and water. If the fire becomes uncontrollable, consider appropriate evacuation.

ALLYL CHLOROFORMATE

DOT Number: UN 1722 *DOT Hazard Class:* Flammable liquid *DOT Guide Number:* 29
Synonyms: ally chlorocarbonate
STCC Number: 4907607 *Reportable Qty:* n/a
Mfg Name: Chemtron Corp. *Phone No:* 1-216-934-6131

Physical Description:

Physical Form: Liquid *Color:* Colorless *Odor:* Pungent
Other Information: Decomposed by water to give hydrochloric acid.

Chemical Properties:

Specific Gravity: 1.1 *Vapor Density:* 4.2 *Boiling Point: 223° F(106.1° C)*
Melting Point: n/a *Vapor Pressure:* n/a *Solubility in water:* No
Other Information: Vapors are heavier than air.

Health Hazards:

Inhalation Hazard: Poisonous if inhaled.
Ingestion Hazard: Poisonous if swallowed.
Absorption Hazard: Poisonous to the skin and eyes.
Hazards to Wildlife: n/a
Decontamination Procedures: Wash away any material with copious amounts of soap and water.
First Aid Procedures: Remove victim to fresh air, call emergency medical care. If not breathing give CPR. If breathing is difficult administer oxygen. Treat for shock.

Fire Hazards:

Flashpoint: 88° F(31.1° C) *Ignition temperature:* n/a
Flammable Explosive High Range: n/a *Low Range:* n/a
Toxic Products of Combustion: Hydrochloric acid, a corrosive material with evolution of heat.
Other Hazards: n/a
Possible extinguishing agents: Use alcohol foam, dry chemical, carbon dioxide, or flooding quantities of water. Do not use water on the material itself.

Reactivity Hazards:

Reactive With: Water *Other Reactions:* n/a

Corrosivity Hazards:

Corrosive With: Metals and tissue *Neutralizing Agent:* Crushed limestone, soda ash, or lime

Radioactivity Hazards:

Radiation Emitted: n/a *Other Hazards:* n/a

Recommended Protection for Response Personnel:

Avoid breathing vapors, keep upwind. Avoid bodily contact with the material itself. Wear a sealed chemical suit, (butyl rubber). Wash away any material which may have come into contact with the body with copious amounts of soap and water. If the fire becomes uncontrollable, consider appropriate evacuation.

ALLYL TRICHLOROSILANE

DOT Number: UN 1724 *DOT Hazard Class:* Corrosive *DOT Guide Number:* 29
Synonyms: allylsilicone trichloride
STCC Number: 4934205 *Reportable Qty:* n/a
Mfg Name: Dow Chemical *Phone No:* 1-517-636-4400

Physical Description:

Physical Form: Liquid *Color:* Colorless *Odor:* Pungent
Other Information: Decomposed by water to give hydrochloric acid.

Chemical Properties:

Specific Gravity: 1.2 *Vapor Density:* 6.05 *Boiling Point: 243° F(117.2° C)*
Melting Point: n/a *Vapor Pressure:* n/a *Solubility in water:* n/a
Other Information: n/a

Health Hazards:

Inhalation Hazard: Coughing, difficulty in breathing.
Ingestion Hazard: Harmful if swallowed.
Absorption Hazard: Will burn the skin and eyes.
Hazards to Wildlife: n/a
Decontamination Procedures: Wash away any material with copious amounts of soap and water.
First Aid Procedures: Remove victim to fresh air, call emergency medical care. If not breathing give CPR. If breathing is difficult administer oxygen. Treat for shock.

Fire Hazards:

Flashpoint: 95° F(35° C) *Ignition temperature:* n/a
Flammable Explosive High Range: n/a *Low Range:* n/a
Toxic Products of Combustion: Hydrochloric acid, a corrosive material with evolution of heat.
Other Hazards: n/a
Possible extinguishing agents: Use dry chemical, dry sand, or carbon dioxide. Do not use water on the material itself.

Reactivity Hazards:

Reactive With: Moisture or water *Other Reactions:* n/a

Corrosivity Hazards:

Corrosive With: Metals and tissues *Neutralizing Agent:* Crushed limestone, soda ash, or lime

Radioactivity Hazards:

Radiation Emitted: n/a *Other Hazards:* n/a

Recommended Protection for Response Personnel:

Avoid breathing vapors, keep upwind. Avoid bodily contact with this material. Wear a sealed chemical suit, (butyl rubber). Wash away any material which may have come into contact with the body with copious amounts of soap and water. Consider appropriate evacuation.

ALUMINUM SULFATE (liquid)

DOT Number: NA 1760 *DOT Hazard Class:* ORM-B *DOT Guide Number:* 60
Synonyms: none given
STCC Number: 4944165 *Reportable Qty:* 5000/2270
Mfg Name: Allied Corp. *Phone No:* 1-201-455-2000

Physical Description:

Physical Form: Liquid *Color:* Colorless *Odor:* Odorless
Other Information: It is used in papermaking, in firefighting foams, in sewage treatment, water purification, and as a fireproofing agent.

Chemical Properties:

Specific Gravity: .837 *Vapor Density:* 3.9 *Boiling Point: n/a*
Melting Point: n/a *Vapor Pressure:* n/a *Solubility in water:* Yes
Other Information: Mixes slowly with water. Weighs 10.9 lbs/4.7 kg per gallon/3.8 l.

Health Hazards:

Inhalation Hazard: Will cause nausea or vomiting.
Ingestion Hazard: Difficulty in breathing.
Absorption Hazard: Irritating to the skin and eyes.
Hazards to Wildlife: Dangerous to Aquatic life.
Decontamination Procedures: Wash away any material with copious amounts of soap and water.
First Aid Procedures: Remove victim to fresh air, call emergency medical care. If not breathing give CPR. If breathing is difficult administer oxygen. Treat for shock.

Fire Hazards:

Flashpoint: n/a *Ignition temperature:* n/a
Flammable Explosive High Range: n/a *Low Range:* n/a
Toxic Products of Combustion: n/a
Other Hazards: n/a
Possible extinguishing agents: Extinguish fire using suitable agent for the type of surrounding fire.

Reactivity Hazards:

Reactive With: n/a *Other Reactions:* n/a

Corrosivity Hazards:

Corrosive With: Aluminum *Neutralizing Agent:* Crushed limestone, soda ash, or lime.

Radioactivity Hazards:

Radiation Emitted: n/a *Other Hazards:* n/a

Recommended Protection for Response Personnel:

Avoid breathing vapors, keep upwind. Wear a sealed chemical suit, (polycarbonate, viton, butyl rubber, chlorinated polyethylene, pvc, nitrile, neoprene). Wash away any material which may have come into contact with the body with copious amounts of soap and water. Consider appropriate evacuation.

ALUMINUM SULFATE (solid)

DOT Number: NA 9078 *DOT Hazard Class:* ORM-E *DOT Guide Number:* 31
Synonyms: cake aluminum, patent aluminum
STCC Number: 4963303 *Reportable Qty:* 5000/2270
Mfg Name: Allied Corp. *Phone No:* 1-201-455-2000

Physical Description:

Physical Form: Solid *Color:* Gray to white *Odor:* Odorless
Other Information: It is used in papermaking, in firefighting foams, in sewage treatment, and water purification.

Chemical Properties:

Specific Gravity: .837 *Vapor Density:* 3.9 *Boiling Point: n/a*
Melting Point: n/a *Vapor Pressure:* n/a *Solubility in water:* Yes
Other Information: Sinks and mixes slowly with water.

Health Hazards:

Inhalation Hazard: Will cause nausea or vomiting.
Ingestion Hazard: Difficulty in breathing.
Absorption Hazard: Irritating to the skin and eyes.
Hazards to Wildlife: Dangerous to Aquatic life.
Decontamination Procedures: Wash away any material with copious amounts of soap and water.
First Aid Procedures: Remove victim to fresh air, call emergency medical care. If not breathing give CPR. If breathing is difficult administer oxygen. Treat for shock.

Fire Hazards:

Flashpoint: n/a *Ignition temperature:* n/a
Flammable Explosive High Range: n/a *Low Range:* n/a
Toxic Products of Combustion: n/a
Other Hazards: n/a
Possible extinguishing agents: Extinguish fire using suitable agent for the type of surrounding fire.

Reactivity Hazards:

Reactive With: Metals in the presence of moisture *Other Reactions:* n/a

Corrosivity Hazards:

Corrosive With: n/a *Neutralizing Agent:* n/a

Radioactivity Hazards:

Radiation Emitted: n/a *Other Hazards:* n/a

Recommended Protection for Response Personnel:

Avoid breathing vapors, keep upwind. Wear a sealed chemical suit (polycarbonate, viton, butyl rubber, chlorinated polyethylene, pvc, nitrile, neoprene). Wash away any material which may have come into contact with the body with copious amounts of soap and water. Consider appropriate evacuation.

ALUMINUM CHLORIDE

DOT Number: UN 1726 *DOT Hazard Class:* Corrosive *DOT Guide Number:* 39
Synonyms: anhydrous aluminum chloride
STCC Number: 4932302 *Reportable Qty:* n/a
Mfg Name: Allied Corp. *Phone No:* 1-201-455-2000

Physical Description:

Physical Form: Solid-powder *Color:* White to gray *Odor:* Pungent
Other Information: Decomposed by water to give hydrochloric acid and aluminum hydroxide with the release of heat.

Chemical Properties:

Specific Gravity: 2.44 *Vapor Density:* 1 *Boiling Point: n/a*
Melting Point: n/a *Vapor Pressure:* n/a *Solubility in water:* n/a
Other Information: n/a

Health Hazards:

Inhalation Hazard: Harmful if inhaled.
Ingestion Hazard: Harmful if swallowed.
Absorption Hazard: Will burn the skin and eyes.
Hazards to Wildlife: n/a
Decontamination Procedures: Wash away any material with copious amounts of soap and water.
First Aid Procedures: Remove victim to fresh air, call emergency medical care. If not breathing give CPR. If breathing is difficult administer oxygen. Treat for shock.

Fire Hazards:

Flashpoint: n/a *Ignition temperature:* n/a
Flammable Explosive High Range: n/a *Low Range:* n/a
Toxic Products of Combustion: n/a
Other Hazards: Flammable hydrogen is formed when wet.
Possible extinguishing agents: Use dry chemical, dry sand or carbon dioxide. Do not use water on the material itself.

Reactivity Hazards:

Reactive With: Moisture or water *Other Reactions:* n/a

Corrosivity Hazards:

Corrosive With: Metals (when wet), and tissues
Neutralizing Agent: Crushed limestone, soda ash, or lime.

Radioactivity Hazards:

Radiation Emitted: n/a *Other Hazards:* n/a

Recommended Protection for Response Personnel:

Avoid breathing vapors, keep upwind. Avoid bodily contact with this material. Wear a sealed chemical suit, (polycarbonate, viton, butyl rubber, nitrile, neoprene, chlorinated polyethylene). Wash away any material which may have come into contact with the body with copious amounts of soap and water. Consider appropriate evacuation.

ALUMINUM NITRATE

DOT Number: UN 1438 *DOT Hazard Class:* Oxidizer *DOT Guide Number:* 35
Synonyms: aluminum salt, nitric acid
STCC Number: 4918701 *Reportable Qty:* n/a *CHEMTREC Phone No:* 1-800-424-9300

Physical Description:

Physical Form: Solid *Color:* White *Odor:* Odorless
Other Information: Used in petroleum refining, dyeing in leather, and numerous other uses.

Chemical Properties:

Specific Gravity: Less than 1 *Vapor Density:* n/a *Boiling Point: n/a*
Melting Point: n/a *Vapor Pressure:* n/a *Solubility in water:* Mixes slowly
Other Information: Toxic oxides of nitrogen are produced in fires involving this material.

Health Hazards:

Inhalation Hazard: Harmful if inhaled.

Ingestion Hazard: Nausea or vomiting.

Absorption Hazard: Irritating to the skin and eyes.

Hazards to Wildlife: n/a

Decontamination Procedures: Wash away any material with copious amounts of soap and water.

First Aid Procedures: Remove victim to fresh air, call emergency medical care. If not breathing give CPR. If breathing is difficult administer oxygen. Treat for shock.

Fire Hazards:

Flashpoint: n/a *Ignition temperature:* n/a

Flammable Explosive High Range: n/a *Low Range:* n/a

Toxic Products of Combustion: Toxic oxides of nitrogen are produced in fires involving this material.

Other Hazards: Material will accelerate combustion of other materials if prolonged exposure to fire in heat may result in an explosion.

Possible extinguishing agents: Flooding quantities of water.

Reactivity Hazards:

Reactive With: Moisture or water *Other Reactions:* Absorbs moisture and dissolves.

Corrosivity Hazards:

Corrosive With: Metals (when wet) *Neutralizing Agent:* Dilute with water.

Radioactivity Hazards:

Radiation Emitted: n/a *Other Hazards:* n/a

Recommended Protection for Response Personnel:

Avoid breathing vapors, keep upwind. Avoid bodily contact with this material. Wear a sealed chemical suit, (pvc). Wash away any material which may have come into contact with the body with copious amounts of soap and water. Consider appropriate evacuation.

2-(2)-AMINOETHOXY(ETHANOL)

DOT Number: NA 1760 *DOT Hazard Class:* Corrosive *DOT Guide Number:* 60
Synonyms: diglycolamine
STCC Number: 4935605 *Reportable Qty:* n/a *CHEMTREC Phone No:* 1-800-424-9300

Physical Description:

Physical Form: Liquid *Color:* Colorless *Odor:* Faint, fish
Other Information: n/a

Chemical Properties:

Specific Gravity: 1.06 *Vapor Density:* n/a *Boiling Point: 430° F(231.1° C)*
Melting Point: n/a *Vapor Pressure:* n/a *Solubility in water:* Yes
Other Information: Sinks slowly and mixes with water.

Health Hazards:

Inhalation Hazard: n/a
Ingestion Hazard: Harmful if swallowed.
Absorption Hazard: Irritating to the skin and eyes.
Hazards to Wildlife: n/a
Decontamination Procedures: Wash away any material with copious amounts of soap and water.
First Aid Procedures: Remove victim to fresh air, call emergency medical care. If not breathing give CPR. If breathing is difficult administer oxygen. Treat for shock.

Fire Hazards:

Flashpoint: 260° F(126.6° C) *Ignition temperature:* n/a
Flammable Explosive High Range: n/a *Low Range:* n/a
Toxic Products of Combustion: Toxic oxides of nitrogen are produced.
Other Hazards: n/a
Possible extinguishing agents: Use water from as far a distance as possible. Use alcohol foam, dry chemical, or carbon dioxide.

Reactivity Hazards:

Reactive With: n/a *Other Reactions:* None

Corrosivity Hazards:

Corrosive With: Tissue *Neutralizing Agent:* n/a

Radioactivity Hazards:

Radiation Emitted: n/a *Other Hazards:* n/a

Recommended Protection for Response Personnel:

Avoid breathing vapors, keep upwind. Wear a sealed chemical suit, (polyethylene). Wash away any material which may have come into contact with the body with copious amounts of soap and water. Consider appropriate evacuation.

AMINOETHYLETHANOLAMINE

DOT Number: UN 1760 *DOT Hazard Class:* Corrosive *DOT Guide Number:* n/a
Synonyms: none given
STCC Number: 4935674 *Reportable Qty:* n/a
Mfg Name: Dow Chemical *Phone No:* 1-517-636-4400

Physical Description:

Physical Form: Liquid *Color:* Colorless *Odor:* Ammonia
Other Information: It is used to make other chemicals.

Chemical Properties:

Specific Gravity: 1.03 *Vapor Density:* 3.6 *Boiling Point: 469° F(242.7° C)*
Melting Point: n/a *Vapor Pressure:* n/a *Solubility in water:* Yes
Other Information: Sinks and mixes with water, vapors are heavier than air.

Health Hazards:

Inhalation Hazard: n/a
Ingestion Hazard: n/a
Absorption Hazard: Will burn the skin and eyes.
Hazards to Wildlife: n/a
Decontamination Procedures: Wash away any material with copious amounts of soap and water.
First Aid Procedures: Remove victim to fresh air, call emergency medical care. If not breathing give CPR. If breathing is difficult administer oxygen. Treat for shock.

Fire Hazards:

Flashpoint: 265° F(129.4° C) *Ignition temperature:* 695° F(368.3° C)
Flammable Explosive High Range: 8 *Low Range:* 1
Toxic Products of Combustion: Toxic oxides of nitrogen
Other Hazards: n/a
Possible extinguishing agents: Use alcohol foam, dry chemical, or carbon dioxide. Do not extinguish the fire unless the flow can be stopped.

Reactivity Hazards:

Reactive With: n/a *Other Reactions:* n/a

Corrosivity Hazards:

Corrosive With: Tissue *Neutralizing Agent:* Water

Radioactivity Hazards:

Radiation Emitted: n/a *Other Hazards:* n/a

Recommended Protection for Response Personnel:

Avoid breathing vapors, keep upwind. Structural protective clothing provides limited protection. Wash away any material which may have come into contact with the body with copious amounts of soap and water. Consider appropriate evacuation.

AMMONIA, ANHYDROUS

DOT Number: UN 1005 *DOT Hazard Class:* Nonflammable gas *DOT Guide Number:* 15
Synonyms: liquid ammonia
STCC Number: 4904210 *Reportable Qty:* 100/45.4
Mfg Name: PPG Industries *Phone No:* 1-412-434-3131

Physical Description:

Physical Form: Gas *Color:* Clear *Odor:* Characteristic ammonia
Other Information: Used as a fertilizer, refrigerant, and the manufacture of other chemicals.

Chemical Properties:

Specific Gravity: .07 *Vapor Density:* .6 *Boiling Point: -28° F(-33.3° C)*
Melting Point: -108° F(-77.7° C) *Vapor Pressure:* Less than 1 atm at 68° F(20° C)
Solubility in water: Yes *Degree of Solubility:* 51%
Other Information: Contact with liquid can cause frostbite. It is soluble in water forming a corrosive liquid. Weighs 6 lbs/2.7 kg per gallon/3.8 l.

Health Hazards:

Inhalation Hazard: Poisonous if inhaled.
Ingestion Hazard: Harmful if swallowed.
Absorption Hazard: Will burn the skin and eyes, cause frostbite.
Hazards to Wildlife: Dangerous to aquatic life.
Decontamination Procedures: Wash away any material with copious amounts of soap and water.
First Aid Procedures: Remove victim to fresh air, call emergency medical care. If not breathing give CPR. If breathing is difficult administer oxygen. Treat for shock.

Fire Hazards:

Flashpoint: n/a *Ignition temperature:* 1204° F(651.1° C)
Flammable Explosive High Range: 28 *Low Range:* 15
Toxic Products of Combustion: n/a
Other Hazards: Presence of oils or other combustibles will increase fire hazard.
Possible extinguishing agents: Use suitable agent for the type of surrounding fire.

Reactivity Hazards:

Reactive With: Strong oxidizers, bleaches, gold, mercury, silver *Other Reactions:* Water forming a corrosive liquid.

Corrosivity Hazards:

Corrosive With: Copper and galvanized surfaces *Neutralizing Agent:* None given

Radioactivity Hazards:

Radiation Emitted: n/a *Other Hazards:* n/a

Recommended Protection for Response Personnel:

Avoid breathing vapors, keep upwind. Avoid bodily contact with this material. Wear a sealed chemical suit, (polycarbonate, butyl rubber, chlorinated Polyethylene, pvc, neoprene). Wash away any material which may have come into contact with the body with copious amounts of soap and water. Consider appropriate evacuation.

AMMONIUM ACETATE

DOT Number: NA 9079 *DOT Hazard Class:* ORM-E *DOT Guide Number:* 31
Synonyms: acetic acid, ammonium salt
STCC Number: 4966708 *Reportable Qty:* 5000/2270
Mfg Name: Allied Corp. *Phone No:* 1-201-455-2000

Physical Description:

Physical Form: Solid *Color:* White *Odor:* Weak ammonia
Other Information: It is used in chemical analysis, pharmaceuticals, preserving foods, and for other uses.

Chemical Properties:

Specific Gravity: 1.17 *Vapor Density:* 2.6 *Boiling Point: n/a*
Melting Point: n/a *Vapor Pressure:* n/a *Solubility in water:* Yes
Other Information: Sinks and mixes with water.

Health Hazards:

Inhalation Hazard: Difficulty in breathing
Ingestion Hazard: Will cause nausea.
Absorption Hazard: Irritating to the skin and eyes.
Hazards to Wildlife: Dangerous to aquatic life.
Decontamination Procedures: Wash away any material with copious amounts of soap and water.
First Aid Procedures: Remove victim to fresh air, call emergency medical care. If not breathing give CPR. If breathing is difficult administer oxygen. Treat for shock.

Fire Hazards:

Flashpoint: n/a *Ignition temperature:* n/a
Flammable Explosive High Range: n/a *Low Range:* n/a
Toxic Products of Combustion: Irritating vapors of ammonia and acetic acid may be formed in fires.
Other Hazards: n/a
Possible extinguishing agents: Extinguish fire using suitable agent for the type of surrounding fire.

Reactivity Hazards:

Reactive With: n/a *Other Reactions:* n/a

Corrosivity Hazards:

Corrosive With: n/a *Neutralizing Agent:* n/a

Radioactivity Hazards:

Radiation Emitted: n/a *Other Hazards:* n/a

Recommended Protection for Response Personnel:

Avoid breathing vapors, keep upwind. Structural protective clothing provides limited protection. Wash away any material which may have come into contact with the body with copious amounts of soap and water. Consider appropriate evacuation.

AMMONIUM BENZOATE

DOT Number: NA 9080 *DOT Hazard Class:* ORM-E *DOT Guide Number:* 31
Synonyms: benzoic acid, ammonium salt
STCC Number: 4966304 *Reportable Qty:* 5000/2270
Mfg Name: Ashland Oil *Phone No:* 1-614-889-3333

Physical Description:

Physical Form: Solid *Color:* White *Odor:* Odorless
Other Information: It is used in medicine, and as a food preservative.

Chemical Properties:

Specific Gravity: 1.26 *Vapor Density:* 2.7 *Boiling Point: n/a*
Melting Point: 388° F(197.7° C) *Vapor Pressure:* n/a *Solubility in water:* Yes
Other Information: Sinks and mixes slowly with water.

Health Hazards:

Inhalation Hazard: Harmful if inhaled.
Ingestion Hazard: Harmful if swallowed.
Absorption Hazard: Irritating to the skin and eyes.
Hazards to Wildlife: n/a
Decontamination Procedures: Wash away any material with copious amounts of soap and water.
First Aid Procedures: Remove victim to fresh air, call emergency medical care. If not breathing give CPR. If breathing is difficult administer oxygen. Treat for shock.

Fire Hazards:

Flashpoint: n/a *Ignition temperature:* n/a
Flammable Explosive High Range: n/a *Low Range:* n/a
Toxic Products of Combustion: Irritating and toxic ammonia gases may be formed in fires.
Other Hazards: n/a
Possible extinguishing agents: Extinguish fire using suitable agent for the type of surrounding fire.

Reactivity Hazards:

Reactive With: Metals and tissue *Other Reactions:* Crushed limestone, soda ash, or lime.

Corrosivity Hazards:

Corrosive With: n/a *Neutralizing Agent:* n/a

Radioactivity Hazards:

Radiation Emitted: n/a *Other Hazards:* n/a

Recommended Protection for Response Personnel:

Avoid breathing vapors, keep upwind. Structural protective clothing provides limited protection. Wash away any material which may have come into contact with the body with copious amounts of soap and water. Consider appropriate evacuation.

AMMONIUM BICARBONATE

DOT Number: NA 9081 *DOT Hazard Class:* ORM-E *DOT Guide Number:* 31
Synonyms: acid ammonia carbonate; ammonium hydrogen carbonate; carbonic acid, monoammonium salt
STCC Number: 4966308 *Reportable Qty:* 5000/2270
Mfg Name: Allied Corp. *Phone No:* 1-201-455-2000

Physical Description:

Physical Form: Solid *Color:* White *Odor:* Weak ammonia
Other Information: It is used to make ammonium compounds, food processing, and for other uses.

Chemical Properties:

Specific Gravity: 1.57 *Vapor Density:* 2.7 *Boiling Point: n/a*
Melting Point: 95° F(35° C) *Vapor Pressure:* n/a *Solubility in water:* Yes
Other Information: Sinks and mixes slowly with water.

Health Hazards:

Inhalation Hazard: Difficulty in breathing.
Ingestion Hazard: Harmful if swallowed.
Absorption Hazard: Irritating to the skin and eyes.
Hazards to Wildlife: n/a
Decontamination Procedures: Wash away any material with copious amounts of soap and water.
First Aid Procedures: Remove victim to fresh air, call emergency medical care. If not breathing give CPR. If breathing is difficult administer oxygen. Treat for shock.

Fire Hazards:

Flashpoint: n/a *Ignition temperature:* n/a
Flammable Explosive High Range: n/a *Low Range:* n/a
Toxic Products of Combustion: Irritating and toxic ammonia gases may be formed in fires.
Other Hazards: Decomposes, but reaction is not explosive.
Possible extinguishing agents: Extinguish fire using suitable agent for the type of surrounding fire.

Reactivity Hazards:

Reactive With: n/a *Other Reactions:* n/a

Corrosivity Hazards:

Corrosive With: n/a *Neutralizing Agent:* n/a

Radioactivity Hazards:

Radiation Emitted: n/a *Other Hazards:* n/a

Recommended Protection for Response Personnel:

Avoid breathing vapors, keep upwind. Structural protective clothing provides limited protection. Wash away any material which may have come into contact with the body with copious amounts of soap and water. Consider appropriate evacuation.

AMMONIUM BIFLUORIDE

DOT Number: UN 1727 *DOT Hazard Class:* Corrosive *DOT Guide Number:* 60
Synonyms: acid ammonium fluoride, ammonium acid fluoride, ammonium hydrogen fluoride
STCC Number: n/a *Reportable Qty:* n/a
Mfg Name: Dow Chemical *Phone No:* 1-517-636-4400

Physical Description:

Physical Form: Solid *Color:* White *Odor:* Odorless
Other Information: n/a

Chemical Properties:

Specific Gravity: 1.5 *Vapor Density:* 2 *Boiling Point: 463° F(239.4° C)*
Melting Point: 258° F(125.5° C) *Vapor Pressure:* n/a *Solubility in water:* Yes
Other Information: Sinks and mixes with water.

Health Hazards:

Inhalation Hazard: Coughing or difficulty in breathing.
Ingestion Hazard: Nausea or vomiting.
Absorption Hazard: Will burn the skin and the eyes.
Hazards to Wildlife: n/a
Decontamination Procedures: Wash away any material with copious amounts of soap and water.
First Aid Procedures: Remove victim to fresh air, call emergency medical care. If not breathing give CPR. If breathing is difficult administer oxygen. Treat for shock.

Fire Hazards:

Flashpoint: n/a *Ignition temperature:* n/a
Flammable Explosive High Range: n/a *Low Range:* n/a
Toxic Products of Combustion: Toxic ammonia and hydrogen fluoride gases may be formed in fire.
Other Hazards: n/a
Possible extinguishing agents: Use water, dry chemical or carbon dioxide. Apply water from as far a distance as possible.

Reactivity Hazards:

Reactive With: Water to form a weak hydrofluoric acid. *Other Reactions:* n/a

Corrosivity Hazards:

Corrosive With: Metals and tissue *Neutralizing Agent:* Crushed limestone, soda ash, or lime.

Radioactivity Hazards:

Radiation Emitted: n/a *Other Hazards:* n/a

Recommended Protection for Response Personnel:

Avoid breathing vapors, keep upwind. Wear a sealed chemical suit, (polycarbonate, butyl rubber). Wash away any material which may have come into contact with the body with copious amounts of soap and water. Consider appropriate evacuation.

AMMONIUM BISULFITE

DOT Number: NA 2693 *DOT Hazard Class:* Corrosive *DOT Guide Number:* 60
Synonyms: ammonium hydrogen sulfite
STCC Number: 4932348 *Reportable Qty:* 5000/2270
Mfg Name: Dow Chemical *Phone No:* 1-517-636-4400

Physical Description:

Physical Form: Liquid *Color:* Colorless to yellow *Odor:* Burning sulfur
Other Information: n/a

Chemical Properties:

Specific Gravity: 1.4 *Vapor Density:* 3.4 *Boiling Point: 302° F(150° C)*
Melting Point: n/a *Vapor Pressure:* n/a *Solubility in water:* Yes
Other Information: Sinks and mixes with water

Health Hazards:

Inhalation Hazard: Exposure data are not available.
Ingestion Hazard: Exposure data are not available.
Absorption Hazard: Exposure data are not available.
Hazards to Wildlife: n/a
Decontamination Procedures: Wash away any material with copious amounts of soap and water.
First Aid Procedures: Remove victim to fresh air, call emergency medical care. If not breathing give CPR. If breathing is difficult administer oxygen. Treat for shock.

Fire Hazards:

Flashpoint: n/a *Ignition temperature:* n/a
Flammable Explosive High Range: n/a *Low Range:* n/a
Toxic Products of Combustion: n/a
Other Hazards: n/a
Possible extinguishing agents: Extinguish fire using suitable agent for the type of surrounding fire.

Reactivity Hazards:

Reactive With: n/a *Other Reactions:* n/a

Corrosivity Hazards:

Corrosive With: Metals and tissue *Neutralizing Agent:* Crushed limestone, soda ash, or lime

Radioactivity Hazards:

Radiation Emitted: n/a *Other Hazards:* n/a

Recommended Protection for Response Personnel:

Avoid breathing vapors, keep upwind. Wear a sealed chemical suit, (polycarbonate, butyl rubber). Wash away any material which may have come into contact with the body with copious amounts of soap and water. Consider appropriate evacuation.

AMMONIUM CARBAMATE

DOT Number: NA 9083 *DOT Hazard Class:* ORM-A *DOT Guide Number:* 31
Synonyms: ammonium aminoformate; carbamic acid, ammonium salt
STCC Number: 4941145 *Reportable Qty:* 5000/2270 *CHEMTREC Phone No:* 1-800-424-9300

Physical Description:

Physical Form: Solid *Color:* White *Odor:* Ammonia
Other Information: It is used as a fertilizer.

Chemical Properties:

Specific Gravity: n/a *Vapor Density:* 2.7 *Boiling Point: 140° F(60° C)*
Melting Point: n/a *Vapor Pressure:* n/a *Solubility in water:* Yes
Other Information: Mixes with water.

Health Hazards:

Inhalation Hazard: Harmful if swallowed.
Ingestion Hazard: Harmful if swallowed.
Absorption Hazard: Irritating to the skin and eyes.
Hazards to Wildlife: Dangerous to aquatic life.
Decontamination Procedures: Wash away any material with copious amounts of soap and water.
First Aid Procedures: Remove victim to fresh air, call emergency medical care. If not breathing give CPR. If breathing is difficult administer oxygen. Treat for shock.

Fire Hazards:

Flashpoint: n/a *Ignition temperature:* n/a
Flammable Explosive High Range: n/a *Low Range:* n/a
Toxic Products of Combustion: n/a
Other Hazards: Moderate fire and explosion hazard when exposed to heat or flame.
Possible extinguishing agents: Extinguish fire using suitable agent for the type of surrounding fire.

Reactivity Hazards:

Reactive With: n/a *Other Reactions:* n/a

Corrosivity Hazards:

Corrosive With: n/a *Neutralizing Agent:* n/a

Radioactivity Hazards:

Radiation Emitted: n/a *Other Hazards:* n/a

Recommended Protection for Response Personnel:

Avoid breathing vapors, keep upwind. Structural protective clothing provides limited protection. Wash away any material which may have come into contact with the body with copious amounts of soap and water. Consider appropriate evacuation.

AMMONIUM CARBONATE

DOT Number: NA 9084 *DOT Hazard Class:* ORM-A *DOT Guide Number:* 31
Synonyms: hartshorn, sal volatile
STCC Number: 4941149 *Reportable Qty:* 5000/2270
Mfg Name: Harshaw Chemical Co. *Phone No:* 1-215-721-8300

Physical Description:

Physical Form: Solid *Color:* White *Odor:* Strong ammonia
Other Information: It is used to make other ammonium compounds, food processing, pharmaceuticals, and for other uses.

Chemical Properties:

Specific Gravity: 1.5 *Vapor Density:* 2.7 *Boiling Point: n/a*
Melting Point: n/a *Vapor Pressure:* n/a *Solubility in water:* Yes
Other Information: Sinks and mixes with water.

Health Hazards:

Inhalation Hazard: Difficulty in breathing.
Ingestion Hazard: Will cause nausea.
Absorption Hazard: Irritating to the skin and eyes.
Hazards to Wildlife: n/a
Decontamination Procedures: Wash away any material with copious amounts of soap and water.
First Aid Procedures: Remove victim to fresh air, call emergency medical care. If not breathing give CPR. If breathing is difficult administer oxygen. Treat for shock.

Fire Hazards:

Flashpoint: n/a *Ignition temperature:* n/a
Flammable Explosive High Range: n/a *Low Range:* n/a
Toxic Products of Combustion: Toxic ammonia gas will be formed in fires.
Other Hazards: Decomposes, but reaction is not explosive.
Possible extinguishing agents: Extinguish fire using suitable agent for the type of surrounding fire.

Reactivity Hazards:

Reactive With: n/a *Other Reactions:* n/a

Corrosivity Hazards:

Corrosive With: n/a *Neutralizing Agent:* n/a

Radioactivity Hazards:

Radiation Emitted: n/a *Other Hazards:* n/a

Recommended Protection for Response Personnel:

Avoid breathing vapors, keep upwind. Structural protective clothing provides limited protection. Wash away any material which may have come into contact with the body with copious amounts of soap and water. Consider appropriate evacuation.

AMMONIUM CHLORIDE

DOT Number: NA 9085 *DOT Hazard Class:* ORM-E *DOT Guide Number:* 31
Synonyms: Amchlor, amchloride, ammoneric, sal ammoniac, salmaic
STCC Number: 4966316 *Reportable Qty:* 5000/2270
Mfg Name: Allied Corp. *Phone No:* 1-201-455-2000

Physical Description:

Physical Form: Solid *Color:* White *Odor:* Odorless
Other Information: It is used to make other ammonium compounds, as a soldering flux, fertilizer, and for many other uses.

Chemical Properties:

Specific Gravity: 1.53 *Vapor Density:* 1.8 *Boiling Point: n/a*
Melting Point: n/a *Vapor Pressure:* n/a *Solubility in water:* Yes
Other Information: Sinks and Mixes slowly with water.

Health Hazards:

Inhalation Hazard: Difficulty in breathing, coughing.
Ingestion Hazard: Will cause nausea.
Absorption Hazard: Irritating to the skin and eyes.
Hazards to Wildlife: n/a
Decontamination Procedures: Wash away any material with copious amounts of soap and water.
First Aid Procedures: Remove victim to fresh air, call emergency medical care. If not breathing give CPR. If breathing is difficult administer oxygen. Treat for shock.

Fire Hazards:

Flashpoint: n/a *Ignition temperature:* n/a
Flammable Explosive High Range: n/a *Low Range:* n/a
Toxic Products of Combustion: Toxic and irritating ammonia and hydrogen chloride gases.
Other Hazards: May volatilize and condense on cool surfaces.
Possible extinguishing agents: Extinguish fire using suitable agent for the type of surrounding fire.

Reactivity Hazards:

Reactive With: n/a *Other Reactions:* n/a

Corrosivity Hazards:

Corrosive With: n/a *Neutralizing Agent:* n/a

Radioactivity Hazards:

Radiation Emitted: n/a *Other Hazards:* n/a

Recommended Protection for Response Personnel:

Avoid breathing vapors, keep upwind. Structural protective clothing provides limited protection. Wash away any material which may have come into contact with the body with copious amounts of soap and water. Consider appropriate evacuation.

AMMONIUM CHROMATE

DOT Number: NA 9086 *DOT Hazard Class:* ORM-E *DOT Guide Number:* 31
Synonyms: diammonium chromate, neutral ammonium chromate
STCC Number: 4966302 *Reportable Qty:* 1000/454
Mfg Name: Allied Corp. *Phone No:* 1-201-455-2000

Physical Description:

Physical Form: Solid *Color:* Yellow *Odor:* Ammonia
Other Information: It is used in dyeing, photography, chemical analysis, and as a corrosion inhibitor.

Chemical Properties:

Specific Gravity: 1.91 *Vapor Density:* 5.2 *Boiling Point: 356° F(180° C)*
Melting Point: n/a *Vapor Pressure:* n/a *Solubility in water:* Yes
Other Information: Sinks and mixes with water.

Health Hazards:

Inhalation Hazard: Poisonous if inhaled.
Ingestion Hazard: Poisonous if swallowed.
Absorption Hazard: Irritating and corrosive to skin, severely irritating to the eyes.
Hazards to Wildlife: Dangerous to aquatic life.
Decontamination Procedures: Wash away any material with copious amounts of soap and water.
First Aid Procedures: Remove victim to fresh air, call emergency medical care. If not breathing give CPR. If breathing is difficult administer oxygen. Treat for shock.

Fire Hazards:

Flashpoint: n/a *Ignition temperature:* n/a
Flammable Explosive High Range: n/a *Low Range:* n/a
Toxic Products of Combustion: Decomposes producing toxic and combustible products.
Other Hazards: Can Explode when Heated or shocked.
Possible extinguishing agents: Extinguish fire using suitable agent for the type of surrounding fire.

Reactivity Hazards:

Reactive With: n/a *Other Reactions:* n/a

Corrosivity Hazards:

Corrosive With: n/a *Neutralizing Agent:* n/a

Radioactivity Hazards:

Radiation Emitted: n/a *Other Hazards:* n/a

Recommended Protection for Response Personnel:

Avoid breathing vapors, keep upwind. Wear a sealed chemical suit. Wash away any material which may have come into contact with the body with copious amounts of soap and water. if the fire becomes uncontrollable, consider appropriate evacuation.

AMMONIUM CITRATE

DOT Number: NA 9087 *DOT Hazard Class:* ORM-E *DOT Guide Number:* 31
Synonyms: citric acid, diammonium citrate, diammonium salt, dibasic
STCC Number: 4966320 *Reportable Qty:* 5000/2270
Mfg Name: Mallinckrodt Inc. *Phone No:* 1-314-895-2000

Physical Description:

Physical Form: Solid *Color:* White *Odor:* Weak ammonia
Other Information: It is used in pharmaceuticals, and in chemical analysis.

Chemical Properties:

Specific Gravity: 1.48 *Vapor Density:* n/a *Boiling Point: n/a*
Melting Point: n/a *Vapor Pressure:* n/a *Solubility in water:* Yes
Other Information: Sinks and mixes with water.

Health Hazards:

Inhalation Hazard: Coughing, difficulty in breathing.
Ingestion Hazard: Will cause nausea.
Absorption Hazard: Irritating to the skin and eyes.
Hazards to Wildlife: n/a
Decontamination Procedures: Wash away any material with copious amounts of soap and water.
First Aid Procedures: Remove victim to fresh air, call emergency medical care. If not breathing give CPR. If breathing is difficult administer oxygen. Treat for shock.

Fire Hazards:

Flashpoint: n/a *Ignition temperature:* n/a
Flammable Explosive High Range: n/a *Low Range:* n/a
Toxic Products of Combustion: Toxic ammonia gas
Other Hazards: n/a
Possible extinguishing agents: Extinguish fire using suitable agent for the type of surrounding fire.

Reactivity Hazards:

Reactive With: n/a *Other Reactions:* n/a

Corrosivity Hazards:

Corrosive With: n/a *Neutralizing Agent:* n/a

Radioactivity Hazards:

Radiation Emitted: n/a *Other Hazards:* n/a

Recommended Protection for Response Personnel:

Avoid breathing vapors, keep upwind. Structural protective clothing provides limited protection. Wash away any material which may have come into contact with the body with copious amounts of soap and water. Consider appropriate evacuation.

AMMONIUM DICHROMATE

DOT Number: UN 1439 *DOT Hazard Class:* Oxidizer *DOT Guide Number:* 35
Synonyms: ammonium dichromate
STCC Number: 4918330 *Reportable Qty:* 1000/454
Mfg Name: Allied Corp. *Phone No:* 1-201-455-2000

Physical Description:

Physical Form: Solid *Color:* Orange to red *Odor:* Odorless
Other Information: n/a

Chemical Properties:

Specific Gravity: 2.13 *Vapor Density:* 8.7 *Boiling Point: n/a*
Melting Point: n/a *Vapor Pressure:* n/a *Solubility in water:* Yes
Other Information: Sinks and mixes with water

Health Hazards:

Inhalation Hazard: Coughing, difficulty in breathing.
Ingestion Hazard: Harmful if swallowed.
Absorption Hazard: Irritating to the skin and eyes.
Hazards to Wildlife: Dangerous to aquatic life.
Decontamination Procedures: Wash away any material with copious amounts of soap and water.
First Aid Procedures: Remove victim to fresh air, call emergency medical care. If not breathing give CPR. If breathing is difficult administer oxygen. Treat for shock.

Fire Hazards:

Flashpoint: n/a *Ignition temperature:* 437° F(225° C)
Flammable Explosive High Range: n/a *Low Range:* n/a
Toxic Products of Combustion: n/a
Other Hazards: May cause fire upon contact with combustible. Containers may explode in a fire.
Possible extinguishing agents: Flood with water.

Reactivity Hazards:

Reactive With: Combustible material e.g. (wood shavings) *Other Reactions:* n/a

Corrosivity Hazards:

Corrosive With: n/a *Neutralizing Agent:* n/a

Radioactivity Hazards:

Radiation Emitted: n/a *Other Hazards:* n/a

Recommended Protection for Response Personnel:

Avoid breathing vapors, keep upwind. Wear a sealed chemical suit, (polycarbonate, butyl rubber). Wash away any material which may have come into contact with the body with copious amounts of soap and water. if the fire becomes uncontrollable, consider appropriate evacuation.

AMMONIUM FLUOBORATE

DOT Number: NA 9088 *DOT Hazard Class:* ORM-B *DOT Guide Number:* 31
Synonyms: ammonium borofluoride, ammonium tetrafluoroborate
STCC Number: 4944125 *Reportable Qty:* 5000/2270
Mfg Name: Allied Corp. *Phone No:* 1-201-455-2000

Physical Description:

Physical Form: Solid *Color:* White *Odor:* Odorless
Other Information: It is used in metal plating and finishing.

Chemical Properties:

Specific Gravity: 1.87 *Vapor Density:* n/a *Boiling Point: 460° F(237.7° C)*
Melting Point: n/a *Vapor Pressure:* n/a *Solubility in water:* Yes
Other Information: Sinks and mixes with water.

Health Hazards:

Inhalation Hazard: Will cause nosebleeds and nausea.
Ingestion Hazard: n/a
Absorption Hazard: Irritating to the skin and eyes.
Hazards to Wildlife: n/a
Decontamination Procedures: Wash away any material with copious amounts of soap and water.
First Aid Procedures: Remove victim to fresh air, call emergency medical care. If not breathing give CPR. If breathing is difficult administer oxygen. Treat for shock.

Fire Hazards:

Flashpoint: n/a *Ignition temperature:* n/a
Flammable Explosive High Range: n/a *Low Range:* n/a
Toxic Products of Combustion: Changes from a solid to a gas at 460° F(237° C), yielding toxic fumes.
Other Hazards: n/a
Possible extinguishing agents: Extinguish fire using suitable agent for the type of surrounding fire.

Reactivity Hazards:

Reactive With: n/a *Other Reactions:* n/a

Corrosivity Hazards:

Corrosive With: Aluminum *Neutralizing Agent:* n/a

Radioactivity Hazards:

Radiation Emitted: n/a *Other Hazards:* n/a

Recommended Protection for Response Personnel:

Avoid breathing vapors, keep upwind. Wear a sealed chemical suit, (viton). Wash away any material which may have come into contact with the body with copious amounts of soap and water. Consider appropriate evacuation.

AMMONIUM FLUORIDE

DOT Number: UN 2505 *DOT Hazard Class:* ORM-B *DOT Guide Number:* 54
Synonyms: natural ammonium fluoride
STCC Number: 4944105 *Reportable Qty:* 100/45.4
Mfg Name: Allied Corp. *Phone No:* 1-201-455-2000

Physical Description:

Physical Form: Solid *Color:* White *Odor:* Odorless
Other Information: It is used in chemical analysis, in brewing, and as a preservative for wood.

Chemical Properties:

Specific Gravity: 1.32 *Vapor Density:* 1.3 *Boiling Point: n/a*
Melting Point: n/a *Vapor Pressure:* n/a *Solubility in water:* Yes
Other Information: Sinks and mixes with water.

Health Hazards:

Inhalation Hazard: Coughing or difficulty in breathing.
Ingestion Hazard: Poisonous if inhaled.
Absorption Hazard: Will burn the skin and eyes.
Hazards to Wildlife: n/a
Decontamination Procedures: Wash away any material with copious amounts of soap and water.
First Aid Procedures: Remove victim to fresh air, call emergency medical care. If not breathing give CPR. If breathing is difficult administer oxygen. Treat for shock.

Fire Hazards:

Flashpoint: n/a *Ignition temperature:* n/a
Flammable Explosive High Range: n/a *Low Range:* n/a
Toxic Products of Combustion: Toxic ammonia and hydrogen fluoride gases may be formed in fires.
Other Hazards: May sublime when hot or condensed on cool surfaces.
Possible extinguishing agents: Extinguish fire using suitable agent for the type of surrounding fire.

Reactivity Hazards:

Reactive With: n/a *Other Reactions:* n/a

Corrosivity Hazards:

Corrosive With: n/a *Neutralizing Agent:* n/a

Radioactivity Hazards:

Radiation Emitted: n/a *Other Hazards:* n/a

Recommended Protection for Response Personnel:

Avoid breathing vapors, keep upwind. Wear a sealed chemical suit, (viton). Wash away any material which may have come into contact with the body with copious amounts of soap and water. Consider appropriate evacuation.

AMMONIUM HYDROXIDE

DOT Number: NA 2672 *DOT Hazard Class:* Corrosive *DOT Guide Number:* 60
Synonyms: aqueous ammonia, ammonia water, household ammonia
STCC Number: 4935234 *Reportable Qty:* 1000/454
Mfg Name: Allied Corp. *Phone No:* 1-201-455-2000

Physical Description:

Physical Form: Liquid *Color:* Colorless *Odor:* Ammonia
Other Information: It is used in cleaning compounds, to make other chemicals, and many other uses.

Chemical Properties:

Specific Gravity: .89 *Vapor Density:* 1 *Boiling Point: n/a*
Melting Point: n/a *Vapor Pressure:* n/a *Solubility in water:* Yes
Other Information: Floats and mixes with water. Irritating vapors are produced. Weighs 7.3 lbs/3.17 kg per gallon/3.8 l.

Health Hazards:

Inhalation Hazard: Nausea, difficulty in breathing, vomiting, loss of consciousness.
Ingestion Hazard: Harmful if swallowed.
Absorption Hazard: Will burn the skin and eyes.
Hazards to Wildlife: Dangerous to aquatic life.
Decontamination Procedures: Wash away any material with copious amounts of soap and water.
First Aid Procedures: Remove victim to fresh air, call emergency medical care. If not breathing give CPR. If breathing is difficult administer oxygen. Treat for shock.

Fire Hazards:

Flashpoint: n/a *Ignition temperature:* n/a
Flammable Explosive High Range: n/a *Low Range:* n/a
Toxic Products of Combustion: n/a
Other Hazards: n/a
Possible extinguishing agents: Extinguish fire using suitable agent for type of surrounding fire.

Reactivity Hazards:

Reactive With: Water to form a mild liberation of heat. *Other Reactions:* n/a

Corrosivity Hazards:

Corrosive With: Copper, copper alloys, aluminum, aluminum alloys, and galvanized surfaces.
Neutralizing Agent: Vinegar or other dilute acids.

Radioactivity Hazards:

Radiation Emitted: n/a *Other Hazards:* n/a

Recommended Protection for Response Personnel:

Avoid breathing vapors, keep upwind. Wear a sealed chemical suit, (neoprene, butyl rubber, pvc, chlorinated polyethylene). Wash away any material which may have come into contact with the body with copious amounts of soap and water. Consider appropriate evacuation.

AMMONIUM NITRATE-PHOSPHATE

DOT Number: UN 2070 *DOT Hazard Class:* Oxidizer *DOT Guide Number:* 43
Synonyms: none given
STCC Number: 4918703 *Reportable Qty:* n/a
Mfg Name: Chevron Chemical *Phone No:* 1-415-233-3737

Physical Description:

Physical Form: Solid *Color:* Grey to white *Odor:* Odorless
Other Information: It is used as a dry fertilizer for formulation.

Chemical Properties:

Specific Gravity: 1.8 *Vapor Density:* 1 *Boiling Point: n/a*
Melting Point: n/a *Vapor Pressure:* n/a *Solubility in water:* Yes
Other Information: Mixes with water

Health Hazards:

Inhalation Hazard: Coughing, difficulty in breathing.
Ingestion Hazard: Harmful if swallowed.
Absorption Hazard: Irritating to the skin and eyes.
Hazards to Wildlife: n/a
Decontamination Procedures: Wash away any material with copious amounts of soap and water.
First Aid Procedures: Remove victim to fresh air, call emergency medical care. If not breathing give CPR. If breathing is difficult administer oxygen. Treat for shock.

Fire Hazards:

Flashpoint: n/a *Ignition temperature:* n/a
Flammable Explosive High Range: n/a *Low Range:* n/a
Toxic Products of Combustion: Toxic oxides of nitrogen are produced during combustion of this material.
Other Hazards: Will increase the intensity of a fire. Containers may explode in a fire.
Possible extinguishing agents: Flood with water.

Reactivity Hazards:

Reactive With: n/a *Other Reactions:* n/a

Corrosivity Hazards:

Corrosive With: Metals *Neutralizing Agent:* n/a

Radioactivity Hazards:

Radiation Emitted: n/a *Other Hazards:* n/a

Recommended Protection for Response Personnel:

Avoid breathing vapors, keep upwind. Wear a sealed chemical suit, (polycarbonate, butyl rubber). Wash away any material which may have come into contact with the body with copious amounts of soap and water. If the fire becomes uncontrollable, consider appropriate evacuation.

AMMONIUM NITRATE

DOT Number: NA 1942 *DOT Hazard Class:* Oxidizer *DOT Guide Number:* 43
Synonyms: nitram
STCC Number: 4918312 *Reportable Qty:* n/a
Mfg Name: Allied Corp. *Phone No:* 1-201-455-2000

Physical Description:

Physical Form: Solid *Color:* White to gray or brown *Odor:* Odorless
Other Information: n/a

Chemical Properties:

Specific Gravity: 1.72 *Vapor Density:* 2.8 *Boiling Point: n/a*
Melting Point: n/a *Vapor Pressure:* n/a *Solubility in water:* Yes
Other Information: Mixes with water

Health Hazards:

Inhalation Hazard: Coughing, difficulty in breathing.
Ingestion Hazard: n/a
Absorption Hazard: n/a
Hazards to Wildlife: n/a
Decontamination Procedures: Wash away any material with copious amounts of soap and water.
First Aid Procedures: Remove victim to fresh air, call emergency medical care. If not breathing give CPR. If breathing is difficult administer oxygen. Treat for shock.

Fire Hazards:

Flashpoint: n/a *Ignition temperature:* n/a
Flammable Explosive High Range: n/a *Low Range:* n/a
Toxic Products of Combustion: Toxic oxides of nitrogen are produced during combustion of this material.
Other Hazards: Containers may explode in a fire.
Possible extinguishing agents: Flood with water.

Reactivity Hazards:

Reactive With: Common organic fuels *Other Reactions:* n/a

Corrosivity Hazards:

Corrosive With: n/a *Neutralizing Agent:* n/a

Radioactivity Hazards:

Radiation Emitted: n/a *Other Hazards:* n/a

Recommended Protection for Response Personnel:

Avoid breathing vapors, keep upwind. Wear a sealed chemical suit, (polycarbonate, butyl rubber). Wash away any material which may have come into contact with the body with copious amounts of soap and water. if the fire becomes uncontrollable, consider appropriate evacuation.

AMMONIUM OXALATE

DOT Number: NA 2449 *DOT Hazard Class:* ORM-A *DOT Guide Number:* 54
Synonyms: diammonium oxalate, diammonium salt; ammonium oxalate hydrate; oxalic acid
STCC Number: 4940350 *Reportable Qty:* 5000/2270
Mfg Name: Allied Corp. *Phone No:* 1-201-455-2000

Physical Description:

Physical Form: Solid *Color:* White *Odor:* Odorless
Other Information: It is used as a regent in chemical analysis, for rust and scale removal, and for many other uses.

Chemical Properties:

Specific Gravity: 1.50 *Vapor Density:* n/a *Boiling Point: n/a*
Melting Point: n/a *Vapor Pressure:* n/a *Solubility in water:* Yes
Other Information: Sinks and mixes slowly with water.

Health Hazards:

Inhalation Hazard: Poisonous if inhaled.
Ingestion Hazard: Poisonous if swallowed.
Absorption Hazard: Poisonous if absorbed.
Hazards to Wildlife: n/a
Decontamination Procedures: Wash away any material with copious amounts of soap and water.
First Aid Procedures: Remove victim to fresh air, call emergency medical care. If not breathing give CPR. If breathing is difficult administer oxygen. Treat for shock.

Fire Hazards:

Flashpoint: n/a *Ignition temperature:* n/a
Flammable Explosive High Range: n/a *Low Range:* n/a
Toxic Products of Combustion: Toxic oxides of nitrogen may be formed in fires.
Other Hazards: n/a
Possible extinguishing agents: Extinguish fire using suitable agent for the type of surrounding fire.

Reactivity Hazards:

Reactive With: n/a *Other Reactions:* n/a

Corrosivity Hazards:

Corrosive With: n/a *Neutralizing Agent:* n/a

Radioactivity Hazards:

Radiation Emitted: n/a *Other Hazards:* n/a

Recommended Protection for Response Personnel:

Avoid breathing vapors, keep upwind. Wear a sealed chemical suit, (polycarbonate, butyl rubber). Wash away any material which may have come into contact with the body with copious amounts of soap and water. Consider appropriate evacuation.

AMMONIUM PERCHLORATE

DOT Number: UN 1442 *DOT Hazard Class:* Oxidizer *DOT Guide Number:* 43
Synonyms: none given
STCC Number: 4918320 *Reportable Qty:* n/a
Mfg Name: Kerr & McGee Chemical *Phone No:* 1-405-270-1313

Physical Description:

Physical Form: Solid *Color:* White *Odor:* Odorless
Other Information: It is used to make rocket propellants, explosives, and other uses.

Chemical Properties:

Specific Gravity: 1.95 *Vapor Density:* 4.1 *Boiling Point: n/a*
Melting Point: n/a *Vapor Pressure:* n/a *Solubility in water:* Yes
Other Information: Sinks and mixes with water.

Health Hazards:

Inhalation Hazard: n/a
Ingestion Hazard: Harmful if swallowed.
Absorption Hazard: Irritating to the skin and eyes.
Hazards to Wildlife: n/a
Decontamination Procedures: Wash away any material with copious amounts of soap and water.
First Aid Procedures: Remove victim to fresh air, call emergency medical care. If not breathing give CPR. If breathing is difficult administer oxygen. Treat for shock.

Fire Hazards:

Flashpoint: n/a *Ignition temperature:* n/a
Flammable Explosive High Range: n/a *Low Range:* n/a
Toxic Products of Combustion: Poisonous gases may be produced in a fire.
Other Hazards: May cause fire and explode upon contact with combustibles.
Possible extinguishing agents: Flood with water.

Reactivity Hazards:

Reactive With: Combustible materials *Other Reactions:* n/a

Corrosivity Hazards:

Corrosive With: n/a *Neutralizing Agent:* n/a

Radioactivity Hazards:

Radiation Emitted: n/a *Other Hazards:* n/a

Recommended Protection for Response Personnel:

Avoid breathing vapors, keep upwind. Wear a sealed chemical suit,(polycarbonate, butyl rubber). Wash away any material which may have come into contact with the body with copious amounts of soap and water. If the fire becomes uncontrollable, consider appropriate evacuation.

AMMONIUM PERSULFATE

DOT Number: UN 1444 *DOT Hazard Class:* Oxidizer *DOT Guide Number:* 35
Synonyms: ammonium peroxydisulfate, diammonium salt; peroxydisulfuric acid
STCC Number: 4918706 *Reportable Qty:* n/a
Mfg Name: Allied Corp. *Phone No:* 1-201-455-2000

Physical Description:

Physical Form: Solid *Color:* Light yellow to colorless *Odor:* Mild unpleasant
Other Information: n/a

Chemical Properties:

Specific Gravity: 1.98 *Vapor Density:* 7.9 *Boiling Point: n/a*
Melting Point: n/a *Vapor Pressure:* n/a *Solubility in water:* Yes
Other Information: Sinks and mixes with water.

Health Hazards:

Inhalation Hazard: Harmful if inhaled.
Ingestion Hazard: Harmful if swallowed.
Absorption Hazard: Irritating to the skin and eyes.
Hazards to Wildlife: Dangerous to aquatic life.
Decontamination Procedures: Wash away any material with copious amounts of soap and water.
First Aid Procedures: Remove victim to fresh air, call emergency medical care. If not breathing give CPR. If breathing is difficult administer oxygen. Treat for shock.

Fire Hazards:

Flashpoint: n/a *Ignition temperature:* n/a
Flammable Explosive High Range: n/a *Low Range:* n/a
Toxic Products of Combustion: Toxic oxides of nitrogen and sulfuric acid fumes may be formed in fires.
Other Hazards: Decomposes with loss of oxygen, and increases intensity of the fire.
Possible extinguishing agents: Extinguish fire using suitable agent for the type of surrounding fire.

Reactivity Hazards:

Reactive With: Grease, wood, and other combustibles *Other Reactions:* n/a

Corrosivity Hazards:

Corrosive With: n/a *Neutralizing Agent:* n/a

Radioactivity Hazards:

Radiation Emitted: n/a *Other Hazards:* n/a

Recommended Protection for Response Personnel:

Avoid breathing vapors, keep upwind. Wear a sealed chemical suit, (polycarbonate, butyl rubber, viton, chlorinated polyethylene, pvc, neoprene). Wash away any material which may have come into contact with the body with copious amounts of soap and water. Consider appropriate evacuation.

AMMONIUM SILICOFLUORIDE

DOT Number: UN 2854   *DOT Hazard Class:* ORM-B   *DOT Guide Number:* 53
Synonyms: ammonia fluosilicate, cryptohalite
STCC Number: n/a   *Reportable Qty:* n/a
Mfg Name: Gallard-Schlesinger   *Phone No:* 1-516-333-5600

Physical Description:

Physical Form: Solid   *Color:* White   *Odor:* Odorless
Other Information: n/a

Chemical Properties:

Specific Gravity: 2   *Vapor Density:* 6.1   *Boiling Point: Decomposes*
Melting Point: n/a   *Vapor Pressure:* n/a   *Solubility in water:* Yes
Other Information: Sinks and mixes slowly with water.

Health Hazards:

Inhalation Hazard: Poisonous if inhaled.
Ingestion Hazard: Poisonous if swallowed.
Absorption Hazard: Poisonous to the skin and eyes.
Hazards to Wildlife: n/a
Decontamination Procedures: Wash away any material with copious amounts of soap and water.
First Aid Procedures: Remove victim to fresh air, call emergency medical care. If not breathing give CPR. If breathing is difficult administer oxygen. Treat for shock.

Fire Hazards:

Flashpoint: n/a   *Ignition temperature:* n/a
Flammable Explosive High Range: n/a   *Low Range:* n/a
Toxic Products of Combustion: Toxic and irritating hydrogen fluoride, silicon, tetrafluoride, and oxides of nitrogen may be formed in fires.
Other Hazards: n/a
Possible extinguishing agents: Water, foam, dry chemical or carbon dioxide.

Reactivity Hazards:

Reactive With: n/a   *Other Reactions:* n/a

Corrosivity Hazards:

Corrosive With: n/a   *Neutralizing Agent:* n/a

Radioactivity Hazards:

Radiation Emitted: n/a   *Other Hazards:* n/a

Recommended Protection for Response Personnel:

Avoid breathing vapors, keep upwind. Wear a sealed chemical suit, (polycarbonate, butyl rubber). Wash away any material which may have come into contact with the body with copious amounts of soap and water. Consider appropriate evacuation.

AMMONIUM SULFAMATE

DOT Number: NA 9188 *DOT Hazard Class:* ORM-E *DOT Guide Number:* 31
Synonyms: ammate, AMS, sulfamic acid
STCC Number: 4966732 *Reportable Qty:* 5000/2270
Mfg Name: Mallinckrodt Inc. *Phone No:* 1-314-895-2000

Physical Description:

Physical Form: Solid *Color:* White to brownish gray *Odor:* Odorless
Other Information: It is used to flameproof fabrics and papers, in weed or brush killing products, and in many other uses.

Chemical Properties:

Specific Gravity: 1 *Vapor Density:* 3.9 *Boiling Point: 392° F (200° C)*
Melting Point: 268° F(131.1° C) *Vapor Pressure:* 0 mm Hg at 68° F (20° C)
Solubility in water: Yes *Degree of Solubility:* 200%
Other Information: Sinks and mixes with water.

Health Hazards:

Inhalation Hazard: Coughing, difficulty in breathing.
Ingestion Hazard: Will cause nausea and vomiting.
Absorption Hazard: Irritating to the skin and eyes.
Hazards to Wildlife: Dangerous to aquatic life.
Decontamination Procedures: Wash away any material with copious amounts of soap and water.
First Aid Procedures: Remove victim to fresh air, call emergency medical care. If not breathing give CPR. If breathing is difficult administer oxygen. Treat for shock.

Fire Hazards:

Flashpoint: n/a *Ignition temperature:* n/a
Flammable Explosive High Range: n/a *Low Range:* n/a
Toxic Products of Combustion: Poisonous gases may be produced in a fire.
Other Hazards: n/a
Possible extinguishing agents: Extinguish fire using suitable agent for type of surrounding fire.

Reactivity Hazards:

Reactive With: n/a *Other Reactions:* n/a

Corrosivity Hazards:

Corrosive With: n/a *Neutralizing Agent:* n/a

Radioactivity Hazards:

Radiation Emitted: n/a *Other Hazards:* n/a

Recommended Protection for Response Personnel:

Avoid breathing vapors, keep upwind. Structural protective clothing will provide limited protection. Wash away any material which may have come into contact with the body with copious amounts of soap and water. Consider appropriate evacuation.

AMMONIUM SULFIDE

DOT Number: UN 2683 *DOT Hazard Class:* Flammable liquid *DOT Guide Number:* 28
Synonyms: ammonium hydrogen sulfide solution
STCC Number: 4909303 *Reportable Qty:* 100/45.4
Mfg Name: J.T. Baker Chemical *Phone No:* 1-201-859-2151

Physical Description:

Physical Form: Liquid *Color:* Colorless to yellow *Odor:* Rotten eggs, ammonia
Other Information: n/a

Chemical Properties:

Specific Gravity: 1 *Vapor Density:* 2.3 *Boiling Point: 104° F (40° C)*
Melting Point: n/a *Vapor Pressure:* n/a *Solubility in water:* Yes
Other Information: Mixes with water, irritating vapor is produced. Weighs 9 lbs/4 kg per gallon/3.8 l.

Health Hazards:

Inhalation Hazard: Coughing, difficulty in breathing, dizziness, Headaches.
Ingestion Hazard: Will cause nausea.
Absorption Hazard: Will burn the skin and eyes.
Hazards to Wildlife: Dangerous to aquatic life.
Decontamination Procedures: Wash away any material with copious amounts of soap and water.
First Aid Procedures: Remove victim to fresh air, call emergency medical care. If not breathing give CPR. If breathing is difficult administer oxygen. Treat for shock.

Fire Hazards:

Flashpoint: 72° F(22.2° C) *Ignition temperature:* n/a
Flammable Explosive High Range: 46 *Low Range:* 4
Toxic Products of Combustion: Toxic hydrogen sulfide. gas and sulfur dioxide gas are produced in a fire.
Other Hazards: n/a
Possible extinguishing agents: Apply water from as far a distance as possible. Use alcohol foam, dry chemical, or carbon dioxide.

Reactivity Hazards:

Reactive With: Moisture giving off hydrogen sulfide. *Other Reactions:* n/a

Corrosivity Hazards:

Corrosive With: Copper, zinc, and their alloys.
Neutralizing Agent: Dilute with water. DO NOT ATTEMPT TO NEUTRALIZE WITH ACID.

Radioactivity Hazards:

Radiation Emitted: n/a *Other Hazards:* n/a

Recommended Protection for Response Personnel:

Avoid breathing vapors, keep upwind. Wear a sealed chemical suit,(polycarbonate, butyl rubber). Wash away any material which may have come into contact with the body with copious amounts of soap and water. Consider appropriate evacuation.

AMMONIUM SULFITE

DOT Number: NA 9090 *DOT Hazard Class:* ORM-E *DOT Guide Number:* 31
Synonyms: none given
STCC Number: 4966332 *Reportable Qty:* 5000/2270
Mfg Name: J.T. Baker Chemical *Phone No:* 1-201-859-2151

Physical Description:

Physical Form: Solid *Color:* White *Odor:* Odorless
Other Information: It is used in the manufacture of other chemicals, medicine, and in photography.

Chemical Properties:

Specific Gravity: 1.1 *Vapor Density:* 4.6 *Boiling Point: n/a*
Melting Point: n/a *Vapor Pressure:* n/a *Solubility in water:* Yes
Other Information: Sinks and mixes slowly with water.

Health Hazards:

Inhalation Hazard: Coughing, difficulty in breathing
Ingestion Hazard: Will cause nausea.
Absorption Hazard: Irritating to the skin and eyes.
Hazards to Wildlife: Dangerous to aquatic life.
Decontamination Procedures: Wash away any material with copious amounts of soap and water.
First Aid Procedures: Remove victim to fresh air, call emergency medical care. If not breathing give CPR. If breathing is difficult administer oxygen. Treat for shock.

Fire Hazards:

Flashpoint: n/a *Ignition temperature:* n/a
Flammable Explosive High Range: n/a *Low Range:* n/a
Toxic Products of Combustion: Toxic sulfur dioxide and oxides of nitrogen may be formed in a fire.
Other Hazards: When dissolved in water or acids, it may liberate flammable sulfide vapors.
Possible extinguishing agents: Extinguish using suitable agent for the type of surrounding fire.

Reactivity Hazards:

Reactive With: Water and acids to form flammable sulfide vapors *Other Reactions:* n/a

Corrosivity Hazards:

Corrosive With: n/a *Neutralizing Agent:* n/a

Radioactivity Hazards:

Radiation Emitted: n/a *Other Hazards:* n/a

Recommended Protection for Response Personnel:

Avoid breathing vapors, keep upwind. Structural protective clothing provides limited protection. Wash away any material which may have come into contact with the body with copious amounts of soap and water. Consider appropriate evacuation.

AMMONIUM TARTRATE

DOT Number: NA 9091 *DOT Hazard Class:* ORM-E *DOT Guide Number:* 31
Synonyms: 1-tartaric acid, ammonium salt
STCC Number: 4966336 *Reportable Qty:* 5000/2270
Mfg Name: Ganes Chemical *Phone No:* 1-212-391-2580

Physical Description:

Physical Form: Solid *Color:* White *Odor:* Odorless
Other Information: It is used to manufacture fabrics, and in medicine.

Chemical Properties:

Specific Gravity: 1.6 *Vapor Density:* 1 *Boiling Point: n/a*
Melting Point: n/a *Vapor Pressure:* n/a *Solubility in water:* Yes
Other Information: Sinks and mixes with water.

Health Hazards:

Inhalation Hazard: Harmful if inhaled
Ingestion Hazard: Harmful if swallowed
Absorption Hazard: Irritating to the skin and eyes
Hazards to Wildlife: Dangerous to aquatic life.
Decontamination Procedures: Wash away any material with copious amounts of soap and water.
First Aid Procedures: Remove victim to fresh air, call emergency medical care. If not breathing give CPR. If breathing is difficult administer oxygen. Treat for shock.

Fire Hazards:

Flashpoint: n/a *Ignition temperature:* n/a
Flammable Explosive High Range: n/a *Low Range:* n/a
Toxic Products of Combustion: Toxic oxides of nitrogen or ammonia gas may be formed in fires.
Other Hazards: Combustible solid.
Possible extinguishing agents: Extinguish fire using suitable agent for the type of surrounding fire.

Reactivity Hazards:

Reactive With: n/a *Other Reactions:* n/a

Corrosivity Hazards:

Corrosive With: n/a *Neutralizing Agent:* n/a

Radioactivity Hazards:

Radiation Emitted: n/a *Other Hazards:* n/a

Recommended Protection for Response Personnel:

Avoid breathing vapors, keep upwind. Wear appropriate chemical clothing, wash away any material which may have come into contact with the body with copious amounts of soap and water. Consider appropriate evacuation.

AMMONIUM THIOCYANATE

DOT Number: NA 9092 *DOT Hazard Class:* ORM-E *DOT Guide Number:* 31
Synonyms: ammonium rhodanate; ammonium rhodanide; ammonium sulfocyanate; thiocyanic acid, ammonium salt
STCC Number: 4966738 *Reportable Qty:* 5000/2270
Mfg Name: J.T. Baker Chemical *Phone No:* 1-201-859-2151

Physical Description:

Physical Form: Solid *Color:* White *Odor:* Odorless
Other Information: It is used in chemical analysis, photography, as a fertilizer, and for many other uses.

Chemical Properties:

Specific Gravity: 1.1 *Vapor Density:* 2.6 *Boiling Point: n/a*
Melting Point: n/a *Vapor Pressure:* n/a *Solubility in water:* Yes
Other Information: Sinks and mixes with water.

Health Hazards:

Inhalation Hazard: Coughing, difficulty in breathing.
Ingestion Hazard: Nausea or vomiting.
Absorption Hazard: Irritating to the skin and eyes
Hazards to Wildlife: Dangerous to aquatic life.
Decontamination Procedures: Wash away any material with copious amounts of soap and water.
First Aid Procedures: Remove victim to fresh air, call emergency medical care. If not breathing give CPR. If breathing is difficult administer oxygen. Treat for shock.

Fire Hazards:

Flashpoint: n/a *Ignition temperature:* n/a
Flammable Explosive High Range: n/a *Low Range:* n/a
Toxic Products of Combustion: Ammonia hydrogen sulfide, nitrogen cyanide, and oxides of nitrogen.
Other Hazards: Solid may be combustible.
Possible extinguishing agents: Extinguish fire using suitable agent for the type of surrounding fire.

Reactivity Hazards:

Reactive With: n/a *Other Reactions:* n/a

Corrosivity Hazards:

Corrosive With: n/a *Neutralizing Agent:* n/a

Radioactivity Hazards:

Radiation Emitted: n/a *Other Hazards:* n/a

Recommended Protection for Response Personnel:

Avoid breathing vapors, keep upwind. Structural protective clothing provides limited protection. Wash away any material which may have come into contact with the body with copious amounts of soap and water. Consider appropriate evacuation.

AMMONIUM THIOCYANATE LIQUOR

DOT Number: NA 9188 *DOT Hazard Class:* ORM-E *DOT Guide Number:* 31
Synonyms: ammonium rhodanate; ammonium rhodanide; ammonium sulfocyanate; thiocyanic acid, ammonium salt
STCC Number: 4966744 *Reportable Qty:* 5000/2270
Mfg Name: J.T. Baker Chemical *Phone No:* 1-201-859-2151

Physical Description:

Physical Form: Liquid *Color:* Colorless *Odor:* n/a
Other Information: It is used to make other chemicals, fertilizers, and for many other uses.

Chemical Properties:

Specific Gravity: 1.1 *Vapor Density:* 2.6 *Boiling Point: 239° F(115° C)*
Melting Point: n/a *Vapor Pressure:* n/a *Solubility in water:* Yes
Other Information: Sinks and mixes with water.

Health Hazards:

Inhalation Hazard: Coughing, difficulty in breathing.
Ingestion Hazard: Nausea or vomiting.
Absorption Hazard: Irritating to the skin and eyes
Hazards to Wildlife: Dangerous to aquatic life.
Decontamination Procedures: Wash away any material with copious amounts of soap and water.
First Aid Procedures: Remove victim to fresh air, call emergency medical care. If not breathing give CPR. If breathing is difficult administer oxygen. Treat for shock.

Fire Hazards:

Flashpoint: n/a *Ignition temperature:* n/a
Flammable Explosive High Range: n/a *Low Range:* n/a
Toxic Products of Combustion: Ammonia hydrogen sulfide, nitrogen cyanide, and oxides of nitrogen.
Other Hazards: n/a
Possible extinguishing agents: Extinguish fire using suitable agent for the type of surrounding fire.

Reactivity Hazards:

Reactive With: n/a *Other Reactions:* n/a

Corrosivity Hazards:

Corrosive With: n/a *Neutralizing Agent:* n/a

Radioactivity Hazards:

Radiation Emitted: n/a *Other Hazards:* n/a

Recommended Protection for Response Personnel:

Avoid breathing vapors, keep upwind. Structural protective clothing provides limited protection. Wash away any material which may have come into contact with the body with copious amounts of soap and water. Consider appropriate evacuation.

AMMONIUM THIOSULFATE (liquid)

DOT Number: NA 9188 *DOT Hazard Class:* ORM-E *DOT Guide Number:* 31
Synonyms: ammonium hyposulfite
STCC Number: 4966756 *Reportable Qty:* 5000/2270
Mfg Name: Allied Corp *Phone No:* 1-201-455-2000

Physical Description:

Physical Form: Liquid *Color:* Colorless *Odor:* Ammonia
Other Information: It is used in photography and for many other uses.

Chemical Properties:

Specific Gravity: 2 *Vapor Density:* 2.6 *Boiling Point: n/a*
Melting Point: n/a *Vapor Pressure:* n/a *Solubility in water:* Yes
Other Information: Sinks and mixes with water.

Health Hazards:

Inhalation Hazard: Difficulty in breathing.
Ingestion Hazard: Harmful if swallowed.
Absorption Hazard: Irritating to the skin and eyes
Hazards to Wildlife: n/a
Decontamination Procedures: Wash away any material with copious amounts of soap and water.
First Aid Procedures: Remove victim to fresh air, call emergency medical care. If not breathing give CPR. If breathing is difficult administer oxygen. Treat for shock.

Fire Hazards:

Flashpoint: n/a *Ignition temperature:* n/a
Flammable Explosive High Range: n/a *Low Range:* n/a
Toxic Products of Combustion: Toxic ammonia, hydrogen sulfide, and oxides of nitrogen and sulfur may be formed in fires.
Other Hazards: n/a
Possible extinguishing agents: Extinguish fire using suitable agent for the type of surrounding fire.

Reactivity Hazards:

Reactive With: n/a *Other Reactions:* n/a

Corrosivity Hazards:

Corrosive With: n/a *Neutralizing Agent:* n/a

Radioactivity Hazards:

Radiation Emitted: n/a *Other Hazards:* n/a

Recommended Protection for Response Personnel:

Avoid breathing vapors, keep upwind. Wear appropriate chemical clothing. Wash away any material which may have come into contact with the body with copious amounts of soap and water. Consider appropriate evacuation.

AMMONIUM THIOSULFATE (solid)

DOT Number: NA 9093 *DOT Hazard Class:* ORM-E *DOT Guide Number:* 31
Synonyms: ammonium hyposulfite
STCC Number: 4966750 *Reportable Qty:* 5000/2270
Mfg Name: Allied Corp. *Phone No:* 1-201-455-2000

Physical Description:

Physical Form: Solid *Color:* White *Odor:* Ammonia
Other Information: It is used in photography, in chemical analysis, and for many other uses.

Chemical Properties:

Specific Gravity: 2 *Vapor Density:* 2.6 *Boiling Point: n/a*
Melting Point: n/a *Vapor Pressure:* n/a *Solubility in water:* Yes
Other Information: Sinks and mixes with water.

Health Hazards:

Inhalation Hazard: Difficulty in breathing.
Ingestion Hazard: Harmful if swallowed.
Absorption Hazard: Irritating to the skin and eyes
Hazards to Wildlife: n/a
Decontamination Procedures: Wash away any material with copious amounts of soap and water.
First Aid Procedures: Remove victim to fresh air, call emergency medical care. If not breathing give CPR. If breathing is difficult administer oxygen. Treat for shock.

Fire Hazards:

Flashpoint: n/a *Ignition temperature:* n/a
Flammable Explosive High Range: n/a *Low Range:* n/a
Toxic Products of Combustion: Toxic ammonia, hydrogen sulfide, and oxides of nitrogen and sulfur may be formed in fires.
Other Hazards: n/a
Possible extinguishing agents: Extinguish fire using suitable agent for the type of surrounding fire.

Reactivity Hazards:

Reactive With: n/a *Other Reactions:* n/a

Corrosivity Hazards:

Corrosive With: n/a *Neutralizing Agent:* n/a

Radioactivity Hazards:

Radiation Emitted: n/a *Other Hazards:* n/a

Recommended Protection for Response Personnel:

Avoid breathing vapors, keep upwind. Wear appropriate chemical clothing. Wash away any material which may have come into contact with the body with copious amounts of soap and water. Consider appropriate evacuation.

AMYL ACETATE

DOT Number: UN 1104
DOT Hazard Class: Flammable liquid
DOT Guide Number: 26
Synonyms: pentanol acetate
STCC Number: 4909111
Reportable Qty: 5000/2270
Mfg Name: Union Carbide
Phone No: 1-203-794-2000

Physical Description:

Physical Form: Liquid
Color: Colorless
Odor: Banana
Other Information: Vapors are heavier than air. Weighs 7.2 lbs/3.1 kg per gallon/3.8 l.

Chemical Properties:

Specific Gravity: .9
Vapor Density: 4.5
Boiling Point: 295° F(146.1° C)
Melting Point: -95° F(-70.5° C)
Vapor Pressure: 7 mm Hg at 68° F (20° C)
Solubility in water: Yes
Degree of Solubility: .2%
Other Information: n/a

Health Hazards:

Inhalation Hazard: Nausea, headache, dizziness.
Ingestion Hazard: n/a
Absorption Hazard: Irritating to the skin and eyes.
Hazards to Wildlife: Dangerous to aquatic life.
Decontamination Procedures: Wash away any material with copious amounts of soap and water.
First Aid Procedures: Remove victim to fresh air, call emergency medical care. If not breathing give CPR. If breathing is difficult administer oxygen. Treat for shock.

Fire Hazards:

Flashpoint: 60° F(15.5° C)
Ignition temperature: 680° F(360° C)
Flammable Explosive High Range: 7.5
Low Range: 1.1
Toxic Products of Combustion: n/a
Other Hazards: Flashback along vapor trail may occur.
Possible extinguishing agents: Use alcohol foam, dry chemical, carbon dioxide, and water from as far a distance as possible.

Reactivity Hazards:

Reactive With: Strong oxidizers, alkalis, and acids.
Other Reaction: n/a

Corrosivity Hazards:

Corrosive With: n/a
Neutralizing Agent: n/a

Radioactivity Hazards:

Radiation Emitted: n/a
Other Hazards: n/a

Recommended Protection for Response Personnel:

Avoid breathing vapors, keep upwind. Avoid bodily contact with this material. Structural protective clothing will provide limited protection. Wash away any material which may have come into contact with the body with copious amounts of soap and water. Consider appropriate evacuation.

AMYL ALCOHOL

DOT Number: UN 1105 *DOT Hazard Class:* Flammable liquid *DOT Guide Number:* 26
Synonyms: Pentanol
STCC Number: n/a *Reportable Qty:* n/a
Mfg Name: Ashland Chemical Co. *Phone No:* 1-614-889-3333

Physical Description:

Physical Form: Liquid *Color:* Colorless *Odor:* Mild, sweet
Other Information: Floats on water.

Chemical Properties:

Specific Gravity: .8 *Vapor Density:* 3.0 *Boiling Point: 280° F(137.7° C)*
Melting Point: n/a *Vapor Pressure:* n/a *Solubility in water:* Slight
Other Information: n/a

Health Hazards:

Inhalation Hazard: Coughing, nausea, headaches, difficulty in breathing
Ingestion Hazard: Harmful if swallowed.
Absorption Hazard: Not an irritant to the skin.
Hazards to Wildlife: n/a
Decontamination Procedures: Wash away any material with copious amounts of soap and water.
First Aid Procedures: Remove victim to fresh air, call emergency medical care. If not breathing give CPR. If breathing is difficult administer oxygen. Treat for shock.

Fire Hazards:

Flashpoint: 91° F(32.7° C) *Ignition temperature:* 572° F(300° C)
Flammable Explosive High Range: 10 *Low Range:* 1.2
Toxic Products of Combustion: n/a
Other Hazards: n/a
Possible extinguishing agents: Use alcohol foam, dry chemical, carbon dioxide, and water from as far a distance as possible.

Reactivity Hazards:

Reactive With: n/a *Other Reaction: n/a*

Corrosivity Hazards:

Corrosive With: n/a *Neutralizing Agent:* n/a

Radioactivity Hazards:

Radiation Emitted: n/a *Other Hazards:* n/a

Recommended Protection for Response Personnel:

Avoid breathing vapors, keep upwind. Avoid bodily contact with this material. Structural protective clothing provides limited protection. Wash away any material which may have come into contact with the body with with copious amounts of soap and water. Consider appropriate evacuation.

AMYL CHLORIDE

DOT Number: UN 1107 *DOT Hazard Class:* Flammable liquid *DOT Guide Number:* 26
Synonyms: chloride of amyl, pentyl chloride
STCC Number: 4909115 *Reportable Qty:* n/a
Mfg Name: Eastman Organic *Phone No:* 1-800-455-6325

Physical Description:

Physical Form: Liquid *Color:* Colorless to purple *Odor:* Pleasant
Other Information: n/a

Chemical Properties:

Specific Gravity: .9 *Vapor Density:* 3.7 *Boiling Point: 223° F(106.1° C)*
Melting Point: n/a *Vapor Pressure:* n/a *Solubility in water:* No
Other Information: n/a

Health Hazards:

Inhalation Hazards: n/a
Ingestion Hazard: n/a
Absorption Hazard: Irritating to the skin and eyes.
Hazards to Wildlife: n/a
Decontamination Procedures: Wash away any material with copious amounts of soap and water.
First Aid Procedures: Remove victim to fresh air, call emergency medical care. If not breathing give CPR. If breathing is difficult administer oxygen. Treat for shock.

Fire Hazards:

Flashpoint: 55° F(12.7° C) *Ignition temperature:* 500° F(260° C)
Flammable Explosive High Range: 8.6 *Low Range:* 1.6
Toxic Products of Combustion: n/a
Other Hazards: Possible flashback
Possible extinguishing agents: Use alcohol foam, dry chemical, carbon dioxide, and water from as far a distance as possible.

Reactivity Hazards:

Reactive With: n/a *Other Reaction:* n/a

Corrosivity Hazards:

Corrosive With: n/a *Neutralizing Agent:* n/a

Radioactivity Hazards:

Radiation Emitted: n/a *Other Hazards:* n/a

Recommended Protection for Response Personnel:

Avoid breathing vapors, keep upwind. Avoid bodily contact with this material. Structural protective clothing provides limited protection. Wash away any material which may have come into contact with the body with copious amounts of soap and water. Consider appropriate evacuation.

AMYL MERCAPTAN

DOT Number: UN 1111 *DOT Hazard Class:* Flammable liquid *DOT Guide Number:* 27
Synonyms: 1-pentanethiol
STCC Number: 4909119 *Reportable Qty:* n/a
Mfg Name: Eastman Organic *Phone No:* 1-800-455-6325

Physical Description:

Physical Form: Liquid *Color:* Colorless to light yellow *Odor:* Offensive
Other Information: n/a

Chemical Properties:

Specific Gravity: .8 *Vapor Density:* 3.59 *Boiling Point: 260° F(126° C)*
Melting Point: n/a *Vapor Pressure:* n/a *Solubility in water:* No
Other Information: Floats on water, flammable vapor is produced.

Health Hazards:

Inhalation Hazard: n/a
Ingestion Hazard: n/a
Absorption Hazard: Irritating to the skin and eyes.
Hazards to Wildlife: n/a
Decontamination Procedures: Wash away any material with copious amounts of soap and water.
First Aid Procedures: Remove victim to fresh air, call emergency medical care. If not breathing give CPR. If breathing is difficult administer oxygen. Treat for shock.

Fire Hazards:

Flashpoint: 65° F(18.3° C) *Ignition temperature:* n/a
Flammable Explosive High Range: n/a *Low Range:* n/a
Toxic Products of Combustion: Sulfur dioxide gas is formed.
Other Hazards: Flammable vapors are produced.
Possible extinguishing agents: Use alcohol foam, dry chemical, carbon dioxide, and water from as far a distance as possible.

Reactivity Hazards:

Reactive With: n/a *Other Reactions:* n/a

Corrosivity Hazards:

Corrosive With: n/a *Neutralizing Agent:* n/a

Radioactivity Hazards:

Radiation Emitted: n/a *Other Hazards:* n/a

Recommended Protection for Response Personnel:

Avoid breathing vapors, keep upwind. Avoid bodily contact with this material. Structural protective clothing provides limited protection. Wash away any material which may have come into contact with the body with copious amounts of soap and water. Consider appropriate evacuation.

AMYL METHYL KETONE

DOT Number: UN 1110 *DOT Hazard Class:* Combustible liquid *DOT Guide Number:* 26
Synonyms: 2-heptanone, 2-ketoheptane, methyl amyl ketone, methyl pentyl ketone, pentyl methyl ketone
STCC Number: n/a *Reportable Qty:* n/a
Mfg Name: Aldrich Chemical *Phone No:* 1-414-273-3850

Physical Description:

Physical Form: Liquid *Color:* White *Odor:* Penetrating, fruity
Other Information: n/a

Chemical Properties:

Specific Gravity: .82 *Vapor Density:* 3.9 *Boiling Point: 304° F(151° C)*
Melting Point: -31° F(-35° C) *Vapor Pressure:* n/a *Solubility in water:* Yes
Other Information: Floats and mixes slowly with water.

Health Hazards:

Inhalation Hazard: Dizziness, headache, difficulty in breathing.
Ingestion Hazard: Nausea or vomiting.
Absorption Hazard: Irritating to the skin and eyes.
Hazards to Wildlife: n/a
Decontamination Procedures: Wash away any material with copious amounts of soap and water.
First Aid Procedures: Remove victim to fresh air, call emergency medical care. If not breathing give CPR. If breathing is difficult administer oxygen. Treat for shock.

Fire Hazards:

Flashpoint: 102° F(38.8° C) *Ignition temperature:* 740° F(393° C)
Flammable Explosive High Range: 7.9 *Low Range:* 1.1
Toxic Products of Combustion: n/a
Other Hazards: n/a
Possible extinguishing agents: Alcohol foam, dry chemical or carbon dioxide.

Reactivity Hazards:

Reactive With: Will attack some forms of plastic. *Other Reactions:* n/a

Corrosivity Hazards:

Corrosive With: n/a *Neutralizing Agent:* n/a

Radioactivity Hazards:

Radiation Emitted: n/a *Other Hazards:* n/a

Recommended Protection for Response Personnel:

Avoid breathing vapors, keep upwind. Structural protective clothing provides limited protection. Wash away any material which may have come into contact with the body with copious amounts of soap and water. Consider appropriate evacuation.

AMYL NITRATE

DOT Number: NA 1112 *DOT Hazard Class:* Combustible liquid *DOT Guide Number:* 26
Synonyms: diesel ignition improver, mixed primary amyl nitrates
STCC Number: 4913107 *Reportable Qty:* n/a
Mfg Name: Ethyl Chemical Corp. *Phone No:* 1-804-788-5000

Physical Description:

Physical Form: Liquid *Color:* Colorless to yellowish *Odor:* Ethereal
Other Information: Used as an additive to diesel fuels.

Chemical Properties:

Specific Gravity: 1 *Vapor Density:* 4.6 *Boiling Point: 300° F(148° C)*
Melting Point: n/a *Vapor Pressure:* n/a *Solubility in water:* No
Other Information: May Floats or sinks in water.

Health Hazards:

Inhalation Hazard: May cause headaches.
Ingestion Hazard: May cause nausea or headache.
Absorption Hazard: Irritating to the skin and the eyes.
Hazards to Wildlife: n/a
Decontamination Procedures: Wash away any material with copious amounts of soap and water.
First Aid Procedures: Remove victim to fresh air, call emergency medical care. If not breathing give CPR. If breathing is difficult administer oxygen. Treat for shock.

Fire Hazards:

Flashpoint: 118° F(47.7° C) *Ignition temperature:* 410° F(210° C)
Flammable Explosive High Range: n/a *Low Range:* n/a
Toxic Products of Combustion: Toxic oxides of nitrogen are produced during combustion of this material.
Other Hazards: May explode if exposed to heat or flames.
Possible extinguishing agents: Do not extinguish the fire unless the flow can be stopped, apply water from as far as possible. Use foam, dry chemical, or carbon dioxide.

Reactivity Hazards:

Reactive With: Wood or combustible materials to form a combustible mixture.
Other Reactions: Liquid will attack some plastics.

Corrosivity Hazards:

Corrosive With: n/a *Neutralizing Agent:* n/a

Radioactivity Hazards:

Radiation Emitted: n/a *Other Hazards:* n/a

Recommended Protection for Response Personnel:

Avoid breathing vapors, keep upwind. Avoid bodily contact with this material. Structural protective clothing will provide limited protection. Wash away any material which may come into contact with copious amounts of soap and water. Consider appropriate evacuation.

AMYL NITRITE

DOT Number: UN 1113 *DOT Hazard Class:* Flammable liquid *DOT Guide Number:* 26
Synonyms: isopentyl nitrite, 3-methylbutyl nitrite
STCC Number: 4909121 *Reportable Qty:* n/a
Mfg Name: James Alexander Corp. *Phone No:* 1-201-362-9266

Physical Description:

Physical Form: Liquid *Color:* Colorless to light yellow *Odor:* Pleasant, fruity
Other Information: Used in medicine, and to make other chemicals.

Chemical Properties:

Specific Gravity: .87 *Vapor Density:* 4 *Boiling Point: 210° F(98.8° C)*
Melting Point: n/a *Vapor Pressure:* n/a *Solubility in water:* n/a
Other Information: Floats on water. Poisonous gas is produced on contact with water.

Health Hazards:

Inhalation Hazard: Poisonous if inhaled
Ingestion Hazard: Will cause dizziness, headache, loss of consciousness.
Absorption Hazard: Will burn the skin and eyes.
Hazards to Wildlife: n/a
Decontamination Procedures: Wash away any material with copious amounts of soap and water.
First Aid Procedures: Remove victim to fresh air, call emergency medical care. If not breathing give CPR. If breathing is difficult administer oxygen. Treat for shock.

Fire Hazards:

Flashpoint: 0° F(-17.7° C) *Ignition temperature:* 410° F(210° C)
Flammable Explosive High Range: n/a *Low Range:* n/a
Toxic Products of Combustion: Toxic oxides of nitrogen are produced during combustion of this material.
Other Hazards: Flashback along vapor trail may occur, vapors may explode if ignited in an enclosed area.
Possible extinguishing agents: Do not extinguish the fire unless the flow can be stopped. Apply water from as far a distance as possible. Use foam, dry chemical, or carbon dioxide.

Reactivity Hazards:

Reactive With: Water, air or light evolving toxic oxides of nitrogen which are orange in color.
Other Reactions: n/a

Corrosivity Hazards:

Corrosive With: May corrode metals if wet.
Neutralizing Agent: Crushed limestone, soda ash, or lime

Radioactivity Hazards:

Radiation Emitted: n/a *Other Hazards:* n/a

Recommended Protection for Response Personnel:

Avoid breathing vapors, keep upwind. Avoid bodily contact with this material. Structural protective clothing will provide limited protection. Wash away any material which may come into contact with the body with copious amounts of soap and water. Consider appropriate evacuation.

AMYLTRICHLOROSILANE

DOT Number: UN 1728 *DOT Hazard Class:* Corrosive *DOT Guide Number:* 29
Synonyms: pentylsilicon trichloride, trichloroamylsilane, trichloropentylsilane
STCC Number: 4934210 *Reportable Qty:* n/a
Mfg Name: Dow Chemical *Phone No:* 1-517-636-4400

Physical Description:

Physical Form: Liquid *Color:* Colorless *Odor:* Sharp, irritating
Other Information: n/a

Chemical Properties:

Specific Gravity: 1.14 *Vapor Density:* 7 *Boiling Point: 320° F(160° C)*
Melting Point: n/a *Vapor Pressure:* n/a *Solubility in water:* No
Other Information: Decomposed by water.

Health Hazards:

Inhalation Hazard: Difficulty in breathing.
Ingestion Hazard: Harmful if swallowed.
Absorption Hazard: Will burn the skin and eyes.
Hazards to Wildlife: n/a
Decontamination Procedures: Wash away any material with copious amounts of soap and water.
First Aid Procedures: Remove victim to fresh air, call emergency medical care. If not breathing give CPR. If breathing is difficult administer oxygen. Treat for shock.

Fire Hazards:

Flashpoint: 145° F(62.7° C) *Ignition temperature:* n/a
Flammable Explosive High Range: n/a *Low Range:* n/a
Toxic Products of Combustion: Irritating hydrogen chloride and toxic phosgene may be formed.
Other Hazards: Reacts violently with water, irritating visible vapor cloud is produced.
Possible extinguishing agents: Use dry chemical, dry sand, and carbon dioxide. Do not use water on the material itself.

Reactivity Hazards:

Reactive With: Water to generate hydrochloric acid. *Other Reactions:* n/a

Corrosivity Hazards:

Corrosive With: Metal and tissue *Neutralizing Agent:* Crushed limestone, soda ash, or lime.

Radioactivity Hazards:

Radiation Emitted: n/a *Other Hazards:* n/a

Recommended Protection for Response Personnel:

Avoid breathing vapors, keep upwind. Wear a sealed chemical suit, (butyl rubber). Wash away any material which may have come into contact with the body with copious amounts of soap and water. Consider appropriate evacuation.

ANILINE

DOT Number: UN 1547 *DOT Hazard Class:* Poison B *DOT Guide Number:* 57
Synonyms: aniline oil, blue oil
STCC Number: 4921410 *Reportable Qty:* 5000/2270
Mfg Name: E.I. Du Pont *Phone No:* 1-800-441-3637

Physical Description:

Physical Form: Oily liquid *Color:* Colorless to brown *Odor:* Weak amine
Other Information: Used to make other chemicals, dyes, photographic chemicals, agricultural chemicals, and others.

Chemical Properties:

Specific Gravity: 1.0 *Vapor Density:* 3.2 *Boiling Point: 364° F(184° C)*
Melting Point: 21° F(-6.1° C) *Vapor Pressure:* .6 mm Hg at 68° F(20° C)
Solubility in water: Slight *Degree of Solubility:* 3.5%
Other Information: Weighs 8.5 lbs/3.8 kg per gallon/3.8 l.

Health Hazards:

Inhalation Hazard: Headaches
Ingestion Hazard: Weak, dizziness
Absorption Hazard: Eye irritant and blue disorder.
Hazards to Wildlife: Dangerous to aquatic life.
Decontamination Procedures: Wash away any material with copious amounts of soap and water.
First Aid Procedures: Remove victim to fresh air, call emergency medical care. If not breathing give CPR. If breathing is difficult administer oxygen. Treat for shock.

Fire Hazards:

Flashpoint: 158° F(70° C) *Ignition temperature:* 1139° F(615° C)
Flammable Explosive High Range: 11 *Low Range:* 1.3
Toxic Products of Combustion: Toxic oxides of nitrogen are produced during combustion of this material.
Other Hazards: It can react with toluene diisocyanate with reaction leading to possible ignition.
Possible extinguishing agents: Use alcohol foam, dry chemical or carbon dioxide. Use water in flooding quantities as a fog.

Reactivity Hazards:

Reactive With: Toluene diisocyanate *Other Reactions:* Strong acids and oxidizers.

Corrosivity Hazards:

Corrosive With: n/a *Neutralizing Agent:* n/a

Radioactivity Hazards:

Radiation Emitted: n/a *Other Hazards:* n/a

Recommended Protection for Response Personnel:

Avoid breathing vapors, keep upwind. Avoid bodily contact with this material. Wear a sealed chemical suit, (butyl rubber, viton, chlorinated polyethylene). Wash away any material which may have come into contact with the body with copious amounts of soap and water. Consider appropriate evacuation.

ANISOYL CHLORIDE

DOT Number: UN 1729 *DOT Hazard Class:* Corrosive *DOT Guide Number:* 60
Synonyms: p-anisoyl chloride
STCC Number: 4931710 *Reportable Qty:* n/a
Mfg Name: Aldrich Chemical *Phone No:* 1-414-273-3850

Physical Description:

Physical Form: Liquid *Color:* Yellow to brown *Odor:* Sharp, penetrating
Other Information: n/a

Chemical Properties:

Specific Gravity: 1.24 *Vapor Density:* 6.2 *Boiling Point: 504° F(262° C)*
Melting Point: n/a *Vapor Pressure:* n/a *Solubility in water:* Reacts
Other Information: Decomposed by water

Health Hazards:

Inhalation Hazard: Difficulty in breathing
Ingestion Hazard: Harmful if swallowed
Absorption Hazard: Irritating to the skin and eyes
Hazards to Wildlife: n/a
Decontamination Procedures: Wash away any material with copious amounts of soap and water.
First Aid Procedures: Remove victim to fresh air, call emergency medical care. If not breathing give CPR. If breathing is difficult administer oxygen. Treat for shock.

Fire Hazards:

Flashpoint: n/a *Ignition temperature:* n/a
Flammable Explosive High Range: n/a *Low Range:* n/a
Toxic Products of Combustion: Irritating hydrogen chloride fumes may be formed.
Other Hazards: Reacts with water, irritating vapor is produced.
Possible extinguishing agents: Use dry chemical or carbon dioxide. Do not use water on the material itself.

Reactivity Hazards:

Reactive With: Water to generate hydrochloric acid. *Other Reactions:* n/a

Corrosivity Hazards:

Corrosive With: Metal and tissue *Neutralizing Agent:* Crushed limestone, soda ash, or lime.

Radioactivity Hazards:

Radiation Emitted: n/a *Other Hazards:* n/a

Recommended Protection for Response Personnel:

Avoid breathing vapors, keep upwind. Wear a sealed chemical suit, (butyl rubber). Wash away any material which may have come into contact with the body with copious amounts of soap and water. Consider appropriate evacuation.

ANTIMONY PENTACHLORIDE

DOT Number: UN 1730 *DOT Hazard Class:* Corrosive *DOT Guide Number:* 60
Synonyms: antimony perchloride
STCC Number: 4932310 *Reportable Qty:* 1000/454
Mfg Name: Mallinckrodt Inc. *Phone No:* 1-314-895-2000

Physical Description:

Physical Form: Liquid *Color:* Colorless to brown *Odor:* Unpleasant
Other Information: Reddish, yellow fuming liquid used to make other chemicals.

Chemical Properties:

Specific Gravity: 2.35 *Vapor Density:* 10 *Boiling Point: 347° F(175° C)*
Melting Point: n/a *Vapor Pressure:* n/a *Solubility in water:* No (sinks)
Other Information: It solidifies at 37° F(2.7° C)

Health Hazards:

Inhalation Hazard: Throat and nose irritation.
Ingestion Hazard: Causes vomiting and severe burns to mouth and stomach.
Absorption Hazard: Causes severe burns to the skin and eyes.
Hazards to Wildlife: n/a
Decontamination Procedures: Wash away any material with copious amounts of soap and water.
First Aid Procedures: Remove victim to fresh air, call emergency medical care. If not breathing give CPR. If breathing is difficult administer oxygen. Treat for shock.

Fire Hazards:

Flashpoint: n/a *Ignition temperature:* n/a
Flammable Explosive High Range: n/a *Low Range:* n/a
Toxic Products of Combustion: n/a
Other Hazards: It is decomposed by water to form hydrochloric acid with releases of heat.
Possible extinguishing agents: Use dry chemical, dry sand, or carbon dioxide. Do not use water on material itself.

Reactivity Hazards:

Reactive With: Water
Other Reaction: n/a

Corrosivity Hazards:

Corrosive With: Metal and tissue *Neutralizing Agent:* Crushed limestone, soda ash, or lime

Radioactivity Hazards:

Radiation Emitted: n/a *Other Hazards:* n/a

Recommended Protection for Response Personnel:

Avoid breathing vapors, keep upwind. Avoid bodily contact with this material. Wear a sealed chemical suit, (butyl rubber, pvc). Wash away any material which may have come into contact with the body with copious amounts of soap and water. Consider appropriate evacuation.

ANTIMONY PENTAFLUORIDE

DOT Number: UN 1732 *DOT Hazard Class:* Corrosive *DOT Guide Number:* 59
Synonyms: none given
STCC Number: 4932005 *Reportable Qty:* n/a
Mfg Name: Allied Corp. *Phone No:* 1-201-455-2000

Physical Description:

Physical Form: Liquid *Color:* Colorless *Odor:* Sharp
Other Information: It is used to make other chemicals and as a catalyst in the manufacture of other chemicals.

Chemical Properties:

Specific Gravity: 2.34 *Vapor Density:* 7.5 *Boiling Point: 289° F(142.7° C)*
Melting Point: n/a *Vapor Pressure:* n/a *Solubility in water:* Reacts
Other Information: Reacts violently with water. The material may only be shipped in cylinders.

Health Hazards:

Inhalation Hazard: Poisonous if inhaled.
Ingestion Hazard: Poisonous if swallowed.
Absorption Hazard: Will burn the skin and eyes.
Hazards to Wildlife: n/a
Decontamination Procedures: Wash away any material with copious amounts of soap and water.
First Aid Procedures: Remove victim to fresh air, call emergency medical care. If not breathing give CPR. If breathing is difficult administer oxygen. Treat for shock.

Fire Hazards:

Flashpoint: n/a *Ignition temperature:* n/a
Flammable Explosive High Range: n/a *Low Range:* n/a
Toxic Products of Combustion: Gives off toxic hydrogen fluoride fumes when water is used to extinguish fires.
Other Hazards: Prolonged exposure of the cylinders to fire or heat may result in their violent rupture and rocketing.
Possible extinguishing agents: Do not use water. Use dry chemical, dry sand, or carbon dioxide.

Reactivity Hazards:

Reactive With: Water to give hydrogen fluoride (hydrofluoric acid)
Other Reactions: May cause fire upon contact with combustibles.

Corrosivity Hazards:

Corrosive With: Metals and tissue *Neutralizing Agent:* Crushed limestone, soda ash, or lime

Radioactivity Hazards:

Radiation Emitted: n/a *Other Hazards:* n/a

Recommended Protection for Response Personnel:

Avoid breathing vapors, keep upwind. Wear a sealed chemical suit, (butyl rubber, polycarbonate). Wash away any material which may have come into contact with the body with copious amounts of soap and water. Consider appropriate evacuation.

ANTIMONY POTASSIUM TARTRATE

DOT Number: UN 1551 *DOT Hazard Class:* ORM-A *DOT Guide Number:* 53
Synonyms: potassium antimonyl tartrate, tartar emetic, tartarized antimony, tartrated antimony
STCC Number: 4941114 *Reportable Qty:* 100/45.4
Mfg Name: Pfizer Chemical *Phone No:* 1-212-546-7721

Physical Description:

Physical Form: Solid *Color:* White *Odor:* Odorless
Other Information: It is used in textile, leather processing, and in insecticides.

Chemical Properties:

Specific Gravity: 2.60 *Vapor Density:* 12 *Boiling Point: n/a*
Melting Point: n/a *Vapor Pressure:* n/a *Solubility in water:* Yes
Other Information: Sinks in water

Health Hazards:

Inhalation Hazard: Poisonous if inhaled.
Ingestion Hazard: Poisonous if swallowed.
Absorption Hazard: Poisonous to the skin and eyes
Hazards to Wildlife: Dangerous to aquatic life.
Decontamination Procedures: Wash away any material with copious amounts of soap and water.
First Aid Procedures: Remove victim to fresh air, call emergency medical care. If not breathing give CPR. If breathing is difficult administer oxygen. Treat for shock.

Fire Hazards:

Flashpoint: n/a *Ignition temperature:* n/a
Flammable Explosive High Range: n/a *Low Range:* n/a
Toxic Products of Combustion: n/a
Other Hazards: n/a
Possible extinguishing agents: Extinguish fire using suitable agent for the type of surrounding fire.

Reactivity Hazards:

Reactive With: n/a *Other Reactions:* n/a

Corrosivity Hazards:

Corrosive With: n/a *Neutralizing Agent:* n/a

Radioactivity Hazards:

Radiation Emitted: n/a *Other Hazards:* n/a

Recommended Protection for Response Personnel:

Avoid breathing vapors, keep upwind. Structural protective clothing provides limited protection. Wash away any material which may have come into contact with the body with copious amounts of soap and water. Consider appropriate evacuation.

ANTIMONY TRIBROMIDE

DOT Number: NA 1549 *DOT Hazard Class:* Corrosive *DOT Guide Number:* 60
Synonyms: antimonous bromide
STCC Number: 4932319 *Reportable Qty:* 1000/454
Mfg Name: Great Western Inorganics *Phone No:* 1-303-423-9770

Physical Description:

Physical Form: Solid *Color:* Colorless to yellow *Odor:* n/a
Other Information: It is used to make other antimonous compounds in chemical analysis, and in dyeing.

Chemical Properties:

Specific Gravity: 4.15 *Vapor Density:* 12 *Boiling Point: 536° F(280° C)*
Melting Point: n/a *Vapor Pressure:* 1 mm Hg at 94° F(34.4° C) *Solubility in water:* Yes
Other Information: Sinks and mixes in water

Health Hazards:

Inhalation Hazard: n/a
Ingestion Hazard: Harmful if swallowed.
Absorption Hazard: Irritating to the skin and eyes.
Hazards to Wildlife: Dangerous to aquatic life.
Decontamination Procedures: Wash away any material with copious amounts of soap and water.
First Aid Procedures: Remove victim to fresh air, call emergency medical care.If not breathing give CPR. If breathing is difficult administer oxygen. Treat for shock.

Fire Hazards:

Flashpoint: n/a *Ignition temperature:* n/a
Flammable Explosive High Range: n/a *Low Range:* n/a
Toxic Products of Combustion: n/a
Other Hazards: n/a
Possible extinguishing agents: Do not use water. Use dry chemical, dry sand, or carbon dioxide.

Reactivity Hazards:

Reactive With: Water to form antimony oxide, and hydrobromic acid. *Other Reactions:* n/a

Corrosivity Hazards:

Corrosive With: Tissue *Neutralizing Agent:* Crushed limestone, soda ash, or lime

Radioactivity Hazards:

Radiation Emitted: n/a *Other Hazards:* n/a

Recommended Protection for Response Personnel:

Avoid breathing vapors, keep upwind. Structural protective clothing will provide limited protection. Wash away any material which may have come into contact with the body with copious amounts of soap and water. Consider appropriate evacuation.

ANTIMONY TRICHLORIDE

DOT Number: UN 1733 *DOT Hazard Class:* Corrosive *DOT Guide Number:* 60
Synonyms: antimony(III) chloride, butter of antimony
STCC Number: 4932316 *Reportable Qty:* 1000/454
Mfg Name: Stauffer Chemical *Phone No:* 1-203-222-3000

Physical Description:

Physical Form: Solid *Color:* White to pale yellow *Odor:* Sharp, unpleasant
Other Information: n/a

Chemical Properties:

Specific Gravity: 3.14 *Vapor Density:* 7.8 *Boiling Point: 433° F(222.7° C)*
Melting Point: n/a *Vapor Pressure:* 1 at 49° F(9.4° C) *Solubility in water:* Yes
Other Information: Sinks and mixes violently with water.

Health Hazards:

Inhalation Hazard: Poisonous if inhaled.
Ingestion Hazard: Nausea, vomiting, loss of consciousness.
Absorption Hazard: Poisonous to the skin and eyes.
Hazards to Wildlife: Dangerous to aquatic life.
Decontamination Procedures: Wash away any material with copious amounts of soap and water.
First Aid Procedures: Remove victim to fresh air, call emergency medical care. If not breathing give CPR. If breathing is difficult administer oxygen. Treat for shock.

Fire Hazards:

Flashpoint: n/a *Ignition temperature:* n/a
Flammable Explosive High Range: n/a *Low Range:* n/a
Toxic Products of Combustion: Poisonous gases may be produced in a fire.
Other Hazards: n/a
Possible extinguishing agents: Do not use water. Use dry chemical, dry sand, or carbon dioxide.

Reactivity Hazards:

Reactive With: Water to form hydrochloric acid. *Other Reactions:* n/a

Corrosivity Hazards:

Corrosive With: Metals and tissue *Neutralizing Agent:* Crushed limestone, soda ash, or lime

Radioactivity Hazards:

Radiation Emitted: n/a *Other Hazards:* n/a

Recommended Protection for Response Personnel:

Avoid breathing vapors, keep upwind. Wear a sealed chemical suit, (butyl rubber, polycarbonate). Wash away any material which may have come into contact with the body with copious amounts of soap and water. Consider appropriate evacuation.

ANTIMONY TRIFLUORIDE

DOT Number: NA 1549 *DOT Hazard Class:* Corrosive *DOT Guide Number:* 60
Synonyms: none given
STCC Number: 4932335 *Reportable Qty:* 1000/454
Mfg Name: A.D. MacKay Inc. *Phone No:* 1-203-655-7401

Physical Description:

Physical Form: Solid *Color:* White *Odor:* Odorless
Other Information: It is used in ceramics, and to make other chemicals.

Chemical Properties:

Specific Gravity: 4.38 *Vapor Density:* 6.2 *Boiling Point: 558° F(292.2° C)*
Melting Point: n/a *Vapor Pressure:* n/a *Solubility in water:* n/a
Other Information: Sinks in water

Health Hazards:

Inhalation Hazard: n/a
Ingestion Hazard: Harmful if swallowed.
Absorption Hazard: Irritating to the skin and eyes.
Hazards to Wildlife: Dangerous to aquatic life.
Decontamination Procedures: Wash away any material with copious amounts of soap and water.
First Aid Procedures: Remove victim to fresh air, call emergency medical care. If not breathing give CPR. If breathing is difficult administer oxygen. Treat for shock.

Fire Hazards:

Flashpoint: n/a *Ignition temperature:* n/a
Flammable Explosive High Range: n/a *Low Range:* n/a
Toxic Products of Combustion: n/a
Other Hazards: n/a
Possible extinguishing agents: Apply water from as far a distance as possible. Extinguish fire using suitable agent for type of surrounding fire.

Reactivity Hazards:

Reactive With: n/a *Other Reactions:* n/a

Corrosivity Hazards:

Corrosive With: Tissue *Neutralizing Agent:* Crushed limestone, soda ash, or lime

Radioactivity Hazards:

Radiation Emitted: n/a *Other Hazards:* n/a

Recommended Protection for Response Personnel:

Avoid breathing vapors, keep upwind. Wear a sealed chemical suit, (polycarbonate, butyl rubber). Wash away any material which may have come into contact with the body with copious amounts of soap and water. Consider appropriate evacuation.

ANTIMONY TRIOXIDE

DOT Number: NA 9201 *DOT Hazard Class:* ORM-E *DOT Guide Number:* 31
Synonyms: diantimony trioxide, exitelite, flowers of antimony, valentinite
STCC Number: 4966905 *Reportable Qty:* 1000/454
Mfg Name: Mallinckrodt Inc. *Phone No:* 1-314-895-2000

Physical Description:

Physical Form: Solid *Color:* White *Odor:* Odorless
Other Information: It is used to fireproof fabrics, paper and plastics, as a paint pigment, and for many other uses.

Chemical Properties:

Specific Gravity: 5.2 *Vapor Density:* 10 *Boiling Point: n/a*
Melting Point: n/a *Vapor Pressure:* 1 mm Hg at 574° F(301° C) *Solubility in water:* No
Other Information: Sinks in water

Health Hazards:

Inhalation Hazard: Poisonous if inhaled.
Ingestion Hazard: Poisonous if swallowed.
Absorption Hazard: Poisonous to the skin and eyes
Hazards to Wildlife: Dangerous to aquatic life.
Decontamination Procedures: Wash away any material with copious amounts of soap and water.
First Aid Procedures: Remove victim to fresh air, call emergency medical care. If not breathing give CPR. If breathing is difficult administer oxygen. Treat for shock.

Fire Hazards:

Flashpoint: n/a *Ignition temperature:* n/a
Flammable Explosive High Range: n/a *Low Range:* n/a
Toxic Products of Combustion: n/a
Other Hazards: n/a
Possible extinguishing agents: Extinguish fire using suitable agent for the type of surrounding fire.

Reactivity Hazards:

Reactive With: n/a *Other Reactions:* n/a

Corrosivity Hazards:

Corrosive With: n/a *Neutralizing Agent:* n/a

Radioactivity Hazards:

Radiation Emitted: n/a *Other Hazards:* n/a

Recommended Protection for Response Personnel:

Avoid breathing vapors, keep upwind. Structural protective clothing provides limited protection. Wash away any material which may have come into contact with the body with copious amounts of soap and water. Consider appropriate evacuation.

ARSENIC ACID (liquid)

DOT Number: UN 1553 *DOT Hazard Class:* Poison B *DOT Guide Number:* 53
Synonyms: arsenic pentoxide, orthoarsenic acid
STCC Number: 4923106 *Reportable Qty:* 1/.454
Mfg Name: Powell Metals & Chemicals *Phone No:* 1-815-963-3595

Physical Description:

Physical Form: Liquid *Color:* Colorless *Odor:* Odorless
Other Information: n/a

Chemical Properties:

Specific Gravity: 2.2 *Vapor Density:* 5.2 *Boiling Point: n/a*
Melting Point: n/a *Vapor Pressure:* n/a *Solubility in water:* Yes
Other Information: Sinks and mixes with water

Health Hazards:

Inhalation Hazard: Poisonous if inhaled.
Ingestion Hazard: Poisonous if swallowed.
Absorption Hazard: Irritating to the skin and eyes.
Hazards to Wildlife: n/a
Decontamination Procedures: Wash away any material with copious amounts of soap and water.
First Aid Procedures: Remove victim to fresh air, call emergency medical care. If not breathing give CPR. If breathing is difficult administer oxygen. Treat for shock.

Fire Hazards:

Flashpoint: n/a *Ignition temperature:* n/a
Flammable Explosive High Range: n/a *Low Range:* n/a
Toxic Products of Combustion: n/a
Other Hazards: n/a
Possible extinguishing agents: Use foam, dry chemical or carbon dioxide. Extinguish fire using suitable agent for type of surrounding fire.

Reactivity Hazards:

Reactive With: n/a *Other Reactions:* n/a

Corrosivity Hazards:

Corrosive With: Metal to give off toxic arsine gas.
Neutralizing Agent: Crushed limestone, soda ash, or lime

Radioactivity Hazards:

Radiation Emitted: n/a *Other Hazards:* n/a

Recommended Protection for Response Personnel:

Avoid breathing vapors, keep upwind. Wear a sealed chemical suit, (nitrile, neoprene, butyl rubber, polycarbonate, chlorinated polyethylene, viton). Wash away any material which may have come into contact with the body with copious amounts of soap and water. Consider appropriate evacuation.

ARSENIC ACID (solid)

DOT Number: UN 1554 *DOT Hazard Class:* Poison B *DOT Guide Number:* 53
Synonyms: arsenic pentoxide, orthoarsenic acid
STCC Number: 4923105 *Reportable Qty:* 1/.454
Mfg Name: Powell Metals & Chemicals *Phone No:* 1-815-963-3595

Physical Description:

Physical Form: Solid *Color:* White or Colorless *Odor:* Odorless
Other Information: n/a

Chemical Properties:

Specific Gravity: 2.2 *Vapor Density:* 5.2 *Boiling Point: n/a*
Melting Point: n/a *Vapor Pressure:* n/a *Solubility in water:* Yes
Other Information: Sinks and mixes with water

Health Hazards:

Inhalation Hazard: Poisonous if inhaled.
Ingestion Hazard: Poisonous if swallowed.
Absorption Hazard: Irritating to the skin and eyes.
Hazards to Wildlife: n/a
Decontamination Procedures: Wash away any material with copious amounts of soap and water.
First Aid Procedures: Remove victim to fresh air, call emergency medical care. If not breathing give CPR. If breathing is difficult administer oxygen. Treat for shock.

Fire Hazards:

Flashpoint: n/a *Ignition temperature:* n/a
Flammable Explosive High Range: n/a *Low Range:* n/a
Toxic Products of Combustion: n/a
Other Hazards: n/a
Possible extinguishing agents: Use alcohol foam, dry chemical or carbon dioxide. Extinguish fire using suitable agent for type of surrounding fire.

Reactivity Hazards:

Reactive With: n/a *Other Reactions:* n/a

Corrosivity Hazards:

Corrosive With: Metal to give off toxic arsine gas.
Neutralizing Agent: Crushed limestone, soda ash, or lime

Radioactivity Hazards:

Radiation Emitted: n/a *Other Hazards:* n/a

Recommended Protection for Response Personnel:

Avoid breathing vapors, keep upwind. Wear a sealed chemical suit, (nitrile, neoprene, butyl rubber, polycarbonate, chlorinated polyethylene, viton). Wash away any material which may have come into contact with the body with copious amounts of soap and water. Consider appropriate evacuation.

ARSENIC DISULFIDE

DOT Number: NA 1557 *DOT Hazard Class:* Poison B *DOT Guide Number:* 53
Synonyms: realgar, red arsenic glass, red orpiment, ruby arsenic
STCC Number: 4923208 *Reportable Qty:* 5000/2270
Mfg Name: Gallard-Schlesinger *Phone No:* 1-516-333-5600

Physical Description:

Physical Form: Solid *Color:* Red brown *Odor:* Odorless
Other Information: n/a

Chemical Properties:

Specific Gravity: 3.5 *Vapor Density:* 7 *Boiling Point: 1049° F(565° C)*
Melting Point: n/a *Vapor Pressure:* n/a *Solubility in water:* n/a
Other Information: Sinks in water

Health Hazards:

Inhalation Hazard: Poisonous if inhaled.
Ingestion Hazard: Poisonous if swallowed.
Absorption Hazard: Will burn the skin and eyes.
Hazards to Wildlife: n/a
Decontamination Procedures: Wash away any material with copious amounts of soap and water.
First Aid Procedures: Remove victim to fresh air, call emergency medical care. If not breathing give CPR. If breathing is difficult administer oxygen. Treat for shock.

Fire Hazards:

Flashpoint: n/a *Ignition temperature:* n/a
Flammable Explosive High Range: n/a *Low Range:* n/a
Toxic Products of Combustion: Poisonous fumes of the compound may be formed in fires. If ignited, will form sulfur dioxide gas.
Other Hazards: May ignite at very high temperatures.
Possible extinguishing agents: Use foam, dry chemical or carbon dioxide.

Reactivity Hazards:

Reactive With: n/a *Other Reactions:* n/a

Corrosivity Hazards:

Corrosive With: n/a *Neutralizing Agent:* n/a

Radioactivity Hazards:

Radiation Emitted: n/a *Other Hazards:* n/a

Recommended Protection for Response Personnel:

Avoid breathing vapors, keep upwind. Structural protective clothing provides limited protection. Wash away any any material which have come into contact with the body with copious amounts of soap and water. Consider appropriate evacuation.

ARSENIC PENTOXIDE

DOT Number: UN 1559 *DOT Hazard Class:* Poison B *DOT Guide Number:* 53
Synonyms: anhydride, arsenic acid,
STCC Number: 4923112 *Reportable Qty:* 5000/2270 *CHEMTREC Phone No:* 1-800-424-9300

Physical Description:

Physical Form: Amorphous solid *Color:* White *Odor:* Odorless
Other Information: n/a

Chemical Properties:

Specific Gravity: .30 *Vapor Density:* 7.9 *Boiling Point: n/a*
Melting Point: n/a *Vapor Pressure:* n/a *Solubility in water:* Yes
Other Information: It is soluble in water to form arsenic acid.

Health Hazards:

Inhalation Hazard: Poisonous if inhaled.
Ingestion Hazard: Poisonous if swallowed.
Absorption Hazard: Poisonous if absorbed.
Hazards to Wildlife: Dangerous to aquatic life.
Decontamination Procedures: Wash away any material with copious amounts of soap and water.
First Aid Procedures: Remove victim to fresh air, call emergency medical care. If not breathing give CPR. If breathing is difficult administer oxygen. Treat for shock.

Fire Hazards:

Flashpoint: n/a *Ignition temperature:* n/a
Flammable Explosive High Range: n/a *Low Range:* n/a
Toxic Products of Combustion: n/a
Other Hazards: n/a
Possible extinguishing agents: Extinguish fire with suitable agent for type of surrounding fire.

Reactivity Hazards:

Reactive With: Water *Other Reaction:* n/a

Corrosivity Hazards:

Corrosive With: Metal *Neutralizing Agent:* Crushed limestone, soda ash, or lime

Radioactivity Hazards:

Radiation Emitted: n/a *Other Hazards:* n/a

Recommended Protection for Response Personnel:

Avoid breathing dust and fumes from burning material, keep upwind. Structural protective clothing provides limited protection. Wash away any material which may have come into contact with the body with copious amounts of soap and water. Consider appropriate evacuation.

ARSENIC TRICHLORIDE

DOT Number: UN 1560 *DOT Hazard Class:* Poison B *DOT Guide Number:* 55
Synonyms: butter of arsenic, fuming liquid arsenic
STCC Number: 4923209 *Reportable Qty:* 5000/2270
Mfg Name: Ventron Corp. *Phone No:* 1-617-774-3100

Physical Description:

Physical Form: Liquid *Color:* Colorless *Odor:* Unpleasant
Other Information: n/a

Chemical Properties:

Specific Gravity: 2.56 *Vapor Density:* 7.9 *Boiling Point: 266.4° F(130.2° C)*
Melting Point: n/a *Vapor Pressure:* 10 mm Hg at 23° F(-5° C) *Solubility in water:* Reacts
Other Information: Sinks and reacts in water, poisonous, visible vapor cloud is produced.

Health Hazards:

Inhalation Hazard: Poisonous if inhaled.
Ingestion Hazard: Poisonous if swallowed.
Absorption Hazard: Irritating to the skin and eyes.
Hazards to Wildlife: n/a
Decontamination Procedures: Wash away any material with copious amounts of soap and water.
First Aid Procedures: Remove victim to fresh air, call emergency medical care. If not breathing give CPR. If breathing is difficult administer oxygen. Treat for shock.

Fire Hazards:

Flashpoint: n/a *Ignition temperature:* n/a
Flammable Explosive High Range: n/a *Low Range:* n/a
Toxic Products of Combustion: Irritating and toxic hydrogen chloride formed when involved in a fire.
Other Hazards: n/a
Possible extinguishing agents: Extinguish fire using suitable agent for type of surrounding fire. Do not use water on the material itself.

Reactivity Hazards:

Reactive With: Water to generate hydrochloric acid. *Other Reactions:* n/a

Corrosivity Hazards:

Corrosive With: Metal *Neutralizing Agent:* Crushed limestone, soda ash, or lime

Radioactivity Hazards:

Radiation Emitted: n/a *Other Hazards:* n/a

Recommended Protection for Response Personnel:

Avoid breathing vapors, keep upwind. Wear a sealed chemical suit, (nitrile, neoprene, viton, chlorinated polyethylene, pvc). Wash away any material which may have come into contact with the body with copious amounts of soap and water. Consider appropriate evacuation.

ARSENIC TRIOXIDE

DOT Number: UN 1561 *DOT Hazard Class:* Poison B *DOT Guide Number:* 53
Synonyms: arsenous acid, white arsenic
STCC Number: 4923115 *Reportable Qty:* 5000/2270
Mfg Name: Ventron Corp. *Phone No:* 1-617-774-3100

Physical Description:

Physical Form: Solid powder *Color:* White/amorphous powder *Odor:* Odorless
Other Information: n/a

Chemical Properties:

Specific Gravity: 3.7 *Vapor Density:* 16 *Boiling Point: 855° F(457.2° C)*
Melting Point: n/a *Vapor Pressure:* n/a *Solubility in water:* Mixes slowly
Other Information: n/a

Health Hazards:

Inhalation Hazard: Poisonous if inhaled.
Ingestion Hazard: Poisonous if swallowed.
Absorption Hazard: Irritating to the skin and eyes.
Hazards to Wildlife: Dangerous to aquatic life.
Decontamination Procedures: Wash away any material with copious amounts of soap and water.
First Aid Procedures: Remove victim to fresh air, call emergency medical care. If not breathing give CPR. If breathing is difficult administer oxygen. Treat for shock.

Fire Hazards:

Flashpoint: n/a *Ignition temperature:* n/a
Flammable Explosive High Range: n/a *Low Range:* n/a
Toxic Products of Combustion: n/a
Other Hazards: n/a
Possible extinguishing agents: Extinguish fire with suitable agent for the type of surrounding fire.

Reactivity Hazards:

Reactive With: Water *Other Reaction:* n/a

Corrosivity Hazards:

Corrosive With: Metal (when wet) *Neutralizing Agent:* Crushed limestone, soda ash, or lime

Radioactivity Hazards:

Radiation Emitted: n/a *Other Hazards:* n/a

Recommended Protection for Response Personnel:

Avoid breathing dust and fumes from burning material, keep upwind. Structural protective clothing provides limited protection. wash away any material which may have come into contact with the body with copious amounts of soap and water. Consider appropriate evacuation.

ARSENIC TRISULFIDE

DOT Number: UN 1557 *DOT Hazard Class:* Poison B *DOT Guide Number:* 53
Synonyms: arsenic yellow, king's gold, king's yellow, orpiment
STCC Number: 4923222 *Reportable Qty:* 5000/2270
Mfg Name: Cerac Inc. *Phone No:* 1-414-289-9800

Physical Description:

Physical Form: Solid *Color:* Yellow orange *Odor:* Odorless
Other Information: n/a

Chemical Properties:

Specific Gravity: 3.43 *Vapor Density:* 16 *Boiling Point: n/a*
Melting Point: n/a *Vapor Pressure:* n/a *Solubility in water:* No
Other Information: Sinks, Insoluble in water

Health Hazards:

Inhalation Hazard: Poisonous if inhaled.
Ingestion Hazard: Poisonous if swallowed.
Absorption Hazard: Will burn the skin and eyes.
Hazards to Wildlife: n/a
Decontamination Procedures: Wash away any material with copious amounts of soap and water.
First Aid Procedures: Remove victim to fresh air, call emergency medical care. If not breathing give CPR. If breathing is difficult administer oxygen. Treat for shock.

Fire Hazards:

Flashpoint: n/a *Ignition temperature:* n/a
Flammable Explosive High Range: n/a *Low Range:* n/a
Toxic Products of Combustion: Poisonous fumes of the compounds may be formed when involved in a fire.
Other Hazards: May ignite at very high temperatures.
Possible extinguishing agents: Use foam, dry chemical, or carbon dioxide.

Reactivity Hazards:

Reactive With: n/a *Other Reactions:* n/a

Corrosivity Hazards:

Corrosive With: n/a *Neutralizing Agent:* n/a

Radioactivity Hazards:

Radiation Emitted: n/a *Other Hazards:* n/a

Recommended Protection for Response Personnel:

Avoid breathing vapors, keep upwind. Wear appropriate chemical clothing, wash away any material which may come into contact with the body with copious amounts of soap and water. Consider appropriate evacuation.

ARSINE

DOT Number: UN 2188 *DOT Hazard Class:* Poison A *DOT Guide Number:* 18
Synonyms: none given
STCC Number: 4920135 *Reportable Qty:* n/a *CHEMTREC Phone No:* 1-800-424-9300

Physical Description:

Physical Form: Gas *Color:* Colorless *Odor:* Garlic like
Other Information: n/a

Chemical Properties:

Specific Gravity: n/a *Vapor Density:* n/a *Boiling Point: n/a*
Melting Point: n/a *Vapor Pressure:* n/a *Solubility in water:* n/a
Other Information: May only be shipped in cylinders that may have to be overpacked in wooden or metal containers.

Health Hazards:

Inhalation Hazard: Poisonous if inhaled.
Ingestion Hazard: Poisonous if swallowed.
Absorption Hazard: Will burn the skin and eyes.
Hazards to Wildlife: n/a
Decontamination Procedures: Wash away any material with copious amounts of soap and water.
First Aid Procedures: Remove victim to fresh air, call emergency medical care. If not breathing give CPR. If breathing is difficult administer oxygen. Treat for shock.

Fire Hazards:

Flashpoint: n/a *Ignition temperature:* n/a
Flammable Explosive High Range: n/a *Low Range:* n/a
Toxic Products of Combustion: n/a
Other Hazards: Prolonged exposure of the cylinders to heat or fire may result in the cylinders violent rupturing and rocketing.
Possible extinguishing agents: Use alcohol foam, dry chemical, or carbon dioxide. Apply water from as far a distance as possible.

Reactivity Hazards:

Reactive With: n/a *Other Reactions:* n/a

Corrosivity Hazards:

Corrosive With: n/a *Neutralizing Agent:* n/a

Radioactivity Hazards:

Radiation Emitted: n/a *Other Hazards:* n/a

Recommended Protection for Response Personnel:

Avoid breathing vapors, keep upwind. Wear appropriate chemical clothing, wash away any material which may come into contact with the body with copious amounts of soap and water. Consider appropriate evacuation.

ASBESTOS, BLUE

DOT Number: UN 2212 *DOT Hazard Class:* ORM-C *DOT Guide Number:* 31
Synonyms: same as above
STCC Number: 4945705 *Reportable Qty:* 1/.454 *CHEMTREC Phone No:* 1-800-424-9300

Physical Description:

Physical Form: Solid, slender, fine, flaxy fibers *Color:* White, gray, green or brown
Odor: Odorless *Other Information:* n/a

Chemical Properties:

Specific Gravity: n/a *Vapor Density:* n/a *Boiling Point: n/a*
Melting Point: n/a *Vapor Pressure:* n/a *Solubility in water:* Insoluble
Other Information: Sinks in water

Health Hazards:

Inhalation Hazard: Hazardous if inhaled (dust)
Ingestion Hazard: n/a
Absorption Hazard: n/a
Hazards to Wildlife: n/a
Decontamination Procedures: Wash away any material with copious amounts of soap and water.
First Aid Procedures: Remove victim to fresh air, call emergency medical care. If not breathing give CPR. If breathing is difficult administer oxygen. Treat for shock.

Fire Hazards:

Flashpoint: n/a *Ignition temperature:* n/a
Flammable Explosive High Range: n/a *Low Range:* n/a
Toxic Products of Combustion: Poisonous gases may be produced when heated.
Other Hazards: n/a
Possible extinguishing agents: Extinguish fire using suitable agent for the type of surrounding fire.

Reactivity Hazards:

Reactive With: n/a *Other Reaction:* n/a

Corrosivity Hazards:

Corrosive With: n/a *Neutralizing Agent:* n/a

Radioactivity Hazards:

Radiation Emitted: n/a *Other Hazards:* n/a

Recommended Protection for Response Personnel:

Avoid breathing dust, wear self contained breathing apparatus and the appropriate chemical clothing. Wash away any material which may have come into contact with the body with copious amounts of soap and water. Consider appropriate evacuation.

ASPHALT

DOT Number: NA 1999 *DOT Hazard Class:* Combustible liquid *DOT Guide Number:* 27
Synonyms: asphalt cement, petroleum asphalt
STCC Number: 4915320 *Reportable Qty:* n/a
Mfg Name: Chevron USA Inc. *Phone No:* 1-415-894-5064

Physical Description:

Physical Form: Thick black, generally heated *Color:* Dark brown to black *Odor:* Tar smell
Other Information: n/a

Chemical Properties:

Specific Gravity: 1.1 *Vapor Density:* 1 *Boiling Point: n/a*
Melting Point: n/a *Vapor Pressure:* n/a *Solubility in water:* Unsoluble
Other Information: May float or sink in water. A rubbery solid is produced when cooled.

Health Hazards:

Inhalation Hazard: Harmful if inhaled.
Ingestion Hazard: Poisonous if swallowed.
Absorption Hazard: Will burn the skin and eyes.
Hazards to Wildlife: n/a
Decontamination Procedures: Wash away any material with copious amounts of soap and water.
First Aid Procedures: Remove victim to fresh air, call emergency medical care. If not breathing give CPR. If breathing is difficult administer oxygen. Treat for shock.

Fire Hazards:

Flashpoint: 75° F(23.8° C) *Ignition temperature:* 905° F(485° C)
Flammable Explosive High Range: n/a *Low Range:* n/a
Toxic Products of Combustion: Vapor
Other Hazards: n/a
Possible extinguishing agents: Use water, dry chemical, foam, or carbon dioxide.

Reactivity Hazards:

Reactive With: n/a *Other Reaction: n/a*

Corrosivity Hazards:

Corrosive With: n/a *Neutralizing Agent:* n/a

Radioactivity Hazards:

Radiation Emitted: n/a *Other Hazards:* n/a

Recommended Protection for Response Personnel:

Avoid breathing vapors from burning material, keep upwind. Structural protective clothing provides limited protection. Wash away any material which may have come into contact with the body with copious amounts of soap and water. Consider appropriate evacuation.

AZINPHOS METHYL

DOT Number: NA 2783 *DOT Hazard Class:* Poison B *DOT Guide Number:* 55
Synonyms: Guthion
STCC Number: 4921528 *Reportable Qty:* 1/.454 *CHEMTREC Phone No:* 1-800-424-9300

Physical Description:

Physical Form: Solid, waxy solid *Color:* Brown, waxy or bright crystalline *Odor:* Odorless
Other Information: Used as a pesticide

Chemical Properties:

Specific Gravity: 2.2 *Vapor Density:* 5.2 *Boiling Point: Decomposes*
Melting Point: 163° F(72.7° C) *Vapor Pressure:* n/a *Solubility in water:* Yes
Degree of Solubility: .003% *Other Information:* In case of damage to, or leaking from containers, contact the Pesticide Safety Team Network at 1-800-424-9300.

Health Hazards:

Inhalation Hazard: Poisonous if inhaled.
Ingestion Hazard: Poisonous if swallowed.
Absorption Hazard: Toxic by skin absorption.
Hazards to Wildlife: Dangerous to aquatic and waterfowl.
Decontamination Procedures: Wash away any material with copious amounts of soap and water.
First Aid Procedures: Remove victim to fresh air, call emergency medical care. If not breathing give CPR. If breathing is difficult administer oxygen. Treat for shock.

Fire Hazards:

Flashpoint: n/a *Ignition temperature:* n/a
Flammable Explosive High Range: n/a *Low Range:* n/a
Toxic Products of Combustion: n/a
Other Hazards: When exposed to high temperatures or flames, the containers may explode.
Possible extinguishing agents: Apply water from as far a distance as possible, use dry chemical, foam, or carbon dioxide.

Reactivity Hazards:

Reactive With: Strong oxidizers *Other Reactions:* n/a

Corrosivity Hazards:

Corrosive With: n/a *Neutralizing Agent:* n/a

Radioactivity Hazards:

Radiation Emitted: n/a *Other Hazards:* n/a

Recommended Protection for Response Personnel:

Do not extinguish fire unless flow can be stopped. Avoid breathing vapors, keep upwind. Wear a sealed chemical suit, (polycarbonate, butyl rubber). Wash away any material which may have come into contact with the body with copious amounts of soap and water. Consider appropriate evacuation.

BARIUM CARBONATE

DOT Number: UN 1564 *DOT Hazard Class:* Poison A *DOT Guide Number:* 55
Synonyms: none given
STCC Number: n/a *Reportable Qty:* n/a
Mfg Name: Chemical Products Division *Phone No:* 1-404-382-2144

Physical Description:

Physical Form: Solid *Color:* White *Odor:* Odorless
Other Information: n/a

Chemical Properties:

Specific Gravity: 4.3 *Vapor Density:* 4.3 *Boiling Point: n/a*
Melting Point: n/a *Vapor Pressure:* 1 mm Hg at 68° F (20° C) *Solubility in water:* n/a
Other Information: Sinks in water.

Health Hazards:

Inhalation Hazard: n/a
Ingestion Hazard: Causes nausea and vomiting.
Absorption Hazard: n/a
Hazards to Wildlife: Dangerous to aquatic life.
Decontamination Procedures: Wash away any material with copious amounts of soap and water.
First Aid Procedures: Remove victim to fresh air, call emergency medical care. If not breathing give CPR. If breathing is difficult administer oxygen. Treat for shock.

Fire Hazards:

Flashpoint: n/a *Ignition temperature:* n/a
Flammable Explosive High Range: n/a *Low Range:* n/a
Toxic Products of Combustion: n/a
Other Hazards: n/a
Possible extinguishing agents: Extinguish fire using suitable agent for the type of surrounding fire.

Reactivity Hazards:

Reactive With: n/a *Other Reactions:* n/a

Corrosivity Hazards:

Corrosive With: n/a *Neutralizing Agent:* n/a

Radioactivity Hazards:

Radiation Emitted: n/a *Other Hazards:* n/a

Recommended Protection for Response Personnel:

Avoid breathing vapors, keep upwind. Structural protective clothing provides limited protection. Wash away any material which may have come into contact with the body with copious amounts of soap and water. Consider appropriate evacuation.

BARIUM CHLORATE

DOT Number: UN 1445 *DOT Hazard Class:* Oxidizer *DOT Guide Number:* 42
Synonyms: barium chlorate, monohydrate
STCC Number: 4918708 *Reportable Qty:* n/a
Mfg Name: Barium & Chemicals Co. *Phone No:* 1-614-282-9776

Physical Description:

Physical Form: Solid *Color:* White *Odor:* Odorless
Other Information: It is used in explosives and pyrotechnics. In dyeing textiles, and to make other chemicals.

Chemical Properties:

Specific Gravity: 3.18 *Vapor Density:* 11 *Boiling Point: n/a*
Melting Point: n/a *Vapor Pressure:* n/a *Solubility in water:* Yes
Other Information: Sinks and mixes with water.

Health Hazards:

Inhalation Hazard: Poisonous if inhaled.
Ingestion Hazard: Poisonous if swallowed.
Absorption Hazard: Irritating to the skin and the eyes.
Hazards to Wildlife: n/a
Decontamination Procedures: Wash away any material with copious amounts of soap and water.
First Aid Procedures: Remove victim to fresh air, call emergency medical care. If not breathing give CPR. If breathing is difficult administer oxygen. Treat for shock.

Fire Hazards:

Flashpoint: n/a *Ignition temperature:* n/a
Flammable Explosive High Range: n/a *Low Range:* n/a
Toxic Products of Combustion: Poisonous gases may be produced in a fire.
Other Hazards: May cause fire upon contact with combustibles. Prolonged exposure to heat or fire may result in an explosion.
Possible extinguishing agents: Flood with water. Apply water from as far a distance as possible.

Reactivity Hazards:

Reactive With: Combustible materials to form explosive mixtures. *Other Reactions:* n/a

Corrosivity Hazards:

Corrosive With: n/a *Neutralizing Agent:* n/a

Radioactivity Hazards:

Radiation Emitted: n/a *Other Hazards:* n/a

Recommended Protection for Response Personnel:

Avoid breathing vapors, keep upwind. Structural protective clothing provides limited protection. Wash away any material which may have come into contact with the body with copious amounts of soap and water. if the fire becomes uncontrollable, consider appropriate evacuation.

BARIUM CYANIDE

DOT Number: UN 1565 *DOT Hazard Class:* Poison B *DOT Guide Number:* 53
Synonyms: barium cyanide solid
STCC Number: 4923410 *Reportable Qty:* 10/4.54
Mfg Name: City Chemical *Phone No:* 1-212-929-2723

Physical Description:

Physical Form: Solid *Color:* White *Odor:* Odorless
Other Information: n/a

Chemical Properties:

Specific Gravity: n/a *Vapor Density:* 6.5 *Boiling Point: n/a*
Melting Point: n/a *Vapor Pressure:* n/a *Solubility in water:* Yes
Other Information: Sinks and mixes with water.

Health Hazards:

Inhalation Hazard: Poisonous if inhaled.
Ingestion Hazard: Poisonous if swallowed.
Absorption Hazard: Irritating to the skin and eyes.
Hazards to Wildlife: Dangerous to aquatic life.
Decontamination Procedures: Wash away any material with copious amounts of soap and water.
First Aid Procedures: Remove victim to fresh air, call emergency medical care. If not breathing give CPR. If breathing is difficult administer oxygen. Treat for shock.

Fire Hazards:

Flashpoint: n/a *Ignition temperature:* n/a
Flammable Explosive High Range: n/a *Low Range:* n/a
Toxic Products of Combustion: Toxic oxides of nitrogen are produced in a fire.
Other Hazards: n/a
Possible extinguishing agents: Extinguish fire using suitable agent for the type of surrounding fire. Do not use water on the material itself.

Reactivity Hazards:

Reactive With: Water to form an acid *Other Reactions:* n/a

Corrosivity Hazards:

Corrosive With: Metals in the presence of moisture.
Neutralizing Agent: Crushed limestone, soda ash, or lime

Radioactivity Hazards:

Radiation Emitted: n/a *Other Hazards:* n/a

Recommended Protection for Response Personnel:

Avoid breathing vapors, keep upwind. Wear appropriate chemical clothing. Wash away any material which may come into contact with the body with copious amounts of soap and water. Consider appropriate evacuation.

BARIUM NITRATE

DOT Number: UN 1446 *DOT Hazard Class:* Oxidizer *DOT Guide Number:* 42
Synonyms: none given
STCC Number: 4918709 *Reportable Qty:* n/a
Mfg Name: J.T. Baker Chemical *Phone No:* 1-201-859-2151

Physical Description:

Physical Form: Solid *Color:* White *Odor:* Odorless
Other Information: n/a

Chemical Properties:

Specific Gravity: 3.24 *Vapor Density:* 9 *Boiling Point: Decomposes*
Melting Point: n/a *Vapor Pressure:* n/a *Solubility in water:* Yes
Other Information: Sinks and mixes with water

Health Hazards:

Inhalation Hazard: Poisonous if inhaled.
Ingestion Hazard: Poisonous if swallowed.
Absorption Hazard: Irritating to the skin and the eyes.
Hazards to Wildlife: Dangerous to aquatic life.
Decontamination Procedures: Wash away any material with copious amounts of soap and water.
First Aid Procedures: Remove victim to fresh air, call emergency medical care. If not breathing give CPR. If breathing is difficult administer oxygen. Treat for shock.

Fire Hazards:

Flashpoint: n/a *Ignition temperature:* n/a
Flammable Explosive High Range: n/a *Low Range:* n/a
Toxic Products of Combustion: Toxic oxides of nitrogen are produced in a fire.
Other Hazards: May cause fire upon contact with combustibles.
Possible extinguishing agents: Apply water from as far a distance as possible. Flood with water.

Reactivity Hazards:

Reactive With: Combustible materials
Other Reactions: Prolonged exposure of this material to fire or heat may result in an explosion.

Corrosivity Hazards:

Corrosive With: n/a *Neutralizing Agent:* n/a

Radioactivity Hazards:

Radiation Emitted: n/a *Other Hazards:* n/a

Recommended Protection for Response Personnel:

Avoid breathing vapors, keep upwind. Structural protective clothing provides limited protection. Wash away any material which may have come into contact with the body with copious amounts of soap and water. If the fire becomes uncontrollable. Consider appropriate evacuation.

BARIUM PERCHLORATE

DOT Number: UN 1447 *DOT Hazard Class:* Oxidizer *DOT Guide Number:* 42
Synonyms: barium perchlorate trihydrate
STCC Number: 4918710 *Reportable Qty:* n/a
Mfg Name: Barium & Chemicals Co. *Phone No:* 1-614-282-9776

Physical Description:

Physical Form: Solid *Color:* White *Odor:* Odorless
Other Information: It is used to make explosives.

Chemical Properties:

Specific Gravity: 3.2 *Vapor Density:* 13 *Boiling Point: Decomposes*
Melting Point: n/a *Vapor Pressure:* n/a *Solubility in water:* Yes
Other Information: Sinks and mixes with water

Health Hazards:

Inhalation Hazard: Poisonous if inhaled.
Ingestion Hazard: Poisonous if swallowed.
Absorption Hazard: Irritating to the skin and eyes.
Hazards to Wildlife: n/a
Decontamination Procedures: Wash away any material with copious amounts of soap and water.
First Aid Procedures: Remove victim to fresh air, call emergency medical care. If not breathing give CPR. If breathing is difficult administer oxygen. Treat for shock.

Fire Hazards:

Flashpoint: n/a *Ignition temperature:* n/a
Flammable Explosive High Range: n/a *Low Range:* n/a
Toxic Products of Combustion: Poisonous gases may be produced when heated.
Other Hazards: Material increases intensity of the fire. May explode upon contact with combustibles and finely divided metals. Containers may explode in a fire.
Possible extinguishing agents: Apply water from as far a distance as possible. Flood with water.

Reactivity Hazards:

Reactive With: Combustible materials
Other Reactions: Prolonged exposure of this material to fire or heat may result in an explosion.

Corrosivity Hazards:

Corrosive With: n/a *Neutralizing Agent:* n/a

Radioactivity Hazards:

Radiation Emitted: n/a *Other Hazards:* n/a

Recommended Protection for Response Personnel:

Avoid breathing vapors, keep upwind. Structural protective clothing provides limited protection. Wash away any material which may have come into contact with the body with copious amounts of soap and water. If the fire becomes uncontrollable, consider appropriate evacuation.

BARIUM PERMANGANATE

DOT Number: UN 1448 *DOT Hazard Class:* Oxidizer *DOT Guide Number:* 42
Synonyms: none given
STCC Number: 4918711 *Reportable Qty:* n/a
Mfg Name: Barium & Chemicals Co. *Phone No:* 1-614-282-9776

Physical Description:

Physical Form: Solid *Color:* Dark purple to black *Odor:* Odorless
Other Information: It is used as a disinfectant, and to make other permanganates.

Chemical Properties:

Specific Gravity: 3.77 *Vapor Density:* 13 *Boiling Point: Decomposes*
Melting Point: n/a *Vapor Pressure:* n/a *Solubility in water:* Yes
Other Information: Sinks and mixes with water

Health Hazards:

Inhalation Hazard: Poisonous if inhaled.
Ingestion Hazard: Poisonous if swallowed.
Absorption Hazard: Irritating to the skin and eyes.
Hazards to Wildlife: Dangerous to aquatic life.
Decontamination Procedures: Wash away any material with copious amounts of soap and water.
First Aid Procedures: Remove victim to fresh air, call emergency medical care. If not breathing give CPR. If breathing is difficult administer oxygen. Treat for shock.

Fire Hazards:

Flashpoint: n/a *Ignition temperature:* n/a
Flammable Explosive High Range: n/a *Low Range:* n/a
Toxic Products of Combustion: n/a
Other Hazards: May cause fire upon contact with combustibles containers may explode in a fire.
Possible extinguishing agents: Apply water from as far a distance as possible. Flood with water.

Reactivity Hazards:

Reactive With: Combustible materials *Other Reactions:* n/a

Corrosivity Hazards:

Corrosive With: n/a *Neutralizing Agent:* n/a

Radioactivity Hazards:

Radiation Emitted: n/a *Other Hazards:* n/a

Recommended Protection for Response Personnel:

Avoid breathing vapors, keep upwind. Structural protective clothing provides limited protection. Wash away any material which may have come into contact with the body with copious amounts of soap and water. If the fire becomes uncontrollable, consider appropriate evacuation.

BARIUM PEROXIDE

DOT Number: UN 1449 *DOT Hazard Class:* Oxidizer *DOT Guide Number:* 42
Synonyms: barium binoxide, barium dioxide, barium super oxide
STCC Number: 4918712 *Reportable Qty:* n/a
Mfg Name: Barium & Chemicals Co. *Phone No:* 1-614-282-9776

Physical Description:

Physical Form: Solid *Color:* Light gray to tan *Odor:* Odorless
Other Information: n/a

Chemical Properties:

Specific Gravity: 4.96 *Vapor Density:* n/a *Boiling Point: Decomposes*
Melting Point: n/a *Vapor Pressure:* n/a *Solubility in water:* Insoluble
Other Information: Sinks, insoluble in water

Health Hazards:

Inhalation Hazard: Poisonous if inhaled.
Ingestion Hazard: Poisonous if swallowed.
Absorption Hazard: Will burn the skin and eyes.
Hazards to Wildlife: n/a
Decontamination Procedures: Wash away any material with copious amounts of soap and water.
First Aid Procedures: Remove victim to fresh air, call emergency medical care. If not breathing give CPR. If breathing is difficult administer oxygen. Treat for shock.

Fire Hazards:

Flashpoint: n/a *Ignition temperature:* n/a
Flammable Explosive High Range: n/a *Low Range:* n/a
Toxic Products of Combustion: n/a
Other Hazards: May cause fire upon contact with combustibles. Containers may explode in a fire.
Possible extinguishing agents: Apply water from as far a distance as possible. Flood with water.

Reactivity Hazards:

Reactive With: Combustible materials *Other Reactions:* n/a

Corrosivity Hazards:

Corrosive With: n/a *Neutralizing Agent:* n/a

Radioactivity Hazards:

Radiation Emitted: n/a *Other Hazards:* n/a

Recommended Protection for Response Personnel:

Avoid breathing vapors, keep upwind. Structural protective clothing provides limited protection. Wash away any material which may have come into contact with the body with copious amounts of soap and water. If the fire becomes uncontrollable, consider appropriate evacuation.

BENZALDEHYDE

DOT Number: UN 1989 *DOT Hazard Class:* Combustible liquid *DOT Guide Number:* 26
Synonyms: benzoic aldehyde, oil of bitter almond
STCC Number: 4913111 *Reportable Qty:* n/a
Mfg Name: Mallinckrodt Inc. *Phone No:* 1-314-895-2000

Physical Description:

Physical Form: Liquid *Color:* Colorless to pale yellow *Odor:* Bitter almond
Other Information: It is used in flavoring, and in perfume making.

Chemical Properties:

Specific Gravity: 1.1 *Vapor Density:* 3.7 *Boiling Point: 355° F(179.4° C)*
Melting Point: n/a *Vapor Pressure:* 1 mm Hg at 68° F (20° C) *Solubility in water:* No
Other Information: May float or sink in water, vapors are heavier than air.

Health Hazards:

Inhalation Hazard: n/a
Ingestion Hazard: Harmful if swallowed.
Absorption Hazard: Irritating to the skin and eyes.
Hazards to Wildlife: n/a
Decontamination Procedures: Wash away any material with copious amounts of soap and water.
First Aid Procedures: Remove victim to fresh air, call emergency medical care. If not breathing give CPR. If breathing is difficult administer oxygen. Treat for shock.

Fire Hazards:

Flashpoint: 145° F(62.7° C) *Ignition temperature:* 377° F(191.6° C)
Flammable Explosive High Range: n/a *Low Range:* n/a
Toxic Products of Combustion: n/a
Other Hazards: n/a
Possible extinguishing agents: Apply water from as far a distance as possible. Use foam, dry chemical, or carbon dioxide.

Reactivity Hazards:

Reactive With: n/a *Other Reactions:* n/a

Corrosivity Hazards:

Corrosive With: n/a *Neutralizing Agent:* n/a

Radioactivity Hazards:

Radiation Emitted: n/a *Other Hazards:* n/a

Recommended Protection for Response Personnel:

Avoid breathing vapors, keep upwind. Wear a sealed chemical suit, (butyl rubber). Wash away any material which may have come into into contact with the body with copious amounts of soap and water. Consider appropriate evacuation.

BENZENE

DOT Number: UN 1114 *DOT Hazard Class:* Flammable liquid *DOT Guide Number:* 27
Synonyms: benzol, benzole
STCC Number: 4908110 *Reportable Qty:* 1000/454
Mfg Name: Witco Chemical Co. *Phone No:* 1-212-605-3800

Physical Description:

Physical Form: Liquid *Color:* Colorless *Odor:* Gasoline like
Other Information: It is used to make other chemicals, as a solvent, and as a gasoline additive.

Chemical Properties:

Specific Gravity: .9 *Vapor Density:* 2.8 *Boiling Point: 176° F(80° C)*
Melting Point: 42° F(5.5° C) *Vapor Pressure:* 75 mm Hg at 68° F (20° C)
Solubility in water: No *Degree of Solubility:* .18%
Other Information: Lighter than and insoluble in water. Vapors are heavier than air. Weighs 7.3lbs/3.17 kg per gallon/3.8 l.

Health Hazards:

Inhalation Hazard: Headache, difficulty in breathing, loss of consciousness.
Ingestion Hazard: Harmful if swallowed.
Absorption Hazard: Irritating to the skin and eyes.
Hazards to Wildlife: Dangerous to aquatic life.
Decontamination Procedures: Wash away any material with copious amounts of soap and water.
First Aid Procedures: Remove victim to fresh air, call emergency medical care. If not breathing give CPR. If breathing is difficult administer oxygen. Treat for shock.

Fire Hazards:

Flashpoint: 12° F(−11.1° C) *Ignition temperature:* 928° F(497.7° C)
Flammable Explosive High Range: 7.9 *Low Range:* 1.3
Toxic Products of Combustion: n/a
Other Hazards: Flashback along vapor trail may occur. Vapors may explode if ignited in an enclosed area.
Possible extinguishing agents: Apply water from as far a distance as possible. Use foam, dry chemical, or carbon dioxide.

Reactivity Hazards:

Reactive With: n/a *Other Reactions:* n/a

Corrosivity Hazards:

Corrosive With: n/a *Neutralizing Agent:* n/a

Radioactivity Hazards:

Radiation Emitted: n/a *Other Hazards:* n/a

Recommended Protection for Response Personnel:

Avoid breathing vapors, keep upwind. Wear a sealed chemical suit, (viton). Wash away any material which may come into contact with the body with copious amounts of soap and water. Consider appropriate evacuation.

BENZENE HEXACHLORIDE

DOT Number: UN 2761 *DOT Hazard Class:* Poison B *DOT Guide Number:* 55
Synonyms: BHC, gammexane, 1,2,3,4,5,6-hexachloro-hexachlorocyclohexane, lindane, tri-6
STCC Number: n/a *Reportable Qty:* n/a
Mfg Name: Hooker Chemical Corp. *Phone No:* 1-716-278-7000

Physical Description:

Physical Form: Solid *Color:* Light to dark brown *Odor:* Musty
Other Information: n/a

Chemical Properties:

Specific Gravity: 1.89 *Vapor Density:* 10 *Boiling Point: n/a*
Melting Point: n/a *Vapor Pressure:* n/a *Solubility in water:* n/a
Other Information: Sinks in water.

Health Hazards:

Inhalation Hazard: Poisonous if inhaled.
Ingestion Hazard: Poisonous if swallowed.
Absorption Hazard: Irritating to the skin and eyes.
Hazards to Wildlife: Dangerous to aquatic life.
Decontamination Procedures: Wash away any material with copious amounts of soap and water.
First Aid Procedures: Remove victim to fresh air, call emergency medical care. If not breathing give CPR. If breathing is difficult administer oxygen. Treat for shock.

Fire Hazards:

Flashpoint: n/a *Ignition temperature:* n/a
Flammable Explosive High Range: n/a *Low Range:* n/a
Toxic Products of Combustion: Toxic gases are generated when solid is heated.
Other Hazards: Solid is not flammable.
Possible extinguishing agents: Extinguish fire using suitable agent for the type of surrounding fire.

Reactivity Hazards:

Reactive With: n/a *Other Reactions:* n/a

Corrosivity Hazards:

Corrosive With: n/a *Neutralizing Agent:* n/a

Radioactivity Hazards:

Radiation Emitted: n/a *Other Hazards:* n/a

Recommended Protection for Response Personnel:

Avoid breathing vapors, keep upwind. Structural protective clothing provides limited protection. Wash away any material which may have come into contact with the body with copious amounts of soap and water. Consider appropriate evacuation.

BENZENE PHOSPHORUS DICHLORIDE

DOT Number: UN 2798 *DOT Hazard Class:* Corrosive *DOT Guide Number:* 39
Synonyms: dichlorophenylphosphine, phenyphosphorus dichloride
STCC Number: 4931715 *Reportable Qty:* n/a
Mfg Name: Aldrich Chemical *Phone No:* 1-414-273-3850

Physical Description:

Physical Form: Liquid *Color:* Colorless *Odor:* Unpleasant
Other Information: n/a

Chemical Properties:

Specific Gravity: 1.14 *Vapor Density:* 6.2 *Boiling Point: 430° F(221° C)*
Melting Point: n/a *Vapor Pressure:* n/a *Solubility in water:* Reacts
Other Information: Reacts violently with water. A poisonous vapor cloud is produced.

Health Hazards:

Inhalation Hazard: Poisonous if inhaled.
Ingestion Hazard: Harmful if swallowed.
Absorption Hazard: Irritating to the skin and eyes.
Hazards to Wildlife: Dangerous to aquatic life.
Decontamination Procedures: Wash away any material with copious amounts of soap and water.
First Aid Procedures: Remove victim to fresh air, call emergency medical care. If not breathing give CPR. If breathing is difficult administer oxygen. Treat for shock.

Fire Hazards:

Flashpoint: 215° F(101.6° C) *Ignition temperature:* 319° F(159.4° C)
Flammable Explosive High Range: n/a *Low Range:* n/a
Toxic Products of Combustion: Toxic fumes include oxides of phosphorus and hydrogen chloride.
Other Hazards: Containers may rupture. Hot liquid is spontaneously flammable because of dissolved phosphorus.
Possible extinguishing agents: Apply water from as far a distance as possible.

Reactivity Hazards:

Reactive With: Water to form hydrochloric acid. *Other Reactions:* n/a

Corrosivity Hazards:

Corrosive With: Metals and tissue *Neutralizing Agent:* Crushed limestone, soda ash, or lime

Radioactivity Hazards:

Radiation Emitted: n/a *Other Hazards:* n/a

Recommended Protection for Response Personnel:

Avoid breathing vapors, keep upwind. Structural protective clothing provides limited protection. Wash away any material which may have come into contact with the body with copious amounts of soap and water. Consider appropriate evacuation.

BENZENE PHOSPHORUS THIODICHLORIDE

DOT Number: UN 2799 *DOT Hazard Class:* Corrosive *DOT Guide Number:* 39
Synonyms: phenylphosphonothioic, dichloride
STCC Number: 4931720 *Reportable Qty:* n/a
Mfg Name: Aldrich Chemical *Phone No:* 1-414-273-3850

Physical Description:

Physical Form: Liquid *Color:* Colorless *Odor:* Unpleasant
Other Information: n/a

Chemical Properties:

Specific Gravity: 1.38 *Vapor Density:* 6.2 *Boiling Point: 518° F (270° C)*
Melting Point: n/a *Vapor Pressure:* n/a *Solubility in water:* Reacts
Other Information: Reacts violently with water. A poisonous vapor cloud is produced.

Health Hazards:

Inhalation Hazard: Poisonous if inhaled.
Ingestion Hazard: Harmful if swallowed.
Absorption Hazard: Irritating to the skin and eyes.
Hazards to Wildlife: n/a
Decontamination Procedures: Wash away any material with copious amounts of soap and water.
First Aid Procedures: Remove victim to fresh air, call emergency medical care. If not breathing give CPR. If breathing is difficult administer oxygen. Treat for shock.

Fire Hazards:

Flashpoint: 252° F(122.2° C) *Ignition temperature:* 338° F(170° C)
Flammable Explosive High Range: n/a *Low Range:* n/a
Toxic Products of Combustion: n/a *Other Hazards:* n/a
Possible *Possible extinguishing agents:* Do not use water, use dry chemical or carbon dioxide.

Reactivity Hazards:

Reactive With: Water to form hydrochloric acid. *Other Reactions:* n/a

Corrosivity Hazards:

Corrosive With: Metals and tissue
Neutralizing Agent: Crushed limestone, soda ash, or lime

Radioactivity Hazards:

Radiation Emitted: n/a *Other Hazards:* n/a

Recommended Protection for Response Personnel:

Avoid breathing vapors, keep upwind. Wear a sealed chemical suit, (butyl rubber). Wash away any material which may have come into contact with the body with copious amounts of soap and water. Consider appropriate evacuation.

BENZOIC ACID

DOT Number: NA 9094 *DOT Hazard Class:* ORM-E *DOT Guide Number:* 31
Synonyms: benzenecarboxylic acid, carboxylbenzene, dracyclic acid
STCC Number: 4966340 *Reportable Qty:* 5000/2270
Mfg Name: Shell Chemical Co. *Phone No:* 1-713-241-6161

Physical Description:

Physical Form: Solid *Color:* White *Odor:* Faint, pleasant
Other Information: It is used to make other chemicals, as a food preservative, and for other uses.

Chemical Properties:

Specific Gravity: 1.27 *Vapor Density:* 4.21 *Boiling Point: 482° F(250° C)*
Melting Point: n/a *Vapor Pressure:* 1 mm Hg at 205° F(96.1° C) *Solubility in water:* Slight
Other Information: Sinks, slightly soluble in water.

Health Hazards:

Inhalation Hazard: n/a
Ingestion Hazard: n/a
Absorption Hazard: Irritating to the skin and eyes.
Hazards to Wildlife: Dangerous to aquatic life.
Decontamination Procedures: Wash away any material with copious amounts of soap and water.
First Aid Procedures: Remove victim to fresh air, call emergency medical care. If not breathing give CPR. If breathing is difficult administer oxygen. Treat for shock.

Fire Hazards:

Flashpoint: 250° F(121° C) *Ignition temperature:* 1058° F(570° C)
Flammable Explosive High Range: n/a *Low Range:* n/a
Toxic Products of Combustion: n/a
Other Hazards: Vapors may explode if ignited in an enclosed area. Dust may form an explosive mixture with air.
Possible extinguishing agents: Apply water from as far a distance as possible. Use foam, dry chemical, or carbon dioxide.

Reactivity Hazards:

Reactive With: n/a *Other Reactions:* n/a

Corrosivity Hazards:

Corrosive With: n/a *Neutralizing Agent:* n/a

Radioactivity Hazards:

Radiation Emitted: n/a *Other Hazards:* n/a

Recommended Protection for Response Personnel:

Avoid breathing vapors, keep upwind. Structural protective clothing provides limited protection. Wash away any material which may have come into contact with the body with copious amounts of soap and water. Consider appropriate evacuation.

BENZONITRILE

DOT Number: UN 2224 *DOT Hazard Class:* Combustible liquid *DOT Guide Number:* 55
Synonyms: benzoic acid nitrile, cyanobenzene, phenylcyanide
STCC Number: 4913134 *Reportable Qty:* 5000/2270
Mfg Name: Aldrich Chemical *Phone No:* 1-414-273-3850

Physical Description:

Physical Form: Liquid *Color:* Colorless *Odor:* Almondlike
Other Information: It is used as a specialty solvent, and to make other chemicals.

Chemical Properties:

Specific Gravity: 1.1 *Vapor Density:* 3.6 *Boiling Point: 376° F(191.1° C)*
Melting Point: n/a *Vapor Pressure:* n/a *Solubility in water:* Slight
Other Information: Heavier, slightly soluble in water. Weighs 8.4 lbs/3.6 kg per gallon/3.8 l.

Health Hazards:

Inhalation Hazard: Headaches, difficulty in breathing, loss of consciousness.
Ingestion Hazard: Headaches, nausea, vomiting, loss of consciousness.
Absorption Hazard: Irritating to the skin and eyes.
Hazards to Wildlife: n/a
Decontamination Procedures: Wash away any material with copious amounts of soap and water.
First Aid Procedures: Remove victim to fresh air, call emergency medical care. If not breathing give CPR. If breathing is difficult administer oxygen. Treat for shock.

Fire Hazards:

Flashpoint: 167° F(75° C) *Ignition temperature:* n/a
Flammable Explosive High Range: n/a *Low Range:* n/a
Toxic Products of Combustion: Toxic hydrogen cyanide and oxides of nitrogen may be formed in fires.
Other Hazards: n/a
Possible extinguishing agents: Apply water from as far a distance as possible. Use alcohol foam, dry chemical, or carbon dioxide. Do not extinguish the fire unless the flow can be stopped.

Reactivity Hazards:

Reactive With: Will attack some plastics. *Other Reactions:* n/a

Corrosivity Hazards:

Corrosive With: n/a *Neutralizing Agent:* n/a

Radioactivity Hazards:

Radiation Emitted: n/a *Other Hazards:* n/a

Recommended Protection for Response Personnel:

Avoid breathing vapors, keep upwind. Wear a sealed chemical suit, (polycarbonate, butyl rubber). Wash away any material which may have come into contact with the body with copious amounts of soap and water. Consider appropriate evacuation.

BENZOYL CHLORIDE

DOT Number: UN 1736 *DOT Hazard Class:* Corrosive *DOT Guide Number:* 39
Synonyms: benzenecarbonyl chloride
STCC Number: 4931725 *Reportable Qty:* 1000/454
Mfg Name: Mallinckrodt Inc. *Phone No:* 1-314-895-2000

Physical Description:

Physical Form: Liquid *Color:* Colorless to slight brown *Odor:* Pungent
Other Information: Fuming liquid used in medicine and the manufacture of other chemicals. Weighs 10.2 lbs/4.4 kg per gallon/3.8 l.

Chemical Properties:

Specific Gravity: 1.2 *Vapor Density:* 4.9 *Boiling Point: 387° F(197.2° C)*
Melting Point: 30.9° F(-.6° C) *Vapor Pressure:* n/a *Solubility in water:* Decomposes
Other Information: Decomposed by water to hydrochloric acid and insoluble benzoic acid, with a release of heat.

Health Hazards:

Inhalation Hazard: Hazardous if inhaled.
Ingestion Hazard: Hazardous if swallowed.
Absorption Hazard: Will burn the skin and eyes.
Hazards to Wildlife: Dangerous to aquatic life and waterfowl in low quantities.
Decontamination Procedures: Wash away any material with copious amounts of soap and water.
First Aid Procedures: Remove victim to fresh air, call emergency medical care. If not breathing give CPR. If breathing is difficult administer oxygen. Treat for shock.

Fire Hazards:

Flashpoint: 162° F(72.2° C) *Ignition temperature:* n/a
Flammable Explosive High Range: n/a *Low Range:* n/a
Toxic Products of Combustion: Poisonous gases are produced in fire and when heated.
Other Hazards: Avoid contact with vapors and liquid.
Possible extinguishing agents: Dry chemical or CO_2

Reactivity Hazards:

Reactive With: Water *Other Reactions:* n/a

Corrosivity Hazards:

Corrosive With: Metal and tissue *Neutralizing Agent:* Crushed limestone, soda ash, or lime

Radioactivity Hazards:

Radiation Emitted: n/a *Other Hazards:* n/a

Recommended Protection for Response Personnel:

Avoid breathing vapors, keep upwind. Structural protective clothing provides limited protection. Wash away any material which may have come into contact with the body with copious amounts of soap and water. Consider appropriate evacuation.

BENZOYL PEROXIDE

DOT Number: UN 2085 *DOT Hazard Class:* Organic peroxide *DOT Guide Number:* 49
Synonyms: benzoic acid, benzoyl superoxide, Epiclear
STCC Number: 4936010 *Reportable Qty:* n/a *CHEMTREC Phone No:* 1-800-424-9300

Physical Description:

Physical Form: Solid *Color:* White *Odor:* Odorless
Other Information: Used to manufacture paints, plastics, and rubber.

Chemical Properties:

Specific Gravity: 1.33 *Vapor Density:* 8.3 *Boiling Point: 217° F(102.7° C)*
Melting Point: n/a *Vapor Pressure:* 1 mm Hg at 68° F (20° C) *Solubility in water:* Yes
Other Information: Can form explosive mixtures with finely divided combustibles.

Health Hazards:

Inhalation Hazard: Irritation to the eyes, nose, and throat.
Ingestion Hazard: Irritating if swallowed.
Absorption Hazard: Irritating to the skin and eyes.
Hazards to Wildlife: n/a
Decontamination Procedures: Wash away any material with copious amounts of soap and water.
First Aid Procedures: Remove victim to fresh air, call emergency medical care. If not breathing give CPR. If breathing is difficult administer oxygen. Treat for shock.

Fire Hazards:

Flashpoint: n/a *Ignition temperature:* 167° F(75° C)
Flammable Explosive High Range: n/a *Low Range:* n/a
Toxic Products of Combustion: Produces suffocating dense white smoke containing benzoic acid.
Other Hazards: Avoid breathing dust and fumes of burning material, dangerously explosive!
Possible extinguishing agents: Apply water from as far a distance as possible. Use alcohol foam, dry chemical, or carbon dioxide.

Reactivity Hazards:

Reactive With: Ordinary combustible materials. *Other Reactions:* Easily oxidizable materials.

Corrosivity Hazards:

Corrosive With: n/a *Neutralizing Agent:* n/a

Radioactivity Hazards:

Radiation Emitted: n/a *Other Hazards:* n/a

Recommended Protection for Response Personnel:

Avoid breathing vapors, keep upwind. Structural protective clothing provides limited protection. Wash away any material which may have come into contact with the body with copious amounts of soap and water. Consider appropriate evacuation.

BENZYL BROMIDE

DOT Number: UN 1737 *DOT Hazard Class:* Corrosive *DOT Guide Number:* 59
Synonyms: α-bromotoluene, omega-bromotoluene
STCC Number: 4936010 *Reportable Qty:* n/a
Mfg Name: Eastman Organic *Phone No:* 1-800-455-6325

Physical Description:

Physical Form: Liquid *Color:* Colorless to yellow *Odor:* Sharp, irritating
Other Information: n/a

Chemical Properties:

Specific Gravity: 1.44 *Vapor Density:* 8.5 *Boiling Point: 388° F(197.7° C)*
Melting Point: n/a *Vapor Pressure:* n/a *Solubility in water:* Slight
Other Information: Heavier, slightly soluble in water.

Health Hazards:

Inhalation Hazard: n/a
Ingestion Hazard: Harmful if swallowed.
Absorption Hazard: Irritating to the skin and eyes.
Hazards to Wildlife: Dangerous to aquatic life.
Decontamination Procedures: Wash away any material with copious amounts of soap and water.
First Aid Procedures: Remove victim to fresh air, call emergency medical care. If not breathing give CPR. If breathing is difficult administer oxygen. Treat for shock.

Fire Hazards:

Flashpoint: 174° F(78.8° C) *Ignition temperature:* n/a
Flammable Explosive High Range: n/a *Low Range:* n/a
Toxic Products of Combustion: Forms vapor that is a powerful tear gas.
Other Hazards: Polymerizes when in contact with all common metals with the exceptions of nickel and lead.
Possible extinguishing agents: Apply water from as far a distance as possible. Use alcohol foam, dry chemical, or carbon dioxide.

Reactivity Hazards:

Reactive With: Water to form hydrobromic acid. *Other Reactions:* n/a

Corrosivity Hazards:

Corrosive With: Metals and tissue *Neutralizing Agent:* Crushed limestone, soda ash, or lime

Radioactivity Hazards:

Radiation Emitted: n/a *Other Hazards:* n/a

Recommended Protection for Response Personnel:

Avoid breathing vapors, keep upwind. Wear a sealed chemical suit, (polycarbonate, butyl rubber). Wash away any material which may have come into contact with the body with copious amounts of soap and water. Consider appropriate evacuation.

BENZYL CHLORIDE

DOT Number: UN 1738 *DOT Hazard Class:* Corrosive *DOT Guide Number:* 59
Synonyms: α-chlorotoluene
STCC Number: 4936012 *Reportable Qty:* 100/45.4
Mfg Name: Stauffer Chemical *Phone No:* 1-201-985-6262

Physical Description:

Physical Form: Liquid *Color:* Colorless to yellow *Odor:* Irritating
Other Information: Weighs 9.2 lbs/4.0 kg per gallon/3.8 l.

Chemical Properties:

Specific Gravity: 1.1 *Vapor Density:* 4.4 *Boiling Point: 354° F(178.8° C)*
Melting Point: −39° F(−39.4° C) *Vapor Pressure:* .9 mm Hg at 68° F (20° C)
Solubility in water: No *Degree of Solubility:* .0033%
Other Information: Undergoes slow hydrolysis, liberating hydrogen chloride.

Health Hazards:

Inhalation Hazard: Severe irritation to the respiratory tract.
Ingestion Hazard: Nausea and vomiting.
Absorption Hazard: Will burn skin and eyes on contact.
Hazards to Wildlife: Dangerous to aquatic life and waterfowl in low quantities.
Decontamination Procedures: Wash away any material with copious amounts of soap and water.
First Aid Procedures: Remove victim to fresh air, call emergency medical care. If not breathing give CPR. If breathing is difficult administer oxygen. Treat for shock.

Fire Hazards:

Flashpoint: 153° F(67.2° C) *Ignition temperature:* 1085° F(585° C)
Flammable Explosive High Range: n/a *Low Range:* 1.1
Toxic Products of Combustion: Irritating gases are produced in fire and when heated.
Other Hazards: Avoid contact with vapors and liquid.
Possible extinguishing agents: Water, foam, dry chemical, and carbon dioxide.

Reactivity Hazards:

Reactive With: Water *Other Reactions:* n/a

Corrosivity Hazards:

Corrosive With: All metals and tissue
Neutralizing Agent: Crushed limestone, soda ash, or lime

Radioactivity Hazards:

Radiation Emitted: n/a *Other Hazards:* n/a

Recommended Protection for Response Personnel:

Avoid breathing vapors, keep upwind. Wear a sealed chemical suit, (polycarbonate, viton). Wash away any material which may have come into contact with the body with copious amounts of soap and water. Consider appropriate evacuation.

BENZYL CHLORFORMATE

DOT Number: UN 1739 *DOT Hazard Class:* Corrosive *DOT Guide Number:* 39
Synonyms: benzyl ester, carbobenzoxy chloride
STCC Number: 4933010 *Reportable Qty:* n/a
Mfg Name: Aldrich Chemical *Phone No:* 1-414-273-3850

Physical Description:

Physical Form: Liquid *Color:* Colorless *Odor:* Sharp, irritating
Other Information: n/a

Chemical Properties:

Specific Gravity: 1.22 *Vapor Density:* 5.4 *Boiling Point: 306° F(152.2° C)*
Melting Point: n/a *Vapor Pressure:* n/a *Solubility in water:* Reacts
Other Information: Sinks and reacts in water.

Health Hazards:

Inhalation Hazard: n/a
Ingestion Hazard: Harmful if swallowed.
Absorption Hazard: Irritating to the skin and eyes.
Hazards to Wildlife: n/a
Decontamination Procedures: Wash away any material with copious amounts of soap and water.
First Aid Procedures: Remove victim to fresh air, call emergency medical care. If not breathing give CPR. If breathing is difficult administer oxygen. Treat for shock.

Fire Hazards:

Flashpoint: 108.3° F *Ignition temperature:* n/a
Flammable Explosive High Range: n/a *Low Range:* n/a
Toxic Products of Combustion: Toxic phosgene hydrogen chloride and benzyl chloride vapors may be formed.
Other Hazards: Containers may explode in a fire.
Possible extinguishing agents: Apply water from as far a distance as possible. Use dry sand, dry chemical, or carbon dioxide.

Reactivity Hazards:

Reactive With: Water to form hydrochloric acid. *Other Reactions:* n/a

Corrosivity Hazards:

Corrosive With: Metals and tissue
Neutralizing Agent: Crushed limestone, soda ash, or lime

Radioactivity Hazards:

Radiation Emitted: n/a *Other Hazards:* n/a

Recommended Protection for Response Personnel:

Avoid breathing vapors, keep upwind. Wear a sealed chemical suit, (pvc, butyl rubber). Wash away any material which may have come into contact with the body with copious amounts of soap and water. Consider appropriate evacuation.

BENZYLAMINE

DOT Number: n/a *DOT Hazard Class:* n/a *DOT Guide Number:* n/a
Synonyms: α-aminotoluene, phenylmethyl amine
STCC Number: n/a *Reportable Qty:* n/a
Mfg Name: Eastman Organic *Phone No:* 1-800-455-6325

Physical Description:

Physical Form: Liquid *Color:* Colorless to light yellow *Odor:* Strong ammonia
Other Information: n/a

Chemical Properties:

Specific Gravity: .98 *Vapor Density:* 3.7 *Boiling Point: 364° F(184.4° C)*
Melting Point: −51° F(−46.1° C) *Vapor Pressure:* n/a *Solubility in water:* Yes
Other Information: Floats and mixes with water.

Health Hazards:

Inhalation Hazard: Dizziness, headache, difficult breathing.
Ingestion Hazard: Harmful if swallowed.
Absorption Hazard: Will burn the skin and the eyes.
Hazards to Wildlife: Dangerous to aquatic life.
Decontamination Procedures: Wash away any material with copious amounts of soap and water.
First Aid Procedures: Remove victim to fresh air, call emergency medical care. If not breathing give CPR. If breathing is difficult administer oxygen. Treat for shock.

Fire Hazards:

Flashpoint: 168° F(75.5° C) *Ignition temperature:* n/a
Flammable Explosive High Range: n/a *Low Range:* n/a
Toxic Products of Combustion: Toxic oxides of nitrogen may be formed in fire.
Other Hazards: n/a
Possible extinguishing agents: Alcohol foam, dry chemical, carbon dioxide. Water may be ineffective!

Reactivity Hazards:

Reactive With: Liquid will attack plastics. *Other Reactions:* n/a

Corrosivity Hazards:

Corrosive With: n/a *Neutralizing Agent:* n/a

Radioactivity Hazards:

Radiation Emitted: n/a *Other Hazards:* n/a

Recommended Protection for Response Personnel:

Avoid breathing vapors, keep upwind. Wear a sealed chemical suit, (butyl rubber). Wash away any material which may have come into contact with the body with copious amounts of soap and water. Consider appropriate evacuation.

BERYLLIUM CHLORIDE

DOT Number: NA 1566 *DOT Hazard Class:* Poison B *DOT Guide Number:* 53
Synonyms: none given
STCC Number: 4923305 *Reportable Qty:* 5000/2270
Mfg Name: Research Organic *Phone No:* 1-201-759-7800

Physical Description:

Physical Form: Solid *Color:* White to green *Odor:* Sharp
Other Information: Absorbs water from the air

Chemical Properties:

Specific Gravity: 1.90 *Vapor Density:* 2 *Boiling Point: 968° F(520° C)*
Melting Point: n/a *Vapor Pressure:* n/a *Solubility in water:* Yes
Other Information: Reacts vigorously with water with evolution of heat, forms beryllium oxide and hydrochloric acid solution.

Health Hazards:

Inhalation Hazard: Poisonous if inhaled.
Ingestion Hazard: Poisonous if swallowed.
Absorption Hazard: Poisonous if absorbed.
Hazards to Wildlife: Dangerous to aquatic life.
Decontamination Procedures: Wash away any material with copious amounts of soap and water.
First Aid Procedures: Remove victim to fresh air, call emergency medical care. If not breathing give CPR. If breathing is difficult administer oxygen. Treat for shock.

Fire Hazards:

Flashpoint: n/a *Ignition temperature:* n/a
Flammable Explosive High Range: n/a *Low Range:* n/a
Toxic Products of Combustion: Irritating gases are produced in fire and when heated.
Other Hazards: Flammable and explosive nitrogen gases may be collected in enclosed spaces.
Possible extinguishing agents: Use suitable agent for type of surrounding fire.

Reactivity Hazards:

Reactive With: Water
Other Reaction: Reacts with common materials.

Corrosivity Hazards:

Corrosive With: Most metals in the presence of moisture.
Neutralizing Agent: Crushed limestone, soda ash, or lime

Radioactivity Hazards:

Radiation Emitted: n/a *Other Hazards:* n/a

Recommended Protection for Response Personnel:

Avoid breathing vapors, keep upwind. Wear a sealed chemical suit, (polycarbonate, butyl rubber). Wash away any material which may have come into contact with the body with copious amounts of soap and water. Consider appropriate evacuation.

BERYLLIUM FLUORIDE

DOT Number: NA 1556 *DOT Hazard Class:* Poison B *DOT Guide Number:* 53
Synonyms: none given
STCC Number: 4923310 *Reportable Qty:* 5000/2270
Mfg Name: Gallard-Schlesinger *Phone No:* 1-516-333-5600

Physical Description:

Physical Form: Solid *Color:* White *Odor:* Odorless
Other Information: n/a

Chemical Properties:

Specific Gravity: 1.99 *Vapor Density:* 2 *Boiling Point: n/a*
Melting Point: n/a *Vapor Pressure:* n/a *Solubility in water:* Yes

Health Hazards:

Inhalation Hazard: Poisonous if inhaled.
Ingestion Hazard: Poisonous if swallowed.
Absorption Hazard: Poisonous to the skin and eyes.
Hazards to Wildlife: Dangerous to aquatic life.
Decontamination Procedures: Wash away any material with copious amounts of soap and water.
First Aid Procedures: Remove victim to fresh air, call emergency medical care. If not breathing give CPR. If breathing is difficult administer oxygen. Treat for shock.

Fire Hazards:

Flashpoint: n/a *Ignition temperature:* n/a
Flammable Explosive High Range: n/a *Low Range:* n/a
Toxic Products of Combustion: Toxic and irritating vapors of unburned material may be formed in a fire.
Other Hazards: n/a
Possible extinguishing agents: Apply water from as far a distance as possible. Use alcohol foam, dry chemical, or carbon dioxide.

Reactivity Hazards:

Reactive With: n/a *Other Reactions:* n/a

Corrosivity Hazards:

Corrosive With: n/a *Neutralizing Agent:* n/a

Radioactivity Hazards:

Radiation Emitted: n/a *Other Hazards:* n/a

Recommended Protection for Response Personnel:

Avoid breathing vapors, keep upwind. Wear a sealed chemical suit, (polycarbonate, butyl rubber). Wash away any material which may have come into contact with the body with copious amounts of soap and water. Consider appropriate evacuation.

BERYLLIUM NITRATE

DOT Number: UN 2464 *DOT Hazard Class:* Oxidizer *DOT Guide Number:* 42
Synonyms: none given
STCC Number: 4918759 *Reportable Qty:* 5000/2270
Mfg Name: Brush Wellman Inc. *Phone No:* 1-216-486-4200

Physical Description:

Physical Form: Solid *Color:* White to pale yellow *Odor:* n/a
Other Information: Used in chemical analysis

Chemical Properties:

Specific Gravity: 1.56 *Vapor Density:* 6.4 *Boiling Point: n/a*
Melting Point: n/a *Vapor Pressure:* n/a *Solubility in water:* Yes
Other Information: n/a

Health Hazards:

Inhalation Hazard: Poisonous if inhaled.
Ingestion Hazard: Poisonous if swallowed.
Absorption Hazard: Poisonous if absorbed.
Hazards to Wildlife: Dangerous to aquatic life.
Decontamination Procedures: Wash away any material with copious amounts of soap and water.
First Aid Procedures: Remove victim to fresh air, call emergency medical care. If not breathing give CPR. If breathing is difficult administer oxygen. Treat for shock.

Fire Hazards:

Flashpoint: n/a *Ignition temperature:* n/a
Flammable Explosive High Range: n/a *Low Range:* n/a
Toxic Products of Combustion: Toxic oxides of nitrogen are produced in fire.
Other Hazards: Will accelerate the burning of other combustible materials. Prolonged exposure to fire and heat will result in an explosion.
Possible extinguishing agents: Flood with water from as far a distance as possible. Cool all affected containers.

Reactivity Hazards:

Reactive With: Water to form a weak solution of nitric acid. *Other Reaction:* n/a

Corrosivity Hazards:

Corrosive With: Wood and metal in the presence of moisture.
Neutralizing Agent: Crushed limestone, soda ash, or lime

Radioactivity Hazards:

Radiation Emitted: n/a *Other Hazards:* n/a

Recommended Protection for Response Personnel:

Avoid breathing dust and fumes of burning material, keep upwind. Wear a sealed chemical suit, (polycarbonate, butyl rubber). Wash away any material which may have come into contact with the body with copious amounts of soap and water. If the fire becomes uncontrollable, consider appropriate evacuation

BERYLLIUM OXIDE

DOT Number: UN 1566 *DOT Hazard Class:* Poison B *DOT Guide Number:* 53
Synonyms: beryllia, bromelite
STCC Number: n/a *Reportable Qty:* n/a
Mfg Name: Brush Wellman Inc. *Phone No:* 1-216-486-4200

Physical Description:

Physical Form: Solid *Color:* White *Odor:* Odorless
Other Information: n/a

Chemical Properties:

Specific Gravity: 3 *Vapor Density:* n/a *Boiling Point: n/a*
Melting Point: n/a *Vapor Pressure:* n/a *Solubility in water:* n/a
Other Information: Sinks in water.

Health Hazards:

Inhalation Hazard: Poisonous if inhaled.
Ingestion Hazard: Poisonous if swallowed.
Absorption Hazard: Poisonous to the skin and eyes.
Hazards to Wildlife: n/a
Decontamination Procedures: Wash away any material with copious amounts of soap and water.
First Aid Procedures: Remove victim to fresh air, call emergency medical care. If not breathing give CPR. If breathing is difficult administer oxygen. Treat for shock.

Fire Hazards:

Flashpoint: n/a *Ignition temperature:* n/a
Flammable Explosive High Range: n/a *Low Range:* n/a
Toxic Products of Combustion: Toxic beryllium oxide fumes may be formed in fire.
Other Hazards: n/a
Possible extinguishing agents: n/a

Reactivity Hazards:

Reactive With: n/a *Other Reactions:* n/a

Corrosivity Hazards:

Corrosive With: n/a *Neutralizing Agent:* n/a

Radioactivity Hazards:

Radiation Emitted: n/a *Other Hazards:* n/a

Recommended Protection for Response Personnel:

Avoid breathing vapors, keep upwind. Wear a sealed chemical suit, (polycarbonate, butyl rubber) Wash away any material which may have come into contact with the body with copious amounts of soap and water. Consider appropriate evacuation.

BERYLLIUM SULFATE

DOT Number: UN 1566 *DOT Hazard Class:* Poison B *DOT Guide Number:* 53
Synonyms: beryllium sulfate tetrahydrate
STCC Number: 4923330 *Reportable Qty:* n/a
Mfg Name: A.D. MacKay Inc. *Phone No:* 1-203-655-7401

Physical Description:

Physical Form: Solid *Color:* White *Odor:* Odorless
Other Information: n/a

Chemical Properties:

Specific Gravity: 1.71 *Vapor Density:* n/a *Boiling Point: n/a*
Melting Point: n/a *Vapor Pressure:* n/a *Solubility in water:* Yes
Other Information: Sinks and mixes in water.

Health Hazards:

Inhalation Hazard: Poisonous if inhaled.
Ingestion Hazard: Harmful if swallowed.
Absorption Hazard: Irritating to the skin and eyes.
Hazards to Wildlife: Dangerous to aquatic life.
Decontamination Procedures: Wash away any material with copious amounts of soap and water.
First Aid Procedures: Remove victim to fresh air, call emergency medical care. If not breathing give CPR. If breathing is difficult administer oxygen. Treat for shock.

Fire Hazards:

Flashpoint: n/a *Ignition temperature:* n/a
Flammable Explosive High Range: n/a *Low Range:* n/a
Toxic Products of Combustion: Toxic beryllium oxide and sulfuric acid fumes may be formed in a fire.
Other Hazards: n/a
Possible extinguishing agents: Apply water from as far a distance as possible. Use alcohol foam, dry chemical, or carbon dioxide.

Reactivity Hazards:

Reactive With: n/a *Other Reactions:* n/a

Corrosivity Hazards:

Corrosive With: n/a *Neutralizing Agent:* n/a

Radioactivity Hazards:

Radiation Emitted: n/a *Other Hazards:* n/a

Recommended Protection for Response Personnel:

Avoid breathing vapors, keep upwind. Wear a sealed chemical suit, (polycarbonate, butyl rubber). Wash away any material which may have come into contact with the body with copious amounts of soap and water. Consider appropriate evacuation.

BORON TRIBROMIDE

DOT Number: UN 2692 *DOT Hazard Class:* Corrosive *DOT Guide Number:* 59
Synonyms: none given
STCC Number: 4932010 *Reportable Qty:* n/a
Mfg Name: Gallard-Schlesinger *Phone No:* 1-516-333-5600

Physical Description:

Physical Form: Liquid *Color:* Colorless *Odor:* Sharp
Other Information: n/a

Chemical Properties:

Specific Gravity: 2.65 *Vapor Density:* 8.6 *Boiling Point: 196° F(91.1° C)*
Melting Point: n/a *Vapor Pressure:* n/a *Solubility in water:* Reacts
Other Information: Reacts violently with water, poisonous vapor is produced.

Health Hazards:

Inhalation Hazard: Coughing, difficulty in breathing.
Ingestion Hazard: Nausea and vomiting.
Absorption Hazard: Will burn the skin and eyes.
Hazards to Wildlife: n/a
Decontamination Procedures: Wash away any material with copious amounts of soap and water.
First Aid Procedures: Remove victim to fresh air, call emergency medical care. If not breathing give CPR. If breathing is difficult administer oxygen. Treat for shock.

Fire Hazards:

Flashpoint: n/a *Ignition temperature:* n/a
Flammable Explosive High Range: n/a *Low Range:* n/a
Toxic Products of Combustion: Toxic fumes of the chemical or hydrogen bromide may be formed in a fire.
Other Hazards: n/a
Possible extinguishing agents: Do not use water, use dry chemical, dry sand, or carbon dioxide.

Reactivity Hazards:

Reactive With: Water to form hydrochromic acid solution and fumes. *Other Reactions:* n/a

Corrosivity Hazards:

Corrosive With: Metal and tissue *Neutralizing Agent:* Crushed limestone, soda ash, or lime.

Radioactivity Hazards:

Radiation Emitted: n/a *Other Hazards:* n/a

Recommended Protection for Response Personnel:

Avoid breathing vapors, keep upwind. Wear a sealed chemical suit (polycarbonate, butyl rubber). Wash away any material which may have come into contact with the body with copious amounts of soap and water. Consider appropriate evacuation.

BORON TRICHLORIDE

DOT Number: UN 1741 *DOT Hazard Class:* Corrosive *DOT Guide Number:* 15
Synonyms: boron chloride
STCC Number: 4932011 *Reportable Qty:* n/a
Mfg Name: Matheson Gas *Phone No:* 1-201-867-4100

Physical Description:

Physical Form: Liquid *Color:* Colorless *Odor:* Irritating
Other Information: It is used as a catalyst in chemical manufacturing, in soldering flux, and many other uses.

Chemical Properties:

Specific Gravity: 1.35 *Vapor Density:* 4.3 *Boiling Point: 54.3° F(12.3° C)*
Melting Point: n/a *Vapor Pressure:* 4.4 psig at 68° F(20° C) *Solubility in water:* Reacts
Other Information: Reacts violently with water. Irritating, visible vapor cloud is produced.

Health Hazards:

Inhalation Hazard: Coughing, difficulty in breathing.
Ingestion Hazard: Poisonous if swallowed.
Absorption Hazard: Will burn the skin and the eyes.
Hazards to Wildlife: n/a
Decontamination Procedures: Wash away any material with copious amounts of soap and water.
First Aid Procedures: Remove victim to fresh air, call emergency medical care. If not breathing give CPR. If breathing is difficult administer oxygen. Treat for shock.

Fire Hazards:

Flashpoint: n/a *Ignition temperature:* n/a
Flammable Explosive High Range: n/a *Low Range:* n/a
Toxic Products of Combustion: Poisonous gases may be produced when heated.
Other Hazards: n/a
Possible extinguishing agents: Do not use water. Use dry chemical, dry sand or carbon dioxide.

Reactivity Hazards:

Reactive With: Water to form hydrochloric acid and boric acid. *Other Reactions:* n/a

Corrosivity Hazards:

Corrosive With: Metal and tissue *Neutralizing Agent:* Crushed limestone, soda ash, or lime.

Radioactivity Hazards:

Radiation Emitted: n/a *Other Hazards:* n/a

Recommended Protection for Response Personnel:

Avoid breathing vapors, keep upwind. Wear a sealed chemical suit, (butyl rubber, pvc). Wash away any material which may have come into contact with the body with copious amounts of soap and water. Consider appropriate evacuation.

BROMINE

DOT Number: UN 1744 *DOT Hazard Class:* Corrosive *DOT Guide Number:* 59
Synonyms: none given
STCC Number: 4936110 *Reportable Qty:* n/a
Mfg Name: Dow Chemical *Phone No:* 1-517-636-4400

Physical Description:

Physical Form: Liquid *Color:* Reddish brown *Odor:* Sharp, irritating
Other Information: This material is also poisonous, and an oxidizer.

Chemical Properties:

Specific Gravity: 3.12 *Vapor Density:* 5.5 *Boiling Point: 138° F(58.5° C)*
Melting Point: 19° F(−7.2° C) *Vapor Pressure:* 175 mm Hg at 68° F (20° C)
Solubility in water: Yes *Degree of Solubility:* 3.5%
Other Information: Not combustible, but a strong oxidizer.

Health Hazards:

Inhalation Hazard: Coughing, difficulty in breathing, loss of consciousness.
Ingestion Hazard: Harmful if swallowed.
Absorption Hazard: Will burn the skin and eyes.
Hazards to Wildlife: Dangerous to aquatic life.
Decontamination Procedures: Wash away any material with copious amounts of soap and water.
First Aid Procedures: Remove victim to fresh air, call emergency medical care. If not breathing give CPR. If breathing is difficult administer oxygen. Treat for shock.

Fire Hazards:

Flashpoint: n/a *Ignition temperature:* n/a
Flammable Explosive High Range: 11.5 *Low Range:* 2
Toxic Products of Combustion: Poisonous gases are produced in fire.
Other Hazards: May cause fire on contact with other combustibles.
Possible extinguishing agents: Flood with water from as far a distance as possible. Cool all affected containers.

Reactivity Hazards:

Reactive With: Reacts violently with aluminum.
Other Reactions: Iron, steel, stainless steel are corroded with bromine.

Corrosivity Hazards:

Corrosive With: Metals and tissue *Neutralizing Agent:* Crushed limestone, soda ash, or lime

Radioactivity Hazards:

Radiation Emitted: n/a *Other Hazards:* n/a

Recommended Protection for Response Personnel:

Avoid breathing dust and fumes of burning material, keep upwind. Wear a sealed chemical suit, (chlorinated polyethylene, viton, pvc). Wash away any material which may have come into contact with the body with copious amounts of soap and water. If the fire becomes uncontrollable, consider appropriate evacuation.

BROMINE PENTAFLUORIDE

DOT Number: UN 1745 *DOT Hazard Class:* Oxidizer *DOT Guide Number:* 44
Synonyms: none given
STCC Number: 4918505 *Reportable Qty:* n/a
Mfg Name: Union Carbide *Phone No:* 1-203-794-2000

Physical Description:

Physical Form: Liquefied gas *Color:* Colorless *Odor:* Irritating
Other Information: It is used to make other chemicals, and in rockets.

Chemical Properties:

Specific Gravity: 2.48 *Vapor Density:* 6.1 *Boiling Point: 106° F(41.1° C)*
Melting Point: n/a *Vapor Pressure:* n/a *Solubility in water:* Reacts
Other Information: Reacts violently with water. Poisonous vapor is produced. Material may only be shipped in containers without a safety relief device.

Health Hazards:

Inhalation Hazard: Coughing, difficulty in breathing.
Ingestion Hazard: Harmful if swallowed.
Absorption Hazard: Will burn the skin and eyes.
Hazards to Wildlife: n/a
Decontamination Procedures: Wash away any material with copious amounts of soap and water.
First Aid Procedures: Remove victim to fresh air, call emergency medical care. If not breathing give CPR. If breathing is difficult administer oxygen. Treat for shock.

Fire Hazards:

Flashpoint: n/a *Ignition temperature:* n/a
Flammable Explosive High Range: n/a *Low Range:* n/a
Toxic Products of Combustion: Poisonous gases may be produced in a fire.
Other Hazards: n/a
Possible extinguishing agents: Do not use water, use dry chemical, dry sand, or carbon dioxide.

Reactivity Hazards:

Reactive With: Water to form hydrogen chloride. *Other Reactions:* n/a

Corrosivity Hazards:

Corrosive With: Metal and tissue *Neutralizing Agent:* Crushed limestone, soda ash, or lime.

Radioactivity Hazards:

Radiation Emitted: n/a *Other Hazards:* n/a

Recommended Protection for Response Personnel:

Avoid breathing vapors, keep upwind. Wear appropriate chemical clothing. Wash away any material which may come into contact with the body with copious amounts of soap and water. If the cylinders are involved in a fire, they may violently rupture and rocket. (BLEVE!). Consider appropriate evacuation.

BROMINE TRIFLUORIDE

DOT Number: UN 1746 *DOT Hazard Class:* Oxidizer *DOT Guide Number:* 44
Synonyms: none given
STCC Number: 4918507 *Reportable Qty:* n/a
Mfg Name: Matheson Gas *Phone No:* 1-201-867-4100

Physical Description:

Physical Form: Liquid *Color:* Colorless *Odor:* Extremely irritating
Other Information: n/a

Chemical Properties:

Specific Gravity: 2.81 *Vapor Density:* 4.7 *Boiling Point: 258° F(125.5° C)*
Melting Point: n/a *Vapor Pressure:* .15 psia at 68° F(20° C) *Solubility in water:* Reacts
Other Information: Reacts violently with water. Poisonous vapor is produced. Material may only be shipped in containers without a safety relief device.

Health Hazards:

Inhalation Hazard: Poisonous if inhaled.
Ingestion Hazard: Poisonous if swallowed.
Absorption Hazard: Will burn the skin and eyes.
Hazards to Wildlife: n/a
Decontamination Procedures: Wash away any material with copious amounts of soap and water.
First Aid Procedures: Remove victim to fresh air, call emergency medical care. If not breathing give CPR. If breathing is difficult administer oxygen. Treat for shock.

Fire Hazards:

Flashpoint: n/a *Ignition temperature:* n/a
Flammable Explosive High Range: n/a *Low Range:* n/a
Toxic Products of Combustion: Poisonous gases may be produced in a fire.
Other Hazards: Will accelerate the burning of combustible material.
Possible extinguishing agents: Do not use water. Use dry chemical, dry sand, or carbon dioxide.

Reactivity Hazards:

Reactive With: Water to form hydrochloric acid. *Other Reactions:* n/a

Corrosivity Hazards:

Corrosive With: Metal and tissue *Neutralizing Agent:* Crushed limestone, soda ash, or lime.

Radioactivity Hazards:

Radiation Emitted: n/a *Other Hazards:* n/a

Recommended Protection for Response Personnel:

Avoid breathing vapors, keep upwind. Wear appropriate chemical clothing. Wash away any material which may come into contact with the body with copious amounts of soap and water. If the cylinders are involved in a fire, they may violently rupture and rocket. (BLEVE!) Consider appropriate evacuation.

BROMOBENZENE

DOT Number: UN 2514 *DOT Hazard Class:* Combustible liquid *DOT Guide Number:* 26
Synonyms: bromobenzol, monobromobenzene, phenyl bromide
STCC Number: 4913112 *Reportable Qty:* n/a
Mfg Name: Mallinckrodt Inc. *Phone No:* 1-314-895-2000

Physical Description:

Physical Form: Liquid *Color:* Colorless *Odor:* Pleasant
Other Information: n/a

Chemical Properties:

Specific Gravity: 1.5 *Vapor Density:* 5.4 *Boiling Point: 313° F(156.1° C)*
Melting Point: n/a *Vapor Pressure:* n/a *Solubility in water:* No
Other Information: Heavier, insoluble in water, vapors are heavier than air.

Health Hazards:

Inhalation Hazard: Harmful if inhaled.
Ingestion Hazard: Harmful if swallowed.
Absorption Hazard: Irritating to the skin and eyes.
Hazards to Wildlife: n/a
Decontamination Procedures: Wash away any material with copious amounts of soap and water.
First Aid Procedures: Remove victim to fresh air, call emergency medical care. If not breathing give CPR. If breathing is difficult administer oxygen. Treat for shock.

Fire Hazards:

Flashpoint: 124° F(51.1c) *Ignition temperature:* 1049° F(565° C)
Flammable Explosive High Range: n/a *Low Range:* n/a
Toxic Products of Combustion: Irritating hydrogen bromide and other irritating gases may be formed in fires.
Other Hazards: Keep exposed containers cool with water.
Possible extinguishing agents: Do not extinguish the fire unless the flow can be stopped. Use foam, dry chemical, or carbon dioxide.

Reactivity Hazards:

Reactive With: n/a *Other Reactions:* n/a

Corrosivity Hazards:

Corrosive With: n/a *Neutralizing Agent:* n/a

Radioactivity Hazards:

Radiation Emitted: n/a *Other Hazards:* n/a

Recommended Protection for Response Personnel:

Avoid breathing vapors, keep upwind. Structural protective clothing provides limited protection. Wash away any material which may have come into contact with the body with copious amounts of soap and water. Consider appropriate evacuation.

BRUCINE

DOT Number: UN 1570 *DOT Hazard Class:* Poison B *DOT Guide Number:* 53
Synonyms: 10,11-dimethoxystrychnine, (-)brucine dihydrate
STCC Number: 4921411 *Reportable Qty:* 100/45.4
Mfg Name: Gallard-Schlesinger *Phone No:* 1-516-333-5600

Physical Description:

Physical Form: Solid *Color:* White *Odor:* Odorless
Other Information: n/a

Chemical Properties:

Specific Gravity: 1 *Vapor Density:* 13 *Boiling Point: Decomposes*
Melting Point: n/a *Vapor Pressure:* n/a *Solubility in water:* Slight
Other Information: Sinks in water.

Health Hazards:

Inhalation Hazard: Poisonous if inhaled.
Ingestion Hazard: Poisonous if swallowed.
Absorption Hazard: Poisonous to the skin and eyes.
Hazards to Wildlife: n/a
Decontamination Procedures: Wash away any material with copious amounts of soap and water.
First Aid Procedures: Remove victim to fresh air, call emergency medical care. If not breathing give CPR. If breathing is difficult administer oxygen. Treat for shock.

Fire Hazards:

Flashpoint: n/a *Ignition temperature:* n/a
Flammable Explosive High Range: n/a *Low Range:* n/a
Toxic Products of Combustion: Toxic oxides of nitrogen may be formed in a fire.
Other Hazards: n/a
Possible extinguishing agents: Extinguish fire using suitable agent for type of surrounding fire. Use foam, dry chemical, or carbon dioxide.

Reactivity Hazards:

Reactive With: n/a *Other Reactions:* n/a

Corrosivity Hazards:

Corrosive With: n/a *Neutralizing Agent:* n/a

Radioactivity Hazards:

Radiation Emitted: n/a *Other Hazards:* n/a

Recommended Protection for Response Personnel:

Avoid breathing vapors, keep upwind. Structural protective clothing provides limited protection. Wash away any material which may have come into contact with the body with copious amounts of soap and water. Consider appropriate evacuation.

BUTADIENE, INHIBITED

DOT Number: UN 1010 *DOT Hazard Class:* Flammable liquid *DOT Guide Number:* 17
Synonyms: bivinyl, divinal
STCC Number: 4905704 *Reportable Qty:* n/a
Mfg Name: Phillips Petroleum *Phone No:* 1-918-661-6600

Physical Description:

Physical Form: Gas *Color:* Colorless *Odor:* Gasoline
Other Information: Used to make synthetic rubber, plastics, and other chemicals.

Chemical Properties:

Specific Gravity: .621 *Vapor Density:* 9.1 *Boiling Point: 24° F(−4.4° C)*
Melting Point: −164° F(−108.8° C) *Vapor Pressure:* 910 mm Hg at 68° F (20° C)
Solubility in water: No *Degree of Solubility:* .05%
Other Information: Liquid can cause frostbite, is shipped in inhibited form as the chemical name butadiene and is liable to polymerize.

Health Hazards:

Inhalation Hazard: Harmful if inhaled.
Ingestion Hazard: Harmful if swallowed.
Absorption Hazard: Will cause frostbite.
Hazards to Wildlife: n/a
Decontamination Procedures: Wash away any material with copious amounts of soap and water.
First Aid Procedures: Remove victim to fresh air, call emergency medical care. If not breathing give CPR. If breathing is difficult administer oxygen. Treat for shock.

Fire Hazards:

Flashpoint: 21° F(−6.1° C) *Ignition temperature:* 788° F(420° C)
Flammable Explosive High Range: 12 *Low Range:* 2
Toxic Products of Combustion: Under fire conditions, the containers may violently rupture and rocket.
Other Hazards: If polymerization occurs, container may violently rupture. Vapors may easily ignite and may flashback.
Possible extinguishing agents: Do not extinguish fire unless flow can be stopped! Use large quantities of water.

Reactivity Hazards:

Reactive With: Heat and air *Other Reactions:* n/a

Corrosivity Hazards:

Corrosive With: n/a *Neutralizing Agent:* n/a

Radioactivity Hazards:

Radiation Emitted: n/a *Other Hazards:* n/a

Recommended Protection for Response Personnel:

Avoid breathing vapors of burning material, keep upwind. Structural protective clothing provides limited protection. Wash away any material which may have come into contact with the body with copious amounts of soap and water. If the fire becomes uncontrollable, consider appropriate evacuation.

BUTANE

DOT Number: UN 1011 *DOT Hazard Class:* Flammable gas *DOT Guide Number:* 22
Synonyms: n-butane
STCC Number: 4905706 *Reportable Qty:* n/a
Mfg Name: Phillips Petroleum *Phone No:* 1-918-661-6600

Physical Description:

Physical Form: Gas *Color:* Colorless *Odor:* Gasoline
Other Information: Used as a fuel, as an aerosol propellant in cigarette lighters, and to make other chemicals.

Chemical Properties:

Specific Gravity: .60 *Vapor Density:* 2 *Boiling Point: 31° F(-.5° C)*
Melting Point: n/a *Vapor Pressure:* 20 mm Hg at 68° F (20° C) *Solubility in water:* No
Other Information: Shipped as a liquefied gas, can cause frostbite.

Health Hazards:

Inhalation Hazard: Dizziness, difficulty in breathing.
Ingestion Hazard: n/a
Absorption Hazard: Will cause frostbite.
Hazards to Wildlife: n/a
Decontamination Procedures: Wash away any material with copious amounts of soap and water.
First Aid Procedures: Remove victim to fresh air, call emergency medical care. If not breathing give CPR. If breathing is difficult administer oxygen. Treat for shock.

Fire Hazards:

Flashpoint: – 100° F(– 73.3° C) *Ignition temperature:* 550° F(287.7c)
Flammable Explosive High Range: 8.4 *Low Range:* 1.6
Toxic Products of Combustion: Easily ignited, can flashback to the source.
Other Hazards: The leak can be either liquid or vapor, under fire conditions, the container may violently rupture and rocket.(BLEVE)
Possible extinguishing agents: Do not extinguish fire unless flow can be stopped! Use large quantities of water.

Reactivity Hazards:

Reactive With: n/a *Other Reactions:* n/a

Corrosivity Hazards:

Corrosive With: n/a *Neutralizing Agent:* n/a

Radioactivity Hazards:

Radiation Emitted: n/a *Other Hazards:* n/a

Recommended Protection for Response Personnel:

Avoid breathing vapors of burning material, keep upwind. Wear a sealed chemical suit, (polycarbonate, viton, nitrile, neoprene). Wash away any material which may have come into contact with the body with copious amounts of soap and water. If the fire becomes uncontrollable, consider appropriate evacuation.

BUTANEDIOL

DOT Number: UN 1987 *DOT Hazard Class:* Flammable liquid *DOT Guide Number:* 26
Synonyms: 1,4-dihydroxybutane, tetramethylene glycol
STCC Number: n/a *Reportable Qty:* n/a
Mfg Name: GAF Corp. *Phone No:* 1-201-628-3000

Physical Description:

Physical Form: Liquid *Color:* Colorless *Odor:* Odorless
Other Information: n/a

Chemical Properties:

Specific Gravity: 1 *Vapor Density:* 3.2 *Boiling Point: 442° F(227.7° C)*
Melting Point: 86° F(30° C) *Vapor Pressure:* 1 mm Hg at 100° F(37.7° C) *Solubility in water:* Yes
Other Information: Sinks and mixes with water.

Health Hazards:

Inhalation Hazard: n/a
Ingestion Hazard: Harmful if swallowed.
Absorption Hazard: Irritating to the skin and eyes.
Hazards to Wildlife: n/a
Decontamination Procedures: Wash away any material with copious amounts of soap and water.
First Aid Procedures: Remove victim to fresh air, call emergency medical care. If not breathing give CPR. If breathing is difficult administer oxygen. Treat for shock.

Fire Hazards:

Flashpoint: 250° F(121.1° C) *Ignition temperature:* 556° F(291.1° C)
Flammable Explosive High Range: n/a *Low Range:* n/a
Toxic Products of Combustion: n/a
Other Hazards: n/a
Possible extinguishing agents: Alcohol foam, dry chemical, carbon dioxide.

Reactivity Hazards:

Reactive With: n/a *Other Reactions:* n/a

Corrosivity Hazards:

Corrosive With: n/a *Neutralizing Agent:* n/a

Radioactivity Hazards:

Radiation Emitted: n/a *Other Hazards:* n/a

Recommended Protection for Response Personnel:

Avoid breathing vapors, keep upwind. Structural protective clothing provides limited protection. Wash away any material which may have come into contact with the body with copious amounts of soap and water. Consider appropriate evacuation.

BUTENE

DOT Number: UN 1012 *DOT Hazard Class:* Flammable gas *DOT Guide Number:* 22
Synonyms: α-butene, 1-butene, butylene, ethylethylene
STCC Number: 4905707 *Reportable Qty:* n/a *CHEMTREC Phone No:* 1-800-424-9300

Physical Description:

Physical Form: Gas *Color:* Colorless *Odor:* Sweet, mild, aromatic
Other Information: It is shipped as a liquefied gas under its own vapor pressure. For transportation, it may be stenched.

Chemical Properties:

Specific Gravity: .59 *Vapor Density:* 1.9 *Boiling Point: 20.7° F(−6.2° C)*
Melting Point: −279° F(−172.7° C) *Vapor Pressure:* 1.36 atm at 68° F(20° C) *Solubility in water:* No
Other Information: Can easily asphyxiate by displacement of air.

Health Hazards:

Inhalation Hazard: May cause dizziness or unconsciousness.
Ingestion Hazard: Will cause frostbite.
Absorption Hazard: Will cause frostbite.
Hazards to Wildlife: n/a
Decontamination Procedures: Wash away any material with copious amounts of soap and water.
First Aid Procedures: Remove victim to fresh air, call emergency medical care. If not breathing give CPR. If breathing is difficult administer oxygen. Treat for shock.

Fire Hazards:

Flashpoint: 0° F(−17.7° C) *Ignition temperature:* 725° F(385° C)
Flammable Explosive High Range: 10 *Low Range:* 1.6
Toxic Products of Combustion: n/a
Other Hazards: Do not extinguish fire unless the flow can be stopped. Under fire conditions, cylinders or tank cars may violently rupture or rocket. (BLEVE)
Possible extinguishing agents: Apply water from as far a distance as possible, use water fog in large quantities.

Reactivity Hazards:

Reactive With: Organic peroxides. *Other Reactions:* n/a

Corrosivity Hazards:

Corrosive With: n/a *Neutralizing Agent:* n/a

Radioactivity Hazards:

Radiation Emitted: n/a *Other Hazards:* n/a

Recommended Protection for Response Personnel:

Avoid breathing vapors, keep upwind. Structural protective clothing provides limited protection. Wash away any material which may have come into contact with the body with copious amounts of soap and water. If the fire becomes uncontrollable, consider appropriate evacuation. (BLEVE)

BUTENEDIOL

DOT Number: UN 1987 *DOT Hazard Class:* Flammable liquid *DOT Guide Number:* 26
Synonyms: 2-butene-1,4-diol, cis-2-butene-1,4-diol, 1,4-dihydroxy-2-butene
STCC Number: n/a *Reportable Qty:* n/a
Mfg Name: GAF Corp. *Phone No:* 1-201-628-3000

Physical Description:

Physical Form: Liquid *Color:* Light yellow *Odor:* Odorless
Other Information: n/a

Chemical Properties:

Specific Gravity: 1.07 *Vapor Density:* n/a *Boiling Point: 453° F(233.8° C)*
Melting Point: 45° F(7.2° C) *Vapor Pressure:* n/a *Solubility in water:* Yes
Other Information: Sinks and mixes with water.

Health Hazards:

Inhalation Hazard: n/a
Ingestion Hazard: Harmful if swallowed.
Absorption Hazard: Irritating to the skin and eyes.
Hazards to Wildlife: n/a
Decontamination Procedures: Wash away any material with copious amounts of soap and water.
First Aid Procedures: Remove victim to fresh air, call emergency medical care. If not breathing give CPR. If breathing is difficult administer oxygen. Treat for shock.

Fire Hazards:

Flashpoint: 263° F(128.3° C) *Ignition temperature:* n/a
Flammable Explosive High Range: n/a *Low Range:* n/a
Toxic Products of Combustion: n/a
Other Hazards: n/a
Possible extinguishing agents: Alcohol foam, dry chemical, carbon dioxide.

Reactivity Hazards:

Reactive With: n/a *Other Reactions:* n/a

Corrosivity Hazards:

Corrosive With: n/a *Neutralizing Agent:* n/a

Radioactivity Hazards:

Radiation Emitted: n/a *Other Hazards:* n/a

Recommended Protection for Response Personnel:

Avoid breathing vapors, keep upwind. Structural protective clothing provides limited protection. Wash away any material which may have come into contact with the body with copious amounts of soap and water. Consider appropriate evacuation.

BUTYL ACETATE

DOT Number: UN 1123 *DOT Hazard Class:* Flammable liquid *DOT Guide Number:* 26
Synonyms: n-butyl acetate
STCC Number: 4909128 *Reportable Qty:* 5000/2270
Mfg Name: Allied Corp. *Phone No:* 1-201-455-2000

Physical Description:

Physical Form: Liquid *Color:* Colorless *Odor:* Fruity
Other Information: Lighter than and very soluble in water. Weighs 7.4 lbs/3.2 kg per gallon/3.8 l.

Chemical Properties:

Specific Gravity: .9 *Vapor Density:* 4 *Boiling Point: 260° F(126.6° C)*
Melting Point: –101° F(–73.8° C) *Vapor Pressure:* 10 mm Hg at 68° F (20° C)
Solubility in water: Slight *Degree of Solubility:* .8%
Other Information: n/a

Health Hazards:

Inhalation Hazard: Headaches, dizziness, difficulty in breathing.
Ingestion Hazard: Harmful if swallowed.
Absorption Hazard: Irritating to the skin and eyes.
Hazards to Wildlife: n/a
Decontamination Procedures: Wash away any material with copious amounts of soap and water.
First Aid Procedures: Remove victim to fresh air, call emergency medical care. If not breathing give CPR. If breathing is difficult administer oxygen. Treat for shock.

Fire Hazards:

Flashpoint: 72° F(22.2° C) *Ignition temperature:* 797° F(425° C)
Flammable Explosive High Range: 7.6 *Low Range:* 1.7
Toxic Products of Combustion: Easily ignited, can flashback to the source.
Other Hazards: The leak can be either liquid or vapor. Under fire conditions, the container may violently rupture and rocket. (BLEVE)
Possible extinguishing agents: Do not extinguish the fire unless the flow can be stopped. Use large quantities of water. Use foam, dry chemical, CO_2.

Reactivity Hazards:

Reactive With: n/a *Other Reactions:* n/a

Corrosivity Hazards:

Corrosive With: n/a *Neutralizing Agent:* n/a

Radioactivity Hazards:

Radiation Emitted: n/a *Other Hazards:* n/a

Recommended Protection for Response Personnel:

Avoid breathing vapors of burning material, keep upwind. Structural protective clothing provides limited protection. Wash away any material which may have come into contact with the body with copious amounts of soap and water. If the fire becomes uncontrollable, consider appropriate evacuation.

sec-BUTYL ACETATE

DOT Number: UN 1124 *DOT Hazard Class:* Flammable liquid *DOT Guide Number:* 26
Synonyms: acetic acid, sec-butyl ester
STCC Number: n/a *Reportable Qty:* n/a
Mfg Name: Jones Chemical *Phone No:* 1-716-538-2311

Physical Description:

Physical Form: Liquid *Color:* Colorless *Odor:* Pleasant fruity
Other Information: n/a

Chemical Properties:

Specific Gravity: .87 *Vapor Density:* 4 *Boiling Point: 234° F(112.2° C)*
Melting Point: –100° F(–73.3° C) *Vapor Pressure:* n/a *Solubility in water:* n/a
Other Information: Floats on water. Flammable, irritating vapor is produced.

Health Hazards:

Inhalation Hazard: Nausea, headache, difficulty in breathing.
Ingestion Hazard: n/a
Absorption Hazard: Irritating to the skin and eyes.
Hazards to Wildlife: n/a
Decontamination Procedures: Wash away any material with copious amounts of soap and water.
First Aid Procedures: Remove victim to fresh air, call emergency medical care. If not breathing give CPR. If breathing is difficult administer oxygen. Treat for shock.

Fire Hazards:

Flashpoint: 62° F(16.6° C) *Ignition temperature:* n/a
Flammable Explosive High Range: 9.8 *Low Range:* 1.7
Toxic Products of Combustion: n/a
Other Hazards: Flashback along vapor trail may occur. Vapors may explode if ignited in an enclosed area.
Possible extinguishing agents: Foam, dry chemical, carbon dioxide. Cool all affected containers with large quantities of water.

Reactivity Hazards:

Reactive With: n/a *Other Reactions:* n/a

Corrosivity Hazards:

Corrosive With: n/a *Neutralizing Agent:* n/a

Radioactivity Hazards:

Radiation Emitted: n/a *Other Hazards:* n/a

Recommended Protection for Response Personnel:

Avoid breathing vapors, keep upwind. Structural protective clothing provides limited protection. Wash away any material which may have come into contact with the body with copious amounts of soap and water. If the fire becomes uncontrollable, consider appropriate evacuation.

BUTYL ACRYLATE

DOT Number: NA 2348 *DOT Hazard Class:* Combustible liquid *DOT Guide Number:* 26
Synonyms: acrylic acid, n-butyl acrylate, n-butyl ester
STCC Number: 4912215 *Reportable Qty:* n/a
Mfg Name: Union Carbide *Phone No:* 1-203-794-2000

Physical Description:

Physical Form: Liquid *Color:* Colorless *Odor:* Sharp fragrant
Other Information: It is used in making paints, coating, caulks, sealants, adhesives, other chemicals, and a variety of other products.

Chemical Properties:

Specific Gravity: .90 *Vapor Density:* 4.4 *Boiling Point: 295° F(146.1° C)*
Melting Point: −84.3° F(−64.6° C) *Vapor Pressure:* 4 mm Hg at 68° F (20° C) *Solubility in water:* No
Other Information: Floats on water

Health Hazards:

Inhalation Hazard: n/a
Ingestion Hazard: Harmful if swallowed.
Absorption Hazard: Irritating to the skin and eyes.
Hazards to Wildlife: n/a
Decontamination Procedures: Wash away any material with copious amounts of soap and water.
First Aid Procedures: Remove victim to fresh air, call emergency medical care. If not breathing give CPR. If breathing is difficult administer oxygen. Treat for shock.

Fire Hazards:

Flashpoint: 105° F(40.5° C) *Ignition temperature:* 540° F(282.2° C)
Flammable Explosive High Range: 9.9 *Low Range:* 1.3
Toxic Products of Combustion: n/a
Other Hazards: Accumulation of vapors if confined in spaces such as buildings and sewers may explode if ignited.
Possible extinguishing agents: Apply water from as far a distance as possible, use foam, dry chemical, or carbon dioxide.

Reactivity Hazards:

Reactive With: Oxidizing agents and sunlight *Other Reactions:* n/a

Corrosivity Hazards:

Corrosive With: n/a *Neutralizing Agent:* n/a

Radioactivity Hazards:

Radiation Emitted: n/a *Other Hazards:* n/a

Recommended Protection for Response Personnel:

Avoid breathing vapors, keep upwind. Structural protective clothing provides limited protection. Wash away any material which may have come into contact with the body with copious amounts of soap and water. If the containers become involved in fire, consider appropriate evacuation.

iso-BUTYL ACRYLATE

DOT Number: UN 2527 *DOT Hazard Class:* n/a *DOT Guide Number:* 27
Synonyms: acrylic acid, isobutyl ester, isobutyl 2-propenoate
STCC Number: n/a *Reportable Qty:* n/a
Mfg Name: Union Carbide *Phone No:* 1-203-794-2000

Physical Description:

Physical Form: Liquid *Color:* Colorless *Odor:* Sharp fragrant
Other Information: n/a

Chemical Properties:

Specific Gravity: .88 *Vapor Density:* 2.5 *Boiling Point: 280° F(137.7° C)*
Melting Point: −78° F(−61.1° C) *Vapor Pressure:* n/a *Solubility in water:* n/a
Other Information: Floats on water. Irritating vapor is produced.

Health Hazards:

Inhalation Hazard: Nausea, headache, difficulty in breathing.
Ingestion Hazard: Harmful if swallowed.
Absorption Hazard: Irritating to the skin and eyes.
Hazards to Wildlife: n/a
Decontamination Procedures: Wash away any material with copious amounts of soap and water.
First Aid Procedures: Remove victim to fresh air, call emergency medical care. If not breathing give CPR. If breathing is difficult administer oxygen. Treat for shock.

Fire Hazards:

Flashpoint: 94° F(34.4° C) *Ignition temperature:* 644° F(340° C)
Flammable Explosive High Range: 8 *Low Range:* 1.9
Toxic Products of Combustion: n/a
Other Hazards: Flashback along vapor trail may occur. Vapors may explode if ignited in an enclosed area. Will polymerize when hot. Polymerization can be explosive!
Possible extinguishing agents: Foam, dry chemical, carbon dioxide. Cool all affected all effected containers with large quantities of water.

Reactivity Hazards:

Reactive With: n/a *Other Reactions:* n/a

Corrosivity Hazards:

Corrosive With: n/a *Neutralizing Agent:* n/a

Radioactivity Hazards:

Radiation Emitted: n/a *Other Hazards:* n/a

Recommended Protection for Response Personnel:

Avoid breathing vapors, keep upwind. Structural protective clothing provides limited protection. Wash away any material which may have come into contact with the body with copious amounts of soap and water. If the fire becomes uncontrollable, consider appropriate evacuation.

n-BUTYL ACRYLATE

DOT Number: UN 2348 *DOT Hazard Class:* n/a *DOT Guide Number:* 26
Synonyms: acrylic acid, butyl acrylate, n-butyl ester, n-butyl-2-propenoate
STCC Number: n/a *Reportable Qty:* n/a
Mfg Name: Union Carbide *Phone No:* 1-203-794-2000

Physical Description:

Physical Form: Liquid *Color:* Colorless *Odor:* Sharp fragrant
Other Information: n/a

Chemical Properties:

Specific Gravity: .89 *Vapor Density:* 4.4 *Boiling Point: 299° F(148.3° C)*
Melting Point: −83° F(−63.8° C) *Vapor Pressure:* n/a *Solubility in water:* n/a
Other Information: Floats on water

Health Hazards:

Inhalation Hazard: Nausea, headache, difficulty in breathing.
Ingestion Hazard: n/a
Absorption Hazard: Irritating to the skin and eyes.
Hazards to Wildlife: n/a
Decontamination Procedures: Wash away any material with copious amounts of soap and water.
First Aid Procedures: Remove victim to fresh air, call emergency medical care. If not breathing give CPR. If breathing is difficult administer oxygen. Treat for shock.

Fire Hazards:

Flashpoint: 118° F(47.7° C) *Ignition temperature:* 534° F(278.8° C)
Flammable Explosive High Range: 1.4 *Low Range:* 1.4
Toxic Products of Combustion: n/a
Other Hazards: Containers may explode in fire. Will polymerize when hot. Polymerization can be explosive!
Possible extinguishing agents: Foam, dry chemical, carbon dioxide. Cool all affected containers with large quantities of water.

Reactivity Hazards:

Reactive With: n/a *Other Reactions:* n/a

Corrosivity Hazards:

Corrosive With: n/a *Neutralizing Agent:* n/a

Radioactivity Hazards:

Radiation Emitted: n/a *Other Hazards:* n/a

Recommended Protection for Response Personnel:

Avoid breathing vapors, keep upwind. Structural protective clothing provides limited protection. Wash away any material which may have come into contact with the body with copious amounts of soap and water. If the fire becomes uncontrollable, consider appropriate evacuation.

BUTYLENE

DOT Number: UN 1012 *DOT Hazard Class:* Flammable gas *DOT Guide Number:* 22
Synonyms: 1-butene
STCC Number: 4905711 *Reportable Qty:* n/a
Mfg Name: Petro-Tex Chemical Corp. *Phone No:* 1-713-477-9211

Physical Description:

Physical Form: Gas *Color:* Colorless *Odor:* Fragrant, gasolinelike
Other Information: It is shipped as a liquefied gas under its own vapor pressure. For transportation, it may be stenched.

Chemical Properties:

Specific Gravity: .59 *Vapor Density:* 1.9 *Boiling Point: 20.7° F(−6.2° C)*
Melting Point: −279° F(−172.7° C) *Vapor Pressure:* 1.36 atm at 68° F (20° C) *Solubility in water:* No
Other Information: Can easily asphyxiatc by displacement of air. Contact with liquid can cause frostbite.

Health Hazards:

Inhalation Hazard: Dizziness and difficulty in breathing.
Ingestion Hazard: Will cause frostbite.
Absorption Hazard: Will cause frostbite.
Hazards to Wildlife: n/a
Decontamination Procedures: Wash away any material with copious amounts of soap and water.
First Aid Procedures: Remove victim to fresh air, call emergency medical care. If not breathing give CPR. If breathing is difficult administer oxygen. Treat for shock.

Fire Hazards:

Flashpoint: 0° F(−17.7° C) *Ignition temperature:* 725° F(385° C)
Flammable Explosive High Range: 10 *Low Range:* 1.6
Toxic Products of Combustion: n/a
Other Hazards: Do not extinguish fire unless the flow can be stopped. Under fire conditions, cylinders or tank cars may violently rupture or rocket. (BLEVE)
Possible extinguishing agents: Apply water from as far a distance as possible. Use water fog in large quantities.

Reactivity Hazards:

Reactive With: Organic peroxides. *Other Reactions:* n/a

Corrosivity Hazards:

Corrosive With: n/a *Neutralizing Agent:* n/a

Radioactivity Hazards:

Radiation Emitted: n/a *Other Hazards:* n/a

Recommended Protection for Response Personnel:

Avoid breathing vapors, keep upwind. Structural protective clothing provides limited protection. Wash away any material which may have come into contact with the body with copious amounts of soap and water. If the fire becomes uncontrollable, consider appropriate evacuation.(BLEVE)

BUTYLENE OXIDE

DOT Number: UN 3022 *DOT Hazard Class:* Flammable liquid *DOT Guide Number:* 26
Synonyms: α-butylene oxide, 1-butene oxide, 1,2-epoxybutane
STCC Number: 4908144 *Reportable Qty:* n/a
Mfg Name: Dow Chemical *Phone No:* 1-517-636-4400

Physical Description:

Physical Form: Liquid *Color:* Colorless *Odor:* Sharp
Other Information: Used as an intermediate to make various polymers.

Chemical Properties:

Specific Gravity: .8 *Vapor Density:* 2.2 *Boiling Point: 145° F(62.7° C)*
Melting Point: n/a *Vapor Pressure:* n/a *Solubility in water:* Yes
Other Information: It is soluble in water and may decompose upon contact with water. Vapors are heavier than air. Weighs 6.9 lbs/3.0 kg per gallon/3.8 l.

Health Hazards:

Inhalation Hazard: Coughing, difficulty in breathing.
Ingestion Hazard: Nausea and vomiting.
Absorption Hazard: Will burn the skin and eyes.
Hazards to Wildlife: n/a
Decontamination Procedures: Wash away any material with copious amounts of soap and water.
First Aid Procedures: Remove victim to fresh air, call emergency medical care. If not breathing give CPR. If breathing is difficult administer oxygen. Treat for shock.

Fire Hazards:

Flashpoint: −20° F(−28.8° C) *Ignition temperature:* 959° F(515° C)
Flammable Explosive High Range: 18.3 *Low Range:* 1.5
Toxic Products of Combustion: n/a
Other Hazards: Containers may explode in a fire. Flashback along vapor trail may occur. Containers may explode if ignited in an enclosed area.
Possible extinguishing agents: Do not extinguish the fire unless the flow can be stopped. Apply water from as far a distance as possible. Use alcohol foam, dry chemical, or carbon dioxide.

Reactivity Hazards:

Reactive With: n/a *Other Reactions:* n/a

Corrosivity Hazards:

Corrosive With: n/a *Neutralizing Agent:* n/a

Radioactivity Hazards:

Radiation Emitted: n/a *Other Hazards:* n/a

Recommended Protection for Response Personnel:

Avoid breathing vapors, keep upwind. Wear a sealed chemical suit, (polycarbonate, butyl rubber). Wash away any material which may have come into contact with the body with copious amounts of soap and water. Consider appropriate evacuation.

BUTYL PHENOL

DOT Number: UN 2229 *DOT Hazard Class:* Flammable liquid *DOT Guide Number:* 53
Synonyms: none given
STCC Number: n/a *Reportable Qty:* n/a
Mfg Name: Dow Chemical *Phone No:* 1-517-636-4400

Physical Description:

Physical Form: Solid *Color:* White *Odor:* Disinfectant
Other Information: n/a

Chemical Properties:

Specific Gravity: 1.03 *Vapor Density:* 5.2 *Boiling Point: 463° F(239.4° C)*
Melting Point: 210° F(98.8° C) *Vapor Pressure:* 1 mm Hg at 158° F(70° C) *Solubility in water:* n/a
Other Information: May float or sink in water.

Health Hazards:

Inhalation Hazard: Difficulty in breathing.
Ingestion Hazard: Cause nausea and vomiting.
Absorption Hazard: Will burn the skin and eyes.
Hazards to Wildlife: n/a
Decontamination Procedures: Wash away any material with copious amounts of soap and water.
First Aid Procedures: Remove victim to fresh air, call emergency medical care. If not breathing give CPR. If breathing is difficult administer oxygen. Treat for shock.

Fire Hazards:

Flashpoint: 235° F(112.7° C) *Ignition temperature:* n/a
Flammable Explosive High Range: n/a *Low Range:* n/a
Toxic Products of Combustion: n/a
Other Hazards: n/a
Possible extinguishing agents: Use foam, dry chemical or carbon dioxide. Water may be ineffective!

Reactivity Hazards:

Reactive With: n/a *Other Reactions:* n/a

Corrosivity Hazards:

Corrosive With: n/a *Neutralizing Agent:* n/a

Radioactivity Hazards:

Radiation Emitted: n/a *Other Hazards:* n/a

Recommended Protection for Response Personnel:

Avoid breathing vapors of burning material, keep upwind. Structural protective clothing provides limited protection. Wash away any material which may have come into contact with the body with copious amounts of soap and water. If the fire becomes uncontrollable, consider appropriate evacuation.

BUTYL TRICHLOROSILANE

DOT Number: UN 1747 *DOT Hazard Class:* Corrosive *DOT Guide Number:* 29
Synonyms: n-butyl trichlorosilane
STCC Number: 4934215 *Reportable Qty:* n/a
Mfg Name: Dow Chemical *Phone No:* 1-517-636-4400

Physical Description:

Physical Form: Liquid *Color:* Colorless *Odor:* Sharp, irritating
Other Information: It is used to make various silicon containing compounds.

Chemical Properties:

Specific Gravity: 1.16 *Vapor Density:* 6.4 *Boiling Point: 288° F(142.2° C)*
Melting Point: n/a *Vapor Pressure:* n/a *Solubility in water:* Reacts
Other Information: Reacts violently with water. A visible irritating vapor cloud is produced.

Health Hazards:

Inhalation Hazard: Harmful if inhaled.
Ingestion Hazard: Harmful if swallowed.
Absorption Hazard: Will burn the skin and eyes.
Hazards to Wildlife: n/a
Decontamination Procedures: Wash away any material with copious amounts of soap and water.
First Aid Procedures: Remove victim to fresh air, call emergency medical care. If not breathing give CPR. If breathing is difficult administer oxygen. Treat for shock.

Fire Hazards:

Flashpoint: 130° F(54.4° C) *Ignition temperature:* n/a
Flammable Explosive High Range: n/a *Low Range:* n/a
Toxic Products of Combustion: Hydrogen chloride, chlorine, or phosgene may be formed.
Other Hazards: Material is difficult to extinguish, reignition may occur!
Possible extinguishing agents: Do not use water, use dry chemical, dry sand, or carbon dioxide.

Reactivity Hazards:

Reactive With: Water to generate hydrochloric acid. *Other Reactions:* n/a

Corrosivity Hazards:

Corrosive With: Metals and tissue *Neutralizing Agent:* Crushed limestone, soda ash, or lime

Radioactivity Hazards:

Radiation Emitted: n/a *Other Hazards:* n/a

Recommended Protection for Response Personnel:

Avoid breathing vapors, keep upwind. Wear a sealed chemical suit, (butyl rubber). Wash away any material which may have come into contact with the body with copious amounts of soap and water. Consider appropriate evacuation.

BUTYRALDEHYDE

DOT Number: UN 1129 *DOT Hazard Class:* Flammable liquid *DOT Guide Number:* 26
Synonyms: isobutylaldehyde, isobutyraldehyde
STCC Number: 4908119 *Reportable Qty:* n/a
Mfg Name: Union Carbide *Phone No:* 1-203-794-2000

Physical Description:

Physical Form: Liquid *Color:* Colorless *Odor:* Gasoline like
Other Information: n/a

Chemical Properties:

Specific Gravity: .8 *Vapor Density:* 2.5 *Boiling Point: 169° F(76.1° C)*
Melting Point: n/a *Vapor Pressure:* n/a *Solubility in water:* No
Other Information: Floats and mixes slowly with water. A flammable irritating vapor is produced.

Health Hazards:

Inhalation Hazard: Difficulty in breathing.
Ingestion Hazard: Harmful if swallowed.
Absorption Hazard: Irritating to the skin, will burn eyes.
Hazards to Wildlife: n/a
Decontamination Procedures: Wash away any material with copious amounts of soap and water.
First Aid Procedures: Remove victim to fresh air, call emergency medical care. If not breathing give CPR. If breathing is difficult administer oxygen. Treat for shock.

Fire Hazards:

Flashpoint: −8° F(−22.2° C) *Ignition temperature:* 425° F(218.3° C)
Flammable Explosive High Range: 12.5 *Low Range:* 1.9
Toxic Products of Combustion: n/a
Other Hazards: Flashback along vapor trail may occur. Vapors may explode if ignited in an enclosed area.
Possible extinguishing agents: Apply water from as far a distance as possible, use foam, dry chemical, or carbon dioxide.

Reactivity Hazards:

Reactive With: n/a *Other Reactions:* n/a

Corrosivity Hazards:

Corrosive With: n/a *Neutralizing Agent:* n/a

Radioactivity Hazards:

Radiation Emitted: n/a *Other Hazards:* n/a

Recommended Protection for Response Personnel:

Avoid breathing vapors, keep upwind. Wear a sealed chemical suit, (butyl rubber, chlorinated polyethylene). Wash away any material which may have come into contact with the body with copious amounts of soap and water. If the fire becomes uncontrollable, or if the containers are exposed to flame, consider appropriate evacuation.

iso-BUTYRALDEHYDE

DOT Number: UN 1129 *DOT Hazard Class:* Flammable liquid *DOT Guide Number:* 26
Synonyms: isobutyraldehyde, isobutyric aldehyde, isobutylaldehyde, 2-methylpropanal
STCC Number: n/a *Reportable Qty:* n/a
Mfg Name: Union Carbide *Phone No:* 1-203-794-2000

Physical Description:

Physical Form: Liquid *Color:* Colorless *Odor:* Pleasant, gasoline like
Other Information: n/a

Chemical Properties:

Specific Gravity: .79 *Vapor Density:* 2.5 *Boiling Point: 147° F(63.8° C)*
Melting Point: −112° F(−80° C) *Vapor Pressure:* n/a *Solubility in water:* Yes
Other Information: Floats and mixes slowly with water. A flammable irritating vapor is produced.

Health Hazards:

Inhalation Hazard: Difficulty in breathing.
Ingestion Hazard: Harmful if swallowed.
Absorption Hazard: Irritating to the skin, will burn eyes.
Hazards to Wildlife: n/a
Decontamination Procedures: Wash away any material with copious amounts of soap and water.
First Aid Procedures: Remove victim to fresh air, call emergency medical care. If not breathing give CPR. If breathing is difficult administer oxygen. Treat for shock.

Fire Hazards:

Flashpoint: −40° F(−40° C) *Ignition temperature:* 385° F(196.1° C)
Flammable Explosive High Range: 10 *Low Range:* 2
Toxic Products of Combustion: n/a
Other Hazards: Flashback along vapor trail may occur. Vapors may explode if ignited in an enclosed area. Fires are difficult to control due to ease of reignition.
Possible extinguishing agents: Apply water from as far a distance as possible. Use foam, dry chemical, or carbon dioxide.

Reactivity Hazards:

Reactive With: n/a *Other Reactions:* n/a

Corrosivity Hazards:

Corrosive With: n/a *Neutralizing Agent:* n/a

Radioactivity Hazards:

Radiation Emitted: n/a *Other Hazards:* n/a

Recommended Protection for Response Personnel:

Avoid breathing vapors, keep upwind. Wear a sealed chemical suit, (butyl rubber, chlorinated polyethylene). Wash away any material which may have come into contact with the body with copious amounts of soap and water. If the fire becomes uncontrollable, or if the containers are exposed to flame, consider appropriate evacuation.

BUTYRIC ACID

DOT Number: UN 2820 *DOT Hazard Class:* Corrosive *DOT Guide Number:* 60
Synonyms: butanic acid, n-butyric acid
STCC Number: 4931414 *Reportable Qty:* 5000/2270
Mfg Name: Mallinckrodt Inc. *Phone No:* 1-341-895-2000

Physical Description:

Physical Form: Liquid *Color:* Colorless *Odor:* Rancid butter
Other Information: Soluble in water. Weighs 8 lbs/3.6 kg per gallon/3.8 l.

Chemical Properties:

Specific Gravity: 1 *Vapor Density:* 3 *Boiling Point: 327° F(163.8° C)*
Melting Point: n/a *Vapor Pressure:* 84 mm Hg at 68° F (20° C) *Solubility in water:* Yes
Other Information: n/a

Health Hazards:

Inhalation Hazard: Will cause coughing, and difficulty in breathing.
Ingestion Hazard: Will cause nausea and vomiting.
Absorption Hazard: Will burn the skin and the eyes.
Hazards to Wildlife: Dangerous to aquatic life.
Decontamination Procedures: Wash away any material with copious amounts of soap and water.
First Aid Procedures: Remove victim to fresh air, call emergency medical care. If not breathing give CPR. If breathing is difficult administer oxygen. Treat for shock.

Fire Hazards:

Flashpoint: 161° F(71.6° C) *Ignition temperature:* 830° F(443.3° C)
Flammable Explosive High Range: 10 *Low Range:* 2
Toxic Products of Combustion: n/a
Other Hazards: Water may be ineffective on fire.
Possible extinguishing agents: Alcohol foam, dry chemical, carbon dioxide.

Reactivity Hazards:

Reactive With: Aluminum or other light metals forming flammable nitrogen gas. *Other Reactions:* n/a

Corrosivity Hazards:

Corrosive With: Metals and tissue *Neutralizing Agent:* Crushed limestone, soda ash, or lime.

Radioactivity Hazards:

Radiation Emitted: n/a *Other Hazards:* n/a

Recommended Protection for Response Personnel:

Avoid breathing vapors, keep upwind. Structural protective clothing provides limited protection. Wash away any material which may have come into contact with the body with copious amounts of soap and water. If the fire becomes uncontrollable, consider appropriate evacuation.

CACODYLIC ACID

DOT Number: UN 1572 *DOT Hazard Class:* Poison B *DOT Guide Number:* 53
Synonyms: Ansar, dimethylarsinic acid, hydroxydimethylarsine oxide, Silvisar-510
STCC Number: n/a *Reportable Qty:* n/a
Mfg Name: Ansul Company *Phone No:* 1-517-735-7411

Physical Description:

Physical Form: Solid *Color:* Colorless or dyed blue *Odor:* n/a
Other Information: n/a

Chemical Properties:

Specific Gravity: 1.1 *Vapor Density:* 4.8 *Boiling Point: 192° F(88.8° C)*
Melting Point: n/a *Vapor Pressure:* n/a *Solubility in water:* Yes
Other Information: Sinks and mixes with water.

Health Hazards:

Inhalation Hazard: Poisonous if inhaled.
Ingestion Hazard: Poisonous if swallowed.
Absorption Hazard: Poisonous to the skin and eyes.
Hazards to Wildlife: n/a
Decontamination Procedures: Wash away any material with copious amounts of soap and water.
First Aid Procedures: Remove victim to fresh air, call emergency medical care. If not breathing give CPR. If breathing is difficult administer oxygen. Treat for shock.

Fire Hazards:

Flashpoint: n/a *Ignition temperature:* n/a
Flammable Explosive High Range: n/a *Low Range:* n/a
Toxic Products of Combustion: May form toxic oxides of arsenic when heated.
Other Hazards: n/a
Possible extinguishing agents: Extinguish fire using suitable agent for the type of surrounding fire.

Reactivity Hazards:

Reactive With: n/a *Other Reactions:* n/a

Corrosivity Hazards:

Corrosive With: n/a *Neutralizing Agent:* n/a

Radioactivity Hazards:

Radiation Emitted: n/a *Other Hazards:* n/a

Recommended Protection for Response Personnel:

Avoid breathing vapors, keep upwind. Structural protective clothing provides limited protection. Wash away any material which may have come into contact with the body with copious amounts of soap and water. Consider appropriate evacuation.

CADMIUM ACETATE

DOT Number: UN 2570 *DOT Hazard Class:* ORM-E *DOT Guide Number:* 53
Synonyms: cadmium acetate dihydrate
STCC Number: 4962303 *Reportable Qty:* 100/45.4
Mfg Name: J.T. Baker Chemical Co. *Phone No:* 1-201-859-2151

Physical Description:

Physical Form: Solid *Color:* Colorless *Odor:* Odorless
Other Information: It is used in ceramics, and in chemical analysis.

Chemical Properties:

Specific Gravity: 2.34 *Vapor Density:* 8 *Boiling Point: n/a*
Melting Point: n/a *Vapor Pressure:* n/a *Solubility in water:* Yes
Other Information: Sinks and mixes with water.

Health Hazards:

Inhalation Hazard: Poisonous if inhaled.
Ingestion Hazard: Poisonous if swallowed.
Absorption Hazard: Poisonous to the skin and eyes.
Hazards to Wildlife: n/a
Decontamination Procedures: Wash away any material with copious amounts of soap and water.
First Aid Procedures: Remove victim to fresh air, call emergency medical care. If not breathing give CPR. If breathing is difficult administer oxygen. Treat for shock.

Fire Hazards:

Flashpoint: n/a *Ignition temperature:* n/a
Flammable Explosive High Range: n/a *Low Range:* n/a
Toxic Products of Combustion: Toxic cadmium oxide fumes may be formed in fires.
Other Hazards: n/a
Possible extinguishing agents: Extinguish fire using suitable agent for the type of surrounding fire.

Reactivity Hazards:

Reactive With: n/a *Other Reactions:* n/a

Corrosivity Hazards:

Corrosive With: n/a *Neutralizing Agent:* n/a

Radioactivity Hazards:

Radiation Emitted: n/a *Other Hazards:* n/a

Recommended Protection for Response Personnel:

Avoid breathing dust and fumes, keep upwind. Wear a sealed chemical suit (polycarbonate, butyl rubber). Wash away any material which may have come into contact with the body with copious amounts of soap and water. Consider appropriate evacuation.

CADMIUM BROMIDE

DOT Number: UN 2570 *DOT Hazard Class:* ORM-E *DOT Guide Number:* 53
Synonyms: cadmium bromide tetrahydrate
STCC Number: 4962305 *Reportable Qty:* 100/45.4
Mfg Name: Gallard-Schlesinger *Phone No:* 1-516-333-5600

Physical Description:

Physical Form: Solid *Color:* White *Odor:* Odorless
Other Information: It is used in photography, printing, and lithography.

Chemical Properties:

Specific Gravity: 1.1 *Vapor Density:* 8 *Boiling Point: n/a*
Melting Point: n/a *Vapor Pressure:* n/a *Solubility in water:* Yes
Other Information: Mixes with water.

Health Hazards:

Inhalation Hazard: Poisonous if inhaled.
Ingestion Hazard: Poisonous if swallowed.
Absorption Hazard: Irritating to the skin and eyes.
Hazards to Wildlife: n/a
Decontamination Procedures: Wash away any material with copious amounts of soap and water.
First Aid Procedures: Remove victim to fresh air, call emergency medical care. If not breathing give CPR. If breathing is difficult administer oxygen. Treat for shock.

Fire Hazards:

Flashpoint: n/a *Ignition temperature:* n/a
Flammable Explosive High Range: n/a *Low Range:* n/a
Toxic Products of Combustion: Toxic cadmium oxide fumes may be formed in fires.
Other Hazards: n/a
Possible extinguishing agents: Extinguish fire using suitable agent for the type of surrounding fire.

Reactivity Hazards:

Reactive With: n/a *Other Reactions:* n/a

Corrosivity Hazards:

Corrosive With: n/a *Neutralizing Agent:* n/a

Radioactivity Hazards:

Radiation Emitted: n/a *Other Hazards:* n/a

Recommended Protection for Response Personnel:

Avoid breathing dust and fumes, keep upwind. Wear a sealed chemical suit (polycarbonate, butyl rubber). Wash away any material which may have come into contact with the body with copious amounts of soap and water. Consider appropriate evacuation.

CADMIUM CHLORIDE

DOT Number: UN 2570 *DOT Hazard Class:* ORM-E *DOT Guide Number:* 53
Synonyms: None given
STCC Number: 4962505 *Reportable Qty:* 100/45.4
Mfg Name: Allied Chemical *Phone No:* 1-212-455-2000

Physical Description:

Physical Form: Solid *Color:* White *Odor:* Odorless
Other Information: It is used in photography, fabric printing, chemical analysis, and in many other uses.

Chemical Properties:

Specific Gravity: 4.05 *Vapor Density:* 6.3 *Boiling Point: n/a*
Melting Point: n/a *Vapor Pressure:* n/a *Solubility in water:* Yes
Other Information: Sinks and mixes with water.

Health Hazards:

Inhalation Hazard: n/a
Ingestion Hazard: Harmful if swallowed.
Absorption Hazard: n/a
Hazards to Wildlife: Dangerous to aquatic life.
Decontamination Procedures: Wash away any material with copious amounts of soap and water.
First Aid Procedures: Remove victim to fresh air, call emergency medical care. If not breathing give CPR. If breathing is difficult administer oxygen. Treat for shock.

Fire Hazards:

Flashpoint: n/a *Ignition temperature:* n/a
Flammable Explosive High Range: n/a *Low Range:* n/a
Toxic Products of Combustion: n/a
Other Hazards: n/a
Possible *Possible extinguishing agents:* Extinguish fire using suitable agent for the type of surrounding fire.

Reactivity Hazards:

Reactive With: n/a *Other Reactions:* n/a

Corrosivity Hazards:

Corrosive With: n/a *Neutralizing Agent:* n/a

Radioactivity Hazards:

Radiation Emitted: n/a *Other Hazards:* n/a

Recommended Protection for Response Personnel:

Avoid breathing dust and fumes, keep upwind. Wear a sealed chemical suit (polycarbonate, butyl rubber). Wash away any material which may have come into contact with the body with copious amounts of soap and water. Consider appropriate evacuation.

CADMIUM FLUOROBORATE

DOT Number: UN 2570 *DOT Hazard Class:* n/a *DOT Guide Number:* 53
Synonyms: cadmium fluoborate, cadmium fluoborate solution
STCC Number: n/a *Reportable Qty:* n/a
Mfg Name: Allied Chemical *Phone No:* 1-212-455-2000

Physical Description:

Physical Form: Liquid *Color:* Colorless *Odor:* Odorless
Other Information: n/a

Chemical Properties:

Specific Gravity: 1.6 *Vapor Density:* n/a *Boiling Point: n/a*
Melting Point: n/a *Vapor Pressure:* n/a *Solubility in water:* Yes
Other Information: Sinks and mixes with water.

Health Hazards:

Inhalation Hazard: Poisonous if inhaled.
Ingestion Hazard: Harmful if swallowed.
Absorption Hazard: Irritating to the skin and eyes.
Hazards to Wildlife: n/a
Decontamination Procedures: Wash away any material with copious amounts of soap and water.
First Aid Procedures: Remove victim to fresh air, call emergency medical care. If not breathing give CPR. If breathing is difficult administer oxygen. Treat for shock.

Fire Hazards:

Flashpoint: n/a *Ignition temperature:* n/a
Flammable Explosive High Range: n/a *Low Range:* n/a
Toxic Products of Combustion: Toxic hydrogen fluoride and cadmium oxides are formed in fires.
Other Hazards: n/a
Possible extinguishing agents: Extinguish fire using suitable agent for the type of surrounding fire.

Reactivity Hazards:

Reactive With: n/a *Other Reactions:* n/a

Corrosivity Hazards:

Corrosive With: n/a *Neutralizing Agent:* n/a

Radioactivity Hazards:

Radiation Emitted: n/a *Other Hazards:* n/a

Recommended Protection for Response Personnel:

Avoid breathing dust and fumes, keep upwind. Wear a sealed chemical suit. Wash away any material which may have come into contact with the body with copious amounts of soap and water. Consider appropriate evacuation.

CADMIUM NITRATE

DOT Number: UN 2570 *DOT Hazard Class:* n/a *DOT Guide Number:* 53
Synonyms: cadmium nitrate tetrahydrate
STCC Number: n/a *Reportable Qty:* n/a
Mfg Name: Hall Chemical Co. *Phone No:* 1-216-944-8500

Physical Description:

Physical Form: Solid *Color:* White *Odor:* Odorless
Other Information: n/a

Chemical Properties:

Specific Gravity: 2.45 *Vapor Density:* n/a *Boiling Point: n/a*
Melting Point: n/a *Vapor Pressure:* n/a *Solubility in water:* Sinks
Other Information: n/a

Health Hazards:

Inhalation Hazard: Poisonous if inhaled.
Ingestion Hazard: Poisonous if swallowed.
Absorption Hazard: Irritating to the skin and eyes.
Hazards to Wildlife: Dangerous to aquatic life.
Decontamination Procedures: Wash away any material with copious amounts of soap and water.
First Aid Procedures: Remove victim to fresh air, call emergency medical care. If not breathing give CPR. If breathing is difficult administer oxygen. Treat for shock.

Fire Hazards:

Flashpoint: n/a *Ignition temperature:* n/a
Flammable Explosive High Range: n/a *Low Range:* n/a
Toxic Products of Combustion: Toxic oxides of nitrogen and cadmium oxide fumes may be formed in fires.
Other Hazards: Will increase the intensity of a fire on contact with combustible materials.
Possible extinguishing agents: Extinguish fire using suitable agent for the type of surrounding fire.

Reactivity Hazards:

Reactive With: n/a *Other Reactions:* n/a

Corrosivity Hazards:

Corrosive With: n/a *Neutralizing Agent:* n/a

Radioactivity Hazards:

Radiation Emitted: n/a *Other Hazards:* n/a

Recommended Protection for Response Personnel:

Avoid breathing dust and fumes, keep upwind. Wear a sealed chemical suit (polycarbonate, butyl rubber). Wash away any material which may have come into contact with the body with copious amounts of soap and water. Consider appropriate evacuation.

CADMIUM OXIDE

DOT Number: UN 2570 *DOT Hazard Class:* n/a *DOT Guide Number:* 53
Synonyms: cadmium fume
STCC Number: n/a *Reportable Qty:* n/a
Mfg Name: J.T. Baker Chemical Co. *Phone No:* 1-201-859-2151

Physical Description:

Physical Form: Solid *Color:* Yellow to brown *Odor:* Odorless
Other Information: n/a

Chemical Properties:

Specific Gravity: 6.95 *Vapor Density:* n/a *Boiling Point: n/a*
Melting Point: n/a *Vapor Pressure:* n/a *Solubility in water:* n/a
Other Information: Sinks in water.

Health Hazards:

Inhalation Hazard: Poisonous if inhaled.
Ingestion Hazard: Nausea and vomiting.
Absorption Hazard: Irritating to the skin and eyes.
Hazards to Wildlife: n/a
Decontamination Procedures: Wash away any material with copious amounts of soap and water.
First Aid Procedures: Remove victim to fresh air, call emergency medical care. If not breathing give CPR. If breathing is difficult administer oxygen. Treat for shock.

Fire Hazards:

Flashpoint: n/a *Ignition temperature:* n/a
Flammable Explosive High Range: n/a *Low Range:* n/a
Toxic Products of Combustion: Toxic cadmium oxide fumes may be formed in fires.
Other Hazards: n/a
Possible extinguishing agents: Extinguish fire using suitable agent for the type of surrounding fire.

Reactivity Hazards:

Reactive With: n/a *Other Reactions:* n/a

Corrosivity Hazards:

Corrosive With: n/a *Neutralizing Agent:* n/a

Radioactivity Hazards:

Radiation Emitted: n/a *Other Hazards:* n/a

Recommended Protection for Response Personnel:

Avoid breathing dust and fumes, keep upwind. Wear a sealed chemical suit (polycarbonate, butyl rubber). Wash away any material which may have come into contact with the body with copious amounts of soap and water. Consider appropriate evacuation.

CADMIUM SULFATE

DOT Number: UN 2570 *DOT Hazard Class:* n/a *DOT Guide Number:* 53
Synonyms: None given
STCC Number: n/a *Reportable Qty:* n/a
Mfg Name: J.T. Baker Chemical Co. *Phone No:* 1-201-859-2151

Physical Description:

Physical Form: Solid *Color:* White *Odor:* Odorless
Other Information: n/a

Chemical Properties:

Specific Gravity: 4.7 *Vapor Density:* n/a *Boiling Point: n/a*
Melting Point: n/a *Vapor Pressure:* n/a *Solubility in water:* Yes
Other Information: Sinks and mixes slowly with water.

Health Hazards:

Inhalation Hazard: Poisonous if inhaled.
Ingestion Hazard: Poisonous if swallowed.
Absorption Hazard: Irritating to the skin and eyes.
Hazards to Wildlife: Dangerous to aquatic life.
Decontamination Procedures: Wash away any material with copious amounts of soap and water.
First Aid Procedures: Remove victim to fresh air, call emergency medical care. If not breathing give CPR. If breathing is difficult administer oxygen. Treat for shock.

Fire Hazards:

Flashpoint: n/a *Ignition temperature:* n/a
Flammable Explosive High Range: n/a *Low Range:* n/a
Toxic Products of Combustion: Toxic cadmium oxide fumes may be formed in fires.
Other Hazards: n/a
Possible extinguishing agents: Extinguish fire using suitable agent for the type of surrounding fire.

Reactivity Hazards:

Reactive With: n/a *Other Reactions:* n/a

Corrosivity Hazards:

Corrosive With: n/a *Neutralizing Agent:* n/a

Radioactivity Hazards:

Radiation Emitted: n/a *Other Hazards:* n/a

Recommended Protection for Response Personnel:

Avoid breathing dust and fumes, keep upwind. Wear a sealed chemical suit (polycarbonate, butyl rubber). Wash away any material which may have come into contact with the body with copious amounts of soap and water. Consider appropriate evacuation.

CALCIUM ARSENATE

DOT Number: UN 1573 *DOT Hazard Class:* Poison B *DOT Guide Number:* 53
Synonyms: cucumber dust
STCC Number: 4923217 *Reportable Qty:* 1000/454
Mfg Name: Los Angeles Chemical Co. *Phone No:* 1-213-583-7461

Physical Description:

Physical Form: Solid *Color:* White *Odor:* None
Other Information: It is slightly soluble in water.

Chemical Properties:

Specific Gravity: 2.62 *Vapor Density:* 14 *Boiling Point: n/a*
Melting Point: n/a *Vapor Pressure:* 0 mm Hg at 68° F(20° C)
Solubility in water: Slight *Degree of Solubility:* .013%
Other Information: n/a

Health Hazards:

Inhalation Hazard: Poisonous if inhaled.
Ingestion Hazard: Poisonous if swallowed.
Absorption Hazard: Irritating to the skin and eyes.
Hazards to Wildlife: Dangerous to aquatic life.
Decontamination Procedures: Wash away any material with copious amounts of soap and water.
First Aid Procedures: Remove victim to fresh air, call emergency medical care. If not breathing give CPR. If breathing is difficult administer oxygen. Treat for shock.

Fire Hazards:

Flashpoint: n/a *Ignition temperature:* n/a
Flammable Explosive High Range: 10 *Low Range:* 2
Toxic Products of Combustion: Poisonous products may be produced in fire.
Other Hazards: Avoid breathing dust
Possible extinguishing agents: Use appropriate extinguishing agent for the type of surrounding fire.
NOTE: The chemical itself does not burn.

Reactivity Hazards:

Reactive With: n/a *Other Reactions:* n/a

Corrosivity Hazards:

Corrosive With: n/a *Neutralizing Agent:* n/a

Radioactivity Hazards:

Radiation Emitted: n/a *Other Hazards:* n/a

Recommended Protection for Response Personnel:

Avoid breathing dust and fumes, keep upwind. Wear a sealed chemical suit (polycarbonate, butyl rubber). Wash away any material which may have come into contact with the body with copious amounts of soap and water. Consider appropriate evacuation.

CALCIUM ARSENITE

DOT Number: NA 1574 *DOT Hazard Class:* Poison B *DOT Guide Number:* 53
Synonyms: calcium salt, arsenous acid
STCC Number: 4923219 *Reportable Qty:* 1000/454
Mfg Name: Great Western Inorgranics *Phone No:* 1-303-423-9770

Physical Description:

Physical Form: Solid *Color:* White *Odor:* Odorless
Other Information: n/a

Chemical Properties:

Specific Gravity: n/a *Vapor Density:* 9.3 *Boiling Point: n/a*
Melting Point: n/a *Vapor Pressure:* n/a *Solubility in water:* No
Other Information: Avoid contact with solid. Keep people away!

Health Hazards:

Inhalation Hazard: n/a
Ingestion Hazard: Poisonous if swallowed.
Absorption Hazard: Irritating to the skin and eyes.
Hazards to Wildlife: Dangerous to aquatic life.
Decontamination Procedures: Wash away any material with copious amounts of soap and water.
First Aid Procedures: Remove victim to fresh air, call emergency medical care. If not breathing give CPR. If breathing is difficult administer oxygen. Treat for shock.

Fire Hazards:

Flashpoint: n/a *Ignition temperature:* n/a
Flammable Explosive High Range: n/a *Low Range:* n/a
Toxic Products of Combustion: n/a
Other Hazards: n/a
Possible extinguishing agents: Use suitable agent for type of surrounding fire, use foam, dry chemical, or carbon dioxide.

Reactivity Hazards:

Reactive With: n/a *Other Reactions:* n/a

Corrosivity Hazards:

Corrosive With: n/a *Neutralizing Agent:* n/a

Radioactivity Hazards:

Radiation Emitted: n/a *Other Hazards:* n/a

Recommended Protection for Response Personnel:

Avoid breathing vapors, keep upwind. Wear a sealed chemical suit (polycarbonate, butyl rubber). Wash away any material which may have come into contact with the body with copious amounts of soap and water. Consider appropriate evacuation.

CALCIUM CARBIDE

DOT Number: UN 1402 *DOT Hazard Class:* Flammable solid *DOT Guide Number:* 40
Synonyms: acetylenogen, carbide
STCC Number: 4916408 *Reportable Qty:* 10/4.54
Mfg Name: Midwest Carbide Corp. *Phone No:* 1-319-524-6510

Physical Description:

Physical Form: Solid *Color:* Gray to bluish black *Odor:* Garlic
Other Information: It is used to make acetylene, and in steel manufacturing.

Chemical Properties:

Specific Gravity: 2.22 *Vapor Density:* 2.2 *Boiling Point: n/a*
Melting Point: n/a *Vapor Pressure:* n/a *Solubility in water:* Yes
Other Information: Sinks in water, and bubbles appear on the surface as a flammable gas is produced.

Health Hazards:

Inhalation Hazard: n/a
Ingestion Hazard: n/a
Absorption Hazard: Irritating to the skin and eyes.
Hazards to Wildlife: n/a
Decontamination Procedures: Wash away any material with copious amounts of soap and water.
First Aid Procedures: Remove victim to fresh air, call emergency medical care. If not breathing give CPR. If breathing is difficult administer oxygen. Treat for shock.

Fire Hazards:

Flashpoint: n/a *Ignition temperature:* n/a
Flammable Explosive High Range: n/a *Low Range:* n/a
Toxic Products of Combustion: n/a
Other Hazards: Flammable explosive gas is produced upon contact with water
Possible extinguishing agents: Do not use water, use graphite, soda ash, powdered sodium chloride, or suitable dry powder.

Reactivity Hazards:

Reactive With: Water to form acetylene gas which may ignite. *Other Reactions:* n/a

Corrosivity Hazards:

Corrosive With: Copper and brass to form an explosive compound. *Neutralizing Agent:* n/a

Radioactivity Hazards:

Radiation Emitted: n/a *Other Hazards:* n/a

Recommended Protection for Response Personnel:

Avoid breathing vapors, keep upwind. Structural protective clothing provides limited protection. Wash away any material which may have come into contact with the body with copious amounts of soap and water. Consider appropriate evacuation.

CALCIUM CHLORATE

DOT Number: UN 1452 *DOT Hazard Class:* Oxidizer *DOT Guide Number:* 35
Synonyms: None given
STCC Number: 4918713 *Reportable Qty:* n/a
Mfg Name: Gallard-Schlesinger *Phone No:* 1-516-333-5600

Physical Description:

Physical Form: Solid *Color:* White *Odor:* Odorless
Other Information: It is used in photographs, in pyrotechnics, and as a herbicide.

Chemical Properties:

Specific Gravity: 2.71 *Vapor Density:* 8.4 *Boiling Point: Decomposes*
Melting Point: n/a *Vapor Pressure:* n/a *Solubility in water:* Yes
Other Information: Sinks and mixes with water.

Health Hazards:

Inhalation Hazard: Irritating
Ingestion Hazard: Harmful if swallowed.
Absorption Hazard: Irritating to the skin and eyes.
Hazards to Wildlife: n/a
Decontamination Procedures: Wash away any material with copious amounts of soap and water.
First Aid Procedures: Remove victim to fresh air, call emergency medical care. If not breathing give CPR. If breathing is difficult administer oxygen. Treat for shock.

Fire Hazards:

Flashpoint: n/a *Ignition temperature:* n/a
Flammable Explosive High Range: n/a *Low Range:* n/a
Toxic Products of Combustion: Irritating gases may be produced when heated.
Other Hazards: May explode upon contact with combustibles.
Possible extinguishing agents: Flood with water. Apply water from as far a distance as possible.

Reactivity Hazards:

Reactive With: Reacts with combustible materials. *Other Reactions:* n/a

Corrosivity Hazards:

Corrosive With: n/a *Neutralizing Agent:* n/a

Radioactivity Hazards:

Radiation Emitted: n/a *Other Hazards:* n/a

Recommended Protection for Response Personnel:

Avoid breathing vapors, keep upwind. Structural protective clothing provides limited protection. Wash away any material which may have come into contact with the body with copious amounts of soap and water. If the fire becomes uncontrollable, consider appropriate evacuation.

CALCIUM CHROMATE

DOT Number: NA 9096 *DOT Hazard Class:* ORM-E *DOT Guide Number:* 31
Synonyms: calcium chromate(VI), calcium chromate dihydrate, gelbin yellow ultra marine, steinbuln yellow
STCC Number: 4963307 *Reportable Qty:* 1000/454
Mfg Name: A.D. MacKay Inc. *Phone No:* 1-203-655-7401

Physical Description:

Physical Form: Solid *Color:* Yellow *Odor:* Odorless
Other Information: It is used to manufacture other chemicals.

Chemical Properties:

Specific Gravity: 1 *Vapor Density:* 5.4 *Boiling Point: n/a*
Melting Point: n/a *Vapor Pressure:* n/a *Solubility in water:* Yes
Other Information: Sinks and mixes slowly with water.

Health Hazards:

Inhalation Hazard: Coughing or difficulty in breathing.
Ingestion Hazard: Harmful if swallowed.
Absorption Hazard: Will burn the skin and eyes.
Hazards to Wildlife: n/a
Decontamination Procedures: Wash away any material with copious amounts of soap and water.
First Aid Procedures: Remove victim to fresh air, call emergency medical care. If not breathing give CPR. If breathing is difficult administer oxygen. Treat for shock.

Fire Hazards:

Flashpoint: n/a *Ignition temperature:* n/a
Flammable Explosive High Range: n/a *Low Range:* n/a
Toxic Products of Combustion: Toxic chromium fumes may be formed in a fire.
Other Hazards: n/a
Possible extinguishing agents: Extinguish fire using suitable agent for the type of surrounding fire.

Reactivity Hazards:

Reactive With: n/a *Other Reactions:* n/a

Corrosivity Hazards:

Corrosive With: n/a *Neutralizing Agent:* n/a

Radioactivity Hazards:

Radiation Emitted: n/a *Other Hazards:* n/a

Recommended Protection for Response Personnel:

Avoid breathing vapors, keep upwind. Wear a sealed chemical suit (polycarbonate, butyl rubber). Wash away any material which may have come into contact with the body with copious amounts of soap and water. Consider appropriate evacuation.

CALCIUM CYANIDE

DOT Number: UN 1575 *DOT Hazard Class:* Poison B *DOT Guide Number:* 55
Synonyms: A-dust, cyanide of calcium, Cyanogas
STCC Number: 4923223 *Reportable Qty:* 10/4.54
Mfg Name: American Cyanamid Co. *Phone No:* 1-201-831-2000

Physical Description:

Physical Form: Solid *Color:* White to gray or black *Odor:* Almond
Other Information: n/a

Chemical Properties:

Specific Gravity: 1.85 *Vapor Density:* 3.2 *Boiling Point: Decomposes*
Melting Point: n/a *Vapor Pressure:* n/a *Solubility in water:* Yes
Other Information: Sinks and mixes with water.

Health Hazards:

Inhalation Hazard: Poisonous if inhaled.
Ingestion Hazard: Poisonous if swallowed.
Absorption Hazard: Irritating to the skin and eyes.
Hazards to Wildlife: Dangerous to aquatic life.
Decontamination Procedures: Wash away any material with copious amounts of soap and water.
First Aid Procedures: Remove victim to fresh air, call emergency medical care. If not breathing give CPR. If breathing is difficult administer oxygen. Treat for shock.

Fire Hazards:

Flashpoint: n/a *Ignition temperature:* n/a
Flammable Explosive High Range: n/a *Low Range:* n/a
Toxic Products of Combustion: Decomposes in fire to give off toxic gases including hydrogen cyanide.
Other Hazards: n/a
Possible extinguishing agents: Extinguish fire using suitable agent for type of surrounding fire, use foam, dry chemical, or carbon dioxide. Do not use water on the material itself.

Reactivity Hazards:

Reactive With: Water to form very poisonous hydrogen cyanide. *Other Reactions:* n/a

Corrosivity Hazards:

Corrosive With: n/a *Neutralizing Agent:* n/a

Radioactivity Hazards:

Radiation Emitted: n/a *Other Hazards:* n/a

Recommended Protection for Response Personnel:

Avoid breathing vapors, keep upwind. Wear a sealed chemical suit (polycarbonate, butyl rubber). Wash away any material which may have come into contact with the body with copious amounts of soap and water. Consider appropriate evacuation.

CALCIUM HYPOCHLORITE

DOT Number: UN 1748 *DOT Hazard Class:* Oxidizer *DOT Guide Number:* 45
Synonyms: HTH, HTH dry chlorine, Sentry
STCC Number: 4918715 *Reportable Qty:* 10/4.54
Mfg Name: PPG Industries Inc. *Phone No:* 1-412-434-3131

Physical Description:

Physical Form: Solid *Color:* White *Odor:* Household bleaching powder
Other Information: It is used in water purification as a swimming pool disinfectant, as a bleaching compound, and for many other uses.

Chemical Properties:

Specific Gravity: 2.35 *Vapor Density:* 6.9 *Boiling Point: n/a*
Melting Point: n/a *Vapor Pressure:* n/a *Solubility in water:* Yes
Other Information: Sinks and mixes with water.

Health Hazards:

Inhalation Hazard: Nausea, vomiting, or loss of consciousness.
Ingestion Hazard: n/a
Absorption Hazard: Irritating to the skin and eyes.
Hazards to Wildlife: Dangerous to aquatic life.
Decontamination Procedures: Wash away any material with copious amounts of soap and water.
First Aid Procedures: Remove victim to fresh air, call emergency medical care. If not breathing give CPR. If breathing is difficult administer oxygen. Treat for shock.

Fire Hazards:

Flashpoint: n/a *Ignition temperature:* n/a
Flammable Explosive High Range: n/a *Low Range:* n/a
Toxic Products of Combustion: Poisonous gases may be produced when heated.
Other Hazards: Will accelerate the burning of other combustible materials.
Possible extinguishing agents: Flood with water. Apply water from as far a distance as possible.

Reactivity Hazards:

Reactive With: Wood or straw to cause fire. *Other Reactions:* n/a

Corrosivity Hazards:

Corrosive With: Most metals *Neutralizing Agent:* n/a

Radioactivity Hazards:

Radiation Emitted: n/a *Other Hazards:* n/a

Recommended Protection for Response Personnel:

Avoid breathing vapors, keep upwind. Structural protective clothing provides limited protection. Wash away any material which may have come into contact with the body with copious amounts of soap and water. If the fire becomes uncontrollable, consider appropriate evacuation.

CALCIUM NITRATE

DOT Number: UN 1454 *DOT Hazard Class:* Oxidizer *DOT Guide Number:* 35
Synonyms: calcium nitrate tetrahydrate
STCC Number: 4918716 *Reportable Qty:* n/a
Mfg Name: J.T. Baker Chemical Co. *Phone No:* 1-201-859-2151

Physical Description:

Physical Form: Solid *Color:* White *Odor:* Odorless
Other Information: It is used in fertilizers, explosives, and pyrotechnics.

Chemical Properties:

Specific Gravity: 2.50 *Vapor Density:* 8.1 *Boiling Point: Decomposes*
Melting Point: n/a *Vapor Pressure:* n/a *Solubility in water:* Yes
Other Information: Sinks and mixes with water.

Health Hazards:

Inhalation Hazard: n/a
Ingestion Hazard: Harmful if swallowed.
Absorption Hazard: Irritating to the skin and eyes.
Hazards to Wildlife: Dangerous to aquatic life.
Decontamination Procedures: Wash away any material with copious amounts of soap and water.
First Aid Procedures: Remove victim to fresh air, call emergency medical care. If not breathing give CPR. If breathing is difficult administer oxygen. Treat for shock.

Fire Hazards:

Flashpoint: n/a *Ignition temperature:* n/a
Flammable Explosive High Range: n/a *Low Range:* n/a
Toxic Products of Combustion: Toxic oxides of nitrogen when involved in a fire.
Other Hazards: Will accelerate the burning of other combustible materials.
Possible extinguishing agents: Flood with water. Apply water from as far a distance as possible.

Reactivity Hazards:

Reactive With: Combustible material *Other Reactions:* n/a

Corrosivity Hazards:

Corrosive With: n/a *Neutralizing Agent:* n/a

Radioactivity Hazards:

Radiation Emitted: n/a *Other Hazards:* n/a

Recommended Protection for Response Personnel:

Avoid breathing vapors, keep upwind. Structural protective clothing provides limited protection. Wash away any material which may have come into contact with the body with copious amounts of soap and water. If the fire becomes uncontrollable, consider appropriate evacuation.

CALCIUM OXIDE

DOT Number: UN 1910 *DOT Hazard Class:* ORM-B *DOT Guide Number:* 53
Synonyms: quicklime, unslaked lime
STCC Number: 4944515 *Reportable Qty:* n/a
Mfg Name: Barium & Chemical Inc. *Phone No:* 1-614-282-9776

Physical Description:

Physical Form: Solid granular *Color:* White to gray *Odor:* Odorless
Other Information: Sinks, reacts violently with water, and appears to boil.

Chemical Properties:

Specific Gravity: 3.3 *Vapor Density:* 1.9 *Boiling Point: 5162° F(2850° C)*
Melting Point: 4658° F(2570° C) *Vapor Pressure:* 0 mm Hg at 68° F(20° C) *Solubility in water:* Reacts
Other Information: Nonflammable but will support combustion by liberation of oxygen especially in the presence of organic materials.

Health Hazards:

Inhalation Hazard: Avoid breathing dust, irritating to nose and throat.
Ingestion Hazard: Harmful if swallowed.
Absorption Hazard: Will burn the skin and eyes.
Hazards to Wildlife: Dangerous to aquatic life.
Decontamination Procedures: Wash away any material with copious amounts of soap and water.
First Aid Procedures: Remove victim to fresh air, call emergency medical care. If not breathing give CPR. If breathing is difficult administer oxygen. Treat for shock.

Fire Hazards:

Flashpoint: n/a *Ignition temperature:* n/a
Flammable Explosive High Range: n/a *Low Range:* n/a
Toxic Products of Combustion: n/a
Other Hazards: Avoid breathing dust
Possible extinguishing agents: Dry chemical or carbon dioxide

Reactivity Hazards:

Reactive With: Water which produces calcium hydroxide, a corrosive alkali and heat!
Other Reactions: n/a

Corrosivity Hazards:

Corrosive With: n/a *Neutralizing Agent:* n/a

Radioactivity Hazards:

Radiation Emitted: n/a *Other Hazards:* n/a

Recommended Protection for Response Personnel:

Avoid breathing dust and fumes, keep upwind. Structural protective clothing provides limited protection. Wash away any material which may have come into contact with the body with copious amounts of soap and water. Consider appropriate evacuation.

CALCIUM PEROXIDE

DOT Number: UN 1457 *DOT Hazard Class:* Oxidizer *DOT Guide Number:* 35
Synonyms: calcium dioxide
STCC Number: 4918717 *Reportable Qty:* n/a
Mfg Name: Barium & Chemical Inc. *Phone No:* 1-614-282-9776

Physical Description:

Physical Form: Solid *Color:* White to yellow *Odor:* Odorless
Other Information: It is used in baking, medicine, in bleaching oils, and in many other uses.

Chemical Properties:

Specific Gravity: 2.92 *Vapor Density:* 2.1 *Boiling Point: Decomposes*
Melting Point: n/a *Vapor Pressure:* n/a *Solubility in water:* No
Other Information: Sinks in water.

Health Hazards:

Inhalation Hazard: n/a
Ingestion Hazard: Harmful if swallowed.
Absorption Hazard: Irritating to the skin and eyes.
Hazards to Wildlife: n/a
Decontamination Procedures: Wash away any material with copious amounts of soap and water.
First Aid Procedures: Remove victim to fresh air, call emergency medical care. If not breathing give CPR. If breathing is difficult administer oxygen. Treat for shock.

Fire Hazards:

Flashpoint: n/a *Ignition temperature:* n/a
Flammable Explosive High Range: n/a *Low Range:* n/a
Toxic Products of Combustion: n/a
Other Hazards: May cause ignition upon contact with combustibles. Containers may explode in a fire.
Possible extinguishing agents: Flood with water. Apply water from as far a distance as possible.

Reactivity Hazards:

Reactive With: Water to form lime water and oxygen gas. *Other Reactions:* n/a

Corrosivity Hazards:

Corrosive With: n/a *Neutralizing Agent:* n/a

Radioactivity Hazards:

Radiation Emitted: n/a *Other Hazards:* n/a

Recommended Protection for Response Personnel:

Avoid breathing vapors, keep upwind. Structural protective clothing provides limited protection. Wash away any material which may have come into contact with the body with copious amounts of soap and water. If the fire becomes uncontrollable, consider appropriate evacuation.

CALCIUM PHOSPHIDE

DOT Number: UN 1360 *DOT Hazard Class:* Flammable solid *DOT Guide Number:* 41
Synonyms: Photophor
STCC Number: 4916310 *Reportable Qty:* n/a
Mfg Name: Ventron Corporation *Phone No:* 1-617-774-3100

Physical Description:

Physical Form: Solid *Color:* Gray *Odor:* Musty
Other Information: n/a

Chemical Properties:

Specific Gravity: 2.5 *Vapor Density:* n/a *Boiling Point: Decomposes*
Melting Point: n/a *Vapor Pressure:* n/a *Solubility in water:* Reacts
Other Information: Reacts violently in water. A poisonous flammable vapor is produced.

Health Hazards:

Inhalation Hazard: Poisonous if inhaled.
Ingestion Hazard: Harmful if swallowed.
Absorption Hazard: Irritating to the skin and eyes.
Hazards to Wildlife: n/a
Decontamination Procedures: Wash away any material with copious amounts of soap and water.
First Aid Procedures: Remove victim to fresh air, call emergency medical care. If not breathing give CPR. If breathing is difficult administer oxygen. Treat for shock.

Fire Hazards:

Flashpoint: n/a *Ignition temperature:* n/a
Flammable Explosive High Range: n/a *Low Range:* n/a
Toxic Products of Combustion: Poisonous gases are produced in a fire.
Other Hazards: Ignites when exposed to moisture.
Possible extinguishing agents: Do not use water. Use dry chemical or carbon dioxide.

Reactivity Hazards:

Reactive With: Water to form phosphine, a poisonous, spontaneously flammable gas.
Other Reactions: Surface moisture

Corrosivity Hazards:

Corrosive With: n/a *Neutralizing Agent:* n/a

Radioactivity Hazards:

Radiation Emitted: n/a *Other Hazards:* n/a

Recommended Protection for Response Personnel:

Avoid breathing vapors, keep upwind. Structural protective clothing provides limited protection. Wash away any material which may have come into contact with the body with copious amounts of soap and water. Consider appropriate evacuation.

CALCIUM RESINATE

DOT Number: UN 1313 *DOT Hazard Class:* Flammable solid *DOT Guide Number:* 32
Synonyms: calcium rosin, metallic resinate
STCC Number: 4916703 *Reportable Qty:* n/a
Mfg Name: Crosby Chemical *Phone No:* 1-504-581-7047

Physical Description:

Physical Form: Solid *Color:* Yellow to dark brown *Odor:* Odorless
Other Information: n/a

Chemical Properties:

Specific Gravity: 1.13 *Vapor Density:* 1 *Boiling Point: 600° F(315.5° C)*
Melting Point: n/a *Vapor Pressure:* n/a *Solubility in water:* Insoluble
Other Information: Sinks in water.

Health Hazards:

Inhalation Hazard: Coughing, difficulty in breathing.
Ingestion Hazard: Harmful if swallowed.
Absorption Hazard: n/a
Hazards to Wildlife: n/a
Decontamination Procedures: Wash away any material with copious amounts of soap and water.
First Aid Procedures: Remove victim to fresh air, call emergency medical care. If not breathing give CPR. If breathing is difficult administer oxygen. Treat for shock.

Fire Hazards:

Flashpoint: n/a *Ignition temperature:* n/a
Flammable Explosive High Range: n/a *Low Range:* n/a
Toxic Products of Combustion: n/a
Other Hazards: May ignite when exposed to air.
Possible extinguishing agents: Use water, foam, dry chemical, or carbon dioxide.

Reactivity Hazards:

Reactive With: n/a *Other Reactions:* n/a

Corrosivity Hazards:

Corrosive With: n/a *Neutralizing Agent:* n/a

Radioactivity Hazards:

Radiation Emitted: n/a *Other Hazards:* n/a

Recommended Protection for Response Personnel:

Avoid breathing vapors, keep upwind. Wear appropriate chemical clothing. Wash away any material which may have come into contact with the body with copious amounts of soap and water. Consider appropriate evacuation.

CAMPHENE

DOT Number: NA 9011 *DOT Hazard Class:* ORM-A *DOT Guide Number:* 58
Synonyms: 2,2-dimethyl-3-methylene-norborane, 3,3-dimethyl-2-methylene norcamphane
STCC Number: 4940308 *Reportable Qty:* n/a
Mfg Name: Hercules Inc. *Phone No:* 1-212-869-4830

Physical Description:

Physical Form: Solid *Color:* White *Odor:* Camphor like
Other Information: It is used in the manufacture of synthetics, and camphor.

Chemical Properties:

Specific Gravity: .87 *Vapor Density:* 4.7 *Boiling Point: 310° F(154.4° C)*
Melting Point: 122° F(50° C) *Vapor Pressure:* n/a *Solubility in water:* n/a
Other Information: Floats on water.

Health Hazards:

Inhalation Hazard: Headaches or difficulty in breathing.
Ingestion Hazard: Harmful if swallowed.
Absorption Hazard: Irritating to the skin and eyes.
Hazards to Wildlife: n/a
Decontamination Procedures: Wash away any material with copious amounts of soap and water.
First Aid Procedures: Remove victim to fresh air, call emergency medical care. If not breathing give CPR. If breathing is difficult administer oxygen. Treat for shock.

Fire Hazards:

Flashpoint: 100° F(37.7° C) *Ignition temperature:* n/a
Flammable Explosive High Range: n/a *Low Range:* n/a
Toxic Products of Combustion: Yields Flammable vapors when heated.
Other Hazards: At higher temperature, will emit acrid smoke and irritating fumes.
Possible extinguishing agents: Apply water from as far a distance as possible. Use foam, dry chemical, or carbon dioxide.

Reactivity Hazards:

Reactive With: n/a *Other Reactions:* n/a

Corrosivity Hazards:

Corrosive With: n/a *Neutralizing Agent:* n/a

Radioactivity Hazards:

Radiation Emitted: n/a *Other Hazards:* n/a

Recommended Protection for Response Personnel:

Avoid breathing vapors, keep upwind. Wear appropriate chemical clothing. Wash away any material which may have come into contact with the body with copious amounts of soap and water. Consider appropriate evacuation.

CAMPHOR OIL

DOT Number: UN 1130 *DOT Hazard Class:* Combustible liquid *DOT Guide Number:* 27
Synonyms: liquid camphor, liquid gum camphor, liquid impure camphor
STCC Number: 4913153 *Reportable Qty:* n/a
Mfg Name: Mutchler Chemical Co. *Phone No:* 1-201-666-7002

Physical Description:

Physical Form: Liquid *Color:* Colorless, brown or blue *Odor:* Penetrating camphor
Other Information: n/a

Chemical Properties:

Specific Gravity: .92 *Vapor Density:* 1 *Boiling Point: 392° F(200° C)*
Melting Point: n/a *Vapor Pressure:* n/a *Solubility in water:* No
Other Information: Generally lighter than water. Vapors are heavier than air.

Health Hazards:

Inhalation Hazard: n/a
Ingestion Hazard: Nausea, vomiting, or loss of consciousness.
Absorption Hazard: Irritating to the skin and eyes.
Hazards to Wildlife: n/a
Decontamination Procedures: Wash away any material with copious amounts of soap and water.
First Aid Procedures: Remove victim to fresh air, call emergency medical care. If not breathing give CPR. If breathing is difficult administer oxygen. Treat for shock.

Fire Hazards:

Flashpoint: 117° F(47.2° C) *Ignition temperature:* 466° F(241.1° C)
Flammable Explosive High Range: n/a *Low Range:* n/a
Toxic Products of Combustion: n/a
Other Hazards: n/a
Possible extinguishing agents: Apply water from as far a distance as possible. Use foam, dry chemical, or carbon dioxide. Do not extinguish the fire unless the flow can be stopped.

Reactivity Hazards:

Reactive With: n/a *Other Reactions:* n/a

Corrosivity Hazards:

Corrosive With: n/a *Neutralizing Agent:* n/a

Radioactivity Hazards:

Radiation Emitted: n/a *Other Hazards:* n/a

Recommended Protection for Response Personnel:

Avoid breathing vapors, keep upwind. Structural protective clothing provides limited protection. Wash away any material which may have come into contact with the body with copious amounts of soap and water. Consider appropriate evacuation.

CAPTAN (liquid)

DOT Number: NA 9099 *DOT Hazard Class:* ORM-E *DOT Guide Number:* 31
Synonyms: Orthocide, Vancide
STCC Number: 4961167 *Reportable Qty:* 10/4.54
Mfg Name: Stauffer Chemical Co. *Phone No:* 1-203-222-3000

Physical Description:

Physical Form: Liquid *Color:* n/a *Odor:* Slight
Other Information: It is used as a fungicide.

Chemical Properties:

Specific Gravity: 1.74 *Vapor Density:* n/a *Boiling Point: n/a*
Melting Point: 338° F(170° C) *Vapor Pressure:* n/a *Solubility in water:* Yes
Other Information: Sinks and mixes in water. In case of damage to or leaking from containers of this material, contact the Pesticide Safety Team Network at 1-800-424-9300.

Health Hazards:

Inhalation Hazard: Poisonous if inhaled.
Ingestion Hazard: Poisonous if swallowed.
Absorption Hazard: Poisonous to the skin and eyes.
Hazards to Wildlife: Dangerous to aquatic life.
Decontamination Procedures: Wash away any material with copious amounts of soap and water.
First Aid Procedures: Remove victim to fresh air, call emergency medical care. If not breathing give CPR. If breathing is difficult administer oxygen. Treat for shock.

Fire Hazards:

Flashpoint: n/a *Ignition temperature:* n/a
Flammable Explosive High Range: n/a *Low Range:* n/a
Toxic Products of Combustion: Irritating toxic gases may be formed in a fire such gases include sulfur dioxide, hydrogen chloride, phosgene and oxides of nitrogen.
Other Hazards: n/a
Possible extinguishing agents: Extinguish fire using suitable agent for the type of surrounding fire.

Reactivity Hazards:

Reactive With: n/a *Other Reactions:* n/a

Corrosivity Hazards:

Corrosive With: n/a *Neutralizing Agent:* n/a

Radioactivity Hazards:

Radiation Emitted: n/a *Other Hazards:* n/a

Recommended Protection for Response Personnel:

Avoid breathing vapors, keep upwind. Wear a sealed chemical suit (polycarbonate, butyl rubber). Wash away any material which may have come into contact with the body with copious amounts of soap and water. Consider appropriate evacuation.

CAPTAN (solid)

DOT Number: NA 9099 *DOT Hazard Class:* ORM-E *DOT Guide Number:* 31
Synonyms: Orthocide, Vancide
STCC Number: 4961164 *Reportable Qty:* 10/4.54
Mfg Name: Stauffer Chemical Co. *Phone No:* 1-203-222-3000

Physical Description:

Physical Form: Solid *Color:* White to brown *Odor:* Slight
Other Information: It is used as a fungicide.

Chemical Properties:

Specific Gravity: 1.74 *Vapor Density:* n/a *Boiling Point: n/a*
Melting Point: 338° F(170° C) *Vapor Pressure:* n/a *Solubility in water:* Yes
Other Information: Sinks and mixes in water. In case of damage to or leaking from containers of this material, contact the Pesticide Safcty Team Network at 1-800-424-9300.

Health Hazards:

Inhalation Hazard: Poisonous if inhaled.

Ingestion Hazard: Poisonous if swallowed.

Absorption Hazard: Poisonous to the skin and eyes.

Hazards to Wildlife: Dangerous to aquatic life.

Decontamination Procedures: Wash away any material with copious amounts of soap and water.

First Aid Procedures: Remove victim to fresh air, call emergency medical care. If not breathing give CPR. If breathing is difficult administer oxygen. Treat for shock.

Fire Hazards:

Flashpoint: n/a *Ignition temperature:* n/a
Flammable Explosive High Range: n/a *Low Range:* n/a
Toxic Products of Combustion: Irritating toxic gases may be formed in a fire such gases include sulfur dioxide, hydrogen chloride, phosgene, and oxides of nitrogen.

Other Hazards: n/a

Possible *Possible extinguishing agents:* Extinguish fire using suitable agent for the type of surrounding fire.

Reactivity Hazards:

Reactive With: n/a *Other Reactions:* n/a

Corrosivity Hazards:

Corrosive With: n/a *Neutralizing Agent:* n/a

Radioactivity Hazards:

Radiation Emitted: n/a *Other Hazards:* n/a

Recommended Protection for Response Personnel:

Avoid breathing vapors, keep upwind. Wear a sealed chemical suit (polycarbonate, butyl rubber). Wash away any material which may have come into contact with the body with copious amounts of soap and water. Consider appropriate evacuation.

CARBARYL (liquid)

DOT Number: NA 2757 *DOT Hazard Class:* ORM-A *DOT Guide Number:* 55
Synonyms: 1-naphthyl n-methyl-carbamate, Sevin
STCC Number: 4941121 *Reportable Qty:* 100/45.4
Mfg Name: Union Carbide *Phone No:* 1-203-794-2000

Physical Description:

Physical Form: Liquid *Color:* n/a *Odor:* Weak
Other Information: It is a water emulsified liquid insecticide.

Chemical Properties:

Specific Gravity: 1.23 *Vapor Density:* 6 *Boiling Point: n/a*
Melting Point: 288° F(142.2° C) *Vapor Pressure:* n/a *Solubility in water:* Yes
Other Information: May float on water. In case of damage to or leaking from containers of this material, contact the Pesticide Safety Team Network at 1-800-424-9300.

Health Hazards:

Inhalation Hazard: n/a
Ingestion Hazard: Harmful if swallowed.
Absorption Hazard: Irritating to the skin and eyes.
Hazards to Wildlife: Dangerous to aquatic life.
Decontamination Procedures: Wash away any material with copious amounts of soap and water.
First Aid Procedures: Remove victim to fresh air, call emergency medical care. If not breathing give CPR. If breathing is difficult administer oxygen. Treat for shock.

Fire Hazards:

Flashpoint: n/a *Ignition temperature:* n/a
Flammable Explosive High Range: n/a *Low Range:* n/a
Toxic Products of Combustion: n/a
Other Hazards: n/a
Possible extinguishing agents: Extinguish fire using suitable agent for the type of surrounding fire.

Reactivity Hazards:

Reactive With: n/a *Other Reactions:* n/a

Corrosivity Hazards:

Corrosive With: n/a *Neutralizing Agent:* n/a

Radioactivity Hazards:

Radiation Emitted: n/a *Other Hazards:* n/a

Recommended Protection for Response Personnel:

Avoid breathing vapors, keep upwind. Wear a sealed chemical suit (polycarbonate, butyl rubber). Wash away any material which may have come into contact with the body with copious amounts of soap and water. Consider appropriate evacuation.

CARBARYL (solid)

DOT Number: NA 2757 *DOT Hazard Class:* ORM-A *DOT Guide Number:* 55
Synonyms: 1-naphthyl n-methyl-carbamate, Sevin
STCC Number: 4941122 *Reportable Qty:* 100/45.4
Mfg Name: Union Carbide *Phone No:* 1-203-794-2000

Physical Description:

Physical Form: Solid *Color:* White to gray *Odor:* Weak
Other Information: It is used as a pesticide.

Chemical Properties:

Specific Gravity: 1.23 *Vapor Density:* 6 *Boiling Point: n/a*
Melting Point: 288° F(142.2° C) *Vapor Pressure:* n/a *Solubility in water:* Yes
Other Information: May float on water. In case of damage to or leaking from containers of this material, contact the Pesticide Safety Team Network at 1-800-424-9300.

Health Hazards:

Inhalation Hazard: n/a
Ingestion Hazard: Harmful if swallowed.
Absorption Hazard: Irritating to the skin and eyes.
Hazards to Wildlife: Dangerous to aquatic life.
Decontamination Procedures: Wash away any material with copious amounts of soap and water.
First Aid Procedures: Remove victim to fresh air, call emergency medical care. If not breathing give CPR. If breathing is difficult administer oxygen. Treat for shock.

Fire Hazards:

Flashpoint: n/a *Ignition temperature:* n/a
Flammable Explosive High Range: n/a *Low Range:* n/a
Toxic Products of Combustion: n/a
Other Hazards: n/a
Possible *Possible extinguishing agents:* Extinguish fire using suitable agent for the type of surrounding fire.

Reactivity Hazards:

Reactive With: n/a *Other Reactions:* n/a

Corrosivity Hazards:

Corrosive With: n/a *Neutralizing Agent:* n/a

Radioactivity Hazards:

Radiation Emitted: n/a *Other Hazards:* n/a

Recommended Protection for Response Personnel:

Avoid breathing vapors, keep upwind. Wear a sealed chemical suit (polycarbonate, butyl rubber). Wash away any material which may have come into contact with the body with copious amounts of soap and water. Consider appropriate evacuation.

CARBOFURAN

DOT Number: NA 2757 *DOT Hazard Class:* Poison B *DOT Guide Number:* 55
Synonyms: Curaterr, Furadan, Niagara 10242
STCC Number: 4921525 *Reportable Qty:* 10/4.54 *CHEMTREC Phone No:* 1-800-424-9300

Physical Description:

Physical Form: Solid *Color:* White *Odor:* Odorless
Other Information: It is used as a pesticide.

Chemical Properties:

Specific Gravity: 1.18 *Vapor Density:* n/a *Boiling Point: 305° F(151.6° C)*
Melting Point: n/a *Vapor Pressure:* n/a *Solubility in water:* Yes
Other Information: In case of damage to, or leaking from containers of this material, contact the Pesticide Safety Team Network at 1-800-424-9300.

Health Hazards:

Inhalation Hazard: Poisonous if inhaled.
Ingestion Hazard: Poisonous if swallowed.
Absorption Hazard: Poisonous if absorbed.
Hazards to Wildlife: Dangerous to both waterfowl and aquatic life.
Decontamination Procedures: Wash away any material with copious amounts of soap and water.
First Aid Procedures: Remove victim to fresh air, call emergency medical care. If not breathing give CPR. If breathing is difficult administer oxygen. Treat for shock.

Fire Hazards:

Flashpoint: n/a *Ignition temperature:* n/a
Flammable Explosive High Range: n/a *Low Range:* n/a
Toxic Products of Combustion: Toxic fumes of nitrogen oxides.
Other Hazards: Will support combustion if ignited.
Possible extinguishing agents: Apply water from as far as distance as possible. Use foam, dry chemical, or carbon dioxide.

Reactivity Hazards:

Reactive With: n/a *Other Reactions:* n/a

Corrosivity Hazards:

Corrosive With: n/a *Neutralizing Agent:* n/a

Radioactivity Hazards:

Radiation Emitted: n/a *Other Hazards:* n/a

Recommended Protection for Response Personnel:

Avoid breathing vapors, keep upwind. Wear appropriate chemical clothing, wash away any material which may have come into contact with the body with copious amounts of soap and water. Consider appropriate evacuation.

CARBOLIC ACID

DOT Number: UN 1671 *DOT Hazard Class:* Poison B *DOT Guide Number:* 55
Synonyms: phenol, phenylic acid
STCC Number: 4921220 *Reportable Qty:* 1000/454
Mfg Name: Allied Corp. *Phone No:* 1-212-455-2000

Physical Description:

Physical Form: Solid or liquid *Color:* Colorless to pink or red *Odor:* Sweet tarry
Other Information: It is used to make plastics, adhesives, and other chemicals. Weighs 8.9 lbs/3.8 kg per gallon/3.8 l.

Chemical Properties:

Specific Gravity: 1.1 *Vapor Density:* 3.2 *Boiling Point: 358° F(181.1° C)*
Melting Point: 109° F(93° C) *Vapor Pressure:* .36 mm Hg at 68° F(20° C)
Solubility in water: Yes *Degree of Solubility:* 8.4%
Other Information: Soluble in water, vapors are heavier than air.

Health Hazards:

Inhalation Hazard: n/a
Ingestion Hazard: Poisonous if swallowed.
Absorption Hazard: Will burn or numb skin and eyes.
Hazards to Wildlife: Dangerous to aquatic life.
Decontamination Procedures: Wash away any material with copious amounts of soap and water.
First Aid Procedures: Remove victim to fresh air, call emergency medical care. If not breathing give CPR. If breathing is difficult administer oxygen. Treat for shock.

Fire Hazards:

Flashpoint: 175° F(79.4° C) *Ignition temperature:* 1319° F(715° C)
Flammable Explosive High Range: 8.6 *Low Range:* 1.8
Toxic Products of Combustion: Poisonous gases are produced in a fire.
Other Hazards: Avoid contact with the liquid or solid.
Possible extinguishing agents: Apply water from as far a distance as possible. Use foam, dry chemical, or carbon dioxide.

Reactivity Hazards:

Reactive With: n/a *Other Reactions:* n/a

Corrosivity Hazards:

Corrosive With: Lead and alloys, certain plastics and rubbers.
Neutralizing Agent: Crushed limestone, soda ash, or lime

Radioactivity Hazards:

Radiation Emitted: n/a *Other Hazards:* n/a

Recommended Protection for Response Personnel:

Avoid breathing vapors, keep upwind. Structural protective clothing provides limited protection. Wash away any material which may have come into contact with the body with copious amounts of soap and water. Consider appropriate evacuation.

CARBOLIC OIL

DOT Number: UN 2821 *DOT Hazard Class:* Poison B *DOT Guide Number:* 55
Synonyms: carbolic acid, liquefied phenol, middle oil
STCC Number: n/a *Reportable Qty:* n/a
Mfg Name: Dow Chemical Co. *Phone No:* 1-517-636-4400

Physical Description:

Physical Form: Liquid *Color:* Colorless *Odor:* Sweet tar
Other Information: Material darkens on exposure to light.

Chemical Properties:

Specific Gravity: 1.04 *Vapor Density:* 3.24 *Boiling Point: 359° F(181.6° C)*
Melting Point: 105° F(40.5° C) *Vapor Pressure:* 1 mm Hg at 104° F(40° C) *Solubility in water:* Yes
Other Information: Sinks and mixes with water.

Health Hazards:

Inhalation Hazard: Poisonous if inhaled.
Ingestion Hazard: Poisonous if swallowed.
Absorption Hazard: Poisonous to the skin and eyes.
Hazards to Wildlife: Dangerous to aquatic life.
Decontamination Procedures: Wash away any material with copious amounts of soap and water.
First Aid Procedures: Remove victim to fresh air, call emergency medical care. If not breathing give CPR. If breathing is difficult administer oxygen. Treat for shock.

Fire Hazards:

Flashpoint: 175° F(79.4° C) *Ignition temperature:* 1319° F(715° C)
Flammable Explosive High Range: 8.6 *Low Range:* 1.7
Toxic Products of Combustion: Poisonous gases are produced in a fire.
Other Hazards: Yields Flammable vapors when heated which form explosive mixtures with air.
Possible extinguishing agents: Apply water from as far a distance as possible. Use foam, dry chemical, or carbon dioxide.

Reactivity Hazards:

Reactive With: n/a *Other Reactions:* n/a

Corrosivity Hazards:

Corrosive With: n/a *Neutralizing Agent:* n/a

Radioactivity Hazards:

Radiation Emitted: n/a *Other Hazards:* n/a

Recommended Protection for Response Personnel:

Avoid breathing vapors, keep upwind. Wear a sealed chemical suit (polycarbonate, butyl rubber, chlorinated polyethylene, viton, pvc). Wash away any material which may have come into contact with the body with copious amounts of soap and water. Consider appropriate evacuation.

CARBON BISULFIDE

DOT Number: UN 1131 *DOT Hazard Class:* Flammable liquid *DOT Guide Number:* 28
Synonyms: carbon disulfide, dithiocarbonic anhydride,
STCC Number: 4908125 *Reportable Qty:* 100/45.4
Mfg Name: PPG Industries Inc. *Phone No:* 1-412-434-3131

Physical Description:

Physical Form: Liquid *Color:* Colorless to light yellow *Odor:* Disagreeable
Other Information: It is used to manufacture rayon, cellophane, and in the manufacture of flotation agents, and as a solvent.

Chemical Properties:

Specific Gravity: 1.3 *Vapor Density:* 2.6 *Boiling Point: 115° F(46.1° C)*
Melting Point: −168.9 f.(−111.6° C) *Vapor Pressure:* 300 mm Hg at 68° F(20° C)
Other Information: Slightly soluble in water, weighs 10.5 lbs/4.7 kg per gallon/3.8 l.

Health Hazards:

Inhalation Hazard: May cause headache, nausea, and dizziness.
Ingestion Hazard: Vomiting, diarrhea, and headache.
Absorption Hazard: Immediate and severe irritations.
Hazards to Wildlife: n/a
Decontamination Procedures: Wash away any material with copious amounts of soap and water.
First Aid Procedures: Remove victim to fresh air, call emergency medical care. If not breathing give CPR. If breathing is difficult administer oxygen. Treat for shock.

Fire Hazards:

Flashpoint: −22° F(−30° C) *Ignition temperature:* 200° F(93.3° C)
Flammable Explosive High Range: 50 *Low Range:* 1
Toxic Products of Combustion: Toxic sulfur oxide, and carbon monoxide gases.
Other Hazards: When exposed to high temperatures or flames, the containers may explode.
Possible extinguishing agents: Apply water from as far a distance as possible. Use foam, dry chemical, or carbon dioxide.

Reactivity Hazards:

Reactive With: Aluminum, strong oxidizers, *Other Reactions:* Chemically active metals

Corrosivity Hazards:

Corrosive With: Some metals, will attack some plastics, rubber and coatings. *Neutralizing Agent:* n/a

Radioactivity Hazards:

Radiation Emitted: n/a *Other Hazards:* n/a

Recommended Protection for Response Personnel:

Avoid breathing vapors, keep upwind. Wear a sealed chemical suit (viton). Wash away any material which may have come into contact with the body with copious amounts of soap and water. If the fire becomes uncontrollable, or if the container is exposed to direct flame, consider appropriate evacuation.

CARBON DIOXIDE

DOT Number: UN 1013 *DOT Hazard Class:* Nonflammable gas *DOT Guide Number:* 21
Synonyms: carbonic acid gas, dry ice
STCC Number: 4904535 *Reportable Qty:* n/a
Mfg Name: Liquid Carbonic *Phone No:* 1-312-855-2642

Physical Description:

Physical Form: Liquefied compressed gas or solid
Color: Colorless gas or white solid *Odor:* Odorless
Other Information: Solid sinks and boils in water, and a visible vapor cloud is produced.

Chemical Properties:

Specific Gravity: 1.56 *Vapor Density:* 1.5 *Boiling Point: – 109° F(42.7° C)*
Melting Point: –109° F(42.7° C) *Vapor Pressure:* 1 atm at 68° F(20° C)
Solubility in water: Slight *Degree of Solubility:* .14%
Other Information: The gas is heavier than air and can asphyxiate by the displacement of air.

Health Hazards:

Inhalation Hazard: Causes headaches and increased respiratory rate.
Ingestion Hazard: Freezing burns to mouth and airway.
Absorption Hazard: Will cause frostbite.
Hazards to Wildlife: No
Decontamination Procedures: Wash away any material with copious amounts of soap and water.
First Aid Procedures: Remove victim to fresh air, call emergency medical care. If not breathing give CPR. If breathing is difficult administer oxygen. Treat for shock.

Fire Hazards:

Flashpoint: n/a *Ignition temperature:* n/a
Flammable Explosive High Range: n/a *Low Range:* n/a
Toxic Products of Combustion: n/a
Other Hazards: Avoid breathing vapor
Possible *Possible extinguishing agents:* n/a

Reactivity Hazards:

Reactive With: Water which forms carbonic acid. *Other Reactions:* n/a

Corrosivity Hazards:

Corrosive With: n/a *Neutralizing Agent:* n/a

Radioactivity Hazards:

Radiation Emitted: n/a *Other Hazards:* n/a

Recommended Protection for Response Personnel:

Avoid breathing vapors, keep upwind. Wear appropriate chemical clothing. Wash away any material which may come into contact with the body with copious amounts of soap and water. Consider appropriate evacuation.

CARBON DISULFIDE

DOT Number: UN 1131 *DOT Hazard Class:* Flammable liquid *DOT Guide Number:* 28
Synonyms: carbon bisulfide
STCC Number: n/a *Reportable Qty:* n/a
Mfg Name: PPG Industries Inc. *Phone No:* 1-412-434-3131

Physical Description:

Physical Form: Liquid *Color:* Colorless to yellow *Odor:* Rotten egg to sweet
Other Information: Sinks in water. Flammable, irritating gas is produced.

Chemical Properties:

Specific Gravity: 1.25 *Vapor Density:* 2.64 *Boiling Point: 115° F(46.1° C)*
Melting Point: −169° F(−111.6° C) *Vapor Pressure:* 300mm at 68° F(20° C)
Solubility in water: Yes *Degree of Solubility: .2%*
Other Information: n/a

Health Hazards:

Inhalation Hazard: Causes headaches
Ingestion Hazard: n/a
Absorption Hazard: Mild to moderate irritation to skin and eyes.
Hazards to Wildlife: No
Decontamination Procedures: Wash away any material with copious amounts of soap and water.
First Aid Procedures: Remove victim to fresh air, call emergency medical care. If not breathing give CPR. If breathing is difficult administer oxygen. Treat for shock.

Fire Hazards:

Flashpoint: −22° F(−30° C) *Ignition temperature:* 212° F(100° C)
Flammable Explosive High Range: 50 *Low Range:* 1.3
Toxic Products of Combustion: Toxic gases are generated.
Other Hazards: n/a
Possible extinguishing agents: Dry chemical or carbon dioxide. Do not use water or foam

Reactivity Hazards:

Reactive With: n/a *Other Reactions:* n/a

Corrosivity Hazards:

Corrosive With: n/a *Neutralizing Agent:* n/a

Radioactivity Hazards:

Radiation Emitted: n/a *Other Hazards:* n/a

Recommended Protection for Response Personnel:

Avoid breathing vapors, keep upwind. Wear a sealed chemical suit (viton). Wash away any material which may come into contact with the body with copious amounts of soap and water. Consider appropriate evacuation.

CARBON MONOXIDE

DOT Number: UN 1016 *DOT Hazard Class:* Flammable gas *DOT Guide Number:* 18
Synonyms: monoxide
STCC Number: 4905709 *Reportable Qty:* n/a
Mfg Name: Matheson Gas Products *Phone No:* 1-201-867-4100

Physical Description:

Physical Form: Gas *Color:* Colorless *Odor:* Odorless
Other Information: Used to make pesticides, dyes, and other chemicals.

Chemical Properties:

Specific Gravity: .8 *Vapor Density:* 1 *Boiling Point: −314° F(−192° C)*
Melting Point: −326° F(−198° C) *Vapor Pressure:* 1atm at 68° F(20° C) *Solubility in water:* Slight
Other Information: Lighter than air, and flame can flashback very easily to the source of ignition.

Health Hazards:

Inhalation Hazard: Causes headaches, nausea, dizziness, weakness of limbs.
Ingestion Hazard: n/a
Absorption Hazard: Frostbite from liquid.
Hazards to Wildlife: No
Decontamination Procedures: Wash away any material with copious amounts of soap and water.
First Aid Procedures: Remove victim to fresh air, call emergency medical care. If not breathing give CPR. If breathing is difficult administer oxygen. Treat for shock.

Fire Hazards:

Flashpoint: n/a *Ignition temperature:* 1128° F(608.8° C)
Flammable Explosive High Range: 74 *Low Range:* 12.5
Toxic Products of Combustion: n/a
Other Hazards: Under fire conditions, a cylinder may violently rupture and rocket. (BLEVE)
Possible extinguishing agents: Do not extinguish fire unless flow can be stopped. Use large quantities of water as a fog with unmanned lines.

Reactivity Hazards:

Reactive With: n/a *Other Reactions:* n/a

Corrosivity Hazards:

Corrosive With: n/a *Neutralizing Agent:* n/a

Radioactivity Hazards:

Radiation Emitted: n/a *Other Hazards:* n/a

Recommended Protection for Response Personnel:

Avoid breathing vapors, keep upwind. Structural protective clothing provides limited protection. Wash away any material which may have come into contact with the body with copious amounts of soap and water. Prolonged exposure to this material can cause asphyxiation. If the fire becomes uncontrollable consider appropriate evacuation.

CARBON TETRACHLORIDE

DOT Number: UN 1846 *DOT Hazard Class:* ORM-A *DOT Guide Number:* 55
Synonyms: benzinoform, carbon tet, necatorina, tetrachloromethane
STCC Number: 4940320 *Reportable Qty:* 5000/2270
Mfg Name: Ashland Chemical *Phone No:* 1-614-889-3333

Physical Description:

Physical Form: Liquid *Color:* Colorless *Odor:* Sweet
Other Information: It is used as a solvent, in the manufacture of other chemicals, and an agricultural fumigant.

Chemical Properties:

Specific Gravity: 1.58 *Vapor Density:* 8.2 *Boiling Point: 170.2° F(76.7° C)*
Melting Point: 8.7° F(−12.4) *Vapor Pressure:* 91.3 mm Hg at 68° F(20° C) *Solubility in water:* No
Other Information: Insoluble in water, weighs 13.2 lbs/5.9 kg per gallon/3.8 l.

Health Hazards:

Inhalation Hazard: Poisonous if inhaled.
Ingestion Hazard: Poisonous if swallowed.
Absorption Hazard: Irritating to the skin and eyes.
Hazards to Wildlife: n/a
Decontamination Procedures: Wash away any material with copious amounts of soap and water.
First Aid Procedures: Remove victim to fresh air, call emergency medical care. If not breathing give CPR. If breathing is difficult administer oxygen. Treat for shock.

Fire Hazards:

Flashpoint: n/a *Ignition temperature:* n/a
Flammable Explosive High Range: n/a *Low Range:* n/a
Toxic Products of Combustion: Hydrogen chloride, chlorine, phosgene, and carbon monoxide.
Other Hazards: Containers may rupture in a fire due to overpressurization
Possible extinguishing agents: Extinguish fire using suitable agent for the type of surrounding fire.

Reactivity Hazards:

Reactive With: Reactions vary with other chemicals *Other Reactions:* n/a

Corrosivity Hazards:

Corrosive With: Will react with aluminum to form explosive mixtures. Will attack some plastics, rubbers, and coatings.
Neutralizing Agent: n/a

Radioactivity Hazards:

Radiation Emitted: n/a *Other Hazards:* n/a

Recommended Protection for Response Personnel:

Avoid breathing vapors, keep upwind. Structural protective clothing provides limited protection. Wash away any material which may have come into contact with the body with copious amounts of soap and water. Consider appropriate evacuation.

CAUSTIC POTASH SOLUTION

DOT Number: UN 1814 *DOT Hazard Class:* Corrosive *DOT Guide Number:* 60
Synonyms: lye, potassium hydroxide solution
STCC Number: 4935245 *Reportable Qty:* 1000/454
Mfg Name: Monsanto Chemical *Phone No:* 1-314-694-1000

Physical Description:

Physical Form: Liquid *Color:* Colorless *Odor:* Odorless
Other Information: n/a

Chemical Properties:

Specific Gravity: 1.45 *Vapor Density:* 1.9 *Boiling Point: 266° F(130° C)*
Melting Point: n/a *Vapor Pressure:* n/a *Solubility in water:* Yes
Other Information: It dissolves in water with releases of heat. Weighs 12.7 lbs/5.7 kg per gallon/3.8 l.

Health Hazards:

Inhalation Hazard: n/a
Ingestion Hazard: Harmful if swallowed.
Absorption Hazard: Will burn the skin and eyes.
Hazards to Wildlife: Dangerous to aquatic life.
Decontamination Procedures: Wash away any material with copious amounts of soap and water.
First Aid Procedures: Remove victim to fresh air, call emergency medical care. If not breathing give CPR. If breathing is difficult administer oxygen. Treat for shock.

Fire Hazards:

Flashpoint: n/a *Ignition temperature:* n/a
Flammable Explosive High Range: n/a *Low Range:* n/a
Toxic Products of Combustion: n/a
Other Hazards: n/a
Possible extinguishing agents: Extinguish fire using suitable agent for type of surrounding fire.

Reactivity Hazards:

Reactive With: n/a *Other Reactions:* n/a

Corrosivity Hazards:

Corrosive With: Metals and tissue *Neutralizing Agent:* Vinegar or other dilute acid

Radioactivity Hazards:

Radiation Emitted: n/a *Other Hazards:* n/a

Recommended Protection for Response Personnel:

Avoid breathing vapors, keep upwind. Wear a sealed chemical suit (nitrile, viton, butyl rubber, chlorinated polyethylene). Wash away any material which may have come into contact with the body with copious amounts of soap and water. Consider appropriate evacuation.

CAUSTIC SODA SOLUTION

DOT Number: UN 1824 *DOT Hazard Class:* Corrosive *DOT Guide Number:* 60
Synonyms: lye, sodium hydroxide solution
STCC Number: 4935206 *Reportable Qty:* 1000/454
Mfg Name: PPG Industries Inc. *Phone No:* 1-412-434-3131

Physical Description:

Physical Form: Liquid *Color:* Colorless *Odor:* Odorless
Other Information: n/a

Chemical Properties:

Specific Gravity: 1.5 *Vapor Density:* 1.4 *Boiling Point: 266° F(130° C)*
Melting Point: n/a *Vapor Pressure:* 1 mm Hg at 1362° F(738.8° C) *Solubility in water:* Yes
Other Information: It is partially soluble in water with releases of heat. Weighs 12.7 lbs/5.7 kg per gallon/3.8 l. This solution may contain other chemicals. Contact the shipper through CHEMTREC at 1-800-424-9300 to obtain commodity specific information.

Health Hazards:

Inhalation Hazard: n/a
Ingestion Hazard: Harmful if swallowed.
Absorption Hazard: Will burn the skin and eyes.
Hazards to Wildlife: Dangerous to aquatic life.
Decontamination Procedures: Wash away any material with copious amounts of soap and water.
First Aid Procedures: Remove victim to fresh air, call emergency medical care. If not breathing give CPR. If breathing is difficult administer oxygen. Treat for shock.

Fire Hazards:

Flashpoint: n/a *Ignition temperature:* n/a
Flammable Explosive High Range: n/a *Low Range:* n/a
Toxic Products of Combustion: n/a
Other Hazards: n/a
Possible extinguishing agents: Extinguish fire using suitable agent for type of surrounding fire. Apply water from as far a distance as possible.

Reactivity Hazards:

Reactive With: n/a *Other Reactions:* n/a

Corrosivity Hazards:

Corrosive With: Metals and tissue *Neutralizing Agent:* Vinegar or other dilute acid

Radioactivity Hazards:

Radiation Emitted: n/a *Other Hazards:* n/a

Recommended Protection for Response Personnel:

Avoid breathing vapors, keep upwind. Wear a sealed chemical suit (butyl rubber). Wash way any material which may have come into contact with the body with copious amounts of soap and water. Consider appropriate evacuation.

CHARCOAL

DOT Number: NA 1361 *DOT Hazard Class:* Flammable solid *DOT Guide Number:* 32
Synonyms: activated, wood, shell, vegetable, animal, charcoal: carbon
STCC Number: 4917322 *Reportable Qty:* n/a
Mfg Name: Mallinckrodt Inc. *Phone No:* 1-314-895-2000

Physical Description:

Physical Form: Solid *Color:* Black *Odor:* Odorless
Other Information: n/a

Chemical Properties:

Specific Gravity: 2 *Vapor Density:* 1 *Boiling Point: Very high*
Melting Point: n/a *Vapor Pressure:* n/a *Solubility in water:* No
Other Information: May float or sink in water.

Health Hazards:

Inhalation Hazard: n/a
Ingestion Hazard: n/a
Absorption Hazard: n/a
Hazards to Wildlife: n/a
Decontamination Procedures: Wash away any material with copious amounts of soap and water.
First Aid Procedures: Remove victim to fresh air, call emergency medical care. If not breathing give CPR. If breathing is difficult administer oxygen. Treat for shock.

Fire Hazards:

Flashpoint: n/a *Ignition temperature:* 600-750° F(315-398° C)
Flammable Explosive High Range: n/a *Low Range:* n/a
Toxic Products of Combustion: Incomplete combustion forms toxic carbon monoxide.
Other Hazards: May spontaneously ignite in air.
Possible extinguishing agents: Remove burning material from load if safe. Use dry chemical and carbon dioxide.

Reactivity Hazards:

Reactive With: n/a *Other Reactions:* n/a

Corrosivity Hazards:

Corrosive With: n/a *Neutralizing Agent:* n/a

Radioactivity Hazards:

Radiation Emitted: n/a *Other Hazards:* n/a

Recommended Protection for Response Personnel:

Avoid breathing vapors, keep upwind. Structural protective clothing provides limited protection. Wash away any material which may have come into contact with the body with copious amounts of soap and water. Consider appropriate evacuation.

CHLORDANE

DOT Number: UN 2762 *DOT Hazard Class:* Flammable liquid *DOT Guide Number:* 28
Synonyms: chlordan, 1,2,4,5,6,7,8,8-octachloro-2,3,3a,4,7,7a-hexahydro-4,7-methanoindene, Octa-klor, Toxichlor, Velsicol 1068
STCC Number: n/a *Reportable Qty:* n/a
Mfg Name: Velsicol Chemical Corp. *Phone No:* 1-312-670-4500

Physical Description:

Physical Form: Liquid *Color:* Brown *Odor:* Sharp
Other Information: n/a

Chemical Properties:

Specific Gravity: 1.6 *Vapor Density:* 1.6 *Boiling Point: Decomposes*
Melting Point: n/a *Vapor Pressure:* n/a *Solubility in water:* n/a
Other Information: Sinks in water

Health Hazards:

Inhalation Hazard: Poisonous if inhaled.
Ingestion Hazard: Poisonous if swallowed.
Absorption Hazard: Poisonous to the skin and eyes.
Hazards to Wildlife: Dangerous to aquatic life.
Decontamination Procedures: Wash away any material with copious amounts of soap and water.
First Aid Procedures: Remove victim to fresh air, call emergency medical care. If not breathing give CPR. If breathing is difficult administer oxygen. Treat for shock.

Fire Hazards:

Flashpoint: 132° F(55.5° C) *Ignition temperature:* 410° F(210° C)
Flammable Explosive High Range: 5.7 *Low Range:* .7
Toxic Products of Combustion: Irritating toxic hydrogen chloride and phosgene gases may be formed when kerosene solution of compound burns.
Other Hazards: n/a
Possible extinguishing agents: Foam, dry chemical or carbon dioxide. Water may be ineffective on solution fire.

Reactivity Hazards:

Reactive With: n/a *Other Reactions:* n/a

Corrosivity Hazards:

Corrosive With: n/a *Neutralizing Agent:* n/a

Radioactivity Hazards:

Radiation Emitted: n/a *Other Hazards:* n/a

Recommended Protection for Response Personnel:

Avoid breathing vapors, keep upwind. Wear a sealed chemical suit (polycarbonate, butyl rubber). Wash away any material which may have come into contact with the body with copious amounts of soap and water. Consider appropriate evacuation.

CHLORINE

DOT Number: UN 1017 *DOT Hazard Class:* Nonflammable gas *DOT Guide Number:* 20
Synonyms: None listed *STCC Number:* 4904120 *Reportable Qty:* 10/4.54
Mfg Name: Dow Chemical Co. *Phone No:* 1-517-636-4400

Physical Description:

Physical Form: Gas *Color:* Greenish yellow *Odor:* Pungent, suffocating
Other Information: Used to purify water, bleach wood pulp and to make other chemicals.

Chemical Properties:

Specific Gravity: 1.42 *Vapor Density:* 2.4 *Boiling Point: −29° F(−33.8° C)*
Melting Point: −150° F(−110.1 C) *Vapor Pressure:* 1 atm at 68° F (20° C)
Solubility in water: Slight *Degree of Solubility: .7%*
Other Information: Liquid readily vaporizes into a gas, vapors are much heavier than air and tend to settle in low lying areas. Weighs 7lbs/3.17 kg per gallon/3.8 l. Contact CHEMTREC to activate the Chlorine Response Team 1-800-424-9300 or 202-483-7616.

Health Hazards:

Inhalation Hazard: Poisonous if inhaled.
Ingestion Hazard: Burning of mouth, irritating throat, nausea, stupor.
Absorption Hazard: Frostbite from liquid, dermatitis. *Hazards to Wildlife:* Dangerous to aquatic life.
Decontamination Procedures: Wash away any material with copious amounts of soap and water.
First Aid Procedures: Remove victim to fresh air, call emergency medical care. If not breathing give CPR. If breathing is difficult administer oxygen. Treat for shock.

Fire Hazards:

Flashpoint: n/a *Ignition temperature:* n/a
Flammable Explosive High Range: n/a *Low Range:* n/a
Toxic Products of Combustion: Poisonous gases are produced when involved in a fire.
Other Hazards: Chlorine does not burn, but will support combustion.
Possible extinguishing agents: Extinguish fire using suitable agent for type of surrounding fire. NOTE: Chlorine itself does not burn.

Reactivity Hazards:

Reactive With: Reacts violently with water to form a corrosive solution.
Other Reactions: It reacts explosively or forms explosive compounds with many common chemicals.

Corrosivity Hazards:

Corrosive With: Most metals at high temperatures *Neutralizing Agent:* n/a

Radioactivity Hazards:

Radiation Emitted: n/a *Other Hazards:* n/a

Recommended Protection for Response Personnel:

Avoid breathing vapors, keep upwind. Wear a sealed chemical suit (polycarbonate, butyl rubber, viton, chlorinated polyethylene). Wash away any material which may have come into contact with the body with copious amounts of soap and water. Prolonged exposure to this material will cause asphyxiation. If the fire becomes uncontrollable, consider appropriate evacuation.

CHLORINE TRIFLUORIDE

DOT Number: UN 1749 *DOT Hazard Class:* Oxidizer *DOT Guide Number:* 44
Synonyms: CTF
STCC Number: 4918210 *Reportable Qty:* n/a
Mfg Name: Atomergic Chemetals Corp. *Phone No:* 1-516-349-8800

Physical Description:

Physical Form: Liquid *Color:* Green *Odor:* Pungent
Other Information: n/a

Chemical Properties:

Specific Gravity: 1.85 *Vapor Density:* 3.1 *Boiling Point: 53° F(11.6° C)*
Melting Point: –105° F(–76.1° C) *Vapor Pressure:* 1atm at 68° F(20° C) *Solubility in water:* Reacts
Other Information: May only be shipped in containers not equipped with safety release devices.

Health Hazards:

Inhalation Hazard: Poisonous if inhaled.
Ingestion Hazard: Poisonous if swallowed.
Absorption Hazard: Will burn the skin and eyes.
Hazards to Wildlife: n/a
Decontamination Procedures: Wash away any material with copious amounts of soap and water.
First Aid Procedures: Remove victim to fresh air, call emergency medical care. If not breathing give CPR. If breathing is difficult administer oxygen. Treat for shock.

Fire Hazards:

Flashpoint: n/a *Ignition temperature:* n/a
Flammable Explosive High Range: n/a *Low Range:* n/a
Toxic Products of Combustion: Poisonous gases are produced when heated.
Other Hazards: May explode upon contact with combustibles. In a fire, the containers may violently rupture or rocket.
Possible extinguishing agents: Do not use water!! Use dry chemical or carbon dioxide.

Reactivity Hazards:

Reactive With: Reacts explosively with water to form hydrofluoric acid and chlorine.
Other Reactions: Causes ignition of all combustible materials even sand or concrete. Very similar to fluorine gas.

Corrosivity Hazards:

Corrosive With: Metals and tissue *Neutralizing Agent:* Crushed limestone, soda ash, or lime

Radioactivity Hazards:

Radiation Emitted: n/a *Other Hazards:* n/a

Recommended Protection for Response Personnel:

Avoid breathing vapors, keep upwind. Wear a sealed chemical suit (chlorinated polyethylene, pvc). Wash away any material which may have come into contact with the body with copious amounts of soap and water. If the fire becomes uncontrollable, consider appropriate evacuation.

CHLOROACETIC ACID

DOT Number: UN 1751 *DOT Hazard Class:* Corrosive *DOT Guide Number:* 60
Synonyms: chloracetic acid
STCC Number: 4931416 *Reportable Qty:* n/a *CHEMTREC Phone No:* 1-800-424-9300

Physical Description:

Physical Form: Liquid *Color:* Cloudy white *Odor:* Vinegar like
Other Information: n/a

Chemical Properties:

Specific Gravity: 1.58 *Vapor Density:* n/a *Boiling Point: 372° F(188.8° C)*
Melting Point: n/a *Vapor Pressure:* n/a *Solubility in water:* Yes
Other Information: It absorbs moisture from air to form a syrup.

Health Hazards:

Inhalation Hazard: n/a

Ingestion Hazard: Harmful if swallowed.

Absorption Hazard: Will burn the skin and eyes.

Hazards to Wildlife: n/a

Decontamination Procedures: Wash away any material with copious amounts of soap and water.

First Aid Procedures: Remove victim to fresh air, call emergency medical care. If not breathing give CPR. If breathing is difficult administer oxygen. Treat for shock.

Fire Hazards:

Flashpoint: 259° F(226.1° C) *Ignition temperature:* n/a

Flammable Explosive High Range: n/a *Low Range:* 8

Toxic Products of Combustion: Hydrogen chloride and phosgene.

Other Hazards: n/a

Possible extinguishing agents: Extinguish fire using suitable agent for type of surrounding fire.

Reactivity Hazards:

Reactive With: n/a *Other Reactions:* n/a

Corrosivity Hazards:

Corrosive With: Metals and tissue

Neutralizing Agent: Crushed limestone, soda ash, or lime

Radioactivity Hazards:

Radiation Emitted: n/a *Other Hazards:* n/a

Recommended Protection for Response Personnel:

Avoid breathing vapors, keep upwind. Wear appropriate chemical clothing, wash away any material which may have come into contact with the body with copious amounts of soap and water. Consider appropriate evacuation.

CHLOROACETOPHENONE

DOT Number: UN 1697 *DOT Hazard Class:* Irritant *DOT Guide Number:* 53
Synonyms: phenyl chloromethylketone, tear gas
STCC Number: 4925220 *Reportable Qty:* n/a
Mfg Name: Federal Laboratories Inc. *Phone No:* 1-412-639-3511

Physical Description:

Physical Form: Solid *Color:* Cloudy white to light yellow *Odor:* Sharp
Other Information: n/a

Chemical Properties:

Specific Gravity: 1.32 *Vapor Density:* 5.32 *Boiling Point: 477° F(247.2° C)*
Melting Point: n/a *Vapor Pressure:* .012 mm Hg at 32° F(.0° C) *Solubility in water:* No
Other Information: n/a

Health Hazards:

Inhalation Hazard: Poisonous if inhaled.
Ingestion Hazard: Poisonous if swallowed.
Absorption Hazard: Irritating to the skin and eyes.
Hazards to Wildlife: n/a
Decontamination Procedures: Wash away any material with copious amounts of soap and water.
First Aid Procedures: Remove victim to fresh air, call emergency medical care. If not breathing give CPR. If breathing is difficult administer oxygen. Treat for shock.

Fire Hazards:

Flashpoint: 259° F (126.1° C) *Ignition temperature:* n/a
Flammable Explosive High Range: n/a *Low Range:* 8
Toxic Products of Combustion: Irritating hydrogen chloride may form.
Other Hazards: Unburned material may become volatile and cause severe eye irritation.
Possible extinguishing agents: Use water in flooding quantities as fog. Use foam, dry chemical, or carbon dioxide.

Reactivity Hazards:

Reactive With: Reacts slowly with water generating hydrogen chloride. *Other Reactions:* n/a

Corrosivity Hazards:

Corrosive With: Metals causing mild corrosion *Neutralizing Agent:* n/a

Radioactivity Hazards:

Radiation Emitted: n/a *Other Hazards:* n/a

Recommended Protection for Response Personnel:

Avoid breathing vapors, keep upwind. Wear a sealed chemical suit (polycarbonate, butyl rubber). Wash away any material which may have come into contact with the body with copious amounts of soap and water. Consider appropriate evacuation.

CHLOROACETYL CHLORIDE

DOT Number: UN 1752 *DOT Hazard Class:* Corrosive *DOT Guide Number:* 59
Synonyms: chloracetyl chloride
STCC Number: 4931210 *Reportable Qty:* n/a
Mfg Name: Dow Chemical Co. *Phone No:* 1-517-636-4400

Physical Description:

Physical Form: Liquid *Color:* Colorless to light yellow *Odor:* Sharp, extremely irritating
Other Information: n/a

Chemical Properties:

Specific Gravity: 1.42 *Vapor Density:* 3.9 *Boiling Point: 221° F(105° C)*
Melting Point: n/a *Vapor Pressure:* n/a *Solubility in water:* Reacts
Other Information: Reacts violently with water, an irritating vapor is produced.

Health Hazards:

Inhalation Hazard: n/a
Ingestion Hazard: Harmful if swallowed.
Absorption Hazard: Will burn the skin and eyes.
Hazards to Wildlife: n/a
Decontamination Procedures: Wash away any material with copious amounts of soap and water.
First Aid Procedures: Remove victim to fresh air, call emergency medical care. If not breathing give CPR. If breathing is difficult administer oxygen. Treat for shock.

Fire Hazards:

Flashpoint: n/a *Ignition temperature:* n/a
Flammable Explosive High Range: n/a *Low Range:* n/a
Toxic Products of Combustion: Highly toxic and irritating hydrogen chloride and phosgene vapors.
Other Hazards: Highly irritating tear gas vapors are released when heated
Possible extinguishing agents: Dry chemical, or carbon dioxide. Do not use water on the material itself.

Reactivity Hazards:

Reactive With: Water to form hydrogen chloride. *Other Reactions:* n/a

Corrosivity Hazards:

Corrosive With: Metals and tissue *Neutralizing Agent:* Crushed limestone, soda ash, or lime

Radioactivity Hazards:

Radiation Emitted: n/a *Other Hazards:* n/a

Recommended Protection for Response Personnel:

Avoid breathing vapors, keep upwind. Wear a sealed chemical suit (butyl rubber). Wash away any material which may have come into contact with the body with copious amounts of soap and water. Consider appropriate evacuation.

CHLOROANILINE

DOT Number: UN 2018 *DOT Hazard Class:* n/a *DOT Guide Number:* 53
Synonyms: 1-amino-4-chlorobenzene, 4-chloroaniline, 4-chlorophenylamine
STCC Number: n/a *Reportable Qty:* n/a
Mfg Name: Aldrich Chemical *Phone No:* 1-414-273-3851

Physical Description:

Physical Form: Solid *Color:* Yellowish white *Odor:* Mild sweet
Other Information: n/a

Chemical Properties:

Specific Gravity: 1.43 *Vapor Density:* n/a *Boiling Point: 446° F(230° C)*
Melting Point: 158° F(70° C) *Vapor Pressure:* 1 mm Hg at 138° F(58.8° C) *Solubility in water:* Yes
Other Information: Sinks and mixes slowly with water.

Health Hazards:

Inhalation Hazard: Poisonous if inhaled.
Ingestion Hazard: Poisonous if swallowed.
Absorption Hazard: Poisonous to the skin and eyes.
Hazards to Wildlife: n/a
Decontamination Procedures: Wash away any material with copious amounts of soap and water.
First Aid Procedures: Remove victim to fresh air, call emergency medical care. If not breathing give CPR. If breathing is difficult administer oxygen. Treat for shock.

Fire Hazards:

Flashpoint: 2201° F (1205° C) *Ignition temperature:* n/a
Flammable Explosive High Range: n/a *Low Range:* n/a
Toxic Products of Combustion: Irritating toxic hydrogen chloride and oxides of nitrogen may be formed in fires.
Other Hazards: n/a
Possible extinguishing agents: Extinguish fire using suitable agent for the type of surrounding fire.

Reactivity Hazards:

Reactive With: n/a *Other Reactions:* n/a

Corrosivity Hazards:

Corrosive With: n/a *Neutralizing Agent:* n/a

Radioactivity Hazards:

Radiation Emitted: n/a *Other Hazards:* n/a

Recommended Protection for Response Personnel:

Avoid breathing vapors, keep upwind. Wear a sealed chemical suit (polycarbonate, butyl rubber). Wash away any material which may have come into contact with the body with copious amounts of soap and water. Consider appropriate evacuation.

CHLOROBENZENE

DOT Number: UN 1134 *DOT Hazard Class:* Flammable liquid *DOT Guide Number:* 27
Synonyms: MCB, phenyl chloride
STCC Number: 4909153 *Reportable Qty:* 100/45.4
Mfg Name: PPG Industries Inc. *Phone No:* 1-412-434-3131

Physical Description:

Physical Form: Liquid *Color:* Colorless *Odor:* Sweet, almond
Other Information: Used to make pesticides, dyes, and other chemicals.

Chemical Properties:

Specific Gravity: 1.1 *Vapor Density:* 3.1 *Boiling Point: 270° F (132.2° C)*
Melting Point: −47° F(−43.8° C) *Vapor Pressure:* 8.8 mm Hg at 68° F (20° C)
Solubility in water: No *Degree of Solubility:* .1%
Other Information: Accumulations of vapors in confined spaces may explode if ignited. Product weighs approximately 9.2 lbs/4.17 kg per gallon/3.8 l.

Health Hazards:

Inhalation Hazard: Damage to lungs, liver, and kidneys.
Ingestion Hazard: Harmful if swallowed.
Absorption Hazard: May cause skin dermatitis.
Hazards to Wildlife: Dangerous to aquatic life.
Decontamination Procedures: Wash away any material with copious amounts of soap and water.
First Aid Procedures: Remove victim to fresh air, call emergency medical care. If not breathing give CPR. If breathing is difficult administer oxygen. Treat for shock.

Fire Hazards:

Flashpoint: 1099° F(592.7° C) *Ignition temperature:* 1099° F(592.7° C)
Flammable Explosive High Range: 9.6 *Low Range:* 1.3
Toxic Products of Combustion: n/a
Other Hazards: Flashback along vapor trail may occur. Vapors may explode if ignited in a closed area.
Possible extinguishing agents: Do not extinguish the fire unless the flow can be topped. Use foam, dry chemical, CO_2

Reactivity Hazards:

Reactive With: n/a *Other Reactions:* n/a

Corrosivity Hazards:

Corrosive With: n/a *Neutralizing Agent:* n/a

Radioactivity Hazards:

Radiation Emitted: n/a *Other Hazards:* n/a

Recommended Protection for Response Personnel:

Avoid breathing vapors, keep upwind. Structural protective clothing provides limited protection. Wash away any material which may have come into contact with the body with copious amounts of soap and water. Consider appropriate evacuation.

CHLOROFORM

DOT Number: UN 1888 *DOT Hazard Class:* ORM-A *DOT Guide Number:* 55
Synonyms: freon 20, R 20, trichloromethane
STCC Number: 4940311 *Reportable Qty:* 5000/2270
Mfg Name: Dow Chemical Co. *Phone No:* 1-517-636-3851

Physical Description:

Physical Form: Liquid *Color:* Colorless *Odor:* Sweet
Other Information: It is used as a solvent, and to make other chemicals, as a fumigant, and in many other uses.

Chemical Properties:

Specific Gravity: 1.49 *Vapor Density:* 4.1 *Boiling Point: 142° F(61° C)*
Melting Point: −82.3° F(−63.5c) *Vapor Pressure:* 155 mm Hg at 68° F(20° C) *Solubility in water:* Yes
Other Information: Slightly soluble in water. Weighs 12.3 lbs/5.5 kg per gallon/3.8 l. May decompose very slowly in the presence of air or light to form toxic phosgene and hydrogen chloride.

Health Hazards:

Inhalation Hazard: Headache, nausea, dizziness, loss of consciousness.
Ingestion Hazard: Harmful if swallowed.
Absorption Hazard: Irritating to the skin and eyes.
Hazards to Wildlife: n/a
Decontamination Procedures: Wash away any material with copious amounts of soap and water.
First Aid Procedures: Remove victim to fresh air, call emergency medical care. If not breathing give CPR. If breathing is difficult administer oxygen. Treat for shock.

Fire Hazards:

Flashpoint: n/a *Ignition temperature:* n/a
Flammable Explosive High Range: n/a *Low Range:* n/a
Toxic Products of Combustion: Poisonous and irritating gases are produced when heated.
Other Hazards: Containers may rupture in a fire due to overpressurization
Possible extinguishing agents: Extinguish fire using suitable agent for the type of surrounding fire.

Reactivity Hazards:

Reactive With: n/a *Other Reactions:* n/a

Corrosivity Hazards:

Corrosive With: n/a *Neutralizing Agent:* n/a

Radioactivity Hazards:

Radiation Emitted: n/a *Other Hazards:* n/a

Recommended Protection for Response Personnel:

Avoid breathing vapors, keep upwind. Wear a sealed chemical suit (viton). Wash away any material which may have come into contact with the body with copious amounts of soap and water. Consider appropriate evacuation.

CHLOROHYDRINS

DOT Number: UN 2023 *DOT Hazard Class:* n/a *DOT Guide Number:* 30
Synonyms: crude epichlorohydrin
STCC Number: n/a *Reportable Qty:* n/a
Mfg Name: Dow Chemical Co. *Phone No:* 1-517-636-3851

Physical Description:

Physical Form: Liquid *Color:* Colorless to yellow *Odor:* Garlic
Other Information: n/a

Chemical Properties:

Specific Gravity: 1.18 *Vapor Density:* n/a *Boiling Point: n/a*
Melting Point: n/a *Vapor Pressure:* n/a *Solubility in water:* Yes
Other Information: Sinks and mixes with water. Poisonous vapor is produced.

Health Hazards:

Inhalation Hazard: Poisonous if inhaled.
Ingestion Hazard: Poisonous if swallowed.
Absorption Hazard: Irritating to the skin and eyes.
Hazards to Wildlife: n/a
Decontamination Procedures: Wash away any material with copious amounts of soap and water.
First Aid Procedures: Remove victim to fresh air, call emergency medical care. If not breathing give CPR. If breathing is difficult administer oxygen. Treat for shock.

Fire Hazards:

Flashpoint: 92° F(33.3° C) *Ignition temperature:* 804° F(428.8° C)
Flammable Explosive High Range: 21 *Low Range:* 3.8
Toxic Products of Combustion: Toxic, irritating vapors are generated when heated.
Other Hazards: Containers may explode if fire because of polymerization.
Possible extinguishing agents: Foam, dry chemical, carbon dioxide.

Reactivity Hazards:

Reactive With: Water, but not likely to be hazardous. *Other Reactions:* n/a

Corrosivity Hazards:

Corrosive With: n/a *Neutralizing Agent:* n/a

Radioactivity Hazards:

Radiation Emitted: n/a *Other Hazards:* n/a

Recommended Protection for Response Personnel:

Avoid breathing vapors, keep upwind. Wear appropriate sealed chemical clothing. Wash away any material which may have come into contact with the body with copious amounts of soap and water. Consider appropriate evacuation.

CHLOROMETHYL METHYL ETHER

DOT Number: UN 1239 *DOT Hazard Class:* Flammable liquid *DOT Guide Number:* 57
Synonyms: monochloromethyl ether; methyl chloromethyl ether, anhydrous
STCC Number: n/a *Reportable Qty:* n/a
Mfg Name: Aldrich Chemical *Phone No:* 1-414-273-3850

Physical Description:

Physical Form: Liquid *Color:* Colorless *Odor:* Irritating
Other Information: n/a

Chemical Properties:

Specific Gravity: 1.07 *Vapor Density:* 1 *Boiling Point: 140° F(60° C)*
Melting Point: −154° F(−103° C) *Vapor Pressure:* n/a *Solubility in water:* n/a
Other Information: May float or sink in water. Poisonous, flammable vapor is produced.

Health Hazards:

Inhalation Hazard: Difficulty in breathing.
Ingestion Hazard: Harmful if swallowed.
Absorption Hazard: Will burn the skin and eyes.
Hazards to Wildlife: n/a
Decontamination Procedures: Wash away any material with copious amounts of soap and water.
First Aid Procedures: Remove victim to fresh air, call emergency medical care. If not breathing give CPR. If breathing is difficult administer oxygen. Treat for shock.

Fire Hazards:

Flashpoint: 0° F(−17.7° C) *Ignition temperature:* n/a
Flammable Explosive High Range: n/a *Low Range:* n/a
Toxic Products of Combustion: Toxic, irritating hydrogen chloride and phosgene gases may be formed in fires.
Other Hazards: Unburned material may form powerful tear gas. When wet, will also form irritating formaldehyde gas.
Possible extinguishing agents: Foam, dry chemical, carbon dioxide. Water may be ineffective.

Reactivity Hazards:

Reactive With: Water to evolve hydrogen chloride and formaldehyde. The reaction is not violent.
Other Reactions: n/a

Corrosivity Hazards:

Corrosive With: Surface moisture to evolve hydrogen chloride which is corrosive to metal.
Neutralizing Agent: Crushed limestone, soda ash, or lime.

Radioactivity Hazards:

Radiation Emitted: n/a *Other Hazards:* n/a

Recommended Protection for Response Personnel:

Avoid breathing vapors, keep upwind. Wear appropriate sealed chemical clothing. Wash away any material which may have come into contact with the body with copious amounts of soap and water. Consider appropriate evacuation.

CHLORONITROBENZENE

DOT Number: UN 1578 *DOT Hazard Class:* Poison B *DOT Guide Number:* 55
Synonyms: nitrochlorobenzene, 1-chloro-2-nitrobenzene
STCC Number: 4921424 *Reportable Qty:* n/a
Mfg Name: Fisher Scientific *Phone No:* 1-412-394-3322

Physical Description:

Physical Form: Solid *Color:* Colorless to yellow *Odor:* Almond
Other Information: n/a

Chemical Properties:

Specific Gravity: 1.36 *Vapor Density:* 5.4 *Boiling Point: 474.8° F(246° C)*
Melting Point: n/a *Vapor Pressure:* n/a *Solubility in water:* Insoluble
Other Information: Sinks in water

Health Hazards:

Inhalation Hazard: Headache, languor, cyanosis, shallow respiration, coma
Ingestion Hazard: Poisonous if swallowed.
Absorption Hazard: Poisonous if absorbed.
Hazards to Wildlife: Dangerous to aquatic life.
Decontamination Procedures: Wash away any material with copious amounts of soap and water.
First Aid Procedures: Remove victim to fresh air, call emergency medical care. If not breathing give CPR. If breathing is difficult administer oxygen. Treat for shock.

Fire Hazards:

Flashpoint: 261° F(127.2° C) *Ignition temperature:* n/a
Flammable Explosive High Range: n/a *Low Range:* n/a
Toxic Products of Combustion: Nitrogen oxide and hydrogen chloride fumes.
Other Hazards: Gives off flammable vapors when heated, forming explosive mixtures with air.
Possible extinguishing agents: Use water in flooding quantities as a fog, use foam, dry chemical, or carbon dioxide.

Reactivity Hazards:

Reactive With: n/a *Other Reactions:* n/a

Corrosivity Hazards:

Corrosive With: n/a *Neutralizing Agent:* n/a

Radioactivity Hazards:

Radiation Emitted: n/a *Other Hazards:* n/a

Recommended Protection for Response Personnel:

Avoid breathing vapors, keep upwind. Wear appropriate chemical clothing, wash away any material which may come into contact with the body with copious amounts of soap and water. Consider appropriate evacuation.

CHLOROPHENOL

DOT Number: UN 2020 *DOT Hazard Class:* n/a *DOT Guide Number:* 53
Synonyms: 4-chlorophenol
STCC Number: n/a *Reportable Qty:* n/a
Mfg Name: Dow Chemical Co. *Phone No:* 1-517-636-4400

Physical Description:

Physical Form: Solid *Color:* White to straw *Odor:* Medicinal
Other Information: n/a

Chemical Properties:

Specific Gravity: 1.4 *Vapor Density:* 4.4 *Boiling Point: 428° F(220° C)*
Melting Point: 109° F(42.7c) *Vapor Pressure:* 1 mm Hg at 122° F(50° C) *Solubility in water:* n/a
Other Information: Sinks in water

Health Hazards:

Inhalation Hazard: Headaches, dizziness
Ingestion Hazard: Nausea and vomiting.
Absorption Hazard: Will burn the skin and eyes.
Hazards to Wildlife: Dangerous to aquatic life.
Decontamination Procedures: Wash away any material with copious amounts of soap and water.
First Aid Procedures: Remove victim to fresh air, call emergency medical care. If not breathing give CPR. If breathing is difficult administer oxygen. Treat for shock.

Fire Hazards:

Flashpoint: 250° F(121.1° C) *Ignition temperature:* n/a
Flammable Explosive High Range: n/a *Low Range:* n/a
Toxic Products of Combustion: Toxic, irritating hydrogen chloride and chlorine gases may be formed in fires.
Other Hazards: Irritating gases may be produced when heated.
Possible extinguishing agents: Water, foam, dry chemical, carbon dioxide.

Reactivity Hazards:

Reactive With: n/a *Other Reactions:* n/a

Corrosivity Hazards:

Corrosive With: n/a *Neutralizing Agent:* n/a

Radioactivity Hazards:

Radiation Emitted: n/a *Other Hazards:* n/a

Recommended Protection for Response Personnel:

Avoid breathing vapors, keep upwind. Wear a sealed chemical suit (polycarbonate, butyl rubber). Wash away any material which may have come into contact with the body with copious amounts of soap and water. Consider appropriate evacuation.

CHLOROPICRIN

DOT Number: UN 1580 *DOT Hazard Class:* Poison B *DOT Guide Number:* 56
Synonyms: nitrochloroform, picfume, trichloronitromethane
STCC Number: 4921414 *Reportable Qty:* n/a
Mfg Name: LCP Chemicals & Plastics *Phone No:* 1-201-225-4840

Physical Description:

Physical Form: Liquid *Color:* Colorless *Odor:* Extremely irritating
Other Information: n/a

Chemical Properties:

Specific Gravity: 1.64 *Vapor Density:* 5.70 *Boiling Point: 234° F(112.2° C)*
Melting Point: n/a *Vapor Pressure:* 44 mm Hg at 92.8° F(33.7° C) *Solubility in water:* Yes
Other Information: In case of damage to, or leaking from this container, call the Pesticide Safety Team Network at 1-800-424-9300.

Health Hazards:

Inhalation Hazard: Poisonous if inhaled.
Ingestion Hazard: Poisonous if swallowed.
Absorption Hazard: Poisonous if absorbed.
Hazards to Wildlife: n/a
Decontamination Procedures: Wash away any material with copious amounts of soap and water.
First Aid Procedures: Remove victim to fresh air, call emergency medical care. If not breathing give CPR. If breathing is difficult administer oxygen. Treat for shock.

Fire Hazards:

Flashpoint: n/a *Ignition temperature:* n/a
Flammable Explosive High Range: n/a *Low Range:* n/a
Toxic Products of Combustion: Poisonous gases may be produced in a fire, containers may explode in a fire.
Other Hazards: Compound forms a powerful tear gas when heated.
Possible extinguishing agents: Extinguish agent using suitable agent for type of surrounding fire.

Reactivity Hazards:

Reactive With: n/a *Other Reactions:* n/a

Corrosivity Hazards:

Corrosive With: n/a *Neutralizing Agent:* n/a

Radioactivity Hazards:

Radiation Emitted: n/a *Other Hazards:* n/a

Recommended Protection for Response Personnel:

Avoid breathing vapors, keep upwind. Wear a sealed chemical suit (polycarbonate, butyl rubber). Wash away any material which may have come into contact with the body with copious amounts of soap and water. Consider appropriate evacuation.

CHLOROPRENE

DOT Number: UN 1991 *DOT Hazard Class:* Flammable liquid *DOT Guide Number:* 30
Synonyms: 2-chloro-1,3-butadiene
STCC Number: 4907223 *Reportable Qty:* n/a
Mfg Name: Hooker Chemical Corp. *Phone No:* 1-716-278-7000

Physical Description:

Physical Form: Liquid *Color:* Colorless *Odor:* Slight, etheric
Other Information: Used to make neoprene rubber.

Chemical Properties:

Specific Gravity: 1 *Vapor Density:* 3 *Boiling Point: 138° F(58.8° C)*
Melting Point: –202° F(–130° C) *Vapor Pressure:* 179 mm Hg at 68° F (20° C) *Solubility in water:* Insoluble
Other Information: If subject to heat material may polymerize.

Health Hazards:

Inhalation Hazard: Fatigue, psychic changes, irritability.
Ingestion Hazard: Harmful if swallowed.
Absorption Hazard: Can cause conjunctivitis, temporary loss of hair, rapid absorption by skin, dermatitis.
Hazards to Wildlife: Dangerous to aquatic life.
Decontamination Procedures: Wash away any material with copious amounts of soap and water.
First Aid Procedures: Remove victim to fresh air, call emergency medical care. If not breathing give CPR. If breathing is difficult administer oxygen. Treat for shock.

Fire Hazards:

Flashpoint: –4° F(–20° C) *Ignition temperature:* n/a
Flammable Explosive High Range: 20 *Low Range:* 4
Toxic Products of Combustion: Poisonous gases are produced in fire.
Other Hazards: Flashback along vapor trail may occur. Containers may explode in fire.
Possible extinguishing agents: Do not extinguish fire unless flow can be stopped, use foam, dry chemical, or CO_2.

Reactivity Hazards:

Reactive With: n/a *Other Reactions:* n/a

Corrosivity Hazards:

Corrosive With: n/a *Neutralizing Agent:* n/a

Radioactivity Hazards:

Radiation Emitted: n/a *Other Hazards:* n/a

Recommended Protection for Response Personnel:

Avoid breathing vapors, keep upwind. Wear appropriate chemical clothing, wash away any material which may come into contact with the body with copious amounts of soap and water. If fire becomes uncontrollable, consider evacuation.

CHLOROSULPHONIC ACID

DOT Number: UN 1754 *DOT Hazard Class:* Corrosive *DOT Guide Number:* 39
Synonyms: chlorsulfonic acid, chlorsulfuric acid
STCC Number: 4930204 *Reportable Qty:* 1000/454
Mfg Name: E.I. Du Pont *Phone No:* 1-800-441-3637

Physical Description:

Physical Form: Liquid *Color:* Colorless to light yellow *Odor:* Sharp, choking
Other Information: n/a

Chemical Properties:

Specific Gravity: 1.75 *Vapor Density:* 4 *Boiling Point: 311° F(155° C)*
Melting Point: n/a *Vapor Pressure:* 1 mm Hg at 89.6° F(32.0° C) *Solubility in water:* Decomposes
Other Information: Weighs 14.7 lbs/6.6 kg per gallon/3.8 l.

Health Hazards:

Inhalation Hazard: Harmful if inhaled.
Ingestion Hazard: Harmful if swallowed.
Absorption Hazard: Will burn the skin and eyes.
Hazards to Wildlife: Dangerous to aquatic life.
Decontamination Procedures: Wash away any material with copious amounts of soap and water.
First Aid Procedures: Remove victim to fresh air, call emergency medical care. If not breathing give CPR. If breathing is difficult administer oxygen. Treat for shock.

Fire Hazards:

Flashpoint: n/a *Ignition temperature:* n/a
Flammable Explosive High Range: n/a *Low Range:* n/a
Toxic Products of Combustion: Decomposes into irritating and toxic gases.
Other Hazards: May cause fire upon contact with combustibles. Flammable explosive gases may be formed on contact with metals and moisture.
Possible extinguishing agents: Do not use water, use dry chemical, or carbon dioxide.

Reactivity Hazards:

Reactive With: Water to form hydrochloric acid (vapor) and sulfuric acid.
Other Reactions: With metal to form hydrogen, a highly explosive gas.

Corrosivity Hazards:

Corrosive With: Skin and tissue *Neutralizing Agent:* Crushed limestone, soda ash, or lime

Radioactivity Hazards:

Radiation Emitted: n/a *Other Hazards:* n/a

Recommended Protection for Response Personnel:

Avoid breathing vapors, keep upwind. Wear a sealed chemical suit (pvc, chlorinated polyethylene). Wash away any material which may have come into contact with the body with copious amounts of soap and water. Consider appropriate evacuation.

CHLOROTOLUENE o-m-p

DOT Number: NA 2239 *DOT Hazard Class:* Combustible liquid *DOT Guide Number:* 27
Synonyms: 1-chloro-4-methylbenzene, 4-chloro-1-methylbenzene, 4-chlorotoluene, p-tolyl chloride
STCC Number: 4913148 *Reportable Qty:* n/a
Mfg Name: Tenneco Chemical *Phone No:* 1-201-981-5000

Physical Description:

Physical Form: Liquid *Color:* Colorless *Odor:* n/a
Other Information: It is used for solvents, and as intermediates for making other chemicals and dyes.

Chemical Properties:

Specific Gravity: 1.07 *Vapor Density:* 4.3 *Boiling Point: 324° F(162.2° C)*
Melting Point: 45.5° F(7.5° C) *Vapor Pressure:* 10 mm Hg at 111° F(43.8° C) *Solubility in water:* Yes
Other Information: Chlorotoluene has 3 isomers: ortho-, meta-, and para-. The vapors from all isomers are irritating and narcotic in high concentrations.

Health Hazards:

Inhalation Hazard: n/a
Ingestion Hazard: Harmful if swallowed.
Absorption Hazard: Irritating to the skin and eyes.
Hazards to Wildlife: Dangerous to aquatic life.
Decontamination Procedures: Wash away any material with copious amounts of soap and water.
First Aid Procedures: Remove victim to fresh air, call emergency medical care. If not breathing give CPR. If breathing is difficult administer oxygen. Treat for shock.

Fire Hazards:

Flashpoint: 120° F(48.8° C) *Ignition temperature:* n/a
Flammable Explosive High Range: n/a *Low Range:* n/a
Toxic Products of Combustion: n/a
Other Hazards: n/a
Possible extinguishing agents: Do not extinguish the fire unless the flow can be stopped. Apply water from as far a distance as possible. Use foam, dry chemical, or carbon dioxide.

Reactivity Hazards:

Reactive With: n/a *Other Reactions:* n/a

Corrosivity Hazards:

Corrosive With: n/a *Neutralizing Agent:* n/a

Radioactivity Hazards:

Radiation Emitted: n/a *Other Hazards:* n/a

Recommended Protection for Response Personnel:

Avoid breathing vapors, keep upwind. Wear a sealed chemical suit. (viton, chlorinated polyethylene). Wash away any material which may have come into contact with the body with copious amounts of soap and water. Consider appropriate evacuation.

CHROMIC ACETATE

DOT Number: NA 9101 *DOT Hazard Class:* ORM-E *DOT Guide Number:* 31
Synonyms: acetic acid, chromium salt; chromic(III) acetate; chromium acetate; chromium triacetate
STCC Number: 4963312 *Reportable Qty:* 1000/454
Mfg Name: McGean-Rohco Inc. *Phone No:* 1-206-621-6425

Physical Description:

Physical Form: Liquid *Color:* Dark green to violet *Odor:* Acetic acid
Other Information: It is used in tanning, and textile dyeing.

Chemical Properties:

Specific Gravity: 1.3 *Vapor Density:* 1 *Boiling Point: 212° F(100° C)*
Melting Point: n/a *Vapor Pressure:* n/a *Solubility in water:* Yes
Other Information: Sinks and mixes with water.

Health Hazards:

Inhalation Hazard: Harmful if inhaled.
Ingestion Hazard: Harmful if swallowed.
Absorption Hazard: Irritating to the skin and eyes.
Hazards to Wildlife: Dangerous to aquatic life.
Decontamination Procedures: Wash away any material with copious amounts of soap and water.
First Aid Procedures: Remove victim to fresh air, call emergency medical care. If not breathing give CPR. If breathing is difficult administer oxygen. Treat for shock.

Fire Hazards:

Flashpoint: n/a *Ignition temperature:* n/a
Flammable Explosive High Range: n/a *Low Range:* n/a
Toxic Products of Combustion: n/a
Possible extinguishing agents: Extinguish fire using suitable agent for the type of surrounding fire.

Reactivity Hazards:

Reactive With: n/a *Other Reactions:* n/a

Corrosivity Hazards:

Corrosive With: n/a *Neutralizing Agent:* n/a

Radioactivity Hazards:

Radiation Emitted: n/a *Other Hazards:* n/a

Recommended Protection for Response Personnel:

Avoid breathing dust, keep upwind. Wear a sealed chemical suit. Wash away any material which may have come into contact with the body with copious amounts of soap and water. Consider appropriate evacuation.

CHROMIC ACID

DOT Number: NA 1463 *DOT Hazard Class:* Oxidizer *DOT Guide Number:* 42
Synonyms: chromic anhydride, chromic oxide
STCC Number: 4918510 *Reportable Qty:* 1000/454
Mfg Name: Allied Corp. *Phone No:* 1-201-455-2000

Physical Description:

Physical Form: Solid *Color:* Dark purplish red *Odor:* Odorless
Other Information: n/a

Chemical Properties:

Specific Gravity: 2.7 *Vapor Density:* 3.4 *Boiling Point: n/a*
Melting Point: 386° F(196.6° C) *Vapor Pressure:* n/a *Solubility in water:* Yes
Other Information: It is soluble in water. Weighs 13.9 lbs/6.3 kg per gallon/3.8 l.

Health Hazards:

Inhalation Hazard: Very irritating to the respiratory tract.
Ingestion Hazard: Harmful if swallowed.
Absorption Hazard: Upon contact, material will burn the skin and eyes.
Hazards to Wildlife: Dangerous to aquatic life.
Decontamination Procedures: Wash away any material with copious amounts of soap and water.
First Aid Procedures: Remove victim to fresh air, call emergency medical care.If not breathing give CPR. If breathing is difficult administer oxygen. Treat for shock.

Fire Hazards:

Flashpoint: n/a *Ignition temperature:* n/a
Flammable Explosive High Range: n/a *Low Range:* n/a
Toxic Products of Combustion: n/a
Other Hazards: May cause fire upon contact with combustibles, containers may explode when heated in a fire.
Possible extinguishing agents: Water to cool affected containers.

Reactivity Hazards:

Reactive With: May react with organic materials.
Other Reactions: Prolonged contact particularly on wood floors may produce a fire hazard.

Corrosivity Hazards:

Corrosive With: Most metals, cloth, leather, and some plastics, possibly causing spontaneous ignition.
Neutralizing Agent: Crushed limestone, soda ash, or lime

Radioactivity Hazards:

Radiation Emitted: n/a *Other Hazards:* n/a

Recommended Protection for Response Personnel:

Avoid breathing dust, keep upwind. Wear a sealed chemical suit (polycarbonate, butyl rubber). Wash away any material which may have come into contact with the body with copious amounts of soap and water. If the fire becomes uncontrollable. Consider appropriate evacuation.

CHROMIC ANHYDRIDE

DOT Number: NA 1463 *DOT Hazard Class:* Oxidizer *DOT Guide Number:* 42
Synonyms: chromic acid, chromic oxide
STCC Number: 4918510 *Reportable Qty:* 1000/454
Mfg Name: Allied Corp. *Phone No:* 1-201-455-2000

Physical Description:

Physical Form: Solid *Color:* Dark purplish red *Odor:* Odorless
Other Information: n/a

Chemical Properties:

Specific Gravity: 2.7 *Vapor Density:* 3.4 *Boiling Point: n/a*
Melting Point: n/a *Vapor Pressure:* n/a *Solubility in water:* Yes
Other Information: It is soluble in water to form a corrosive solution.

Health Hazards:

Inhalation Hazard: Very irritating to the respiratory tract.
Ingestion Hazard: Harmful if swallowed.
Absorption Hazard: Upon contact, material will burn the skin and eyes.
Hazards to Wildlife: Dangerous to aquatic life.
Decontamination Procedures: Wash away any material with copious amounts of soap and water.
First Aid Procedures: Remove victim to fresh air, call emergency medical care. If not breathing give CPR. If breathing is difficult administer oxygen. Treat for shock.

Fire Hazards:

Flashpoint: n/a *Ignition temperature:* n/a
Flammable Explosive High Range: n/a *Low Range:* n/a
Toxic Products of Combustion: n/a
Other Hazards: May cause fire upon contact with combustibles, the containers may explode when heated in a fire.
Possible extinguishing agents: Water to cool the affected containers.

Reactivity Hazards:

Reactive With: Organic materials *Other Reactions:* n/a

Corrosivity Hazards:

Corrosive With: No material given *Neutralizing Agent:* Crushed limestone, soda ash, or lime

Radioactivity Hazards:

Radiation Emitted: n/a *Other Hazards:* n/a

Recommended Protection for Response Personnel:

Avoid breathing dust, keep upwind. Wear a sealed chemical suit (polycarbonate, butyl rubber). Wash away any material which may have come into contact with the body with copious amounts of soap and water. If the fire becomes uncontrollable. Consider appropriate evacuation.

CHROMIC SULFATE

DOT Number: NA 9100 *DOT Hazard Class:* ORM-E *DOT Guide Number:* 31
Synonyms: chromium sulfate, dichromium sulfate
STCC Number: 4963314 *Reportable Qty:* 1000/454
Mfg Name: Mallinckrodt Inc. *Phone No:* 1-314-895-2000

Physical Description:

Physical Form: Solid *Color:* Peach, violet, dark green *Odor:* Odorless
Other Information: It is used in ceramics, as a pigment for paints, and in textile dyeing.

Chemical Properties:

Specific Gravity: 3.012 *Vapor Density:* 1 *Boiling Point: n/a*
Melting Point: n/a *Vapor Pressure:* n/a *Solubility in water:* Yes
Other Information: Sinks and mixes with water.

Health Hazards:

Inhalation Hazard: Harmful if inhaled.
Ingestion Hazard: Harmful if swallowed.
Absorption Hazard: Irritating to the skin and eyes.
Hazards to Wildlife: Dangerous to aquatic life.
Decontamination Procedures: Wash away any material with copious amounts of soap and water.
First Aid Procedures: Remove victim to fresh air, call emergency medical care. If not breathing give CPR. If breathing is difficult administer oxygen. Treat for shock.

Fire Hazards:

Flashpoint: n/a *Ignition temperature:* n/a
Flammable Explosive High Range: n/a *Low Range:* n/a
Toxic Products of Combustion: Decomposes to chromic acid when heated.
Other Hazards: n/a
Possible extinguishing agents: Extinguish fire using suitable agent for type of surrounding fire.

Reactivity Hazards:

Reactive With: n/a *Other Reactions:* n/a

Corrosivity Hazards:

Corrosive With: n/a *Neutralizing Agent:* n/a

Radioactivity Hazards:

Radiation Emitted: n/a *Other Hazards:* n/a

Recommended Protection for Response Personnel:

Avoid breathing vapors, keep upwind. Wear a sealed chemical suit (butyl rubber). Wash away any material which may have come into contact with the body with copious amounts of soap and water. Consider appropriate evacuation.

CHROMOUS CHLORIDE

DOT Number: NA 9102 *DOT Hazard Class:* ORM-E *DOT Guide Number:* 31
Synonyms: chromium chloride, chromium dichloride
STCC Number: 4963322 *Reportable Qty:* 1000/454
Mfg Name: A.D. MacKay Inc. *Phone No:* 1-203-655-3000

Physical Description:

Physical Form: Solid *Color:* White to blue *Odor:* n/a
Other Information: It is used to make other chemicals, and as an oxygen absorbent.

Chemical Properties:

Specific Gravity: 2.75 *Vapor Density:* 4.2 *Boiling Point: n/a*
Melting Point: n/a *Vapor Pressure:* n/a *Solubility in water:* Yes
Other Information: Sinks and mixes with water.

Health Hazards:

Inhalation Hazard: Harmful if inhaled.
Ingestion Hazard: Harmful if swallowed.
Absorption Hazard: Irritating to the skin and eyes.
Hazards to Wildlife: Dangerous to aquatic life.
Decontamination Procedures: Wash away any material with copious amounts of soap and water.
First Aid Procedures: Remove victim to fresh air, call emergency medical care. If not breathing give CPR. If breathing is difficult administer oxygen. Treat for shock.

Fire Hazards:

Flashpoint: n/a *Ignition temperature:* n/a
Flammable Explosive High Range: n/a *Low Range:* n/a
Toxic Products of Combustion: n/a
Possible extinguishing agents: Extinguish fire using suitable agent for the type of surrounding fire.

Reactivity Hazards:

Reactive With: n/a *Other Reactions:* n/a

Corrosivity Hazards:

Corrosive With: n/a *Neutralizing Agent:* n/a

Radioactivity Hazards:

Radiation Emitted: n/a *Other Hazards:* n/a

Recommended Protection for Response Personnel:

Avoid breathing dust, keep upwind. Wear appropriate chemical clothing. Wash away any material which may have come into contact with the body with copious amounts of soap and water. Consider appropriate evacuation.

CHROMYL CHLORIDE

DOT Number: UN 1758 *DOT Hazard Class:* Corrosive *DOT Guide Number:* 31
Synonyms: chromium(VI) dioxychloride, chromium oxychloride
STCC Number: 4930203 *Reportable Qty:* n/a
Mfg Name: Gallard-Schlesinger *Phone No:* 1-516-333-5600

Physical Description:

Physical Form: Liquid *Color:* Dark red *Odor:* Pleasant
Other Information: n/a

Chemical Properties:

Specific Gravity: 1.96 *Vapor Density:* 5.3 *Boiling Point: 241° F(116.1° C)*
Melting Point: −141.7deg f(−96.5° C) *Vapor Pressure:* 20 mm Hg at 68° F(20° C)
Other Information: Reacts violently with water. Irritating visible vapor cloud is produced.

Health Hazards:

Inhalation Hazard: Difficulty in breathing.
Ingestion Hazard: Poisonous if swallowed.
Absorption Hazard: Will burn the skin and eyes.
Hazards to Wildlife: n/a
Decontamination Procedures: Wash away any material with copious amounts of soap and water.
First Aid Procedures: Remove victim to fresh air, call emergency medical care. If not breathing give CPR. If breathing is difficult administer oxygen. Treat for shock.

Fire Hazards:

Flashpoint: n/a *Ignition temperature:* n/a
Flammable Explosive High Range: n/a *Low Range:* n/a
Toxic Products of Combustion: Irritating gases are produced when heated.
Other Hazards: May cause fire upon contact with combustibles. Containers may explode in fire.
Possible extinguishing agents: Extinguish fire using suitable agent for the type of surrounding fire. Do not use water on the material itself

Reactivity Hazards:

Reactive With: Water to form chlorine gases, chromic acid, and hydrochloric acid. *Other Reactions:* n/a

Corrosivity Hazards:

Corrosive With: Will cause severe corrosion of common metals.
Neutralizing Agent: Crushed limestone, soda ash, or lime.

Radioactivity Hazards:

Radiation Emitted: n/a *Other Hazards:* n/a

Recommended Protection for Response Personnel:

Avoid breathing dust, keep upwind. Wear appropriate chemical clothing. Wash away any material which may have come into contact with the body with copious amounts of soap and water. Consider appropriate evacuation.

COLLODION

DOT Number: NA 2059 *DOT Hazard Class:* Flammable liquid *DOT Guide Number:* 26
Synonyms: box toe gum, nitrocellulose gum, pyroxylin solution
STCC Number: 4910144 *Reportable Qty:* n/a
Mfg Name: Mallinckrodt Inc. *Phone No:* 1-314-895-2000

Physical Description:

Physical Form: Liquid *Color:* Colorless *Odor:* Ether like
Other Information: n/a

Chemical Properties:

Specific Gravity: .772 *Vapor Density:* 1 *Boiling Point: 93° F(33.8° C)*
Melting Point: n/a *Vapor Pressure:* n/a *Solubility in water:* No
Other Information: Floats on water

Health Hazards:

Inhalation Hazard: Dizziness, difficulty in breathing, loss of consciousness.
Ingestion Hazard: n/a
Absorption Hazard: n/a
Hazards to Wildlife: n/a
Decontamination Procedures: Wash away any material with copious amounts of soap and water.
First Aid Procedures: Remove victim to fresh air, call emergency medical care. If not breathing give CPR. If breathing is difficult administer oxygen. Treat for shock.

Fire Hazards:

Flashpoint: –49° F(–45° C) *Ignition temperature:* 359° F(181.6° C)
Flammable Explosive High Range: 36 *Low Range:* 1.9
Toxic Products of Combustion: Forms extremely toxic gases, oxides of nitrogen, hydrogen cyanide, and carbon dioxide.
Other Hazards: Containers may explode in a fire. Flashback along vapor trail may occur. Vapors may explode if ignited in an enclosed area.
Possible extinguishing agents: Apply water from as far a distance as possible. Use alcohol foam, dry chemical, carbon dioxide.

Reactivity Hazards:

Reactive With: n/a *Other Reactions:* n/a

Corrosivity Hazards:

Corrosive With: n/a *Neutralizing Agent:* n/a

Radioactivity Hazards:

Radiation Emitted: n/a *Other Hazards:* n/a

Recommended Protection for Response Personnel:

Avoid breathing vapors, keep upwind. Structural protective clothing provides limited protection. Wash away any material which may have come into contact with the body with copious amounts of soap and water. Consider appropriate evacuation.

COPPER ACETATE

DOT Number: NA 9106 *DOT Hazard Class:* ORM-E *DOT Guide Number:* 31
Synonyms: acetic acid, cupric salt; crystallized verdigris; cupric acetate monohydrate; neutral verdigris
STCC Number: 4962310 *Reportable Qty:* 100/45.4
Mfg Name: J.T. Baker Chemical Co. *Phone No:* 1-201-859-2151

Physical Description:

Physical Form: Solid *Color:* Bluish green *Odor:* Odorless
Other Information: n/a

Chemical Properties:

Specific Gravity: 1.9 *Vapor Density:* 1 *Boiling Point: n/a*
Melting Point: 239° F(115° C) *Vapor Pressure:* n/a *Solubility in water:* No
Other Information: n/a

Health Hazards:

Inhalation Hazard: Coughing or difficulty in breathing.
Ingestion Hazard: Nausea, vomiting, or loss of consciousness.
Absorption Hazard: Irritating to the skin, will burn the eyes.
Hazards to Wildlife: n/a
Decontamination Procedures: Wash away any material with copious amounts of soap and water.
First Aid Procedures: Remove victim to fresh air, call emergency medical care. If not breathing give CPR. If breathing is difficult administer oxygen. Treat for shock.

Fire Hazards:

Flashpoint: n/a *Ignition temperature:* n/a
Flammable Explosive High Range: n/a *Low Range:* n/a
Toxic Products of Combustion: Irritating vapors of acetic acid may be formed in fires.
Other Hazards: n/a
Possible extinguishing agents: Use suitable agent for the type of surrounding fire.

Reactivity Hazards:

Reactive With: n/a *Other Reactions:* n/a

Corrosivity Hazards:

Corrosive With: n/a *Neutralizing Agent:* n/a

Radioactivity Hazards:

Radiation Emitted: n/a *Other Hazards:* n/a

Recommended Protection for Response Personnel:

Avoid breathing dust, keep upwind. Wear a sealed chemical suit (polycarbonate, butyl rubber). Wash away any material which may have come into contact with the body with copious amounts of soap and water. Consider appropriate evacuation.

COPPER ACETOARSENITE

DOT Number: UN 1585 *DOT Hazard Class:* Poison B *DOT Guide Number:* 53
Synonyms: imperial green, Paris green
STCC Number: 4923220 *Reportable Qty:* 100/45.4
Mfg Name: Los Angeles Chemical Co. *Phone No:* 1-213-583-4761

Physical Description:

Physical Form: Solid *Color:* Emerald green powder *Odor:* Odorless
Other Information: n/a

Chemical Properties:

Specific Gravity: 1.1 *Vapor Density:* 1 *Boiling Point: n/a*
Melting Point: n/a *Vapor Pressure:* n/a *Solubility in water:* No
Other Information: n/a

Health Hazards:

Inhalation Hazard: Poisonous if inhaled.
Ingestion Hazard: Poisonous if swallowed.
Absorption Hazard: Irritating to the skin and eyes.
Hazards to Wildlife: Dangerous to waterfowl.
Decontamination Procedures: Wash away any material with copious amounts of soap and water.
First Aid Procedures: Remove victim to fresh air, call emergency medical care. If not breathing give CPR. If breathing is difficult administer oxygen. Treat for shock.

Fire Hazards:

Flashpoint: n/a *Ignition temperature:* n/a
Flammable Explosive High Range: n/a *Low Range:* n/a
Toxic Products of Combustion: Poisonous gases may be produced when heated
Other Hazards: n/a
Possible extinguishing agents: Use suitable agent for the type of surrounding fire. Use foam, dry chemical, carbon dioxide.

Reactivity Hazards:

Reactive With: n/a *Other Reactions:* n/a

Corrosivity Hazards:

Corrosive With: n/a *Neutralizing Agent:* n/a

Radioactivity Hazards:

Radiation Emitted: n/a *Other Hazards:* n/a

Recommended Protection for Response Personnel:

Avoid breathing dust, keep upwind. Wear a sealed chemical suit (polycarbonate, butyl rubber). Wash away any material which may have come into contact with the body with copious amounts of soap and water. Consider appropriate evacuation.

COPPER ARSENITE

DOT Number: UN 1586 *DOT Hazard Class:* Poison B *DOT Guide Number:* 53
Synonyms: cupric arsenite, Swedish green
STCC Number: 4923221 *Reportable Qty:* n/a
Mfg Name: Gallard-Schlesinger *Phone No:* 1-516-333-5600

Physical Description:

Physical Form: Solid *Color:* Light green powder *Odor:* Odorless
Other Information: n/a

Chemical Properties:

Specific Gravity: 1.1 *Vapor Density:* 6.5 *Boiling Point: n/a*
Melting Point: n/a *Vapor Pressure:* n/a *Solubility in water:* No
Other Information: Sinks in water

Health Hazards:

Inhalation Hazard: Poisonous if inhaled.
Ingestion Hazard: Poisonous if swallowed.
Absorption Hazard: Irritating to the skin and eyes.
Hazards to Wildlife: n/a
Decontamination Procedures: Wash away any material with copious amounts of soap and water.
First Aid Procedures: Remove victim to fresh air, call emergency medical care. If not breathing give CPR. If breathing is difficult administer oxygen. Treat for shock.

Fire Hazards:

Flashpoint: n/a *Ignition temperature:* n/a
Flammable Explosive High Range: n/a *Low Range:* n/a
Toxic Products of Combustion: Poisonous gases may be produced when heated
Other Hazards: n/a
Possible extinguishing agents: Use suitable agent for the type of surrounding fire. Use foam, dry chemical, carbon dioxide.

Reactivitity Hazards:

Reactive With: n/a *Other Reactions:* n/a

Corrosivity Hazards:

Corrosive With: n/a *Neutralizing Agent:* n/a

Radioactivity Hazards:

Radiation Emitted: n/a *Other Hazards:* n/a

Recommended Protection for Response Personnel:

Avoid breathing dust, keep upwind. Wear a sealed chemical suit (polycarbonate, butyl rubber). Wash away any material which may have come into contact with the body with copious amounts of soap and water. Consider appropriate evacuation.

COPPER CHLORIDE

DOT Number: UN 2802 *DOT Hazard Class:* ORM-B *DOT Guide Number:* 60
Synonyms: cupric chloride dihydrate, eriocholcite (anhydrous)
STCC Number: 4944173 *Reportable Qty:* 10/4.54
Mfg Name: J.T. Baker Chemical Co. *Phone No:* 1-201-859-2151

Physical Description:

Physical Form: Solid *Color:* Blue green *Odor:* Odorless
Other Information: It is used to manufacture other chemicals, in dyeing, printing, fungicides, as a wood preservative, and in many other uses.

Chemical Properties:

Specific Gravity: 2.54 *Vapor Density:* 4.6 *Boiling Point: n/a*
Melting Point: n/a *Vapor Pressure:* n/a *Solubility in water:* Yes
Other Information: Sinks and mixes with water.

Health Hazards:

Inhalation Hazard: Coughing or difficulty in breathing.
Ingestion Hazard: Nausea and vomiting.
Absorption Hazard: Irritating to the skin, will burn the eyes.
Hazards to Wildlife: Dangerous to aquatic life.
Decontamination Procedures: Wash away any material with copious amounts of soap and water.
First Aid Procedures: Remove victim to fresh air, call emergency medical care. If not breathing give CPR. If breathing is difficult administer oxygen. Treat for shock.

Fire Hazards:

Flashpoint: n/a *Ignition temperature:* n/a
Flammable Explosive High Range: n/a *Low Range:* n/a
Toxic Products of Combustion: Irritating hydrogen chloride gas may be formed in fires.
Other Hazards: n/a
Possible extinguishing agents: Use suitable agent for the type of surrounding fire.

Reactivity Hazards:

Reactive With: n/a *Other Reactions:* n/a

Corrosivity Hazards:

Corrosive With: Aluminum *Neutralizing Agent:* Crushed limestone, soda ash, or lime.

Radioactivity Hazards:

Radiation Emitted: n/a *Other Hazards:* n/a

Recommended Protection for Response Personnel:

Avoid breathing dust, keep upwind. Structural protective clothing provides limited protection. Wash away any material which may have come into contact with the body with copious amounts of soap and water. Consider appropriate evacuation.

COPPER CYANIDE

DOT Number: UN 1587 *DOT Hazard Class:* Poison B *DOT Guide Number:* 53
Synonyms: cupricin cyanide, cuprous cyanide
STCC Number: 4923418 *Reportable Qty:* 10/4.54
Mfg Name: Ashland Chemical *Phone No:* 1-614-889-3333

Physical Description:

Physical Form: Solid *Color:* Green powder *Odor:* n/a
Other Information: Decomposed by acids to give off hydrogen cyanide, a flammable poisonous gas.

Chemical Properties:

Specific Gravity: 2.92 *Vapor Density:* 4.0 *Boiling Point: n/a*
Melting Point: n/a *Vapor Pressure:* n/a *Solubility in water:* Insoluble
Other Information: n/a

Health Hazards:

Inhalation Hazard: Will cause dizziness or loss of consciousness.
Ingestion Hazard: Will cause dizziness and loss of consciousness.
Absorption Hazard: Irritating to the skin and eyes.
Hazards to Wildlife: n/a
Decontamination Procedures: Wash away any material with copious amounts of soap and water.
First Aid Procedures: Remove victim to fresh air, call emergency medical care. If not breathing give CPR. If breathing is difficult administer oxygen. Treat for shock.

Fire Hazards:

Flashpoint: n/a *Ignition temperature:* n/a
Flammable Explosive High Range: n/a *Low Range:* n/a
Toxic Products of Combustion: Poisonous gases may be produced when heated
Other Hazards: Toxic hydrogen cyanide gases may be produced in fire.
Possible extinguishing agents: Do not use water on the material itself. Use foam, dry chemical, or carbon dioxide.

Reactivity Hazards:

Reactive With: Water *Other Reactions:* n/a

Corrosivity Hazards:

Corrosive With: n/a *Neutralizing Agent:* n/a

Radioactivity Hazards:

Radiation Emitted: n/a *Other Hazards:* n/a

Recommended Protection for Response Personnel:

Avoid breathing vapors or dust, keep upwind. Wear a sealed chemical suit (polycarbonate, butyl rubber). Wash away any material which may have come into contact with the body with copious amounts of soap and water. Consider appropriate evacuation.

COPPER NAPHTHENATE

DOT Number: UN 1168 *DOT Hazard Class:* Flammable liquid *DOT Guide Number:* 26
Synonyms: paint dryer
STCC Number: 4910516 *Reportable Qty:* n/a
Mfg Name: Ferro Corp. *Phone No:* 1-216-641-8580

Physical Description:

Physical Form: Liquid *Color:* Dark green *Odor:* Gasoline like
Other Information: n/a

Chemical Properties:

Specific Gravity: 1.05-.93 *Vapor Density:* 7.7 *Boiling Point: 310-395° F(154-201° C)*
Melting Point: n/a *Vapor Pressure:* n/a *Solubility in water:* No
Other Information: May float or sink in water. Vapors are heavier than air. lighter, insoluble in water.

Health Hazards:

Inhalation Hazard: n/a
Ingestion Hazard: Harmful if swallowed.
Absorption Hazard: Irritating to the skin and the eyes.
Hazards to Wildlife: Dangerous to aquatic life.
Decontamination Procedures: Wash away any material with copious amounts of soap and water.
First Aid Procedures: Remove victim to fresh air, call emergency medical care. If not breathing give CPR. If breathing is difficult administer oxygen. Treat for shock.

Fire Hazards:

Flashpoint: 100° F(37.7° C) *Ignition temperature:* 500° F(260° C)
Flammable Explosive High Range: 5 *Low Range:* .8
Toxic Products of Combustion: n/a
Other Hazards: n/a
Possible extinguishing agents: Do not extinguish the fire unless the flow can be stopped. Apply water from as far a distance as possible. Use foam, dry chemical, or carbon dioxide.

Reactivity Hazards:

Reactive With: n/a *Other Reactions:* n/a

Corrosivity Hazards:

Corrosive With: n/a *Neutralizing Agent:* n/a

Radioactivity Hazards:

Radiation Emitted: n/a *Other Hazards:* n/a

Recommended Protection for Response Personnel:

Avoid breathing dust, keep upwind. Structural protective clothing provides limited protection. Wash away any material which may have come into contact with the body with copious amounts of soap and water. Consider appropriate evacuation.

COPPER NITRATE

DOT Number: NA 1479 *DOT Hazard Class:* Oxidizer *DOT Guide Number:* 35
Synonyms: cupric nitrate trihydrate, gerhardite
STCC Number: 4918745 *Reportable Qty:* 100/45.4
Mfg Name: Allied Corp. *Phone No:* 1-201-455-2000

Physical Description:

Physical Form: Solid *Color:* Blue *Odor:* Odorless
Other Information: It is used in medicine, as an insecticide, in chemical analysis, making light sensitive papers, and for many other uses.

Chemical Properties:

Specific Gravity: 2.32 *Vapor Density:* 8.3 *Boiling Point: n/a*
Melting Point: 238° F(114.4° C) *Vapor Pressure:* n/a *Solubility in water:* Yes
Other Information: Sinks and mixes with water.

Health Hazards:

Inhalation Hazard: Coughing or difficulty in breathing.
Ingestion Hazard: Nausea, vomiting, or loss of consciousness.
Absorption Hazard: Irritating to the skin and the eyes.
Hazards to Wildlife: Dangerous to aquatic life.
Decontamination Procedures: Wash away any material with copious amounts of soap and water.
First Aid Procedures: Remove victim to fresh air, call emergency medical care. If not breathing give CPR. If breathing is difficult administer oxygen. Treat for shock.

Fire Hazards:

Flashpoint: n/a *Ignition temperature:* n/a
Flammable Explosive High Range: n/a *Low Range:* n/a
Toxic Products of Combustion: Toxic and irritating oxides of nitrogen are formed in fires.
Other Hazards: Can increase the intensity of a fire upon contact with combustible materials.
Possible extinguishing agents: Flood with water. Apply water from as far a distance as possible.

Reactivity Hazards:

Reactive With: Wood, paper and other combustibles. *Other Reactions:* n/a

Corrosivity Hazards:

Corrosive With: n/a *Neutralizing Agent:* n/a

Radioactivity Hazards:

Radiation Emitted: n/a *Other Hazards:* n/a

Recommended Protection for Response Personnel:

Avoid breathing dust, keep upwind. Structural protective clothing provides limited protection. Wash away any material which may have come into contact with the body with copious amounts of soap and water. Consider appropriate evacuation.

COPPER OXALATE

DOT Number: UN 2449 *DOT Hazard Class:* n/a *DOT Guide Number:* 54
Synonyms: cupric oxalate, hemihydrate
STCC Number: n/a *Reportable Qty:* n/a
Mfg Name: Atomergic Chemetals Corp. *Phone No:* 1-516-349-8800

Physical Description:

Physical Form: Solid *Color:* Bluish white *Odor:* Odorless
Other Information: n/a

Chemical Properties:

Specific Gravity: 1 *Vapor Density:* n/a *Boiling Point: Decomposes*
Melting Point: n/a *Vapor Pressure:* n/a *Solubility in water:* n/a
Other Information: Sinks in water

Health Hazards:

Inhalation Hazard: Coughing, difficulty in breathing.
Ingestion Hazard: Nausea, vomiting, or loss of consciousness.
Absorption Hazard: Irritating to the skin and eyes.
Hazards to Wildlife: n/a
Decontamination Procedures: Wash away any material with copious amounts of soap and water.
First Aid Procedures: Remove victim to fresh air, call emergency medical care. If not breathing give CPR. If breathing is difficult administer oxygen. Treat for shock.

Fire Hazards:

Flashpoint: n/a *Ignition temperature:* n/a
Flammable Explosive High Range: n/a *Low Range:* n/a
Toxic Products of Combustion: Toxic carbon monoxide gas may be formed in fires
Other Hazards: n/a
Possible extinguishing agents: Extinguish fire using suitable agent for the type of surrounding fire.

Reactivity Hazards:

Reactive With: n/a *Other Reactions:* n/a

Corrosivity Hazards:

Corrosive With: n/a *Neutralizing Agent:* n/a

Radioactivity Hazards:

Radiation Emitted: n/a *Other Hazards:* n/a

Recommended Protection for Response Personnel:

Avoid breathing vapors, keep upwind. Wear a sealed chemical suit (polycarbonate, butyl rubber). Wash away any material which may have come into contact with the body with copious amounts of soap and water. Consider appropriate evacuation.

COPPER SULFATE

DOT Number: NA 9109 *DOT Hazard Class:* ORM-E *DOT Guide Number:* 31
Synonyms: blue vitol, cupric sulfate, copper sulfate pentahydrate, sulfate of copper
STCC Number: 4961317 *Reportable Qty:* 10/4.54
Mfg Name: Mallinckrodt Inc. *Phone No:* 1-314-895-2000

Physical Description:

Physical Form: Solid *Color:* White to blue *Odor:* Odorless
Other Information: It is used as an agricultural soil conditioner.

Chemical Properties:

Specific Gravity: 2.92 *Vapor Density:* n/a *Boiling Point: n/a*
Melting Point: n/a *Vapor Pressure:* n/a *Solubility in water:* Yes
Other Information: Sinks and mixes with water.

Health Hazards:

Inhalation Hazard: n/a
Ingestion Hazard: Nausea, vomiting, or loss of consciousness.
Absorption Hazard: n/a
Hazards to Wildlife: Dangerous to aquatic life.
Decontamination Procedures: Wash away any material with copious amounts of soap and water.
First Aid Procedures: Remove victim to fresh air, call emergency medical care. If not breathing give CPR. If breathing is difficult administer oxygen. Treat for shock.

Fire Hazards:

Flashpoint: n/a *Ignition temperature:* n/a
Flammable Explosive High Range: n/a *Low Range:* n/a
Toxic Products of Combustion: n/a
Other Hazards: n/a
Possible extinguishing agents: Extinguish fire using suitable agent for the type of surrounding fire.

Reactivity Hazards:

Reactive With: n/a *Other Reactions:* n/a

Corrosivity Hazards:

Corrosive With: n/a *Neutralizing Agent:* n/a

Radioactivity Hazards:

Radiation Emitted: n/a *Other Hazards:* n/a

Recommended Protection for Response Personnel:

Avoid breathing dust, keep upwind. Structural protective clothing provides limited protection. Wash away any material which may have come into contact with the body with copious amounts of soap and water. Consider appropriate evacuation.

COUMAPHOS

DOT Number: NA 2783 *DOT Hazard Class:* Poison B *DOT Guide Number:* 55
Synonyms: Co-ral, O,O-diethyl-O-(3-chloro-4-methyl-7-coumarinyl)phosphorothioate
STCC Number: 4921505 *Reportable Qty:* 10/4.54
Mfg Name: Bayvet Corp. *Phone No:* 1-913-631-4800

Physical Description:

Physical Form: Solid *Color:* White *Odor:* Weak, sulfurous
Other Information: Sinks in water

Chemical Properties:

Specific Gravity: 1.47 *Vapor Density:* 1 *Boiling Point: n/a*
Melting Point: n/a *Vapor Pressure:* n/a *Solubility in water:* Insoluble
Other Information: In case of damage to or leaking from containers of this material, contact the Pesticide Safety Team Network at 1-800-424-9300.

Health Hazards:

Inhalation Hazard: Harmful if inhaled.
Ingestion Hazard: Harmful if swallowed.
Absorption Hazard: Irritating to the skin and eyes.
Hazards to Wildlife: Dangerous to aquatic life and waterfowl.
Decontamination Procedures: Wash away any material with copious amounts of soap and water.
First Aid Procedures: Remove victim to fresh air, call emergency medical care. If not breathing give CPR. If breathing is difficult administer oxygen. Treat for shock.

Fire Hazards:

Flashpoint: n/a *Ignition temperature:* n/a
Flammable Explosive High Range: n/a *Low Range:* n/a
Toxic Products of Combustion: Decomposes to release toxic and irritating oxides of sulfur, phosphorus, and chlorides.
Other Hazards: n/a
Possible extinguishing agents: Extinguish fire using suitable agent for type of surrounding fire.

Reactivity Hazards:

Reactive With: n/a *Other Reactions:* n/a

Corrosivity Hazards:

Corrosive With: n/a *Neutralizing Agent:* n/a

Radioactivity Hazards:

Radiation Emitted: n/a *Other Hazards:* n/a

Recommended Protection for Response Personnel:

Avoid breathing vapors or dust, keep upwind. Wear appropriate chemical clothing. Wash away any material which may come into contact with the body with copious amounts of soap and water. Consider appropriate evacuation.

CREOSOTE, COAL TAR

DOT Number: UN 2076 *DOT Hazard Class:* Combustible liquid *DOT Guide Number:* 27
Synonyms: coal tar, creosote oil, dead oil, liquid pitch, tar oil
STCC Number: 4915133 *Reportable Qty:* n/a
Mfg Name: Allied Corp. *Phone No:* 1-201-455-2000

Physical Description:

Physical Form: Liquid *Color:* Yellow to black *Odor:* Tarry
Other Information: May float or sink in water. Weighs 9 lbs/4.0 kg per gallon/3.8 l.

Chemical Properties:

Specific Gravity: 1.08 *Vapor Density:* 1 *Boiling Point: 365 + ° F(185° C)*
Melting Point: n/a *Vapor Pressure:* 80 mm Hg at 212° F(100° C) *Solubility in water:* No
Other Information: n/a

Health Hazards:

Inhalation Hazard: n/a
Ingestion Hazard: Harmful if swallowed.
Absorption Hazard: Irritating to the skin and eyes.
Hazards to Wildlife: n/a
Decontamination Procedures: Wash away any material with copious amounts of soap and water.
First Aid Procedures: Remove victim to fresh air, call emergency medical care. If not breathing give CPR. If breathing is difficult administer oxygen. Treat for shock.

Fire Hazards:

Flashpoint: 165° F(73.8° C) *Ignition temperature:* 637° F(336.1° C)
Flammable Explosive High Range: n/a *Low Range:* n/a
Toxic Products of Combustion: Heavy toxic and irritating black smoke is formed in a fire.
Other Hazards: n/a
Possible extinguishing agents: Use foam, dry chemical, or carbon dioxide. Water may be ineffective.

Reactivity Hazards:

Reactive With: Strong oxidizers *Other Reactions:* n/a

Corrosivity Hazards:

Corrosive With: n/a *Neutralizing Agent:* n/a

Radioactivity Hazards:

Radiation Emitted: n/a *Other Hazards:* n/a

Recommended Protection for Response Personnel:

Avoid breathing dust, keep upwind. Wear a sealed chemical suit (polycarbonate, butyl rubber). Wash away any material which may have come into contact with the body with copious amounts of soap and water. If the fire becomes uncontrollable, consider appropriate evacuation.

CRESOL (o-m-p)

DOT Number: UN 2067 *DOT Hazard Class:* Corrosive *DOT Guide Number:* 55
Synonyms: 3-cresol, (o-m-p) cresylic acid
STCC Number: 4931417 *Reportable Qty:* 1000/454
Mfg Name: U.S.S. Chemical *Phone No:* 1-412-433-1121

Physical Description:

Physical Form: Liquid *Color:* Colorless *Odor:* Sweet, tarry
Other Information: n/a

Chemical Properties:

Specific Gravity: 1.03 *Vapor Density:* 3.72 *Boiling Point: 397° F(202.7° C)*
Melting Point: 95° F(35° C) *Vapor Pressure:* .25 mm Hg at 68° F(20° C)
Solubility in water: Slowly mixes *Degree of Solubility:* 2.2%
Other Information: Sinks in water. Weighs 8.7 lbs/3.9 kg per gallon/3.8 l.

Health Hazards:

Inhalation Hazard: Headaches, coughing.
Ingestion Hazard: Burning sensation of mouth and esophagus, vomiting.
Absorption Hazard: Will burn the skin and eyes.
Hazards to Wildlife: Dangerous to aquatic life.
Decontamination Procedures: Wash away any material with copious amounts of soap and water.
First Aid Procedures: Remove victim to fresh air, call emergency medical care. If not breathing give CPR. If breathing is difficult administer oxygen. Treat for shock.

Fire Hazards:

Flashpoint: 178° F(81.1° C) *Ignition temperature:* 1038° F(558.8° C)
Flammable Explosive High Range: 1.35 *Low Range:* 1.06
Toxic Products of Combustion: Poisonous flammable gases may be produced in fire
Other Hazards: Vapors may form explosive mixtures with air.
Possible extinguishing agents: Use foam, dry chemical, carbon dioxide, and water to knock down the vapors.

Reactivity Hazards:

Reactive With: n/a *Other Reactions:* n/a

Corrosivity Hazards:

Corrosive With: Tissue *Neutralizing Agent:* Crushed limestone, soda ash, or lime.

Radioactivity Hazards:

Radiation Emitted: n/a *Other Hazards:* n/a

Recommended Protection for Response Personnel:

Avoid breathing vapors or dust, keep upwind. Structural protective clothing provides limited protection. Wash away any material which may have come into contact with the body with copious amounts of soap and water. Consider appropriate evacuation.

CROTONALDEHYDE

DOT Number: UN 1143 *DOT Hazard Class:* Flammable liquid *DOT Guide Number:* 28
Synonyms: crotenaldehyde, crotonic aldehyde
STCC Number: 4909137 *Reportable Qty:* 100/45.4
Mfg Name: Union Carbide *Phone No:* 1-213-794-2000

Physical Description:

Physical Form: Liquid *Color:* Yellow *Odor:* Tar
Other Information: If subject to heat, the material may polymerize. Weighs 7.1 lbs/3.2 kg per gallon/3.8 l.

Chemical Properties:

Specific Gravity: .9 *Vapor Density:* 2.4 *Boiling Point: 216° F(102.2° C)*
Melting Point: −103° F(−75° C) *Vapor Pressure:* 30 mm Hg at 68° F(20° C)
Solubility in water: Yes *Degree of Solubility:* 15.5%
Other Information: If polymerization occurs, the container may violently rupture.

Health Hazards:

Inhalation Hazard: Coughing, nausea, loss of consciousness.
Ingestion Hazard: Harmful if swallowed.
Absorption Hazard: Will burn the skin and eyes.
Hazards to Wildlife: n/a
Decontamination Procedures: Wash away any material with copious amounts of soap and water.
First Aid Procedures: Remove victim to fresh air, call emergency medical care. If not breathing give CPR. If breathing is difficult administer oxygen. Treat for shock.

Fire Hazards:

Flashpoint: 55° F(12.7° C) *Ignition temperature:* 450° F(232.2° C)
Flammable Explosive High Range: 15.5 *Low Range:* 2.1
Toxic Products of Combustion: n/a
Other Hazards: Flashback along vapor trail may occur.
Possible extinguishing agents: Do not extinguish the fire unless the flow can be stopped. Use alcohol foam, dry chemical, or carbon dioxide.

Reactivity Hazards:

Reactive With: n/a *Other Reactions:* n/a

Corrosivity Hazards:

Corrosive With: n/a *Neutralizing Agent:* n/a

Radioactivity Hazards:

Radiation Emitted: n/a *Other Hazards:* n/a

Recommended Protection for Response Personnel:

Avoid breathing vapors, keep upwind. Wear a sealed chemical suit (butyl rubber). Wash away any material which may have come into contact with the body with copious amounts of soap and water. Consider appropriate evacuation.

CUMENE

DOT Number: NA 1918 *DOT Hazard Class:* Combustible liquid *DOT Guide Number:* 28
Synonyms: cumol, isopropylbenzene
STCC Number: 4913125 *Reportable Qty:* 5000/2270
Mfg Name: Marathon Oil *Phone No:* 1-419-422-2121

Physical Description:

Physical Form: Liquid *Color:* Clear *Odor:* Gasoline like
Other Information: n/a

Chemical Properties:

Specific Gravity: .9 *Vapor Density:* 4.1 *Boiling Point: 306° F(152.2° C)*
Melting Point: −141° F(−96.1° C) *Vapor Pressure:* 8 mm Hg at 68° F(20° C) *Solubility in water:* Insoluble
Other Information: n/a

Health Hazards:

Inhalation Hazard: Harmful if inhaled.
Ingestion Hazard: Harmful if swallowed.
Absorption Hazard: Irritating to the skin and eyes.
Hazards to Wildlife: Dangerous to aquatic life.
Decontamination Procedures: Wash away any material with copious amounts of soap and water.
First Aid Procedures: Remove victim to fresh air, call emergency medical care. If not breathing give CPR. If breathing is difficult administer oxygen. Treat for shock.

Fire Hazards:

Flashpoint: 96° F(35.5° C) *Ignition temperature:* 795° F(423.8° C)
Flammable Explosive High Range: 6.5 *Low Range:* .9
Toxic Products of Combustion: n/a
Other Hazards: Flashback along vapor trail may occur.
Possible extinguishing agents: Do not extinguish the fire unless the flow can be stopped. Use water, foam, dry chemical, or carbon dioxide.

Reactivity Hazards:

Reactive With: n/a *Other Reactions:* n/a

Corrosivity Hazards:

Corrosive With: n/a *Neutralizing Agent:* n/a

Radioactivity Hazards:

Radiation Emitted: n/a *Other Hazards:* n/a

Recommended Protection for Response Personnel:

Avoid breathing vapors, keep upwind. Wear a sealed chemical suit (polycarbonate, butyl rubber). Wash away any material which may have come into contact with the body with copious amounts of soap and water. Consider appropriate evacuation.

CUMENE HYDROPEROXIDE

DOT Number: UN 2116 *DOT Hazard Class:* Organic peroxide *DOT Guide Number:* 51
Synonyms: α-CHP, cumyl hydroperoxide
STCC Number: 4919525 *Reportable Qty:* n/a
Mfg Name: Allied Corp. *Phone No:* 1-201-455-2000

Physical Description:

Physical Form: Liquid *Color:* Colorless to light yellow *Odor:* Sharp, irritating
Other Information: n/a

Chemical Properties:

Specific Gravity: 1.3 *Vapor Density:* 5.2 *Boiling Point: 16° F(−8.8° C)*
Melting Point: n/a *Vapor Pressure:* n/a *Solubility in water:* n/a
Other Information: Sinks in water

Health Hazards:

Inhalation Hazard: Headaches, coughing.
Ingestion Hazard: Harmful if swallowed.
Absorption Hazard: Irritating to the skin and eyes.
Hazards to Wildlife: n/a
Decontamination Procedures: Wash away any material with copious amounts of soap and water.
First Aid Procedures: Remove victim to fresh air, call emergency medical care. If not breathing give CPR. If breathing is difficult administer oxygen. Treat for shock.

Fire Hazards:

Flashpoint: 147° F(63.8° C) *Ignition temperature:* 300° F(148.8° C)
Flammable Explosive High Range: 6.5 *Low Range:* .9
Toxic Products of Combustion: Toxic phenol vapors may form from hot material.
Other Hazards: Containers may explode in a fire.
Possible extinguishing agents: Apply water from as far a distance as possible, use alcohol foam, dry chemical, or carbon dioxide.

Reactivity Hazards:

Reactive With: n/a *Other Reactions:* n/a

Corrosivity Hazards:

Corrosive With: Metal *Neutralizing Agent:* n/a

Radioactivity Hazards:

Radiation Emitted: n/a *Other Hazards:* n/a

Recommended Protection for Response Personnel:

Avoid breathing vapors, keep upwind. Wear a sealed chemical suit (polycarbonate, butyl rubber). Wash away any material which may have come into contact with the body with copious amounts of soap and water. If the fire becomes uncontrollable. Consider appropriate evacuation.

CUPRIC NITRATE

DOT Number: NA 1479 *DOT Hazard Class:* Oxidizer *DOT Guide Number:* 35
Synonyms: copper nitrate trihydrate, gerhardite
STCC Number: 4918744 *Reportable Qty:* 100/45.4
Mfg Name: Allied Corp. *Phone No:* 1-201-455-2000

Physical Description:

Physical Form: Solid *Color:* Blue *Odor:* Odorless
Other Information: It is used in medicine, as an insecticide, in chemical analysis, making light sensitive papers, and for many other uses.

Chemical Properties:

Specific Gravity: 2.32 *Vapor Density:* 8.3 *Boiling Point: n/a*
Melting Point: 238° F(144.4° C) *Vapor Pressure:* n/a *Solubility in water:* Yes
Other Information: Sinks and mixes with water.

Health Hazards:

Inhalation Hazard: Coughing or difficulty in breathing.
Ingestion Hazard: Nausea, vomiting, or loss of consciousness.
Absorption Hazard: Irritating to the skin and the eyes.
Hazards to Wildlife: Dangerous to aquatic life.
Decontamination Procedures: Wash away any material with copious amounts of soap and water.
First Aid Procedures: Remove victim to fresh air, call emergency medical care. If not breathing give CPR. If breathing is difficult administer oxygen. Treat for shock.

Fire Hazards:

Flashpoint: n/a *Ignition temperature:* n/a
Flammable Explosive High Range: n/a *Low Range:* n/a
Toxic Products of Combustion: Toxic and irritating oxides of nitrogen are formed in fires.
Other Hazards: Can increase the intensity of a fire upon contact with combustible materials.
Possible extinguishing agents: Flood with water. Apply water from as far a distance as possible.

Reactivity Hazards:

Reactive With: Wood, paper and other combustibles. *Other Reactions:* n/a

Corrosivity Hazards:

Corrosive With: n/a *Neutralizing Agent:* n/a

Radioactivity Hazards:

Radiation Emitted: n/a *Other Hazards:* n/a

Recommended Protection for Response Personnel:

Avoid breathing dust, keep upwind. Structural protective clothing provides limited protection. Wash away any material which may have come into contact with the body with copious amounts of soap and water. Consider appropriate evacuation.

CUPRIC SULFATE

DOT Number: NA 9109 *DOT Hazard Class:* ORM-E *DOT Guide Number:* 31
Synonyms: blue vitol, copper sulfate, copper sulfate pentahydrate, sulfate of copper
STCC Number: 4961316 *Reportable Qty:* 10/4.54
Mfg Name: Mallinckrodt Inc. *Phone No:* 1-314-895-2000

Physical Description:

Physical Form: Solid *Color:* White to blue *Odor:* Odorless
Other Information: It is used as an agricultural soil conditioner.

Chemical Properties:

Specific Gravity: 2.92 *Vapor Density:* n/a *Boiling Point: n/a*
Melting Point: n/a *Vapor Pressure:* n/a *Solubility in water:* Yes
Other Information: Sinks and mixes with water.

Health Hazards:

Inhalation Hazard: n/a
Ingestion Hazard: Nausea, vomiting, or loss of consciousness.
Absorption Hazard: n/a
Hazards to Wildlife: Dangerous to aquatic life.
Decontamination Procedures: Wash away any material with copious amounts of soap and water.
First Aid Procedures: Remove victim to fresh air, call emergency medical care. If not breathing give CPR. If breathing is difficult administer oxygen. Treat for shock.

Fire Hazards:

Flashpoint: n/a *Ignition temperature:* n/a
Flammable Explosive High Range: n/a *Low Range:* n/a
Toxic Products of Combustion: n/a
Other Hazards: n/a
Possible extinguishing agents: Extinguish fire using suitable agent for the type of surrounding fire.

Reactivity Hazards:

Reactive With: n/a *Other Reactions:* n/a

Corrosivity Hazards:

Corrosive With: n/a *Neutralizing Agent:* n/a

Radioactivity Hazards:

Radiation Emitted: n/a *Other Hazards:* n/a

Recommended Protection for Response Personnel:

Avoid breathing dust, keep upwind. Structural protective clothing provides limited protection. Wash away any material which may have come into contact with the body with copious amounts of soap and water. Consider appropriate evacuation.

CUPRIETHYLENEDIAMINE SOLUTION

DOT Number: UN 1761 *DOT Hazard Class:* Corrosive *DOT Guide Number:* 59
Synonyms: cypriethylendiamine hydroxide solution
STCC Number: 4935630 *Reportable Qty:* n/a
Mfg Name: Olin Corp. *Phone No:* 1-203-356-2000

Physical Description:

Physical Form: Liquid *Color:* Blue to dark purple *Odor:* Fishy
Other Information: n/a

Chemical Properties:

Specific Gravity: 1.1 *Vapor Density:* n/a *Boiling Point: 212° F(100° C)*
Melting Point: n/a *Vapor Pressure:* n/a *Solubility in water:* Yes
Other Information: Mixes and sinks in water. Irritating vapor is produced.

Health Hazards:

Inhalation Hazard: Will cause difficulty in breathing.
Ingestion Hazard: Poisonous if swallowed.
Absorption Hazard: Poisonous if absorbed.
Hazards to Wildlife: n/a
Decontamination Procedures: Wash away any material with copious amounts of soap and water.
First Aid Procedures: Remove victim to fresh air, call emergency medical care. If not breathing give CPR. If breathing is difficult administer oxygen. Treat for shock.

Fire Hazards:

Flashpoint: n/a *Ignition temperature:* n/a
Flammable Explosive High Range: n/a *Low Range:* n/a
Toxic Products of Combustion: Irritating vapors of ethylenediamine may be produced when heated.
Other Hazards: n/a
Possible extinguishing agents: Apply water from as far a distance as possible. Use alcohol foam, dry chemical, or carbon dioxide.

Reactivity Hazards:

Reactive With: n/a *Other Reactions:* n/a

Corrosivity Hazards:

Corrosive With: Metal and tissue *Neutralizing Agent:* Flush with water

Radioactivity Hazards:

Radiation Emitted: n/a *Other Hazards:* n/a

Recommended Protection for Response Personnel:

Avoid breathing vapors, keep upwind. Wear appropriate chemical clothing, wash away any material which may come into contact with the body with copious amounts of soap and water. Consider appropriate evacuation.

CYANOACETIC ACID

DOT Number: UN 1935 *DOT Hazard Class:* n/a *DOT Guide Number:* 55
Synonyms: cyanacetic acid, malonic mononitrile
STCC Number: n/a *Reportable Qty:* n/a
Mfg Name: M.G. Chemical Co. *Phone No:* 1-212-269-5533

Physical Description:

Physical Form: Liquid *Color:* Yellowish brown *Odor:* Unpleasant
Other Information: n/a

Chemical Properties:

Specific Gravity: 1.1 *Vapor Density:* 2.9 *Boiling Point: Decomposes*
Melting Point: 151° F(66.1° C) *Vapor Pressure:* n/a *Solubility in water:* Yes
Other Information: Sinks and mixes with water.

Health Hazards:

Inhalation Hazard: Harmful if inhaled.
Ingestion Hazard: Harmful if swallowed.
Absorption Hazard: Irritating to the skin and eyes.
Hazards to Wildlife: Dangerous to aquatic life.
Decontamination Procedures: Wash away any material with copious amounts of soap and water.
First Aid Procedures: Remove victim to fresh air, call emergency medical care. If not breathing give CPR. If breathing is difficult administer oxygen. Treat for shock.

Fire Hazards:

Flashpoint: n/a *Ignition temperature:* n/a
Flammable Explosive High Range: n/a *Low Range:* n/a
Toxic Products of Combustion: Toxic oxides of nitrogen, and toxic flammable acetonitrile vapors may form in fires.
Other Hazards: n/a
Possible extinguishing agents: Extinguish fire using suitable agent for the type of surrounding fire.

Reactivity Hazards:

Reactive With: n/a *Other Reactions:* n/a

Corrosivity Hazards:

Corrosive With: n/a *Neutralizing Agent:* n/a

Radioactivity Hazards:

Radiation Emitted: n/a *Other Hazards:* n/a

Recommended Protection for Response Personnel:

Avoid breathing vapors, keep upwind. Wear a sealed chemical suit (polycarbonate, butyl rubber). Wash away any material which may have come into contact with the body with copious amounts of soap and water. Consider appropriate evacuation.

CYANOGEN

DOT Number: UN 1026 *DOT Hazard Class:* Poison A *DOT Guide Number:* 18
Synonyms: Dicyan, ethane dinitrile
STCC Number: 4920115 *Reportable Qty:* 100/45.4
Mfg Name: Matheson Gas Products *Phone No:* 1-201-867-4100

Physical Description:

Physical Form: Gas *Color:* Colorless *Odor:* Almond
Other Information: It is used to make other chemicals, as a fumigant, and as rocket propellant.

Chemical Properties:

Specific Gravity: .59 *Vapor Density:* 1.8 *Boiling Point: −6° F(−21.1° C)*
Melting Point: n/a *Vapor Pressure:* n/a *Solubility in water:* n/a
Other Information: Floats and boils on water. A poisonous, flammable, visible vapor cloud is produced.

Health Hazards:

Inhalation Hazard: Poisonous if inhaled.
Ingestion Hazard: Poisonous if swallowed.
Absorption Hazard: Will cause frostbite.
Hazards to Wildlife: n/a
Decontamination Procedures: Wash away any material with copious amounts of soap and water.
First Aid Procedures: Remove victim to fresh air, call emergency medical care. If not breathing give CPR. If breathing is difficult administer oxygen. Treat for shock.

Fire Hazards:

Flashpoint: n/a *Ignition temperature:* n/a
Flammable Explosive High Range: 32 *Low Range:* 6.6
Toxic Products of Combustion: Unburned vapors are highly toxic.
Other Hazards: Containers may explode in fire. Flashback along vapor trail may occur. Vapor may explode if ignited in an enclosed area.
Possible extinguishing agents: Do not extinguish the fire unless the flow can be stopped. Apply water from as far a distance as possible. Use alcohol foam, dry chemical, or carbon dioxide.

Reactivity Hazards:

Reactive With: n/a *Other Reactions:* n/a

Corrosivity Hazards:

Corrosive With: n/a *Neutralizing Agent:* n/a

Radioactivity Hazards:

Radiation Emitted: n/a *Other Hazards:* n/a

Recommended Protection for Response Personnel:

Avoid breathing vapors, keep upwind. Wear a sealed chemical suit (polycarbonate, butyl rubber). Wash away any material which may have come into contact with the body with copious amounts of soap and water. Prolonged exposure to heat or flame may cause the cylinders to violently rupture/rocket. (BLEVE) Consider appropriate evacuation.

CYANOGEN BROMIDE

DOT Number: UN 1889 *DOT Hazard Class:* Poison B *DOT Guide Number:* 55
Synonyms: None given
STCC Number: 4923229 *Reportable Qty:* 1000/454
Mfg Name: Eastman Organic *Phone No:* 1-800-455-6325

Physical Description:

Physical Form: Solid *Color:* Colorless *Odor:* Penetrating
Other Information: It is used in gold extraction, to make other chemicals, and as a fumigant.

Chemical Properties:

Specific Gravity: 2.01 *Vapor Density:* n/a *Boiling Point: n/a*
Melting Point: n/a *Vapor Pressure:* 100 mm Hg at 74° F(23.3° C) *Solubility in water:* Yes
Other Information: Sinks and mixes with water.

Health Hazards:

Inhalation Hazard: Poisonous if inhaled.
Ingestion Hazard: Poisonous if swallowed.
Absorption Hazard: Poisonous if absorbed.
Hazards to Wildlife: n/a
Decontamination Procedures: Wash away any material with copious amounts of soap and water.
First Aid Procedures: Remove victim to fresh air, call emergency medical care. If not breathing give CPR. If breathing is difficult administer oxygen. Treat for shock.

Fire Hazards:

Flashpoint: n/a *Ignition temperature:* n/a
Flammable Explosive High Range: n/a *Low Range:* n/a
Toxic Products of Combustion: Poisonous gases are produced when heated. Toxic oxides of nitrogen are produced.
Other Hazards: n/a
Possible extinguishing agents: Extinguish fire using suitable agent for type of surrounding fire.

Reactivity Hazards:

Reactive With: n/a *Other Reactions:* n/a

Corrosivity Hazards:

Corrosive With: n/a *Neutralizing Agent:* n/a

Radioactivity Hazards:

Radiation Emitted: n/a *Other Hazards:* n/a

Recommended Protection for Response Personnel:

Avoid breathing vapors, keep upwind. Wear a sealed chemical suit (polycarbonate, butyl rubber). Wash away any material which may have come into contact with the body with copious amounts of soap and water. Consider appropriate evacuation.

CYANOGEN CHLORIDE

DOT Number: UN 1589 *DOT Hazard Class:* Poison A *DOT Guide Number:* 15
Synonyms: None given
STCC Number: 4920506 *Reportable Qty:*.54
Mfg Name: Atomergic Chemetals Corp. *Phone No:* 1-516-349-8800

Physical Description:

Physical Form: Liquid *Color:* Colorless *Odor:* Strong
Other Information: It weighs 10 lbs/4.5 kg per gallon/3.8 l. A poisonous vapor cloud is produced.

Chemical Properties:

Specific Gravity: 1.22 *Vapor Density:* 2 *Boiling Point: 55.6° F(13.1° C)*
Melting Point: n/a *Vapor Pressure:* 1010 mm Hg at 68° F (20° C) *Solubility in water:* n/a
Other Information: May only be shipped in cylinders.

Health Hazards:

Inhalation Hazard: Poisonous if inhaled.
Ingestion Hazard: Poisonous if swallowed.
Absorption Hazard: Poisonous if absorbed.
Hazards to Wildlife: Dangerous to aquatic life.
Decontamination Procedures: Wash away any material with copious amounts of soap and water.
First Aid Procedures: Remove victim to fresh air, call emergency medical care. If not breathing give CPR. If breathing is difficult administer oxygen. Treat for shock.

Fire Hazards:

Flashpoint: n/a *Ignition temperature:* n/a
Flammable Explosive High Range: n/a *Low Range:* n/a
Toxic Products of Combustion: Poisonous gases are produced when heated, and when involved in a fire.
Other Hazards: Overheated containers may explode.
Possible extinguishing agents: Extinguish fire using suitable agent for type of surrounding fire. Apply water from as far a distance as possible. Use alcohol foam, dry chemical, or carbon dioxide.

Reactivity Hazards:

Reactive With: n/a *Other Reactions:* n/a

Corrosivity Hazards:

Corrosive With: n/a *Neutralizing Agent:* n/a

Radioactivity Hazards:

Radiation Emitted: n/a *Other Hazards:* n/a

Recommended Protection for Response Personnel:

Avoid breathing vapors, keep upwind. Wear a sealed chemical suit (polycarbonate, butyl rubber). Wash away any material which may have come into contact with the body with copious amounts of soap and water. Prolonged exposure to heat or flame may cause the cylinders to violently rupture/rocket. (BLEVE) Consider appropriate evacuation.

CYCLOHEXANE

DOT Number: UN 1145 *DOT Hazard Class:* Flammable liquid *DOT Guide Number:* 26
Synonyms: hexahydrobenzene
STCC Number: 4908132 *Reportable Qty:* 1000/454
Mfg Name: Phillips Petroleum *Phone No:* 1-918-661-6600

Physical Description:

Physical Form: Liquid *Color:* Colorless *Odor:* Gasolinelike
Other Information: It is used to make nylon, as a solvent, paint remover, and to make other chemicals.

Chemical Properties:

Specific Gravity: .8 *Vapor Density:* 29 *Boiling Point: 179° F(81.6° C)*
Melting Point: 44° F(6.6° C) *Vapor Pressure:* 95 mm Hg at 68° F(20° C) *Solubility in water:* No
Other Information: Lighter than water. Vapors are heavier than air. Weighs 6.1 lbs/2.7 kg per gallon/3.8 l.

Health Hazards:

Inhalation Hazard: Dizziness, nausea, loss of consciousness.
Ingestion Hazard: Harmful if swallowed.
Absorption Hazard: Irritating to the skin and eyes.
Hazards to Wildlife: Dangerous to aquatic life.
Decontamination Procedures: Wash away any material with copious amounts of soap and water.
First Aid Procedures: Remove victim to fresh air, call emergency medical care. If not breathing give CPR. If breathing is difficult administer oxygen. Treat for shock.

Fire Hazards:

Flashpoint: −4° F(−20° C) *Ignition temperature:* 473° F(245° C)
Flammable Explosive High Range: 8 *Low Range:* 1.3
Toxic Products of Combustion: n/a
Other Hazards: Flashback along vapor trail may occur. Vapors may explode if ignited in an enclosed area.
Possible extinguishing agents: Do not extinguish the fire unless the flow can be stopped apply water from as far a distance as possible. Use dry chemical or carbon dioxide.

Reactivity Hazards:

Reactive With: n/a *Other Reactions:* n/a

Corrosivity Hazards:

Corrosive With: n/a *Neutralizing Agent:* n/a

Radioactivity Hazards:

Radiation Emitted: n/a *Other Hazards:* n/a

Recommended Protection for Response Personnel:

Avoid breathing vapors, keep upwind. Structural protective clothing provides limited protection. Wash away any material which may have come into contact with the body with copious amounts of soap and water. If the fire becomes uncontrollable, or if the containers are exposed to direct flame, consider appropriate evacuation.

CYCLOHEXANOL

DOT Number: NA 1993 *DOT Hazard Class:* Combustible liquid *DOT Guide Number:* n/a
Synonyms: Adronal, cyclohexy alcohol, hexalinanol, hexahydrophenor, hydroxycyclohexane
STCC Number: 4915518 *Reportable Qty:* n/a
Mfg Name: Allied Corp. *Phone No:* 1-201-455-2000

Physical Description:

Physical Form: Liquid *Color:* Colorless to light yellow *Odor:* Alcohol
Other Information: It is used in soap making, lacquers, and in plastics.

Chemical Properties:

Specific Gravity: .94 *Vapor Density:* 3.45 *Boiling Point: 322° F(161.1° C)*
Melting Point: 74.5° F(23.6° C) *Vapor Pressure:* 1 mm Hg at 70° F(21.1° C) *Solubility in water:* Yes
Other Information: Floats and mixes slowly with water. May solidify.

Health Hazards:

Inhalation Hazard: n/a
Ingestion Hazard: Harmful if swallowed.
Absorption Hazard: Will burn the skin and eyes.
Hazards to Wildlife: n/a
Decontamination Procedures: Wash away any material with copious amounts of soap and water.
First Aid Procedures: Remove victim to fresh air, call emergency medical care. If not breathing give CPR. If breathing is difficult administer oxygen. Treat for shock.

Fire Hazards:

Flashpoint: 160° F(71.1° C) *Ignition temperature:* 572° F(300° C)
Flammable Explosive High Range: n/a *Low Range:* n/a
Toxic Products of Combustion: n/a
Other Hazards: Cool exposed containers with water.
Possible extinguishing agents: Do not extinguish the fire unless the flow can be stopped apply water from as far a distance as possible. Use alcohol foam, dry chemical, or carbon dioxide.

Reactivity Hazards:

Reactive With: n/a *Other Reactions:* n/a

Corrosivity Hazards:

Corrosive With: n/a *Neutralizing Agent:* n/a

Radioactivity Hazards:

Radiation Emitted: n/a *Other Hazards:* n/a

Recommended Protection for Response Personnel:

Avoid breathing vapors, keep upwind. Wear a sealed chemical suit (polycarbonate, butyl rubber). Wash away any material which may have come into contact with the body with copious amounts of soap and water. Consider appropriate evacuation.

CYCLOHEXANONE

DOT Number: NA 1915 *DOT Hazard Class:* Combustible liquid *DOT Guide Number:* 26
Synonyms: Anonehytrol O, Nadone, pimelic ketone, Sextone
STCC Number: 4913179 *Reportable Qty:* 5000/2270
Mfg Name: Allied Corp. *Phone No:* 1-201-455-2000

Physical Description:

Physical Form: Liquid *Color:* Colorless to light yellow *Odor:* Sweet peppermint
Other Information: It is used to make nylon, as a chemical reaction medium, and as a solvent.

Chemical Properties:

Specific Gravity: .946 *Vapor Density:* 3.4 *Boiling Point: 313° F(156.1° C)*
Melting Point: −26° F(−32.2° C) *Vapor Pressure:* 2 to 4 mm Hg at 68° F(20° C)
Solubility in water: Yes *Degree of Solubility:* 2.3%
Other Information: Slightly soluble in water. Vapors are heavier than air.

Health Hazards:

Inhalation Hazard: n/a
Ingestion Hazard: Harmful if swallowed.
Absorption Hazard: Will burn the skin and eyes.
Hazards to Wildlife: n/a
Decontamination Procedures: Wash away any material with copious amounts of soap and water.
First Aid Procedures: Remove victim to fresh air. Call emergency medical care. If not breathing give CPR. If breathing is difficult administer oxygen. Treat for shock.

Fire Hazards:

Flashpoint: 111° F (43.8° C) *Ignition temperature:* 788° F(420° C)
Flammable Explosive High Range: 9.8 *Low Range:* 1.1
Toxic Products of Combustion: Acrid smoke and irritating fumes
Other Hazards: There is a possibility that the containers may violently rupture in a fire.
Possible extinguishing agents: Use alcohol foam, dry chemical or carbon dioxide. Apply water from as far a distance as possible.

Reactivity Hazards:

Reactive With: Oxidizing agents *Other Reactions:* Oxidizers may cause fire or explosion.

Corrosivity Hazards:

Corrosive With: Paints and various plastics *Neutralizing Agent:* n/a

Radioactivity Hazards:

Radiation Emitted: n/a *Other Hazards:* n/a

Recommended Protection for Response Personnel:

Avoid breathing dust, keep upwind. Structural protective clothing provides limited protection. Wash away any material which may have come into contact with the body with copious amounts of soap and water. If the fire becomes uncontrollable, consider appropriate evacuation.

CYCLOHEXANONE PEROXIDE

DOT Number: UN 2118 *DOT Hazard Class:* Organic peroxide *DOT Guide Number:* 51
Synonyms: Cadox HDP, dicyclohexanone diperoxide, Luperco JDB-50-T
STCC Number: n/a *Reportable Qty:* n/a
Mfg Name: Penwald Corp. *Phone No:* 1-215-587-7000

Physical Description:

Physical Form: Liquid *Color:* White *Odor:* Odorless
Other Information: n/a

Chemical Properties:

Specific Gravity: 1.05 *Vapor Density:* 1 *Boiling Point: Decomposes*
Melting Point: n/a *Vapor Pressure:* n/a *Solubility in water:* n/a
Other Information: Sinks in water

Health Hazards:

Inhalation Hazard: Harmful if inhaled.
Ingestion Hazard: Harmful if swallowed.
Absorption Hazard: Irritating to the skin and eyes.
Hazards to Wildlife: n/a
Decontamination Procedures: Wash away any material with copious amounts of soap and water.
First Aid Procedures: Remove victim to fresh air, call emergency medical care. If not breathing give CPR. If breathing is difficult administer oxygen. Treat for shock.

Fire Hazards:

Flashpoint: 315° F(157.2° C) *Ignition temperature:* 757° F(402.7° C)
Flammable Explosive High Range: n/a *Low Range:* n/a
Toxic Products of Combustion: n/a
Other Hazards: Containers may explode in fire.
Possible extinguishing agents: Extinguish fire using suitable agent for the type of surrounding fire.

Reactivity Hazards:

Reactive With: n/a *Other Reactions:* n/a

Corrosivity Hazards:

Corrosive With: n/a *Neutralizing Agent:* n/a

Radioactivity Hazards:

Radiation Emitted: n/a *Other Hazards:* n/a

Recommended Protection for Response Personnel:

Avoid breathing dust, keep upwind. Wear a sealed chemical suit (polycarbonate, butyl rubber). Wash away any material which may have come into contact with the body with copious amounts of soap and water. Consider appropriate evacuation.

CYCLOHEXENYL TRICHLOROSILANE

DOT Number: UN 1762 *DOT Hazard Class:* Corrosive *DOT Guide Number:* 29
Synonyms: None given
STCC Number: 4934217 *Reportable Qty:* n/a *CHEMTREC Phone No:* 1-800-424-9300

Physical Description:

Physical Form: Liquid *Color:* Colorless *Odor:* Sharp
Other Information: It is used to make various silicone containing compounds.

Chemical Properties:

Specific Gravity: 1.23 *Vapor Density:* 7.4 *Boiling Point: 300° F(148.8° C)*
Melting Point: n/a *Vapor Pressure:* n/a *Solubility in water:* Reacts
Other Information: Poisonous gas is produced upon contact with water.

Health Hazards:

Inhalation Hazard: Poisonous if inhaled
Ingestion Hazard: Harmful if swallowed.
Absorption Hazard: Will burn the skin and eyes.
Hazards to Wildlife: n/a
Decontamination Procedures: Wash away any material with copious amounts of soap and water.
First Aid Procedures: Remove victim to fresh air, call emergency medical care. If not breathing give CPR. If breathing is difficult administer oxygen. Treat for shock.

Fire Hazards:

Flashpoint: 150° F(65.5° C) *Ignition temperature:* n/a
Flammable Explosive High Range: n/a *Low Range:* n/a
Toxic Products of Combustion: Irritating toxic hydrogen chloride and phosgene
Other Hazards: Difficult to extinguish, reignition may occur.
Possible *Possible extinguishing agents:* Do not use water on the material itself. Use dry chemical, dry sand, or carbon dioxide.

Reactivity Hazards:

Reactive With: Water to form hydrochloric acid. *Other Reactions:* n/a

Corrosivity Hazards:

Corrosive With: Metals and tissue *Neutralizing Agent:* Crushed limestone, soda ash, or lime

Radioactivity Hazards:

Radiation Emitted: n/a *Other Hazards:* n/a

Recommended Protection for Response Personnel:

Avoid breathing dust, keep upwind. Wear appropriate chemical clothing. Wash away any material which may have come into contact with the body with copious amounts of soap and water. Consider appropriate evacuation.

CYCLOHEXENYLAMINE

DOT Number: UN 2357 *DOT Hazard Class:* Flammable liquid *DOT Guide Number:* 68
Synonyms: aminocyclohexane, hexahydroaniline
STCC Number: 4909139 *Reportable Qty:* n/a
Mfg Name: Monsanto Chemical *Phone No:* 1-314-694-1000

Physical Description:

Physical Form: Liquid *Color:* Colorless *Odor:* Strong, fishy
Other Information: n/a

Chemical Properties:

Specific Gravity: .9 *Vapor Density:* 3.4 *Boiling Point: 274° F(134.4° C)*
Melting Point: n/a *Vapor Pressure:* n/a *Solubility in water:* Yes
Other Information: Floats and mixes with water. Vapors are heavier than air.

Health Hazards:

Inhalation Hazard: n/a
Ingestion Hazard: Harmful if swallowed.
Absorption Hazard: Will burn the skin and eyes.
Hazards to Wildlife: n/a
Decontamination Procedures: Wash away any material with copious amounts of soap and water.
First Aid Procedures: Remove victim to fresh air, call emergency medical care. If not breathing give CPR. If breathing is difficult administer oxygen. Treat for shock.

Fire Hazards:

Flashpoint: 90° F(32.2° C) *Ignition temperature:* 560° F(293.3° C)
Flammable Explosive High Range: n/a *Low Range:* n/a
Toxic Products of Combustion: n/a
Other Hazards: Flashback along vapor trail may occur. Vapors may explode if ignited in an enclosed area.
Possible extinguishing agents: Do not extinguish the fire unless the flow can be stopped. Apply water from as far a distance as possible. Use foam, dry chemical, or carbon dioxide.

Reactivity Hazards:

Reactive With: n/a *Other Reactions:* n/a

Corrosivity Hazards:

Corrosive With: n/a *Neutralizing Agent:* n/a

Radioactivity Hazards:

Radiation Emitted: n/a *Other Hazards:* n/a

Recommended Protection for Response Personnel:

Avoid breathing dust, keep upwind. Wear a sealed chemical suit (butyl rubber). Wash away any material which may have come into contact with the body with copious amounts of soap and water. Consider appropriate evacuation.

CYCLOPENTANE

DOT Number: UN 1146 *DOT Hazard Class:* Flammable liquid *DOT Guide Number:* 27
Synonyms: pentamethylene
STCC Number: 4908135 *Reportable Qty:* n/a
Mfg Name: Phillips Petroleum *Phone No:* 1-918-661-6600

Physical Description:

Physical Form: Liquid *Color:* Colorless *Odor:* Mild, sweet
Other Information: n/a

Chemical Properties:

Specific Gravity: .8 *Vapor Density:* 2.35 *Boiling Point: 111° F(43.8° C)*
Melting Point: n/a *Vapor Pressure:* 400 mm Hg at 88° F(31.1° C) *Solubility in water:* No
Other Information: Floats on water, flammable, irritating vapor is produced. Vapors are heavier than air.

Health Hazards:

Inhalation Hazard: Dizziness, nausea, vomiting, difficulty in breathing, loss of consciousness
Ingestion Hazard: Harmful if swallowed.
Absorption Hazard: Irritating to the skin and eyes.
Hazards to Wildlife: n/a
Decontamination Procedures: Wash away any material with copious amounts of soap and water.
First Aid Procedures: Remove victim to fresh air, call emergency medical care. If not breathing give CPR. If breathing is difficult administer oxygen. Treat for shock.

Fire Hazards:

Flashpoint: –20° F(–28.8° C) *Ignition temperature:* 743° F(395° C)
Flammable Explosive High Range: 8.7 *Low Range:* 1.1
Toxic Products of Combustion: n/a
Other Hazards: Flashback along vapor trail may occur, vapors may explode if confined in an enclosed area. Containers may explode in a fire.
Possible extinguishing agents: Do not extinguish fire unless the flow can be stopped. Apply water from as far a distance as possible. Use foam dry chemical, or carbon dioxide.

Reactivity Hazards:

Reactive With: n/a *Other Reactions:* n/a

Corrosivity Hazards:

Corrosive With: n/a *Neutralizing Agent:* n/a

Radioactivity Hazards:

Radiation Emitted: n/a *Other Hazards:* n/a

Recommended Protection for Response Personnel:

Avoid breathing vapors, keep upwind. Structural protective clothing provides limited protection. Wash away any material which may have come into contact with the body with copious amounts of soap and water. If the fire becomes uncontrollable, or if the containers are exposed to direct flame, consider appropriate evacuation.

CYCLOPROPANE

DOT Number: UN 1027 *DOT Hazard Class:* Liquefied gas *DOT Guide Number:* 22
Synonyms: trimethylene
STCC Number: 4905713 *Reportable Qty:* n/a
Mfg Name: Airco Industrial Gases *Phone No:* 1-201-464-8100

Physical Description:

Physical Form: Gas *Color:* Colorless *Odor:* Mild, sweet
Other Information: n/a

Chemical Properties:

Specific Gravity: .67 *Vapor Density:* 1.5 *Boiling Point: – 29° F(– 33.8° C)*
Melting Point: n/a *Vapor Pressure:* n/a *Solubility in water:* No
Other Information: Floats and boils on water. Flammable, visible vapor cloud is produced.

Health Hazards:

Inhalation Hazard: Will cause difficulty in breathing.
Ingestion Hazard: n/a
Absorption Hazard: Will cause frostbite.
Hazards to Wildlife: n/a
Decontamination Procedures: Wash away any material with copious amounts of soap and water.
First Aid Procedures: Remove victim to fresh air, call emergency medical care. If not breathing give CPR. If breathing is difficult administer oxygen. Treat for shock.

Fire Hazards:

Flashpoint: n/a *Ignition temperature:* 928° F(497.7° C)
Flammable Explosive High Range: 10.4 *Low Range:* 2.4
Toxic Products of Combustion: n/a
Other Hazards: Flashback along vapor trail may occur, vapors may ignite if confined in an enclosed area. Containers may explode in fire
Possible extinguishing agents: Do not extinguish fire unless the flow can be stopped. Apply water from as far a distance as possible.

Reactivity Hazards:

Reactive With: n/a *Other Reactions:* n/a

Corrosivity Hazards:

Corrosive With: n/a *Neutralizing Agent:* n/a

Radioactivity Hazards:

Radiation Emitted: n/a *Other Hazards:* n/a

Recommended Protection for Response Personnel:

Avoid breathing vapors, keep upwind. Structural protective clothing provides limited protection. Wash away any material which may have come into contact with the body with copious amounts of soap and water. Under fire conditions, the containers may violently rupture or rocket. (BLEVE) Consider appropriate evacuation.

CYMENE

DOT Number: UN 2046 *DOT Hazard Class:* n/a *DOT Guide Number:* 27
Synonyms: cymol, p-isopropyltoluene, isopropyltoluol, 1-methyl-4-isopropylbenzene, methyl propyl benzene
STCC Number: n/a *Reportable Qty:* n/a
Mfg Name: Aldrich Chemical *Phone No:* 1-414-273-3850

Physical Description:

Physical Form: Liquid *Color:* Colorless *Odor:* Mild pleasant
Other Information: n/a

Chemical Properties:

Specific Gravity: .857 *Vapor Density:* 4.62 *Boiling Point: 351° F(177.2° C)*
Melting Point: −90° F(−67.7° C) *Vapor Pressure:* 1 mm Hg at 63° F(17.2° C) *Solubility in water:* n/a
Other Information: Floats on water

Health Hazards:

Inhalation Hazard: n/a
Ingestion Hazard: n/a
Absorption Hazard: Irritating to the skin and eyes.
Hazards to Wildlife: n/a
Decontamination Procedures: Wash away any material with copious amounts of soap and water.
First Aid Procedures: Remove victim to fresh air, call emergency medical care. If not breathing give CPR. If breathing is difficult administer oxygen. Treat for shock.

Fire Hazards:

Flashpoint: 117° F (47.2° C) *Ignition temperature:* 817° F(436.1° C)
Flammable Explosive High Range: 5.6 *Low Range:* .7
Toxic Products of Combustion: n/a
Other Hazards: n/a
Possible extinguishing agents: Foam, dry chemical or carbon dioxide. Cool all exposed containers with water. Water may be ineffective on fire. Apply water from as far a distance as possible.

Reactivity Hazards:

Reactive With: n/a *Other Reactions:* n/a

Corrosivity Hazards:

Corrosive With: n/a *Neutralizing Agent:* n/a

Radioactivity Hazards:

Radiation Emitted: n/a *Other Hazards:* n/a

Recommended Protection for Response Personnel:

Avoid breathing dust, keep upwind. Structural protective clothing provides limited protection. Wash away any material which may have come into contact with the body with copious amounts of soap and water. Consider appropriate evacuation.

2,4-D ESTERS

DOT Number: UN 2765 *DOT Hazard Class:* ORM-E *DOT Guide Number:* 55

Synonyms: butyl 2,4-dichlorophenoxyacetate; 2,4-dichlorophenoxy acetic acid, butoxyethyl ester; isopropyl 2,4-dichlorophenoxy acetate

STCC Number: n/a *Reportable Qty:* n/a

Mfg Name: Dow Chemical *Phone No:* 1-517-636-4400

Physical Description:

Physical Form: Liquid *Color:* Yellowish brown *Odor:* Fuel oil

Other Information: n/a

Chemical Properties:

Specific Gravity: 1.2 *Vapor Density:* 7.63 *Boiling Point: Very high*

Melting Point: n/a *Vapor Pressure:* n/a *Solubility in water:* n/a

Other Information: Sinks in water.

Health Hazards:

Inhalation Hazard: n/a

Ingestion Hazard: Harmful if swallowed.

Absorption Hazard: Irritating to the skin and eyes.

Hazards to Wildlife: Dangerous to aquatic life.

Decontamination Procedures: Wash away any material with copious amounts of soap and water.

First Aid Procedures: Remove victim to fresh air, call emergency medical care. If not breathing give CPR. If breathing is difficult administer oxygen. Treat for shock.

Fire Hazards:

Flashpoint: 175° F(79.4° C) *Ignition temperature:* n/a

Flammable Explosive High Range: n/a *Low Range:* n/a

Toxic Products of Combustion: Irritating hydrogen chloride vapors may be formed in fires.

Other Hazards: n/a

Possible extinguishing agents: Foam, dry chemical, carbon dioxide. Water may be ineffective on fires.

Reactivity Hazards:

Reactive With: n/a *Other Reactions:* n/a

Corrosivity Hazards:

Corrosive With: May attack some forms of plastic. *Neutralizing Agent:* n/a

Radioactivity Hazards:

Radiation Emitted: n/a *Other Hazards:* n/a

Recommended Protection for Response Personnel:

Avoid breathing vapors, keep upwind. Wear appropriate sealed chemical clothing. Wash away any material which may have come into contact with the body with copious amounts of soap and water. Consider appropriate evacuation.

DALAPON

DOT Number: UN 1760 *DOT Hazard Class:* Corrosive *DOT Guide Number:* 60
Synonyms: 2,2-dichloropropanic acid, 2,2-dichloropropanoic acid
STCC Number: n/a *Reportable Qty:* n/a
Mfg Name: Dow Chemical *Phone No:* 1-517-636-4100

Physical Description:

Physical Form: Liquid *Color:* Colorless *Odor:* Acrid
Other Information: n/a

Chemical Properties:

Specific Gravity: 1.39 *Vapor Density:* n/a *Boiling Point: 374° F(190° C)*
Melting Point: 46° F7.7° C) *Vapor Pressure:* n/a *Solubility in water:* Yes
Other Information: Sinks and mixes with water.

Health Hazards:

Inhalation Hazard: n/a
Ingestion Hazard: Harmful if swallowed.
Absorption Hazard: Will burn the skin and the eyes.
Hazards to Wildlife: Dangerous to aquatic life.
Decontamination Procedures: Wash away any material with copious amounts of soap and water.
First Aid Procedures: Remove victim to fresh air, call emergency medical care. If not breathing give CPR. If breathing is difficult administer oxygen. Treat for shock.

Fire Hazards:

Flashpoint: n/a *Ignition temperature:* n/a
Flammable Explosive High Range: n/a *Low Range:* n/a
Toxic Products of Combustion: Irritating fumes of hydrochloric acid may be formed in fires.
Other Hazards: n/a
Possible extinguishing agents: Use alcohol foam, dry chemical or carbon dioxide. Water may be ineffective on fires.

Reactivity Hazards:

Reactive With: Water to form hydrochloric and pyruvic acids. The reaction is not hazardous.
Other Reactions: n/a

Corrosivity Hazards:

Corrosive With: Very corrosive to aluminum and copper alloys. Flammable and corrosive hydrogen gas may be formed in spaces. *Neutralizing Agent:* Crushed limestone, soda ash, or lime.

Radioactivity Hazards:

Radiation Emitted: n/a *Other Hazards:* n/a

Recommended Protection for Response Personnel:

Avoid breathing vapors, keep upwind. Structural protective clothing provides limited protection. Wash away any material which may have come into contact with the body with copious amounts of soap and water. Consider appropriate evacuation.

DDD

DOT Number: UN 2761 *DOT Hazard Class:* n/a *DOT Guide Number:* 55
Synonyms: 1,1-dichloro-2,2-bis(p-chlorophenyl) ethane, dichlorodiphenyldichloroethane, ENT 4225, rhothane, TDE
STCC Number: n/a *Reportable Qty:* n/a
Mfg Name: Aldrich Chemical *Phone No:* 1-414-273-3850

Physical Description:

Physical Form: Solid *Color:* White *Odor:* n/a
Other Information: n/a

Chemical Properties:

Specific Gravity: 1.476 *Vapor Density:* 11 *Boiling Point: Decomposes*
Melting Point: 234° F(112.2° C) *Vapor Pressure:* n/a *Solubility in water:* Yes
Other Information: Sinks in water.

Health Hazards:

Inhalation Hazard: n/a
Ingestion Hazard: Harmful if swallowed.
Absorption Hazard: Irritating to the skin and eyes.
Hazards to Wildlife: Dangerous to aquatic life.
Decontamination Procedures: Wash away any material with copious amounts of soap and water.
First Aid Procedures: Remove victim to fresh air, call emergency medical care. If not breathing give CPR. If breathing is difficult administer oxygen. Treat for shock.

Fire Hazards:

Flashpoint: n/a *Ignition temperature:* n/a
Flammable Explosive High Range: n/a *Low Range:* n/a
Toxic Products of Combustion: Irritating fumes of hydrogen chloride may be formed in fires.
Other Hazards: n/a
Possible extinguishing agents: Extinguish fire using suitable agent for the type of surrounding fire.

Reactivity Hazards:

Reactive With: n/a *Other Reactions:* n/a

Corrosivity Hazards:

Corrosive With: n/a *Neutralizing Agent:* n/a

Radioactivity Hazards:

Radiation Emitted: n/a *Other Hazards:* n/a

Recommended Protection for Response Personnel:

Avoid breathing vapors, keep upwind. Structural protective clothing provides limited protection. Wash away any material which may have come into contact with the body with copious amounts of soap and water. Consider appropriate evacuation.

DDT

DOT Number: NA 2761 *DOT Hazard Class:* ORM-A *DOT Guide Number:* 55
Synonyms: dichlorodiphenyltrichloroethane
STCC Number: 4941129 *Reportable Qty:* 1/.454
Mfg Name: Olin Corp. *Phone No:* 1-203-356-2000

Physical Description:

Physical Form: Solid *Color:* White *Odor:* Odorless
Other Information: Used as a pesticide. Is a wettable powder or a water emulsified liquid.

Chemical Properties:

Specific Gravity: 1.56 *Vapor Density:* 12 *Boiling Point: n/a*
Melting Point: 228° F(108.8° C) *Vapor Pressure:* 0 mm Hg at 68° F(20° C)
Solubility in water: Sinks in water *Degree of Solubility:* .00001%
Other Information: In case of damage to or leaking from containers, contact the Pesticide Safety Team Network at 1-800-424-9300.

Health Hazards:

Inhalation Hazard: Will cause dizziness, nausea, and loss of consciousness
Ingestion Hazard: Harmful if swallowed.
Absorption Hazard: Irritating to the skin and eyes.
Hazards to Wildlife: Dangerous to aquatic life and waterfowl.
Decontamination Procedures: Wash away any material with copious amounts of soap and water.
First Aid Procedures: Remove victim to fresh air, call emergency medical care. If not breathing give CPR. If breathing is difficult administer oxygen. Treat for shock.

Fire Hazards:

Flashpoint: 162° F(72.2° C) *Ignition temperature:* n/a
Flammable Explosive High Range: n/a *Low Range:* n/a
Toxic Products of Combustion: Poisonous gases are produced in fire.
Other Hazards: n/a
Possible extinguishing agents: Extinguish fire using suitable agent for the type of surrounding fire.

Reactivity Hazards:

Reactive With: n/a *Other Reactions:* n/a

Corrosivity Hazards:

Corrosive With: n/a *Neutralizing Agent:* n/a

Radioactivity Hazards:

Radiation Emitted: n/a *Other Hazards:* n/a

Recommended Protection for Response Personnel:

Avoid breathing vapors, keep upwind. Structural protective clothing provides limited protection. Wash away any material which may have come into contact with the body with copious amounts of soap and water. If the fire becomes uncontrollable, consider appropriate evacuation.

DECABORANE

DOT Number: UN 1868 *DOT Hazard Class:* Flammable solid *DOT Guide Number:* 34
Synonyms: none given
STCC Number: 4916610 *Reportable Qty:* n/a
Mfg Name: Callery Chemical Co. *Phone No:* 1-412-538-3510

Physical Description:

Physical Form: Solid *Color:* White *Odor:* Persistent, disagreeable
Other Information: Readily ignited. Once burning it emits a green flame. Forms shock sensitive compounds when dissolved in oxygenated or halogenated solvents.

Chemical Properties:

Specific Gravity: .94 *Vapor Density:* 4.2 *Boiling Point: 416° F(213.3° C)*
Melting Point: 212° F(100° C) *Vapor Pressure:* .05 mm Hg at 77° F(25° C) *Solubility in water:* Insoluble
Other Information: It is slightly soluble in cold water. Slowly hydrolyzed by hot water to give off a flammable gas.

Health Hazards:

Inhalation Hazard: Poisonous if inhaled.
Ingestion Hazard: Poisonous if swallowed.
Absorption Hazard: Poisonous if the skin is exposed.
Hazards to Wildlife: n/a
Decontamination Procedures: Wash away any material with copious amounts of soap and water.
First Aid Procedures: Remove victim to fresh air, call emergency medical care. If not breathing give CPR. If breathing is difficult administer oxygen. Treat for shock.

Fire Hazards:

Flashpoint: 176° F(80° C) *Ignition temperature:* 300° F(148° C)
Flammable Explosive High Range: n/a *Low Range:* n/a
Toxic Products of Combustion: Poisonous gases are produced in fire.
Other Hazards: It may spontaneously ignite upon exposure to air.
Possible extinguishing agents: Use water in flooding quantities as a fog. Use dry chemical or carbon dioxide.

Reactivity Hazards:

Reactive With: Water, natural, synthetic rubber, greases and other lubricants. *Other Reactions:* n/a

Corrosivity Hazards:

Corrosive With: n/a *Neutralizing Agent:* n/a

Radioactivity Hazards:

Radiation Emitted: n/a *Other Hazards:* n/a

Recommended Protection for Response Personnel:

Avoid breathing vapors, keep upwind. Wear a sealed chemical suit (polycarbonate, butyl rubber). Wash away any material which may have come into contact with the body with copious amounts of soap and water. Consider appropriate evacuation.

DECAHYDRONAPHTHALENE

DOT Number: UN 1147 *DOT Hazard Class:* Combustible liquid *DOT Guide Number:* 27
Synonyms: bicyclo(4.4.0)decane, de-kalin, naphthane, napthalane, perhydronapthalene,
STCC Number: 4913167 *Reportable Qty:* n/a
Mfg Name: E.I. Du Pont *Phone No:* 1-800-441-3637

Physical Description:

Physical Form: Liquid *Color:* Colorless *Odor:* Turpentine
Other Information: n/a

Chemical Properties:

Specific Gravity: .89 *Vapor Density:* 4.77 *Boiling Point: 383° F(195° C)*
Melting Point: −44° F(−42.2° C) *Vapor Pressure:* n/a *Solubility in water:* No
Other Information: It is lighter than and insoluble in water. Vapors are heavier than air.

Health Hazards:

Inhalation Hazard: n/a
Ingestion Hazard: Headache, nausea or vomiting.
Absorption Hazard: Irritating to the skin and eyes.
Hazards to Wildlife: n/a
Decontamination Procedures: Wash away any material with copious amounts of soap and water.
First Aid Procedures: Remove victim to fresh air, call emergency medical care. If not breathing give CPR. If breathing is difficult administer oxygen. Treat for shock.

Fire Hazards:

Flashpoint: 134° F *Ignition temperature:* 482° F
Flammable Explosive High Range: 5.4 *Low Range:* .7
Toxic Products of Combustion: n/a
Other Hazards: n/a
Possible extinguishing agents: Do not extinguish the fire unless the flow can be stopped. Apply water from as far a distance as possible. Use foam, dry chemical, or carbon dioxide.

Reactivity Hazards:

Reactive With: n/a *Other Reactions:* n/a

Corrosivity Hazards:

Corrosive With: n/a *Neutralizing Agent:* n/a

Radioactivity Hazards:

Radiation Emitted: n/a *Other Hazards:* n/a

Recommended Protection for Response Personnel:

Avoid breathing vapors, keep upwind. Structural protective clothing provides limited protection. Wash away any material which may have come into contact with the body with copious amounts of soap and water. Consider appropriate evacuation.

DECYL ALCOHOL

DOT Number: UN 1987 *DOT Hazard Class:* Combustible liquid *DOT Guide Number:* 26
Synonyms: alcohol C-10, capric alcohol, Dytol S-91, 1-decanol, Lorol-22, nonylcarbinol
STCC Number: 4913121 *Reportable Qty:* n/a
Mfg Name: Proctor and Gamble *Phone No:* 1-513-562-1100

Physical Description:

Physical Form: Liquid *Color:* Colorless to light yellow *Odor:* Faint alcohol
Other Information: n/a

Chemical Properties:

Specific Gravity: .84 *Vapor Density:* 5.3 *Boiling Point: 446° F(230° C)*
Melting Point: 44° F(6.6° C) *Vapor Pressure:* 1 mm Hg at 157° F(69.4° C) *Solubility in water:* No
Other Information: It is lighter than and insoluble in water. Vapors are heavier than air.

Health Hazards:

Inhalation Hazard: n/a
Ingestion Hazard: n/a
Absorption Hazard: Irritating to the eyes.
Hazards to Wildlife: n/a
Decontamination Procedures: Wash away any material with copious amounts of soap and water.
First Aid Procedures: Remove victim to fresh air, call emergency medical care. If not breathing give CPR. If breathing is difficult administer oxygen. Treat for shock.

Fire Hazards:

Flashpoint: 180° F(82.2° C) *Ignition temperature:* 550° F(287.7° C)
Flammable Explosive High Range: n/a *Low Range:* n/a
Toxic Products of Combustion: n/a
Other Hazards: n/a
Possible extinguishing agents: Do not extinguish the fire unless the flow can be stopped. Apply water from as far a distance as possible. Use foam, dry chemical, or carbon dioxide.

Reactivity Hazards:

Reactive With: n/a *Other Reactions:* n/a

Corrosivity Hazards:

Corrosive With: n/a *Neutralizing Agent:* n/a

Radioactivity Hazards:

Radiation Emitted: n/a *Other Hazards:* n/a

Recommended Protection for Response Personnel:

Avoid breathing vapors, keep upwind. Structural protective clothing provides limited protection. Wash away any material which may have come into contact with the body with copious amounts of soap and water. Consider appropriate evacuation.

DI-(2-ETHYLHEXYL) PHOSPHORIC ACID

DOT Number: UN 1902 *DOT Hazard Class:* N/A *DOT Guide Number:* 60
Synonyms: bis(2-ethylhexyl)hydrogen phosphate, DEHPA, di-(2-ethylhexyl)phosphate, 2-ethyl-1-hexanol hydrogen phosphate
STCC Number: n/a *Reportable Qty:* n/a
Mfg Name: Union Carbide *Phone No:* 1-203-794-2000

Physical Description:

Physical Form: Liquid *Color:* Light yellow *Odor:* Odorless
Other Information: n/a

Chemical Properties:

Specific Gravity: .97 *Vapor Density:* 11 *Boiling Point: Decomposes*
Melting Point: −76° F(−60° C) *Vapor Pressure:* n/a *Solubility in water:* n/a
Other Information: Floats on water

Health Hazards:

Inhalation Hazard: n/a
Ingestion Hazard: Harmful if swallowed.
Absorption Hazard: Irritating to the skin and eyes.
Hazards to Wildlife: n/a
Decontamination Procedures: Wash away any material with copious amounts of soap and water.
First Aid Procedures: Remove victim to fresh air, call emergency medical care. If not breathing give CPR. If breathing is difficult administer oxygen. Treat for shock.

Fire Hazards:

Flashpoint: 385° F(196.1° C) *Ignition temperature:* n/a
Flammable Explosive High Range: n/a *Low Range:* n/a
Toxic Products of Combustion: Irritating phosphorus oxides may be released.
Other Hazards: n/a
Possible extinguishing agents: Alcohol foam, dry chemical or carbon dioxide. Water or foam may cause frothing.

Reactivity Hazards:

Reactive With: n/a *Other Reactions:* n/a

Corrosivity Hazards:

Corrosive With: Mildly corrosive to most metals. May form flammable hydrogen gas.
Neutralizing Agent: Crushed limestone, soda ash, or lime.

Radioactivity Hazards:

Radiation Emitted: n/a *Other Hazards:* n/a

Recommended Protection for Response Personnel:

Avoid breathing vapors, keep upwind. Wear a sealed chemical suit (polycarbonate, butyl rubber). Wash away any material which may have come into contact with the body with copious amounts of soap and water. If the fire becomes uncontrollable, consider appropriate evacuation.

DI-BUTYL ETHER

DOT Number: UN 1149 *DOT Hazard Class:* Flammable liquid *DOT Guide Number:* 26
Synonyms: 1-butoxy butane, butyl ether, n-butyl ether, dibutyl ether, n-dibutyl ether, dibutyl oxide
STCC Number: n/a *Reportable Qty:* n/a
Mfg Name: Union Carbide *Phone No:* 1-203-794-2000

Physical Description:

Physical Form: Liquid *Color:* Colorless *Odor:* Mild pleasant
Other Information: n/a

Chemical Properties:

Specific Gravity: .76 *Vapor Density:* 4.5 *Boiling Point: 288° F(142.2° C)*
Melting Point: 139° F(59.4° C) *Vapor Pressure:* n/a *Solubility in water:* n/a
Other Information: Floats on water. Flammable, irritating vapor is produced.

Health Hazards:

Inhalation Hazard: n/a
Ingestion Hazard: n/a
Absorption Hazard: Irritating to the skin and eyes.
Hazards to Wildlife: n/a
Decontamination Procedures: Wash away any material with copious amounts of soap and water.
First Aid Procedures: Remove victim to fresh air, call emergency medical care. If not breathing give CPR. If breathing is difficult administer oxygen. Treat for shock.

Fire Hazards:

Flashpoint: 92° F(33.3° C) *Ignition temperature:* 382° F(194.4° C)
Flammable Explosive High Range: 7.6 *Low Range:* 1.5
Toxic Products of Combustion: n/a
Other Hazards: Containers may explode in fire. Flashback along vapor trail may occur. Vapor may explode if ignited in an enclosed area.
Possible extinguishing agents: Water may be ineffective on fires. Use alcohol foam, dry chemical, carbon dioxide. Cool all affected containers with water.

Reactivity Hazards:

Reactive With: n/a Other reaction: n/a

Corrosivity Hazards:

Corrosive With: n/a *Neutralizing Agent:* n/a

Radioactivity Hazards:

Radiation Emitted: n/a *Other Hazards:* n/a

Recommended Protection for Response Personnel:

Avoid breathing vapors, keep upwind. Structural protective clothing provides limited protection. Wash away any material which may have come into contact with the body with copious amounts of soap and water. Consider appropriate evacuation.

DIACETONE ALCOHOL

DOT Number: UN 1148 *DOT Hazard Class:* Flammable liquid *DOT Guide Number:* 26
Synonyms: 4-hydroxy-4-methyl-2-pentanone
STCC Number: 4915325 *Reportable Qty:* n/a
Mfg Name: Shell Chemical *Phone No:* 1-713-241-6161

Physical Description:

Physical Form: Liquid *Color:* Colorless to light yellow *Odor:* Mild, pleasant
Other Information: Used as a solvent, in antifreeze, and hydraulic fluid.

Chemical Properties:

Specific Gravity: .9 *Vapor Density:* 4 *Boiling Point: 228° F(108.8° C)*
Melting Point: −45° F(−42.7° C) *Vapor Pressure:* .8 mm Hg at 68° F(20° C) *Solubility in water:* Yes
Other Information: It may emit acrid smoke and fumes when on fire.

Health Hazards:

Inhalation Hazard: Irritating to mucous membranes.
Ingestion Hazard: Harmful if swallowed
Absorption Hazard: Irritating to the skin and eyes.
Hazards to Wildlife: n/a
Decontamination Procedures: Wash away any material with copious amounts of soap and water.
First Aid Procedures: Remove victim to fresh air, call emergency medical care. If not breathing give CPR. If breathing is difficult administer oxygen. Treat for shock.

Fire Hazards:

Flashpoint: 148° F(64.4° C) *Ignition temperature:* 1118° F(603.3° C)
Flammable Explosive High Range: 6.9 *Low Range:* 1.8
Toxic Products of Combustion: n/a
Other Hazards: Flashback along vapor trail may occur. Vapors may ignite in an enclosed area.
Possible extinguishing agents: Use alcohol foam, dry chemical, or carbon dioxide.

Reactivity Hazards:

Reactive With: n/a *Other Reactions:* n/a

Corrosivity Hazards:

Corrosive With: n/a *Neutralizing Agent:* n/a

Radioactivity Hazards:

Radiation Emitted: n/a *Other Hazards:* n/a

Recommended Protection for Response Personnel:

Avoid breathing vapors, keep upwind. Structural protective clothing provides limited protection. Wash away any material which may have come into contact with the body with copious amounts of soap and water. If the fire becomes uncontrollable, consider appropriate evacuation.

DIAZINON

DOT Number: NA 2783 *DOT Hazard Class:* ORM-A *DOT Guide Number:* 55
Synonyms: Alfa-Tox, Saralex, Spectracide
STCC Number: 4941140 *Reportable Qty:* 1/.454
Mfg Name: Ciba-Geigy Agricultural *Phone No:* 1-914-478-3131

Physical Description:

Physical Form: Liquid *Color:* Light to dark brown *Odor:* Odorless
Other Information: It is used as a pesticide.

Chemical Properties:

Specific Gravity: 1.117 *Vapor Density:* 10 *Boiling Point: Decomposes*
Melting Point: n/a *Vapor Pressure:* n/a *Solubility in water:* Yes
Other Information: In case of damage to, or leaking from the containers, contact the Pesticide Safety Network at 1-800-424-9300.

Health Hazards:

Inhalation Hazard: n/a
Ingestion Hazard: Poisonous if swallowed.
Absorption Hazard: Irritating to the skin and eyes.
Hazards to Wildlife: Dangerous to both aquatic life and waterfowl.
Decontamination Procedures: Wash away any material with copious amounts of soap and water.
First Aid Procedures: Remove victim to fresh air, call emergency medical care. If not breathing give CPR. If breathing is difficult administer oxygen. Treat for shock.

Fire Hazards:

Flashpoint: 82-105° F(27.7-40.5° C) *Ignition temperature:* n/a
Flammable Explosive High Range: *Low Range:*
Toxic Products of Combustion: Oxides of sulfur and of phosphorus are generated in a fire.
Other Hazards: n/a
Possible extinguishing agents: Extinguish fire using suitable agent for type of surrounding fire.

Reactivity Hazards:

Reactive With: n/a *Other Reactions:* n/a

Corrosivity Hazards:

Corrosive With: n/a *Neutralizing Agent:* n/a

Radioactivity Hazards:

Radiation Emitted: n/a *Other Hazards:* n/a

Recommended Protection for Response Personnel:

Avoid breathing vapors, keep upwind. Wear a sealed chemical suit (polycarbonate, butyl rubber). Wash away any material which may have come into contact with the body with copious amounts of soap and water. Consider appropriate evacuation.

DIBENZOYL PEROXIDE

DOT Number: UN 2085 *DOT Hazard Class:* Organic peroxide *DOT Guide Number:* 49
Synonyms: benzoyl peroxide, benzoyl superoxide, BPO, BP, lucidol oxalate
STCC Number: n/a *Reportable Qty:* n/a
Mfg Name: Pennwald Corp. *Phone No:* 1-215-587-7000

Physical Description:

Physical Form: Solid *Color:* White *Odor:* Odorless
Other Information: n/a

Chemical Properties:

Specific Gravity: 1.33 *Vapor Density:* 1 *Boiling Point: n/a*
Melting Point: 217° F(102.7° C) *Vapor Pressure:* n/a *Solubility in water:* n/a
Other Information: Can form explosive mixtures with finely divided combustibles.

Health Hazards:

Inhalation Hazard: Irritation to the eyes, nose, and throat.
Ingestion Hazard: Harmful if swallowed.
Absorption Hazard: Irritating to the skin and eyes.
Hazards to Wildlife: n/a
Decontamination Procedures: Wash away any material with copious amounts of soap and water.
First Aid Procedures: Remove victim to fresh air, call emergency medical care. If not breathing give CPR. If breathing is difficult administer oxygen. Treat for shock.

Fire Hazards:

Flashpoint: n/a (highly Flammable solid) *Ignition temperature:* n/a
Flammable Explosive High Range: n/a *Low Range:* n/a
Toxic Products of Combustion: Produces suffocating dense white smoke containing benzoic acid.
Other Hazards: May explode if subject to heat, shock or friction. May cause fire and explode on contact with combustibles.
Possible extinguishing agents: Apply water from as far a distance as possible. Use alcohol foam, dry chemical, or carbon dioxide.

Reactivity Hazards:

Reactive With: Ordinary combustible materials. *Other Reactions:* Easily oxidizable materials.

Corrosivity Hazards:

Corrosive With: n/a *Neutralizing Agent:* n/a

Radioactivity Hazards:

Radiation Emitted: n/a *Other Hazards:* n/a

Recommended Protection for Response Personnel:

Avoid breathing vapors, keep upwind. Structural protective clothing provides limited protection. Wash away any material which may have come into contact with the body with copious amounts of soap and water. Consider appropriate evacuation.

DIBORANE

DOT Number: UN 1911 *DOT Hazard Class:* Flammable gas *DOT Guide Number:* 18
Synonyms: boroethane
STCC Number: 4905425 *Reportable Qty:* n/a *CHEMTREC Phone No:* 1-800-424-9300

Physical Description:

Physical Form: Gas *Color:* Colorless *Odor:* Repulsive, sweet
Other Information: Used in electronics.

Chemical Properties:

Specific Gravity: n/a *Vapor Density:* 1 *Boiling Point: – 134° F(– 92.2° C)*
Melting Point: –265° F(–165° C) *Vapor Pressure:* 1 atm at 68° F(20° C)
Solubility in water: Reacts with water.
Other Information: Decomposed by water to give off hydrogen, another flammable gas, and boric acid.

Health Hazards:

Inhalation Hazard: Hazardous by inhalation.
Ingestion Hazard: Harmful if swallowed.
Absorption Hazard: Harmful to the skin and eyes.
Hazards to Wildlife: n/a
Decontamination Procedures: Wash away any material with copious amounts of soap and water.
First Aid Procedures: Remove victim to fresh air, call emergency medical care. If not breathing give CPR. If breathing is difficult administer oxygen. Treat for shock.

Fire Hazards:

Flashpoint: n/a *Ignition temperature:* 100° F(37.7° C)
Flammable Explosive High Range: 98 *Low Range:* .8
Toxic Products of Combustion: n/a
Other Hazards: n/a
Possible extinguishing agents: Do not extinguish fire unless flow can be stopped. Apply water from as far a distance as possible. Use water spray to knock down the vapors.

Reactivity Hazards:

Reactive With: Water *Other Reactions:* n/a

Corrosivity Hazards:

Corrosive With: n/a *Neutralizing Agent:* n/a

Radioactivity Hazards:

Radiation Emitted: n/a *Other Hazards:* n/a

Recommended Protection for Response Personnel:

Avoid breathing vapors, keep upwind. Wear appropriate chemical clothing. Wash away any material which may have come into contact with the body with copious amounts of soap and water. Consider appropriate evacuation.

DIBUTYL PHTHALATE

DOT Number: NA 9095 *DOT Hazard Class:* n/a *DOT Guide Number:* 31
Synonyms: butyl phthalate; DBP; phthalic acid, dibutyl ester; RC plasticizer DBP; Witicizer 300
STCC Number: n/a *Reportable Qty:* n/a
Mfg Name: Ashland Chemical *Phone No:* 1-614-889-3333

Physical Description:

Physical Form: Liquid *Color:* Colorless *Odor:* Odorless
Other Information: n/a

Chemical Properties:

Specific Gravity: 1.05 *Vapor Density:* 9.58 *Boiling Point: 635° F(335° C)*
Melting Point: −31 deg F (−35° C) *Vapor Pressure:* n/a *Solubility in water:* n/a
Other Information: Sinks slowly in water.

Health Hazards:

Inhalation Hazard: n/a
Ingestion Hazard: n/a
Absorption Hazard: n/a
Hazards to Wildlife: n/a
Decontamination Procedures: Wash away any material with copious amounts of soap and water.
First Aid Procedures: Remove victim to fresh air, call emergency medical care. If not breathing give CPR. If breathing is difficult administer oxygen. Treat for shock.

Fire Hazards:

Flashpoint: 315° F(157.2° C) *Ignition temperature:* 757° F(402.7° C)
Flammable Explosive High Range: 2.5 *Low Range:* .5
Toxic Products of Combustion: n/a
Other Hazards: Water and foam may cause frothing.
Possible extinguishing agents: Foam, dry chemical, carbon dioxide.

Reactivity Hazards:

Reactive With: n/a Other reaction: n/a

Corrosivity Hazards:

Corrosive With: n/a *Neutralizing Agent:* n/a

Radioactivity Hazards:

Radiation Emitted: n/a *Other Hazards:* n/a

Recommended Protection for Response Personnel:

Avoid breathing vapors, keep upwind. Wear a sealed chemical suit (viton, butyl rubber). Wash away any material which may have come into contact with the body with copious amounts of soap and water. Consider appropriate evacuation.

DICAMBA (liquid)

DOT Number: NA 2769 *DOT Hazard Class:* ORM-E *DOT Guide Number:* 55
Synonyms: Banvil D, 3,6-dichloro-o-anisic acid, Midiben
STCC Number: 4963334 *Reportable Qty:* 1000/454
Mfg Name: Veliscol Chemical *Phone No:* 1-312-670-4500

Physical Description:

Physical Form: Liquid *Color:* n/a *Odor:* Odorless
Other Information: It is used as a herbicide. It is also a water emulsified liquid.

Chemical Properties:

Specific Gravity: n/a *Vapor Density:* 7.6 *Boiling Point: n/a*
Melting Point: 240° F(115.5° C) *Vapor Pressure:* n/a *Solubility in water:* Yes
Other Information: In case of damage to or leaking from containers of this material, contact the Pesticide Safety Team Network at 1-800-424-9300.

Health Hazards:

Inhalation Hazard: Harmful if inhaled.
Ingestion Hazard: Harmful if swallowed.
Absorption Hazard: Harmful to the skin and eyes.
Hazards to Wildlife: Dangerous to aquatic life.
Decontamination Procedures: Wash away any material with copious amounts of soap and water.
First Aid Procedures: Remove victim to fresh air, call emergency medical care. If not breathing give CPR. If breathing is difficult administer oxygen. Treat for shock.

Fire Hazards:

Flashpoint: n/a *Ignition temperature:* n/a
Flammable Explosive High Range: n/a *Low Range:* n/a
Toxic Products of Combustion: n/a
Other Hazards: n/a
Possible extinguishing agents: Extinguish fire using suitable agent for the type of surrounding fire.

Reactivity Hazards:

Reactive With: n/a *Other Reactions:* n/a

Corrosivity Hazards:

Corrosive With: n/a *Neutralizing Agent:* n/a

Radioactivity Hazards:

Radiation Emitted: n/a *Other Hazards:* n/a

Recommended Protection for Response Personnel:

Avoid breathing vapors, keep upwind. Structural protective clothing provides limited protection. Wash away any material which may have come into contact with the body with copious amounts of soap and water. Consider appropriate evacuation.

DICAMBA (solid)

DOT Number: NA 2769 *DOT Hazard Class:* ORM-E *DOT Guide Number:* 55
Synonyms: Banvil D, 3,6-dichloro-o-anisic acid, Midiben
STCC Number: 4963337 *Reportable Qty:* 1000/454
Mfg Name: Veliscol Chemical *Phone No:* 1-312-670-4500

Physical Description:

Physical Form: Solid *Color:* White or brown *Odor:* Odorless
Other Information: It is used as a herbicide.

Chemical Properties:

Specific Gravity: n/a *Vapor Density:* 7.6 *Boiling Point: n/a*
Melting Point: 240° F(115.5° C) *Vapor Pressure:* n/a *Solubility in water:* Yes
Other Information: In case of damage to or leaking from containers of this material, contact the Pesticide Safety Team Network at 1-800-424-9300.

Health Hazards:

Inhalation Hazard: Harmful if inhaled.
Ingestion Hazard: Harmful if swallowed.
Absorption Hazard: Harmful to the skin and eyes.
Hazards to Wildlife: Dangerous to aquatic life.
Decontamination Procedures: Wash away any material with copious amounts of soap and water.
First Aid Procedures: Remove victim to fresh air, call emergency medical care. If not breathing give CPR. If breathing is difficult administer oxygen. Treat for shock.

Fire Hazards:

Flashpoint: n/a *Ignition temperature:* n/a
Flammable Explosive High Range: n/a *Low Range:* n/a
Toxic Products of Combustion: n/a
Other Hazards: n/a
Possible extinguishing agents: Extinguish fire using suitable agent for the type of surrounding fire.

Reactivity Hazards:

Reactive With: n/a *Other Reactions:* n/a

Corrosivity Hazards:

Corrosive With: n/a *Neutralizing Agent:* n/a

Radioactivity Hazards:

Radiation Emitted: n/a *Other Hazards:* n/a

Recommended Protection for Response Personnel:

Avoid breathing vapors, keep upwind. Structural protective clothing provides limited protection. Wash away any material which may have come into contact with the body with copious amounts of soap and water. Consider appropriate evacuation.

DICHLOBENIL (liquid)

DOT Number: NA 2769 *DOT Hazard Class:* ORM-E *DOT Guide Number:* 55
Synonyms: Casoron, 2,6-DBN, 2,6-dichlorobenzonitrile, Du-Sprex, Nia 5996
STCC Number: 4963809 *Reportable Qty:* 100/45.4
Mfg Name: Aldrich Chemical *Phone No:* 1-414-273-3850

Physical Description:

Physical Form: Liquid *Color:* n/a *Odor:* Aromatic
Other Information: It is used as a herbicide. It is a water emulsified liquid.

Chemical Properties:

Specific Gravity: n/a *Vapor Density:* 5.9 *Boiling Point: 518° F(270° C)*
Melting Point: 294° F(145.5° C) *Vapor Pressure:* n/a *Solubility in water:* Yes
Other Information: In case of damage to or leaking from containers of this material, contact the Pesticide Safety Team Network at 1-800-424-9300.

Health Hazards:

Inhalation Hazard: Harmful if inhaled.
Ingestion Hazard: Harmful if swallowed.
Absorption Hazard: Harmful to the skin and eyes.
Hazards to Wildlife: Dangerous to aquatic life.
Decontamination Procedures: Wash away any material with copious amounts of soap and water.
First Aid Procedures: Remove victim to fresh air, call emergency medical care. If not breathing give CPR. If breathing is difficult administer oxygen. Treat for shock.

Fire Hazards:

Flashpoint: n/a *Ignition temperature:* n/a
Flammable Explosive High Range: n/a *Low Range:* n/a
Toxic Products of Combustion: n/a
Other Hazards: n/a
Possible extinguishing agents: Extinguish fire using suitable agent for the type of surrounding fire.

Reactivity Hazards:

Reactive With: n/a *Other Reactions:* n/a

Corrosivity Hazards:

Corrosive With: n/a *Neutralizing Agent:* n/a

Radioactivity Hazards:

Radiation Emitted: n/a *Other Hazards:* n/a

Recommended Protection for Response Personnel:

Avoid breathing vapors, keep upwind. Structural protective clothing provides limited protection. Wash away any material which may have come into contact with the body with copious amounts of soap and water. Consider appropriate evacuation.

DICHLOBENIL (solid)

DOT Number: NA 2769 *DOT Hazard Class:* ORM-E *DOT Guide Number:* 55
Synonyms: Casoron, 2,6-DBN, 2,6-dichlorobenzonitrile, Du-Sprex, Nia 5996
STCC Number: 4963814 *Reportable Qty:* 100/45.4
Mfg Name: Aldrich Chemical *Phone No:* 1-414-273-3850

Physical Description:

Physical Form: Solid *Color:* n/a *Odor:* Aromatic
Other Information: It is used as a herbicide.

Chemical Properties:

Specific Gravity: n/a *Vapor Density:* 5.9 *Boiling Point: 518° F(270° C)*
Melting Point: 294° F(145.5° C) *Vapor Pressure:* n/a *Solubility in water:* Yes
Other Information: In case of damage to or leaking from containers of this material, contact the Pesticide Safety Team Network at 1-800-424-9300.

Health Hazards:

Inhalation Hazard: Harmful if inhaled.
Ingestion Hazard: Harmful if swallowed.
Absorption Hazard: Harmful to the skin and eyes.
Hazards to Wildlife: Dangerous to aquatic life.
Decontamination Procedures: Wash away any material with copious amounts of soap and water.
First Aid Procedures: Remove victim to fresh air, call emergency medical care. If not breathing give CPR. If breathing is difficult administer oxygen. Treat for shock.

Fire Hazards:

Flashpoint: n/a *Ignition temperature:* n/a
Flammable Explosive High Range: n/a *Low Range:* n/a
Toxic Products of Combustion: n/a
Other Hazards: n/a
Possible extinguishing agents: Extinguish fire using suitable agent for the type of surrounding fire.

Reactivity Hazards:

Reactive With: n/a *Other Reactions:* n/a

Corrosivity Hazards:

Corrosive With: n/a *Neutralizing Agent:* n/a

Radioactivity Hazards:

Radiation Emitted: n/a *Other Hazards:* n/a

Recommended Protection for Response Personnel:

Avoid breathing vapors, keep upwind. Structural protective clothing provides limited protection. Wash away any material which may have come into contact with the body with copious amounts of soap and water. Consider appropriate evacuation.

DICHLONE (liquid)

DOT Number: NA 2761 *DOT Hazard Class:* ORM-E *DOT Guide Number:* 55
Synonyms: 2,3-dichloro-1,4-naphtho-quinone, Phygon, Phygon-XL
STCC Number: 4960617 *Reportable Qty:* 1/.454 *CHEMTREC Phone No:* 1-800-424-9300

Physical Description:

Physical Form: Liquid *Color:* n/a *Odor:* n/a
Other Information: It is used as a fungicide. It is a water emulsified liquid.

Chemical Properties:

Specific Gravity: n/a *Vapor Density:* 7.8 *Boiling Point: 527° F(275° C)*
Melting Point: 379° F(192.7° C) *Vapor Pressure:* n/a *Solubility in water:* Yes
Other Information: In case of damage to or leaking from containers of this material, contact the Pesticide Safety Team Network at 1-800-424-9300.

Health Hazards:

Inhalation Hazard: Harmful if inhaled.
Ingestion Hazard: Harmful if swallowed.
Absorption Hazard: Harmful to the skin and eyes.
Hazards to Wildlife: Dangerous to aquatic life.
Decontamination Procedures: Wash away any material with copious amounts of soap and water.
First Aid Procedures: Remove victim to fresh air, call emergency medical care. If not breathing give CPR. If breathing is difficult administer oxygen. Treat for shock.

Fire Hazards:

Flashpoint: n/a *Ignition temperature:* n/a
Flammable Explosive High Range: n/a *Low Range:* n/a
Toxic Products of Combustion: Highly toxic fumes are Emitted.
Other Hazards: n/a
Possible extinguishing agents: Extinguish fire using suitable agent for the type of surrounding fire.

Reactivity Hazards:

Reactive With: n/a *Other Reactions:* n/a

Corrosivity Hazards:

Corrosive With: n/a *Neutralizing Agent:* n/a

Radioactivity Hazards:

Radiation Emitted: n/a *Other Hazards:* n/a

Recommended Protection for Response Personnel:

Avoid breathing vapors, keep upwind. Structural protective clothing provides limited protection. Wash away any material which may have come into contact with the body with copious amounts of soap and water. Consider appropriate evacuation.

DICHLONE (solid)

DOT Number: NA 2761 *DOT Hazard Class:* ORM-E *DOT Guide Number:* 55
Synonyms: 2,3-dichloro-1,4-naphtho-quinone, Phygon, Phygon-XL
STCC Number: 4960616 *Reportable Qty:* 1/.454 *CHEMTREC Phone No:* 1-800-424-9300

Physical Description:

Physical Form: Solid *Color:* Yellow *Odor:* n/a
Other Information: It is used as a fungicide. Sinks and mixes with water.

Chemical Properties:

Specific Gravity: n/a *Vapor Density:* 7.8 *Boiling Point: 527° F(275° C)*
Melting Point: 379° F(192.7° C) *Vapor Pressure:* n/a *Solubility in water:* Yes
Other Information: In case of damage to or leaking from containers of this material, contact the Pesticide Safety Team Network at 1-800-424-9300.

Health Hazards:

Inhalation Hazard: Harmful if inhaled.
Ingestion Hazard: Harmful if swallowed.
Absorption Hazard: Harmful to the skin and eyes.
Hazards to Wildlife: Dangerous to aquatic life.
Decontamination Procedures: Wash away any material with copious amounts of soap and water.
First Aid Procedures: Remove victim to fresh air, call emergency medical care. If not breathing give CPR. If breathing is difficult administer oxygen. Treat for shock.

Fire Hazards:

Flashpoint: n/a *Ignition temperature:* n/a
Flammable Explosive High Range: n/a *Low Range:* n/a
Toxic Products of Combustion: Highly toxic fumes are emitted.
Other Hazards: n/a
Possible extinguishing agents: Extinguish fire using suitable agent for the type of surrounding fire.

Reactivity Hazards:

Reactive With: n/a *Other Reactions:* n/a

Corrosivity Hazards:

Corrosive With: n/a *Neutralizing Agent:* n/a

Radioactivity Hazards:

Radiation Emitted: n/a *Other Hazards:* n/a

Recommended Protection for Response Personnel:

Avoid breathing vapors, keep upwind. Structural protective clothing provides limited protection. Wash away any material which may have come into contact with the body with copious amounts of soap and water. Consider appropriate evacuation.

1,2-DICHLOROETHYLENE

DOT Number: UN 1150 *DOT Hazard Class:* Flammable liquid *DOT Guide Number:* 29
Synonyms: cis-acetylene, Dioform
STCC Number: 4909145 *Reportable Qty:* 5000/2270
Mfg Name: Eastman Organic *Phone No:* 1-800-445-6325

Physical Description:

Physical Form: Liquid *Color:* Colorless *Odor:* Sweet, pleasant
Other Information: Used in making perfumes. It is heavier and insoluble in water.

Chemical Properties:

Specific Gravity: 1.25 *Vapor Density:* 3.42 *Boiling Point: 113-140° F(45-60° C)*
Melting Point: −56 to −115 ° F (−48/−81 C) *Vapor Pressure:* 180 to 265 mm Hg at 68° F(20° C)
Solubility in water: n/a *Degree of Solubility:* .35% to .63%
Other Information: n/a

Health Hazards:

Inhalation Hazard: Will cause dizziness, nausea, difficulty in breathing.
Ingestion Hazard: Harmful if swallowed.
Absorption Hazard: Irritating to the skin and eyes.
Hazards to Wildlife: n/a
Decontamination Procedures: Wash away any material with copious amounts of soap and water.
First Aid Procedures: Remove victim to fresh air, call emergency medical care. If not breathing give CPR. If breathing is difficult administer oxygen. Treat for shock.

Fire Hazards:

Flashpoint: 36-39° F(2.2 to 3.8° C) *Ignition temperature:* 860° F
Flammable Explosive High Range: 12.8 *Low Range:* 9.7
Toxic Products of Combustion: Phosgene and nitrogen chloride fumes may be formed in fires.
Other Hazards: Containers may explode in fire. Flashback along vapor trail may occur. (BLEVE)
Possible extinguishing agents: Water may be ineffective. Use dry chemical, foam, and carbon dioxide.

Reactivity Hazards:

Reactive With: Water *Other Reactions:* Reacts with heat

Corrosivity Hazards:

Corrosive With: n/a *Neutralizing Agent:* n/a

Radioactivity Hazards:

Radiation Emitted: n/a *Other Hazards:* n/a

Recommended Protection for Response Personnel:

Avoid breathing vapors, keep upwind. Wear appropriate chemical clothing. Wash away any material which may have come into contact with the body with copious amounts of soap and water. If the fire becomes uncontrollable, consider appropriate evacuation (BLEVE)

o-DICHLOROBENZENE

DOT Number: UN 1591 *DOT Hazard Class:* ORM-A *DOT Guide Number:* 58
Synonyms: ortho-dichlorbenzene
STCC Number: 4941127 *Reportable Qty:* 100/45.4
Mfg Name: Monsanto Chemical *Phone No:* 1-314-694-1000

Physical Description:

Physical Form: Liquid *Color:* Colorless *Odor:* Pleasant
Other Information: Used to make other chemicals, solvent, fumigant, pesticides, and many other uses.

Chemical Properties:

Specific Gravity: 1.3 *Vapor Density:* 5.1 *Boiling Point: 356° F(180° C)*
Melting Point: .5° F(−17.5° C) *Vapor Pressure:* 1.2 mm Hg at 68° F(20° C)
Solubility in water: No *Degree of Solubility:* .008%
Other Information: It may burn if heated to high temperatures.

Health Hazards:

Inhalation Hazard: Damage to lungs, liver and kidneys.
Ingestion Hazard: Poisonous if swallowed. (No known antidote)!!
Absorption Hazard: Irritating to the skin, eyes, and mucous membranes.
Hazards to Wildlife: Dangerous to aquatic life.
Decontamination Procedures: Wash away any material with copious amounts of soap and water.
First Aid Procedures: Remove victim to fresh air, call emergency medical care. If not breathing give CPR. If breathing is difficult administer oxygen. Induce vomiting if swallowed. Treat for shock.

Fire Hazards:

Flashpoint: 151° F(66.1° C) *Ignition temperature:* 1198° F(647.7° C)
Flammable Explosive High Range: 5.1 *Low Range:* 1.3
Toxic Products of Combustion: Irritating vapors including hydrogen chloride gas, chlorocarbons, and chlorine gas.
Other Hazards: n/a
Possible extinguishing agents: Extinguish fire using suitable agent for type of surrounding fire. Use water in flooding quantities as fog, use foam, dry chemical or carbon dioxide.

Reactivity Hazards:

Reactive With: n/a *Other Reactions:* n/a

Corrosivity Hazards:

Corrosive With: n/a *Neutralizing Agent:* n/a

Radioactivity Hazards:

Radiation Emitted: n/a *Other Hazards:* n/a

Recommended Protection for Response Personnel:

Avoid breathing vapors or dust,keep upwind. Wear a sealed chemical suit (viton). Wash away any material which may have come into contact with the body with copious amounts of soap and water. Consider appropriate evacuation.

p-DICHLOROBENZENE

DOT Number: UN 1592 *DOT Hazard Class:* ORM-A *DOT Guide Number:* 58
Synonyms: Paradi, Paradow, Parimoth
STCC Number: 4941128 *Reportable Qty:* 100/45.4
Mfg Name: Dow Chemical *Phone No:* 1-517-636-4400

Physical Description:

Physical Form: Solid crystals *Color:* Clear to white *Odor:* Mothball
Other Information: Used to make moth repellant, to make other chemicals, as a fumigant, and many other uses.

Chemical Properties:

Specific Gravity: 1.5 *Vapor Density:* 5.1 *Boiling Point: 345° F(173.8° C)*
Melting Point: 127° F(52.7° C) *Vapor Pressure:* .4 mm Hg at 68° F(20° C)
Solubility in water: No *Degree of Solubility:* .015%
Other Information: It is heavier than and insoluble in water. Material may burn, but is difficult to ignite.

Health Hazards:

Inhalation Hazard: Irritating to the upper respiratory tract.
Ingestion Hazard: Harmful if swallowed.
Absorption Hazard: Irritating to the skin and eyes.
Hazards to Wildlife: Dangerous to aquatic life.
Decontamination Procedures: Wash away any material with copious amounts of soap and water.
First Aid Procedures: Remove victim to fresh air, call emergency medical care. If not breathing give CPR. If breathing is difficult administer oxygen. Treat for shock.

Fire Hazards:

Flashpoint: 150° F(65.5° C) *Ignition temperature:* n/a
Flammable Explosive High Range: n/a *Low Range:* 2.5
Toxic Products of Combustion: Toxic chlorine, hydrogen chloride, and phosgene gases.
Other Hazards: n/a
Possible extinguishing agents: Extinguish fire using suitable agent for type of surrounding fire. Use water in flooding quantities as fog. Use foam, dry chemical or carbon dioxide.

Reactivity Hazards:

Reactive With: n/a *Other Reactions:* n/a

Corrosivity Hazards:

Corrosive With: n/a *Neutralizing Agent:* n/a

Radioactivity Hazards:

Radiation Emitted: n/a *Other Hazards:* n/a

Recommended Protection for Response Personnel:

Avoid breathing vapors or dust, keep upwind. Structural protective clothing provides limited protection. Wash away any material which may have come into contact with the body with copious amounts of soap and water. Consider appropriate evacuation.

DICHLOROBENZOYL PEROXIDE-2,4

DOT Number: UN 2138 *DOT Hazard Class:* Organic peroxide *DOT Guide Number:* 48
Synonyms: Cadox-PS
STCC Number: 4919243 *Reportable Qty:* n/a *CHEMTREC Phone No:* 1-800-424-9300

Physical Description:

Physical Form: Solid or paste *Color:* White *Odor:* Odorless
Other Information: It is used to manufacture paints, plastics, and rubber.

Chemical Properties:

Specific Gravity: 1.1 *Vapor Density:* 1 *Boiling Point: n/a*
Melting Point: n/a *Vapor Pressure:* n/a *Solubility in water:* No
Other Information: Insoluble in water.

Health Hazards:

Inhalation Hazard: Harmful if inhaled.

Ingestion Hazard: Harmful if swallowed.

Absorption Hazard: Irritating to the skin and eyes.

Hazards to Wildlife: n/a

Decontamination Procedures: Wash away any material with copious amounts of soap and water.

First Aid Procedures: Remove victim to fresh air, call emergency medical care. If not breathing give CPR. If breathing is difficult administer oxygen. Treat for shock.

Fire Hazards:

Flashpoint: n/a *Ignition temperature:* n/a

Flammable Explosive High Range: n/a *Low Range:* n/a

Toxic Products of Combustion: Toxic chlorinated biphenyls are formed in fires.

Other Hazards: Solids may explode. Burns very rapidly when ignited.

Possible extinguishing agents: Apply water from as far a distance as possible. Use alcohol foam, dry chemical, or carbon dioxide.

Reactivity Hazards:

Reactive With: Combustible materials to form explosive materials if finely divided. *Other Reactions:* n/a

Corrosivity Hazards:

Corrosive With: n/a *Neutralizing Agent:* n/a

Radioactivity Hazards:

Radiation Emitted: n/a *Other Hazards:* n/a

Recommended Protection for Response Personnel:

Avoid breathing vapors, keep upwind. Wear appropriate chemical clothing. Wash away any material which may have come into contact with the body with copious amounts of soap and water. Consider appropriate evacuation. Material is dangerously explosive!!

DICHLOROBUTENE

DOT Number: NA 2924 *DOT Hazard Class:* Flammable liquid *DOT Guide Number:* 29
Synonyms: 2-butylene dichloride, 1,4-dichloro-2-butene
STCC Number: 4907605 *Reportable Qty:* 1/.454
Mfg Name: Petro-Tex Chemical Corp. *Phone No:* 1-713-477-9211

Physical Description:

Physical Form: Liquid *Color:* Colorless *Odor:* Sweet
Other Information: n/a

Chemical Properties:

Specific Gravity: 1.11 *Vapor Density:* 4.3 *Boiling Point: 258° F (125.5° C)*
Melting Point: n/a *Vapor Pressure:* n/a *Solubility in water:* No
Other Information: Heavier, insoluble in water. Vapors are heavier than air.

Health Hazards:

Inhalation Hazard: n/a
Ingestion Hazard: Harmful if swallowed.
Absorption Hazard: Irritating to the skin and eyes.
Hazards to Wildlife: n/a
Decontamination Procedures: Wash away any material with copious amounts of soap and water.
First Aid Procedures: Remove victim to fresh air, call emergency medical care. If not breathing give CPR. If breathing is difficult administer oxygen. Treat for shock.

Fire Hazards:

Flashpoint: 80° F(26.6° C) *Ignition temperature:* n/a
Flammable Explosive High Range: 4 *Low Range:* 1.5
Toxic Products of Combustion: Decomposition vapors contain phosgene and hydrogen chloride gases. Both are toxic and irritating.
Other Hazards: Flashback along vapor trail may occur, vapors may explode if ignited in an enclosed area.
Possible extinguishing agents: Do not extinguish fire unless the flow can be stopped. Apply water from as far a distance as possible. Use foam, dry chemical, or carbon dioxide.

Reactivity Hazards:

Reactive With: Slowly with water to form hydrochloric acid. *Other Reactions:* n/a

Corrosivity Hazards:

Corrosive With: Metal and tissue *Neutralizing Agent:* Crushed limestone, soda ash, or lime.

Radioactivity Hazards:

Radiation Emitted: n/a *Other Hazards:* n/a

Recommended Protection for Response Personnel:

Avoid breathing vapors, keep upwind. Wear appropriate chemical clothing. Wash away any material which may have come into contact with the body with copious amounts of soap and water. If the fire becomes uncontrollable, or if the are exposed to direct flame, consider appropriate evacuation.

DICHLORODIFLUOROMETHANE

DOT Number: UN 1028 *DOT Hazard Class:* Nonflammable gas *DOT Guide Number:* 12
Synonyms: Freon 12, Halon 122, Yukon 12
STCC Number: 4904516 *Reportable Qty:* 1/.454
Mfg Name: Allied Corp. *Phone No:* 1-201-455-2000

Physical Description:

Physical Form: Gas *Color:* Colorless *Odor:* Faint Ethereal
Other Information: Shipped as a liquid under its own vapor pressure. Heavier than air and can cause asphyxiation.

Chemical Properties:

Specific Gravity: 135 *Vapor Density:* 2.4 *Boiling Point: −21.6° F(−29.7° C)*
Melting Point: −252° F(−157.7° C) *Vapor Pressure:* 5.7 atm at 68° F(20° C)
Solubility in water: No *Degree of Solubility:* .008%
Other Information: n/a

Health Hazards:

Inhalation Hazard: Will cause dizziness, difficulty in breathing.
Ingestion Hazard: n/a
Absorption Hazard: Contact with liquid can cause frostbite.
Hazards to Wildlife: n/a
Decontamination Procedures: Wash away any material with copious amounts of soap and water.
First Aid Procedures: Remove victim to fresh air, call emergency medical care. If not breathing give CPR. If breathing is difficult administer oxygen. Treat for shock.

Fire Hazards:

Flashpoint: n/a *Ignition temperature:* n/a
Flammable Explosive High Range: n/a *Low Range:* n/a
Toxic Products of Combustion: Toxic gases can be produced.
Other Hazards: n/a
Possible extinguishing agents: Extinguish fire using suitable agent for type of surrounding fire. Use water in flooding quantities as fog. Apply water from as far a distance as possible.

Reactivity Hazards:

Reactive With: Water *Other Reactions:* n/a

Corrosivity Hazards:

Corrosive With: n/a *Neutralizing Agent:* n/a

Radioactivity Hazards:

Radiation Emitted: n/a *Other Hazards:* n/a

Recommended Protection for Response Personnel:

Avoid breathing vapors, keep upwind. Structural protective clothing provides limited protection. Wash away any material which may have come into contact with the body with copious amounts of soap and water. Consider appropriate evacuation.

DICHLOROETHANE

DOT Number: UN 2362 *DOT Hazard Class:* n/a *DOT Guide Number:* 27
Synonyms: ethylidene chloride, ethylidene dichloride, chlorinated hydrochloric ether
STCC Number: n/a *Reportable Qty:* n/a
Mfg Name: Dow Chemical *Phone No:* 1-517-636-4400

Physical Description:

Physical Form: Liquid *Color:* Colorless *Odor:* Chloroform
Other Information: n/a

Chemical Properties:

Specific Gravity: 1.17 *Vapor Density:* 3.44 *Boiling Point: 135° F(57.2° C)*
Melting Point: −143° F(97.2° C) *Vapor Pressure:* 230 mm Hg at 77° F(25° C) *Solubility in water:* Yes
Other Information: Sinks and mixes with water.

Health Hazards:

Inhalation Hazard: n/a
Ingestion Hazard: Nausea, vomiting, and faintness.
Absorption Hazard: Irritating to the skin and eyes.
Hazards to Wildlife: n/a
Decontamination Procedures: Wash away any material with copious amounts of soap and water.
First Aid Procedures: Remove victim to fresh air, call emergency medical care. If not breathing give CPR. If breathing is difficult administer oxygen. Treat for shock.

Fire Hazards:

Flashpoint: 22° F(−5.5° C) *Ignition temperature:* 856° F(457.7° C)
Flammable Explosive High Range: 11.4 *Low Range:* 5.6
Toxic Products of Combustion: When heated to decomposition, material emits highly toxic fumes to phosgene.
Other Hazards: Containers may explode in fire. Flashback along vapor trail may occur.
Possible extinguishing agents: Water may be ineffective. Use alcohol foam, dry chemical, carbon dioxide, and carbon tetrachloride.

Reactivity Hazards:

Reactive With: n/a *Other Reactions:* n/a

Corrosivity Hazards:

Corrosive With: n/a *Neutralizing Agent:* n/a

Radioactivity Hazards:

Radiation Emitted: n/a *Other Hazards:* n/a

Recommended Protection for Response Personnel:

Avoid breathing vapors, keep upwind. Structural protective clothing provides limited protection. Wash away any material which may have come into contact with the body with copious amounts of soap and water. If the fire becomes uncontrollable, consider appropriate evacuation.

DICHLOROETHYL ETHER

DOT Number: UN 1916 *DOT Hazard Class:* Poison B *DOT Guide Number:* 57

Synonyms: Chlorex, DCEE, dichlorodiethyl ether, di-(2-chloroethyl) ether, dichloroether, bis(2-chloroethyl) ether

STCC Number: 4921550 *Reportable Qty:* 1/.454

Mfg Name: Aldrich Chemical *Phone No:* 1-414-273-3850

Physical Description:

Physical Form: Liquid *Color:* Colorless *Odor:* Sweet, pleasant

Other Information: It is used in cleaning compounds, paints, textile finishing, and as a general solvent.

Chemical Properties:

Specific Gravity: 1.22 *Vapor Density:* 4.93 *Boiling Point: 353° F(178.3° C)*

Melting Point: −62° F(−52.2° C) *Vapor Pressure:* .7 mm Hg at 68° F(20° C) *Solubility in water:* No

Other Information: Insoluble and heavier than water.

Health Hazards:

Inhalation Hazard: Poisonous if inhaled.

Ingestion Hazard: Poisonous if swallowed.

Absorption Hazard: Poisonous to the skin and eyes.

Hazards to Wildlife: n/a

Decontamination Procedures: Wash away any material with copious amounts of soap and water.

First Aid Procedures: Remove victim to fresh air, call emergency medical care. If not breathing give CPR. If breathing is difficult administer oxygen. Treat for shock.

Fire Hazards:

Flashpoint: 131° F(55.0° C) *Ignition temperature:* 696° F(368.8° C)

Flammable Explosive High Range: n/a *Low Range:* n/a

Toxic Products of Combustion: Phosgene or hydrogen chloride are formed in fires.

Other Hazards: Containers may explode if exposed to heat or flame.

Possible extinguishing agents: Do not extinguish the fire unless the flow can be stopped. Apply water from as far a distance as possible. Cool all affected containers. Use foam, dry chemical, or carbon dioxide.

Reactivity Hazards:

Reactive With: n/a *Other Reactions:* n/a

Corrosivity Hazards:

Corrosive With: n/a *Neutralizing Agent:* n/a

Radioactivity Hazards:

Radiation Emitted: n/a *Other Hazards:* n/a

Recommended Protection for Response Personnel:

Avoid breathing vapors, keep upwind. Wear appropriate chemical clothing. Wash away any material which may have come into contact with the body with copious amounts of soap and water. Consider appropriate evacuation.

DICHLOROMETHANE

DOT Number: UN 1593 *DOT Hazard Class:* ORM-A *DOT Guide Number:* 74
Synonyms: methylene chloride, methylene dichloride
STCC Number: 4941132 *Reportable Qty:* 1000/454
Mfg Name: Mallinckrodt Inc. *Phone No:* 1-314-895-2000

Physical Description:

Physical Form: Liquid *Color:* Colorless *Odor:* Sweet, pleasant
Other Information: It is used as a solvent, and as a paint remover.

Chemical Properties:

Specific Gravity: 1.3 *Vapor Density:* 2.9 *Boiling Point: 104° F(40° C)*
Melting Point: n/a *Vapor Pressure:* n/a *Solubility in water:* Slight
Other Information: Dichloromethane vapors are narcotic in high concentrations.

Health Hazards:

Inhalation Hazard: Will cause nausea and dizziness.
Ingestion Hazard: Harmful if swallowed.
Absorption Hazard: Irritating to the skin and eyes.
Hazards to Wildlife: n/a
Decontamination Procedures: Wash away any material with copious amounts of soap and water.
First Aid Procedures: Remove victim to fresh air, call emergency medical care. If not breathing give CPR. If breathing is difficult administer oxygen. Treat for shock. No known antidote!

Fire Hazards:

Flashpoint: n/a *Ignition temperature:* 1033° F(556.1° C)
Flammable Explosive High Range: 22 *Low Range:* 14
Toxic Products of Combustion: Poisonous gases are produced when heated.
Other Hazards: n/a
Possible extinguishing agents: Extinguish fire using suitable agent for the type of surrounding fire.

Reactivity Hazards:

Reactive With: n/a *Other Reactions:* n/a

Corrosivity Hazards:

Corrosive With: n/a *Neutralizing Agent:* n/a

Radioactivity Hazards:

Radiation Emitted: n/a *Other Hazards:* n/a

Recommended Protection for Response Personnel:

Avoid breathing vapors, keep upwind. Structural protective clothing provides limited protection. Wash away any material which may have come into contact with the body with copious amounts of soap and water. Consider appropriate evacuation.

DICHLOROPHENOL

DOT Number: UN 2020 *DOT Hazard Class:* Poison A *DOT Guide Number:* 53
Synonyms: none given
STCC Number: n/a *Reportable Qty:* n/a
Mfg Name: Dow Chemical *Phone No:* 1-517-636-4400

Physical Description:

Physical Form: Solid *Color:* Colorless *Odor:* Medicinal
Other Information: n/a

Chemical Properties:

Specific Gravity: 1.4 *Vapor Density:* 5.62 *Boiling Point: 421° F(216.1° C)*
Melting Point: 110° F(43.3° C) *Vapor Pressure:* 1 mm Hg at 127° F(52.7° C) *Solubility in water:* n/a
Other Information: Sinks in water.

Health Hazards:

Inhalation Hazard: n/a
Ingestion Hazard: Poisonous if swallowed.
Absorption Hazard: Will burn the skin and eyes.
Hazards to Wildlife: n/a
Decontamination Procedures: Wash away any material with copious amounts of soap and water.
First Aid Procedures: Remove victim to fresh air, call emergency medical care. If not breathing give CPR. If breathing is difficult administer oxygen. Treat for shock.

Fire Hazards:

Flashpoint: 200° F(93.3° C) *Ignition temperature:* n/a
Flammable Explosive High Range: n/a *Low Range:* n/a
Toxic Products of Combustion: Poisonous gases are produced in fire.
Other Hazards: Solid melts and burns.
Possible extinguishing agents: Water or foam may cause frothing!! Use foam, dry chemical, or carbon dioxide.

Reactivity Hazards:

Reactive With: May react vigorously with oxidizing materials. *Other Reactions:* n/a

Corrosivity Hazards:

Corrosive With: n/a *Neutralizing Agent:* n/a

Radioactivity Hazards:

Radiation Emitted: n/a *Other Hazards:* n/a

Recommended Protection for Response Personnel:

Avoid breathing vapors, keep upwind. Wear a sealed chemical suit (polycarbonate, butyl rubber). Wash away any material which may have come into contact with the body with copious amounts of soap and water. If the fire becomes uncontrollable, consider appropriate evacuation.

DICHLOROPHENOXYACETIC ACID

DOT Number: UN 2765 *DOT Hazard Class:* ORM-A *DOT Guide Number:* 55
Synonyms: 2,4-D
STCC Number: n/a *Reportable Qty:* n/a
Mfg Name: Dow Chemical *Phone No:* 1-517-636-4400

Physical Description:

Physical Form: Solid *Color:* White to tan *Odor:* Odorless
Other Information: n/a

Chemical Properties:

Specific Gravity: 1.56 *Vapor Density:* 7.63 *Boiling Point: Very high*
Melting Point: 286° F(141.1° C) *Vapor Pressure:* n/a *Solubility in water:* n/a
Other Information: Sinks in water

Health Hazards:

Inhalation Hazard: Poisonous if inhaled.
Ingestion Hazard: Poisonous if swallowed.
Absorption Hazard: Poisonous to the skin and eyes.
Hazards to Wildlife: Dangerous to aquatic life.
Decontamination Procedures: Wash away any material with copious amounts of soap and water.
First Aid Procedures: Remove victim to fresh air, call emergency medical care. If not breathing give CPR. If breathing is difficult administer oxygen. Treat for shock.

Fire Hazards:

Flashpoint: n/a (combustible solid) *Ignition temperature:* n/a
Flammable Explosive High Range: n/a *Low Range:* n/a
Toxic Products of Combustion: Toxic and irritating hydrogen chloride and phosgene gases may be formed.
Other Hazards: n/a
Possible extinguishing agents: Extinguish fire using suitable agent for the type of surrounding fire.

Reactivity Hazards:

Reactive With: n/a *Other Reactions:* n/a

Corrosivity Hazards:

Corrosive With: n/a *Neutralizing Agent:* n/a

Radioactivity Hazards:

Radiation Emitted: n/a *Other Hazards:* n/a

Recommended Protection for Response Personnel:

Avoid breathing vapors, keep upwind. Structural protective clothing provides limited protection. Wash away any material which may have come into contact with the body with copious amounts of soap and water. If the fire becomes uncontrollable, consider appropriate evacuation.

1,2-DICHLOROPROPANE

DOT Number: UN 1279 *DOT Hazard Class:* Flammable liquid *DOT Guide Number:* 27
Synonyms: dichloropropane
STCC Number: 4909269 *Reportable Qty:* 1000/454
Mfg Name: Dow Chemical *Phone No:* 1-517-636-4400

Physical Description:

Physical Form: Liquid *Color:* Colorless *Odor:* Chloroform like
Other Information: Heavier and insoluble in water. Vapors are heavier than air. Weighs 9.6 lbs/4.3 kg per gallon/3.8 l.

Chemical Properties:

Specific Gravity: 1.15 *Vapor Density:* 3.5 *Boiling Point: 2.6° F(– 16.3° C)*
Melting Point: n/a *Vapor Pressure:* 1.9 psia at 68° F(20° C) *Solubility in water:* No
Other Information: n/a

Health Hazards:

Inhalation Hazard: n/a
Ingestion Hazard: Harmful if swallowed.
Absorption Hazard: Irritating to the skin and eyes.
Hazards to Wildlife: Dangerous to aquatic life.
Decontamination Procedures: Wash away any material with copious amounts of soap and water.
First Aid Procedures: Remove victim to fresh air, call emergency medical care. If not breathing give CPR. If breathing is difficult administer oxygen. Treat for shock.

Fire Hazards:

Flashpoint: 60° F(15.5° C) *Ignition temperature:* 1035° F(557.2° C)
Flammable Explosive High Range: 14.5 *Low Range:* 3.4
Toxic Products of Combustion: Poisonous gases may be produced in fire.
Other Hazards: n/a
Possible extinguishing agents: Do not extinguish fire unless flow can be stopped. Use foam, dry chemical, carbon dioxide.

Reactivity Hazards:

Reactive With: n/a *Other Reactions:* n/a

Corrosivity Hazards:

Corrosive With: n/a *Neutralizing Agent:* n/a

Radioactivity Hazards:

Radiation Emitted: n/a *Other Hazards:* n/a

Recommended Protection for Response Personnel:

Avoid breathing vapors, keep upwind. Wear a sealed chemical suit (butyl rubber). Wash away any material which may have come into contact with the body with copious amounts of soap and water. If the fire becomes uncontrollable, consider appropriate evacuation.

1,3-DICHLOROPROPENE

DOT Number: UN 2047 *DOT Hazard Class:* Flammable liquid *DOT Guide Number:* 29
Synonyms: dichloropropene
STCC Number: 4909255 *Reportable Qty:* 100/45.4
Mfg Name: Dow Chemical *Phone No:* 1-517-636-4400

Physical Description:

Physical Form: Liquid *Color:* Colorless *Odor:* Sweet
Other Information: Used to make other chemicals, and a soil fumigant.

Chemical Properties:

Specific Gravity: 1.2 *Vapor Density:* 3.8 *Boiling Point: 2.9° F(– 16.1° C)*
Melting Point: n/a *Vapor Pressure:* n/a *Solubility in water:* No
Other Information: It is heavier and insoluble in water. Its vapors are heavier than air.

Health Hazards:

Inhalation Hazard: n/a
Ingestion Hazard: Harmful if swallowed.
Absorption Hazard: Irritating to nose and throat. Will burn skin and eyes.
Hazards to Wildlife: Dangerous to aquatic life.
Decontamination Procedures: Wash away any material with copious amounts of soap and water.
First Aid Procedures: Remove victim to fresh air, call emergency medical care. If not breathing give CPR. If breathing is difficult administer oxygen. Treat for shock.

Fire Hazards:

Flashpoint: 95° F(35.0° C) *Ignition temperature:* n/a
Flammable Explosive High Range: 14.5 *Low Range:* 5.3
Toxic Products of Combustion: Poisonous gases are produced in fire.
Other Hazards: Flashback along vapor trail may occur.
Possible extinguishing agents: Do not extinguish fire unless flow can be stopped. Use foam, dry chemical, CO_2. Apply water from as far a distance as possible.

Reactivity Hazards:

Reactive With: n/a *Other Reactions:* n/a

Corrosivity Hazards:

Corrosive With: n/a *Neutralizing Agent:* n/a

Radioactivity Hazards:

Radiation Emitted: n/a *Other Hazards:* n/a

Recommended Protection for Response Personnel:

Avoid breathing vapors, keep upwind. Structural protective clothing provides limited protection. Wash away any material which may have come into contact with the body with copious amounts of soap and water. If the fire becomes uncontrollable, consider appropriate evacuation.

DICHLORVOS (liquid)

DOT Number: NA 2783 *DOT Hazard Class:* Poison B *DOT Guide Number:* 55
Synonyms: DDVP, Nerkol, Vapona
STCC Number: 4921534 *Reportable Qty:* 10/4.54
Mfg Name: Shell Chemical *Phone No:* 1-713-241-6161

Physical Description:

Physical Form: Liquid *Color:* Clear to amber *Odor:* Aromatic
Other Information: Used as a pesticide. Is heavier than and slightly soluble in water.

Chemical Properties:

Specific Gravity: 1.41 *Vapor Density:* 7.6 *Boiling Point: 284° F(140° C)*
Melting Point: n/a *Vapor Pressure:* .032 mm Hg at 90° F(32.2° C)
Solubility in water: Slightly *Degree of Solubility:* 1%
Other Information: In case of damage to or leaking from containers, contact the Pesticide Safety Team Network at 1-800-424-9300

Health Hazards:

Inhalation Hazard: Poisonous if inhaled.
Ingestion Hazard: Poisonous if swallowed.
Absorption Hazard: Poisonous if the skin is exposed.
Hazards to Wildlife: Dangerous to aquatic life and waterfowl.
Decontamination Procedures: Wash away any material with copious amounts of soap and water.
First Aid Procedures: Remove victim to fresh air, call emergency medical care. If not breathing give CPR. If breathing is difficult administer oxygen. Treat for shock.

Fire Hazards:

Flashpoint: n/a *Ignition temperature:* n/a
Flammable Explosive High Range: n/a *Low Range:* n/a
Toxic Products of Combustion: Chloride fumes and phosgene gas may be emitted in high temperatures.
Other Hazards: n/a
Possible extinguishing agents: Do not extinguish the fire unless the flow can be stopped. Apply foam, dry chemical, carbon dioxide. Apply water from as far a distance as possible.

Reactivity Hazards:

Reactive With: n/a *Other Reactions:* n/a

Corrosivity Hazards:

Corrosive With: n/a *Neutralizing Agent:* n/a

Radioactivity Hazards:

Radiation Emitted: n/a *Other Hazards:* n/a

Recommended Protection for Response Personnel:

Avoid breathing vapors, keep upwind. Wear appropriate chemical clothing. Wash away any material which may have come into contact with the body with copious amounts of soap and water. Consider appropriate evacuation.

DICHLORVOS (solid)

DOT Number: NA 2783 *DOT Hazard Class:* Poison B *DOT Guide Number:* 55
Synonyms: DDVP, Nerkol, Vapona
STCC Number: 4921537 *Reportable Qty:* 10/4.54
Mfg Name: Shell Chemical *Phone No:* 1-713-241-6161

Physical Description:

Physical Form: Solid *Color:* n/a *Odor:* Chemical
Other Information: It is absorbed in a dry carrier. Used as a pesticide, and is heavier than and slightly soluble in water.

Chemical Properties:

Specific Gravity: 1.41 *Vapor Density:* 6.7 *Boiling Point: 284° F(140° C)*
Melting Point: n/a *Vapor Pressure:* .032 mm Hg at 90° F(32.2° C)
Solubility in water: Slightly *Degree of Solubility:* 1%
Other Information: In case of damage to or leaking from containers, contact the Pesticide Safety Team Network at 1-800-424-9300

Health Hazards:

Inhalation Hazard: Poisonous if inhaled.
Ingestion Hazard: Poisonous if swallowed.
Absorption Hazard: Poisonous if the skin is exposed.
Hazards to Wildlife: Dangerous to aquatic life and waterfowl.
Decontamination Procedures: Wash away any material with copious amounts of soap and water.
First Aid Procedures: Remove victim to fresh air, call emergency medical care. If not breathing give CPR. If breathing is difficult administer oxygen. Treat for shock.

Fire Hazards:

Flashpoint: n/a *Ignition temperature:* n/a
Flammable Explosive High Range: n/a *Low Range:* n/a
Toxic Products of Combustion: Chloride fumes and phosgene gas may be emitted at high temperatures.
Other Hazards: n/a
Possible extinguishing agents: Use foam, dry chemical, or carbon dioxide. Apply water from as far a distance as possible.

Reactivity Hazards:

Reactive With: n/a *Other Reactions:* n/a

Corrosivity Hazards:

Corrosive With: n/a *Neutralizing Agent:* n/a

Radioactivity Hazards:

Radiation Emitted: n/a *Other Hazards:* n/a

Recommended Protection for Response Personnel:

Avoid breathing vapors, keep upwind. Wear appropriate chemical clothing. Wash away any material which may have come into contact with the body with copious amounts of soap and water. Consider appropriate evacuation.

DICYCLOPENTADIENE

DOT Number: UN 2048 *DOT Hazard Class:* Flammable liquid *DOT Guide Number:* 26
Synonyms: Dicy, 3a,4,7,7a-tetrahydro-4,7-methanoindene
STCC Number: 4907219 *Reportable Qty:* n/a
Mfg Name: Exxon Chemical America *Phone No:* 1-713-870-6000

Physical Description:

Physical Form: Liquid *Color:* Colorless *Odor:* Camphor
Other Information: It is used in paints, varnishes, as an intermediate in insecticides, and as a flame retardant in plastics.

Chemical Properties:

Specific Gravity: .98 *Vapor Density:* 4.6 *Boiling Point: 338° F(170° C)*
Melting Point: 41° F(50° C) *Vapor Pressure:* 10 mm Hg at 118° F(47.7° C) *Solubility in water:* No
Other Information: Floats, insoluble in water. Weighs 8.2 lbs/3.7 kg per gallon/3.8 l.

Health Hazards:

Inhalation Hazard: n/a
Ingestion Hazard: n/a
Absorption Hazard: Irritating to the skin and eyes.
Hazards to Wildlife: n/a
Decontamination Procedures: Wash away any material with copious amounts of soap and water.
First Aid Procedures: Remove victim to fresh air, call emergency medical care. If not breathing give CPR. If breathing is difficult administer oxygen. Treat for shock.

Fire Hazards:

Flashpoint: 90° F(32.2° C) *Ignition temperature:* 941° F(505° C)
Flammable Explosive High Range: 6.3 *Low Range:* .8
Toxic Products of Combustion: n/a
Other Hazards: Flashback along vapor trail may occur. Vapor may explode if ignited in an enclosed area.
Possible extinguishing agents: Do not extinguish the fire unless the flow can be stopped apply water from as far a distance as possible. Cool all affected containers. Use foam, dry chemical, or carbon dioxide.

Reactivity Hazards:

Reactive With: n/a *Other Reactions:* n/a

Corrosivity Hazards:

Corrosive With: n/a *Neutralizing Agent:* n/a

Radioactivity Hazards:

Radiation Emitted: n/a *Other Hazards:* n/a

Recommended Protection for Response Personnel:

Avoid breathing vapors, keep upwind. Wear a sealed chemical suit (polycarbonate, butyl rubber). Wash away any material which may have come into contact with the body with copious amounts of soap and water. Consider appropriate evacuation.

DIELDRIN (liquid)

DOT Number: NA 2761 *DOT Hazard Class:* ORM-A *DOT Guide Number:* 55
Synonyms: ENDO, HEOD
STCC Number: 4941134 *Reportable Qty:* 1/.454
Mfg Name: Triangle Chemical Co. *Phone No:* 1-912-743-1548

Physical Description:

Physical Form: Liquid *Color:* Light tan *Odor:* Chemical
Other Information: It is dissolved in a liquid carrier. Is used as a pesticide, and a water emulsified liquid.

Chemical Properties:

Specific Gravity: 1.75 *Vapor Density:* 13.2 *Boiling Point: n/a*
Melting Point: 349° F(176.1° C) *Vapor Pressure:* 0 mm Hg at 68° F(20° C)
Solubility in water: Sinks in water *Degree of Solubility: .2%*
Other Information: In case of damage to or leaking from containers, contact the Pesticide Safety Team Network at 1-800-424-9300.

Health Hazards:

Inhalation Hazard: Headaches, dizziness, loss of consciousness.
Ingestion Hazard: Poisonous if swallowed.
Absorption Hazard: Poisonous if the skin is exposed.
Hazards to Wildlife: Dangerous to aquatic life and waterfowl.
Decontamination Procedures: Wash away any material with copious amounts of soap and water.
First Aid Procedures: Remove victim to fresh air, call emergency medical care. If not breathing give CPR. If breathing is difficult administer oxygen. Treat for shock.

Fire Hazards:

Flashpoint: n/a *Ignition temperature:* n/a
Flammable Explosive High Range: n/a *Low Range:* n/a
Toxic Products of Combustion: Poisonous gases may be produced when heated.
Other Hazards: n/a
Possible extinguishing agents: The material itself does not burn. Extinguish fire using suitable agent for the type of surrounding fire.

Reactivity Hazards:

Reactive With: n/a *Other Reactions:* n/a

Corrosivity Hazards:

Corrosive With: n/a *Neutralizing Agent:* n/a

Radioactivity Hazards:

Radiation Emitted: n/a *Other Hazards:* n/a

Recommended Protection for Response Personnel:

Avoid breathing vapors, keep upwind. Wear a sealed chemical suit (polycarbonate, butyl rubber). Wash away any material which may have come into contact with the body with copious amounts of soap and water. Consider appropriate evacuation.

DIELDRIN (solid)

DOT Number: NA 2761 *DOT Hazard Class:* ORM-A *DOT Guide Number:* 55
Synonyms: ENDO, HEOD
STCC Number: 4941135 *Reportable Qty:* 1/.454
Mfg Name: Triangle Chemical Co. *Phone No:* 1-912-743-1548

Physical Description:

Physical Form: Solid *Color:* Light tan *Odor:* Chemical
Other Information: It is used as a pesticide. It is a wettable powder.

Chemical Properties:

Specific Gravity: 1.75 *Vapor Density:* 13.2 *Boiling Point: n/a*
Melting Point: 349° F(176.1° C) *Vapor Pressure:* 0 mm Hg at 68° F(20° C)
Solubility in water: Sinks in water. *Degree of Solubility: .2%*
Other Information: In case of damage to or leaking from containers, contact the Pesticide Safety Team Network at 1-800-424-9300.

Health Hazards:

Inhalation Hazard: Will cause headaches, dizziness, loss consciousness.
Ingestion Hazard: Poisonous if swallowed.
Absorption Hazard: Poisonous if the skin is exposed.
Hazards to Wildlife: Dangerous to aquatic life and waterfowl.
Decontamination Procedures: Wash away any material with copious amounts of soap and water.
First Aid Procedures: Remove victim to fresh air, call emergency medical care. If not breathing give CPR. If breathing is difficult administer oxygen. Treat for shock.

Fire Hazards:

Flashpoint: n/a *Ignition temperature:* n/a
Flammable Explosive High Range: n/a *Low Range:* n/a
Toxic Products of Combustion: Poisonous gases may be produced when heated.
Other Hazards: n/a
Possible extinguishing agents: The material itself does not burn. Extinguish fire using suitable agent for the type of surrounding fire.

Reactivity Hazards:

Reactive With: n/a *Other Reactions:* n/a

Corrosivity Hazards:

Corrosive With: n/a *Neutralizing Agent:* n/a

Radioactivity Hazards:

Radiation Emitted: n/a *Other Hazards:* n/a

Recommended Protection for Response Personnel:

Avoid breathing vapors and dust, keep upwind. Wear a sealed chemical suit (polycarbonate, butyl rubber). Wash away any material which may have come into into contact with the body with copious amounts of soap and water. Consider appropriate evacuation.

DIETHYL CARBONATE

DOT Number: UN 2366 *DOT Hazard Class:* Flammable liquid *DOT Guide Number:* 26
Synonyms: carbonic acid, diethyl ester; ethyl carbonate; Eufin
STCC Number: 4909218 *Reportable Qty:* n/a
Mfg Name: FMC Corp. *Phone No:* 1-312-861-5900

Physical Description:

Physical Form: Liquid *Color:* Colorless *Odor:* Pleasant
Other Information: It is used as a solvent.

Chemical Properties:

Specific Gravity: .975 *Vapor Density:* 4.7 *Boiling Point: 260° F(126.6° C)*
Melting Point: −45° F(−42.7° C) *Vapor Pressure:* 10 mm Hg at 73° F(22.7° C) *Solubility in water:* No
Other Information: Lighter than and insoluble in water.

Health Hazards:

Inhalation Hazard: Headache, dizziness, nausea, loss of consciousness.
Ingestion Hazard: Harmful if swallowed.
Absorption Hazard: Irritating to the skin and eyes.
Hazards to Wildlife: n/a
Decontamination Procedures: Wash away any material with copious amounts of soap and water.
First Aid Procedures: Remove victim to fresh air, call emergency medical care. If not breathing give CPR. If breathing is difficult administer oxygen. Treat for shock.

Fire Hazards:

Flashpoint: 77° F(25° C) *Ignition temperature:* n/a
Flammable Explosive High Range: n/a *Low Range:* n/a
Toxic Products of Combustion: n/a
Other Hazards: Flashback along vapor trail may occur. Vapor may explode if ignited in an enclosed area.
Possible extinguishing agents: Do not extinguish the fire unless the flow can be stopped. Apply water from as far a distance as possible. Cool all affected containers. Use foam, dry chemical, or carbon dioxide.

Reactivity Hazards:

Reactive With: n/a *Other Reactions:* n/a

Corrosivity Hazards:

Corrosive With: n/a *Neutralizing Agent:* n/a

Radioactivity Hazards:

Radiation Emitted: n/a *Other Hazards:* n/a

Recommended Protection for Response Personnel:

Avoid breathing vapors, keep upwind. Structural protective clothing provides limited protection. Wash away any material which may have come into contact with the body with copious amounts of soap and water. Consider appropriate evacuation.

DIETHYLAMINE

DOT Number: UN 1154 *DOT Hazard Class:* Flammable liquid *DOT Guide Number:* 68
Synonyms: DEN
STCC Number: 4907815 *Reportable Qty:* 100/45.4
Mfg Name: Ashland Chemical *Phone No:* 1-614-889-3333

Physical Description:

Physical Form: Liquid *Color:* Colorless *Odor:* Fishy, ammonia
Other Information: It is lighter than and soluble in water. Vapors are heavier than air. Weighs 5.9 lbs/2.6 kg per gallon/3.8 l.

Chemical Properties:

Specific Gravity: .7 *Vapor Density:* 2.5 *Boiling Point: 134° F(56.6° C)*
Melting Point: −58° F(−50° C) *Vapor Pressure:* 195 mm Hg at 68° F(20° C) *Solubility in water:* Yes
Other Information: n/a

Health Hazards:

Inhalation Hazard: Harmful if inhaled.
Ingestion Hazard: Harmful if swallowed.
Absorption Hazard: Harmful to nose, throat, and eyes.
Hazards to Wildlife: Dangerous to aquatic life.
Decontamination Procedures: Wash away any material with copious amounts of soap and water.
First Aid Procedures: Remove victim to fresh air, call emergency medical care. If not breathing give CPR. If breathing is difficult administer oxygen. Treat for shock.

Fire Hazards:

Flashpoint: −9° F(−22.7° C) *Ignition temperature:* 594° F(312.2° C)
Flammable Explosive High Range: 1001 *Low Range:* 1.8
Toxic Products of Combustion: Vapors are irritating.
Other Hazards: Flashback along vapor trail may occur. Vapors may explode if ignited in an enclosed area.
Possible extinguishing agents: Use alcohol foam, dry chemical, or carbon dioxide. Use water spray to knock down the vapors.

Reactivity Hazards:

Reactive With: n/a *Other Reactions:* n/a

Corrosivity Hazards:

Corrosive With: Tissue *Neutralizing Agent:* Crushed limestone, soda ash, or lime.

Radioactivity Hazards:

Radiation Emitted: n/a *Other Hazards:* n/a

Recommended Protection for Response Personnel:

Avoid breathing vapors, keep upwind. Structural protective clothing provides limited protection. Wash away any material which may have come into contact with the body with copious amounts of soap and water. Consider appropriate evacuation.

DIETHYLBENZENE

DOT Number: NA 2049 *DOT Hazard Class:* Combustible liquid *DOT Guide Number:* 29
Synonyms: none given
STCC Number: 4913135 *Reportable Qty:* n/a
Mfg Name: Dow Chemical *Phone No:* 1-517-636-4400

Physical Description:

Physical Form: Liquid *Color:* Colorless *Odor:* Sweet, gasoline
Other Information: n/a

Chemical Properties:

Specific Gravity: .86 *Vapor Density:* 4.6 *Boiling Point: 356° F(180° C)*
Melting Point: 160° F(71.1° C) *Vapor Pressure:* 1 mm Hg at 68° F(20° C) *Solubility in water:* No
Other Information: Lighter than and insoluble in water. Vapors are heavier than air.

Health Hazards:

Inhalation Hazard: n/a
Ingestion Hazard: n/a
Absorption Hazard: Irritating to the skin and eyes.
Hazards to Wildlife: n/a
Decontamination Procedures: Wash away any material with copious amounts of soap and water.
First Aid Procedures: Remove victim to fresh air, call emergency medical care. If not breathing give CPR. If breathing is difficult administer oxygen. Treat for shock.

Fire Hazards:

Flashpoint: 135° F(57.2° C) *Ignition temperature:* 743° F(395° C)
Flammable Explosive High Range: n/a *Low Range:* n/a
Toxic Products of Combustion: n/a
Other Hazards: n/a
Possible extinguishing agents: Do not extinguish the fire unless the flow can be stopped. Apply water from as far a distance as possible. cool all affected containers. Use foam, dry chemical, or carbon dioxide.

Reactivity Hazards:

Reactive With: n/a *Other Reactions:* n/a

Corrosivity Hazards:

Corrosive With: n/a *Neutralizing Agent:* n/a

Radioactivity Hazards:

Radiation Emitted: n/a *Other Hazards:* n/a

Recommended Protection for Response Personnel:

Avoid breathing vapors, keep upwind. Structural protective clothing provides limited protection. Wash away any material which may have come into contact with the body with copious amounts of soap and water. Consider appropriate evacuation.

DIETHYLENETRIAMINE

DOT Number: UN 2079 *DOT Hazard Class:* n/a *DOT Guide Number:* 29
Synonyms: bis(2-aminoethyl)amine, 2,2′-diaminodiethylamine
STCC Number: n/a *Reportable Qty:* n/a
Mfg Name: Dow Chemical *Phone No:* 1-517-636-4400

Physical Description:

Physical Form: Liquid *Color:* Colorless to yellow *Odor:* Ammonia
Other Information: n/a

Chemical Properties:

Specific Gravity: .95 *Vapor Density:* 3.5 *Boiling Point: 405° F(207.2° C)*
Melting Point: −38° F(−38.8° C) *Vapor Pressure:* .22 mm Hg at 68° F(20° C) *Solubility in water:* Yes
Other Information: Floats and mixes with water.

Health Hazards:

Inhalation Hazard: n/a
Ingestion Hazard: Harmful if swallowed.
Absorption Hazard: Will burn the skin and eyes.
Hazards to Wildlife: n/a
Decontamination Procedures: Wash away any material with copious amounts of soap and water.
First Aid Procedures: Remove victim to fresh air, call emergency medical care. If not breathing give CPR. If breathing is difficult administer oxygen. Treat for shock.

Fire Hazards:

Flashpoint: 210° F(98.8° C) *Ignition temperature:* 676° F(357.7° C)
Flammable Explosive High Range: 10 *Low Range:* 1
Toxic Products of Combustion: Irritating vapors are generated when heated.
Other Hazards: n/a
Possible extinguishing agents: Alcohol foam, dry chemical or carbon dioxide. Water may cause frothing!!

Reactivity Hazards:

Reactive With: n/a *Other Reactions:* n/a

Corrosivity Hazards:

Corrosive With: n/a *Neutralizing Agent:* n/a

Radioactivity Hazards:

Radiation Emitted: n/a *Other Hazards:* n/a

Recommended Protection for Response Personnel:

Avoid breathing vapors, keep upwind. Structural protective clothing provides limited protection. Wash away any material which may have come into contact with the body with copious amounts of soap and water. If the fire becomes uncontrollable, consider appropriate evacuation.

DIETHYLZINC

DOT Number: UN 1366 *DOT Hazard Class:* Flammable liquid *DOT Guide Number:* 40
Synonyms: ethylzinc, zinc diethyl, zinc ethyl
STCC Number: 4906069 *Reportable Qty:* n/a
Mfg Name: Ventron Corp. *Phone No:* 1-617-774-3100

Physical Description:

Physical Form: Liquid *Color:* Colorless *Odor:* Garlic
Other Information: Pyrophoric, is stable when shipped in sealed tubes with carbon dioxide. Used as an aircraft fuel.

Chemical Properties:

Specific Gravity: 1.2 *Vapor Density:* 1 *Boiling Point: 255° F(123.8° C)*
Melting Point: –18° F(–27.7° C) *Vapor Pressure:* n/a *Solubility in water:* Decomposes violently.
Other Information: Ignites when exposed to air. Flammable, irritating vapor is produced.

Health Hazards:

Inhalation Hazard: Headache, difficulty in breathing.

Ingestion Hazard: Harmful if swallowed.

Absorption Hazard: Will burn the skin and eyes.

Hazards to Wildlife: n/a

Decontamination Procedures: Wash away any material with copious amounts of soap and water.

First Aid Procedures: Remove victim to fresh air, call emergency medical care. If not breathing give CPR. If breathing is difficult administer oxygen. Treat for shock.

Fire Hazards:

Flashpoint: Ignites spontaneously *Ignition temperature:* Below zero (–17.7° C)

Flammable Explosive High Range: n/a *Low Range:* n/a

Toxic Products of Combustion: Yields zinc oxide fumes when burning which can cause metal fume fever.

Other Hazards: Reacts violently with water evolving flammable ethane gas.

Possible extinguishing agents: Do not extinguish the fire unless the flow can be stopped. Do not use water!! Use graphite, soda ash, or powdered sodium chloride.

Reactivity Hazards:

Reactive With: Water to form flammable ethane gas. *Other Reactions:* Air to form heat, flames, and fire.

Corrosivity Hazards:

Corrosive With: n/a *Neutralizing Agent:* n/a

Radioactivity Hazards:

Radiation Emitted: n/a *Other Hazards:* n/a

Recommended Protection for Response Personnel:

Avoid breathing vapors, keep upwind. Structural protective clothing provides limited protection. Wash away any material which may have come into contact with the body with copious amounts of soap and water. Consider appropriate evacuation.

1,1-DIFLUOROETHANE

DOT Number: UN 1030 *DOT Hazard Class:* Flammable gas *DOT Guide Number:* 22
Synonyms: Refrigerant 52A
STCC Number: 4905716 *Reportable Qty:* n/a
Mfg Name: E.I. Du Pont *Phone No:* 1-800-441-3637

Physical Description:

Physical Form: Gas *Color:* Colorless *Odor:* Odorless
Other Information: A liquefied gas shipped under its own pressure. Contact with liquid can cause frostbite.

Chemical Properties:

Specific Gravity: .95 *Vapor Density:* 2.3 *Boiling Point: 52.3° F(11.2° C)*
Melting Point: n/a *Vapor Pressure:* n/a *Solubility in water:* n/a
Other Information: It can asphyxiate by the displacement of air.

Health Hazards:

Inhalation Hazard: Harmful if inhaled.
Ingestion Hazard: Irritating to eyes, nose, and throat.
Absorption Hazard: Will cause frostbite.
Hazards to Wildlife: No
Decontamination Procedures: Wash away any material with copious amounts of soap and water.
First Aid Procedures: Remove victim to fresh air, call emergency medical care. If not breathing give CPR. If breathing is difficult administer oxygen. Treat for shock.

Fire Hazards:

Flashpoint: n/a *Ignition temperature:* n/a
Flammable Explosive High Range: 18 *Low Range:* 3.7
Toxic Products of Combustion: Irritating hydrogen fluoride fumes.
Other Hazards: Flashback along vapor trail may occur. Containers may violently explode and rocket. (BLEVE)
Possible extinguishing agents: Do not extinguish fire unless flow can be stopped. Apply water from as far a distance as possible.

Reactivity Hazards:

Reactive With: n/a *Other Reactions:* n/a

Corrosivity Hazards:

Corrosive With: n/a *Neutralizing Agent:* n/a

Radioactivity Hazards:

Radiation Emitted: n/a *Other Hazards:* n/a

Recommended Protection for Response Personnel:

Avoid breathing vapors, keep upwind. Structural protective clothing provides limited protection. Wash away any material which may have come into contact with the body with copious amounts of soap and water. If the fire becomes uncontrollable, consider appropriate evacuation.

DIFLUOROPHOSPHORIC ACID

DOT Number: UN 1768 *DOT Hazard Class:* Corrosive *DOT Guide Number:* 59
Synonyms: difluorophosphorus acid
STCC Number: 4930214 *Reportable Qty:* n/a
Mfg Name: Ozark-Mahoning Co. *Phone No:* 1-918-585-2661

Physical Description:

Physical Form: Liquid *Color:* Colorless *Odor:* Sharp, irritating
Other Information: A fuming liquid that is soluble in water.

Chemical Properties:

Specific Gravity: 1.58 *Vapor Density:* 3.5 *Boiling Point: 241° F(116.1° C)*
Melting Point: n/a *Vapor Pressure:* n/a *Solubility in water:* Yes
Other Information: Reacts violently with water. An irritating gas is produced on contact with water.

Health Hazards:

Inhalation Hazard: Irritating to skin, nose, and throat.
Ingestion Hazard: Harmful if swallowed.
Absorption Hazard: Will burn the skin and eyes.
Hazards to Wildlife: n/a
Decontamination Procedures: Wash away any material with copious amounts of soap and water.
First Aid Procedures: Remove victim to fresh air, call emergency medical care. If not breathing give CPR. If breathing is difficult administer oxygen. Treat for shock.

Fire Hazards:

Flashpoint: n/a *Ignition temperature:* n/a
Flammable Explosive High Range: n/a *Low Range:* n/a
Toxic Products of Combustion: Poisonous gases may be produced when heated.
Other Hazards: n/a
Possible extinguishing agents: Extinguish fire using suitable agent for type of surrounding fire. Apply water from as far a distance as possible.

Reactivity Hazards:

Reactive With: Water and heat *Other Reactions:* n/a

Corrosivity Hazards:

Corrosive With: Metals and tissue *Neutralizing Agent:* Crushed limestone, soda ash, or lime.

Radioactivity Hazards:

Radiation Emitted: n/a *Other Hazards:* n/a

Recommended Protection for Response Personnel:

Avoid breathing vapors, keep upwind. Wear a sealed chemical suit (butyl rubber). Wash away any material which may have come contact with the body with copious amounts of soap and water. Consider appropriate evacuation.

DIISOBUTYL KETONE

DOT Number: UN 1157 *DOT Hazard Class:* Combustible liquid *DOT Guide Number:* 26
Synonyms: DIBK, Isovalerone, Valerone
STCC Number: 4909177 *Reportable Qty:* n/a
Mfg Name: Union Carbide *Phone No:* 1-203-794-2000

Physical Description:

Physical Form: Liquid *Color:* Colorless *Odor:* Fish like
Other Information: Lighter than and insoluble in water. Vapors are heavier than air.

Chemical Properties:

Specific Gravity: .8 *Vapor Density:* 4.9 *Boiling Point: 335° F(168.3° C)*
Melting Point: −43° F(−41.6° C) *Vapor Pressure:* 1.7 mm Hg at 68° F(20° C)
Solubility in water: No *Degree of Solubility:* .05%
Other Information: Floats on water.

Health Hazards:

Inhalation Hazard: Will cause coughing and difficulty in breathing.
Ingestion Hazard: Will cause nausea and vomiting.
Absorption Hazard: Irritating to the skin and eyes.
Hazards to Wildlife: Dangerous to aquatic life.
Decontamination Procedures: Wash away any material with copious amounts of soap and water.
First Aid Procedures: Remove victim to fresh air, call emergency medical care. If not breathing give CPR. If breathing is difficult administer oxygen. Treat for shock.

Fire Hazards:

Flashpoint: 120° F(48.8° C) *Ignition temperature:* 745° F(396.1° C)
Flammable Explosive High Range: 7.1% at 200° F *Low Range:* .8
Toxic Products of Combustion: n/a
Other Hazards: n/a
Possible extinguishing agents: Water may be ineffective on fire. Use foam, dry chemical, or carbon dioxide. Do not extinguish the fire unless the flow can be stopped.

Reactivity Hazards:

Reactive With: n/a *Other Reactions:* n/a

Corrosivity Hazards:

Corrosive With: n/a *Neutralizing Agent:* n/a

Radioactivity Hazards:

Radiation Emitted: n/a *Other Hazards:* n/a

Recommended Protection for Response Personnel:

Avoid breathing vapors, keep upwind. Structural protective clothing provides limited protection. Wash away any material which may have come into contact with the body with copious amounts of soap and water. Consider appropriate evacuation.

DIISOBUTYLCARBINOL

DOT Number: NA 1993 *DOT Hazard Class:* Combustible liquid *DOT Guide Number:* 45
Synonyms: 2,6-dimethyl-4-heptanol
STCC Number: 4913119 *Reportable Qty:* n/a
Mfg Name: Union Carbide *Phone No:* 1-203-794-2000

Physical Description:

Physical Form: Liquid *Color:* Colorless *Odor:* n/a
Other Information: n/a

Chemical Properties:

Specific Gravity: .18 *Vapor Density:* 5 *Boiling Point: 352° F(177.7° C)*
Melting Point: −85° F(−65° C) *Vapor Pressure:* .3 mm Hg at 68° F(20° C) *Solubility in water:* No
Other Information: Lighter than and insoluble in water. Vapors are heavier than air.

Health Hazards:

Inhalation Hazard: n/a
Ingestion Hazard: Harmful if swallowed.
Absorption Hazard: n/a
Hazards to Wildlife: n/a
Decontamination Procedures: Wash away any material with copious amounts of soap and water.
First Aid Procedures: Remove victim to fresh air, call emergency medical care. If not breathing give CPR. If breathing is difficult administer oxygen. Treat for shock.

Fire Hazards:

Flashpoint: 162° F(72.2° C) *Ignition temperature:* 494° F(256.6° C)
Flammable Explosive High Range: 6.1 *Low Range:* .8
Toxic Products of Combustion: n/a
Other Hazards: n/a
Possible extinguishing agents: Do not extinguish the fire unless the flow can be stopped. Cool all affected containers with water. Use foam, dry chemical, or carbon dioxide.

Reactivity Hazards:

Reactive With: n/a *Other Reactions:* n/a

Corrosivity Hazards:

Corrosive With: n/a *Neutralizing Agent:* n/a

Radioactivity Hazards:

Radiation Emitted: n/a *Other Hazards:* n/a

Recommended Protection for Response Personnel:

Avoid breathing vapors, keep upwind. Structural protective clothing provides limited protection. Wash away any material which may have come into contact with the body with copious amounts of soap and water. Consider appropriate evacuation.

DIISOBUTYLENE

DOT Number: UN 2050 *DOT Hazard Class:* Flammable liquid *DOT Guide Number:* 26
Synonyms: 2,4,4-trimethyl-1-pentene
STCC Number: 4909209 *Reportable Qty:* n/a
Mfg Name: Petrol-Tex Chemical Corp. *Phone No:* 1-713-477-2911

Physical Description:

Physical Form: Liquid *Color:* Colorless *Odor:* Gasoline
Other Information: It is used in the manufacture of other chemicals.

Chemical Properties:

Specific Gravity: .7 *Vapor Density:* 4.9 *Boiling Point: 335° F(168.3° C)*
Melting Point: −150° F(−101.1° C) *Vapor Pressure:* 75 mm Hg at 100° F(37.7° C) *Solubility in water:* No
Other Information: Floats on water. A flammable, irritating vapor is produced.

Health Hazards:

Inhalation Hazard: Dizziness, headache, difficulty in breathing.
Ingestion Hazard: Cause nausea or vomiting.
Absorption Hazard: Irritating to the skin and eyes.
Hazards to Wildlife: n/a
Decontamination Procedures: Wash away any material with copious amounts of soap and water.
First Aid Procedures: Remove victim to fresh air, call emergency medical care. If not breathing give CPR. If breathing is difficult administer oxygen. Treat for shock.

Fire Hazards:

Flashpoint: 20° F(−6.6° C) *Ignition temperature:* 736° F(391.1° C)
Flammable Explosive High Range: 4.8 *Low Range:* .8
Toxic Products of Combustion: n/a
Other Hazards: Flashback along vapor trail may occur. Vapors may explode if ignited in an enclosed area.
Possible extinguishing agents: Use alcohol foam, dry chemical or carbon dioxide. Apply water from as far a distance as possible.

Reactivity Hazards:

Reactive With: n/a *Other Reactions:* n/a

Corrosivity Hazards:

Corrosive With: n/a *Neutralizing Agent:* n/a

Radioactivity Hazards:

Radiation Emitted: n/a *Other Hazards:* n/a

Recommended Protection for Response Personnel:

Avoid breathing dust, keep upwind. Structural protective clothing provides limited protection. Wash away any material which may have come into contact with the body with copious amounts of soap and water. If the fire becomes uncontrollable, consider appropriate evacuation.

DIISOPROPYLAMINE

DOT Number: UN 1158 *DOT Hazard Class:* Flammable liquid *DOT Guide Number:* 68
Synonyms: none given
STCC Number: 4909148 *Reportable Qty:* n/a
Mfg Name: Pennwald Corp. *Phone No:* 1-215-587-7000

Physical Description:

Physical Form: Liquid *Color:* Colorless *Odor:* Fishy
Other Information: It is used to make other chemicals. It is lighter than water and soluble in water.

Chemical Properties:

Specific Gravity: .7 *Vapor Density:* 3.5 *Boiling Point: 183° F(83.3° C)*
Melting Point: −141° F(−96.1° C) *Vapor Pressure:* 60 mm Hg at 68° F(20° C) *Solubility in water:* Yes
Other Information: n/a

Health Hazards:

Inhalation Hazard: Will cause coughing or difficulty in breathing.
Ingestion Hazard: Will cause nausea and vomiting.
Absorption Hazard: Irritating to eyes, nose, and throat.
Hazards to Wildlife: Dangerous to aquatic life.
Decontamination Procedures: Wash away any material with copious amounts of soap and water.
First Aid Procedures: Remove victim to fresh air, call emergency medical care. If not breathing give CPR. If breathing is difficult administer oxygen. Treat for shock.

Fire Hazards:

Flashpoint: 30° F(−1.1° C) *Ignition temperature:* 600° F(315.5° C)
Flammable Explosive High Range: 7.1 *Low Range:* 1.1
Toxic Products of Combustion: Toxic oxides of nitrogen are produced in fire.
Other Hazards: n/a
Possible extinguishing agents: Water may be ineffective on fire. Use alcohol foam, dry chemical or carbon dioxide.

Reactivity Hazards:

Reactive With: n/a *Other Reactions:* n/a

Corrosivity Hazards:

Corrosive With: n/a *Neutralizing Agent:* n/a

Radioactivity Hazards:

Radiation Emitted: n/a *Other Hazards:* n/a

Recommended Protection for Response Personnel:

Avoid breathing vapors, keep upwind. Structural protective clothing provides limited protection. Wash away any material which may have come into contact with the body with copious amounts of soap and water. If the material is leaking and not on fire, consider appropriate evacuation.

DIISOPROPYLBENZENE HYDROPEROXIDE

DOT Number: UN 2171 *DOT Hazard Class:* Organic peroxide *DOT Guide Number:* 48
Synonyms: isopropylcumly hydroperoxide
STCC Number: 4919130 *Reportable Qty:* n/a
Mfg Name: Hercules Inc. *Phone No:* 1-302-594-5000

Physical Description:

Physical Form: Liquid *Color:* Colorless to pale yellow *Odor:* Sharp, unpleasant
Other Information: Must be shipped in a nonviolent solution not exceeding 72% concentrate.

Chemical Properties:

Specific Gravity: .95 *Vapor Density:* 1 *Boiling Point: n/a*
Melting Point: n/a *Vapor Pressure:* n/a *Solubility in water:* n/a
Other Information: Strongly supports the combustion of burning materials.

Health Hazards:

Inhalation Hazard: Will cause coughing and difficulty in breathing.
Ingestion Hazard: Harmful if swallowed.
Absorption Hazard: Irritating to the skin and eyes.
Hazards to Wildlife: n/a
Decontamination Procedures: Wash away any material with copious amounts of soap and water.
First Aid Procedures: Remove victim to fresh air, call emergency medical care. If not breathing give CPR. If breathing is difficult administer oxygen. Treat for shock.

Fire Hazards:

Flashpoint: 175° F(79.4° C) *Ignition temperature:* n/a
Flammable Explosive High Range: n/a *Low Range:* n/a
Toxic Products of Combustion: Flammable alcohol and ketone gases are formed in a fire.
Other Hazards: Will increase the intensity of a fire. May cause fire upon contact with combustibles. The containers may explode in a fire.
Possible extinguishing agents: Apply water from as far a distance as possible. Use alcohol foam, dry chemical, or carbon dioxide.

Reactivity Hazards:

Reactive With: Aluminum, copper, brass, lead, zinc salts, mineral acids, oxidizing or reducing agents can cause rapid decomposition. *Other Reactions:* n/a

Corrosivity Hazards:

Corrosive With: n/a *Neutralizing Agent:* n/a

Radioactivity Hazards:

Radiation Emitted: n/a *Other Hazards:* n/a

Recommended Protection for Response Personnel:

Avoid breathing vapors, keep upwind. Wear appropriate chemical clothing, Wash away any material which may have into contact with the body with copious amounts of soap and water. If the material is involved in a fire or on fire, consider appropriate evacuation. Dangerously explosive!!!

DIMETHYL ETHER

DOT Number: UN 1033 *DOT Hazard Class:* Flammable gas *DOT Guide Number:* 22
Synonyms: methyl ether, wood ether
STCC Number: 4905725 *Reportable Qty:* n/a
Mfg Name: Union Carbide *Phone No:* 1-203-794-2000

Physical Description:

Physical Form: Gas *Color:* Colorless *Odor:* Pleasant
Other Information: It is shipped as a liquefied gas under its own vapor pressure.

Chemical Properties:

Specific Gravity: .724 *Vapor Density:* 1.6 *Boiling Point: – 11° F(–23.8° C)*
Melting Point: n/a *Vapor Pressure:* n/a *Solubility in water:* Yes
Other Information: Floats and boils on water. A flammable, irritating vapor is produced.

Health Hazards:

Inhalation Hazard: Headache, dizziness, loss of consciousness.
Ingestion Hazard: n/a
Absorption Hazard: Irritating to the skin and eyes. Will cause frostbite.
Hazards to Wildlife: n/a
Decontamination Procedures: Wash away any material with copious amounts of soap and water.
First Aid Procedures: Remove victim to fresh air, call emergency medical care. If not breathing give CPR. If breathing is difficult administer oxygen. Treat for shock.

Fire Hazards:

Flashpoint: n/a *Ignition temperature:* 662° F(350° C)
Flammable Explosive High Range: 50 *Low Range:* 2
Toxic Products of Combustion: n/a
Other Hazards: Containers may explode in a fire. Flashback along vapor trail may occur. Containers may explode if ignited in an enclosed area.
Possible extinguishing agents: Apply water from as far a distance as possible. Do not extinguish the fire unless the flow can be stopped.

Reactivity Hazards:

Reactive With: n/a *Other Reactions:* n/a

Corrosivity Hazards:

Corrosive With: n/a *Neutralizing Agent:* n/a

Radioactivity Hazards:

Radiation Emitted: n/a *Other Hazards:* n/a

Recommended Protection for Response Personnel:

Avoid breathing vapors, keep upwind. Structural protective clothing provides limited protection. Wash away any material which may have come into contact with the body with copious amounts of soap and water. If the material is involved in a fire or on fire, consider appropriate evacuation. (BLEVE)

DIMETHYL SULFATE

DOT Number: UN 1595 *DOT Hazard Class:* Corrosive *DOT Guide Number:* 57
Synonyms: methyl sulfate
STCC Number: 4933322 *Reportable Qty:* 1/.454
Mfg Name: E.I. Du Pont *Phone No:* 1-800-441-3637

Physical Description:

Physical Form: Liquid *Color:* Colorless *Odor:* Mild, onion
Other Information: n/a

Chemical Properties:

Specific Gravity: 1.3 *Vapor Density:* 4.4 *Boiling Point: 370° F(187.7° C)*
Melting Point: –25° F(–31.6° C) *Vapor Pressure:* .5 mm Hg at 68° F(20° C)
Solubility in water: Slight *Degree of Solubility:* 2.8%
Other Information: Sinks and mixes slowly with water.

Health Hazards:

Inhalation Hazard: n/a
Ingestion Hazard: Poisonous if swallowed.
Absorption Hazard: Poisonous if absorbed.
Hazards to Wildlife: n/a
Decontamination Procedures: Wash away any material with copious amounts of soap and water.
First Aid Procedures: Remove victim to fresh air, call emergency medical care. If not breathing give CPR. If breathing is difficult administer oxygen. Treat for shock.

Fire Hazards:

Flashpoint: 182° F(83.3° C) *Ignition temperature:* 370° F(187.7° C)
Flammable Explosive High Range: n/a *Low Range:* n/a
Toxic Products of Combustion: Flammable, toxic vapors are generated.
Other Hazards: n/a
Possible extinguishing agents: Use dry chemical, dry sand, or carbon dioxide. Do not use water on the material itself.

Reactivity Hazards:

Reactive With: n/a *Other Reactions:* n/a

Corrosivity Hazards:

Corrosive With: Metal, skin, and tissue
Neutralizing Agent: Crushed limestone, soda ash, or lime.

Radioactivity Hazards:

Radiation Emitted: n/a *Other Hazards:* n/a

Recommended Protection for Response Personnel:

Avoid breathing vapors, keep upwind. Structural protective clothing provides limited protection. Wash away any material which may have come into contact with the body with copious amounts of soap and water. Consider appropriate evacuation.

DIMETHYL SULFIDE

DOT Number: UN 1164 *DOT Hazard Class:* Flammable liquid *DOT Guide Number:* 27
Synonyms: MDMS, methyl sulfide, 2-thiapropane
STCC Number: 4908151 *Reportable Qty:* n/a
Mfg Name: Phillips Petroleum *Phone No:* 1-918-661-6600

Physical Description:

Physical Form: Liquid *Color:* Colorless to light yellow *Odor:* Unpleasant
Other Information: n/a

Chemical Properties:

Specific Gravity: .8 *Vapor Density:* 2.1 *Boiling Point: 99° F(37.7° C)*
Melting Point: n/a *Vapor Pressure:* n/a *Solubility in water:* Slight
Other Information: Lighter than and slightly soluble in water. Vapors are heavier than air.

Health Hazards:

Inhalation Hazard: n/a
Ingestion Hazard: Harmful if swallowed.
Absorption Hazard: Irritating to the skin and eyes.
Hazards to Wildlife: n/a
Decontamination Procedures: Wash away any material with copious amounts of soap and water.
First Aid Procedures: Remove victim to fresh air, call emergency medical care. If not breathing give CPR. If breathing is difficult administer oxygen. Treat for shock.

Fire Hazards:

Flashpoint: −36° F(−37.7° C) *Ignition temperature:* 403° F(206.1° C)
Flammable Explosive High Range: 19.7 *Low Range:* 2.2
Toxic Products of Combustion: Toxic and irritating sulfur dioxide is formed.
Other Hazards: Containers may explode in a fire, flashback along vapor trail may occur. Containers may explode if ignited in an enclosed area
Possible extinguishing agents: Apply water from as far a distance as possible. Use foam, dry chemical, or carbon dioxide.

Reactivity Hazards:

Reactive With: n/a *Other Reactions:* n/a

Corrosivity Hazards:

Corrosive With: n/a *Neutralizing Agent:* n/a

Radioactivity Hazards:

Radiation Emitted: n/a *Other Hazards:* n/a

Recommended Protection for Response Personnel:

Avoid breathing vapors, keep upwind. Wear a sealed chemical suit. (polycarbonate, butyl rubber). Wash away any material which may have come into contact with the body with copious amounts of soap and water. If the fire becomes uncontrollable, or if the containers are exposed to direct flame, consider appropriate evacuation.

DIMETHYLAMINE

DOT Number: UN 1032 *DOT Hazard Class:* Flammable gas *DOT Guide Number:* 19
Synonyms: dimethylamine anhydrous
STCC Number: 4905510 *Reportable Qty:* 1000/454
Mfg Name: GAF Corp. *Phone No:* 1-201-628-3000

Physical Description:

Physical Form: Gas *Color:* Colorless *Odor:* Dead fish or ammonia
Other Information: It is used to make other chemicals, as a solvent, and for many other uses.

Chemical Properties:

Specific Gravity: .67 *Vapor Density:* 1.6 *Boiling Point: 44° F(6.6° C)*
Melting Point: –134° F(–92.2° C) *Vapor Pressure:* 1.7 atm at 68° F(20° C) *Solubility in water:* Yes
Other Information: It is an asphyxiant and weighs 5.5 lbs/2.4 kg per gallon/3.8 l.

Health Hazards:

Inhalation Hazard: Will cause difficulty in breathing.
Ingestion Hazard: Harmful if swallowed.
Absorption Hazard: Frostbite or chemical type burns.
Hazards to Wildlife: Dangerous to aquatic life.
Decontamination Procedures: Wash away any material with copious amounts of soap and water.
First Aid Procedures: Remove victim to fresh air, call emergency medical care. If not breathing give CPR. If breathing is difficult administer oxygen. Treat for shock.

Fire Hazards:

Flashpoint: 20° F(–6.6° C) *Ignition temperature:* 752° F(400° C)
Flammable Explosive High Range: 14.4 *Low Range:* 2.8
Toxic Products of Combustion: Toxic oxides of nitrogen are produced.
Other Hazards: Vapors are heavier than air. Flames can flashback to the source. under fire conditions, containers may rupture or rocket (BLEVE)
Possible extinguishing agents: Do not extinguish fire unless flow can be stopped. Apply water from as far a distance as possible.

Reactivity Hazards:

Reactive With: n/a *Other Reactions:* n/a

Corrosivity Hazards:

Corrosive With: The gas itself is corrosive. *Neutralizing Agent:* n/a

Radioactivity Hazards:

Radiation Emitted: n/a *Other Hazards:* n/a

Recommended Protection for Response Personnel:

Avoid breathing vapors, keep upwind. Structural protective clothing provides limited protection. Wash away any material which may have come into contact with the body with copious amounts of soap and water. If the material is leaking or on fire, consider appropriate evacuation.

DIMETHYLDICHLOROSILANE

DOT Number: UN 1162 *DOT Hazard Class:* Flammable liquid *DOT Guide Number:* 29
Synonyms: None given
STCC Number: 4907610 *Reportable Qty:* n/a
Mfg Name: Union Carbide *Phone No:* 1-203-794-2000

Physical Description:

Physical Form: Liquid *Color:* Colorless *Odor:* Sharp, irritating
Other Information: n/a

Chemical Properties:

Specific Gravity: 1.1 *Vapor Density:* 4.4 *Boiling Point: 158° F(70° C)*
Melting Point: n/a *Vapor Pressure:* n/a *Solubility in water:* Decomposes
Other Information: Reacts violently with water. An irritating gas is produced upon contact with water.

Health Hazards:

Inhalation Hazard: n/a
Ingestion Hazard: Harmful if swallowed.
Absorption Hazard: Will burn the skin and eyes.
Hazards to Wildlife: n/a
Decontamination Procedures: Wash away any material with copious amounts of soap and water.
First Aid Procedures: Remove victim to fresh air, call emergency medical care. If not breathing give CPR. If breathing is difficult administer oxygen. Treat for shock.

Fire Hazards:

Flashpoint: 15° F(−9.4° C) *Ignition temperature:* 750° F(398.8° C)
Flammable Explosive High Range: 9.5 *Low Range:* 3.4
Toxic Products of Combustion: Hydrogen chloride and phosgene gases are formed which are toxic and irritating.
Other Hazards: Flashback along vapor trail may occur. Containers may explode if ignited in an enclosed area.
Possible extinguishing agents: Apply water from as far a distance as possible. Do not use water on the material itself. Use alcohol foam, dry chemical, or carbon dioxide.

Reactivity Hazards:

Reactive With: Water to form hydrochloric acid *Other Reactions:* n/a

Corrosivity Hazards:

Corrosive With: Metal and tissue *Neutralizing Agent:* Crushed limestone, soda ash, or lime.

Radioactivity Hazards:

Radiation Emitted: n/a *Other Hazards:* n/a

Recommended Protection for Response Personnel:

Avoid breathing vapors, keep upwind. Wear a sealed chemical suit (butyl rubber). Wash away any material which may have come into contact with the body with copious amounts of soap and water. If the fire becomes uncontrollable, or if the containers are exposed to direct flame, consider appropriate evacuation.

DIMETHYLFORMAMIDE

DOT Number: NA 2265 *DOT Hazard Class:* Combustible liquid *DOT Guide Number:* 26
Synonyms: N,N-dimethylformamide, DMF
STCC Number: 4913157 *Reportable Qty:* n/a
Mfg Name: E.I. Du Pont *Phone No:* 1-800-441-3637

Physical Description:

Physical Form: Liquid *Color:* Colorless *Odor:* Slight ammonia
Other Information: It is used as a solvent.

Chemical Properties:

Specific Gravity: 9.5 *Vapor Density:* n/a *Boiling Point: 307° F(152.7° C)*
Melting Point: −78° F(−61.1° C) *Vapor Pressure:* 3.5 mm Hg at 77° F(25° C) *Solubility in water:* Yes
Other Information: Floats and mixes with water. Vapors are heavier than air.

Health Hazards:

Inhalation Hazard: n/a
Ingestion Hazard: n/a
Absorption Hazard: Will burn the skin and eyes.
Hazards to Wildlife: n/a
Decontamination Procedures: Wash away any material with copious amounts of soap and water.
First Aid Procedures: Remove victim to fresh air, call emergency medical care. If not breathing give CPR. If breathing is difficult administer oxygen. Treat for shock.

Fire Hazards:

Flashpoint: 136° F(57.7° C) *Ignition temperature:* 833° F(445° C)
Flammable Explosive High Range: 15.2 *Low Range:* 2.2
Toxic Products of Combustion: Irritating vapors
Other Hazards: n/a
Possible extinguishing agents: Do not extinguish the fire unless the flow can be stopped. Cool all affected containers with water. Apply water from as far a distance as possible. Use alcohol foam, dry chemical or carbon dioxide.

Reactivity Hazards:

Reactive With: n/a *Other Reactions:* n/a

Corrosivity Hazards:

Corrosive With: n/a *Neutralizing Agent:* n/a

Radioactivity Hazards:

Radiation Emitted: n/a *Other Hazards:* n/a

Recommended Protection for Response Personnel:

Avoid breathing vapors, keep upwind. Wear a sealed chemical suit (butyl rubber). Wash away any material which may have come into contact with the body with copious amounts of soap and water. Consider appropriate evacuation.

DIMETHYLHEXANE DIHYDROPEROXIDE

DOT Number: UN 2174 *DOT Hazard Class:* Organic peroxide *DOT Guide Number:* 49
Synonyms: 2,5-dihydroperoxy-2,5-dimethylhexane; 2,5-dimethyhexane-2,5-dihydroperoxide
STCC Number: 4919140 *Reportable Qty:* n/a
Mfg Name: Pennwald Corp. *Phone No:* 1-215-587-7000

Physical Description:

Physical Form: Wet solid *Color:* White *Odor:* n/a
Other Information: n/a

Chemical Properties:

Specific Gravity: 1 *Vapor Density:* 6.5 *Boiling Point: n/a*
Melting Point: n/a *Vapor Pressure:* n/a *Solubility in water:* n/a
Other Information: May float or sink in water.

Health Hazards:

Inhalation Hazard: Coughing or difficulty in breathing.
Ingestion Hazard: Harmful if swallowed.
Absorption Hazard: Irritating to the skin and eyes.
Hazards to Wildlife: n/a
Decontamination Procedures: Wash away any material with copious amounts of soap and water.
First Aid Procedures: Remove victim to fresh air, call emergency medical care. If not breathing give CPR. If breathing is difficult administer oxygen. Treat for shock.

Fire Hazards:

Flashpoint: n/a *Ignition temperature:* n/a
Flammable Explosive High Range: n/a *Low Range:* n/a
Toxic Products of Combustion: n/a
Other Hazards: Can increase the intensity of fire when in contact with combustible materials. Containers may explode in fires.
Possible extinguishing agents: Use alcohol foam, dry chemical or carbon dioxide.

Reactivity Hazards:

Reactive With: n/a *Other Reactions:* n/a

Corrosivity Hazards:

Corrosive With: n/a *Neutralizing Agent:* n/a

Radioactivity Hazards:

Radiation Emitted: n/a *Other Hazards:* n/a

Recommended Protection for Response Personnel:

Avoid breathing vapors, keep upwind. Wear appropriate chemical clothing. Wash away any material which may have come into contact with the body with copious amounts of soap and water. Dangerously explosive!! Consider appropriate evacuation.

DIMETHYLHYDRAZINE

DOT Number: UN 1163 *DOT Hazard Class:* Flammable liquid *DOT Guide Number:* 57
Synonyms: dimazine, UDMH
STCC Number: n/a *Reportable Qty:* n/a
Mfg Name: FMC Corp. *Phone No:* 1-312-861-5900

Physical Description:

Physical Form: Liquid *Color:* Colorless *Odor:* Fishy or ammonia like
Other Information: Floats and mixes in water.

Chemical Properties:

Specific Gravity: .79 *Vapor Density:* 2.1 *Boiling Point: 146° F(63.3° C)*
Melting Point: −71° F(−57.2° C) *Vapor Pressure:* 103 mm Hg at 68° F(20° C)] *Solubility in water:* Yes
Other Information: A fuming liquid.

Health Hazards:

Inhalation Hazard: Poisonous if inhaled.
Ingestion Hazard: Poisonous if swallowed.
Absorption Hazard: Poisonous if the skin is exposed.
Hazards to Wildlife: n/a
Decontamination Procedures: Wash away any material with copious amounts of soap and water.
First Aid Procedures: Remove victim to fresh air, call emergency medical care, If not breathing give CPR. If breathing is difficult administer oxygen. Treat for shock.

Fire Hazards:

Flashpoint: 5° F(−15° C) *Ignition temperature:* 452-482° F(233-250° C)
Flammable Explosive High Range: 95 *Low Range:* 2
Toxic Products of Combustion: Poisonous gases are produced when heated.
Other Hazards: Tends to reignite unless diluted with much water. flashback along vapor trail may occur. Vapors may explode if ignited in an enclosed area.
Possible extinguishing agents: Flood with water.

Reactivity Hazards:

Reactive With: Plastics *Other Reactions:* n/a

Corrosivity Hazards:

Corrosive With: n/a *Neutralizing Agent:* n/a

Radioactivity Hazards:

Radiation Emitted: n/a *Other Hazards:* n/a

Recommended Protection for Response Personnel:

Avoid breathing vapors, keep upwind. Wear a sealed chemical suit (butyl rubber). Wash away any material which may have into contact with the body with copious amounts of soap and water. If the material is leaking or on fire, consider appropriate evacuation.

DIMETHYLZINC

DOT Number: UN 1370 *DOT Hazard Class:* Flammable liquid *DOT Guide Number:* 40
Synonyms: methylzinc, zinc dimethyl, zinc methyl
STCC Number: n/a *Reportable Qty:* n/a
Mfg Name: Ventron Corp. *Phone No:* 1-617-774-3100

Physical Description:

Physical Form: Liquid *Color:* Colorless *Odor:* n/a
Other Information: n/a

Chemical Properties:

Specific Gravity: 1.39 *Vapor Density:* 1 *Boiling Point: 113° F(45° C)*
Melting Point: −44° F(−42.2° C) *Vapor Pressure:* n/a *Solubility in water:* n/a
Other Information: Ignites when exposed to air. Reacts violently with water to produce flammable vapor.

Health Hazards:

Inhalation Hazard: n/a
Ingestion Hazard: Nausea or vomiting.
Absorption Hazard: Will burn the skin and eyes.
Hazards to Wildlife: n/a
Decontamination Procedures: Wash away any material with copious amounts of soap and water.
First Aid Procedures: Remove victim to fresh air, call emergency medical care. If not breathing give CPR. If breathing is difficult administer oxygen. Treat for shock.

Fire Hazards:

Flashpoint: Spontaneously ignites *Ignition temperature:* Below zero (−17.7° C)
Flammable Explosive High Range: n/a *Low Range:* n/a
Toxic Products of Combustion: Smoke contains zinc oxide which can irritate the lungs to cause metal fume fever.
Other Hazards: Ignites when exposed to air!!
Possible extinguishing agents: Dry chemical, sand or powdered limestone. Do not use water, foam, halogenated agent or carbon dioxide.

Reactivity Hazards:

Reactive With: Water to generate flammable methane gas. *Other Reactions:* n/a

Corrosivity Hazards:

Corrosive With: n/a *Neutralizing Agent:* n/a

Radioactivity Hazards:

Radiation Emitted: n/a *Other Hazards:* n/a

Recommended Protection for Response Personnel:

Avoid breathing vapors, keep upwind. Structural protective clothing provides limited protection. Dangerously flammable!! Avoid contact with this material! Consider appropriate evacuation.

DINITROANILINE

DOT Number: UN 1596 *DOT Hazard Class:* Poison *DOT Guide Number:* 56
Synonyms: 2,4-dinitroaniline
STCC Number: n/a *Reportable Qty:* n/a
Mfg Name: American Hoechst Corp. *Phone No:* 1-201-231-2000

Physical Description:

Physical Form: Solid *Color:* Yellow *Odor:* Musty
Other Information: n/a

Chemical Properties:

Specific Gravity: 1.61 *Vapor Density:* 6.3 *Boiling Point: n/a*
Melting Point: 368° F(186.6° C) *Vapor Pressure:* n/a *Solubility in water:* n/a
Other Information: Sinks in water

Health Hazards:

Inhalation Hazard: Poisonous if inhaled.
Ingestion Hazard: Poisonous if swallowed.
Absorption Hazard: Poisonous to the skin and eyes.
Hazards to Wildlife: n/a
Decontamination Procedures: Wash away any material with copious amounts of soap and water.
First Aid Procedures: Remove victim to fresh air, call emergency medical care. If not breathing give CPR. If breathing is difficult administer oxygen. Treat for shock.

Fire Hazards:

Flashpoint: 435° F(223.8° C) *Ignition temperature:* n/a
Flammable Explosive High Range: n/a *Low Range:* n/a
Toxic Products of Combustion: Vapors and combustion gases are irritating. poisonous gas is produced when heated.
Other Hazards: May explode if subjected to heat or flame.
Possible extinguishing agents: For small fires, use water, alcohol foam, dry chemical, or carbon dioxide.

Reactivity Hazards:

Reactive With: Oxidizing materials *Other Reactions:* May detonate when heated under confinement.

Corrosivity Hazards:

Corrosive With: n/a *Neutralizing Agent:* n/a

Radioactivity Hazards:

Radiation Emitted: n/a *Other Hazards:* n/a

Recommended Protection for Response Personnel:

Avoid breathing vapors, keep upwind. Wear appropriate chemical clothing. Wash away any material which may have come into contact with the body with copious amounts of soap and water. If the fire becomes uncontrollable, consider appropriate evacuation.

DINITROBENZENE (liquid)

DOT Number: UN 1597 *DOT Hazard Class:* Poison B *DOT Guide Number:* 56
Synonyms: none given
STCC Number: 4921422 *Reportable Qty:* 100/454
Mfg Name: Mallinckrodt Inc. *Phone No:* 1-314-895-2000

Physical Description:

Physical Form: Liquid *Color:* Colorless to yellow *Odor:* n/a
Other Information: n/a

Chemical Properties:

Specific Gravity: 1.3 *Vapor Density:* 5.8 *Boiling Point: 607° F(319.4° C)*
Melting Point: 109° F(42.7° C) *Vapor Pressure:* n/a *Solubility in water:* No
Other Information: Mixes and sinks slowly with water.

Health Hazards:

Inhalation Hazard: Poisonous if inhaled.
Ingestion Hazard: Poisonous if swallowed.
Absorption Hazard: Poisonous if absorbed.
Hazards to Wildlife: Dangerous to aquatic life.
Decontamination Procedures: Wash away any material with copious amounts of soap and water.
First Aid Procedures: Remove victim to fresh air, call emergency medical care. If not breathing give CPR. If breathing is difficult administer oxygen. Treat for shock.

Fire Hazards:

Flashpoint: 302° F(150° C) *Ignition temperature:* n/a
Flammable Explosive High Range: 22 *Low Range:* 2
Toxic Products of Combustion: Emits highly toxic fumes of nitrogen. May explode.
Other Hazards: May explode if subjected to heat and shock.
Possible extinguishing agents: Apply water from as far a distance as possible. Use alcohol foam, dry chemical or carbon dioxide.

Reactivity Hazards:

Reactive With: Oxidizing materials. *Other Reactions:* n/a

Corrosivity Hazards:

Corrosive With: n/a *Neutralizing Agent:* n/a

Radioactivity Hazards:

Radiation Emitted: n/a *Other Hazards:* n/a

Recommended Protection for Response Personnel:

Avoid breathing vapors, keep upwind. Wear appropriate chemical clothing, wash away any material which may have come into contact with the body with copious amounts of soap and water. If the fire becomes uncontrollable, consider appropriate evacuation.

DINITROBENZENE (solid)

DOT Number: UN 1597 *DOT Hazard Class:* Poison B *DOT Guide Number:* 56
Synonyms: 1,2-nitrobenzene, o-nitrobenzoyl
STCC Number: 4921421 *Reportable Qty:* 100/454
Mfg Name: Mallinckrodt Inc. *Phone No:* 1-314-895-2000

Physical Description:

Physical Form: Solid *Color:* Colorless to yellow *Odor:* n/a
Other Information: n/a

Chemical Properties:

Specific Gravity: 1.3 *Vapor Density:* 5.8 *Boiling Point: 607° F(319.4° C)*
Melting Point: 109° F(42.7° C) *Vapor Pressure:* n/a *Solubility in water:* No
Other Information: Mixes and sinks slowly with water.

Health Hazards:

Inhalation Hazard: Poisonous if inhaled.
Ingestion Hazard: Poisonous if swallowed.
Absorption Hazard: Poisonous if absorbed.
Hazards to Wildlife: Dangerous to aquatic life.
Decontamination Procedures: Wash away any material with copious amounts of soap and water.
First Aid Procedures: Remove victim to fresh air, call emergency medical care. If not breathing give CPR. If breathing is difficult administer oxygen. Treat for shock.

Fire Hazards:

Flashpoint: 302° F(150° C) *Ignition temperature:* n/a
Flammable Explosive High Range: 22 *Low Range:* 2
Toxic Products of Combustion: Emits highly toxic fumes of nitrogen. may explode.
Other Hazards: May explode if subjected to heat, shock, or friction.
Possible extinguishing agents: Apply water from as far a distance as possible. Use alcohol foam, dry chemical or carbon dioxide.

Reactivity Hazards:

Reactive With: Oxidizing materials. *Other Reactions:* n/a

Corrosivity Hazards:

Corrosive With: n/a *Neutralizing Agent:* n/a

Radioactivity Hazards:

Radiation Emitted: n/a *Other Hazards:* n/a

Recommended Protection for Response Personnel:

Avoid breathing vapors, keep upwind. Wear appropriate chemical clothing. Wash away any material which may have come into contact with the body with copious amounts of soap and water. If the fire becomes uncontrollable, consider appropriate evacuation.

DINITROCRESOL

DOT Number: UN 1598 *DOT Hazard Class:* Poison A *DOT Guide Number:* 53
Synonyms: 2,6-dinitro-o-cresol, 3,5-dinitro-o-cresol, 4,6-dinitro-o-cresol
STCC Number: n/a *Reportable Qty:* n/a
Mfg Name: J.T. Baker Chemical *Phone No:* 1-201-859-2151

Physical Description:

Physical Form: Solid *Color:* Yellow *Odor:* None
Other Information: n/a

Chemical Properties:

Specific Gravity: 1.1 *Vapor Density:* 6.8 *Boiling Point: n/a*
Melting Point: 180° F(82.2° C) *Vapor Pressure:* n/a *Solubility in water:* n/a
Other Information: Sinks in water

Health Hazards:

Inhalation Hazard: Poisonous if inhaled.
Ingestion Hazard: Poisonous if swallowed.
Absorption Hazard: Poisonous to the skin and eyes.
Hazards to Wildlife: Dangerous to aquatic life.
Decontamination Procedures: Wash away any material with copious amounts of soap and water.
First Aid Procedures: Remove victim to fresh air, call emergency medical care. If not breathing give CPR. If breathing is difficult administer oxygen. Treat for shock.

Fire Hazards:

Flashpoint: n/a *Ignition temperature:* n/a
Flammable Explosive High Range: n/a *Low Range:* n/a
Toxic Products of Combustion: Toxic oxides of nitrogen may be formed in fires.
Other Hazards: Containers may explode in fire.
Possible extinguishing agents: Water, foam, dry chemical, carbon dioxide. Cool all affected containers with large quantities of water.

Reactivity Hazards:

Reactive With: n/a *Other Reactions:* n/a

Corrosivity Hazards:

Corrosive With: n/a *Neutralizing Agent:* n/a

Radioactivity Hazards:

Radiation Emitted: n/a *Other Hazards:* n/a

Recommended Protection for Response Personnel:

Avoid breathing vapors, keep upwind. Wear a sealed chemical suit (polycarbonate, butyl rubber). Wash away any material which may have come into contact with the body with copious amounts of soap and water. If the fire becomes uncontrollable, consider appropriate evacuation.

DINITROPHENOL

DOT Number: UN 1320 *DOT Hazard Class:* Poison B *DOT Guide Number:* 36
Synonyms: Aldifen, α-dinitrophenol
STCC Number: 4921425 *Reportable Qty:* 10/4.54
Mfg Name: Martin Marietta Corp. *Phone No:* 1-301-897-6000

Physical Description:

Physical Form: Liquid *Color:* Yellow colored *Odor:* Sweet, musty
Other Information: n/a

Chemical Properties:

Specific Gravity: 1.68 *Vapor Density:* 6.3 *Boiling Point: n/a*
Melting Point: n/a *Vapor Pressure:* n/a *Solubility in water:* Slight
Other Information: Sinks in water

Health Hazards:

Inhalation Hazard: Poisonous if inhaled.
Ingestion Hazard: Poisonous if swallowed.
Absorption Hazard: Poisonous if absorbed.
Hazards to Wildlife: Dangerous to aquatic life.
Decontamination Procedures: Wash away any material with copious amounts of soap and water.
First Aid Procedures: Remove victim to fresh air, call emergency medical care. If not breathing give CPR. If breathing is difficult administer oxygen. Treat for shock.

Fire Hazards:

Flashpoint: n/a *Ignition temperature:* n/a
Flammable Explosive High Range: n/a *Low Range:* n/a
Toxic Products of Combustion: Toxic oxides of nitrogen are produced during combustion of this material.
Other Hazards: May explode if subject to heat or flame.
Possible extinguishing agents: Apply water from as far a distance as possible. Use foam, dry chemical, or carbon dioxide.

Reactivity Hazards:

Reactive With: Oxidizing materials and combustibles. *Other Reactions:* n/a

Corrosivity Hazards:

Corrosive With: n/a *Neutralizing Agent:* n/a

Radioactivity Hazards:

Radiation Emitted: n/a *Other Hazards:* n/a

Recommended Protection for Response Personnel:

Avoid breathing vapors, keep upwind. Wear a sealed chemical suit (polycarbonate, butyl rubber). Wash away any material which may have come into contact with the body with copious amounts of soap and water. If the fire becomes uncontrollable, consider appropriate evacuation.

DINITROTOLUENE (liquid)

DOT Number: NA 1600 *DOT Hazard Class:* ORM-E *DOT Guide Number:* 56
Synonyms: 2,4-dinitrotoluol, DNT
STCC Number: 4963120 *Reportable Qty:* 1000/454
Mfg Name: E.I. Du Pont *Phone No:* 1-800-441-3637

Physical Description:

Physical Form: Liquid *Color:* Yellow *Odor:* Slight
Other Information: It is used to make dyes, explosives, and to make other chemicals.

Chemical Properties:

Specific Gravity: 1.38 *Vapor Density:* 6.27 *Boiling Point: 572° F(300° C)*
Melting Point: 158° F(70° C) *Vapor Pressure:* 1 mm Hg at 68° F(20° C)
Solubility in water: No *Degree of Solubility:* .03%
Other Information: Solidifies and sinks in water.

Health Hazards:

Inhalation Hazard: n/a
Ingestion Hazard: Poisonous if swallowed.
Absorption Hazard: Poisonous if the skin is exposed.
Hazards to Wildlife: n/a
Decontamination Procedures: Wash away any material with copious amounts of soap and water.
First Aid Procedures: Remove victim to fresh air, call emergency medical care. If not breathing give CPR. If breathing is difficult administer oxygen. Treat for shock.

Fire Hazards:

Flashpoint: 404° F(206.6° C) *Ignition temperature:* n/a
Flammable Explosive High Range: n/a *Low Range:* n/a
Toxic Products of Combustion: Toxic oxides and dense black smoke are produced in a fire.
Other Hazards: Containers may explode in a fire.
Possible extinguishing agents: Apply water from as far a distance as possible. Use foam, dry chemical, or carbon dioxide.

Reactivity Hazards:

Reactive With: n/a *Other Reactions:* n/a

Corrosivity Hazards:

Corrosive With: n/a *Neutralizing Agent:* n/a

Radioactivity Hazards:

Radiation Emitted: n/a *Other Hazards:* n/a

Recommended Protection for Response Personnel:

Avoid breathing vapors, keep upwind. Wear a sealed chemical suit (polycarbonate, butyl rubber). Wash away any material which may have come into contact with the body with copious amounts of soap and water. If the material is involved in a fire, consider appropriate evacuation.

DINITROTOLUENE (solid)

DOT Number: NA 2038 *DOT Hazard Class:* ORM-E *DOT Guide Number:* 56
Synonyms: 2,4-dinitrotoluol, DNT
STCC Number: 4963115 *Reportable Qty:* 1000/454
Mfg Name: E.I. Du Pont *Phone No:* 1-800-441-3637

Physical Description:

Physical Form: Solid *Color:* Yellow to red *Odor:* Slight
Other Information: It is used to make dyes, explosives, and to make other chemicals.

Chemical Properties:

Specific Gravity: 1.38 *Vapor Density:* 6.27 *Boiling Point: 572° F(300° C)*
Melting Point: 158° F(70° C) *Vapor Pressure:* 1 mm Hg at 68° F(20° C)
Solubility in water: No *Degree of Solubility:* .03%
Other Information: Sinks in water.

Health Hazards:

Inhalation Hazard: n/a
Ingestion Hazard: Poisonous if swallowed.
Absorption Hazard: Poisonous if the skin is exposed.
Hazards to Wildlife: n/a
Decontamination Procedures: Wash away any material with copious amounts of soap and water.
First Aid Procedures: Remove victim to fresh air, call emergency medical care. If not breathing give CPR. If breathing is difficult administer oxygen. Treat for shock.

Fire Hazards:

Flashpoint: 404° F(206.6° C) *Ignition temperature:* n/a
Flammable Explosive High Range: n/a *Low Range:* n/a
Toxic Products of Combustion: Toxic oxides and dense black smoke are produced in a fire.
Other Hazards: Containers may explode in a fire.
Possible extinguishing agents: Apply water from as far a distance as possible. Use foam, dry chemical, or carbon dioxide.

Reactivity Hazards:

Reactive With: n/a *Other Reactions:* n/a

Corrosivity Hazards:

Corrosive With: n/a *Neutralizing Agent:* n/a

Radioactivity Hazards:

Radiation Emitted: n/a *Other Hazards:* n/a

Recommended Protection for Response Personnel:

Avoid breathing vapors, keep upwind. Wear a sealed chemical suit (polycarbonate, butyl rubber). Wash away any material which may have come into contact with the body with copious amounts of soap and water. If the material is involved in a fire, consider appropriate evacuation.

DIOXANE

DOT Number: UN 1165 *DOT Hazard Class:* Flammable liquid *DOT Guide Number:* 26
Synonyms: 1,4-dioxane, p-dioxane
STCC Number: 4909155 *Reportable Qty:* 1/.454
Mfg Name: Dow Chemical *Phone No:* 1-517-636-4400

Physical Description:

Physical Form: Liquid *Color:* Clear *Odor:* Slight alcohol
Other Information: Slightly heavier than and soluble in water. Vapors are heavier than air.

Chemical Properties:

Specific Gravity: 1 *Vapor Density:* 3 *Boiling Point: 214° F(101.1° C)*
Melting Point: 53° F(11.6° C) *Vapor Pressure:* 29 mm Hg at 68° F(20° C) *Solubility in water:* Yes
Other Information: n/a

Health Hazards:

Inhalation Hazard: Harmful if inhaled.
Ingestion Hazard: Harmful if swallowed.
Absorption Hazard: Irritating to the skin and eyes.
Hazards to Wildlife: n/a
Decontamination Procedures: Wash away any material with copious amounts of soap and water.
First Aid Procedures: Remove victim to fresh air, call emergency medical care. If not breathing give CPR. If breathing is difficult administer oxygen. Treat for shock.

Fire Hazards:

Flashpoint: 54° F(12.2° C) *Ignition temperature:* 356° F(180° C)
Flammable Explosive High Range: 22 *Low Range:* 2
Toxic Products of Combustion: Toxic vapors are generated when heated.
Other Hazards: Flashback along vapor trail may occur, vapor may explode if ignited in an enclosed area.
Possible extinguishing agents: Do not extinguish fire unless flow can be stopped. Use water from as far a distance as possible. Use alcohol foam, dry chemical or carbon dioxide.

Reactivity Hazards:

Reactive With: n/a *Other Reactions:* n/a

Corrosivity Hazards:

Corrosive With: n/a *Neutralizing Agent:* n/a

Radioactivity Hazards:

Radiation Emitted: n/a *Other Hazards:* n/a

Recommended Protection for Response Personnel:

Avoid breathing vapors, keep upwind. Structural protective clothing provides limited protection. Wash away any material which may have come into contact with the body with copious amounts of soap and water. Consider appropriate evacuation.

DIPENTENE

DOT Number: NA 2052 *DOT Hazard Class:* Combustible liquid *DOT Guide Number:* 27
Synonyms: limonene, terpenene
STCC Number: 4913149 *Reportable Qty:* n/a
Mfg Name: Hercules Inc. *Phone No:* 1-302-594-5000

Physical Description:

Physical Form: Liquid *Color:* Colorless to light yellow *Odor:* Pleasant, lemon like
Other Information: It is used as a solvent, and a dispersing agent for oils, paints, and lacquers.

Chemical Properties:

Specific Gravity: .9 *Vapor Density:* 4.7 *Boiling Point: 339° F(170.5° C)*
Melting Point: n/a *Vapor Pressure:* 1 mm Hg at 57° F(13.8° C) *Solubility in water:* No
Other Information: Vapors are heavier than air. Floats on water. Weighs 7.2 lbs/3.2 kg per gallon/3.8 l.

Health Hazards:

Inhalation Hazard: n/a
Ingestion Hazard: Harmful if swallowed.
Absorption Hazard: Irritating to the skin and eyes.
Hazards to Wildlife: n/a
Decontamination Procedures: Wash away any material with copious amounts of soap and water.
First Aid Procedures: Remove victim to fresh air, call emergency medical care. If not breathing give CPR. If breathing is difficult administer oxygen. Treat for shock.

Fire Hazards:

Flashpoint: 113° F(45° C) *Ignition temperature:* 458° F(236.6° C)
Flammable Explosive High Range: 6.1 *Low Range:* .7
Toxic Products of Combustion: n/a
Other Hazards: Containers may explode in a fire.
Possible extinguishing agents: Apply water from as far a distance as possible. Use foam, dry chemical, or carbon dioxide. carbon dioxide.

Reactivity Hazards:

Reactive With: n/a *Other Reactions:* n/a

Corrosivity Hazards:

Corrosive With: n/a *Neutralizing Agent:* n/a

Radioactivity Hazards:

Radiation Emitted: n/a *Other Hazards:* n/a

Recommended Protection for Response Personnel:

Avoid breathing vapors, keep upwind. Structural protective clothing provides limited protection. Wash away any material which may have come into contact with the body with copious amounts of soap and water. Consider appropriate evacuation.

DIPHENYLAMINE

DOT Number: n/a *DOT Hazard Class:* n/a *DOT Guide Number:* n/a
Synonyms: anilinobenzene, n-phenylaniline
STCC Number: n/a *Reportable Qty:* n/a
Mfg Name: Crowley Chemical Co. *Phone No:* 1-203-794-2000

Physical Description:

Physical Form: Solid *Color:* Tan to brown *Odor:* Pleasant
Other Information: n/a

Chemical Properties:

Specific Gravity: 1.06 *Vapor Density:* 8.5 *Boiling Point: 576° F(302.2° C)*
Melting Point: 127° F(52.7° C) *Vapor Pressure:* 1 mm Hg at 236° F(113.3° C) *Solubility in water:* n/a
Other Information: Sinks in water.

Health Hazards:

Inhalation Hazard: Harmful if inhaled.
Ingestion Hazard: Harmful if swallowed.
Absorption Hazard: Irritating to the skin and eyes.
Hazards to Wildlife: n/a
Decontamination Procedures: Wash away any material with copious amounts of soap and water.
First Aid Procedures: Remove victim to fresh air, call emergency medical care. If not breathing give CPR. If breathing is difficult administer oxygen. Treat for shock.

Fire Hazards:

Flashpoint: 302° F(150° C) *Ignition temperature:* 1175° F(635° C)
Flammable Explosive High Range: n/a *Low Range:* n/a
Toxic Products of Combustion: Toxic oxides of nitrogen may be formed in fire.
Other Hazards: Dust may be explosive if mixed with air in critical proportions and in the presence of a source of ignition.
Possible extinguishing agents: Apply water from as far a distance as possible. Water or foam may cause frothing. Use foam, dry chemical, or carbon dioxide.

Reactivity Hazards:

Reactive With: n/a *Other Reactions:* n/a

Corrosivity Hazards:

Corrosive With: n/a *Neutralizing Agent:* n/a

Radioactivity Hazards:

Radiation Emitted: n/a *Other Hazards:* n/a

Recommended Protection for Response Personnel:

Avoid breathing vapors, keep upwind. Wear a sealed chemical suit (polycarbonate, butyl rubber). Wash away any material which may have come into contact with the body with copious amounts of soap and water. Consider appropriate evacuation.

DIPHENYLDICHLOROSILANE

DOT Number: UN 1769 *DOT Hazard Class:* Corrosive *DOT Guide Number:* 29
Synonyms: dichlorodiphenylsilane
STCC Number: 4934235 *Reportable Qty:* n/a
Mfg Name: Mallinckrodt Inc. *Phone No:* 1-314-895-2000

Physical Description:

Physical Form: Liquid *Color:* Colorless *Odor:* Sharp, irritating
Other Information: n/a

Chemical Properties:

Specific Gravity: 1.22 *Vapor Density:* 8.4 *Boiling Point: 581° F(305° C)*
Melting Point: n/a *Vapor Pressure:* n/a *Solubility in water:* Reacts
Other Information: Reacts with water. An irritating vapor is produced.

Health Hazards:

Inhalation Hazard: n/a
Ingestion Hazard: Harmful if swallowed.
Absorption Hazard: Will burn the skin and eyes.
Hazards to Wildlife: n/a
Decontamination Procedures: Wash away any material with copious amounts of soap and water.
First Aid Procedures: Remove victim to fresh air, call emergency medical care. If not breathing give CPR. If breathing is difficult administer oxygen. Treat for shock.

Fire Hazards:

Flashpoint: 288° F(142.2° C) *Ignition temperature:* n/a
Flammable Explosive High Range: n/a *Low Range:* n/a
Toxic Products of Combustion: Hydrochloric acid and phosgene fumes may be formed.
Other Hazards: Difficult to extinguish, reignition may occur.
Possible extinguishing agents: Apply water from as far a distance as possible. Do not use water on the material itself. Use dry chemical, dry sand, or carbon dioxide.

Reactivity Hazards:

Reactive With: Water to form hydrochloric acid *Other Reactions:* n/a

Corrosivity Hazards:

Corrosive With: Metals and tissue *Neutralizing Agent:* Crushed limestone, soda ash, or lime.

Radioactivity Hazards:

Radiation Emitted: n/a *Other Hazards:* n/a

Recommended Protection for Response Personnel:

Avoid breathing vapors, keep upwind. Wear appropriate chemical clothing. Wash away any material which may have come into contact with the body with copious amounts of soap and water. Consider appropriate evacuation.

DIPHENYLMETHYL DIISOCYANATE

DOT Number: UN 2489 *DOT Hazard Class:* N/A *DOT Guide Number:* 53
Synonyms: Carwinate 125 M, Hylene-M50, MDI, Multrathane M, Nacconate 300, Vilrathane 4300
STCC Number: n/a *Reportable Qty:* n/a
Mfg Name: Fisher Scientific *Phone No:* 1-412-349-3322

Physical Description:

Physical Form: Solid *Color:* White to light yellow *Odor:* n/a
Other Information: n/a

Chemical Properties:

Specific Gravity: 1.2 *Vapor Density:* 8.6 *Boiling Point: 738° F(392.2° C)*
Melting Point: 100° F(37.7° C) *Vapor Pressure:* n/a *Solubility in water:* n/a
Other Information: Sinks in water

Health Hazards:

Inhalation Hazard: n/a
Ingestion Hazard: n/a
Absorption Hazard: Irritating to the skin and eyes.
Hazards to Wildlife: n/a
Decontamination Procedures: Wash away any material with copious amounts of soap and water.
First Aid Procedures: Remove victim to fresh air, call emergency medical care. If not breathing give CPR. If breathing is difficult administer oxygen. Treat for shock.

Fire Hazards:

Flashpoint: 425° F(218.3° C) *Ignition temperature:* n/a
Flammable Explosive High Range: n/a *Low Range:* n/a
Toxic Products of Combustion: Toxic vapors are generated when heated.
Other Hazards: Solid melts and burns.
Possible extinguishing agents: Use foam, dry chemical, carbon dioxide.

Reactivity Hazards:

Reactive With: Water, a slow reaction forming nonhazardous carbon dioxide gas. *Other Reactions:* n/a

Corrosivity Hazards:

Corrosive With: n/a *Neutralizing Agent:* n/a

Radioactivity Hazards:

Radiation Emitted: n/a *Other Hazards:* n/a

Recommended Protection for Response Personnel:

Avoid breathing vapors, keep upwind. Wear a sealed chemical suit (polycarbonate, butyl rubber). Wash away any material which may have come into contact with the body with copious amounts of soap and water. If the fire becomes uncontrollable, consider appropriate evacuation.

DIPROPYLAMINE

DOT Number: UN 2383 *DOT Hazard Class:* Flammable liquid *DOT Guide Number:* 68
Synonyms: dipropylamine, n-dipropylamine, DNPA, n-propyl-1-propanamine
STCC Number: 4909157 *Reportable Qty:* n/a
Mfg Name: Pennwald Corp. *Phone No:* 1-215-587-7000

Physical Description:

Physical Form: Liquid *Color:* Colorless *Odor:* Ammonia like
Other Information: It is used for making other chemicals.

Chemical Properties:

Specific Gravity: .739 *Vapor Density:* 1 *Boiling Point: 126° F(52.2° C)*
Melting Point: −39.3° F(−39.6° C) *Vapor Pressure:* 21-22 mm Hg at 68° F(20° C) *Solubility in water:* Yes
Other Information: It is moderately soluble in water, and lighter than water. May be expected to form a floating slick that quickly dissolves.

Health Hazards:

Inhalation Hazard: Irritating
Ingestion Hazard: Irritating and burning, will cause nausea & vomiting
Absorption Hazard: Severe irritation to the skin and eyes.
Hazards to Wildlife: n/a
Decontamination Procedures: Wash away any material with copious amounts of soap and water.
First Aid Procedures: Remove victim to fresh air, call emergency medical care. If not breathing give CPR. If breathing is difficult administer oxygen. Treat for shock.

Fire Hazards:

Flashpoint: 53° F(11.6° C) *Ignition temperature:* 570° F(298.8° C)
Flammable Explosive High Range: n/a *Low Range:* 2.8
Toxic Products of Combustion: Carbon monoxide and nitrogen oxide.
Other Hazards: Flashback along vapor trail may occur. Vapors may explode if ignited in an enclosed area.
Possible extinguishing agents: Use alcohol foam, dry chemical or carbon dioxide. Apply water from as far a distance as possible.

Reactivity Hazards:

Reactive With: Strong oxidizers *Other Reactions:* n/a

Corrosivity Hazards:

Corrosive With: Copper, aluminum, zinc, and galvanized surfaces. *Neutralizing Agent:* n/a

Radioactivity Hazards:

Radiation Emitted: n/a *Other Hazards:* n/a

Recommended Protection for Response Personnel:

Avoid breathing dust, keep upwind. Wear a sealed chemical suit (butyl rubber). Wash away any material which may have come into contact with the body with copious amounts of soap and water. If the fire becomes uncontrollable, consider appropriate evacuation.

DIQUAT (liquid)

DOT Number: NA 2781 *DOT Hazard Class:* ORM-E *DOT Guide Number:* 55
Synonyms: Dextrone, Aquacide, Reglone
STCC Number: 4963339 *Reportable Qty:* 1000/454
Mfg Name: Chevron Chemical Co. *Phone No:* 1-415-233-3737

Physical Description:

Physical Form: Liquid *Color:* n/a *Odor:* n/a
Other Information: It is used as a herbicide. It is a wettable powder, or water emulsifiable liquid.

Chemical Properties:

Specific Gravity: 1.25 *Vapor Density:* 1 *Boiling Point: n/a*
Melting Point: n/a *Vapor Pressure:* n/a *Solubility in water:* Mixes in water
Other Information: In case of damage to or leaking from containers, contact the Pesticide Safety Team Network at 1-800-424-9300.

Health Hazards:

Inhalation Hazard: Poisonous if inhaled.
Ingestion Hazard: Poisonous if swallowed.
Absorption Hazard: Irritating to the skin and eyes.
Hazards to Wildlife: Dangerous to aquatic life and waterfowl.
Decontamination Procedures: Wash away any material with copious amounts of soap and water.
First Aid Procedures: Remove victim to fresh air, call emergency medical care. If not breathing give CPR. If breathing is difficult administer oxygen. Treat for shock.

Fire Hazards:

Flashpoint: n/a *Ignition temperature:* n/a
Flammable Explosive High Range: n/a *Low Range:* n/a
Toxic Products of Combustion: n/a
Other Hazards: n/a
Possible extinguishing agents: Extinguish fire using suitable agent for the type of surrounding fire.

Reactivity Hazards:

Reactive With: n/a *Other Reactions:* n/a

Corrosivity Hazards:

Corrosive With: n/a *Neutralizing Agent:* n/a

Radioactivity Hazards:

Radiation Emitted: n/a *Other Hazards:* n/a

Recommended Protection for Response Personnel:

Avoid breathing vapors, keep upwind. Wear a sealed chemical suit (polycarbonate, butyl rubber). Wash away any material which may have come into contact with the body with copious amounts of soap and water. Consider appropriate evacuation.

DIQUAT (solid)

DOT Number: NA 2781 *DOT Hazard Class:* ORM-E *DOT Guide Number:* 55
Synonyms: Aquacide, Dextrone, Reglone
STCC Number: 4963344 *Reportable Qty:* 1000/454
Mfg Name: Chevron Chemical Co. *Phone No:* 1-415-233-3737

Physical Description:

Physical Form: Solid *Color:* Yellow to reddish brown *Odor:* n/a
Other Information: Used as a herbicide, and is a wettable powder.

Chemical Properties:

Specific Gravity: 1.25 *Vapor Density:* 1 *Boiling Point: n/a*
Melting Point: n/a *Vapor Pressure:* n/a *Solubility in water:* Mixes in water
Other Information: In case of damage to or leaking from containers, contact the Pesticide Safety Team Network at 1-800-424-9300.

Health Hazards:

Inhalation Hazard: Poisonous if inhaled.
Ingestion Hazard: Poisonous if swallowed.
Absorption Hazard: Irritating to the skin and eyes.
Hazards to Wildlife: Dangerous to aquatic life and waterfowl.
Decontamination Procedures: Wash away any material with copious amounts of soap and water.
First Aid Procedures: Remove victim to fresh air, call emergency medical care. If not breathing give CPR. If breathing is difficult administer oxygen. Treat for shock.

Fire Hazards:

Flashpoint: n/a *Ignition temperature:* n/a
Flammable Explosive High Range: n/a *Low Range:* n/a
Toxic Products of Combustion: n/a
Other Hazards: n/a
Possible extinguishing agents: Extinguish fire using suitable agent for the type of surrounding fire.

Reactivity Hazards:

Reactive With: n/a *Other Reactions:* n/a

Corrosivity Hazards:

Corrosive With: n/a *Neutralizing Agent:* n/a

Radioactivity Hazards:

Radiation Emitted: n/a *Other Hazards:* n/a

Recommended Protection for Response Personnel:

Avoid breathing vapors, keep upwind. Wear a sealed chemical suit (polycarbonate, butyl rubber). Wash away any material which may have come into contact with the body with copious amounts of soap and water. Consider appropriate evacuation.

DISULFOTON (liquid)

DOT Number: NA 2783 *DOT Hazard Class:* Poison B *DOT Guide Number:* 55
Synonyms: Di-Syston, Thiodemeton
STCC Number: 4921511 *Reportable Qty:* 1/.454
Mfg Name: Mobay Chemical Corp. *Phone No:* 1-412-777-2000

Physical Description:

Physical Form: Liquid *Color:* Yellow *Odor:* Sulfur
Other Information: Used as a pesticide.

Chemical Properties:

Specific Gravity: 1.14 *Vapor Density:* 9.45 *Boiling Point: 143° F(61.6° C)*
Melting Point: n/a *Vapor Pressure:* n/a *Solubility in water:* Mixes slowly
Other Information: In case of damage to or leaking from containers, contact the Pesticide Safety Team Network at 1-800-424-9300.

Health Hazards:

Inhalation Hazard: Poisonous if inhaled.
Ingestion Hazard: Poisonous if swallowed.
Absorption Hazard: Poisonous if the skin is exposed.
Hazards to Wildlife: Dangerous to aquatic life and waterfowl.
Decontamination Procedures: Wash away any material with copious amounts of soap and water.
First Aid Procedures: Remove victim to fresh air, call emergency medical care. If not breathing give CPR. If breathing is difficult administer oxygen. Treat for shock.

Fire Hazards:

Flashpoint: n/a *Ignition temperature:* n/a
Flammable Explosive High Range: n/a *Low Range:* n/a
Toxic Products of Combustion: Material may ignite and emit toxic fumes. containers may explode.
Other Hazards: n/a
Possible extinguishing agents: Extinguish fire using suitable agent for the type of surrounding fire. Use foam, dry chemical and carbon dioxide.

Reactivity Hazards:

Reactive With: n/a *Other Reactions:* n/a

Corrosivity Hazards:

Corrosive With: n/a *Neutralizing Agent:* n/a

Radioactivity Hazards:

Radiation Emitted: n/a *Other Hazards:* n/a

Recommended Protection for Response Personnel:

Avoid breathing vapors, keep upwind. Wear appropriate chemical clothing. Wash away any material which may have come into contact with the body with copious amounts of soap and water. Consider appropriate evacuation.

DISULFOTON (solid)

DOT Number: NA 2783 *DOT Hazard Class:* Poison B *DOT Guide Number:* 55
Synonyms: Di-Syston, Thiodemeton
STCC Number: 4921512 *Reportable Qty:* 1/.454
Mfg Name: Mobay Chemical Corp. *Phone No:* 1-412-777-2000

Physical Description:

Physical Form: Solid *Color:* Dark yellow *Odor:* Sulfur
Other Information: Used as a pesticide. Liquid is absorbed on a dry carrier as a wettable powder.

Chemical Properties:

Specific Gravity: 1.14 *Vapor Density:* 9.45 *Boiling Point: 143° F(61.6° C)*
Melting Point: n/a *Vapor Pressure:* n/a *Solubility in water:* Yes
Other Information: In case of damage to or leaking from containers, contact the Pesticide Safety Team Network at 1-800-424-9300.

Health Hazards:

Inhalation Hazard: Poisonous if inhaled.
Ingestion Hazard: Poisonous if swallowed.
Absorption Hazard: Poisonous if the skin is exposed.
Hazards to Wildlife: Dangerous to aquatic life and waterfowl.
Decontamination Procedures: Wash away any material with copious amounts of soap and water.
First Aid Procedures: Remove victim to fresh air, call emergency medical care. If not breathing give CPR. If breathing is difficult administer oxygen. Treat for shock.

Fire Hazards:

Flashpoint: n/a *Ignition temperature:* n/a
Flammable Explosive High Range: n/a *Low Range:* n/a
Toxic Products of Combustion: Material may ignite and emit toxic fumes. containers may explode.
Other Hazards: n/a
Possible extinguishing agents: Extinguish fire using suitable agent for the type of surrounding fire. Use foam, dry chemical and carbon dioxide.

Reactivity Hazards:

Reactive With: n/a *Other Reactions:* n/a

Corrosivity Hazards:

Corrosive With: n/a *Neutralizing Agent:* n/a

Radioactivity Hazards:

Radiation Emitted: n/a *Other Hazards:* n/a

Recommended Protection for Response Personnel:

Avoid breathing vapors, keep upwind. Wear appropriate chemical clothing, Wash away any material which may have come into contact with the body with copious amounts of soap and water. Consider appropriate evacuation.

DIURON

DOT Number: NA 2767 *DOT Hazard Class:* ORM-E *DOT Guide Number:* 55
Synonyms: Di-on, Diurex, Karmex, Marmer
STCC Number: 4962622 *Reportable Qty:* 100/45.4
Mfg Name: E.I. Du Pont *Phone No:* 1-800-441-3637

Physical Description:

Physical Form: Solid *Color:* White *Odor:* Odorless
Other Information: In case of damage to, or leaking from these containers, contact the Pesticide Safety Team Network at 1-800-424-9300.

Chemical Properties:

Specific Gravity: n/a *Vapor Density:* 8 *Boiling Point: n/a*
Melting Point: n/a *Vapor Pressure:* n/a *Solubility in water:* Yes
Other Information: Avoid contact with solid and dust. Keep people away.

Health Hazards:

Inhalation Hazard: Harmful if inhaled.
Ingestion Hazard: Harmful if swallowed.
Absorption Hazard: Irritating to the skin and eyes.
Hazards to Wildlife: Dangerous to aquatic life.
Decontamination Procedures: Wash away any material with copious amounts of soap and water.
First Aid Procedures: Remove victim to fresh air, call emergency medical care. If not breathing give CPR. If breathing is difficult administer oxygen. Treat for shock.

Fire Hazards:

Flashpoint: n/a *Ignition temperature:* n/a
Flammable Explosive High Range: n/a *Low Range:* n/a
Toxic Products of Combustion: Highly toxic fumes are emitted.
Other Hazards: Decomposes at 356° F(180° C)
Possible extinguishing agents: Extinguish fire using suitable agent for the type of surrounding fire.

Reactivity Hazards:

Reactive With: n/a *Other Reactions:* n/a

Corrosivity Hazards:

Corrosive With: n/a *Neutralizing Agent:* n/a

Radioactivity Hazards:

Radiation Emitted: n/a *Other Hazards:* n/a

Recommended Protection for Response Personnel:

Avoid breathing vapors, keep upwind. Wear appropriate chemical clothing. Wash away any material which may have come into contact with the body with copious amounts of soap and water. Consider appropriate evacuation.

DODECYLBENZENESULFONIC ACID

DOT Number: NA 2584 *DOT Hazard Class:* Corrosive *DOT Guide Number:* 60
Synonyms: Conoco SA597, Lacconol 988 A
STCC Number: 4931426 *Reportable Qty:* 1000/454
Mfg Name: Monsanto Chemical *Phone No:* 1-314-694-1000

Physical Description:

Physical Form: Liquid *Color:* Light yellow to brown *Odor:* Possible odor of SO_2
Other Information: It is used in making detergents.

Chemical Properties:

Specific Gravity: 1 *Vapor Density:* 1 *Boiling Point: 440° F(226.6° C)*
Melting Point: n/a *Vapor Pressure:* n/a *Solubility in water:* Yes
Other Information: n/a

Health Hazards:

Inhalation Hazard: n/a
Ingestion Hazard: Harmful if swallowed.
Absorption Hazard: Will burn the skin and eyes.
Hazards to Wildlife: Dangerous to aquatic life.
Decontamination Procedures: Wash away any material with copious amounts of soap and water.
First Aid Procedures: Remove victim to fresh air, call emergency medical care. If not breathing give CPR. If breathing is difficult administer oxygen. Treat for shock.

Fire Hazards:

Flashpoint: 300° F(148.8° C) *Ignition temperature:* n/a
Flammable Explosive High Range: n/a *Low Range:* n/a
Toxic Products of Combustion: May give off SO_3, SO_2, and H_2S.
Other Hazards: n/a
Possible extinguishing agents: Apply water from as far a distance as possible. Use alcohol foam, dry chemical, or carbon dioxide.

Reactivity Hazards:

Reactive With: n/a *Other Reactions:* n/a

Corrosivity Hazards:

Corrosive With: Metal and tissue *Neutralizing Agent:* Crushed limestone, soda ash, or lime.

Radioactivity Hazards:

Radiation Emitted: n/a *Other Hazards:* n/a

Recommended Protection for Response Personnel:

Avoid breathing vapors, keep upwind. Wear appropriate chemical clothing. Wash away any material which may have come into contact with the body with copious amounts of soap and water. Consider appropriate evacuation.

DODECYL TRICHLOROSILANE

DOT Number: UN 1771 *DOT Hazard Class:* Corrosive *DOT Guide Number:* 60
Synonyms: none given
STCC Number: 4934240 *Reportable Qty:* n/a
Mfg Name: Dow Chemical *Phone No:* 1-517-636-4400

Physical Description:

Physical Form: Liquid *Color:* Colorless *Odor:* Sharp, irritating
Other Information: Reacts with water. An irritating gas is produced upon contact with water.

Chemical Properties:

Specific Gravity: 1.03 *Vapor Density:* 10 *Boiling Point: 300° F(148.8° C)*
Melting Point: n/a *Vapor Pressure:* n/a *Solubility in water:* Reacts with water
Other Information: n/a

Health Hazards:

Inhalation Hazard: Irritates mucous membranes.
Ingestion Hazard: Causes severe burns to the mouth and stomach.
Absorption Hazard: Causes severe burns to the skin and eyes.
Hazards to Wildlife: n/a
Decontamination Procedures: Wash away any material with copious amounts of soap and water.
First Aid Procedures: Remove victim to fresh air, call emergency medical care. If not breathing give CPR. If breathing is difficult administer oxygen. Treat for shock.

Fire Hazards:

Flashpoint: 150° F(65.5° C) *Ignition temperature:* n/a
Flammable Explosive High Range: n/a *Low Range:* n/a
Toxic Products of Combustion: Hydrochloric and phosgene fumes may be formed in fires.
Other Hazards: n/a
Possible extinguishing agents: Do not use water or foam. Use dry chemical, carbon dioxide or sand.

Reactivity Hazards:

Reactive With: Surface moisture *Other Reactions:* n/a

Corrosivity Hazards:

Corrosive With: Metals and tissue *Neutralizing Agent:* Crushed limestone, soda ash, or lime.

Radioactivity Hazards:

Radiation Emitted: n/a *Other Hazards:* n/a

Recommended Protection for Response Personnel:

Avoid breathing vapors, keep upwind. Wear a sealed chemical suit (butyl rubber). Wash away any material which may have come into contact with the body with copious amounts of soap and water. Consider appropriate evacuation.

DURSBAN

DOT Number: UN 1615 *DOT Hazard Class:* Poison B *DOT Guide Number:* n/a
Synonyms: Chlorpyrifos, Dowco 179, ENT 27,311, Killmaster, Lorsban
STCC Number: n/a *Reportable Qty:* n/a
Mfg Name: Dow Chemical *Phone No:* 1-517-636-4400

Physical Description:

Physical Form: Solid *Color:* White *Odor:* Mercaptan
Other Information: n/a

Chemical Properties:

Specific Gravity: n/a *Vapor Density:* 1 *Boiling Point: n/a*
Melting Point: 108° F(42.2° C) *Vapor Pressure:* n/a *Solubility in water:* n/a
Other Information: Sinks in water

Health Hazards:

Inhalation Hazard: Poisonous if inhaled.
Ingestion Hazard: Poisonous if swallowed.
Absorption Hazard: Poisonous to the skin and eyes.
Hazards to Wildlife: Dangerous to aquatic life.
Decontamination Procedures: Wash away any material with copious amounts of soap and water.
First Aid Procedures: Remove victim to fresh air, call emergency medical care. If not breathing give CPR. If breathing is difficult administer oxygen. Treat for shock.

Fire Hazards:

Data are not available. Use extreme caution. Treat as though flammable. Material is more of a health hazard than a fire hazard!!
Toxic Products of Combustion: n/a
Other Hazards: n/a
Possible extinguishing agents: Use foam, dry chemical, carbon dioxide, and water.

Reactivity Hazards:

Reactive With: n/a *Other Reactions:* n/a

Corrosivity Hazards:

Corrosive With: n/a *Neutralizing Agent:* n/a

Radioactivity Hazards:

Radiation Emitted: n/a *Other Hazards:* n/a

Recommended Protection for Response Personnel:

Avoid breathing vapors, keep upwind. Wear appropriate sealed chemical suit. Wash away any material which may have come into contact with the body with copious amounts of soap and water. If the fire becomes uncontrollable, consider appropriate evacuation.

ENDOSULFAN (liquid)

DOT Number: NA 2761 *DOT Hazard Class:* Poison B *DOT Guide Number:* 55
Synonyms: Cyclodan, Malix, Thiodan
STCC Number: 4921517 *Reportable Qty:* 1/.454
Mfg Name: Hooker Chemical Co. *Phone No:* 1-716-278-7000

Physical Description:

Physical Form: Liquid *Color:* Brown *Odor:* Sulfur dioxide
Other Information: It is used as a pesticide, herbicide, and fungicide.

Chemical Properties:

Specific Gravity: 1.74 *Vapor Density:* 14 *Boiling Point: n/a*
Melting Point: n/a *Vapor Pressure:* n/a *Solubility in water:* Yes
Other Information: In case of damage to, or leaking from these containers, contact the Pesticide Safety Network at 1-800-424-9300.

Health Hazards:

Inhalation Hazard: Poisonous if inhaled.
Ingestion Hazard: Poisonous if swallowed.
Absorption Hazard: Poisonous if the skin is exposed.
Hazards to Wildlife: Dangerous to both aquatic life and waterfowl.
Decontamination Procedures: Wash away any material with copious amounts of soap and water.
First Aid Procedures: Remove victim to fresh air, call emergency medical care. If not breathing give CPR. If breathing is difficult administer oxygen. Treat for shock.

Fire Hazards:

Flashpoint: n/a *Ignition temperature:* n/a
Flammable Explosive High Range: n/a *Low Range:* n/a
Toxic Products of Combustion: Decomposes to liberate SO_2 (sulfur dioxide).
Other Hazards: Containers may explode if exposed to heat or flame.
Possible extinguishing agents: Extinguish fire using suitable agent for the type of surrounding fire. Use foam, dry chemical, or carbon dioxide.

Reactivity Hazards:

Reactive With: n/a *Other Reactions:* n/a

Corrosivity Hazards:

Corrosive With: n/a *Neutralizing Agent:* n/a

Radioactivity Hazards:

Radiation Emitted: n/a *Other Hazards:* n/a

Recommended Protection for Response Personnel:

Avoid breathing vapors, keep upwind. Wear a sealed chemical suit (polycarbonate, butyl rubber). Wash away any material which may have come into contact with the body with copious amounts of soap and water. Consider appropriate evacuation.

ENDOSULFAN (solid)

DOT Number: NA 2761 *DOT Hazard Class:* Poison B *DOT Guide Number:* 55
Synonyms: Cyclodan, Malix, Thiodan
STCC Number: 4921518 *Reportable Qty:* 1/.454
Mfg Name: Hooker Chemical Co. *Phone No:* 1-716-278-7000

Physical Description:

Physical Form: Solid *Color:* Brown *Odor:* Sulfur dioxide
Other Information: It is used as a pesticide, herbicide, and fungicide.

Chemical Properties:

Specific Gravity: 1.74 *Vapor Density:* 14 *Boiling Point: n/a*
Melting Point: n/a *Vapor Pressure:* n/a *Solubility in water:* Yes
Other Information: In case of damage to, or leaking from these containers, contact the Pesticide Safety Network at 1-800-424-9300.

Health Hazards:

Inhalation Hazard: Poisonous if inhaled.
Ingestion Hazard: Poisonous if swallowed.
Absorption Hazard: Poisonous if the skin is exposed.
Hazards to Wildlife: Dangerous to both aquatic life and waterfowl.
Decontamination Procedures: Wash away any material with copious amounts of soap and water.
First Aid Procedures: Remove victim to fresh air, call emergency medical care. If not breathing give CPR. If breathing is difficult administer oxygen. Treat for shock.

Fire Hazards:

Flashpoint: n/a *Ignition temperature:* n/a
Flammable Explosive High Range: n/a *Low Range:* n/a
Toxic Products of Combustion: Decomposes to liberate SO_2 (sulfur dioxide).
Other Hazards: Containers may explode if exposed to heat or flame.
Possible extinguishing agents: Extinguish fire using suitable agent for the type of surrounding fire. Use alcohol foam, dry chemical, or carbon dioxide.

Reactivity Hazards:

Reactive With: n/a *Other Reactions:* n/a

Corrosivity Hazards:

Corrosive With: n/a *Neutralizing Agent:* n/a

Radioactivity Hazards:

Radiation Emitted: n/a *Other Hazards:* n/a

Recommended Protection for Response Personnel:

Avoid breathing vapors, keep upwind. Wear a sealed chemical suit (polycarbonate, butyl rubber). Wash away any material which may have come into contact with the body with copious amounts of soap and water. Consider appropriate evacuation.

ENDRIN (liquid)

DOT Number: NA 2761 *DOT Hazard Class:* Poison B *DOT Guide Number:* 55
Synonyms: Hexadrin, Mendrin
STCC Number: 4921521 *Reportable Qty:* 1/.454
Mfg Name: Shell Chemical *Phone No:* 1-713-241-6161

Physical Description:

Physical Form: Liquid *Color:* Colorless to tan *Odor:* Odorless
Other Information: In case of damage to, or leaking from these containers, contact the Pesticide Safety Network at 1-800-424-9300.

Chemical Properties:

Specific Gravity: 1.65 *Vapor Density:* 13 *Boiling Point: Decomposes*
Melting Point: n/a *Vapor Pressure:* 0 mm Hg at 68° F(20° C) *Solubility in water:* Yes
Other Information: It is dissolved in a liquid carrier.

Health Hazards:

Inhalation Hazard: Poisonous if inhaled.
Ingestion Hazard: Poisonous if swallowed.
Absorption Hazard: Poisonous if the skin is exposed.
Hazards to Wildlife: Dangerous to both aquatic life and waterfowl.
Decontamination Procedures: Wash away any material with copious amounts of soap and water.
First Aid Procedures: Remove victim to fresh air, call emergency medical care. If not breathing give CPR. If breathing is difficult administer oxygen. Treat for shock.

Fire Hazards:

Flashpoint: 80° F(26.6° C) *Ignition temperature:* n/a
Flammable Explosive High Range: 7 *Low Range:* 1.1
Toxic Products of Combustion: Toxic hydrogen chloride and phosgene gases are generated when the solution burns.
Other Hazards: Water may be ineffective on a solution fire.
Possible extinguishing agents: Apply water from as far a distance as possible. Use foam, dry chemical, or carbon dioxide.

Reactivity Hazards:

Reactive With: n/a *Other Reactions:* n/a

Corrosivity Hazards:

Corrosive With: n/a *Neutralizing Agent:* n/a

Radioactivity Hazards:

Radiation Emitted: n/a *Other Hazards:* n/a

Recommended Protection for Response Personnel:

Avoid breathing vapors, keep upwind. Wear a sealed chemical suit (polycarbonate, butyl rubber). Wash away any material which may have come into contact with the body with copious amounts of soap and water. Consider appropriate evacuation.

ENDRIN (solid)

DOT Number: NA 2761 *DOT Hazard Class:* Poison B *DOT Guide Number:* 55
Synonyms: Hexadrin, Mendrin
STCC Number: 4921522 *Reportable Qty:* 1/.454
Mfg Name: Shell Chemical *Phone No:* 1-713-241-6161

Physical Description:

Physical Form: Solid *Color:* Colorless to tan *Odor:* Odorless
Other Information: In case of damage to, or leaking from these containers, contact the Pesticide Safety Network at 1-800-424-9300.

Chemical Properties:

Specific Gravity: 1.65 *Vapor Density:* 13 *Boiling Point: Decomposes*
Melting Point: n/a *Vapor Pressure:* 0 mm Hg at 68° F(20° C) *Solubility in water:* Insoluble
Other Information: Sinks in water.

Health Hazards:

Inhalation Hazard: Poisonous if inhaled.
Ingestion Hazard: Poisonous if swallowed.
Absorption Hazard: Poisonous if the skin is exposed.
Hazards to Wildlife: Dangerous to both aquatic life and waterfowl.
Decontamination Procedures: Wash away any material with copious amounts of soap and water.
First Aid Procedures: Remove victim to fresh air, call emergency medical care. If not breathing give CPR. If breathing is difficult administer oxygen. Treat for shock.

Fire Hazards:

Flashpoint: n/a *Ignition temperature:* n/a
Flammable Explosive High Range: 7 *Low Range:* 1.1
Toxic Products of Combustion: Poisonous gases are produced in a fire.
Other Hazards: n/a
Possible extinguishing agents: Apply water from as far a distance as possible. Use foam, dry chemical, or carbon dioxide.

Reactivity Hazards:

Reactive With: n/a *Other Reactions:* n/a

Corrosivity Hazards:

Corrosive With: n/a *Neutralizing Agent:* n/a

Radioactivity Hazards:

Radiation Emitted: n/a *Other Hazards:* n/a

Recommended Protection for Response Personnel:

Avoid breathing vapors, keep upwind. Wear a sealed chemical suit (polycarbonate, butyl rubber). Wash away any material which may have come into contact with the body with copious amounts of soap and water. Consider appropriate evacuation.

EPICHLORHYDRIN

DOT Number: UN 2023 *DOT Hazard Class:* Flammable liquid *DOT Guide Number:* 30
Synonyms: 1-chloro-2,3-epoxy-propane
STCC Number: 4907420 *Reportable Qty:* 1000/454
Mfg Name: Shell Chemical *Phone No:* 1-713-241-6161

Physical Description:

Physical Form: Liquid *Color:* Colorless *Odor:* Sweet, garlic odor
Other Information: Used to make plastics, and as a solvent.

Chemical Properties:

Specific Gravity: 1.2 *Vapor Density:* 3.2 *Boiling Point: 239° F(115° C)*
Melting Point: −72° F(−57.7° C) *Vapor Pressure:* 13 mm Hg at 68° F(20° C)
Solubility in water: Yes *Degree of Solubility:* 6.4%
Other Information: If contaminated, material is subject to polymerization, with evolution of heat.

Health Hazards:

Inhalation Hazard: Poisonous if inhaled.
Ingestion Hazard: Poisonous if swallowed.
Absorption Hazard: Will burn the skin and eyes.
Hazards to Wildlife: n/a
Decontamination Procedures: Wash away any material with copious amounts of soap and water.
First Aid Procedures: Remove victim to fresh air, call emergency medical care. If not breathing give CPR. If breathing is difficult administer oxygen. Treat for shock.

Fire Hazards:

Flashpoint: 88° F(31.1° C) *Ignition temperature:* 722° F(383.3° C)
Flammable Explosive High Range: 21 *Low Range:* 3.8
Toxic Products of Combustion: Poisonous gases are produced in a fire.
Other Hazards: Vapors may explode and ignite in an enclosed area.
Possible extinguishing agents: Use alcohol foam, dry chemical, or carbon dioxide.

Reactivity Hazards:

Reactive With: Water *Other Reactions:* Heat

Corrosivity Hazards:

Corrosive With: n/a *Neutralizing Agent:* n/a

Radioactivity Hazards:

Radiation Emitted: n/a *Other Hazards:* n/a

Recommended Protection for Response Personnel:

Avoid breathing vapors, keep upwind. Wear a sealed chemical suit (polycarbonate, butyl rubber). Wash away any material which may have come into contact with the body with copious amounts of soap and water. If the cylinders are under direct flame, consider appropriate evacuation.

ETHANE

DOT Number: UN 1035 *DOT Hazard Class:* Flammable gas *DOT Guide Number:* 22
Synonyms: methylmethane
STCC Number: 4905731 *Reportable Qty:* None
Mfg Name: Phillips Petroleum *Phone No:* 1-918-661-6600

Physical Description:

Physical Form: Gas *Color:* Colorless *Odor:* Odorless
Other Information: May only be shipped in cylinders.

Chemical Properties:

Specific Gravity: .546 *Vapor Density:* 1 *Boiling Point: – 128° F(–88.8° C)*
Melting Point: n/a *Vapor Pressure:* n/a *Solubility in water:* No
Other Information: Can asphyxiate by the displacement of air.

Health Hazards:

Inhalation Hazard: Will cause difficulty in breathing.
Ingestion Hazard: n/a
Absorption Hazard: Will cause frostbite.
Hazards to Wildlife: n/a
Decontamination Procedures: Wash away any material with copious amounts of soap and water.
First Aid Procedures: Remove victim to fresh air, call emergency medical care. If not breathing give CPR. If breathing is difficult administer oxygen. Treat for shock.

Fire Hazards:

Flashpoint: –211° F(–135° C) *Ignition temperature:* 882° F(472.2° C)
Flammable Explosive High Range: 12.5 *Low Range:* 3
Toxic Products of Combustion: n/a
Other Hazards: Flashback along vapor trail may occur. Vapors may explode if ignited in an enclosed area.
Possible extinguishing agents: Water

Reactivity Hazards:

Reactive With: n/a *Other Reactions:* n/a

Corrosivity Hazards:

Corrosive With: n/a *Neutralizing Agent:* n/a

Radioactivity Hazards:

Radiation Emitted: n/a *Other Hazards:* n/a

Recommended Protection for Response Personnel:

Avoid breathing vapors, keep upwind. Structural protective clothing provides limited protection. Wash away any material which may have come into contact with the body with copious amounts of soap and water. under fire conditions, the cylinders may violently rupture and rocket. (BLEVE) Consider appropriate evacuation.

ETHION

DOT Number: NA 2783 *DOT Hazard Class:* Poison B *DOT Guide Number:* 55
Synonyms: Diethion, Nialate, Nia 12 40
STCC Number: 4921565 *Reportable Qty:* 10/4.54
Mfg Name: FMC Corporation *Phone No:* 1-312-861-5900

Physical Description:

Physical Form: Liquid *Color:* Clear to amber *Odor:* Disagreeable
Other Information: Used as a pesticide.

Chemical Properties:

Specific Gravity: 1.22 *Vapor Density:* 13 *Boiling Point: n/a*
Melting Point: n/a *Vapor Pressure:* n/a *Solubility in water:* Yes
Other Information: In case of damage to or leaking from containers, contact the Pesticide Safety Team Network at 1-800-424-9300.

Health Hazards:

Inhalation Hazard: Toxic by inhalation.
Ingestion Hazard: Poisonous if swallowed.
Absorption Hazard: Poisonous if the skin is exposed.
Hazards to Wildlife: Dangerous to both aquatic life and waterfowl.
Decontamination Procedures: Wash away any material with copious amounts of soap and water.
First Aid Procedures: Remove victim to fresh air, call emergency medical care. If not breathing give CPR. If breathing is difficult administer oxygen. Treat for shock.

Fire Hazards:

Flashpoint: n/a *Ignition temperature:* n/a
Flammable Explosive High Range: n/a *Low Range:* n/a
Toxic Products of Combustion: Fire may produce toxic and irritating fumes.
Other Hazards: May decompose rapidly with violence at 302° F(150° C)
Possible extinguishing agents: Use suitable agent for the type of surrounding fire.

Reactivity Hazards:

Reactive With: n/a *Other Reactions:* n/a

Corrosivity Hazards:

Corrosive With: n/a *Neutralizing Agent:* n/a

Radioactivity Hazards:

Radiation Emitted: n/a *Other Hazards:* n/a

Recommended Protection for Response Personnel:

Avoid breathing vapors, keep upwind. Wear a sealed chemical suit (polycarbonate, butyl rubber). Wash away any material which may have come into contact with the body with copious amounts of soap and water. Consider appropriate evacuation.

ETHYL ACETATE

DOT Number: UN 1173 *DOT Hazard Class:* Flammable liquid *DOT Guide Number:* 26
Synonyms: acetic ester, acetic ether
STCC Number: 4909160 *Reportable Qty:* 5000/2270
Mfg Name: Mallinckrodt Inc. *Phone No:* 1-314-895-2000

Physical Description:

Physical Form: Liquid *Color:* Colorless *Odor:* Fruity
Other Information: n/a

Chemical Properties:

Specific Gravity: .9 *Vapor Density:* 3 *Boiling Point: 171° F(77.2° C)*
Melting Point: −117° F(−82.7° C) *Vapor Pressure:* 76 mm Hg at 68° F(20° C)
Solubility in water: Slight *Degree of Solubility:* 8.7%
Other Information: Lighter, slightly soluble in water. Vapors are heavier than air.

Health Hazards:

Inhalation Hazard: Irritating to eyes, ears, and throat.
Ingestion Hazard: Harmful if swallowed.
Absorption Hazard: Irritating to the skin and eyes.
Hazards to Wildlife: n/a
Decontamination Procedures: Wash away any material with copious amounts of soap and water.
First Aid Procedures: Remove victim to fresh air, call emergency medical care. If not breathing give CPR. If breathing is difficult administer oxygen. Treat for shock.

Fire Hazards:

Flashpoint: 24° F(−4.4° C) *Ignition temperature:* 800° F(426.6° C)
Flammable Explosive High Range: 11.5 *Low Range:* 2
Toxic Products of Combustion: n/a
Other Hazards: May flashback along vapor trail. Vapors may ignite in an enclosed area.
Possible extinguishing agents: Apply water from as far a distance as possible. Use alcohol foam, dry chemical, or carbon dioxide.

Reactivity Hazards:

Reactive With: n/a *Other Reactions:* n/a

Corrosivity Hazards:

Corrosive With: n/a *Neutralizing Agent:* n/a

Radioactivity Hazards:

Radiation Emitted: n/a *Other Hazards:* n/a

Recommended Protection for Response Personnel:

Avoid breathing vapors, keep upwind. Structural protective clothing provides limited protection. Wash away any material which may have come into contact with the body with copious amounts of soap and water. Consider appropriate evacuation.

ETHYL ACETOACETATE

DOT Number: NA 1993 *DOT Hazard Class:* Combustible liquid *DOT Guide Number:* 27
Synonyms: acetoacetic acid, ethyl ester; EAA; acetoacetic ester; diacetic ether; ethyl 3-oxobutanoate
STCC Number: 4907215 *Reportable Qty:* n/a
Mfg Name: Mallinckrodt Inc. *Phone No:* 1-314-895-2000

Physical Description:

Physical Form: Liquid *Color:* Colorless *Odor:* Pleasant, fruity
Other Information: It is used in organic synthetics, and in lacquers and paints.

Chemical Properties:

Specific Gravity: 1.03 *Vapor Density:* 4.5 *Boiling Point: 356° F(180° C)*
Melting Point: −112° F(−80° C) *Vapor Pressure:* 1 mm Hg at 83° F(−63.8° C) *Solubility in water:* Yes
Other Information: Mixes with water.

Health Hazards:

Inhalation Hazard: Harmful if inhaled.
Ingestion Hazard: Harmful if swallowed.
Absorption Hazard: Irritating to the skin and the eyes.
Hazards to Wildlife: n/a
Decontamination Procedures: Wash away any material with copious amounts of soap and water.
First Aid Procedures: Remove victim to fresh air, call emergency medical care. If not breathing give CPR. If breathing is difficult administer oxygen. Treat for shock.

Fire Hazards:

Flashpoint: 135 ° F (57.2° C) *Ignition temperature:* 563° F(295° C)
Flammable Explosive High Range: 9.5 *Low Range:* 1.4
Toxic Products of Combustion: n/a
Other Hazards: Water may not be effective.
Possible extinguishing agents: Do not extinguish the fire unless the flow can be stopped. Apply water from as far a distance as possible. Cool all affected containers. Use alcohol foam, dry chemical, or carbon dioxide.

Reactivity Hazards:

Reactive With: n/a *Other Reactions:* n/a

Corrosivity Hazards:

Corrosive With: n/a *Neutralizing Agent:* n/a

Radioactivity Hazards:

Radiation Emitted: n/a *Other Hazards:* n/a

Recommended Protection for Response Personnel:

Avoid breathing vapors, keep upwind. Structural protective clothing provides limited protection. Wash away any material which may have come into contact with the body with copious amounts of soap and water. Consider appropriate evacuation.

ETHYL ACRYLATE, INHIBITED

DOT Number: UN 1917 *DOT Hazard Class:* Flammable liquid *DOT Guide Number:* 26
Synonyms: acrylic acid, ethyl ester
STCC Number: 4909167 *Reportable Qty:* 1000/454
Mfg Name: Rohm and Haas Co. *Phone No:* 1-215-592-3000

Physical Description:

Physical Form: Liquid *Color:* Colorless *Odor:* Sharp, fruity
Other Information: Used as a monomer for acrylic resins.

Chemical Properties:

Specific Gravity: .9 *Vapor Density:* 3.5 *Boiling Point: 211° F(99.4° C)*
Melting Point: −103° F(−75° C) *Vapor Pressure:* 29.5 mm Hg at 68° F(20° C)
Solubility in water: Slight *Degree of Solubility:* 1.5%
Other Information: Liquid is lighter than and immiscible in water. Vapors arc heavier than air.

Health Hazards:

Inhalation Hazard: Cause headache and nausea.
Ingestion Hazard: Harmful if swallowed.
Absorption Hazard: Will burn the skin and eyes.
Hazards to Wildlife: Dangerous to aquatic life.
Decontamination Procedures: Wash away any material with copious amounts of soap and water.
First Aid Procedures: Remove victim to fresh air, call emergency medical care. If not breathing give CPR. If breathing is difficult administer oxygen. Treat for shock.

Fire Hazards:

Flashpoint: 50° F(10° C) *Ignition temperature:* 702° F(372.2° C)
Flammable Explosive High Range: 14 *Low Range:* 1.4
Toxic Products of Combustion: Toxic vapors are generated when this material is heated to high temperatures.
Other Hazards: n/a
Possible extinguishing agents: Apply water from as far a distance as possible. Use foam, dry chemical, or carbon dioxide.

Reactivity Hazards:

Reactive With: n/a *Other Reactions:* n/a

Corrosivity Hazards:

Corrosive With: n/a *Neutralizing Agent:* n/a

Radioactivity Hazards:

Radiation Emitted: n/a *Other Hazards:* n/a

Recommended Protection for Response Personnel:

Avoid breathing vapors, keep upwind. Wear a sealed chemical suit (chlorinated polyethylene, butyl rubber). Wash away any material which may have come into contact with the body with copious amounts of soap and water. Consider appropriate evacuation.

ETHYL ACRYLATE, (liquid)

DOT Number: UN 1173 *DOT Hazard Class:* Flammable liquid *DOT Guide Number:* 26
Synonyms: acrylic acid, ethyl ester
STCC Number: 4907215 *Reportable Qty:* 1000/454
Mfg Name: Rohm and Haas Co. *Phone No:* 1-215-592-3000

Physical Description:

Physical Form: Liquid *Color:* Clear *Odor:* Acrid
Other Information: Used to make paints and plastics.

Chemical Properties:

Specific Gravity: .9 *Vapor Density:* 3.5 *Boiling Point: 211° F(99.4° C)*
Melting Point: –103° F(–75° C) *Vapor Pressure:* 29.5 mm Hg at 68° F(20° C)
Solubility in water: Slight *Degree of Solubility:* 1.5%
Other Information: Liquid is lighter than and immiscible in water. Vapors are heavier than air.

Health Hazards:

Inhalation Hazard: Causes headache and nausea.
Ingestion Hazard: Harmful if swallowed.
Absorption Hazard: Will burn the skin and eyes.
Hazards to Wildlife: Dangerous to aquatic life.
Decontamination Procedures: Wash away any material with copious amounts of soap and water.
First Aid Procedures: Remove victim to fresh air, call emergency medical care. If not breathing give CPR. If breathing is difficult administer oxygen. Treat for shock.

Fire Hazards:

Flashpoint: 60° F(15.5° C) *Ignition temperature:* 702° F(372.2° C)
Flammable Explosive High Range: 14 *Low Range:* 1.4
Toxic Products of Combustion: Toxic vapors are generated when this material is heated to high temperatures.
Other Hazards: Material is polymerizable if contaminated or heated. Container may give off heat and violently rupture.
Possible extinguishing agents: Apply water from as far a distance as possible. Use foam, dry chemical, or carbon dioxide.

Reactivity Hazards:

Reactive With: n/a *Other Reactions:* n/a

Corrosivity Hazards:

Corrosive With: n/a *Neutralizing Agent:* n/a

Radioactivity Hazards:

Radiation Emitted: n/a *Other Hazards:* n/a

Recommended Protection for Response Personnel:

Avoid breathing vapors, keep upwind. Wear a sealed chemical suit (polychlorinated ethylene, butyl rubber). Wash away any material which may have come into contact with the body with copious amounts of soap and water. Consider appropriate evacuation.

ETHYL ALCOHOL

DOT Number: UN 1170 *DOT Hazard Class:* Flammable liquid *DOT Guide Number:* 26
Synonyms: ethanol (international name)
STCC Number: 4909146 *Reportable Qty:* n/a
Mfg Name: Georgia Pacific Co. *Phone No:* 1-404-521-4000

Physical Description:

Physical Form: Liquid *Color:* Colorless *Odor:* Alcohol
Other Information: Weighs 6.5 lbs/2.9 kg per gallon/3.8 l.

Chemical Properties:

Specific Gravity: .8 *Vapor Density:* 1.6 *Boiling Point: 173° F(78.3° C)*
Melting Point: n/a *Vapor Pressure:* n/a *Solubility in water:* Yes
Other Information: Vapors are heavier than air.

Health Hazards:

Inhalation Hazard: Irritating to eyes, nose, and throat.
Ingestion Hazard: Liquid is not harmful if swallowed.
Absorption Hazard: Not harmful.
Hazards to Wildlife: Dangerous to aquatic life.
Decontamination Procedures: Wash away any material with copious amounts of soap and water.
First Aid Procedures: Remove victim to fresh air, call emergency medical care. If not breathing give CPR. If breathing is difficult administer oxygen. Treat for shock.

Fire Hazards:

Flashpoint: 55° F(12.7° C) *Ignition temperature:* 685° F(362.7° C)
Flammable Explosive High Range: 19 *Low Range:* 3.3
Toxic Products of Combustion: n/a
Other Hazards: Material burns without a flame.
Possible extinguishing agents: Apply water from as far a distance as possible. Use alcohol foam, dry chemical or carbon dioxide.

Reactivity Hazards:

Reactive With: n/a *Other Reactions:* n/a

Corrosivity Hazards:

Corrosive With: n/a *Neutralizing Agent:* n/a

Radioactivity Hazards:

Radiation Emitted: n/a *Other Hazards:* n/a

Recommended Protection for Response Personnel:

Avoid breathing vapors, keep upwind. Structural protective clothing provides limited protection. Wash away any material which may have come into contact with the body with copious amounts of soap and water. Consider appropriate evacuation.

ETHYL ALUMINUM SESQUICHLORIDE

DOT Number: UN 1925 *DOT Hazard Class:* Flammable liquid *DOT Guide Number:* 40
Synonyms: EASE
STCC Number: 4906063 *Reportable Qty:* n/a
Mfg Name: Texas Alkyls *Phone No:* 1-713-479-8411

Physical Description:

Physical Form: Liquid *Color:* Colorless to yellow *Odor:* n/a
Other Information: It is pyrophoric. Used to make other chemicals.

Chemical Properties:

Specific Gravity: 1.09 *Vapor Density:* 1.56 *Boiling Point: 297° F(147.2° C)*
Melting Point: n/a *Vapor Pressure:* 2.2 mm Hg at 68° F(20° C) *Solubility in water:* Reacts
Other Information: Ignites when exposed to air. Reacts violently with water. Poisonous and flammable gases are produced on contact with water.

Health Hazards:

Inhalation Hazard: Harmful if inhaled.
Ingestion Hazard: Harmful if swallowed.
Absorption Hazard: Will burn the skin if it is exposed.
Hazards to Wildlife: n/a
Decontamination Procedures: Wash away any material with copious amounts of soap and water.
First Aid Procedures: Remove victim to fresh air, call emergency medical care. If not breathing give CPR. If breathing is difficult administer oxygen. Treat for shock.

Fire Hazards:

Flashpoint: −4° F(−20° C) *Ignition temperature:* Spontaneous
Flammable Explosive High Range: n/a *Low Range:* n/a
Toxic Products of Combustion: Intense smoke may cause metal fume fever. Irritating hydrogen chloride also formed.
Other Hazards: Ignites when exposed to air.
Possible extinguishing agents: Use graphite, soda ash, or powdered sodium chloride.

Reactivity Hazards:

Reactive With: Water to form hydrogen chloride and flammable ethane gas. *Other Reactions:* n/a

Corrosivity Hazards:

Corrosive With: Common metals *Neutralizing Agent:* n/a

Radioactivity Hazards:

Radiation Emitted: n/a *Other Hazards:* n/a

Recommended Protection for Response Personnel:

Avoid breathing vapors, keep upwind. Wear a sealed chemical suit. (butyl rubber) Wash away any material which may come into contact with the body with copious amounts of soap and water. If the fire becomes uncontrollable, or if the containers are exposed to direct flame, consider appropriate evacuation.

ETHYLAMINE

DOT Number: UN 1036 *DOT Hazard Class:* Flammable liquid *DOT Guide Number:* 68
Synonyms: aminoethane, monoethylamine
STCC Number: n/a *Reportable Qty:* n/a
Mfg Name: Penwald Corp. *Phone No:* 1-215-587-7000

Physical Description:

Physical Form: Liquid *Color:* Colorless *Odor:* Strong ammonialike
Other Information: n/a

Chemical Properties:

Specific Gravity: .8 *Vapor Density:* 1.6 *Boiling Point: 62° F(16.6° C)*
Melting Point: −114° F(−81.1° C) *Vapor Pressure:* 1.18 atm at 68° F(20° C) *Solubility in water:* Yes
Other Information: n/a

Health Hazards:

Inhalation Hazard: Harmful if inhaled.
Ingestion Hazard: Harmful if swallowed.
Absorption Hazard: Irritating to the skin and eyes.
Hazards to Wildlife: Dangerous to aquatic life.
Decontamination Procedures: Wash away any material with copious amounts of soap and water.
First Aid Procedures: Remove victim to fresh air, call emergency medical care. If not breathing give CPR. If breathing is difficult administer oxygen. Treat for shock.

Fire Hazards:

Flashpoint: 0° F(−17.7° C) *Ignition temperature:* 725° F(385° C)
Flammable Explosive High Range: 14 *Low Range:* 3.5
Toxic Products of Combustion: Irritating and toxic oxides of nitrogen may be formed.
Other Hazards: Containers may explode in a fire. Flashback along vapor trail may occur. Containers may explode if ignited in an enclosed area.
Possible *Possible extinguishing agents:* Use alcohol foam, dry chemical, or carbon dioxide.

Reactivity Hazards:

Reactive With: n/a *Other Reactions:* n/a

Corrosivity Hazards:

Corrosive With: Paint, plastics, and rubber *Neutralizing Agent:* Flush with water

Radioactivity Hazards:

Radiation Emitted: n/a *Other Hazards:* n/a

Recommended Protection for Response Personnel:

Avoid breathing vapors, keep upwind. Wear a sealed chemical suit (butyl rubber). Wash away any material which may have come into contact with the body with copious amounts of soap and water. If the fire becomes uncontrollable, or if the containers are exposed to direct flame, consider appropriate evacuation.

ETHYLBENZENE

DOT Number: UN 1175 *DOT Hazard Class:* Flammable liquid *DOT Guide Number:* 26
Synonyms: ethylbenzol, penylethane
STCC Number: 4909163 *Reportable Qty:* 1000/454
Mfg Name: Dow Chemical *Phone No:* 1-517-636-4400

Physical Description:

Physical Form: Liquid *Color:* Colorless *Odor:* Aromatic
Other Information: It is used as a solvent, and to make other chemicals.

Chemical Properties:

Specific Gravity: .9 *Vapor Density:* 3.7 *Boiling Point: 277° F(136.1° C)*
Melting Point: –139° F(–95° C) *Vapor Pressure:* 7.1 mm Hg at 68° F(20° C)
Solubility in water: No *Degree of Solubility:* .015%
Other Information: Lighter, insoluble in water. Vapors are heavier than air. Weighs 7.2 lbs/3.26 kg per gallon/3.8 l.

Health Hazards:

Inhalation Hazard: Hazardous if inhaled.
Ingestion Hazard: Hazardous if swallowed.
Absorption Hazard: Irritating to the skin and eyes.
Hazards to Wildlife: n/a
Decontamination Procedures: Wash away any material with copious amounts of soap and water.
First Aid Procedures: Remove victim to fresh air, call emergency medical care. If not breathing give CPR. If breathing is difficult administer oxygen. Treat for shock.

Fire Hazards:

Flashpoint: 20° F(–6.6° C) *Ignition temperature:* 810° F(432.2° C)
Flammable Explosive High Range: 6.7 *Low Range:* 1
Toxic Products of Combustion: n/a
Other Hazards: Containers may explode in a fire. Flashback along vapor trail may occur. Containers may explode if ignited in an enclosed area.
Possible extinguishing agents: Use foam, dry chemical, or carbon dioxide. Apply water from as far a distance as possible.

Reactivity Hazards:

Reactive With: n/a *Other Reactions:* n/a

Corrosivity Hazards:

Corrosive With: n/a *Neutralizing Agent:* n/a

Radioactivity Hazards:

Radiation Emitted: n/a *Other Hazards:* n/a

Recommended Protection for Response Personnel:

Avoid breathing vapors, keep upwind. Structural protective clothing provides limited protection. Wash away any material which may have come into contact with the body with copious amounts of soap and water. Consider appropriate evacuation.

ETHYL BUTANOL

DOT Number: UN 2275 *DOT Hazard Class:* Flammable liquid *DOT Guide Number:* 26
Synonyms: 2-ethyl-1-butanol, 2-ethylbutyl alcohol, sec-hexyl alcohol, sec-pentylcarbinol, pseudohexyl alcohol
STCC Number: n/a *Reportable Qty:* n/a
Mfg Name: Union Carbide *Phone No:* 1-203-794-2000

Physical Description:

Physical Form: Liquid *Color:* Colorless *Odor:* Mild alcohol
Other Information: n/a

Chemical Properties:

Specific Gravity: .834 *Vapor Density:* 3.4 *Boiling Point: 293° F(145° C)*
Melting Point: –173° F(–113.8° C) *Vapor Pressure:* .9 mm Hg at 68° F(20° C) *Solubility in water:* No
Other Information: Floats on water.

Health Hazards:

Inhalation Hazard: Harmful if inhaled.
Ingestion Hazard: Harmful if swallowed.
Absorption Hazard: Irritating to the skin, will burn the eyes.
Hazards to Wildlife: n/a
Decontamination Procedures: Wash away any material with copious amounts of soap and water.
First Aid Procedures: Remove victim to fresh air, call emergency medical care. If not breathing give CPR. If breathing is difficult administer oxygen. Treat for shock.

Fire Hazards:

Flashpoint: 128° F(53.3° C) *Ignition temperature:* 580° F(304.4° C)
Flammable Explosive High Range: 8.8 *Low Range:* 1.9
Toxic Products of Combustion: n/a
Other Hazards: n/a
Possible extinguishing agents: Do not extinguish the fire unless the flow can be stopped apply water from as far a distance as possible. Cool all affected containers. Use alcohol foam, dry chemical, or carbon dioxide.

Reactivity Hazards:

Reactive With: n/a *Other Reactions:* n/a

Corrosivity Hazards:

Corrosive With: n/a *Neutralizing Agent:* n/a

Radioactivity Hazards:

Radiation Emitted: n/a *Other Hazards:* n/a

Recommended Protection for Response Personnel:

Avoid breathing vapors, keep upwind. Structural protective clothing provides limited protection. Wash away any material which may have come into contact with the body with copious amounts of soap and water. Consider appropriate evacuation.

ETHYL BUTYRATE

DOT Number: UN 1180 *DOT Hazard Class:* Flammable liquid *DOT Guide Number:* 26
Synonyms: butyric acid, ethyl butanoate
STCC Number: 4909132 *Reportable Qty:* n/a
Mfg Name: Witco Chemical Corp. *Phone No:* 1-212-605-3800

Physical Description:

Physical Form: Liquid *Color:* Colorless *Odor:* Pineapple
Other Information: n/a

Chemical Properties:

Specific Gravity: .9 *Vapor Density:* 4 *Boiling Point: 248° F(120° C)*
Melting Point: n/a *Vapor Pressure:* 10 mm Hg at 60° F(15.5° C) *Solubility in water:* No
Other Information: Vapors are heavier than air.

Health Hazards:

Inhalation Hazard: Cause headaches and dizziness.
Ingestion Hazard: Cause nausea and vomiting.
Absorption Hazard: Irritating to the skin and eyes.
Hazards to Wildlife: n/a
Decontamination Procedures: Wash away any material with copious amounts of soap and water.
First Aid Procedures: Remove victim to fresh air, call emergency medical care. If not breathing give CPR. If breathing is difficult administer oxygen. Treat for shock.

Fire Hazards:

Flashpoint: 75° F(23.8° C) *Ignition temperature:* 865° F(462.7° C)
Flammable Explosive High Range: n/a *Low Range:* n/a
Toxic Products of Combustion: n/a
Other Hazards: Containers may explode in fire. Flashback along vapor trail may occur. May ignite or explode in an enclosed area.
Possible extinguishing agents: Apply water from as far a distance as possible. Use foam, dry chemical, or carbon dioxide.

Reactivity Hazards:

Reactive With: Some forms of plastics. *Other Reactions:* n/a

Corrosivity Hazards:

Corrosive With: n/a *Neutralizing Agent:* n/a

Radioactivity Hazards:

Radiation Emitted: n/a *Other Hazards:* n/a

Recommended Protection for Response Personnel:

Avoid breathing vapors, keep upwind. Structural protective clothing will provide limited protection. Wash away any material which may have come into contact with the body with copious amounts of soap and water. Consider appropriate evacuation.

ETHYL CHLORIDE

DOT Number: UN 1073 *DOT Hazard Class:* Flammable liquid *DOT Guide Number:* 27
Synonyms: choroethane, muriatic ether
STCC Number: 4908162 *Reportable Qty:* 100/45.4
Mfg Name: Ethyl Corp *Phone No:* 1-804-788-5000

Physical Description:

Physical Form: Liquid *Color:* Colorless *Odor:* Pungent
Other Information: n/a

Chemical Properties:

Specific Gravity: .9 *Vapor Density:* 2.2 *Boiling Point: 54° F(12.2° C)*
Melting Point: −218° F(−138.8° C) *Vapor Pressure:* 5 mm Hg at 68° F(20° C)
Solubility in water: Slight *Degree of Solubility:* .6%
Other Information: Vapors are heavier than air.

Health Hazards:

Inhalation Hazard: Cause dizziness of loss of consciousness.
Ingestion Hazard: n/a
Absorption Hazard: Will cause frostbite.
Hazards to Wildlife: n/a
Decontamination Procedures: Wash away any material with copious amounts of soap and water.
First Aid Procedures: Remove victim to fresh air, call emergency medical care. If not breathing give CPR. If breathing is difficult administer oxygen. Treat for shock.

Fire Hazards:

Flashpoint: −58° F(−50° C) *Ignition temperature:* 966° F(518.8° C)
Flammable Explosive High Range: 15.4 *Low Range:* 3.8
Toxic Products of Combustion: Toxic and irritating gases are produced in a fire.
Other Hazards: Vapors may explode in fire. Flashback along vapor trail may occur. May ignite or explode in an enclosed area.
Possible extinguishing agents: Apply water from as far a distance as possible. Use foam, dry chemical, or carbon dioxide.

Reactivity Hazards:

Reactive With: n/a *Other Reactions:* n/a

Corrosivity Hazards:

Corrosive With: n/a *Neutralizing Agent:* n/a

Radioactivity Hazards:

Radiation Emitted: n/a *Other Hazards:* n/a

Recommended Protection for Response Personnel:

Avoid breathing vapors, keep upwind. Structural protective clothing provides limited protection. Wash away any material which may have come into contact with the body with copious amounts of soap and water. Consider appropriate evacuation.

ETHYL CHLOROACETATE

DOT Number: UN 1181 *DOT Hazard Class:* Combustible liquid *DOT Guide Number:* 55
Synonyms: chloroacetic acid
STCC Number: 4913114 *Reportable Qty:* n/a
Mfg Name: Dow Chemical *Phone No:* 1-517-636-4400

Physical Description:

Physical Form: Liquid *Color:* Colorless to light brown *Odor:* Irritating or fruity
Other Information: n/a

Chemical Properties:

Specific Gravity: 1.15 *Vapor Density:* 4.3 *Boiling Point: 295° F(146.1° C)*
Melting Point: n/a *Vapor Pressure:* 10 mm Hg at 100° F(37.7° C) *Solubility in water:* No
Other Information: Heavier, insoluble in water. Vapors are heavier than air.

Health Hazards:

Inhalation Hazard: Cause headaches/nausea.
Ingestion Hazard: Poisonous if swallowed.
Absorption Hazard: Irritating to the skin and eyes.
Hazards to Wildlife: n/a
Decontamination Procedures: Wash away any material with copious amounts of soap and water.
First Aid Procedures: Remove victim to fresh air, call emergency medical care. If not breathing give CPR. If breathing is difficult administer oxygen. Treat for shock.

Fire Hazards:

Flashpoint: 100° F(37.7° C) *Ignition temperature:* n/a
Flammable Explosive High Range: n/a *Low Range:* n/a
Toxic Products of Combustion: Irritating toxic hydrogen chloride and phosgene gases may be generated in a fire.
Other Hazards: n/a
Possible extinguishing agents: Use foam, dry chemical, or carbon dioxide. Apply water from as far a distance as possible.

Reactivity Hazards:

Reactive With: n/a *Other Reactions:* n/a

Corrosivity Hazards:

Corrosive With: Metals *Neutralizing Agent:* n/a

Radioactivity Hazards:

Radiation Emitted: n/a *Other Hazards:* n/a

Recommended Protection for Response Personnel:

Avoid breathing vapors, keep upwind. Structural protective clothing provides limited protection. Wash away any material which may have come into contact with the body with copious amounts of soap and water. Consider appropriate evacuation.

ETHYL CHLOROFORMATE

DOT Number: UN 1182 *DOT Hazard Class:* Flammable liquid *DOT Guide Number:* 57
Synonyms: chloroformic acid, ethyl ester
STCC Number: 4907617 *Reportable Qty:* n/a
Mfg Name: Fisher Scientific *Phone No:* 1-412-349-3322

Physical Description:

Physical Form: Liquid *Color:* Colorless *Odor:* Pungent
Other Information: n/a

Chemical Properties:

Specific Gravity: 1.1 *Vapor Density:* 3.7 *Boiling Point: 201° F(93.8° C)*
Melting Point: n/a *Vapor Pressure:* n/a *Solubility in water:* Decomposes
Other Information: Vapors are heavier than air. Decomposes in water to give off hydrochloric acid.

Health Hazards:

Inhalation Hazard: Poisonous if inhaled.
Ingestion Hazard: Poisonous if swallowed.
Absorption Hazard: Poisonous if absorbed.
Hazards to Wildlife: n/a
Decontamination Procedures: Wash away any material with copious amounts of soap and water.
First Aid Procedures: Remove victim to fresh air, call emergency medical care. If not breathing give CPR. If breathing is difficult administer oxygen. Treat for shock.

Fire Hazards:

Flashpoint: 61° F(16.1° C) *Ignition temperature:* 932° F(500° C)
Flammable Explosive High Range: n/a *Low Range:* n/a
Toxic Products of Combustion: Poisonous gases are produced in a fire.
Other Hazards: Vapors may explode in fire. Flashback along vapor trail may occur. May ignite or explode in an enclosed area.
Possible extinguishing agents: Apply water from as far a distance as possible. Use alcohol foam, dry chemical, or carbon dioxide.

Reactivity Hazards:

Reactive With: Water producing hydrochloric acid. *Other Reactions:* n/a

Corrosivity Hazards:

Corrosive With: Metals and tissue *Neutralizing Agent:* Crushed limestone, soda ash, or lime.

Radioactivity Hazards:

Radiation Emitted: n/a *Other Hazards:* n/a

Recommended Protection for Response Personnel:

Avoid breathing vapors, keep upwind. Wear a sealed chemical suit (butyl rubber). Wash away any material which may have come into contact with the body with copious amounts of soap and water. If the fire becomes uncontrollable, consider appropriate evacuation.

ETHYL CYANOHYDRIN

DOT Number: n/a *DOT Hazard Class:* n/a *DOT Guide Number:* n/a
Synonyms: 2-cyanoethanol, glycol cyanohydrin, hydhracrylonitrile, 3-hydroxypropanenitrile
STCC Number: n/a *Reportable Qty:* n/a
Mfg Name: Union Carbide *Phone No:* 1-203-794-2000

Physical Description:

Physical Form: Liquid *Color:* Colorless to yellow brown *Odor:* Weak
Other Information: n/a

Chemical Properties:

Specific Gravity: 1.05 *Vapor Density:* 2.45 *Boiling Point: 445° F(229.4° C)*
Melting Point: 551° F(288.3° C) *Vapor Pressure:* .08 mm Hg at 77° F(25° C) *Solubility in water:* Yes
Other Information: Sinks and mixes with water.

Health Hazards:

Inhalation Hazard: n/a
Ingestion Hazard: Harmful if swallowed.
Absorption Hazard: Irritating to the skin and eyes.
Hazards to Wildlife: n/a
Decontamination Procedures: Wash away any material with copious amounts of soap and water.
First Aid Procedures: Remove victim to fresh air, call emergency medical care. If not breathing give CPR. If breathing is difficult administer oxygen. Treat for shock.

Fire Hazards:

Flashpoint: 265° F(129.4° C) *Ignition temperature:* 922° F(494.4° C)
Flammable Explosive High Range: 12.1 *Low Range:* 2.3
Toxic Products of Combustion: Poisonous gases may be produced when heated.
Other Hazards: n/a
Possible extinguishing agents: Carbon dioxide and dry chemical for small fires. alcohol foam for large fires.

Reactivity Hazards:

Reactive With: n/a *Other Reactions:* n/a

Corrosivity Hazards:

Corrosive With: n/a *Neutralizing Agent:* n/a

Radioactivity Hazards:

Radiation Emitted: n/a *Other Hazards:* n/a

Recommended Protection for Response Personnel:

Avoid breathing vapors, keep upwind. Wear a sealed chemical suit (polycarbonate, butyl rubber). Wash away any material which may have come into contact with the body with copious amounts of soap and water. Consider appropriate evacuation.

ETHYLDICHLOROSILANE

DOT Number: UN 1183 *DOT Hazard Class:* Flammable liquid *DOT Guide Number:* 29
Synonyms: none given
STCC Number: 4907615 *Reportable Qty:* n/a
Mfg Name: Dow Chemical *Phone No:* 1-517-636-4400

Physical Description:

Physical Form: Liquid *Color:* Colorless *Odor:* Sharp, irritating
Other Information: n/a

Chemical Properties:

Specific Gravity: 1.1 *Vapor Density:* 4.45 *Boiling Point: 168° F(75.5° C)*
Melting Point: n/a *Vapor Pressure:* n/a *Solubility in water:* Yes
Other Information: Heavier and decomposed by water. Vapors are heavier than air.

Health Hazards:

Inhalation Hazard: Irritating to the nose and throat.
Ingestion Hazard: Harmful if swallowed.
Absorption Hazard: Will burn the skin and eyes.
Hazards to Wildlife: n/a
Decontamination Procedures: Wash away any material with copious amounts of soap and water.
First Aid Procedures: Remove victim to fresh air, call emergency medical care. If not breathing give CPR. If breathing is difficult administer oxygen. Treat for shock.

Fire Hazards:

Flashpoint: 30° F(−1.1° C) *Ignition temperature:* n/a
Flammable Explosive High Range: n/a *Low Range:* 2.9
Toxic Products of Combustion: Toxic hydrogen chloride and phosgene gases may be formed in a fire.
Other Hazards: Flashback along vapor trail may occur. Vapors may explode if ignited in an enclosed area.
Possible extinguishing agents: Use dry chemical, dry sand, or carbon dioxide. Do not use water on the material itself.

Reactivity Hazards:

Reactive With: Water to form hydrochloric acid. *Other Reactions:* n/a

Corrosivity Hazards:

Corrosive With: Metal *Neutralizing Agent:* Crushed limestone, soda ash, or lime.

Radioactivity Hazards:

Radiation Emitted: n/a *Other Hazards:* n/a

Recommended Protection for Response Personnel:

Avoid breathing vapors, keep upwind. Wear a sealed chemical suit (butyl rubber). Wash away any material which may have come into contact with the body with copious amounts of soap and water. If the fire becomes uncontrollable, or if the containers are exposed to direct flame, consider appropriate evacuation.

ETHYLENE

DOT Number: NA 1962 *DOT Hazard Class:* Flammable gas *DOT Guide Number:* 22
Synonyms: ethene, ole faint gas
STCC Number: 4905734 *Reportable Qty:* n/a *CHEMTREC Phone No:* 1-800-424-9300

Physical Description:

Physical Form: Gas *Color:* Colorless *Odor:* Sweet
Other Information: n/a

Chemical Properties:

Specific Gravity: .57 *Vapor Density:* 1 *Boiling Point: – 155° F(– 103.8° C)*
Melting Point: n/a *Vapor Pressure:* n/a *Solubility in water:* Yes
Other Information: Floats and boils on water. Flammable, visible vapor cloud is produced.

Health Hazards:

Inhalation Hazard: Headache, dizziness, loss of consciousness.

Ingestion Hazard: n/a

Absorption Hazard: Will cause frostbite.

Hazards to Wildlife: Dangerous to aquatic life.

Decontamination Procedures: Wash away any material with copious amounts of soap and water.

First Aid Procedures: Remove victim to fresh air, call emergency medical care. If not breathing give CPR. If breathing is difficult administer oxygen. Treat for shock.

Fire Hazards:

Flashpoint: 213° F(100.5° C) *Ignition temperature:* 842° F(450° C)
Flammable Explosive High Range: 36 *Low Range:* 2.7

Toxic Products of Combustion: Vapors are anesthetic.

Other Hazards: Flashback along vapor trail may occur. Vapors may explode if ignited in an enclosed area. Containers may explode in a fire.

Possible extinguishing agents: Do not extinguish the fire unless the flow can be stopped. Apply water from as far a distance as possible.

Reactivity Hazards:

Reactive With: n/a *Other Reactions:* n/a

Corrosivity Hazards:

Corrosive With: n/a *Neutralizing Agent:* n/a

Radioactivity Hazards:

Radiation Emitted: n/a *Other Hazards:* n/a

Recommended Protection for Response Personnel:

Avoid breathing vapors, keep upwind. Wear appropriate chemical clothing. Wash away any material which may come into contact with the body with copious amounts of soap and water. If the fire becomes uncontrollable, or if the containers are exposed to direct flame, consider appropriate evacuation. under fire conditions, the cylinders may violently rupture or rocket. (BLEVE)

ETHYLENE CHLOROHYDRIN

DOT Number: UN 1135 *DOT Hazard Class:* Poison B *DOT Guide Number:* 55
Synonyms: 2-chloroethanol, glycol chlorohydrin
STCC Number: n/a *Reportable Qty:* n/a
Mfg Name: Diamond Shamrock *Phone No:* 1-214-745-2000

Physical Description:

Physical Form: Liquid *Color:* Colorless *Odor:* Faint, sweet, pleasant
Other Information: n/a

Chemical Properties:

Specific Gravity: 1.2 *Vapor Density:* 2.8 *Boiling Point: 266° F(130° C)*
Melting Point: −81° F(−62.7° C) *Vapor Pressure:* 5 mm Hg at 68° F(20° C) *Solubility in water:* Yes
Other Information: Mixes with water. Irritating vapor is produced.

Health Hazards:

Inhalation Hazard: Coughing, difficulty in breathing.
Ingestion Hazard: Harmful if swallowed.
Absorption Hazard: Irritating to the skin and eyes.
Hazards to Wildlife: n/a
Decontamination Procedures: Wash away any material with copious amounts of soap and water.
First Aid Procedures: Remove victim to fresh air, call emergency medical care. If not breathing give CPR. If breathing is difficult administer oxygen. Treat for shock.

Fire Hazards:

Flashpoint: 139° F(59.4° C) *Ignition temperature:* 797° F(425° C)
Flammable Explosive High Range: 15.9 *Low Range:* 4.9
Toxic Products of Combustion: Toxic hydrogen chloride and phosgene gases may be formed in a fire.
Other Hazards: Vapors may flashback to the source.
Possible extinguishing agents: Apply water from as far a distance as possible. Use alcohol foam, dry chemical, or carbon dioxide.

Reactivity Hazards:

Reactive With: n/a *Other Reactions:* n/a

Corrosivity Hazards:

Corrosive With: n/a *Neutralizing Agent:* n/a

Radioactivity Hazards:

Radiation Emitted: n/a *Other Hazards:* n/a

Recommended Protection for Response Personnel:

Avoid breathing vapors, keep upwind. Wear a sealed chemical suit (butyl rubber, viton). Wash away any material which may have come into contact with the body with copious amounts of soap and water. Consider appropriate evacuation.

ETHYLENEDIAMINE

DOT Number: UN 1604 *DOT Hazard Class:* Corrosive *DOT Guide Number:* 29
Synonyms: 1,2-diaminoethane, 1,2-ethylenediamine
STCC Number: 4935628 *Reportable Qty:* 5000/2270
Mfg Name: Dow Chemical *Phone No:* 1-517-636-4400

Physical Description:

Physical Form: Liquid *Color:* Colorless *Odor:* Mild ammonia
Other Information: It is used to make other chemicals, and as a fumigant.

Chemical Properties:

Specific Gravity: .9 *Vapor Density:* 2.1 *Boiling Point: 242° F(116.6° C)*
Melting Point: 52° F(11.1° C) *Vapor Pressure:* 10 mm Hg at 68° F(20° C) *Solubility in water:* Yes
Other Information: Floats and mixes with water. Irritating vapor is produced. Weighs 7.5 lbs/3.4 kg per gallon/3.8 l.

Health Hazards:

Inhalation Hazard: Harmful if inhaled.
Ingestion Hazard: Harmful if swallowed.
Absorption Hazard: Will burn the skin and eyes.
Hazards to Wildlife: Dangerous to aquatic life.
Decontamination Procedures: Wash away any material with copious amounts of soap and water.
First Aid Procedures: Remove victim to fresh air, call emergency medical care. If not breathing give CPR. If breathing is difficult administer oxygen. Treat for shock.

Fire Hazards:

Flashpoint: 99° F(37.2° C) *Ignition temperature:* 715° F(379.4° C)
Flammable Explosive High Range: 14.4 *Low Range:* 4.2
Toxic Products of Combustion: Toxic oxides of nitrogen are produced during combustion of this material.
Other Hazards: n/a
Possible extinguishing agents: Do not extinguish the fire unless the flow can be stopped. Apply water from as far a distance as possible. Use alcohol foam, dry chemical, or carbon dioxide.

Reactivity Hazards:

Reactive With: n/a *Other Reactions:* n/a

Corrosivity Hazards:

Corrosive With: Tissue *Neutralizing Agent:* Flush with water

Radioactivity Hazards:

Radiation Emitted: n/a *Other Hazards:* n/a

Recommended Protection for Response Personnel:

Avoid breathing vapors, keep upwind. Structural protective clothing provides limited protection. Wash away any material which may have come into contact with the body with copious amounts of soap and water. Consider appropriate evacuation.

ETHYLENEDIAMINE TETRACETIC ACID

DOT Number: NA 9117 *DOT Hazard Class:* ORM-E *DOT Guide Number:* 31
Synonyms: EDTA, Endrate, Versene acid, Tetrine acid
STCC Number: n/a *Reportable Qty:* n/a
Mfg Name: Dow Chemical *Phone No:* 1-517-636-4400

Physical Description:

Physical Form: Solid *Color:* White *Odor:* Odorless
Other Information: n/a

Chemical Properties:

Specific Gravity: .86 *Vapor Density:* 1 *Boiling Point: n/a*
Melting Point: n/a *Vapor Pressure:* n/a *Solubility in water:* n/a
Other Information: Floats on water

Health Hazards:

Inhalation Hazard: n/a
Ingestion Hazard: n/a
Absorption Hazard: Irritating to the skin and eyes.
Hazards to Wildlife: n/a
Decontamination Procedures: Wash away any material with copious amounts of soap and water.
First Aid Procedures: Remove victim to fresh air, call emergency medical care. If not breathing give CPR. If breathing is difficult administer oxygen. Treat for shock.

Fire Hazards:

Flashpoint: n/a *Ignition temperature:* n/a
Flammable Explosive High Range: n/a *Low Range:* n/a
Toxic Products of Combustion: n/a
Other Hazards: n/a
Possible extinguishing agents: Extinguish fire using suitable agent for the type of surrounding fire.

Reactivity Hazards:

Reactive With: n/a *Other Reactions:* n/a

Corrosivity Hazards:

Corrosive With: n/a *Neutralizing Agent:* n/a

Radioactivity Hazards:

Radiation Emitted: n/a *Other Hazards:* n/a

Recommended Protection for Response Personnel:

Avoid breathing vapors, keep upwind. Structural protective clothing provides limited protection. Wash away any material which may have come into contact with the body with copious amounts of soap and water. Consider appropriate evacuation.

ETHYLENE DIBROMIDE

DOT Number: UN 1605 *DOT Hazard Class:* ORM-A *DOT Guide Number:* 55
Synonyms: Bromofume, 1,2-dibromoethane, Dow-fume 40, W-10, W-15, W-40
STCC Number: 4940335 *Reportable Qty:* 1000/454
Mfg Name: Dow Chemical *Phone No:* 1-517-636-4400

Physical Description:

Physical Form: Liquid *Color:* Colorless *Odor:* Sweet
Other Information: It is used as a solvent, as a grain fumigant, to make other chemicals, and for many other uses.

Chemical Properties:

Specific Gravity: 2.18 *Vapor Density:* 6.5 *Boiling Point: 268° F(131.1° C)*
Melting Point: 50° F(10° C) *Vapor Pressure:* 11 mm Hg at 68° F(20° C)
Solubility in water: Slight *Degree of Solubility:* .4%
Other Information: Heavier than and slightly soluble in water. Weighs 18.1 lbs/8.2 kg per gallon/3.8 l.

Health Hazards:

Inhalation Hazard: Poisonous if inhaled.
Ingestion Hazard: Poisonous if swallowed.
Absorption Hazard: Poisonous if the skin is exposed.
Hazards to Wildlife: n/a
Decontamination Procedures: Wash away any material with copious amounts of soap and water.
First Aid Procedures: Remove victim to fresh air, call emergency medical care. If not breathing give CPR. If breathing is difficult administer oxygen. Treat for shock.

Fire Hazards:

Flashpoint: n/a *Ignition temperature:* n/a
Flammable Explosive High Range: n/a *Low Range:* n/a
Toxic Products of Combustion: Poisonous gases are produced when heated.
Other Hazards: n/a
Possible extinguishing agents: Extinguish fire using suitable agent for the type of surrounding fire.

Reactivity Hazards:

Reactive With: Hot metals *Other Reactions:* n/a

Corrosivity Hazards:

Corrosive With: n/a *Neutralizing Agent:* n/a

Radioactivity Hazards:

Radiation Emitted: n/a *Other Hazards:* n/a

Recommended Protection for Response Personnel:

Avoid breathing vapors, keep upwind. Wear a sealed chemical suit (polycarbonate, butyl rubber, viton, chlorinated polyethylene). Wash away any material which may have come into contact with the body with copious amounts of soap and water. Consider appropriate evacuation.

ETHYLENE DICHLORIDE

DOT Number: UN 1184 *DOT Hazard Class:* Flammable liquid *DOT Guide Number:* 26
Synonyms: Brocide, Dutch Liquid, EDC, glycol dichloride
STCC Number: 4909166 *Reportable Qty:* 5000/2270
Mfg Name: Dow Chemical *Phone No:* 1-517-636-4400

Physical Description:

Physical Form: Liquid *Color:* Colorless *Odor:* Sweet
Other Information: n/a

Chemical Properties:

Specific Gravity: 1.3 *Vapor Density:* 3.4 *Boiling Point: 183° F(83.8° C)*
Melting Point: −32° F(−35.5° C) *Vapor Pressure:* 62 mm Hg at 68° F(20° C)
Solubility in water: No *Degree of Solubility: .8%*
Other Information: Sinks in water. Flammable, irritating vapor is produced.

Health Hazards:

Inhalation Hazard: Nausea, dizziness, difficulty in breathing.
Ingestion Hazard: Harmful if swallowed.
Absorption Hazard: Will burn the skin and eyes.
Hazards to Wildlife: n/a
Decontamination Procedures: Wash away any material with copious amounts of soap and water.
First Aid Procedures: Remove victim to fresh air, call emergency medical care. If not breathing give CPR. If breathing is difficult administer oxygen. Treat for shock.

Fire Hazards:

Flashpoint: 56° F(13.3° C) *Ignition temperature:* 775° F(412.7° C)
Flammable Explosive High Range: 16 *Low Range:* 6.2
Toxic Products of Combustion: Toxic and irritating hydrogen chloride and phosgene gases are generated in a fire.
Other Hazards: Flashback along vapor trail may occur. Vapors may explode if ignited in an enclosed area.
Possible extinguishing agents: Apply water from as far a distance as possible. Use foam, dry chemical, or carbon dioxide.

Reactivity Hazards:

Reactive With: n/a *Other Reactions:* n/a

Corrosivity Hazards:

Corrosive With: n/a *Neutralizing Agent:* n/a

Radioactivity Hazards:

Radiation Emitted: n/a *Other Hazards:* n/a

Recommended Protection for Response Personnel:

Avoid breathing vapors, keep upwind. Wear a sealed chemical suit (butyl rubber, viton). Wash away any material which may have come into contact with the body with copious amounts of soap and water. Consider appropriate evacuation.

ETHYLENE GLYCOL DIETHYL ETHER

DOT Number: UN 1153 *DOT Hazard Class:* Combustible liquid *DOT Guide Number:* 26
Synonyms: 1,2-diethoxyethane, diethyl "cellosolve"
STCC Number: 4913115 *Reportable Qty:* n/a
Mfg Name: Eastman Chemical *Phone No:* 1-615-229-2000

Physical Description:

Physical Form: Liquid *Color:* Colorless *Odor:* Mild, pleasant
Other Information: n/a

Chemical Properties:

Specific Gravity: .85 *Vapor Density:* 6.56 *Boiling Point: 252° F(122.2° C)*
Melting Point: –101° F(–73.8° C) *Vapor Pressure:* 9.4 mm Hg at 68° F(20° C) *Solubility in water:* No
Other Information: It is lighter than and insoluble in water. Vapors are heavier than air. Irritating vapor is produced.

Health Hazards:

Inhalation Hazard: Harmful if inhaled.
Ingestion Hazard: Harmful if swallowed.
Absorption Hazard: Irritating to the skin and eyes.
Hazards to Wildlife: n/a
Decontamination Procedures: Wash away any material with copious amounts of soap and water.
First Aid Procedures: Remove victim to fresh air, call emergency medical care. If not breathing give CPR. If breathing is difficult administer oxygen. Treat for shock.

Fire Hazards:

Flashpoint: 90° F(32.2° C) *Ignition temperature:* 406° F(207.7° C)
Flammable Explosive High Range: n/a *Low Range:* n/a
Toxic Products of Combustion: n/a
Other Hazards: Water may be ineffective.
Possible extinguishing agents: Do not extinguish the fire unless the flow can be stopped flow can be stopped. Apply water from as far a distance as possible. Use alcohol foam, dry chemical, or carbon dioxide. Cool all affected containers.

Reactivity Hazards:

Reactive With: n/a *Other Reactions:* n/a

Corrosivity Hazards:

Corrosive With: n/a *Neutralizing Agent:* n/a

Radioactivity Hazards:

Radiation Emitted: n/a *Other Hazards:* n/a

Recommended Protection for Response Personnel:

Avoid breathing vapors, keep upwind. Structural protective clothing provides limited protection. Wash away any material which may have come into contact with the body with copious amounts of soap and water. Consider appropriate evacuation.

ETHYLENE GLYCOL MONOETHYL ETHER

DOT Number: UN 1171 *DOT Hazard Class:* Combustible liquid *DOT Guide Number:* 26
Synonyms: Dowanol EM, glycol monomethylether, methyl "cellosolve", Poly-solv EM
STCC Number: 4913116 *Reportable Qty:* n/a
Mfg Name: Union Carbide *Phone No:* 1-203-794-2000

Physical Description:

Physical Form: Liquid *Color:* Colorless *Odor:* Odorless
Other Information: It is used as a solvent for resins, in lacquers, for dyeing leathers and textiles, and in formulating cleaners and varnish removers.

Chemical Properties:

Specific Gravity: .931 *Vapor Density:* 3.1 *Boiling Point: 275° F(135° C)*
Melting Point: −94° F(−70° C) *Vapor Pressure:* 4 mm Hg at 68° F(20° C) *Solubility in water:* Yes
Other Information: It is fully soluble in water. Weighs 7.8 lbs/3.5 kg per gallon/3.8 l

Health Hazards:

Inhalation Hazard: n/a
Ingestion Hazard: Harmful if swallowed.
Absorption Hazard: Irritating to the skin and eyes.
Hazards to Wildlife: n/a
Decontamination Procedures: Wash away any material with copious amounts of soap and water.
First Aid Procedures: Remove victim to fresh air, call emergency medical care. If not breathing give CPR. If breathing is difficult administer oxygen. Treat for shock.

Fire Hazards:

Flashpoint: 115° F(46.1° C) *Ignition temperature:* 455° F(235° C)
Flammable Explosive High Range: 15.6 *Low Range:* 1.7
Toxic Products of Combustion: n/a
Other Hazards: Flashback along vapor trail may occur. Vapors may explode if ignited in an enclosed area. Containers may explode in a fire.
Possible extinguishing agents: Do not extinguish the fire unless the flow can be stopped. Apply water from as far a distance as possible. Use alcohol foam, dry chemical, or carbon dioxide.

Reactivity Hazards:

Reactive With: Oxidizing materials *Other Reactions:* n/a

Corrosivity Hazards:

Corrosive With: n/a *Neutralizing Agent:* n/a

Radioactivity Hazards:

Radiation Emitted: n/a *Other Hazards:* n/a

Recommended Protection for Response Personnel:

Avoid breathing vapors, keep upwind. Structural protective clothing provides limited protection. Wash away any material which may have come into contact with the body with copious amounts of soap and water. If the fire becomes uncontrollable, or if the containers are exposed to direct flame, consider appropriate evacuation. Under fire conditions, the cylinders may violently rupture or rocket.

ETHYLENE GLYCOL MONOETHYL ETHER ACETATE

DOT Number: UN 1172 *DOT Hazard Class:* Combustible liquid *DOT Guide Number:* 26
Synonyms: "cellosolve" acetate, 2-ethoxyethyl acetate, glycol monoethyl ether acetate, Poly-solv EE acetate
STCC Number: 4913117 *Reportable Qty:* n/a
Mfg Name: Union Carbide *Phone No:* 1-203-794-2000

Physical Description:

Physical Form: Liquid *Color:* Colorless *Odor:* Pleasant
Other Information: n/a

Chemical Properties:

Specific Gravity: .97 *Vapor Density:* 4.1 *Boiling Point: 313° F(156.1° C)*
Melting Point: −79° F(−61.6° C) *Vapor Pressure:* 6.2 mm Hg at 68° F(20° C) *Solubility in water:* Yes
Other Information: It is lighter, slightly soluble in water. Vapors are heavier than air.

Health Hazards:

Inhalation Hazard: Harmful if inhaled.
Ingestion Hazard: Harmful if swallowed.
Absorption Hazard: Irritating to the skin and eyes.
Hazards to Wildlife: n/a
Decontamination Procedures: Wash away any material with copious amounts of soap and water.
First Aid Procedures: Remove victim to fresh air, call emergency medical care. If not breathing give CPR. If breathing is difficult administer oxygen. Treat for shock.

Fire Hazards:

Flashpoint: 200° F(93.3° C) *Ignition temperature:* 715° F(379.4° C)
Flammable Explosive High Range: 6.7 *Low Range:* 1.7
Toxic Products of Combustion: n/a
Other Hazards: n/a
Possible extinguishing agents: Do not extinguish the fire unless the flow can be stopped. Apply water from as far a distance as possible. Use alcohol foam, dry chemical, or carbon dioxide. Cool all affected containers.

Reactivity Hazards:

Reactive With: n/a *Other Reactions:* n/a

Corrosivity Hazards:

Corrosive With: n/a *Neutralizing Agent:* n/a

Radioactivity Hazards:

Radiation Emitted: n/a *Other Hazards:* n/a

Recommended Protection for Response Personnel:

Avoid breathing vapors, keep upwind. Structural protective clothing provides limited protection. Wash away any material which may have come into contact with the body with copious amounts of soap and water. Consider appropriate evacuation.

ETHYLENE IMINE

DOT Number: UN 1185 *DOT Hazard Class:* Flammable liquid *DOT Guide Number:* 30
Synonyms: azirane, aziridine
STCC Number: 4906220 *Reportable Qty:* 1/.454
Mfg Name: Dow Chemical *Phone No:* 1-517-636-4400

Physical Description:

Physical Form: Liquid *Color:* Colorless *Odor:* Ammonia
Other Information: n/a

Chemical Properties:

Specific Gravity: .832 *Vapor Density:* 1.5 *Boiling Point: 131° F(55° C)*
Melting Point: 108° F(42.2° C) *Vapor Pressure:* 160 mm Hg at 68° F(20° C) *Solubility in water:* Yes
Other Information: Floats and mixes with water. Poisonous, flammable vapor is produced.

Health Hazards:

Inhalation Hazard: Poisonous if inhaled.
Ingestion Hazard: Poisonous if swallowed.
Absorption Hazard: Poisonous if the skin is exposed.
Hazards to Wildlife: n/a
Decontamination Procedures: Wash away any material with copious amounts of soap and water.
First Aid Procedures: Remove victim to fresh air, call emergency medical care. If not breathing give CPR. If breathing is difficult administer oxygen. Treat for shock.

Fire Hazards:

Flashpoint: 1° F(−17.2° C) *Ignition temperature:* 608° F(320° C)
Flammable Explosive High Range: 54.8 *Low Range:* 3.3
Toxic Products of Combustion: Irritating vapor is generated when heated.
Other Hazards: Flashback along vapor trail may occur. Vapors may explode if ignited in an enclosed area. Containers may explode when heated.
Possible extinguishing agents: Do not extinguish the fire unless the flow can be stopped apply water from as far a distance as possible. Use alcohol foam, dry chemical, or carbon dioxide.

Reactivity Hazards:

Reactive With: n/a *Other Reactions:* n/a

Corrosivity Hazards:

Corrosive With: n/a *Neutralizing Agent:* n/a

Radioactivity Hazards:

Radiation Emitted: n/a *Other Hazards:* n/a

Recommended Protection for Response Personnel:

Avoid breathing vapors, keep upwind. Wear a sealed chemical suit (butyl rubber). Wash away any material which may have come into contact with the body with copious amounts of soap and water. If the fire becomes uncontrollable, consider appropriate evacuation. Prolonged exposure to this material to fire or heat may result in its polymerization and a violent rupturing of the container.

ETHYLENE OXIDE

DOT Number: UN 1040 *DOT Hazard Class:* Flammable liquid *DOT Guide Number:* 69
Synonyms: 1,2-epoxyethane, oxirane
STCC Number: 4906610 *Reportable Qty:* 1/.454
Mfg Name: Dow Chemical *Phone No:* 1-517-636-4400

Physical Description:

Physical Form: Liquid *Color:* Colorless *Odor:* Sweet
Other Information: It is used to make other chemicals, as a fumigant, and in an industrial sterilant.

Chemical Properties:

Specific Gravity: .9 *Vapor Density:* 1.5 *Boiling Point: 51° F(10.5° C)*
Melting Point: 171° F(77.2° C) *Vapor Pressure:* 1095 mm Hg at 68° F(20° C) *Solubility in water:* Yes
Other Information: Must be diluted on the order of 24 to 1 with water before the liquid loses its flammability.

Health Hazards:

Inhalation Hazard: Nausea, vomiting, difficulty in breathing.
Ingestion Hazard: Harmful if swallowed.
Absorption Hazard: Will burn the skin and the eyes.
Hazards to Wildlife: n/a
Decontamination Procedures: Wash away any material with copious amounts of soap and water.
First Aid Procedures: Remove victim to fresh air, call emergency medical care. If not breathing give CPR. If breathing is difficult administer oxygen. Treat for shock.

Fire Hazards:

Flashpoint: 0° F(−17.7° C) *Ignition temperature:* 804° F(428.8° C)
Flammable Explosive High Range: 100 *Low Range:* 3
Toxic Products of Combustion: Irritating vapors are generated when heated.
Other Hazards: Containers may explode in a fire. Flashback along vapor trail may occur. Vapors may explode if ignited in an enclosed area.
Possible extinguishing agents: Apply water from as far a distance as possible. Use alcohol foam, dry chemical, or carbon dioxide.

Reactivity Hazards:

Reactive With: n/a *Other Reactions:* n/a

Corrosivity Hazards:

Corrosive With: n/a *Neutralizing Agent:* n/a

Radioactivity Hazards:

Radiation Emitted: n/a *Other Hazards:* n/a

Recommended Protection for Response Personnel:

Avoid breathing vapors, keep upwind. Wear a sealed chemical suit (polycarbonate, butyl rubber, chlorinated polyethylene). Wash away any material which may have come into contact with the body with copious amounts of soap and water. Consider appropriate evacuation.

ETHYLENETRICHLOROSILANE

DOT Number: UN 1196 *DOT Hazard Class:* Flammable liquid *DOT Guide Number:* 29
Synonyms: ethylsilicon trichloride, trichloroetyl silane, trichloroethyl silicone
STCC Number: n/a *Reportable Qty:* n/a
Mfg Name: Union Carbide *Phone No:* 1-203-794-2000

Physical Description:

Physical Form: Liquid *Color:* Colorless *Odor:* Sharp irritating
Other Information: n/a

Chemical Properties:

Specific Gravity: 1.25 *Vapor Density:* 5.6 *Boiling Point: 210° F(98.8° C)*
Melting Point: n/a *Vapor Pressure:* n/a *Solubility in water:* Reacts
Other Information: Reacts violently with water. Irritating gas is produced on contact with water.

Health Hazards:

Inhalation Hazard: Difficulty in breathing.
Ingestion Hazard: Harmful if swallowed.
Absorption Hazard: Irritating to the skin and eyes.
Hazards to Wildlife: n/a
Decontamination Procedures: Wash away any material with copious amounts of soap and water.
First Aid Procedures: Remove victim to fresh air, call emergency medical care. If not breathing give CPR. If breathing is difficult administer oxygen. Treat for shock.

Fire Hazards:

Flashpoint: 57° F(13.8° C) *Ignition temperature:* n/a
Flammable Explosive High Range: n/a *Low Range:* n/a
Toxic Products of Combustion: Toxic hydrogen chloride and phosgene gas may be formed in fires.
Other Hazards: Flashback along vapor trail may occur. Vapors may explode if ignited in an enclosed area. Containers may explode in a fire.
Possible extinguishing agents: Do not extinguish the fire unless the flow can be stopped. Apply water from as far a distance as possible. Use alcohol foam, dry chemical, carbon dioxide.

Reactivity Hazards:

Reactive With: Reacts vigorously with water evolving hydrochloric acid. *Other Reactions:* n/a

Corrosivity Hazards:

Corrosive With: Surface moisture to form hydrochloric acid which is corrosive to common metals.
Neutralizing Agent: Flood with water. Crushed limestone, soda ash, or lime.

Radioactivity Hazards:

Radiation Emitted: n/a *Other Hazards:* n/a

Recommended Protection for Response Personnel:

Avoid breathing vapors, keep upwind. Wear a sealed chemical suit (butyl rubber). Wash away any material which may have come into contact with the body with copious amounts of soap and water. If the fire becomes uncontrollable, or if the containers are exposed to direct flame, consider appropriate evacuation.

ETHYL ETHER

DOT Number: UN 1155 *DOT Hazard Class:* Flammable liquid *DOT Guide Number:* 26
Synonyms: diethyl ether; ethyl oxide, ether
STCC Number: 4908157 *Reportable Qty:* 100/45.4
Mfg Name: Ashland Chemical *Phone No:* 1-614-889-3333

Physical Description:

Physical Form: Liquid *Color:* Colorless *Odor:* Anesthetic
Other Information: Used as an anesthetic, solvent, and to make other chemicals.

Chemical Properties:

Specific Gravity: .7 *Vapor Density:* 2.6 *Boiling Point: 95° F(35° C)*
Melting Point: –190° F(–123.3° C) *Vapor Pressure:* 442 mm Hg at 68° F(20° C)
Solubility in water: Slight *Degree of Solubility:* 7.5%
Other Information: Vapors are heavier than air. Material is lighter than water.

Health Hazards:

Inhalation Hazard: Cause nausea, loss of consciousness.
Ingestion Hazard: Harmful if swallowed.
Absorption Hazard: Irritating to the skin.
Hazards to Wildlife: n/a
Decontamination Procedures: Wash away any material with copious amounts of soap and water.
First Aid Procedures: Remove victim to fresh air, call emergency medical care. If not breathing give CPR. If breathing is difficult administer oxygen. Treat for shock.

Fire Hazards:

Flashpoint: –49° F(–45° C) *Ignition temperature:* 356° F(180° C)
Flammable Explosive High Range: 36 *Low Range:* 1.9
Toxic Products of Combustion: n/a
Other Hazards: Vapors may explode in fire. Flashback along vapor trail may occur. May ignite or explode in an enclosed area.
Possible extinguishing agents: Apply water from as far a distance as possible. Use foam, dry chemical, or carbon dioxide.

Reactivity Hazards:

Reactive With: n/a *Other Reactions:* n/a

Corrosivity Hazards:

Corrosive With: n/a *Neutralizing Agent:* n/a

Radioactivity Hazards:

Radiation Emitted: n/a *Other Hazards:* n/a

Recommended Protection for Response Personnel:

Avoid breathing vapors, keep upwind. Structural protective clothing provides limited protection. Wash away any material which may have come into contact with the body with copious amounts of soap and water. If the fire becomes uncontrollable, consider appropriate evacuation.

ETHYL FORMATE

DOT Number: UN 1190 *DOT Hazard Class:* Flammable liquid *DOT Guide Number:* 26
Synonyms: formic acid, ethyl ester
STCC Number: 4908167 *Reportable Qty:* n/a
Mfg Name: Florasynth Inc. *Phone No:* 1-212-371-7700

Physical Description:

Physical Form: Liquid *Color:* Colorless *Odor:* Pleasant
Other Information: n/a

Chemical Properties:

Specific Gravity: .9 *Vapor Density:* 2.6 *Boiling Point: 130° F(54.4° C)*
Melting Point: –110° F(–78.8° C) *Vapor Pressure:* 194 mm Hg at 68° F(20° C)
Solubility in water: Yes *Degree of Solubility:* 13.6%
Other Information: Slowly decomposes in water to form formic acid, a corrosive material and ethyl alcohol. Vapors are heavier than air.

Health Hazards:

Inhalation Hazard: Cause difficulty in breathing.
Ingestion Hazard: Harmful if swallowed.
Absorption Hazard: Irritating to the skin and eyes.
Hazards to Wildlife: n/a
Decontamination Procedures: Wash away any material with copious amounts of soap and water.
First Aid Procedures: Remove victim to fresh air, call emergency medical care. If not breathing give CPR. If breathing is difficult administer oxygen. Treat for shock.

Fire Hazards:

Flashpoint: –4° F(–20° C) *Ignition temperature:* 851° F(455° C)
Flammable Explosive High Range: 16 *Low Range:* 2.8
Toxic Products of Combustion: n/a
Other Hazards: Vapors may explode in fire. Flashback along vapor trail may occur. May ignite or explode in an enclosed area. Containers may explode in fire.
Possible extinguishing agents: Apply water from as far a distance as possible. Use alcohol foam, dry chemical, or carbon dioxide.

Reactivity Hazards:

Reactive With: n/a *Other Reactions:* n/a

Corrosivity Hazards:

Corrosive With: n/a *Neutralizing Agent:* n/a

Radioactivity Hazards:

Radiation Emitted: n/a *Other Hazards:* n/a

Recommended Protection for Response Personnel:

Avoid breathing vapors, keep upwind. Structural protective clothing provides limited protection. Wash away any material which may have come into contact with the body with copious amounts of soap and water. If the fire becomes uncontrollable, consider appropriate evacuation.

ETHYL HEXALDEHYDE

DOT Number: UN 1191 *DOT Hazard Class:* Combustible liquid *DOT Guide Number:* 26
Synonyms: butylethylacetaldehyde, 2-ethyl hexaldehyde, 2-ethylhexanal, octylaldehyde
STCC Number: 4913164 *Reportable Qty:* n/a
Mfg Name: Eastman Chemical *Phone No:* 1-615-229-2000

Physical Description:

Physical Form: Liquid *Color:* White *Odor:* Mild
Other Information: n/a

Chemical Properties:

Specific Gravity: .82 *Vapor Density:* 4.42 *Boiling Point: 327° F(163.8° C)*
Melting Point: n/a *Vapor Pressure:* 1.8 mm Hg at 68° F(20° C) *Solubility in water:* Yes
Other Information: It is lighter, slightly soluble in water. Vapors are heavier than air.

Health Hazards:

Inhalation Hazard: Coughing, difficulty in breathing.
Ingestion Hazard: Nausea and vomiting.
Absorption Hazard: Irritating to the skin and eyes.
Hazards to Wildlife: n/a
Decontamination Procedures: Wash away any material with copious amounts of soap and water.
First Aid Procedures: Remove victim to fresh air, call emergency medical care. If not breathing give CPR. If breathing is difficult administer oxygen. Treat for shock.

Fire Hazards:

Flashpoint: 125° F(51.6° C) *Ignition temperature:* 387° F(197.2° C)
Flammable Explosive High Range: n/a *Low Range:* n/a
Toxic Products of Combustion: n/a
Other Hazards: Water may be ineffective.
Possible extinguishing agents: Do not extinguish the fire unless the flow can be stopped. Cool all affected containers. Apply water from as far a distance as possible. Use alcohol foam, dry chemical, or carbon dioxide.

Reactivity Hazards:

Reactive With: Clothing, paper, or other absorbent materials. May spontaneously ignite.
Other Reactions: n/a

Corrosivity Hazards:

Corrosive With: n/a *Neutralizing Agent:* n/a

Radioactivity Hazards:

Radiation Emitted: n/a *Other Hazards:* n/a

Recommended Protection for Response Personnel:

Avoid breathing vapors, keep upwind. Wear appropriate chemical clothing, Wash away any material which may have come into contact with the body with copious amounts of soap and water. Consider appropriate evacuation.

ETHYL LACTATE

DOT Number: UN 1192 *DOT Hazard Class:* Combustible liquid *DOT Guide Number:* 26
Synonyms: lactic acid, ethyl ester; ethyl 2-hydroxypropanoate; ethyl 2-hydroxypropionate
STCC Number: 4913165 *Reportable Qty:* n/a
Mfg Name: Gallard-Schlesinger *Phone No:* 1-516-333-5600

Physical Description:

Physical Form: Liquid *Color:* Colorless *Odor:* Mild
Other Information: n/a

Chemical Properties:

Specific Gravity: 1.03 *Vapor Density:* 4.1 *Boiling Point: 309° F(153.8° C)*
Melting Point: n/a *Vapor Pressure:* n/a *Solubility in water:* Yes
Other Information: It is heavier, slightly soluble in water. Vapors are heavier than air.

Health Hazards:

Inhalation Hazard: n/a
Ingestion Hazard: Harmful if swallowed.
Absorption Hazard: n/a
Hazards to Wildlife: n/a
Decontamination Procedures: Wash away any material with copious amounts of soap and water.
First Aid Procedures: Remove victim to fresh air, call emergency medical care. If not breathing give CPR. If breathing is difficult administer oxygen. Treat for shock.

Fire Hazards:

Flashpoint: 115 ° F (46.1° C) *Ignition temperature:* 752° F(400° C)
Flammable Explosive High Range: 11.4 *Low Range:* 1.5
Toxic Products of Combustion: n/a
Other Hazards: n/a
Possible extinguishing agents: Do not extinguish the fire unless the flow can be stopped. Cool all affected containers. Apply water from as far a distance as possible. Use alcohol foam, dry chemical, or carbon dioxide.

Reactivity Hazards:

Reactive With: n/a *Other Reactions:* n/a

Corrosivity Hazards:

Corrosive With: n/a *Neutralizing Agent:* n/a

Radioactivity Hazards:

Radiation Emitted: n/a *Other Hazards:* n/a

Recommended Protection for Response Personnel:

Avoid breathing vapors, keep upwind. Structural protective clothing provides limited protection. Wash away any material which may have come into contact with the body with copious amounts of soap and water. Consider appropriate evacuation.

ETHYL MERCAPTAN

DOT Number: UN 2363 *DOT Hazard Class:* Flammable liquid *DOT Guide Number:* 27
Synonyms: ethanethiol, ethyl sulfhydrate
STCC Number: 4908169 *Reportable Qty:* n/a
Mfg Name: Penwald Corp. *Phone No:* 1-215-587-5000

Physical Description:

Physical Form: Liquid *Color:* Colorless, clear
Odor: Overpowering garliclike, also smells like a skunk. *Other Information:* Used as a stabilizer for adhesives.

Chemical Properties:

Specific Gravity: .8 *Vapor Density:* 2.1 *Boiling Point: 95° F(35° C)*
Melting Point: −234° F(−147.7° C) *Vapor Pressure:* 442 mm Hg at 68° F(20° C)
Solubility in water: No *Degree of Solubility:* 1.3%
Other Information: n/a

Health Hazards:

Inhalation Hazard: Poisonous if inhaled.
Ingestion Hazard: Poisonous if swallowed.
Absorption Hazard: Poisonous if exposed to the skin.
Hazards to Wildlife: n/a
Decontamination Procedures: Wash away any material with copious amounts of soap and water.
First Aid Procedures: Remove victim to fresh air, call emergency medical care. If not breathing give CPR. If breathing is difficult administer oxygen. Treat for shock.

Fire Hazards:

Flashpoint: −55° F(−48.3° C) *Ignition temperature:* 527° F(275° C)
Flammable Explosive High Range: 18 *Low Range:* 2.8
Toxic Products of Combustion: Poisonous gases are produced in fire.
Other Hazards: Vapors may explode in fire. Flashback along vapor trail may occur. May ignite or explode in an enclosed area. Containers may explode in fire.
Possible extinguishing agents: Apply water from as far a distance as possible. Use alcohol foam, dry chemical, or carbon dioxide.

Reactivity Hazards:

Reactive With: n/a *Other Reactions:* n/a

Corrosivity Hazards:

Corrosive With: n/a *Neutralizing Agent:* n/a

Radioactivity Hazards:

Radiation Emitted: n/a *Other Hazards:* n/a

Recommended Protection for Response Personnel:

Avoid breathing vapors, keep upwind. Structural protective clothing provides limited protection. Wash away any material which may have come into contact with the body with copious amounts of soap and water. If the fire becomes uncontrollable, consider appropriate evacuation.

ETHYL METHACRYLATE

DOT Number: UN 2227 *DOT Hazard Class:* Flammable liquid *DOT Guide Number:* 26
Synonyms: ethyl methacrylate (inhibited); ethyl 2-methacrylate; methacrylic acid, ethyl ester
STCC Number: 4907232 *Reportable Qty:* 1000/454
Mfg Name: E.I. Du Pont *Phone No:* 1-800-441-3637

Physical Description:

Physical Form: Liquid *Color:* Colorless *Odor:* Sharp, unpleasant
Other Information: It is used to make polymers, and other chemicals.

Chemical Properties:

Specific Gravity: .915 *Vapor Density:* 4 *Boiling Point: 243° F(117.2° C)*
Melting Point: −58° F(−50° C) *Vapor Pressure:* n/a *Solubility in water:* No
Other Information: It is lighter than and insoluble in water.

Health Hazards:

Inhalation Hazard: Coughing, difficulty in breathing.
Ingestion Hazard: Nausea and vomiting.
Absorption Hazard: Irritating to the skin and eyes.
Hazards to Wildlife: n/a
Decontamination Procedures: Wash away any material with copious amounts of soap and water.
First Aid Procedures: Remove victim to fresh air, call emergency medical care. If not breathing give CPR. If breathing is difficult administer oxygen. Treat for shock.

Fire Hazards:

Flashpoint: 70° F(21.1° C) *Ignition temperature:* 740° F(393.3° C)
Flammable Explosive High Range: n/a *Low Range:* 1.8
Toxic Products of Combustion: n/a
Other Hazards: Sealed containers may rupture explosively if hot. Flashback along vapor trail may occur. Vapors may explode if ignited in an enclosed area.
Possible extinguishing agents: Do not extinguish the fire unless the flow can be stopped. Cool all affected containers. Apply water from as far a distance as possible. Use foam, dry chemical, or carbon dioxide.

Reactivity Hazards:

Reactive With: Heat causing polymerization with rapid release of energy. *Other Reactions:* n/a

Corrosivity Hazards:

Corrosive With: n/a *Neutralizing Agent:* n/a

Radioactivity Hazards:

Radiation Emitted: n/a *Other Hazards:* n/a

Recommended Protection for Response Personnel:

Avoid breathing vapors, keep upwind. Wear a sealed chemical suit (polycarbonate, butyl rubber). Wash away any material which may have come into contact with the body with copious amounts of soap and water. Consider appropriate evacuation.

ETHYL NITRITE

DOT Number: NA 1993 *DOT Hazard Class:* Flammable liquid *DOT Guide Number:* 27
Synonyms: nitrous ether, spirit of ether nitrate
STCC Number: 4907020 *Reportable Qty:* n/a
Mfg Name: Mallinckrodt Inc. *Phone No:* 1-314-895-2000

Physical Description:

Physical Form: Liquid *Color:* Colorless to yellow *Odor:* Pleasant
Other Information: n/a

Chemical Properties:

Specific Gravity: .9 *Vapor Density:* 2.6 *Boiling Point: 63° F(17.2° C)*
Melting Point: n/a *Vapor Pressure:* n/a *Solubility in water:* No
Other Information: Vapors are heavier than air. Lighter than water.

Health Hazards:

Inhalation Hazard: Headache, dizziness, loss of consciousness.
Ingestion Hazard: Same as inhalation.
Absorption Hazard: n/a
Hazards to Wildlife: n/a
Decontamination Procedures: Wash away any material with copious amounts of soap and water.
First Aid Procedures: Remove victim to fresh air, call emergency medical care. If not breathing give CPR. If breathing is difficult administer oxygen. Treat for shock.

Fire Hazards:

Flashpoint: −31° F(−35° C) *Ignition temperature:* 194° F(90° C)
Flammable Explosive High Range: 50 *Low Range:* 4
Toxic Products of Combustion: Toxic oxides of nitrogen are produced.
Other Hazards: Vapors may ignite spontaneously at 194° F(90° C) it is decomposed by heat or light and may explode violently. thermally unstable.
Possible extinguishing agents: Apply water from as far a distance as possible. Use foam, dry chemical, or carbon dioxide.

Reactivity Hazards:

Reactive With: n/a *Other Reactions:* n/a

Corrosivity Hazards:

Corrosive With: n/a *Neutralizing Agent:* n/a

Radioactivity Hazards:

Radiation Emitted: n/a *Other Hazards:* n/a

Recommended Protection for Response Personnel:

Avoid breathing vapors, keep upwind. Wear appropriate chemical clothing. Wash away any material which may come into contact with the body with copious amounts of soap and water. If the fire becomes uncontrollable, consider appropriate evacuation.

ETHYL PHENYL DICHLOROSILANE

DOT Number: UN 2435 *DOT Hazard Class:* Corrosive *DOT Guide Number:* 39
Synonyms: none given
STCC Number: 4934245 *Reportable Qty:* n/a
Mfg Name: Dow Chemical *Phone No:* 1-517-636-4400

Physical Description:

Physical Form: Liquid *Color:* Colorless *Odor:* Sharp, irritating
Other Information: n/a

Chemical Properties:

Specific Gravity: 1.16 *Vapor Density:* 7.1 *Boiling Point: 300° F(148.8° C)*
Melting Point: n/a *Vapor Pressure:* n/a *Solubility in water:* Reacts
Other Information: Reacts with water. Poisonous gas is produced on contact with water.

Health Hazards:

Inhalation Hazard: Poisonous if inhaled.
Ingestion Hazard: Harmful if swallowed.
Absorption Hazard: Will burn the skin and eyes.
Hazards to Wildlife: n/a
Decontamination Procedures: Wash away any material with copious amounts of soap and water.
First Aid Procedures: Remove victim to fresh air, call emergency medical care. If not breathing give CPR. If breathing is difficult administer oxygen. Treat for shock.

Fire Hazards:

Flashpoint: 150° F(65.5° C) *Ignition temperature:* n/a
Flammable Explosive High Range: n/a *Low Range:* n/a
Toxic Products of Combustion: Toxic hydrogen chloride and phosgene fumes may be formed.
Other Hazards: Difficult to extinguish, reignition may occur.
Possible extinguishing agents: Use dry chemical, dry sand, or carbon dioxide. Do not use water on the material itself. Apply water from as far a distance as possible.

Reactivity Hazards:

Reactive With: Water to form hydrochloric acid. *Other Reactions:* n/a

Corrosivity Hazards:

Corrosive With: Metals and tissue *Neutralizing Agent:* Crushed limestone, soda ash, or lime.

Radioactivity Hazards:

Radiation Emitted: n/a *Other Hazards:* n/a

Recommended Protection for Response Personnel:

Avoid breathing vapors, keep upwind. Wear a sealed chemical suit (butyl rubber). Wash away any material which may have come into contact with the body with copious amounts of soap and water. Consider appropriate evacuation.

ETHYL PHOSPHONOTHIOIC DICHLORIDE

DOT Number: NA 1760 *DOT Hazard Class:* Corrosive *DOT Guide Number:* 60
Synonyms: ethyl thionophosphoryl dichloride
STCC Number: 4933355 *Reportable Qty:* n/a
Mfg Name: Dow Chemical *Phone No:* 1-517-636-4400

Physical Description:

Physical Form: Liquid *Color:* Colorless *Odor:* Choking
Other Information: n/a

Chemical Properties:

Specific Gravity: 1.35 *Vapor Density:* 7.1 *Boiling Point: 342° F(172.2° C)*
Melting Point: n/a *Vapor Pressure:* n/a *Solubility in water:* Reacts
Other Information: Reacts with water. Poisonous gas is produced on contact with water.

Health Hazards:

Inhalation Hazard: Poisonous if inhaled.

Ingestion Hazard: Harmful if swallowed.

Absorption Hazard: Irritating to the skin and eyes.

Hazards to Wildlife: n/a

Decontamination Procedures: Wash away any material with copious amounts of soap and water.

First Aid Procedures: Remove victim to fresh air, call emergency medical care. If not breathing give CPR. If breathing is difficult administer oxygen. Treat for shock.

Fire Hazards:

Flashpoint: 203° F(95° C) *Ignition temperature:* n/a

Flammable Explosive High Range: n/a *Low Range:* n/a

Toxic Products of Combustion: Oxides of sulfur, phosphorus, hydrogen chloride, and phosgene.

Other Hazards: n/a

Possible extinguishing agents: Do not use water on the material itself. Use alcohol foam, dry chemical, or carbon dioxide.

Reactivity Hazards:

Reactive With: Water to form hydrochloric acid. *Other Reactions:* n/a

Corrosivity Hazards:

Corrosive With: Metals and tissue *Neutralizing Agent:* Crushed limestone, soda ash, or lime.

Radioactivity Hazards:

Radiation Emitted: n/a *Other Hazards:* n/a

Recommended Protection for Response Personnel:

Avoid breathing vapors, keep upwind. Wear a sealed chemical suit (butyl rubber). Wash away any material which may have come into contact with the body with copious amounts of soap and water. Consider appropriate evacuation.

ETHYL PHOSPHORODICHLORIDATE

DOT Number: NA 1760 *DOT Hazard Class:* Corrosive *DOT Guide Number:* 60
Synonyms: ethyl dichlorophosphate
STCC Number: 4933333 *Reportable Qty:* n/a
Mfg Name: Aldrich Chemical *Phone No:* 1-414-273-3850

Physical Description:

Physical Form: Liquid *Color:* Colorless *Odor:* Choking
Other Information: n/a

Chemical Properties:

Specific Gravity: 1.35 *Vapor Density:* 5.4 *Boiling Point: n/a*
Melting Point: n/a *Vapor Pressure:* n/a *Solubility in water:* Reacts
Other Information: Reacts with water. Irritating gas is produced on contact with water.

Health Hazards:

Inhalation Hazard: Harmful if inhaled.
Ingestion Hazard: Harmful if swallowed.
Absorption Hazard: Will burn the skin and eyes.
Hazards to Wildlife: n/a
Decontamination Procedures: Wash away any material with copious amounts of soap and water.
First Aid Procedures: Remove victim to fresh air, call emergency medical care. If not breathing give CPR. If breathing is difficult administer oxygen. Treat for shock.

Fire Hazards:

Flashpoint: n/a *Ignition temperature:* n/a
Flammable Explosive High Range: n/a *Low Range:* n/a
Toxic Products of Combustion: Irritating fumes of hydrogen chloride and phosphoric acid are formed.
Other Hazards: n/a
Possible extinguishing agents: Apply water from as far a distance as possible. Use alcohol foam, dry chemical, or carbon dioxide.

Reactivity Hazards:

Reactive With: Water to form hydrochloric acid. *Other Reactions:* n/a

Corrosivity Hazards:

Corrosive With: Metals and tissue *Neutralizing Agent:* Crushed limestone, soda ash, or lime.

Radioactivity Hazards:

Radiation Emitted: n/a *Other Hazards:* n/a

Recommended Protection for Response Personnel:

Avoid breathing vapors, keep upwind. Wear a sealed chemical suit (butyl rubber). Wash away any material which may have come into contact with the body with copious amounts of soap and water. Consider appropriate evacuation.

ETHYL PROPIONATE

DOT Number: UN 1195 *DOT Hazard Class:* Flammable liquid *DOT Guide Number:* 26
Synonyms: none given
STCC Number: 4909170 *Reportable Qty:* n/a *CHEMTREC Phone No:* 1-800-424-9300

Physical Description:

Physical Form: Liquid *Color:* Colorless *Odor:* Pineapple
Other Information: n/a

Chemical Properties:

Specific Gravity: .9 *Vapor Density:* 3.5 *Boiling Point: 210° F(98.8° C)*
Melting Point: n/a *Vapor Pressure:* n/a *Solubility in water:* No
Other Information: Vapors are heavier than air. Lighter than water.

Health Hazards:

Inhalation Hazard: n/a
Ingestion Hazard: n/a
Absorption Hazard: n/a
Hazards to Wildlife: n/a
Decontamination Procedures: Wash away any material with copious amounts of soap and water.
First Aid Procedures: Remove victim to fresh air, call emergency medical care. If not breathing give CPR. If breathing is difficult administer oxygen. Treat for shock.

Fire Hazards:

Flashpoint: 54° F(12.2° C) *Ignition temperature:* 824° F(440° C)
Flammable Explosive High Range: 11 *Low Range:* 1.9
Toxic Products of Combustion: n/a
Other Hazards: n/a
Possible extinguishing agents: Apply water from as far a distance as possible. Use foam, dry chemical, or carbon dioxide.

Reactivity Hazards:

Reactive With: n/a *Other Reactions:* n/a

Corrosivity Hazards:

Corrosive With: n/a *Neutralizing Agent:* n/a

Radioactivity Hazards:

Radiation Emitted: n/a *Other Hazards:* n/a

Recommended Protection for Response Personnel:

Avoid breathing vapors, keep upwind. Wear appropriate chemical clothing. Wash away any material which may come into contact with the body with copious amounts of soap and water. Consider appropriate evacuation.

ETHYL PROPYLACROLENE

DOT Number: UN 1191 *DOT Hazard Class:* Flammable liquid *DOT Guide Number:* 26
Synonyms: 2-ethyl-2-hexenal, 2-ethyl-3-propylacrylaldehyde
STCC Number: n/a *Reportable Qty:* n/a
Mfg Name: Union Carbide *Phone No:* 1-203-794-2000

Physical Description:

Physical Form: Liquid *Color:* Yellow *Odor:* n/a
Other Information: n/a

Chemical Properties:

Specific Gravity: .86 *Vapor Density:* 4.4 *Boiling Point: 283° F(139.4° C)*
Melting Point: n/a *Vapor Pressure:* n/a *Solubility in water:* n/a
Other Information: Floats on water

Health Hazards:

Inhalation Hazard: n/a
Ingestion Hazard: Harmful if swallowed.
Absorption Hazard: Will burn the skin and eyes.
Hazards to Wildlife: n/a
Decontamination Procedures: Wash away any material with copious amounts of soap and water.
First Aid Procedures: Remove victim to fresh air, call emergency medical care. If not breathing give CPR. If breathing is difficult administer oxygen. Treat for shock.

Fire Hazards:

Flashpoint: 155° F(68.3° C) *Ignition temperature:* 200° F(93.3° C)
Flammable Explosive High Range: n/a *Low Range:* n/a
Toxic Products of Combustion: n/a
Other Hazards: n/a
Possible extinguishing agents: Extinguish fire using suitable agent for the type of surrounding fire.

Reactivity Hazards:

Reactive With: n/a *Other Reactions:* n/a

Corrosivity Hazards:

Corrosive With: n/a *Neutralizing Agent:* n/a

Radioactivity Hazards:

Radiation Emitted: n/a *Other Hazards:* n/a

Recommended Protection for Response Personnel:

Avoid breathing vapors, keep upwind. Structural protective clothing provides limited protection. Wash away any material which may have come into contact contact with the body with copious amounts of soap and water. Consider appropriate evacuation.

ETHYL SILICATE

DOT Number: UN 1292 *DOT Hazard Class:* Combustible liquid *DOT Guide Number:* 29
Synonyms: tetraethyl silicate
STCC Number: 4901316 *Reportable Qty:* n/a
Mfg Name: Stauffer Chemical *Phone No:* 1-203-222-3000

Physical Description:

Physical Form: Liquid *Color:* Colorless *Odor:* Faint
Other Information: n/a

Chemical Properties:

Specific Gravity: .9 *Vapor Density:* 7.2 *Boiling Point: 334° F(167.7° C)*
Melting Point: –121° F(–85° C) *Vapor Pressure:* 2 mm Hg at 68° F(20° C) *Solubility in water:* Reacts
Other Information: Vapors are heavier than air, lighter than water.

Health Hazards:

Inhalation Hazard: Irritating to the eyes, nose and throat.
Ingestion Hazard: n/a
Absorption Hazard: n/a
Hazards to Wildlife: n/a
Decontamination Procedures: Wash away any material with copious amounts of soap and water.
First Aid Procedures: Remove victim to fresh air, call emergency medical care. If not breathing give CPR. If breathing is difficult administer oxygen. Treat for shock.

Fire Hazards:

Flashpoint: 125° F(51.6° C) *Ignition temperature:* n/a
Flammable Explosive High Range: 23 *Low Range:* 1.3
Toxic Products of Combustion: n/a
Other Hazards: n/a
Possible extinguishing agents: Apply water from as far a distance as possible. Use alcohol foam, dry chemical, or carbon dioxide.

Reactivity Hazards:

Reactive With: Water forming a silicone adhesive. *Other Reactions:* n/a

Corrosivity Hazards:

Corrosive With: n/a *Neutralizing Agent:* n/a

Radioactivity Hazards:

Radiation Emitted: n/a *Other Hazards:* n/a

Recommended Protection for Response Personnel:

Avoid breathing vapors, keep upwind. Structural protective clothing provides limited protection. Wash away any material which may have come into contact contact with the body with copious amounts of soap and water. Consider appropriate evacuation.

ETIOLOGIC AGENT

DOT Number: NA 2814 *DOT Hazard Class:* Etiologic agent *DOT Guide Number:* 24
Synonyms: N.O.S. Infectious substance, humans
STCC Number: 4925999 *Reportable Qty:* n/a *CHEMTREC Phone No:* 1-800-424-9300

Physical Description:

Physical Form: None *Color:* None *Odor:* None
Other Information: Etiologic agent is a very viable microorganism. In case of damage to or leaking from a container, contact the Centers for Disease Control at 404-633-5313 (24 hrs).

Chemical Properties:

Specific Gravity: n/a *Vapor Density:* n/a *Boiling Point: n/a*
Melting Point: n/a *Vapor Pressure:* n/a *Solubility in water:* n/a
Other Information: n/a

Health Hazards:

Inhalation Hazard: n/a
Ingestion Hazard: n/a
Absorption Hazard: n/a
Hazards to Wildlife: n/a
Decontamination Procedures: Wash away any material with copious amounts of soap and water.
First Aid Procedures: Remove victim to fresh air, call emergency medical care. If not breathing give CPR. If breathing is difficult administer oxygen. Treat for shock.

Fire Hazards:

Flashpoint: n/a *Ignition temperature:* n/a
Flammable Explosive High Range: n/a *Low Range:* n/a
Toxic Products of Combustion: n/a
Other Hazards: May be ignited if carrier agent is flammable. Damage to outer container may not affect inner container. Remove container from fire area if possible.
Possible extinguishing agents: Use appropriate extinguishing agent for the type of surrounding fire.

Reactivity Hazards:

Reactive With: n/a *Other Reactions:* n/a

Corrosivity Hazards:

Corrosive With: n/a *Neutralizing Agent:* n/a

Radioactivity Hazards:

Radiation Emitted: n/a *Other Hazards:* n/a

Recommended Protection for Response Personnel:

Avoid breathing vapors, keep upwind. Wear appropriate chemical clothing, Wash away any material which may come into contact with the body with copious amounts of soap and water. Do not allow this substance to come into contact with any unprotected parts of the body! Consider appropriate evacuation.

EXPLOSIVES A

DOT Number: n/a *DOT Hazard Class:* Explosives A *DOT Guide Number:* 46
Synonyms: blasting cap, bomb, mine, projectile, torpedo
STCC Number: 4901526 *Reportable Qty:* n/a *CHEMTREC Phone No:* 1-800-424-9300

Physical Description:

Physical Form: Solid *Color:* Any *Odor:* n/a
Other Information: Class A explosives are explosives that decompose by detonation.

Chemical Properties:

Specific Gravity: n/a *Vapor Density:* n/a *Boiling Point: n/a*
Melting Point: n/a *Vapor Pressure:* n/a *Solubility in water:* n/a
Other Information: n/a

Health Hazards:

Inhalation Hazard: n/a
Ingestion Hazard: n/a
Absorption Hazard: n/a
Hazards to Wildlife: n/a
Decontamination Procedures: Wash away any material with copious amounts of soap and water.
First Aid Procedures: Remove victim to fresh air, call emergency medical care. If not breathing give CPR. If breathing is difficult administer oxygen. Treat for shock.

Fire Hazards:

Flashpoint: n/a *Ignition temperature:* n/a
Flammable Explosive High Range: n/a *Low Range:* n/a
Toxic Products of Combustion: Shock wave and fragmentation.
Other Hazards: n/a
Possible extinguishing agents: Do not fight fires in a cargo with explosives. Evacuate area and let burn.

Reactivity Hazards:

Reactive With: Sudden shock, high temperature, or a combination of the two. *Other Reactions:* n/a

Corrosivity Hazards:

Corrosive With: n/a *Neutralizing Agent:* n/a

Radioactivity Hazards:

Radiation Emitted: n/a *Other Hazards:* n/a

Recommended Protection for Response Personnel:

Avoid breathing vapors, keep upwind. Wear appropriate chemical clothing, Wash away any material which may come into contact with the body with copious amounts of soap and water. If the material is on fire, or involved in a fire, consider evacuation to a one mile radius. Detonation occurs instantaneously and is violent!

FERRIC AMMONIUM CITRATE

DOT Number: NA 9118 *DOT Hazard Class:* ORM-E *DOT Guide Number:* 31
Synonyms: ammonium ferric citrate, ferric ammonium citrate (green),(brown)
STCC Number: 4963349 *Reportable Qty:* 1000/454
Mfg Name: U.S. Chemical Corp. *Phone No:* 1-617-237-4877

Physical Description:

Physical Form: Solid *Color:* Red, green, or brown *Odor:* Odorless
Other Information: It is used in making medicine, in making blueprints, and as a feed additive.

Chemical Properties:

Specific Gravity: 1.8 *Vapor Density:* n/a *Boiling Point: Decomposes*
Melting Point: n/a *Vapor Pressure:* n/a *Solubility in water:* Yes
Other Information: Sinks and mixes slowly with water.

Health Hazards:

Inhalation Hazard: Will cause coughing, difficulty in breathing.
Ingestion Hazard: Will cause nausea and vomiting.
Absorption Hazard: Irritating to the skin and the eyes.
Hazards to Wildlife: n/a
Decontamination Procedures: Wash away any material with copious amounts of soap and water.
First Aid Procedures: Remove victim to fresh air, call emergency medical care. If not breathing give CPR. If breathing is difficult administer oxygen. Treat for shock.

Fire Hazards:

Flashpoint: n/a *Ignition temperature:* n/a
Flammable Explosive High Range: n/a *Low Range:* n/a
Toxic Products of Combustion: Toxic oxides of nitrogen or ammonia gas may be formed in fires.
Other Hazards: n/a
Possible extinguishing agents: Extinguish fire using suitable agent for the type of surrounding fire.

Reactivity Hazards:

Reactive With: n/a *Other Reactions:* n/a

Corrosivity Hazards:

Corrosive With: n/a *Neutralizing Agent:* n/a

Radioactivity Hazards:

Radiation Emitted: n/a *Other Hazards:* n/a

Recommended Protection for Response Personnel:

Avoid breathing vapors, keep upwind. Wear appropriate chemical clothing. Wash away any material which may come into contact with the body with copious amounts of soap and water. Consider appropriate evacuation.

FERRIC AMMONIUM OXALATE

DOT Number: NA 9119 *DOT Hazard Class:* ORM-E *DOT Guide Number:* 31
Synonyms: ammonium ferric oxalate trihydride
STCC Number: 4963352 *Reportable Qty:* 1000/454
Mfg Name: U.S. Chemical Corp. *Phone No:* 1-617-237-4877

Physical Description:

Physical Form: Solid *Color:* Yellowish green *Odor:* Light burnt sugar
Other Information: It is used in making blueprints.

Chemical Properties:

Specific Gravity: 1.78 *Vapor Density:* n/a *Boiling Point: Decomposes*
Melting Point: n/a *Vapor Pressure:* n/a *Solubility in water:* Yes
Other Information: Sinks and mixes with water.

Health Hazards:

Inhalation Hazard: Will cause coughing, difficulty in breathing.
Ingestion Hazard: Will cause nausea and vomiting.
Absorption Hazard: Will burn the skin and the eyes.
Hazards to Wildlife: n/a
Decontamination Procedures: Wash away any material with copious amounts of soap and water.
First Aid Procedures: Remove victim to fresh air, call emergency medical care. If not breathing give CPR. If breathing is difficult administer oxygen. Treat for shock.

Fire Hazards:

Flashpoint: n/a *Ignition temperature:* n/a
Flammable Explosive High Range: n/a *Low Range:* n/a
Toxic Products of Combustion: Toxic oxides of nitrogen, ammonia, or carbon dioxide may be formed in fires.
Other Hazards: n/a
Possible extinguishing agents: Extinguish fire using suitable agent for the type of surrounding fire.

Reactivity Hazards:

Reactive With: n/a *Other Reactions:* n/a

Corrosivity Hazards:

Corrosive With: n/a *Neutralizing Agent:* n/a

Radioactivity Hazards:

Radiation Emitted: n/a *Other Hazards:* n/a

Recommended Protection for Response Personnel:

Avoid breathing vapors, keep upwind. Wear a sealed chemical suit (polycarbonate, butyl rubber). Wash away any material which may have come into contact with the body with copious amounts of soap and water. Consider appropriate evacuation.

FERRIC CHLORIDE (liquid)

DOT Number: UN 2582 *DOT Hazard Class:* Corrosive *DOT Guide Number:* 31
Synonyms: ferric chloride anhydrous, ferric chloride hexahydrate
STCC Number: 4932342 *Reportable Qty:* 1000/454
Mfg Name: E.I. Du Pont *Phone No:* 1-800-441-3637

Physical Description:

Physical Form: Liquid *Color:* Greenish black *Odor:* Hydrochloric acid
Other Information: n/a

Chemical Properties:

Specific Gravity: 2.8 *Vapor Density:* 5.6 *Boiling Point: Decomposes*
Melting Point: n/a *Vapor Pressure:* 40 mm Hg at 95° F(35° C)
Solubility in water: Yes *Other Information:* Sinks and mixes with water.

Health Hazards:

Inhalation Hazard: Will cause coughing, difficulty in breathing.
Ingestion Hazard: Will cause nausea and vomiting.
Absorption Hazard: Will burn the skin and the eyes.
Hazards to Wildlife: Dangerous to aquatic life.
Decontamination Procedures: Wash away any material with copious amounts of soap and water.
First Aid Procedures: Remove victim to fresh air, call emergency medical care. If not breathing give CPR. If breathing is difficult administer oxygen. Treat for shock.

Fire Hazards:

Flashpoint: n/a *Ignition temperature:* n/a
Flammable Explosive High Range: n/a *Low Range:* n/a
Toxic Products of Combustion: Irritating hydrogen chloride fumes may be formed in a fire.
Other Hazards: n/a
Possible extinguishing agents: Extinguish fire using suitable agent for the type of surrounding fire.

Reactivity Hazards:

Reactive With: n/a *Other Reactions:* n/a

Corrosivity Hazards:

Corrosive With: Most metals in the presence of moisture and tissue
Neutralizing Agent: Crushed limestone, soda ash, or lime.

Radioactivity Hazards:

Radiation Emitted: n/a *Other Hazards:* n/a

Recommended Protection for Response Personnel:

Avoid breathing vapors, keep upwind. Wear a sealed chemical suit (polycarbonate, butyl rubber). Wash away any material which may have come into contact with the body with copious amounts of soap and water. Consider appropriate evacuation.

FERRIC CHLORIDE (solid)

DOT Number: UN 1773 *DOT Hazard Class:* ORM-B *DOT Guide Number:* 31
Synonyms: iron(III) chloride, iron perchloride, iron trichloride
STCC Number: 4944138 *Reportable Qty:* 1000/454
Mfg Name: E.I. Du Pont *Phone No:* 1-800-441-3637

Physical Description:

Physical Form: Solid *Color:* Greenish black *Odor:* Odorless
Other Information: It is used to treat sewage, industrial waste, purify water, as an etching agent for engraving circuit boards, and in the manufacturing of other chemicals.

Chemical Properties:

Specific Gravity: 2.8 *Vapor Density:* 5.6 *Boiling Point: Decomposes*
Melting Point: n/a *Vapor Pressure:* 40 mm Hg at 95° F(35° C)
Solubility in water: Yes *Other Information:* Sinks and mixes with water.

Health Hazards:

Inhalation Hazard: Will cause coughing, difficulty in breathing.
Ingestion Hazard: Will cause nausea and vomiting.
Absorption Hazard: Will burn the skin and the eyes.
Hazards to Wildlife: Dangerous to aquatic life.
Decontamination Procedures: Wash away any material with copious amounts of soap and water.
First Aid Procedures: Remove victim to fresh air, call emergency medical care. If not breathing give CPR. If breathing is difficult administer oxygen. Treat for shock.

Fire Hazards:

Flashpoint: n/a *Ignition temperature:* n/a
Flammable Explosive High Range: n/a *Low Range:* n/a
Toxic Products of Combustion: Irritating hydrogen chloride fumes may be formed in a fire.
Other Hazards: n/a
Possible extinguishing agents: Extinguish fire using suitable agent for the type of surrounding fire.

Reactivity Hazards:

Reactive With: n/a *Other Reactions:* n/a

Corrosivity Hazards:

Corrosive With: Most metals in the presence of moisture.
Neutralizing Agent: Crushed limestone, soda ash, or lime.

Radioactivity Hazards:

Radiation Emitted: n/a *Other Hazards:* n/a

Recommended Protection for Response Personnel:

Avoid breathing vapors, keep upwind. Wear a sealed chemical suit (polycarbonate, butyl rubber). Wash away any material which may have come into contact with the body with copious amounts of soap and water. Consider appropriate evacuation.

FERRIC FLUORIDE

DOT Number: NA 9120 *DOT Hazard Class:* ORM-E *DOT Guide Number:* 31
Synonyms: iron fluoride
STCC Number: 4962626 *Reportable Qty:* 100/45.4
Mfg Name: Ozark-Mahoning *Phone No:* 1-918-585-2661

Physical Description:

Physical Form: Solid *Color:* Green *Odor:* n/a
Other Information: It is used in ceramics.

Chemical Properties:

Specific Gravity: 3.87 *Vapor Density:* 1 *Boiling Point: Sublimes*
Melting Point: n/a *Vapor Pressure:* n/a *Solubility in water:* Yes
Other Information: Sinks and mixes slowly with water.

Health Hazards:

Inhalation Hazard: Irritating to the eyes, nose, and the throat.
Ingestion Hazard: Will cause lethargy, nausea, and vomiting.
Absorption Hazard: n/a
Hazards to Wildlife: Dangerous to aquatic life.
Decontamination Procedures: Wash away any material with copious amounts of soap and water.
First Aid Procedures: Remove victim to fresh air, call emergency medical care. If not breathing give CPR. If breathing is difficult administer oxygen. Treat for shock.

Fire Hazards:

Flashpoint: n/a *Ignition temperature:* n/a
Flammable Explosive High Range: n/a *Low Range:* n/a
Toxic Products of Combustion: May give off fumes or vapors of fluorides and hydrofluoric acid.
Other Hazards: n/a
Possible extinguishing agents: Extinguish fire using suitable agent for the type of surrounding fire.

Reactivity Hazards:

Reactive With: n/a *Other Reactions:* n/a

Corrosivity Hazards:

Corrosive With: n/a *Neutralizing Agent:* n/a

Radioactivity Hazards:

Radiation Emitted: n/a *Other Hazards:* n/a

Recommended Protection for Response Personnel:

Avoid breathing vapors, keep upwind. Wear appropriate chemical clothing. Wash away any material which may come into contact with the body with copious amounts of soap and water. Consider appropriate evacuation.

FERRIC GYLCEROPHOSPHATE

DOT Number: n/a *DOT Hazard Class:* n/a *DOT Guide Number:* n/a
Synonyms: none given
STCC Number: n/a *Reportable Qty:* n/a
Mfg Name: Shepherd Chemical Co. *Phone No:* 1-513-731-1110

Physical Description:

Physical Form: Solid *Color:* Greenish brown to greenish yellow *Odor:* n/a
Other Information: n/a

Chemical Properties:

Specific Gravity: 1.5 *Vapor Density:* n/a *Boiling Point: n/a*
Melting Point: n/a *Vapor Pressure:* n/a *Solubility in water:* Yes
Other Information: Sinks and mixes with water.

Health Hazards:

Inhalation Hazard: Coughing, difficulty in breathing.
Ingestion Hazard: Harmful if swallowed.
Absorption Hazard: Irritating to the skin and the eyes.
Hazards to Wildlife: n/a
Decontamination Procedures: Wash away any material with copious amounts of soap and water.
First Aid Procedures: Remove victim to fresh air, call emergency medical care. If not breathing give CPR. If breathing is difficult administer oxygen. Treat for shock.

Fire Hazards:

Flashpoint: n/a *Ignition temperature:* n/a
Flammable Explosive High Range: n/a *Low Range:* n/a
Toxic Products of Combustion: n/a
Other Hazards: n/a
Possible extinguishing agents: Extinguish fire using suitable agent for the type of surrounding fire.

Reactivity Hazards:

Reactive With: n/a *Other Reactions:* n/a

Corrosivity Hazards:

Corrosive With: n/a *Neutralizing Agent:* n/a

Radioactivity Hazards:

Radiation Emitted: n/a *Other Hazards:* n/a

Recommended Protection for Response Personnel:

Avoid breathing vapors, keep upwind. Structural protective clothing provides limited protection. Wash away any material which may have come into contact with the body with copious amounts of soap and water. Consider appropriate evacuation.

FERRIC NITRATE

DOT Number: UN 1466 *DOT Hazard Class:* Oxidizer *DOT Guide Number:* 35
Synonyms: nitric acid, iron(II) salt
STCC Number: 4918725 *Reportable Qty:* 1000/454
Mfg Name: Allied Corp. *Phone No:* 1-201-455-2000

Physical Description:

Physical Form: Solid *Color:* Green to pale violet *Odor:* Odorless
Other Information: Used for dyeing and tanning, for chemical analysis, and in medicine.

Chemical Properties:

Specific Gravity: 1.7 *Vapor Density:* n/a *Boiling Point: n/a*
Melting Point: n/a *Vapor Pressure:* n/a *Solubility in water:* Yes
Other Information: n/a

Health Hazards:

Inhalation Hazard: Will cause coughing, difficulty in breathing.
Ingestion Hazard: Harmful if swallowed.
Absorption Hazard: Irritating to the skin and eyes.
Hazards to Wildlife: n/a
Decontamination Procedures: Wash away any material with copious amounts of soap and water.
First Aid Procedures: Remove victim to fresh air, call emergency medical care. If not breathing give CPR. If breathing is difficult administer oxygen. Treat for shock.

Fire Hazards:

Flashpoint: n/a *Ignition temperature:* n/a
Flammable Explosive High Range: n/a *Low Range:* n/a
Toxic Products of Combustion: Toxic oxides of nitrogen are produced when this material is involved in fire.
Other Hazards: Will accelerate the burning of combustible materials. if large quantities are involved in fire, or the combustible material is divided, explosion may result.
Possible extinguishing agents: Apply water from as far a distance as possible.

Reactivity Hazards:

Reactive With: Most metals. Contact of solid with wood or paper may cause fire. *Other Reactions:* n/a

Corrosivity Hazards:

Corrosive With: n/a *Neutralizing Agent:* n/a

Radioactivity Hazards:

Radiation Emitted: n/a *Other Hazards:* n/a

Recommended Protection for Response Personnel:

Avoid breathing vapors, keep upwind. Structural protective clothing provides limited protection. Wash away any material which may have come into contact with the body with copious amounts of soap and water. If the fire becomes uncontrollable, consider appropriate evacuation.

FERRIC SULFATE

DOT Number: NA 9121 *DOT Hazard Class:* ORM-E *DOT Guide Number:* 31
Synonyms: iron(III) sulfate, iron sesquisulfate, iron trisulfate
STCC Number: n/a *Reportable Qty:* n/a
Mfg Name: J.T. Baker Chemical *Phone No:* 1-201-859-2151

Physical Description:

Physical Form: Solid *Color:* White to gray *Odor:* Odorless
Other Information: n/a

Chemical Properties:

Specific Gravity: 3.1 *Vapor Density:* 1 *Boiling Point: Decomposes*
Melting Point: n/a *Vapor Pressure:* n/a *Solubility in water:* Yes
Other Information: Sinks and mixes slowly with water.

Health Hazards:

Inhalation Hazard: Coughing, difficulty in breathing.
Ingestion Hazard: Nausea and vomiting.
Absorption Hazard: Irritating to the skin and the eyes.
Hazards to Wildlife: Dangerous to aquatic life.
Decontamination Procedures: Wash away any material with copious amounts of soap and water.
First Aid Procedures: Remove victim to fresh air, call emergency medical care. If not breathing give CPR. If breathing is difficult administer oxygen. Treat for shock.

Fire Hazards:

Flashpoint: n/a *Ignition temperature:* n/a
Flammable Explosive High Range: n/a *Low Range:* n/a
Toxic Products of Combustion: n/a
Other Hazards: n/a
Possible *Possible extinguishing agents:* Extinguish fire using suitable agent for the type of surrounding fire.

Reactivity Hazards:

Reactive With: n/a *Other Reactions:* n/a

Corrosivity Hazards:

Corrosive With: Copper, copper alloys, mild steel, and galvanized steel.
Neutralizing Agent: Flush with water.

Radioactivity Hazards:

Radiation Emitted: n/a *Other Hazards:* n/a

Recommended Protection for Response Personnel:

Avoid breathing vapors, keep upwind. Structural protective clothing provides limited protection. Wash away any material which may have come into contact with the body with copious amounts of soap and water. Consider appropriate evacuation.

FERROUS AMMONIUM SULFATE

DOT Number: NA 9122 *DOT Hazard Class:* ORM-E *DOT Guide Number:* 31
Synonyms: iron ammonium sulfate, Mohr's salt
STCC Number: 4963357 *Reportable Qty:* 1000/454
Mfg Name: Allied Corp. *Phone No:* 1-201-455-2000

Physical Description:

Physical Form: Solid *Color:* Pale blue green *Odor:* Odorless
Other Information: It is used as an ingredient to mix fertilizers, or as a soil additive by itself.

Chemical Properties:

Specific Gravity: 1.86 *Vapor Density:* 1 *Boiling Point: Decomposes*
Melting Point: n/a *Vapor Pressure:* n/a *Solubility in water:* Yes
Other Information: Sinks and mixes slowly with water.

Health Hazards:

Inhalation Hazard: Will cause coughing, difficulty in breathing.
Ingestion Hazard: Will cause nausea and vomiting.
Absorption Hazard: Irritating to the skin and eyes.
Hazards to Wildlife: n/a
Decontamination Procedures: Wash away any material with copious amounts of soap and water.
First Aid Procedures: Remove victim to fresh air, call emergency medical care. If not breathing give CPR. If breathing is difficult administer oxygen. Treat for shock.

Fire Hazards:

Flashpoint: n/a *Ignition temperature:* n/a
Flammable Explosive High Range: n/a *Low Range:* n/a
Toxic Products of Combustion: Irritating and toxic ammonia and oxides of nitrogen may be formed in fires.
Other Hazards: n/a
Possible extinguishing agents: Extinguish fire using suitable agent for the type of surrounding fire.

Reactivity Hazards:

Reactive With: n/a *Other Reactions:* n/a

Corrosivity Hazards:

Corrosive With: n/a *Neutralizing Agent:* n/a

Radioactivity Hazards:

Radiation Emitted: n/a *Other Hazards:* n/a

Recommended Protection for Response Personnel:

Avoid breathing vapors, keep upwind. Structural protective clothing provides limited protection. Wash away any material which may have come into contact with the body with copious amounts of soap and water. Consider appropriate evacuation.

FERROUS CHLORIDE (liquid)

DOT Number: NA 1760 *DOT Hazard Class:* Corrosive *DOT Guide Number:* 60
Synonyms: iron dichloride, iron protochloride
STCC Number: 4932329 *Reportable Qty:* 100/45.4
Mfg Name: E.I. Du Pont *Phone No:* 1-800-441-3637

Physical Description:

Physical Form: Liquid *Color:* Greenish white *Odor:* Odorless
Other Information: It is used in dyeing, medicine, and in sewage treatment.

Chemical Properties:

Specific Gravity: 1.93 *Vapor Density:* 4.4 *Boiling Point: n/a*
Melting Point: n/a *Vapor Pressure:* n/a *Solubility in water:* Yes
Other Information: Sinks and mixes slowly with water.

Health Hazards:

Inhalation Hazard: Will cause difficulty in breathing.
Ingestion Hazard: Will cause nausea and vomiting.
Absorption Hazard: Irritating to the skin and eyes.
Hazards to Wildlife: Dangerous to aquatic life.
Decontamination Procedures: Wash away any material with copious amounts of soap and water.
First Aid Procedures: Remove victim to fresh air, call emergency medical care. If not breathing give CPR. If breathing is difficult administer oxygen. Treat for shock.

Fire Hazards:

Flashpoint: n/a *Ignition temperature:* n/a
Flammable Explosive High Range: n/a *Low Range:* n/a
Toxic Products of Combustion: Irritating hydrogen chloride fumes may be formed in fires.
Other Hazards: n/a
Possible extinguishing agents: Extinguish fire using suitable agent for the type of surrounding fire.

Reactivity Hazards:

Reactive With: n/a *Other Reactions:* n/a

Corrosivity Hazards:

Corrosive With: Metals and tissue *Neutralizing Agent:* Crushed limestone, soda ash, or lime.

Radioactivity Hazards:

Radiation Emitted: n/a *Other Hazards:* n/a

Recommended Protection for Response Personnel:

Avoid breathing vapors, keep upwind. Structural protective clothing provides limited protection. Wash away any material which may have come into contact with the body with copious amounts of soap and water. Consider appropriate evacuation.

FERROUS CHLORIDE (solid)

DOT Number: NA 1759 *DOT Hazard Class:* ORM-A *DOT Guide Number:* 60
Synonyms: iron dichloride, iron protochloride
STCC Number: 4941131 *Reportable Qty:* 100/45.4
Mfg Name: E.I. Du Pont *Phone No:* 1-800-441-3637

Physical Description:

Physical Form: Solid *Color:* Pale green *Odor:* Odorless
Other Information: It is used in sewage treatment, for dyeing of fabrics, and for many other uses.

Chemical Properties:

Specific Gravity: 1.93 *Vapor Density:* 4.4 *Boiling Point: n/a*
Melting Point: n/a *Vapor Pressure:* 10 mm Hg at 1292° F(700° C)
Solubility in water: Yes *Other Information:* Sinks and mixes slowly with water.

Health Hazards:

Inhalation Hazard: Will cause difficulty in breathing.
Ingestion Hazard: Will cause nausea and vomiting.
Absorption Hazard: Irritating to the skin and eyes.
Hazards to Wildlife: Dangerous to aquatic life.
Decontamination Procedures: Wash away any material with copious amounts of soap and water.
First Aid Procedures: Remove victim to fresh air, call emergency medical care. If not breathing give CPR. If breathing is difficult administer oxygen. Treat for shock.

Fire Hazards:

Flashpoint: n/a *Ignition temperature:* n/a
Flammable Explosive High Range: n/a *Low Range:* n/a
Toxic Products of Combustion: Irritating hydrogen chloride fumes may be formed in fires.
Other Hazards: n/a
Possible extinguishing agents: Extinguish fire using suitable agent for the type of surrounding fire.

Reactivity Hazards:

Reactive With: n/a *Other Reactions:* n/a

Corrosivity Hazards:

Corrosive With: Metals in the presence of moisture.
Neutralizing Agent: Crushed limestone, soda ash, or lime.

Radioactivity Hazards:

Radiation Emitted: n/a *Other Hazards:* n/a

Recommended Protection for Response Personnel:

Avoid breathing vapors, keep upwind. Structural protective clothing provides limited protection. Wash away any material which may have come into contact with the body with copious amounts of soap and water. Consider appropriate evacuation.

FERROUS FLUOROBORATE

DOT Number: n/a *DOT Hazard Class:* n/a *DOT Guide Number:* n/a
Synonyms: ferrous borofluoride
STCC Number: n/a *Reportable Qty:* n/a
Mfg Name: Harstan Chemical Co. *Phone No:* 1-718-435-8225

Physical Description:

Physical Form: Liquid *Color:* Yellow green *Odor:* n/a
Other Information: n/a

Chemical Properties:

Specific Gravity: 1.1 *Vapor Density:* n/a *Boiling Point: Decomposes*
Melting Point: n/a *Vapor Pressure:* n/a *Solubility in water:* Yes
Other Information: Sinks and mixes with water.

Health Hazards:

Inhalation Hazard: Poisonous if inhaled.
Ingestion Hazard: Poisonous if swallowed.
Absorption Hazard: Irritating to the skin and eyes.
Hazards to Wildlife: n/a
Decontamination Procedures: Wash away any material with copious amounts of soap and water.
First Aid Procedures: Remove victim to fresh air, call emergency medical care. If not breathing give CPR. If breathing is difficult administer oxygen. Treat for shock.

Fire Hazards:

Flashpoint: n/a *Ignition temperature:* n/a
Flammable Explosive High Range: n/a *Low Range:* n/a
Toxic Products of Combustion: Irritating gases may be produced when heated.
Other Hazards: n/a
Possible extinguishing agents: Extinguish fire using suitable agent for the type of surrounding fire.

Reactivity Hazards:

Reactive With: n/a *Other Reactions:* n/a

Corrosivity Hazards:

Corrosive With: n/a *Neutralizing Agent:* n/a

Radioactivity Hazards:

Radiation Emitted: n/a *Other Hazards:* n/a

Recommended Protection for Response Personnel:

Avoid breathing vapors, keep upwind. Wear appropriate sealed chemical suit. Wash away any material which may have come into contact with the body with copious amounts of soap and water. Consider appropriate evacuation.

FERROUS OXALATE

DOT Number: n/a *DOT Hazard Class:* n/a *DOT Guide Number:* n/a
Synonyms: ferrous oxalate dihydrate; ferrox; iron protoxalate; oxalic acid, ferrous salt;
STCC Number: n/a *Reportable Qty:* n/a
Mfg Name: Tennessee Chemical Co. *Phone No:* 1-404-233-6811

Physical Description:

Physical Form: Solid *Color:* Yellow *Odor:* Odorless
Other Information: n/a

Chemical Properties:

Specific Gravity: 2.3 *Vapor Density:* n/a *Boiling Point: Decomposes*
Melting Point: n/a *Vapor Pressure:* n/a *Solubility in water:* n/a
Other Information: Sinks in water

Health Hazards:

Inhalation Hazard: Coughing or difficulty in breathing.
Ingestion Hazard: Nausea, vomiting, loss of consciousness.
Absorption Hazard: Irritating to the skin and eyes.
Hazards to Wildlife: n/a
Decontamination Procedures: Wash away any material with copious amounts of soap and water.
First Aid Procedures: Remove victim to fresh air, call emergency medical care. If not breathing give CPR. If breathing is difficult administer oxygen. Treat for shock.

Fire Hazards:

Flashpoint: n/a *Ignition temperature:* n/a
Flammable Explosive High Range: n/a *Low Range:* n/a
Toxic Products of Combustion: Iron fume or iron oxide fume may be formed in fire.
Other Hazards: n/a
Possible extinguishing agents: Extinguish fire using suitable agent for the type of surrounding fire.

Reactivity Hazards:

Reactive With: n/a *Other Reactions:* n/a

Corrosivity Hazards:

Corrosive With: n/a *Neutralizing Agent:* n/a

Radioactivity Hazards:

Radiation Emitted: n/a *Other Hazards:* n/a

Recommended Protection for Response Personnel:

Avoid breathing vapors, keep upwind. Wear a sealed chemical suit (polycarbonate, butyl rubber). Wash away any material which may have come into contact with the body with copious amounts of soap and water. Consider appropriate evacuation.

FERROUS SULFATE

DOT Number: NA 9125 *DOT Hazard Class:* ORM-E *DOT Guide Number:* 31
Synonyms: copperas, green vitriol, iron(ous) sulfate, iron vitriol
STCC Number: 4963832 *Reportable Qty:* 1000/454
Mfg Name: Pfizer Chemical *Phone No:* 1-201-546-7721

Physical Description:

Physical Form: Solid *Color:* Green *Odor:* Odorless
Other Information: It is used to make other iron compounds, in water and sewage treatment, as a fertilizer, feed additive, and for many other uses.

Chemical Properties:

Specific Gravity: 1.90 *Vapor Density:* 5.2 *Boiling Point: n/a*
Melting Point: n/a *Vapor Pressure:* n/a *Solubility in water:* Slightly
Other Information: Sinks and mixes with water.

Health Hazards:

Inhalation Hazard: n/a
Ingestion Hazard: Will cause nausea, vomiting, loss of consciousness.
Absorption Hazard: n/a
Hazards to Wildlife: Dangerous to aquatic life.
Decontamination Procedures: Wash away any material with copious amounts of soap and water.
First Aid Procedures: Remove victim to fresh air, call emergency medical care. If not breathing give CPR. If breathing is difficult administer oxygen. Treat for shock.

Fire Hazards:

Flashpoint: n/a *Ignition temperature:* n/a
Flammable Explosive High Range: n/a *Low Range:* n/a
Toxic Products of Combustion: n/a
Other Hazards: n/a
Possible extinguishing agents: Extinguish fire using suitable agent for the type of surrounding fire.

Reactivity Hazards:

Reactive With: n/a *Other Reactions:* n/a

Corrosivity Hazards:

Corrosive With: n/a *Neutralizing Agent:* n/a

Radioactivity Hazards:

Radiation Emitted: n/a *Other Hazards:* n/a

Recommended Protection for Response Personnel:

Avoid breathing vapors, keep upwind. Structural protective clothing provides limited protection. Wash away any material which may have come into contact with the body with copious amounts of soap and water. Consider appropriate evacuation.

FLUORINE

DOT Number: UN 1045 *DOT Hazard Class:* Nonflammable gas *DOT Guide Number:* 20
Synonyms: none given
STCC Number: 4904030 *Reportable Qty:* 10/4.54
Mfg Name: Allied Corp. *Phone No:* 1-201-455-2000

Physical Description:

Physical Form: Gas *Color:* Clear to yellow *Odor:* Pungent
Other Information: Must be shipped in cylinders. (No more than 6 lbs/2.7 kg per cylinder.)

Chemical Properties:

Specific Gravity: 1.5 *Vapor Density:* 1.31
Boiling Point: −307° F(188.3° C) *Melting Point:* −363° F(−219.4 C)
Vapor Pressure: n/a *Solubility in water:* Reacts
Other Information: Forms hydrofluoric acid and oxygen.

Health Hazards:

Inhalation Hazard: Poisonous if inhaled.
Ingestion Hazard: n/a
Absorption Hazard: Will cause frostbite, and burns the skin and eyes.
Hazards to Wildlife: Dangerous to aquatic life.
Decontamination Procedures: Wash away any material with copious amounts of soap and water.
First Aid Procedures: Remove victim to fresh air, call emergency medical care. If not breathing give CPR. If breathing is difficult administer oxygen. Treat for shock.

Fire Hazards:

Flashpoint: n/a *Ignition temperature:* n/a
Flammable Explosive High Range: n/a *Low Range:* n/a
Toxic Products of Combustion: n/a
Other Hazards: Reacts with most combustible materials to the point that combustion often occurs.
Possible extinguishing agents: Use appropriate extinguishing agent for type of surrounding fire.

Reactivity Hazards:

Reactive With: Water to form hydrofluoric acid and oxygen.
Other Reactions: Reacts violently with all combustible materials.

Corrosivity Hazards:

Corrosive With: Most common metals *Neutralizing Agent:* n/a

Radioactivity Hazards:

Radiation Emitted: n/a *Other Hazards:* n/a

Recommended Protection for Response Personnel:

Avoid breathing vapors, keep upwind. Wear a sealed chemical suit (viton). Wash away any material which may come into contact with the body with copious amounts of soap and water. If fire becomes uncontrollable, consider appropriate evacuation. Under fire conditions, cylinders are subject to rocket (BLEVE!!)

FLUOROSILICIC ACID

DOT Number: UN 1778 *DOT Hazard Class:* Corrosive *DOT Guide Number:* 60
Synonyms: fluosilicic acid, hexafluosilicic acid, hydrogen hexafluorosilicate, sand acid, silicofluoric acid
STCC Number: n/a *Reportable Qty:* n/a
Mfg Name: W.R. Grace and Co. *Phone No:* 1-901-522-2000

Physical Description:

Physical Form: Liquid *Color:* Colorless *Odor:* Sharp unpleasant
Other Information: n/a

Chemical Properties:

Specific Gravity: 1.3 *Vapor Density:* 5
Boiling Point: 212° F(100° C) *Melting Point:* −24° F(−31.1° C)
Vapor Pressure: n/a *Solubility in water:* Yes
Other Information: Sinks and mixes with water.

Health Hazards:

Inhalation Hazard: Coughing or difficulty in breathing.
Ingestion Hazard: Causes nausea.
Absorption Hazard: Will burn the skin and eyes.
Hazards to Wildlife: n/a
Decontamination Procedures: Wash away any material with copious amounts of soap and water.
First Aid Procedures: Remove victim to fresh air, call emergency medical care. If not breathing give CPR. If breathing is difficult administer oxygen. Treat for shock.

Fire Hazards:

Flashpoint: n/a *Ignition temperature:* n/a
Flammable Explosive High Range: n/a *Low Range:* n/a
Toxic Products of Combustion: Irritating fumes of hydrogen fluoride may be formed in fires.
Other Hazards: n/a
Possible extinguishing agents: Extinguish fire using suitable agent for the type of surrounding fire.

Reactivity Hazards:

Reactive With: n/a *Other Reactions:* n/a

Corrosivity Hazards:

Corrosive With: Metals producing flammable hydrogen gas.
Neutralizing Agent: Crushed limestone, soda ash, or lime. Flush with water!

Radioactivity Hazards:

Radiation Emitted: n/a *Other Hazards:* n/a

Recommended Protection for Response Personnel:

Avoid breathing vapors, keep upwind. Wear a sealed chemical suit (nitrile, neoprene, chlorinated polyethylene). Wash away any material which may have come into contact with the body with copious amounts of soap and water. Consider appropriate evacuation.

FLUOROSULPHONIC ACID

DOT Number: UN 1777 *DOT Hazard Class:* Corrosive *DOT Guide Number:* 39
Synonyms: fluorosulfonic acid, fluorosulfuric acid
STCC Number: 4930010 *Reportable Qty:* n/a
Mfg Name: Allied Corp. *Phone No:* 1-201-455-2000

Physical Description:

Physical Form: Liquid *Color:* Colorless to light yellow *Odor:* Choking
Other Information: n/a

Chemical Properties:

Specific Gravity: 1.73 *Vapor Density:* 3.4 *Boiling Point: 324° F(162.2° C)*
Melting Point: n/a *Vapor Pressure:* n/a *Solubility in water:* Yes
Other Information: Reacts violently with water, irritating mist and gases are produced on contact with water.

Health Hazards:

Inhalation Hazard: Harmful if inhaled.
Ingestion Hazard: Harmful if swallowed.
Absorption Hazard: Will burn the skin and the eyes.
Hazards to Wildlife: n/a
Decontamination Procedures: Wash away any material with copious amounts of soap and water.
First Aid Procedures: Remove victim to fresh air, call emergency medical care. If not breathing give CPR. If breathing is difficult administer oxygen. Treat for shock.

Fire Hazards:

Flashpoint: n/a *Ignition temperature:* n/a
Flammable Explosive High Range: n/a *Low Range:* n/a
Toxic Products of Combustion: Toxic and irritating fumes of hydrogen fluoride and sulfuric acid may be formed in fires.
Other Hazards: Do not use water or foam on adjacent fires.
Possible extinguishing agents: Extinguish fire using suitable agent for the type of surrounding fire.

Reactivity Hazards:

Reactive With: Water to generate hydrogen fluoride and sulfuric acid mists.
Other Reactions: Reacts with metals to form flammable nitrogen.

Corrosivity Hazards:

Corrosive With: Metals and tissue *Neutralizing Agent:* Crushed limestone, soda ash, or lime.

Radioactivity Hazards:

Radiation Emitted: n/a *Other Hazards:* n/a

Recommended Protection for Response Personnel:

Avoid breathing vapors, keep upwind. Structural protective clothing provides limited protection. Wash away any material which may have come into contact with the body with copious amounts of soap and water. Consider appropriate evacuation.

FORMALDEHYDE

DOT Number: UN 1198 *DOT Hazard Class:* Combustible liquid *DOT Guide Number:* 29
Synonyms: formalin, formalith, hyde
STCC Number: 4913168 *Reportable Qty:* 1000/454
Mfg Name: E.I. Du Pont *Phone No:* 1-800-441-3637

Physical Description:

Physical Form: Liquid *Color:* Colorless *Odor:* Irritating
Other Information: Used to make plastics, other chemicals, fertilizers as a preservative and a corrosion inhibitor.

Chemical Properties:

Specific Gravity: 1.2 *Vapor Density:* 3.3
Boiling Point: 322° F(161.1° C) *Melting Point:* −180° F(−117.7° C)
Vapor Pressure: 19 mm Hg at 77° F(25° C) *Solubility in water:* Yes
Other Information: Vapors are heavier than air. Weighs 9.4 lbs/4.2 kg per gallon/3.8 l.

Health Hazards:

Inhalation Hazard: n/a
Ingestion Hazard: Will cause nausea, loss of consciousness.
Absorption Hazard: Will burn the skin and eyes.
Hazards to Wildlife: Dangerous to aquatic life.
Decontamination Procedures: Wash away any material with copious amounts of soap and water.
First Aid Procedures: Remove victim to fresh air, call emergency medical care. If not breathing give CPR. If breathing is difficult administer oxygen. Treat for shock.

Fire Hazards:

Flashpoint: 182° F(83.3° C) *Ignition temperature:* 806° F(430° C)
Flammable Explosive High Range: 73 *Low Range:* 7
Toxic Products of Combustion: n/a
Other Hazards: The gas readily vaporizes from solution, and is flammable over a wide vapor air concentration range.
Possible extinguishing agents: Apply water from as far a distance as possible. Use alcohol foam, dry chemical, or carbon dioxide.

Reactivity Hazards:

Reactive With: n/a *Other Reactions:* n/a

Corrosivity Hazards:

Corrosive With: n/a *Neutralizing Agent:* n/a

Radioactivity Hazards:

Radiation Emitted: n/a *Other Hazards:* n/a

Recommended Protection for Response Personnel:

Avoid breathing vapors, keep upwind. Wear a sealed chemical suit (polycarbonate, PVC, chlorinated polyethylene, butyl rubber, viton). Wash away any material which may have come into contact with the body with copious amounts of soap and water. Consider appropriate evacuation.

FORMIC ACID

DOT Number: UN 1779 *DOT Hazard Class:* Corrosive *DOT Guide Number:* 60
Synonyms: formylic acid, methanoic acid
STCC Number: 4931320 *Reportable Qty:* 5000/2270
Mfg Name: Mallinckrodt Inc. *Phone No:* 1-314-895-2000

Physical Description:

Physical Form: Liquid *Color:* Colorless *Odor:* Penetrating
Other Information: n/a

Chemical Properties:

Specific Gravity: 1.22 *Vapor Density:* 1.59 *Boiling Point: 225° F(107.2° C)*
Melting Point: 20° F(−6.6° C) *Vapor Pressure:* 23-33 mm Hg at 68° F(20° C) *Solubility in water:* Yes
Other Information: Weighs 10.2 lbs/4.6 kg per gallon/3.8 l.

Health Hazards:

Inhalation Hazard: Harmful if inhaled.
Ingestion Hazard: Harmful if swallowed.
Absorption Hazard: Will burn the skin and eyes.
Hazards to Wildlife: Dangerous to aquatic life.
Decontamination Procedures: Wash away any material with copious amounts of soap and water.
First Aid Procedures: Remove victim to fresh air, call emergency medical care. If not breathing give CPR. If breathing is difficult administer oxygen. Treat for shock.

Fire Hazards:

Flashpoint: 138° F(58.8° C) *Ignition temperature:* 1114° F(601.1° C)
Flammable Explosive High Range: 57 *Low Range:* 18
Toxic Products of Combustion: n/a
Other Hazards: n/a
Possible extinguishing agents: Apply water from as far a distance as possible. Use alcohol foam, dry chemical, or carbon dioxide.

Reactivity Hazards:

Reactive With: n/a *Other Reactions:* n/a

Corrosivity Hazards:

Corrosive With: Metals and tissues *Neutralizing Agent:* Crushed limestone, soda ash, or lime.

Radioactivity Hazards:

Radiation Emitted: n/a *Other Hazards:* n/a

Recommended Protection for Response Personnel:

Avoid breathing vapors, keep upwind. Structural protective clothing provides limited protection. Wash away any material which may have come into contact with the body with copious amounts of soap and water. Consider appropriate evacuation.

FUEL OIL NUMBER 1

DOT Number: NA 1993 *DOT Hazard Class:* Combustible liquid *DOT Guide Number:* 27
Synonyms: coal oil, kerosene, JP-1, range oil
STCC Number: 4915112 *Reportable Qty:* n/a
Mfg Name: Shell Oil Co. *Phone No:* 1-713-241-6161

Physical Description:

Physical Form: Liquid *Color:* Colorless to light brown *Odor:* Petroleum like
Other Information: It is used for jet fuel and for domestic heating.

Chemical Properties:

Specific Gravity: .83 *Vapor Density:* 4.5 *Boiling Point: 380-560° F(193-293° C)*
Melting Point: –45/–55° F(–42/–48 C) *Vapor Pressure:* 2 mm Hg at 68° F(20° C)
Solubility in water: No
Other Information: Insoluble and lighter than water. Weighs 7 lbs/3.1 kg per gallon/3.8 l.

Health Hazards:

Inhalation Hazard: Dizziness, drowsiness
Ingestion Hazard: Nausea, vomiting, and cramping.
Absorption Hazard: Mild irritation
Hazards to Wildlife: n/a
Decontamination Procedures: Wash away any material with copious amounts of soap and water.
First Aid Procedures: Remove victim to fresh air, call emergency medical care. If not breathing give CPR. If breathing is difficult administer oxygen. Treat for shock.

Fire Hazards:

Flashpoint: 100-162° F(37-72° C) *Ignition temperature:* 410-444° F(210-228° C)
Flammable Explosive High Range: 5 *Low Range:* .7
Toxic Products of Combustion: n/a
Other Hazards: Will burn, but difficult to ignite unless warmed, there is a possibility that the containers may violently rupture in a fire.
Possible extinguishing agents: Do not extinguish the fire unless the flow can be stopped. Apply water from as far a distance as possible, use foam, dry chemical, or carbon dioxide.

Reactivity Hazards:

Reactive With: n/a *Other Reactions:* n/a

Corrosivity Hazards:

Corrosive With: n/a *Neutralizing Agent:* n/a

Radioactivity Hazards:

Radiation Emitted: n/a *Other Hazards:* n/a

Recommended Protection for Response Personnel:

Avoid breathing vapors, keep upwind. Structural protective clothing provides limited protection. Wash away any material which may have come into contact with the body with copious amounts of soap and water. Consider appropriate evacuation.

FUEL OIL NUMBER 2

DOT Number: NA 1993 *DOT Hazard Class:* Combustible liquid *DOT Guide Number:* 27
Synonyms: diesel oil, home heating
STCC Number: 4915112 *Reportable Qty:* n/a
Mfg Name: Shell Oil Co. *Phone No:* 1-713-241-6161

Physical Description:

Physical Form: Liquid *Color:* Colorless to light brown *Odor:* Petroleum like
Other Information: It is used for domestic and commercial fuel oil.

Chemical Properties:

Specific Gravity: .88 *Vapor Density:* 2 *Boiling Point: 540-640° F(282-337° C)*
Melting Point: −20° F(−28.8° C) *Vapor Pressure:* 2.6 mm Hg at 122° F(50° C) *Solubility in water:* No
Other Information: Insoluble and lighter than water. Weighs 7.3 lbs/3.3 kg per gallon/3.8 l.

Health Hazards:

Inhalation Hazard: Headache, slight giddiness.
Ingestion Hazard: Nausea, vomiting, and cramping.
Absorption Hazard: Mild irritation
Hazards to Wildlife: n/a
Decontamination Procedures: Wash away any material with copious amounts of soap and water.
First Aid Procedures: Remove victim to fresh air, call emergency medical care. If not breathing give CPR. If breathing is difficult administer oxygen. Treat for shock.

Fire Hazards:

Flashpoint: 126-204° F(52-95° C) *Ignition temperature:* 494° F(256.6° C)
Flammable Explosive High Range: 6 *Low Range:* 1.3
Toxic Products of Combustion: n/a
Other Hazards: Will burn, but difficult to ignite unless warmed, there is a possibility that the containers may violently rupture in a fire.
Possible extinguishing agents: Do not extinguish the fire unless the flow can be stopped. Apply water from as far a distance as possible, use foam, dry chemical, or carbon dioxide.

Reactivity Hazards:

Reactive With: n/a *Other Reactions:* n/a

Corrosivity Hazards:

Corrosive With: n/a *Neutralizing Agent:* n/a

Radioactivity Hazards:

Radiation Emitted: n/a *Other Hazards:* n/a

Recommended Protection for Response Personnel:

Avoid breathing vapors, keep upwind. Structural protective clothing provides limited protection. Wash away any material which may have come into contact with the body with copious amounts of soap and water. Consider appropriate evacuation.

FUEL OIL NUMBER 4

DOT Number: NA 1993 *DOT Hazard Class:* Combustible liquid *DOT Guide Number:* 27
Synonyms: cat cracker feedstock, residential fuel oil
STCC Number: 4915112 *Reportable Qty:* n/a
Mfg Name: Shell Oil Co. *Phone No:* 1-713-241-6161

Physical Description:

Physical Form: Liquid *Color:* Brown *Odor:* Petroleum like
Other Information: It is used for commercial and industrial burner fuel.

Chemical Properties:

Specific Gravity: .904 *Vapor Density:* 2 *Boiling Point: 214-1092° F(101-588° C)*
Melting Point: −15/−20° F(−26/−28 C) *Vapor Pressure:* 2 mm Hg at 68° F(20° C)
Solubility in water: No
Other Information: Insoluble and lighter than water. Weighs 7.5 lbs/3.4 kg per gallon/3.8 l.

Health Hazards:

Inhalation Hazard: Headache, slight giddiness.
Ingestion Hazard: Nausea, vomiting, and cramping.
Absorption Hazard: Mild irritation
Hazards to Wildlife: n/a
Decontamination Procedures: Wash away any material with copious amounts of soap and water.
First Aid Procedures: Remove victim to fresh air, call emergency medical care. If not breathing give CPR. If breathing is difficult administer oxygen. Treat for shock.

Fire Hazards:

Flashpoint: 142-240° F(61-115° C) *Ignition temperature:* 505° F(262.7° C)
Flammable Explosive High Range: 5 *Low Range:* 1
Toxic Products of Combustion: n/a
Other Hazards: Will burn, but difficult to ignite unless warmed, there is a possibility that the containers may violently rupture in a fire.
Possible extinguishing agents: Do not extinguish the fire unless the flow can be stopped. Apply water from as far a distance as possible, use foam, dry chemical, or carbon dioxide.

Reactivity Hazards:

Reactive With: n/a *Other Reactions:* n/a

Corrosivity Hazards:

Corrosive With: n/a *Neutralizing Agent:* n/a

Radioactivity Hazards:

Radiation Emitted: n/a *Other Hazards:* n/a

Recommended Protection for Response Personnel:

Avoid breathing vapors, keep upwind. Structural protective clothing provides limited protection. Wash away any material which may have come into contact with the body with copious amounts of soap and water. Consider appropriate evacuation.

FUEL OIL NUMBER 5

DOT Number: NA 1993 *DOT Hazard Class:* Combustible liquid *DOT Guide Number:* 27
Synonyms: bunker C, Navy special fuel oil, NSFO, residential fuel oil #5
STCC Number: 4915112 *Reportable Qty:* n/a
Mfg Name: Shell Oil Co. *Phone No:* 1-713-241-6161

Physical Description:

Physical Form: Liquid *Color:* Brown *Odor:* Petroleum like
Other Information: It is used for power plants, ships, locomotives, and metallurgical operations.

Chemical Properties:

Specific Gravity: .936 *Vapor Density:* 2 *Boiling Point: 426-1062° F(218-572° C)*
Melting Point: 0° F(−17.7° C) *Vapor Pressure:* 2 mm Hg at 68° F(20° C) *Solubility in water:* No
Other Information: Insoluble and lighter than water. Weighs 7.8 lbs/3.5 kg per gallon/3.8 l.

Health Hazards:

Inhalation Hazard: Headache, slight giddiness.
Ingestion Hazard: Nausea, vomiting, and cramping.
Absorption Hazard: Mild irritation
Hazards to Wildlife: n/a
Decontamination Procedures: Wash away any material with copious amounts of soap and water.
First Aid Procedures: Remove victim to fresh air, call emergency medical care. If not breathing give CPR. If breathing is difficult administer oxygen. Treat for shock.

Fire Hazards:

Flashpoint: 156-336° F(68-168° C) *Ignition temperature:* 765° F(407.2° C)
Flammable Explosive High Range: 5 *Low Range:* 1
Toxic Products of Combustion: n/a
Other Hazards: Will burn, but difficult to ignite unless warmed, there is a possibility that the containers may violently rupture in a fire.
Possible extinguishing agents: Do not extinguish the fire unless the flow can be stopped. Apply water from as far a distance as possible, use foam, dry chemical, or carbon dioxide.

Reactivity Hazards:

Reactive With: n/a *Other Reactions:* n/a

Corrosivity Hazards:

Corrosive With: n/a *Neutralizing Agent:* n/a

Radioactivity Hazards:

Radiation Emitted: n/a *Other Hazards:* n/a

Recommended Protection for Response Personnel:

Avoid breathing vapors, keep upwind. Structural protective clothing provides limited protection. Wash away any material which may have come into contact with the body with copious amounts of soap and water. Consider appropriate evacuation.

FUMARIC ACID

DOT Number: NA 9126 *DOT Hazard Class:* ORM-E *DOT Guide Number:* 31
Synonyms: allomaleic acid, boletic acid, lichenic acid
STCC Number: 4966352 *Reportable Qty:* 5000/2270
Mfg Name: Monsanto Chemical *Phone No:* 1-314-694-1000

Physical Description:

Physical Form: Solid *Color:* White *Odor:* Odorless
Other Information: It is used to make paints and plastics, in food processing, and in preservation, and for many other uses.

Chemical Properties:

Specific Gravity: 1.63 *Vapor Density:* 3.3 *Boiling Point: Very high*
Melting Point: n/a *Vapor Pressure:* n/a *Solubility in water:* Yes
Other Information: Sinks and mixes with water.

Health Hazards:

Inhalation Hazard: Will cause coughing, difficulty in breathing.
Ingestion Hazard: n/a
Absorption Hazard: Irritating to the skin and eyes.
Hazards to Wildlife: n/a
Decontamination Procedures: Wash away any material with copious amounts of soap and water.
First Aid Procedures: Remove victim to fresh air, call emergency medical care. If not breathing give CPR. If breathing is difficult administer oxygen. Treat for shock.

Fire Hazards:

Flashpoint: n/a *Ignition temperature:* 1364° F(740° C)
Flammable Explosive High Range: n/a *Low Range:* n/a
Toxic Products of Combustion: Irritating fumes of maleic anhydride may be formed in fires.
Other Hazards: Dust presents an explosion hazard.
Possible extinguishing agents: Apply water from as far a distance as possible. Use foam, dry chemical, or carbon dioxide.

Reactivity Hazards:

Reactive With: n/a *Other Reactions:* n/a

Corrosivity Hazards:

Corrosive With: n/a *Neutralizing Agent:* n/a

Radioactivity Hazards:

Radiation Emitted: n/a *Other Hazards:* n/a

Recommended Protection for Response Personnel:

Avoid breathing vapors, keep upwind. Structural protective clothing provides limited protection. Wash away any material which may have come into contact with the body with copious amounts of soap and water. Consider appropriate evacuation.

FURFURAL

DOT Number: UN 1199 *DOT Hazard Class:* Combustible liquid *DOT Guide Number:* 29
Synonyms: fural, furfurole, quakeral
STCC Number: 4913146 *Reportable Qty:* 5000/2270
Mfg Name: Quaker Oats Chemical *Phone No:* 1-312-222-7300

Physical Description:

Physical Form: Liquid *Color:* Colorless to reddish brown *Odor:* Almond
Other Information: n/a

Chemical Properties:

Specific Gravity: 1.16 *Vapor Density:* 3.3 *Boiling Point: 323° F(161.6° C)*
Melting Point: −34° F(−36.6° C) *Vapor Pressure:* 2 mm Hg at 68° F(20° C)
Solubility in water: n/a *Degree of Solubility:* 8.3%
Other Information: Weighs 9.7 lbs/4.3 kg per gallon/3.8 l.

Health Hazards:

Inhalation Hazard: Harmful if inhaled.
Ingestion Hazard: Harmful if swallowed.
Absorption Hazard: Will burn the skin and eyes.
Hazards to Wildlife: Dangerous to aquatic life.
Decontamination Procedures: Wash away any material with copious amounts of soap and water.
First Aid Procedures: Remove victim to fresh air, call emergency medical care. If not breathing give CPR. If breathing is difficult administer oxygen. Treat for shock.

Fire Hazards:

Flashpoint: 140° F(60° C) *Ignition temperature:* 739° F(392.7° C)
Flammable Explosive High Range: 19.3 *Low Range:* 2.1
Toxic Products of Combustion: Irritating vapors are generated when heated.
Other Hazards: n/a
Possible extinguishing agents: Apply water from as far a distance as possible. Use alcohol foam, dry chemical, or carbon dioxide.

Reactivity Hazards:

Reactive With: n/a *Other Reactions:* n/a

Corrosivity Hazards:

Corrosive With: n/a *Neutralizing Agent:* n/a

Radioactivity Hazards:

Radiation Emitted: n/a *Other Hazards:* n/a

Recommended Protection for Response Personnel:

Avoid breathing vapors, keep upwind. Structural protective clothing provides limited protection. Wash away any material which may have come into contact with the body with copious amounts of soap and water. Consider appropriate evacuation.

FURFURYL ALCOHOL

DOT Number: UN 2874 *DOT Hazard Class:* Combustible liquid *DOT Guide Number:* 55
Synonyms: furfuralcohol
STCC Number: 4913124 *Reportable Qty:* n/a
Mfg Name: Quaker Oats Chemical *Phone No:* 1-312-222-7300

Physical Description:

Physical Form: Liquid *Color:* Colorless to light yellow *Odor:* Mild, irritating
Other Information: n/a

Chemical Properties:

Specific Gravity: 1.1 *Vapor Density:* 3.4 *Boiling Point: 340° F(171.1° C)*
Melting Point: 6° F(–14.4° C) *Vapor Pressure:* 1 mm Hg at 68° F(20° C) *Solubility in water:* Yes
Other Information: Reacts vigorously with most acids forming a black solid.

Health Hazards:

Inhalation Hazard: Harmful if inhaled.
Ingestion Hazard: Harmful if swallowed.
Absorption Hazard: Irritating to the skin and eyes.
Hazards to Wildlife: n/a
Decontamination Procedures: Wash away any material with copious amounts of soap and water.
First Aid Procedures: Remove victim to fresh air, call emergency medical care. If not breathing give CPR. If breathing is difficult administer oxygen. Treat for shock.

Fire Hazards:

Flashpoint: 167° F(75° C) *Ignition temperature:* 736° F(391.1° C)
Flammable Explosive High Range: 16.3 *Low Range:* 1.8
Toxic Products of Combustion: n/a
Other Hazards: n/a
Possible extinguishing agents: Apply water from as far a distance as possible. Use alcohol foam, dry chemical, or carbon dioxide.

Reactivity Hazards:

Reactive With: n/a *Other Reactions:* n/a

Corrosivity Hazards:

Corrosive With: n/a *Neutralizing Agent:* n/a

Radioactivity Hazards:

Radiation Emitted: n/a *Other Hazards:* n/a

Recommended Protection for Response Personnel:

Avoid breathing vapors, keep upwind. Structural protective clothing provides limited protection. Wash away any material which may have come into contact with the body with copious amounts of soap and water. Consider appropriate evacuation.

GALLIC ACID

DOT Number: n/a *DOT Hazard Class:* n/a *DOT Guide Number:* n/a
Synonyms: gallic acid monohydrate, 3,4,5-trihydroxybenzoic acid
STCC Number: n/a *Reportable Qty:* n/a
Mfg Name: Mallinckrodt Inc. *Phone No:* 1-314-895-2000

Physical Description:

Physical Form: Solid *Color:* White *Odor:* Odorless
Other Information: n/a

Chemical Properties:

Specific Gravity: 1.7 *Vapor Density:* n/a *Boiling Point: n/a*
Melting Point: n/a *Vapor Pressure:* n/a *Solubility in water:* n/a
Other Information: Sinks in watcr

Health Hazards:

Inhalation Hazard: Coughing, difficulty in breathing.
Ingestion Hazard: Harmful if swallowed.
Absorption Hazard: Irritating to the skin and eyes.
Hazards to Wildlife: Dangerous to aquatic life.
Decontamination Procedures: Wash away any material with copious amounts of soap and water.
First Aid Procedures: Remove victim to fresh air, call emergency medical care. If not breathing give CPR. If breathing is difficult administer oxygen. Treat for shock.

Fire Hazards:

Flashpoint: n/a *Ignition temperature:* n/a
Flammable Explosive High Range: n/a *Low Range:* n/a
Toxic Products of Combustion: n/a
Other Hazards: Combustible
Possible extinguishing agents: Extinguish fire using suitable agent for the type of surrounding fire.

Reactivity Hazards:

Reactive With: n/a *Other Reactions:* n/a

Corrosivity Hazards:

Corrosive With: n/a *Neutralizing Agent:* n/a

Radioactivity Hazards:

Radiation Emitted: n/a *Other Hazards:* n/a

Recommended Protection for Response Personnel:

Avoid breathing vapors, keep upwind. Structural protective clothing provides limited protection. Wash away any material which may have come into contact with the body with copious amounts of soap and water. Consider appropriate evacuation.

GAS OIL

DOT Number: UN 1202
DOT Hazard Class: Flammable liquid
DOT Guide Number: 27
Synonyms: none given
STCC Number: n/a
Reportable Qty: n/a
Mfg Name: Shell Oil Co.
Phone No: 1-713-241-6161

Physical Description:

Physical Form: Liquid
Color: Colorless
Odor: Gasolinelike
Other Information: n/a

Chemical Properties:

Specific Gravity: .85
Vapor Density: n/a
Boiling Point: 375-750° F(190-398° C)
Melting Point: n/a
Vapor Pressure: n/a
Solubility in water: No
Other Information: Floats on water

Health Hazards:

Inhalation Hazard: n/a
Ingestion Hazard: Harmful if swallowed.
Absorption Hazard: n/a
Hazards to Wildlife: Dangerous to aquatic life.
Decontamination Procedures: Wash away any material with copious amounts of soap and water.
First Aid Procedures: Remove victim to fresh air, call emergency medical care. If not breathing give CPR. If breathing is difficult administer oxygen. Treat for shock.

Fire Hazards:

Flashpoint: 150° F(65.5° C)
Ignition temperature: 640° F(337.7° C)
Flammable Explosive High Range: 13.5
Low Range: 6
Toxic Products of Combustion: n/a
Other Hazards: Flashback along vapor trail may occur. Vapors may explode if ignited in an enclosed area.
Possible extinguishing agents: Extinguish fire using suitable agent for the type of surrounding fire.

Reactivity Hazards:

Reactive With: n/a
Other Reactions: n/a

Corrosivity Hazards:

Corrosive With: n/a
Neutralizing Agent: n/a

Radioactivity Hazards:

Radiation Emitted: n/a
Other Hazards: n/a

Recommended Protection for Response Personnel:

Avoid breathing vapors, keep upwind. Structural protective clothing provides limited protection. Wash away any material which may have come into contact with the body with copious amounts of soap and water. Consider appropriate evacuation.

GASOLINE

DOT Number: UN 1203 *DOT Hazard Class:* Flammable liquid *DOT Guide Number:* 27
Synonyms: motor spirit, petrol
STCC Number: 4908178 *Reportable Qty:* n/a
Mfg Name: Shell Oil Co. *Phone No:* 1-713-241-6161

Physical Description:

Physical Form: Liquid *Color:* Colorless to pale brown or pink *Odor:* Gasoline like
Other Information: n/a

Chemical Properties:

Specific Gravity: .8 *Vapor Density:* 3-4 *Boiling Point: 100-400° F(37.7-204.4° C)*
Melting Point: n/a *Vapor Pressure:* n/a *Solubility in water:* No
Other Information: n/a

Health Hazards:

Inhalation Hazard: Will cause headache, dizziness.
Ingestion Hazard: Will cause nausea and vomiting.
Absorption Hazard: Irritating to the skin and eyes.
Hazards to Wildlife: Dangerous to aquatic life.
Decontamination Procedures: Wash away any material with copious amounts of soap and water.
First Aid Procedures: Remove victim to fresh air, call emergency medical care. If not breathing give CPR. If breathing is difficult administer oxygen. Treat for shock.

Fire Hazards:

Flashpoint: −45° F(−42.7° C) *Ignition temperature:* 536° F(280° C)
Flammable Explosive High Range: 7.8 *Low Range:* 1.4
Toxic Products of Combustion: n/a
Other Hazards: Flashback along vapor trail may occur. Vapors may explode if ignited in a small area.
Possible extinguishing agents: Water may be ineffective on fire. Use foam, dry chemical, or carbon dioxide

Reactivity Hazards:

Reactive With: n/a *Other Reactions:* n/a

Corrosivity Hazards:

Corrosive With: n/a *Neutralizing Agent:* n/a

Radioactivity Hazards:

Radiation Emitted: n/a *Other Hazards:* n/a

Recommended Protection for Response Personnel:

Avoid breathing vapors, keep upwind. Structural protective clothing provides limited protection. Wash away any material which may have come into contact with the body with copious amounts of soap and water. Consider appropriate evacuation.

GASOLINE, AVIATION

DOT Number: UN 1203 *DOT Hazard Class:* Flammable liquid *DOT Guide Number:* 27
Synonyms: av gas
STCC Number: n/a *Reportable Qty:* n/a
Mfg Name: Shell Oil Co. *Phone No:* 1-713-241-6161

Physical Description:

Physical Form: Liquid *Color:* Red, blue, green, brown or purple *Odor:* Gasoline like
Other Information: Used as an aircraft fuel.

Chemical Properties:

Specific Gravity: .71 *Vapor Density:* 3.4 *Boiling Point: 160 – 340° F(71 – 171° C)*
Melting Point: 76° F(24.4° C) *Vapor Pressure:* n/a *Solubility in water:* No
Other Information: Floats on water. Flammable, irritating vapor is produced.

Health Hazards:

Inhalation Hazard: Dizziness, headache, difficulty in breathing, loss of consciousness.
Ingestion Hazard: Nausea and vomiting.
Absorption Hazard: Irritating to the skin and eyes.
Hazards to Wildlife: Dangerous to aquatic life.
Decontamination Procedures: Wash away any material with copious amounts of soap and water.
First Aid Procedures: Remove victim to fresh air, call emergency medical care. If not breathing give CPR. If breathing is difficult administer oxygen. Treat for shock.

Fire Hazards:

Flashpoint: –50° F(–45.5° C) *Ignition temperature:* 824° F(440° C)
Flammable Explosive High Range: 7.1 *Low Range:* 1.2
Toxic Products of Combustion: n/a
Other Hazards: Flashback along vapor trail may occur. Vapors may explode if ignited in an enclosed area.
Possible extinguishing agents: Use foam, dry chemical or carbon dioxide.

Reactivity Hazards:

Reactive With: n/a *Other Reactions:* n/a

Corrosivity Hazards:

Corrosive With: n/a *Neutralizing Agent:* n/a

Radioactivity Hazards:

Radiation Emitted: n/a *Other Hazards:* n/a

Recommended Protection for Response Personnel:

Avoid breathing vapors, keep upwind. Structural protective clothing provides limited protection. Wash away any material which may have come into contact with the body with copious amounts of soap and water. Consider appropriate evacuation.

GASOLINE, CASINGHEAD

DOT Number: UN 1203 *DOT Hazard Class:* Flammable liquid *DOT Guide Number:* 27
Synonyms: motor spirit, petrol
STCC Number: 4908176 *Reportable Qty:* n/a
Mfg Name: Shell Oil Co. *Phone No:* 1-713-241-6161

Physical Description:

Physical Form: Liquid *Color:* Colorless to amber *Odor:* Gasolinelike
Other Information: n/a

Chemical Properties:

Specific Gravity: .67 *Vapor Density:* n/a *Boiling Point: 58-275° F(14-135° C)*
Melting Point: n/a *Vapor Pressure:* n/a *Solubility in water:* No
Other Information: n/a

Health Hazards:

Inhalation Hazard: Will cause headache, dizziness.
Ingestion Hazard: Will cause nausea and vomiting.
Absorption Hazard: Irritating to the skin and eyes.
Hazards to Wildlife: Dangerous to aquatic life.
Decontamination Procedures: Wash away any material with copious amounts of soap and water.
First Aid Procedures: Remove victim to fresh air, call emergency medical care. If not breathing give CPR. If breathing is difficult administer oxygen. Treat for shock.

Fire Hazards:

Flashpoint: 0° F(−17.7° C) *Ignition temperature:* n/a
Flammable Explosive High Range: 7.1 *Low Range:* 1.3
Toxic Products of Combustion: n/a
Other Hazards: Flashback along vapor trail may occur. Vapors may explode if ignited in a small area.
Possible extinguishing agents: Water may be ineffective on fire. Use foam, dry chemical, or carbon dioxide

Reactivity Hazards:

Reactive With: n/a *Other Reactions:* n/a

Corrosivity Hazards:

Corrosive With: n/a *Neutralizing Agent:* n/a

Radioactivity Hazards:

Radiation Emitted: n/a *Other Hazards:* n/a

Recommended Protection for Response Personnel:

Avoid breathing vapors, keep upwind. Structural protective clothing provides limited protection. Wash away any material which may have come into contact with the body with copious amounts of soap and water. Consider appropriate evacuation.

GELATINE DYNAMITE

DOT Number: n/a *DOT Hazard Class:* Explosives a *DOT Guide Number:* n/a
Synonyms: C-4
STCC Number: 4901549 *Reportable Qty:* n/a *CHEMTREC Phone No:* 1-800-424-9300

Physical Description:

Physical Form: Solid *Color:* Pale yellow *Odor:* n/a
Other Information: It is elastic in nature, usually composed of nitroglycerin or ammonium nitrate.

Chemical Properties:

Specific Gravity: n/a *Vapor Density:* n/a *Boiling Point: n/a*
Melting Point: n/a *Vapor Pressure:* n/a *Solubility in water:* No
Other Information: It is sensitive to heat and shock. It is used in rocket propellants and commercial explosives.

Health Hazards:

Inhalation Hazard: n/a
Ingestion Hazard: n/a
Absorption Hazard: n/a
Hazards to Wildlife: n/a
Decontamination Procedures: Wash away any material with copious amounts of soap and water.
First Aid Procedures: Remove victim to fresh air, call emergency medical care. If not breathing give CPR. If breathing is difficult administer oxygen. Treat for shock.

Fire Hazards:

Flashpoint: n/a *Ignition temperature:* n/a
Flammable Explosive High Range: n/a *Low Range:* n/a
Toxic Products of Combustion: When heated to decomposition, it emits highly toxic nitrogen oxides and carbon monoxide.
Other Hazards: Flashback along vapor trail may occur. Vapors may explode if ignited in a small area.
Possible extinguishing agents: Dangerously explosive! Do not fight fires in a cargo of explosives. Evacuate and let burn.

Reactivity Hazards:

Reactive With: Heat and shock. *Other Reactions:* n/a

Corrosivity Hazards:

Corrosive With: n/a *Neutralizing Agent:* n/a

Radioactivity Hazards:

Radiation Emitted: n/a *Other Hazards:* n/a

Recommended Protection for Response Personnel:

Avoid breathing vapors, keep upwind. Wear appropriate chemical clothing. Wash away any material which may have come into contact with the body with copious amounts of soap and water. Consider appropriate evacuation.

GLYCERINE

DOT Number: n/a *DOT Hazard Class:* n/a *DOT Guide Number:* n/a
Synonyms: gylcerol, 1,2,3-propanetriol, 1,2,3-trihydroxypropane
STCC Number: n/a *Reportable Qty:* n/a
Mfg Name: Dow Chemical *Phone No:* 1-517-636-4400

Physical Description:

Physical Form: Liquid *Color:* Colorless *Odor:* Odorless
Other Information: n/a

Chemical Properties:

Specific Gravity: 1.26 *Vapor Density:* 3.17 *Boiling Point: 554° F(290° C)*
Melting Point: 64° F(17.7° C) *Vapor Pressure:* .0025 mm Hg at 122° F(50° C)
Solubility in water: Yes *Other Information:* Sinks and mixes with water.

Health Hazards:

Inhalation Hazard: n/a
Ingestion Hazard: n/a
Absorption Hazard: n/a
Hazards to Wildlife: n/a
Decontamination Procedures: Wash away any material with copious amounts of soap and water.
First Aid Procedures: Remove victim to fresh air, call emergency medical care. If not breathing give CPR. If breathing is difficult administer oxygen. Treat for shock.

Fire Hazards:

Flashpoint: 320° F(160° C) *Ignition temperature:* 698° F(370° C)
Flammable Explosive High Range: n/a *Low Range:* n/a
Toxic Products of Combustion: n/a
Other Hazards: n/a
Possible extinguishing agents: Use alcohol foam, dry chemical, carbon dioxide. Cool all affected containers with water. Water may be ineffective on fire!

Reactivity Hazards:

Reactive With: n/a *Other Reactions:* n/a

Corrosivity Hazards:

Corrosive With: n/a *Neutralizing Agent:* n/a

Radioactivity Hazards:

Radiation Emitted: n/a *Other Hazards:* n/a

Recommended Protection for Response Personnel:

Avoid breathing vapors, keep upwind. Structural protective clothing provides limited protection. Wash away any material which may have come into contact with the body with copious amounts of soap and water. Consider appropriate evacuation.

GLYCIDALDEHYDE

DOT Number: UN 2622 *DOT Hazard Class:* n/a *DOT Guide Number:* 28
Synonyms: 1,5-pentanedial
STCC Number: n/a *Reportable Qty:* n/a
Mfg Name: Union Carbide *Phone No:* 1-203-794-2000

Physical Description:

Physical Form: Liquid *Color:* Light yellow *Odor:* n/a
Other Information: n/a

Chemical Properties:

Specific Gravity: 1.1 *Vapor Density:* n/a *Boiling Point: 212° F(100° C)*
Melting Point: 20° F(−6.6° C) *Vapor Pressure:* n/a *Solubility in water:* Yes
Other Information: Mixes with water.

Health Hazards:

Inhalation Hazard: n/a
Ingestion Hazard: Harmful if swallowed.
Absorption Hazard: Irritating to the skin and eyes.
Hazards to Wildlife: n/a
Decontamination Procedures: Wash away any material with copious amounts of soap and water.
First Aid Procedures: Remove victim to fresh air, call emergency medical care. If not breathing give CPR. If breathing is difficult administer oxygen. Treat for shock.

Fire Hazards:

Flashpoint: n/a *Ignition temperature:* n/a
Flammable Explosive High Range: n/a *Low Range:* n/a
Toxic Products of Combustion: n/a
Other Hazards: n/a
Possible extinguishing agents: Extinguish fire using suitable agent for the type of surrounding fire.

Reactivity Hazards:

Reactive With: n/a *Other Reactions:* n/a

Corrosivity Hazards:

Corrosive With: n/a *Neutralizing Agent:* n/a

Radioactivity Hazards:

Radiation Emitted: n/a *Other Hazards:* n/a

Recommended Protection for Response Personnel:

Avoid breathing vapors, keep upwind. Wear appropriate sealed chemical suit. Wash away any material which may have come into contact with the body with copious amounts of soap and water. Consider appropriate evacuation.

GLYCIDYL METHACRYLATE

DOT Number: n/a *DOT Hazard Class:* n/a *DOT Guide Number:* n/a
Synonyms: glycidyl α-methyl-acrylate, methacrylic acid-2,3-epoxy propyl ester
STCC Number: n/a *Reportable Qty:* n/a
Mfg Name: Alcoalac Inc. *Phone No:* 1-301-355-2600

Physical Description:

Physical Form: Liquid *Color:* Colorless *Odor:* Fruity
Other Information: n/a

Chemical Properties:

Specific Gravity: 1.07 *Vapor Density:* n/a *Boiling Point: Very high*
Melting Point: n/a *Vapor Pressure:* n/a *Solubility in water:* No
Other Information: Floats on water.

Health Hazards:

Inhalation Hazard: n/a
Ingestion Hazard: Harmful if swallowed.
Absorption Hazard: Will burn the skin and eyes.
Hazards to Wildlife: n/a
Decontamination Procedures: Wash away any material with copious amounts of soap and water.
First Aid Procedures: Remove victim to fresh air, call emergency medical care. If not breathing give CPR. If breathing is difficult administer oxygen. Treat for shock.

Fire Hazards:

Flashpoint: 183° F(83.8° C) *Ignition temperature:* n/a
Flammable Explosive High Range: n/a *Low Range:* n/a
Toxic Products of Combustion: Irritating vapors are generated when heated.
Other Hazards: Combustible
Possible extinguishing agents: Extinguish fire using suitable agent for the type of surrounding fire.

Reactivity Hazards:

Reactive With: Heat causes polymerization. The reaction is not hazardous. *Other Reactions:* n/a

Corrosivity Hazards:

Corrosive With: n/a *Neutralizing Agent:* n/a

Radioactivity Hazards:

Radiation Emitted: n/a *Other Hazards:* n/a

Recommended Protection for Response Personnel:

Avoid breathing vapors, keep upwind. Wear a sealed chemical suit (polycarbonate, butyl rubber). Wash away any material which may have come into contact with the body with copious amounts of soap and water. Consider appropriate evacuation.

GLYOXAL

DOT Number: n/a *DOT Hazard Class:* n/a *DOT Guide Number:* n/a
Synonyms: biformal, biformyl, diformyl, ethanedial, oxal, oxaladehyde
STCC Number: n/a *Reportable Qty:* n/a
Mfg Name: Union Carbide *Phone No:* 1-203-794-2000

Physical Description:

Physical Form: Liquid *Color:* Light yellow *Odor:* Weak sour
Other Information: n/a

Chemical Properties:

Specific Gravity: 1.29 *Vapor Density:* n/a *Boiling Point: n/a*
Melting Point: 5° F(−15° C) *Vapor Pressure:* n/a *Solubility in water:* Yes
Other Information: Mixes with water

Health Hazards:

Inhalation Hazard: n/a
Ingestion Hazard: Harmful if swallowed.
Absorption Hazard: Irritating to the skin and eyes.
Hazards to Wildlife: n/a
Decontamination Procedures: Wash away any material with copious amounts of soap and water.
First Aid Procedures: Remove victim to fresh air, call emergency medical care. If not breathing give CPR. If breathing is difficult administer oxygen. Treat for shock.

Fire Hazards:

Flashpoint: n/a *Ignition temperature:* n/a
Flammable Explosive High Range: n/a *Low Range:* n/a
Toxic Products of Combustion: n/a
Other Hazards: Heat may cause polymerization.
Possible extinguishing agents: Extinguish fire using suitable agent for the type of surrounding fire.

Reactivity Hazards:

Reactive With: Heat may cause polymerization. *Other Reactions:* n/a

Corrosivity Hazards:

Corrosive With: Most metals *Neutralizing Agent:* n/a

Radioactivity Hazards:

Radiation Emitted: n/a *Other Hazards:* n/a

Recommended Protection for Response Personnel:

Avoid breathing vapors, keep upwind. Structural protective clothing provides limited protection. Wash away any material which may have come into contact with the body with copious amounts of soap and water. Consider appropriate evacuation.

GRENADE (WITH POISON B CHARGE)

DOT Number: NA 2016 *DOT Hazard Class:* Poison B *DOT Guide Number:* 15
Synonyms: none given
STCC Number: 4923929 *Reportable Qty:* n/a *CHEMTREC Phone No:* 1-800-424-9300

Physical Description:

Physical Form: Solid *Color:* n/a *Odor:* n/a
Other Information: No bursting charge. A small metal plastic device. Upon impact, will burst and release poison gas.

Chemical Properties:

Specific Gravity: n/a *Vapor Density:* n/a *Boiling Point: n/a*
Melting Point: n/a *Vapor Pressure:* n/a *Solubility in water:* No
Other Information: n/a

Health Hazards:

Inhalation Hazard: Poisonous if inhaled.
Ingestion Hazard: Poisonous if swallowed.
Absorption Hazard: Poisonous to the skin and the eyes.
Hazards to Wildlife: n/a
Decontamination Procedures: Wash away any material with copious amounts of soap and water.
First Aid Procedures: Remove victim to fresh air, call emergency medical care. If not breathing give CPR. If breathing is difficult administer oxygen. Treat for shock.

Fire Hazards:

Flashpoint: n/a *Ignition temperature:* n/a
Flammable Explosive High Range: n/a *Low Range:* n/a
Toxic Products of Combustion: n/a
Other Hazards: Poison B material may be flammable.
Possible extinguishing agents: Apply water from as far a distance as possible. Cool all affected containers with water. Use alcohol foam, dry chemical or carbon dioxide.

Reactivity Hazards:

Reactive With: n/a *Other Reactions:* n/a

Corrosivity Hazards:

Corrosive With: n/a *Neutralizing Agent:* n/a

Radioactivity Hazards:

Radiation Emitted: n/a *Other Hazards:* n/a

Recommended Protection for Response Personnel:

Avoid breathing vapors, keep upwind. Wear appropriate chemical clothing. Wash away any material which may have come into contact with the body with copious amounts of soap and water. Consider appropriate evacuation.

HAZARDOUS WASTE

DOT Number: NA 9189 *DOT Hazard Class:* ORM-E *DOT Guide Number:* 31
Synonyms: none given
STCC Number: 4906131 *Reportable Qty:* Varies!! *CHEMTREC Phone No:* 1-800-424-9300

Physical Description:

Physical Form: Liquid or solid *Color:* Not specified *Odor:* Not specified
Other Information: Material which has been shipped to an EPA designated chemical disposal area.

Chemical Properties:

Specific Gravity: Varies *Vapor Density:* Varies *Boiling Point: Varies*
Melting Point: Varies *Vapor Pressure:* Varies *Solubility in water:* Varies
Other Information: Contact CHEMTREC after location of shipping papers.

Health Hazards: Varies with type of chemical!

Inhalation Hazard: Varies
Ingestion Hazard: Varies
Absorption Hazard: Varies
Hazards to Wildlife: Varies
Decontamination Procedures: Wash away any material with copious amounts of soap and water.
First Aid Procedures: Remove victim to fresh air, call emergency medical care. If not breathing give CPR. If breathing is difficult administer oxygen. Treat for shock.

Fire Hazards: Varies with type of chemical!

Flashpoint: n/a *Ignition temperature:* n/a
Flammable Explosive High Range: n/a *Low Range:* n/a
Toxic Products of Combustion: Varies
Other Hazards: Varies
Possible *Possible extinguishing agents:* Will vary with type of chemical.

Reactivity Hazards

Reactive With: Will vary with type of chemical! *Other Reactions:* Varies

Corrosivity Hazards

Corrosive With: Will vary with type of chemical! *Neutralizing Agent:* Varies

Radioactivity Hazard

Radiation Emitted: Will vary with type of chemical! *Other Hazards:* Varies

Recommended Protection for Response Personnel:

Avoid breathing vapors, keep upwind. Wear appropriate chemical clothing. Wash away any material which may come into contact with the body with copious amounts of soap and water. Type of chemical involved will dictate any other actions taken by response personnel. Consider appropriate evacuation.

HEPTACHLOR

DOT Number: NA 2761 *DOT Hazard Class:* ORM-E *DOT Guide Number:* 55
Synonyms: Delsicol, E3314
STCC Number: 4960630 *Reportable Qty:* 1/.454
Mfg Name: Velsicol Chemical Corp. *Phone No:* 1-312-670-4500

Physical Description:

Physical Form: Solid *Color:* White to tan *Odor:* Camphorlike
Other Information: It is used as an insecticide.

Chemical Properties:

Specific Gravity: 1.66 *Vapor Density:* 13 *Boiling Point: n/a*
Melting Point: 114-165° F(45-57° C) *Vapor Pressure:* .0003 mm Hg at 68° F(20° C)
Solubility in water: Insoluble
Other Information: In case of damage to or leaking from containers, contact the Pesticide Safety Team Network at 1-800-424-9300.

Health Hazards:

Inhalation Hazard: Poisonous if inhaled.
Ingestion Hazard: Poisonous if swallowed.
Absorption Hazard: Poisonous if absorbed.
Hazards to Wildlife: Dangerous to aquatic life and waterfowl.
Decontamination Procedures: Wash away any material with copious amounts of soap and water.
First Aid Procedures: Remove victim to fresh air, call emergency medical care. If not breathing give CPR. If breathing is difficult administer oxygen. Treat for shock.

Fire Hazards:

Flashpoint: n/a *Ignition temperature:* n/a
Flammable Explosive High Range: n/a *Low Range:* n/a
Toxic Products of Combustion: Irritating gases may be produced when heated.
Other Hazards: n/a
Possible extinguishing agents: Use suitable agent for the type of surrounding fire.

Reactivity Hazards:

Reactive With: n/a *Other Reactions:* n/a

Corrosivity Hazards:

Corrosive With: n/a *Neutralizing Agent:* n/a

Radioactivity Hazards:

Radiation Emitted: n/a *Other Hazards:* n/a

Recommended Protection for Response Personnel:

Avoid breathing vapors, keep upwind. Wear a sealed chemical suit (polycarbonate, butyl rubber). Wash away any material which may have come into contact with the body with copious amounts of soap and water. Consider appropriate evacuation.

HEPTANE

DOT Number: UN 1206 *DOT Hazard Class:* Flammable liquid *DOT Guide Number:* 27
Synonyms: n-heptane
STCC Number: 4909190 *Reportable Qty:* n/a
Mfg Name: Phillips Petroleum *Phone No:* 1-981-661-6600

Physical Description:

Physical Form: Liquid *Color:* Colorless *Odor:* Gasoline
Other Information: Floats on water. Flammable vapors are produced.

Chemical Properties:

Specific Gravity: .7 *Vapor Density:* 3.5 *Boiling Point: 209° F(98.3° C)*
Melting Point: −132° F(−91.1° C) *Vapor Pressure:* 40 mm Hg at 68° F(20° C)
Solubility in water: No *Degree of Solubility:* .005%
Other Information: Lighter than water. Vapors are heavier than air.

Health Hazards:

Inhalation Hazard: Will cause coughing, difficulty in breathing.
Ingestion Hazard: Will cause nausea and vomiting.
Absorption Hazard: Irritating to the skin and eyes.
Hazards to Wildlife: Dangerous to aquatic life.
Decontamination Procedures: Wash away any material with copious amounts of soap and water.
First Aid Procedures: Remove victim to fresh air, call emergency medical care. If not breathing give CPR. If breathing is difficult administer oxygen. Treat for shock.

Fire Hazards:

Flashpoint: 25° F(−3.8° C) *Ignition temperature:* 399° F(203.8° C)
Flammable Explosive High Range: 6.7 *Low Range:* 1.05
Toxic Products of Combustion: n/a
Other Hazards: Flashback along vapor trail may occur. May ignite or explode in an enclosed area.
Possible extinguishing agents: Apply water from as far a distance as possible. Use foam, dry chemical or carbon dioxide.

Reactivity Hazards:

Reactive With: n/a *Other Reactions:* n/a

Corrosivity Hazards:

Corrosive With: n/a *Neutralizing Agent:* n/a

Radioactivity Hazards:

Radiation Emitted: n/a *Other Hazards:* n/a

Recommended Protection for Response Personnel:

Avoid breathing vapors, keep upwind. Structural protective clothing provides limited protection. Wash away any material which may have come into contact with the body with copious amounts of soap and water. Consider appropriate evacuation.

HEPTANOL

DOT Number: n/a *DOT Hazard Class:* n/a *DOT Guide Number:* n/a
Synonyms: enanthic alcohol, 1-heptanol, heptyl alcohol, 1-hydroxyheptane
STCC Number: n/a *Reportable Qty:* n/a
Mfg Name: Intsel Corp. *Phone No:* 1-212-758-5880

Physical Description:

Physical Form: Liquid *Color:* Colorless *Odor:* Weak alcohol
Other Information: n/a

Chemical Properties:

Specific Gravity: .82 *Vapor Density:* n/a *Boiling Point: 349° F(176.1° C)*
Melting Point: –29° F(–33.8° C)
Vapor Pressure: n/a *Solubility in water:* n/a
Other Information: Floats on water.

Health Hazards:

Inhalation Hazard: n/a
Ingestion Hazard: n/a
Absorption Hazard: n/a
Hazards to Wildlife: n/a
Decontamination Procedures: Wash away any material with copious amounts of soap and water.
First Aid Procedures: Remove victim to fresh air, call emergency medical care. If not breathing give CPR. If breathing is difficult administer oxygen. Treat for shock.

Fire Hazards:

Flashpoint: 170° F *Ignition temperature:* n/a
Flammable Explosive High Range: n/a *Low Range:* n/a
Toxic Products of Combustion: n/a
Other Hazards: n/a
Possible extinguishing agents: Use alcohol foam, dry chemical or carbon dioxide. Cool all affected containers with water. Water may be ineffective on fires!

Reactivity Hazards:

Reactive With: Strong oxidizers. May cause fire and explosions. *Other Reactions:* n/a

Corrosivity Hazards:

Corrosive With: n/a *Neutralizing Agent:* n/a

Radioactivity Hazards:

Radiation Emitted: n/a *Other Hazards:* n/a

Recommended Protection for Response Personnel:

Avoid breathing vapors, keep upwind. Structural protective clothing provides limited protection. Wash away any material which may have come into contact with the body with copious amounts of soap and water. Consider appropriate evacuation.

HEPTENE

DOT Number: UN 2278 *DOT Hazard Class:* Flammable liquid *DOT Guide Number:* 27
Synonyms: heptylene
STCC Number: n/a *Reportable Qty:* n/a
Mfg Name: Intsel Corp. *Phone No:* 1-212-758-5880

Physical Description:

Physical Form: Liquid *Color:* Colorless *Odor:* Gasolinelike
Other Information: n/a

Chemical Properties:

Specific Gravity: .7 *Vapor Density:* 3.4 *Boiling Point: 200° F(93.3° C)*
Melting Point: −182° F(−118.8 C)
Vapor Pressure: 420 psia *Solubility in water:* n/a
Other Information: Floats on water. Flammable, irritating vapor is produced.

Health Hazards:

Inhalation Hazard: Dizziness, difficulty in breathing.
Ingestion Hazard: Harmful if swallowed.
Absorption Hazard: Irritating to the skin and eyes.
Hazards to Wildlife: n/a
Decontamination Procedures: Wash away any material with copious amounts of soap and water.
First Aid Procedures: Remove victim to fresh air, call emergency medical care. If not breathing give CPR. If breathing is difficult administer oxygen. Treat for shock.

Fire Hazards:

Flashpoint: 25° F(−3.8° C) *Ignition temperature:* 500° F(260° C)
Flammable Explosive High Range: n/a *Low Range:* 1
Toxic Products of Combustion: n/a
Other Hazards: Flashback along vapor trail may occur. Vapors may explode if ignited in an enclosed area.
Possible extinguishing agents: Use foam, dry chemical or carbon dioxide. Cool all affected containers with water. Water may be ineffective on fires.

Reactivity Hazards:

Reactive With: Strong oxidizers. May cause fire and explosions. *Other Reactions:* n/a

Corrosivity Hazards:

Corrosive With: n/a *Neutralizing Agent:* n/a

Radioactivity Hazards:

Radiation Emitted: n/a *Other Hazards:* n/a

Recommended Protection for Response Personnel:

Avoid breathing vapors, keep upwind. Structural protective clothing provides limited protection. Wash away any material which may have come into contact with the body with copious amounts of soap and water. Consider appropriate evacuation.

HEXACHLOROCYCLOPENTADIENE

DOT Number: UN 2646 *DOT Hazard Class:* Corrosive *DOT Guide Number:* 55
Synonyms: C-56, HCCPD, HEX, perchlorocyclopentadiene
STCC Number: n/a *Reportable Qty:* n/a
Mfg Name: Velsicol Chemical Corp. *Phone No:* 1-312-670-4500

Physical Description:

Physical Form: Liquid *Color:* Greenish yellow *Odor:* Harsh unpleasant
Other Information: n/a

Chemical Properties:

Specific Gravity: 1.71 *Vapor Density:* 9.42 *Boiling Point: 462° F(238.8° C)*
Melting Point: 50° F(10° C) *Vapor Pressure:* n/a *Solubility in water:* n/a
Other Information: Sinks in water

Health Hazards:

Inhalation Hazard: Poisonous if inhaled.
Ingestion Hazard: Nausea and vomiting.
Absorption Hazard: Will burn the skin and eyes.
Hazards to Wildlife: Dangerous to aquatic life.
Decontamination Procedures: Wash away any material with copious amounts of soap and water.
First Aid Procedures: Remove victim to fresh air, call emergency medical care. If not breathing give CPR. If breathing is difficult administer oxygen. Treat for shock.

Fire Hazards:

Flashpoint: n/a *Ignition temperature:* n/a
Flammable Explosive High Range: n/a *Low Range:* n/a
Toxic Products of Combustion: Toxic hydrogen chloride, chlorine and phosgene gases may be formed in fire.
Other Hazards: n/a
Possible extinguishing agents: Use alcohol foam, dry chemical or carbon dioxide.

Reactivity Hazards:

Reactive With: Water to form hydrochloric acid. *Other Reactions:* n/a

Corrosivity Hazards:

Corrosive With: In the presence of moisture, will corrode iron and other metals with the release of flammable and explosive hydrogen gas. *Neutralizing Agent:* Crushed limestone, soda ash, or lime.

Radioactivity Hazards:

Radiation Emitted: n/a *Other Hazards:* n/a

Recommended Protection for Response Personnel:

Avoid breathing vapors, keep upwind. Structural protective clothing provides limited protection. Wash away any material which may have come into contact with the body with copious amounts of soap and water. Consider appropriate evacuation.

HEXACHLOROETHANE

DOT Number: NA 9037 *DOT Hazard Class:* ORM-A *DOT Guide Number:* 53
Synonyms: carbon hexachloride, HCE, perchloroethane
STCC Number: 4941225 *Reportable Qty:* 1/.454
Mfg Name: Intsel Corp. *Phone No:* 1-212-758-5880

Physical Description:

Physical Form: Solid *Color:* Colorless *Odor:* Camphorlike
Other Information: It is used to make other chemicals.

Chemical Properties:

Specific Gravity: n/a *Vapor Density:* 8.2 *Boiling Point: n/a*
Melting Point: n/a *Vapor Pressure:* 1 mm Hg at 91° F(32.7° C)
Solubility in water: n/a *Other Information:* n/a

Health Hazards:

Inhalation Hazard: Hazardous if inhaled.
Ingestion Hazard: Hazardous if swallowed.
Absorption Hazard: Irritating to the skin and eyes.
Hazards to Wildlife: n/a
Decontamination Procedures: Wash away any material with copious amounts of soap and water.
First Aid Procedures: Remove victim to fresh air, call emergency medical care. If not breathing give CPR. If breathing is difficult administer oxygen. Treat for shock.

Fire Hazards:

Flashpoint: n/a *Ignition temperature:* n/a
Flammable Explosive High Range: n/a *Low Range:* n/a
Toxic Products of Combustion: When heated to high temperatures, material may emit toxic fumes.
Other Hazards: n/a
Possible extinguishing agents: Extinguish fire using suitable agent for the type of surrounding fire. Cool all affected containers with water. Apply water from as far a distance as possible.

Reactivity Hazards:

Reactive With: n/a *Other Reactions:* n/a

Corrosivity Hazards:

Corrosive With: n/a *Neutralizing Agent:* n/a

Radioactivity Hazards:

Radiation Emitted: n/a *Other Hazards:* n/a

Recommended Protection for Response Personnel:

Avoid breathing vapors, keep upwind. Wear appropriate chemical clothing. Wash away any material which may come into contact with the body with copious amounts of soap and water. If fire becomes uncontrollable, or if containers are exposed to fire, consider appropriate evacuation.

HEXADECYLTRIMETHYLAMMONIUM CHLORIDE

DOT Number: n/a *DOT Hazard Class:* Flammable liquid *DOT Guide Number:* n/a
Synonyms: cetyltrimethylammonium chloride solution
STCC Number: n/a *Reportable Qty:* n/a
Mfg Name: Ashland Chemical *Phone No:* 1-614-889-3333

Physical Description:

Physical Form: Liquid *Color:* Clear to pale yellow *Odor:* Rubbing alcohol
Other Information: n/a

Chemical Properties:

Specific Gravity: .9 *Vapor Density:* n/a *Boiling Point: 180° F(82.2° C)*
Melting Point: n/a *Vapor Pressure:* n/a *Solubility in water:* n/a
Other Information: Floats or sinks in water.

Health Hazards:

Inhalation Hazard: Harmful if inhaled.
Ingestion Hazard: Harmful if swallowed.
Absorption Hazard: Irritating to the skin and eyes.
Hazards to Wildlife: n/a
Decontamination Procedures: Wash away any material with copious amounts of soap and water.
First Aid Procedures: Remove victim to fresh air, call emergency medical care. If not breathing give CPR. If breathing is difficult administer oxygen. Treat for shock.

Fire Hazards:

Flashpoint: 69° F(20.5° C) *Ignition temperature:* 750° F(398.8° C)
Flammable Explosive High Range: 12 *Low Range:* 2
Toxic Products of Combustion: Irritating fumes of hydrogen chloride may be formed in fires.
Other Hazards: Flashback along vapor trail may occur. Vapors may explode if ignited in an enclosed area.
Possible *Possible extinguishing agents:* Use alcohol foam, dry chemical or carbon dioxide. Water may be ineffective on fires. cool all affected containers with water.

Reactivity Hazards:

Reactive With: n/a *Other Reactions:* n/a

Corrosivity Hazards:

Corrosive With: n/a *Neutralizing Agent:* n/a

Radioactivity Hazards:

Radiation Emitted: n/a *Other Hazards:* n/a

Recommended Protection for Response Personnel:

Avoid breathing vapors, keep upwind. Wear appropriate chemical suit. Wash away any material which may have come into contact with the body with copious amounts of soap and water. If the fire becomes uncontrollable, consider appropriate evacuation.

HEXALDEHYDE

DOT Number: UN 1207
DOT Hazard Class: Flammable liquid
DOT Guide Number: 26
Synonyms: hexanal
STCC Number: 4909185
Reportable Qty: n/a
Mfg Name: Aldrich Chemical
Phone No: 1-414-273-3850

Physical Description:

Physical Form: Liquid
Color: Colorless
Odor: Pungent
Other Information: n/a

Chemical Properties:

Specific Gravity: .8
Vapor Density: 3.6
Boiling Point: 268° F(131.1° C)
Melting Point: n/a
Vapor Pressure: 10 mm Hg at 68° F(20° C)
Solubility in water: No
Other Information: Lighter than water. Vapors are heavier than air.

Health Hazards:

Inhalation Hazard: Harmful if inhaled.
Ingestion Hazard: Will cause nausea and vomiting.
Absorption Hazard: Irritating to the skin and eyes.
Hazards to Wildlife: n/a
Decontamination Procedures: Wash away any material with copious amounts of soap and water.
First Aid Procedures: Remove victim to fresh air, call emergency medical care. If not breathing give CPR. If breathing is difficult administer oxygen. Treat for shock.

Fire Hazards:

Flashpoint: 90° F(32.2° C)
Ignition temperature: n/a
Flammable Explosive High Range: n/a
Low Range: n/a
Toxic Products of Combustion: n/a
Other Hazards: Flashback along vapor trail may occur. May ignite or explode in an enclosed area.
Possible extinguishing agents: Apply water from as far a distance as possible. Use foam, dry chemical or carbon dioxide.

Reactivity Hazards:

Reactive With: n/a
Other Reactions: n/a

Corrosivity Hazards:

Corrosive With: n/a
Neutralizing Agent: n/a

Radioactivity Hazards:

Radiation Emitted: n/a
Other Hazards: n/a

Recommended Protection for Response Personnel:

Avoid breathing vapors, keep upwind. Wear a sealed chemical suit (butyl rubber). Wash away any material which may have come into contact with the body with copious amounts of soap and water. Consider appropriate evacuation.

HEXAMETHYLENE DIAMINE

DOT Number: UN 2280 *DOT Hazard Class:* Corrosive *DOT Guide Number:* 60
Synonyms: 1,6-hexanediamine, HMDA
STCC Number: 4935645 *Reportable Qty:* n/a
Mfg Name: E.I. Du Pont *Phone No:* 1-800-441-3637

Physical Description:

Physical Form: Solid *Color:* Clear, colorless *Odor:* Weak ammonia
Other Information: Used to make nylon.

Chemical Properties:

Specific Gravity: .993 *Vapor Density:* n/a *Boiling Point: 401° F(205° C)*
Melting Point: n/a *Vapor Pressure:* n/a *Solubility in water:* Yes
Other Information: Floats and mixes with water.

Health Hazards:

Inhalation Hazard: Poisonous if inhaled.
Ingestion Hazard: Poisonous if swallowed.
Absorption Hazard: Poisonous if the skin is exposed.
Hazards to Wildlife: n/a
Decontamination Procedures: Wash away any material with copious amounts of soap and water.
First Aid Procedures: Remove victim to fresh air, call emergency medical care. If not breathing give CPR. If breathing is difficult administer oxygen. Treat for shock.

Fire Hazards:

Flashpoint: 160° F(71.1° C) *Ignition temperature:* n/a
Flammable Explosive High Range: 6.3 *Low Range:* .7
Toxic Products of Combustion: Toxic oxides of nitrogen are produced.
Other Hazards: n/a
Possible extinguishing agents: Apply water from as far a distance as possible. Use alcohol foam, dry chemical, or carbon dioxide.

Reactivity Hazards:

Reactive With: n/a *Other Reactions:* n/a

Corrosivity Hazards:

Corrosive With: Metals and tissue *Neutralizing Agent:* n/a

Radioactivity Hazards:

Radiation Emitted: n/a *Other Hazards:* n/a

Recommended Protection for Response Personnel:

Avoid breathing vapors, keep upwind. Wear a sealed chemical suit (butyl rubber). Wash away any material which may have come into contact with the body with copious amounts of soap and water. Consider appropriate evacuation.

HEXAMETHYLENEIMINE

DOT Number: UN 2493 *DOT Hazard Class:* Corrosive *DOT Guide Number:* 29
Synonyms: hexahydroazepine
STCC Number: 4907872 *Reportable Qty:* n/a
Mfg Name: Aldrich Chemical *Phone No:* 1-414-273-3850

Physical Description:

Physical Form: Liquid *Color:* Colorless to light yellow *Odor:* Ammonialike
Other Information: n/a

Chemical Properties:

Specific Gravity: .880 *Vapor Density:* 3.4 *Boiling Point: 132° F(55.5° C)*
Melting Point: n/a *Vapor Pressure:* n/a *Solubility in water:* Yes
Other Information: Floats and mixes slowly with water. Irritating vapors are produced.

Health Hazards:

Inhalation Hazard: Will cause coughing, difficulty in breathing.
Ingestion Hazard: Poisonous if swallowed.
Absorption Hazard: Will burn the skin and eyes.
Hazards to Wildlife: n/a
Decontamination Procedures: Wash away any material with copious amounts of soap and water.
First Aid Procedures: Remove victim to fresh air, call emergency medical care. If not breathing give CPR. If breathing is difficult administer oxygen. Treat for shock.

Fire Hazards:

Flashpoint: 65 F (18.3 C) *Ignition temperature:* n/a
Flammable Explosive High Range: 2.3 *Low Range:* 1.6
Toxic Products of Combustion: Toxic oxides of nitrogen are produced.
Other Hazards: Vapors may flash back along vapor trail.
Possible extinguishing agents: Apply water from as far a distance as possible. Use alcohol foam, dry chemical, or carbon dioxide.

Reactivity Hazards:

Reactive With: n/a *Other Reactions:* n/a

Corrosivity Hazards:

Corrosive With: Metals and tissue *Neutralizing Agent:* n/a

Radioactivity Hazards:

Radiation Emitted: n/a *Other Hazards:* n/a

Recommended Protection for Response Personnel:

Avoid breathing vapors, keep upwind. Wear appropriate chemical clothing. Wash away any material which may have come into contact with the body with copious amounts of soap and water. Consider appropriate evacuation.

HEXAMETHYLENETETRAMINE

DOT Number: UN 1328 *DOT Hazard Class:* n/a *DOT Guide Number:* 32
Synonyms: amminoform, ammonioformaldehyde, hexa, hexamine, methaneamine, urotropin
STCC Number: n/a *Reportable Qty:* n/a
Mfg Name: Mallinckrodt Inc. *Phone No:* 1-314-895-2000

Physical Description:

Physical Form: Solid *Color:* White *Odor:* Mild ammonia
Other Information: n/a

Chemical Properties:

Specific Gravity: 1.35 *Vapor Density:* 4.8 *Boiling Point: 180° F(82.2° C)*
Melting Point: n/a *Vapor Pressure:* n/a *Solubility in water:* Yes
Other Information: Sinks and mixes with water.

Health Hazards:

Inhalation Hazard: n/a
Ingestion Hazard: n/a
Absorption Hazard: Irritating to the skin and eyes.
Hazards to Wildlife: n/a
Decontamination Procedures: Wash away any material with copious amounts of soap and water.
First Aid Procedures: Remove victim to fresh air, call emergency medical care. If not breathing give CPR. If breathing is difficult administer oxygen. Treat for shock.

Fire Hazards:

Flashpoint: 482° F(250° C) *Ignition temperature:* 700° F(371.1° C)
Flammable Explosive High Range: n/a *Low Range:* n/a
Toxic Products of Combustion: Formaldehyde gas and ammonia may be given off when hot.
Other Hazards: n/a
Possible *Possible extinguishing agents:* Extinguish fire using suitable agent for the type of surrounding fire.

Reactivity Hazards:

Reactive With: n/a *Other Reactions:* n/a

Corrosivity Hazards:

Corrosive With: n/a *Neutralizing Agent:* n/a

Radioactivity Hazards:

Radiation Emitted: n/a *Other Hazards:* n/a

Recommended Protection for Response Personnel:

Avoid breathing vapors, keep upwind. Wear a sealed chemical suit (polycarbonate, butyl rubber). Wash away any material which may have come into contact with the body with copious amounts of soap and water. Consider appropriate evacuation.

HEXANE

DOT Number: UN 1208 *DOT Hazard Class:* Flammable liquid *DOT Guide Number:* 27
Synonyms: n-hexane, hexyldyride
STCC Number: 4908183 *Reportable Qty:* n/a
Mfg Name: Phillips Petroleum *Phone No:* 1-918-661-6600

Physical Description:

Physical Form: Liquid *Color:* Colorless *Odor:* Gasoline
Other Information: It is used as a solvent and a paint thinner.

Chemical Properties:

Specific Gravity: .7 *Vapor Density:* 3 *Boiling Point: 156° F(68.8° C)*
Melting Point: −139° F(−95° C) *Vapor Pressure:* 124 mm Hg at 68° F(20° C)
Solubility in water: No *Degree of Solubility:* .014%
Other Information: Lighter than and soluble in water. Vapors are heavier than air.

Health Hazards:

Inhalation Hazard: Will cause coughing, dizziness.
Ingestion Hazard: Will cause nausea and vomiting.
Absorption Hazard: Irritating to the skin and eyes.
Hazards to Wildlife: n/a
Decontamination Procedures: Wash away any material with copious amounts of soap and water.
First Aid Procedures: Remove victim to fresh air, call emergency medical care. If not breathing give CPR. If breathing is difficult administer oxygen. Treat for shock.

Fire Hazards:

Flashpoint: −7° F(−21.6° C) *Ignition temperature:* 437° F(225° C)
Flammable Explosive High Range: 7.5 *Low Range:* 1.1
Toxic Products of Combustion: n/a
Other Hazards: Flashback along vapor trail may occur. Vapors may explode if ignited in a enclosed area.
Possible extinguishing agents: Apply water from as far a distance as possible. Use foam, dry chemical, or carbon dioxide.

Reactivity Hazards:

Reactive With: n/a *Other Reactions:* n/a

Corrosivity Hazards:

Corrosive With: n/a *Neutralizing Agent:* n/a

Radioactivity Hazards:

Radiation Emitted: n/a *Other Hazards:* n/a

Recommended Protection for Response Personnel:

Avoid breathing vapors, keep upwind. Structural protective clothing will provide limited protection. Wash away any material which may have come into contact with the body with copious amounts of soap and water. Consider appropriate evacuation.

HEXANOL

DOT Number: UN 2282 *DOT Hazard Class:* Flammable liquid *DOT Guide Number:* 26
Synonyms: amyl carbinol, n-amyl carbinol, 1-hexanol, n-hexyl alcohol, 1-hydroxyhexane
STCC Number: n/a *Reportable Qty:* n/a
Mfg Name: Exxon Chemical America *Phone No:* 1-713-870-6000

Physical Description:

Physical Form: Liquid *Color:* Colorless *Odor:* Sweet
Other Information: n/a

Chemical Properties:

Specific Gravity: .85 *Vapor Density:* 3.52 *Boiling Point: 315° F(157.2° C)*
Melting Point: −84° F(−64.4° C) Vapor Pressure: 1 mm Hg at 75° F(23.8° C) *Solubility in water:* n/a
Other Information: Floats on water

Health Hazards:

Inhalation Hazard: n/a
Ingestion Hazard: Harmful if swallowed.
Absorption Hazard: Will burn the skin and eyes.
Hazards to Wildlife: n/a
Decontamination Procedures: Wash away any material with copious amounts of soap and water.
First Aid Procedures: Remove victim to fresh air, call emergency medical care. If not breathing give CPR. If breathing is difficult administer oxygen. Treat for shock.

Fire Hazards:

Flashpoint: 145° F(62.7° C) *Ignition temperature:* 580° F(304.4° C)
Flammable Explosive High Range: 7.7 *Low Range:* 1.2
Toxic Products of Combustion: n/a
Other Hazards: n/a
Possible extinguishing agents: Alcohol foam, dry chemical and carbon dioxide. Water may be ineffective on fire.

Reactivity Hazards:

Reactive With: n/a *Other Reactions:* n/a

Corrosivity Hazards:

Corrosive With: n/a *Neutralizing Agent:* n/a

Radioactivity Hazards:

Radiation Emitted: n/a *Other Hazards:* n/a

Recommended Protection for Response Personnel:

Avoid breathing vapors, keep upwind. Structural protective clothing provides limited protection. Wash away any material which may have come into contact with the body with copious amounts of soap and water. If the fire becomes uncontrollable, consider appropriate evacuation.

HEXENE

DOT Number: UN 2370 *DOT Hazard Class:* Flammable liquid *DOT Guide Number:* 27
Synonyms: butyl ethylene, α-hexene, 1-hexene, hexylene
STCC Number: n/a *Reportable Qty:* n/a
Mfg Name: Phillips Petroleum *Phone No:* 1-918-661-6600

Physical Description:

Physical Form: Liquid *Color:* Colorless *Odor:* Mild pleasant
Other Information: n/a

Chemical Properties:

Specific Gravity: .67 *Vapor Density:* 3 *Boiling Point: 146° F(63.3° C)*
Melting Point: −219° F(−139.4° C) *Vapor Pressure:* 310 mm Hg at 100° F(37.7° C)
Solubility in water: n/a
Other Information: Floats on water. Flammable irritating vapor is produced.

Health Hazards:

Inhalation Hazard: Dizziness, difficulty in breathing, loss of consciousness.
Ingestion Hazard: Harmful if swallowed.
Absorption Hazard: Irritating to the skin and eyes.
Hazards to Wildlife: n/a
Decontamination Procedures: Wash away any material with copious amounts of soap and water.
First Aid Procedures: Remove victim to fresh air, call emergency medical care. If not breathing give CPR. If breathing is difficult administer oxygen. Treat for shock.

Fire Hazards:

Flashpoint: −15° F(−26.1° C) *Ignition temperature:* 521° F(271.6° C)
Flammable Explosive High Range: n/a *Low Range:* 1.2
Toxic Products of Combustion: n/a
Other Hazards: Flashback along vapor trail may occur. Vapors may explode if ignited in an enclosed area.
Possible extinguishing agents: Alcohol foam, dry chemical and carbon dioxide. Water may be ineffective on fire. Cool all affected containers with water.

Reactivity Hazards:

Reactive With: n/a *Other Reactions:* n/a

Corrosivity Hazards:

Corrosive With: n/a *Neutralizing Agent:* n/a

Radioactivity Hazards:

Radiation Emitted: n/a *Other Hazards:* n/a

Recommended Protection for Response Personnel:

Avoid breathing vapors, keep upwind. Structural protective clothing provides limited protection. Wash away any material which may have come into contact with the body with copious amounts of soap and water. If the fire becomes uncontrollable, consider appropriate evacuation.

HEXYLENE GLYCOL

DOT Number: UN 2030 *DOT Hazard Class:* Corrosive *DOT Guide Number:* 59
Synonyms: 2-methyl-2,4-pentanediol
STCC Number: n/a *Reportable Qty:* n/a
Mfg Name: Ashland Chemical *Phone No:* 1-614-889-3333

Physical Description:

Physical Form: Liquid *Color:* Colorless *Odor:* Mild sweet
Other Information: n/a

Chemical Properties:

Specific Gravity: 1.0 *Vapor Density:* 4 *Boiling Point: 236° F(113.3° C)*
Melting Point: 35° F(1.6° C) *Vapor Pressure:* .05 mm Hg at 68° F(20° C) *Solubility in water:* Yes
Other Information: Floats and mixes slowly with water.

Health Hazards:

Inhalation Hazard: n/a
Ingestion Hazard: n/a
Absorption Hazard: Irritating to the skin and eyes.
Hazards to Wildlife: n/a
Decontamination Procedures: Wash away any material with copious amounts of soap and water.
First Aid Procedures: Remove victim to fresh air, call emergency medical care. If not breathing give CPR. If breathing is difficult administer oxygen. Treat for shock.

Fire Hazards:

Flashpoint: 126° F(52.2° C) *Ignition temperature:* 518° F(270° C)
Flammable Explosive High Range: 100 *Low Range:* 4.7
Toxic Products of Combustion: Toxic vapor is generated when heated.
Other Hazards: May ignite spontaneously. May explode if confined.
Possible extinguishing agents: Extinguish fire using suitable agent for the type of surrounding fire.

Reactivity Hazards:

Reactive With: May ignite on contact with porous materials such as wood, asbestos, cloth, earth, and rusty metals.
Other Reactions: Heat with the release of nitrogen and ammonia gases.

Corrosivity Hazards:

Corrosive With: n/a *Neutralizing Agent:* HTH 7 lbs/3.1 kg per gallon/3.8 l of hydrazine spilled.

Radioactivity Hazards:

Radiation Emitted: n/a *Other Hazards:* n/a

Recommended Protection for Response Personnel:

Avoid breathing vapors, keep upwind. Structural protective clothing provides limited protection. Wash away any material which may have come into contact with the body with copious amounts of soap and water. If the fire becomes uncontrollable, consider appropriate evacuation.

HYDROCHLORIC ACID

DOT Number: UN 1789 *DOT Hazard Class:* Corrosive *DOT Guide Number:* 60
Synonyms: muriatic acid
STCC Number: 4930228 *Reportable Qty:* 5000/2270
Mfg Name: Allied Corp. *Phone No:* 1-201-455-2000

Physical Description:

Physical Form: Liquid *Color:* Colorless to yellow *Odor:* Pungent
Other Information: It is used for cleaning of masonry and metals, in the manufacture of chemicals, and in petroleum production.

Chemical Properties:

Specific Gravity: 1.19 *Vapor Density:* 1.3 *Boiling Point: 123° F(50.5° C)*
Melting Point: n/a *Vapor Pressure:* 4 mm Hg at 63° F(17.2° C) *Solubility in water:* Yes
Other Information: Soluble in water with releases of heat. Weighs 10.1 lbs/4.5 kg per gallon/3.8 l.

Health Hazards:

Inhalation Hazard: Will cause coughing, difficulty in breathing.
Ingestion Hazard: Harmful if swallowed.
Absorption Hazard: Will burn the skin and eyes.
Hazards to Wildlife: Dangerous to aquatic life.
Decontamination Procedures: Wash away any material with copious amounts of soap and water.
First Aid Procedures: Remove victim to fresh air, call emergency medical care. If not breathing give CPR. If breathing is difficult administer oxygen. Treat for shock.

Fire Hazards:

Flashpoint: n/a *Ignition temperature:* n/a
Flammable Explosive High Range: n/a *Low Range:* n/a
Toxic Products of Combustion: n/a
Other Hazards: n/a
Possible extinguishing agents: Extinguish fire using suitable agent for type of surrounding fire.

Reactivity Hazards:

Reactive With: n/a *Other Reactions:* n/a

Corrosivity Hazards:

Corrosive With: Metal and tissue *Neutralizing Agent:* Crushed limestone, soda ash, or lime

Radioactivity Hazards:

Radiation Emitted: n/a *Other Hazards:* n/a

Recommended Protection for Response Personnel:

Avoid breathing vapors, keep upwind. Wear a sealed chemical suit (polycarbonate, butyl rubber, viton, chlorinated polyethylene, PVC). Wash away any material which may have come into contact with the body with copious amounts of soap and water. Consider appropriate evacuation.

HYDRAZINE

DOT Number: UN 2029 *DOT Hazard Class:* Flammable liquid *DOT Guide Number:* 28
Synonyms: hydrazine anhydrous
STCC Number: 4935030 *Reportable Qty:* 1/.454
Mfg Name: Olin Corp. *Phone No:* 1-201-356-2000

Physical Description:

Physical Form: Liquid *Color:* Colorless *Odor:* Ammonia
Other Information: It is used as a rocket propellant and in fuel cells.

Chemical Properties:

Specific Gravity: 1 *Vapor Density:* 1.1 *Boiling Point: 236° F(113.3° C)*
Melting Point: 36° F(2.2° C) *Vapor Pressure:* 10 mm Hg at 68° F(20° C) *Solubility in water:* Yes
Other Information: May ignite spontaneously with porous materials such as earth, wood, or cloth.

Health Hazards:

Inhalation Hazard: Poisonous if inhaled.
Ingestion Hazard: Poisonous if swallowed.
Absorption Hazard: Poisonous on contact with the skin.
Hazards to Wildlife: Dangerous to aquatic life and waterfowl.
Decontamination Procedures: Wash away any material with copious amounts of soap and water.
First Aid Procedures: Remove victim to fresh air, call emergency medical care. If not breathing give CPR. If breathing is difficult administer oxygen. Treat for shock.

Fire Hazards:

Flashpoint: 100° F(37.7° C) *Ignition temperature:* 518° F(270° C)
Flammable Explosive High Range: 100 *Low Range:* 4.7
Toxic Products of Combustion: Toxic oxides of nitrogen are produced.
Other Hazards: Flashback along vapor trail may occur. Vapors may explode if ignited in an enclosed area.
Possible extinguishing agents: Apply water from as far a distance as possible. Use alcohol foam, dry chemical, or carbon dioxide.

Reactivity Hazards:

Reactive With: May ignite on contact with porous materials such as wood, asbestos, cloth, earth, and rusty metals.
Other Reactions: Heat with the release of nitrogen and ammonia gases.

Corrosivity Hazards:

Corrosive With: n/a
Neutralizing Agent: Flush with water, HTH 7 lbs/3.1 kg per gallon/3.8 l of hydrazine spilled.

Radioactivity Hazards:

Radiation Emitted: n/a *Other Hazards:* n/a

Recommended Protection for Response Personnel:

Avoid breathing vapors, keep upwind. Wear a sealed chemical suit (butyl rubber, polycarbonate). Wash away any material which may have come into contact with the body with copious amounts of soap and water. Consider appropriate evacuation.

HYDROCYANIC ACID

DOT Number: NA 1051 *DOT Hazard Class:* Poison A *DOT Guide Number:* 13
Synonyms: none given
STCC Number: 4920125 *Reportable Qty:* 10/4.54 *CHEMTREC Phone No:* 1-800-424-9300

Physical Description:

Physical Form: Liquefied gas *Color:* Colorless *Odor:* Faint aromatic
Other Information: It weighs 5.7 lbs/2.5 kg per gallon/3.8 l.

Chemical Properties:

Specific Gravity: .7 *Vapor Density:* .9 *Boiling Point: 79° F(26.1° C)*
Melting Point: n/a *Vapor Pressure:* n/a *Solubility in water:* Yes
Other Information: It is shipped in cylinders or tank cars.

Health Hazards:

Inhalation Hazard: Lethal.
Ingestion Hazard: Lethal.
Absorption Hazard: Lethal.
Hazards to Wildlife: Dangerous to aquatic life.
Decontamination Procedures: Wash away any material with copious amounts of soap and water.
First Aid Procedures: Remove victim to fresh air, call emergency medical care. If not breathing give CPR. If breathing is difficult administer oxygen. Treat for shock.

Fire Hazards:

Flashpoint: 0° F(–17.7° C) *Ignition temperature:* 1000° F(537.7° C)
Flammable Explosive High Range: 40 *Low Range:* 5.6
Toxic Products of Combustion: n/a
Other Hazards: It is shipped as a liquefied gas under its own pressure, it must be stabilized to avoid polymerization. Prolonged exposure to heat may cause cylinders to rupture/rocket.
Possible extinguishing agents: Apply water from as far a distance as possible. Use alcohol foam, dry chemical, or carbon dioxide.

Reactivity Hazards:

Reactive With: n/a *Other Reactions:* n/a

Corrosivity Hazards:

Corrosive With: n/a *Neutralizing Agent:* n/a

Radioactivity Hazards:

Radiation Emitted: n/a *Other Hazards:* n/a

Recommended Protection for Response Personnel:

Avoid breathing vapors, keep upwind. Wear appropriate chemical clothing. Wash away any material which may have come into contact with the body with copious amounts of soap and water. If fire becomes uncontrollable, or if containers are exposed to fire, consider appropriate evacuation.

HYDROFLUORIC ACID

DOT Number: UN 1790 *DOT Hazard Class:* Corrosive *DOT Guide Number:* 59
Synonyms: none given
STCC Number: 4930022 *Reportable Qty:* 100/45.4
Mfg Name: Allied Corp. *Phone No:* 1-201-455-2000

Physical Description:

Physical Form: Liquid *Color:* Colorless *Odor:* Pungent
Other Information: It weighs 9.6 lbs/4.3 kg per gallon/3.8 l.

Chemical Properties:

Specific Gravity: 1.26 *Vapor Density:* .7 *Boiling Point: 152° F(66.6° C)*
Melting Point: n/a *Vapor Pressure:* 400 mm Hg at 37° F(2.7° C) *Solubility in water:* Yes
Other Information: It is soluble in water with release of heat.

Health Hazards:

Inhalation Hazard: Harmful if inhaled.
Ingestion Hazard: Harmful if swallowed.
Absorption Hazard: Will burn the skin and eyes.
Hazards to Wildlife: Dangerous to aquatic life.
Decontamination Procedures: Wash away any material with copious amounts of soap and water.
First Aid Procedures: Remove victim to fresh air, call emergency medical care. If not breathing give CPR. If breathing is difficult administer oxygen. Treat for shock.

Fire Hazards:

Flashpoint: n/a *Ignition temperature:* n/a
Flammable Explosive High Range: n/a *Low Range:* n/a
Toxic Products of Combustion: n/a
Other Hazards: n/a
Possible extinguishing agents: Use suitable agent for the type of surrounding fire.

Reactivity Hazards:

Reactive With: Glass, concrete, metal, rubber, iron, leather, and many organic materials.
Other Reactions: n/a

Corrosivity Hazards:

Corrosive With: Metals and tissue *Neutralizing Agent:* Crushed limestone, soda ash, or lime

Radioactivity Hazards:

Radiation Emitted: n/a *Other Hazards:* n/a

Recommended Protection for Response Personnel:

Avoid breathing vapors, keep upwind. Wear a sealed chemical suit (viton, PVC, chlorinated polyethylene). Wash away any material which may have come into contact with the body with copious amounts of soap and water. Consider appropriate evacuation.

HYDROGEN

DOT Number: UN 1049 *DOT Hazard Class:* Flammable liquid *DOT Guide Number:* 22
Synonyms: liquid hydrogen, para hydrogen
STCC Number: 4905746 *Reportable Qty:* n/a *CHEMTREC Phone No:* 1-800-424-9300

Physical Description:

Physical Form: Gas *Color:* Colorless *Odor:* Odorless
Other Information: It is used to make other chemicals and in oxy-hydrogen welding.

Chemical Properties:

Specific Gravity: n/a *Vapor Density:* .1 *Boiling Point: −422° F(−252.2° C)*
Melting Point: n/a *Vapor Pressure:* .067 mm Hg at 68° F(20° C) *Solubility in water:* Slight
Other Information: Once ignited, it emits a pale blue flame, almost invisible.

Health Hazards:

Inhalation Hazard: Displaces the oxygen in the air.
Ingestion Hazard: n/a
Absorption Hazard: Will cause frostbite (liquid)
Hazards to Wildlife: No
Decontamination Procedures: Wash away any material with copious amounts of soap and water.
First Aid Procedures: Remove victim to fresh air, call emergency medical care. If not breathing give CPR. If breathing is difficult administer oxygen. Treat for shock.

Fire Hazards:

Flashpoint: n/a *Ignition temperature:* 932° F
Flammable Explosive High Range: 75 *Low Range:* 4
Toxic Products of Combustion: n/a
Other Hazards: It is shipped in cylinders and in special tank cars. Under fire conditions, the cylinders may violently rupture and rocket. (BLEVE)
Possible extinguishing agents: Water

Reactivity Hazards:

Reactive With: n/a *Other Reactions:* n/a

Corrosivity Hazards:

Corrosive With: n/a *Neutralizing Agent:* n/a

Radioactivity Hazards:

Radiation Emitted: n/a *Other Hazards:* n/a

Recommended Protection for Response Personnel:

Avoid breathing vapors, keep upwind. Wear appropriate chemical clothing. Wash away any material which may have come into contact with the body with copious amounts of soap and water. If fire becomes uncontrollable, consider appropriate evacuation.

HYDROGEN BROMIDE

DOT Number: UN 1048 *DOT Hazard Class:* Nonflammable gas *DOT Guide Number:* 15
Synonyms: hydrobromic acid anhydrous, hydrogen bromide anhydrous
STCC Number: 4904260 *Reportable Qty:* n/a
Mfg Name: Dow Chemical *Phone No:* 1-517-636-4400

Physical Description:

Physical Form: Gas *Color:* Colorless *Odor:* Irritating
Other Information: It is used to make other chemicals and as a catalyst in the manufacture of chemicals. Material may be only be shipped in cylinders.

Chemical Properties:

Specific Gravity: 2.14 *Vapor Density:* 2.71 *Boiling Point:* *−88° F(−66.6° C)*
Melting Point: n/a *Vapor Pressure:* 320 psig at 70° F(21.1° C) *Solubility in water:* Yes
Other Information: Sinks and mixes with water. Poisonous, visible vapor cloud is produced.

Health Hazards:

Inhalation Hazard: Poisonous if inhaled.
Ingestion Hazard: Poisonous if swallowed.
Absorption Hazard: Will burn the skin and eyes. Will cause frostbite.
Hazards to Wildlife: Dangerous to aquatic life.
Decontamination Procedures: Wash away any material with copious amounts of soap and water.
First Aid Procedures: Remove victim to fresh air, call emergency medical care. If not breathing give CPR. If breathing is difficult administer oxygen. Treat for shock.

Fire Hazards:

Flashpoint: n/a *Ignition temperature:* n/a
Flammable Explosive High Range: n/a *Low Range:* n/a
Toxic Products of Combustion: n/a
Other Hazards: Pressurized containers may explode and rocket releasing toxic and irritating vapors.
Possible extinguishing agents: Extinguish fire using suitable agent for the type of surrounding fire. Cool all affected containers with flooding quantities of water. Apply water from as far a distance as possible. Do not use water on the material itself!

Reactivity Hazards:

Reactive With: Water with evolution of heat forming hydrobromic acid, highly corrosive to most metals with evolution of flammable hydrogen gas. *Other Reactions:* n/a

Corrosivity Hazards:

Corrosive With: Most metals *Neutralizing Agent:* Crushed limestone, soda ash, or lime.

Radioactivity Hazards:

Radiation Emitted: n/a *Other Hazards:* n/a

Recommended Protection for Response Personnel:

Avoid breathing vapors, keep upwind. Wear a sealed chemical suit (butyl rubber, PVC). Wash away any material which may have come into contact with the body with copious amounts of soap and water. If the fire becomes uncontrollable, consider appropriate evacuation. Under fire conditions, the cylinders may violently rupture and rocket.

HYDROGEN CHLORIDE

DOT Number: UN 1050 *DOT Hazard Class:* Nonflammable gas *DOT Guide Number:* 15
Synonyms: hydrochloric acid
STCC Number: 4904270 *Reportable Qty:* 5000/2270
Mfg Name: Vulcan Materials Co. *Phone No:* 1-205-877-3000

Physical Description:

Physical Form: Gas *Color:* Colorless *Odor:* Sharp, pungent
Other Information: It is used in the manufacture of rubber, pharmaceuticals, chemicals, and in gasoline refining. Weighs 10 lbs/4.5 kg per gallon/3.8 l.

Chemical Properties:

Specific Gravity: 1.3 *Vapor Density:* 1.3
Boiling Point: – 121° F(–85° C) *Melting Point:* –173° F(–113.8° C)
Vapor Pressure: 1 atm at 68° F(20° C)
Solubility in water: Yes *Degree of Solubility:* 62%
Other Information: n/a

Health Hazards:

Inhalation Hazard: Poisonous if inhaled. *Ingestion Hazard:* Poisonous if swallowed.
Absorption Hazard: Will burn the skin and eyes.
Hazards to Wildlife: Dangerous to aquatic life.
Decontamination Procedures: Wash away any material with copious amounts of soap and water.
First Aid Procedures: Remove victim to fresh air, call emergency medical care. If not breathing give CPR. If breathing is difficult administer oxygen. Treat for shock.

Fire Hazards:

Flashpoint: n/a *Ignition temperature:* n/a
Flammable Explosive High Range: n/a *Low Range:* n/a
Toxic Products of Combustion: n/a
Other Hazards: n/a
Possible extinguishing agents: Apply water from as far a distance as possible. Use appropriate agent for type of surrounding fire.

Reactivity Hazards:

Reactive With: Water to form hydrochloric acid. *Other Reactions:* n/a

Corrosivity Hazards:

Corrosive With: Metals with the evolution of flammable hydrogen gas.
Neutralizing Agent: Crushed limestone, soda ash, or lime.

Radioactivity Hazards:

Radiation Emitted: n/a *Other Hazards:* n/a

Recommended Protection for Response Personnel:

Avoid breathing vapors, keep upwind. Wear a sealed chemical suit (polycarbonate, butyl rubber, PVC, viton, chlorinated polyethylene). Wash away any material which may have come into contact with the body with copious amounts of soap and water. If the fire becomes uncontrollable, consider appropriate evacuation.

HYDROGEN CYANIDE

DOT Number: UN 1051 *DOT Hazard Class:* Poison A *DOT Guide Number:* 13
Synonyms: hydrocyanic acid, prussic acid
STCC Number: n/a *Reportable Qty:* n/a
Mfg Name: E.I. Du Pont *Phone No:* 1-800-441-3637

Physical Description:

Physical Form: Liquid or gas *Color:* Colorless *Odor:* Bitter almond
Other Information: n/a

Chemical Properties:

Specific Gravity: .7 *Vapor Density:* 1.3 *Boiling Point: 78° F(25.5° C)*
Melting Point: 8° F(−13.3° C) *Vapor Pressure:* 750 mm Hg at 77° F(25° C) *Solubility in water:* Yes
Other Information: Sinks and mixes with water. Poisonous, flammable vapor is produced and rises.

Health Hazards:

Inhalation Hazard: Lethal.
Ingestion Hazard: Lethal.
Absorption Hazard: Lethal.
Hazards to Wildlife: Dangerous to aquatic life.
Decontamination Procedures: Wash away any material with copious amounts of soap and water.
First Aid Procedures: Remove victim to fresh air, call emergency medical care. If not breathing give CPR. If breathing is difficult administer oxygen. Treat for shock.

Fire Hazards:

Flashpoint: 0° F(−17.7° C) *Ignition temperature:* 1004° F(540° C)
Flammable Explosive High Range: 40 *Low Range:* 5.6
Toxic Products of Combustion: Extremely toxic vapors are generated even at ambient temperatures.
Other Hazards: Containers may explode with ignition of contents.
Possible extinguishing agents: Apply water from as far a distance as possible. Use alcohol foam, dry chemical, or carbon dioxide.

Reactivity Hazards:

Reactive With: n/a *Other Reactions:* n/a

Corrosivity Hazards:

Corrosive With: n/a *Neutralizing Agent:* n/a

Radioactivity Hazards:

Radiation Emitted: n/a *Other Hazards:* n/a

Recommended Protection for Response Personnel:

Avoid breathing vapors, keep upwind. Wear a sealed chemical suit (butyl rubber, PVC). Wash away any material which may have come into contact with the body with copious amounts of soap and water. If the fire becomes uncontrollable, or if the containers are exposed to heat, consider appropriate evacuation.

HYDROGEN FLUORIDE

DOT Number: UN 1052 *DOT Hazard Class:* Corrosive *DOT Guide Number:* 15
Synonyms: fluorohydric acid, hydrofluoric acid anhydrous, hydrogen fluoride anhydrous
STCC Number: 4930024 *Reportable Qty:* 100/45.4
Mfg Name: Allied Corp. *Phone No:* 1-201-455-2000

Physical Description:

Physical Form: Liquid *Color:* Colorless *Odor:* Pungent
Other Information: It is used as a catalyst and raw material in chemical manufacture. Weighs 8.2 lbs/3.7 kg per gallon/3.8 l. It is shipped as a liquefied gas under its own vapor pressure.

Chemical Properties:

Specific Gravity: .99 *Vapor Density:* .7 *Boiling Point: 67° F(19.4° C)*
Melting Point: −134° F(−92.2° C) Vapor Pressure: .3 psig at 68° F(20° C)
Solubility in water: Yes
Other Information: It is soluble in water with release of heat. Poisonous vapor is produced and slowly rises.

Health Hazards:

Inhalation Hazard: Poisonous if inhaled.
Ingestion Hazard: Poisonous if swallowed.
Absorption Hazard: Will burn the skin and eyes.
Hazards to Wildlife: Dangerous to aquatic life.
Decontamination Procedures: Wash away any material with copious amounts of soap and water.
First Aid Procedures: Remove victim to fresh air, call emergency medical care. If not breathing give CPR. If breathing is difficult administer oxygen. Treat for shock.

Fire Hazards:

Flashpoint: n/a *Ignition temperature:* n/a
Flammable Explosive High Range: n/a *Low Range:* n/a
Toxic Products of Combustion: Toxic and irritating vapors are generated when heated.
Other Hazards: n/a
Possible extinguishing agents: Use suitable agent for the type of surrounding fire.

Reactivity Hazards:

Reactive With: Water with liberation of heat.
Other Reactions: Metals forming flammable hydrogen gas.

Corrosivity Hazards:

Corrosive With: Metals and tissue. *Neutralizing Agent:* Crushed limestone, soda ash, or lime.

Radioactivity Hazards:

Radiation Emitted: n/a *Other Hazards:* n/a

Recommended Protection for Response Personnel:

Avoid breathing vapors, keep upwind. Wear a sealed chemical suit (viton, PVC, chlorinated polyethylene). Wash away any material which may have come into contact with the body with copious amounts of soap and water. Consider appropriate evacuation.

HYDROGEN PEROXIDE (52-100%)

DOT Number: UN 2015 *DOT Hazard Class:* Oxidizer *DOT Guide Number:* 47
Synonyms: albone, hydrogen dioxide, peroxide, Superoxyl, t-stuff
STCC Number: 4918335 *Reportable Qty:* n/a
Mfg Name: Allied Corp. *Phone No:* 1-201-445-2000

Physical Description:

Physical Form: Liquid *Color:* Colorless *Odor:* Slightly sharp
Other Information: It is used to bleach textiles and wood pulp. In chemical manufacturing and food processing, and in water purification.

Chemical Properties:

Specific Gravity: 1.29 *Vapor Density:* 1.2 *Boiling Point: 257° F(125° C)*
Melting Point: –40° F(–40° C) *Vapor Pressure:* 1 mm Hg at 60° F(15.5° C) *Solubility in water:* Yes
Other Information: Sinks and mixes with water. Irritating vapor is produced.

Health Hazards:

Inhalation Hazard: Harmful if inhaled.
Ingestion Hazard: Harmful if swallowed.
Absorption Hazard: Will burn the skin and eyes.
Hazards to Wildlife: n/a
Decontamination Procedures: Wash away any material with copious amounts of soap and water.
First Aid Procedures: Remove victim to fresh air, call emergency medical care. If not breathing give CPR. If breathing is difficult administer oxygen. Treat for shock.

Fire Hazards:

Flashpoint: n/a *Ignition temperature:* n/a
Flammable Explosive High Range: n/a *Low Range:* n/a
Toxic Products of Combustion: n/a
Other Hazards: May cause fire and explode on contact with combustibles and metals. Containers may explode when heated.
Possible extinguishing agents: Use suitable agent for the type of surrounding fire.

Reactivity Hazards:

Reactive With: Combustible materials and metals causing a rapid decomposition with liberation of oxygen gas.
Other Reactions: Contamination with metals and dirt can cause rapid decomposition.

Corrosivity Hazards:

Corrosive With: n/a *Neutralizing Agent:* n/a

Radioactivity Hazards:

Radiation Emitted: n/a *Other Hazards:* n/a

Recommended Protection for Response Personnel:

Avoid breathing vapors, keep upwind. Wear a sealed chemical suit (PVC, chlorinated polyethylene, polycarbonate, butyl rubber, viton). Wash away any material which may have come into contact with the body with copious amounts of soap and water. Consider appropriate evacuation. Under exposure to fire or heat, containers of the material can violently rupture due to decomposition of the material.

HYDROGEN SULFIDE

DOT Number: UN 1053 *DOT Hazard Class:* Flammable gas *DOT Guide Number:* 13
Synonyms: hepatic gas, hydrosulfuric acid, sulfuretted hydrogen
STCC Number: 4905410 *Reportable Qty:* 100/45.4
Mfg Name: Matheson Gas Products *Phone No:* 1-201-867-4100

Physical Description:

Physical Form: Gas *Color:* Colorless *Odor:* Rotten egg
Other Information: Material will be odorless at poisonous concentrations. It is shipped as a liquefied gas under its own vapor pressure.

Chemical Properties:

Specific Gravity: .916 *Vapor Density:* 1.19 *Boiling Point: −76° F(−60° C)*
Melting Point: −117° F(−82.7° C) Vapor Pressure: 252 psig at 70° F(21.1° C)
Solubility in water: Yes
Other Information: Mixes, sinks and boils in water. Poisonous, flammable, visible vapor cloud is produced.

Health Hazards:

Inhalation Hazard: Poisonous if inhaled.
Ingestion Hazard: Poisonous if swallowed.
Absorption Hazard: Poisonous to the skin and eyes.
Hazards to Wildlife: Dangerous to aquatic life.
Decontamination Procedures: Wash away any material with copious amounts of soap and water.
First Aid Procedures: Remove victim to fresh air, call emergency medical care. If not breathing give CPR. If breathing is difficult administer oxygen. Treat for shock.

Fire Hazards:

Flashpoint: Gas *Ignition temperature:* 500° F(260° C)
Flammable Explosive High Range: 45 *Low Range:* 4.3
Toxic Products of Combustion: Toxic and irritating vapors are generated when heated.
Other Hazards: Flashback along vapor trail may occur. Vapors may explode if ignited in an enclosed area.
Possible extinguishing agents: Use suitable agent for the type of surrounding fire.

Reactivity Hazards:

Reactive With: n/a *Other Reactions:* n/a

Corrosivity Hazards:

Corrosive With: n/a *Neutralizing Agent:* n/a

Radioactivity Hazards:

Radiation Emitted: n/a *Other Hazards:* n/a

Recommended Protection for Response Personnel:

Avoid breathing vapors, keep upwind. Wear a sealed chemical suit (PVC, chlorinated polyethylene, polycarbonate, butyl rubber). Wash away any material which may have come into contact with the body with copious amounts of soap and water. Consider appropriate evacuation. Under fire conditions, the cylinders and tank cars may violently rupture and rocket.

HYDROQUINONE

DOT Number: UN 2662 *DOT Hazard Class:* n/a *DOT Guide Number:* 53
Synonyms: 1,4-benzenediol, p-dihydroxybenzene, hydroquinol, pyrogentisic acid, quinol
STCC Number: n/a *Reportable Qty:* n/a
Mfg Name: Atomergic Chemetals Corp. *Phone No:* 1-516-349-8800

Physical Description:

Physical Form: Solid *Color:* White, light tan to gray *Odor:* n/a
Other Information: n/a

Chemical Properties:

Specific Gravity: 1.33 *Vapor Density:* 3.81 *Boiling Point: 545° F(285° C)*
Melting Point: 338° F(170° C) *Vapor Pressure:* 1 mm Hg at 270° F(132.2° C) *Solubility in water:* Yes
Other Information: Sinks and mixes with water.

Health Hazards:

Inhalation Hazard: Harmful if inhaled.

Ingestion Hazard: Headache, dizziness, nausea, vomiting, loss of consciousness.

Absorption Hazard: Irritating and burning of the skin and eyes.

Hazards to Wildlife: n/a

Decontamination Procedures: Wash away any material with copious amounts of soap and water.

First Aid Procedures: Remove victim to fresh air, call emergency medical care. If not breathing give CPR. If breathing is difficult administer oxygen. Treat for shock.

Fire Hazards:

Flashpoint: 350° F(176.6° C) *Ignition temperature:* 960° F(515.5° C)

Flammable Explosive High Range: n/a *Low Range:* n/a

Toxic Products of Combustion: n/a

Other Hazards: Dust explosion is possible.

Possible extinguishing agents: Use suitable agent for the type of surrounding fire.

Reactivity Hazards:

Reactive With: n/a *Other Reactions:* n/a

Corrosivity Hazards:

Corrosive With: n/a *Neutralizing Agent:* n/a

Radioactivity Hazards:

Radiation Emitted: n/a *Other Hazards:* n/a

Recommended Protection for Response Personnel:

Avoid breathing vapors, keep upwind. Structural protective clothing provides limited protection. Wash away any material which may have come into contact with the body with copious amounts of soap and water. Consider appropriate evacuation.

HYDROSILICOFLUORIC ACID

DOT Number: NA 1778 *DOT Hazard Class:* Corrosive *DOT Guide Number:* 60
Synonyms: sand acid, silicofluoric acid
STCC Number: 4930026 *Reportable Qty:* n/a *CHEMTREC Phone No:* 1-800-424-9300

Physical Description:

Physical Form: Liquid *Color:* Colorless to straw *Odor:* Acrid, sour
Other Information: It is used for electropolishing metals, to harden cement and ceramics, as a wood preservative, and for water fluoridation.

Chemical Properties:

Specific Gravity: 1.3 *Vapor Density:* 5 *Boiling Point: 112° F(44.4° C)*
Melting Point: −24/−4° F(−31/−20° C) *Vapor Pressure:* n/a
Solubility in water: Yes
Other Information: It is soluble in water.

Health Hazards:

Inhalation Hazard: Severely corrosive to mucous membranes.
Ingestion Hazard: Severe burns to the mouth and stomach.
Absorption Hazard: Will burn the skin and eyes.
Hazards to Wildlife: n/a
Decontamination Procedures: Wash away any material with copious amounts of soap and water.
First Aid Procedures: Remove victim to fresh air, call emergency medical care. If not breathing give CPR. If breathing is difficult administer oxygen. Treat for shock.

Fire Hazards:

Flashpoint: n/a *Ignition temperature:* n/a
Flammable Explosive High Range: n/a *Low Range:* n/a
Toxic Products of Combustion: May decompose in fire to produce toxic and corrosive fumes of hydrogen fluoride.
Other Hazards: n/a
Possible extinguishing agents: Extinguish fire using suitable agent for the type of surrounding fire.

Reactivity Hazards:

Reactive With: Oxidizers *Other Reactions:* n/a

Corrosivity Hazards:

Corrosive With: Metals and tissue with evolution of flammable hydrogen gas.
Neutralizing Agent: Crushed limestone, soda ash, or lime.

Radioactivity Hazards:

Radiation Emitted: n/a *Other Hazards:* n/a

Recommended Protection for Response Personnel:

Avoid breathing vapors, keep upwind. Wear a sealed chemical suit (butyl rubber). Wash away any material which may have come into contact with the body with copious amounts of soap and water. Consider appropriate evacuation.

HYDROXYETHYL ACRYLATE

DOT Number: n/a *DOT Hazard Class:* n/a *DOT Guide Number:* n/a
Synonyms: β-hydroxyethyl acrylate, 2-hydroxyethyl-2-propenoate
STCC Number: n/a *Reportable Qty:* n/a
Mfg Name: Dow Chemical *Phone No:* 1-517-636-4400

Physical Description:

Physical Form: Liquid *Color:* Colorless *Odor:* Sweet unpleasant
Other Information: n/a

Chemical Properties:

Specific Gravity: 1.1 *Vapor Density:* n/a *Boiling Point: 346° F(174.4° C)*
Melting Point: −76° F(−60° C) *Vapor Pressure:* n/a *Solubility in water:* Yes
Other Information: Sinks and mixes with water.

Health Hazards:

Inhalation Hazard: n/a
Ingestion Hazard: Harmful if swallowed.
Absorption Hazard: Will burn the skin and eyes.
Hazards to Wildlife: n/a
Decontamination Procedures: Wash away any material with copious amounts of soap and water.
First Aid Procedures: Remove victim to fresh air, call emergency medical care. If not breathing give CPR. If breathing is difficult administer oxygen. Treat for shock.

Fire Hazards:

Flashpoint: 220° F(104.4° C) *Ignition temperature:* n/a
Flammable Explosive High Range: n/a *Low Range:* n/a
Toxic Products of Combustion: n/a
Other Hazards: Containers may explode in fire.
Possible extinguishing agents: Use suitable agent for the type of surrounding fire.

Reactivity Hazards:

Reactive With: n/a *Other Reactions:* n/a

Corrosivity Hazards:

Corrosive With: n/a *Neutralizing Agent:* n/a

Radioactivity Hazards:

Radiation Emitted: n/a *Other Hazards:* n/a

Recommended Protection for Response Personnel:

Avoid breathing vapors, keep upwind. Wear appropriate sealed chemical suit. Wash away any material which may have come into contact with the body with copious amounts of soap and water. Consider appropriate evacuation.

HYDROXYLAMINE

DOT Number: n/a *DOT Hazard Class:* n/a *DOT Guide Number:* n/a
Synonyms: oxammonium
STCC Number: n/a *Reportable Qty:* n/a
Mfg Name: Allied Corp. *Phone No:* 1-201-455-2000

Physical Description:

Physical Form: Solid *Color:* White *Odor:* Odorless
Other Information: n/a

Chemical Properties:

Specific Gravity: 1.23 *Vapor Density:* 1.1 *Boiling Point: 133° F(56.1° C)*
Melting Point: 91° F(32.7° C) *Vapor Pressure:* 10 mm Hg at 117° F(47.2° C) *Solubility in water:* Yes
Other Information: Sinks and mixes with water.

Health Hazards:

Inhalation Hazard: Headaches, dizziness, ringing in ears, labored breathing, nausea and vomiting.
Absorption Hazard: Irritating to the skin and eyes.
Hazards to Wildlife: Dangerous to aquatic life.
Decontamination Procedures: Wash away any material with copious amounts of soap and water.
First Aid Procedures: Remove victim to fresh air, call emergency medical care. If not breathing give CPR. If breathing is difficult administer oxygen. Treat for shock.

Fire Hazards:

Flashpoint: 265° F(129c) *Ignition temperature:* 265° F(129.4° C)
Flammable Explosive High Range: n/a *Low Range:* n/a
Toxic Products of Combustion: Nitrogen oxides.
Other Hazards: May explode if exposed to heat or flame.
Possible extinguishing agents: Use suitable agent for the type of surrounding fire. Use extreme caution, material may explode. Use remote extinguishing equipment or unmanned fixed turrets and fixed nozzles. Evacuate the area!!

Reactivity Hazards:

Reactive With: Water to form alkaline solutions. *Other Reactions:* n/a

Corrosivity Hazards:

Corrosive With: n/a *Neutralizing Agent:* n/a

Radioactivity Hazards:

Radiation Emitted: n/a *Other Hazards:* n/a

Recommended Protection for Response Personnel:

Avoid breathing vapors, keep upwind. Structural protective clothing provides limited protection. Wash away any material which may have come into contact with the body with copious amounts of soap and water. If the fire becomes uncontrollable, consider appropriate evacuation.

HYDROXYLAMINE SULFATE

DOT Number: UN 2865 *DOT Hazard Class:* n/a *DOT Guide Number:* 60
Synonyms: hydroxyl ammonium sulfate, oxammonium sulfate
STCC Number: n/a *Reportable Qty:* n/a
Mfg Name: Allied Corp. *Phone No:* 1-201-455-2000

Physical Description:

Physical Form: Solid *Color:* White *Odor:* Odorless
Other Information: n/a

Chemical Properties:

Specific Gravity: 1 *Vapor Density:* 5.6 *Boiling Point: Decomposes*
Melting Point: n/a *Vapor Pressure:* n/a *Solubility in water:* Yes
Other Information: Sinks and mixes with water.

Health Hazards:

Inhalation Hazard: Difficulty in breathing, loss of consciousness.
Ingestion Hazard: Poisonous if swallowed.
Absorption Hazard: Irritating to the skin and eyes.
Hazards to Wildlife: n/a
Decontamination Procedures: Wash away any material with copious amounts of soap and water.
First Aid Procedures: Remove victim to fresh air, call emergency medical care. If not breathing give CPR. If breathing is difficult administer oxygen. Treat for shock.

Fire Hazards:

Flashpoint: n/a *Ignition temperature:* n/a
Flammable Explosive High Range: n/a *Low Range:* n/a
Toxic Products of Combustion: Sulfuric acid fumes may be formed in fire.
Other Hazards: n/a
Possible extinguishing agents: Alcohol foam, dry chemical, carbon dioxide. Apply water from as far a distance as possible. Extinguish fire using suitable agent for the type of surrounding fire.

Reactivity Hazards:

Reactive With: n/a *Other Reactions:* n/a

Corrosivity Hazards:

Corrosive With: Metals in the presence of moisture.
Neutralizing Agent: Crushed limestone, soda ash, or lime.

Radioactivity Hazards:

Radiation Emitted: n/a *Other Hazards:* n/a

Recommended Protection for Response Personnel:

Avoid breathing vapors, keep upwind. Wear a sealed chemical suit (PVC). Wash away any material which may have come into contact with the body with copious amounts of soap and water. Consider appropriate evacuation.

HYDROXYPROPYL ACRYLATE

DOT Number: UN 1760 *DOT Hazard Class:* Corrosive *DOT Guide Number:* 60
Synonyms: 1,2-propanediol-1-acrylate, propylene glycol monoacrylate
STCC Number: 4935510 *Reportable Qty:* n/a
Mfg Name: Dow Chemical *Phone No:* 1-517-636-4400

Physical Description:

Physical Form: Liquid *Color:* Colorless *Odor:* Faint, unpleasant
Other Information: It is used to make plastics.

Chemical Properties:

Specific Gravity: 1.06 *Vapor Density:* 4.5 *Boiling Point: n/a*
Melting Point: n/a *Vapor Pressure:* n/a *Solubility in water:* Yes
Other Information: It is slightly heavier than and soluble in water. Vapors are heavier than air.

Health Hazards:

Inhalation Hazard: Coughing, difficulty in breathing.
Ingestion Hazard: Will cause nausea.
Absorption Hazard: Will burn the skin and eyes.
Hazards to Wildlife: n/a
Decontamination Procedures: Wash away any material with copious amounts of soap and water.
First Aid Procedures: Remove victim to fresh air, call emergency medical care. If not breathing give CPR. If breathing is difficult administer oxygen. Treat for shock.

Fire Hazards:

Flashpoint: 136° F(57.7° C) *Ignition temperature:* n/a
Flammable Explosive High Range: n/a *Low Range:* 1.8
Toxic Products of Combustion: n/a
Other Hazards: If polymerization takes place, containers may violently rupture.
Possible extinguishing agents: Apply water from as far a distance as possible. Cool all affected containers with water. Use alcohol foam, dry chemical, or carbon dioxide.

Reactivity Hazards:

Reactive With: n/a *Other Reactions:* n/a

Corrosivity Hazards:

Corrosive With: Tissue *Neutralizing Agent:* Crushed limestone, soda ash, or lime

Radioactivity Hazards:

Radiation Emitted: n/a *Other Hazards:* n/a

Recommended Protection for Response Personnel:

Avoid breathing vapors, keep upwind. Wear a sealed chemical suit (polycarbonate, butyl rubber). Wash away any material which may have come into contact with the body with copious amounts of soap and water. Consider appropriate evacuation.

HYDROXYPROPYL METHACRYLATE

DOT Number: n/a *DOT Hazard Class:* n/a *DOT Guide Number:* n/a
Synonyms: 1,2-propanediol 1-methacrylate, propylene glycol monomethacrylate
STCC Number: n/a *Reportable Qty:* n/a
Mfg Name: Rohm and Haas *Phone No:* 1-215-592-3000

Physical Description:

Physical Form: Liquid *Color:* White *Odor:* Slight unpleasant
Other Information: n/a

Chemical Properties:

Specific Gravity: 1.06 *Vapor Density:* n/a *Boiling Point: Decomposes*
Melting Point: n/a *Vapor Pressure:* n/a *Solubility in water:* n/a
Other Information: May float or sink in water.

Health Hazards:

Inhalation Hazard: Difficulty in breathing, coughing.
Ingestion Hazard: Nausea or vomiting
Absorption Hazard: Irritating and burning of the skin and eyes.
Hazards to Wildlife: n/a
Decontamination Procedures: Wash away any material with copious amounts of soap and water.
First Aid Procedures: Remove victim to fresh air, call emergency medical care. If not breathing give CPR. If breathing is difficult administer oxygen. Treat for shock.

Fire Hazards:

Flashpoint: 250° F(121.1° C) *Ignition temperature:* n/a
Flammable Explosive High Range: n/a *Low Range:* n/a
Toxic Products of Combustion: n/a
Other Hazards: Containers may explode in fire.
Possible extinguishing agents: Alcohol foam, dry chemical, carbon dioxide. Apply water from as far a distance as possible. Water may be ineffective on fire.

Reactivity Hazards:

Reactive With: When hot or exposed to ultraviolet light, material may polymerize. *Other Reactions:* n/a

Corrosivity Hazards:

Corrosive With: n/a *Neutralizing Agent:* n/a

Radioactivity Hazards:

Radiation Emitted: n/a *Other Hazards:* n/a

Recommended Protection for Response Personnel:

Avoid breathing vapors, keep upwind. Structural protective clothing provides limited protection. Wash away any material which may have come into contact with the body with copious amounts of soap and water. Consider appropriate evacuation.

ISOAMYL ALCOHOL

DOT Number: UN 1105 *DOT Hazard Class:* Flammable liquid *DOT Guide Number:* 26
Synonyms: fermentation amyl alcohol, fusel oil, isobutylcarbinol, isopentyl alcohol, 3-methyl-1-butanol, potato spirit oil
STCC Number: n/a *Reportable Qty:* n/a
Mfg Name: Union Carbide *Phone No:* 1-203-794-2000

Physical Description:

Physical Form: Liquid *Color:* Colorless *Odor:* Choking alcohol
Other Information: n/a

Chemical Properties:

Specific Gravity: .81 *Vapor Density:* 3.04 *Boiling Point: 270° F(132.2° C)*
Melting Point: n/a *Vapor Pressure:* n/a *Solubility in water:* Yes
Other Information: Floats and mixes with water. Irritating vapor is produced.

Health Hazards:

Inhalation Hazard: Harmful if inhaled.
Ingestion Hazard: n/a
Absorption Hazard: Irritating to the skin and eyes.
Hazards to Wildlife: Dangerous to aquatic life.
Decontamination Procedures: Wash away any material with copious amounts of soap and water.
First Aid Procedures: Remove victim to fresh air, call emergency medical care. If not breathing give CPR. If breathing is difficult administer oxygen. Treat for shock.

Fire Hazards:

Flashpoint: 114° F(45.5c) *Ignition temperature:* 662° F(350° C)
Flammable Explosive High Range: 9 *Low Range:* 1.2
Toxic Products of Combustion: n/a
Other Hazards: n/a
Possible extinguishing agents: Cool all affected containers with water. Use alcohol foam, dry chemical, carbon dioxide.

Reactivity Hazards:

Reactive With: n/a *Other Reactions:* n/a

Corrosivity Hazards:

Corrosive With: n/a *Neutralizing Agent:* n/a

Radioactivity Hazards:

Radiation Emitted: n/a *Other Hazards:* n/a

Recommended Protection for Response Personnel:

Avoid breathing vapors, keep upwind. Structural protective clothing provides limited protection. Wash away any material which may have come into contact with the body with copious amounts of soap and water. Consider appropriate evacuation.

ISOBUTANE

DOT Number: UN 1969 *DOT Hazard Class:* Gas *DOT Guide Number:* 22
Synonyms: 2-methyl propane
STCC Number: 4905747 *Reportable Qty:* n/a
Mfg Name: Phillips Petroleum *Phone No:* 1-918-661-6600

Physical Description:

Physical Form: Gas *Color:* Colorless *Odor:* Odorless
Other Information: Floats and boils on water.

Chemical Properties:

Specific Gravity: .557 *Vapor Density:* 2 *Boiling Point: 11° F(– 11.6° C)*
Melting Point: n/a *Vapor Pressure:* 30.7 psig at 70° F(21.1° C) *Solubility in water:* No
Other Information: Flammable, visible vapor cloud is produced.

Health Hazards:

Inhalation Hazard: Will cause dizziness, difficulty in breathing.
Ingestion Hazard: n/a
Absorption Hazard: May cause frostbite.
Hazards to Wildlife: No
Decontamination Procedures: Wash away any material with copious amounts of soap and water.
First Aid Procedures: Remove victim to fresh air, call emergency medical care. If not breathing give CPR. If breathing is difficult administer oxygen. Treat for shock.

Fire Hazards:

Flashpoint: n/a *Ignition temperature:* 860° F(460° C)
Flammable Explosive High Range: 8.4 *Low Range:* 1.8
Toxic Products of Combustion: n/a
Other Hazards: Flashback along vapor trail may occur. May ignite if in an enclosed area. Containers exposed to fire may violently rupture/rocket. (BLEVE)
Possible extinguishing agents: Apply water from as far a distance as possible. Use foam, dry chemical or carbon dioxide.

Reactivity Hazards:

Reactive With: n/a *Other Reactions:* n/a

Corrosivity Hazards:

Corrosive With: n/a *Neutralizing Agent:* n/a

Radioactivity Hazards:

Radiation Emitted: n/a *Other Hazards:* n/a

Recommended Protection for Response Personnel:

Avoid breathing vapors, keep upwind. Structural protective clothing provides limited protection. Wash away any material which may have come into contact with the body with copious amounts of soap and water. If the fire becomes uncontrollable, consider appropriate evacuation. (BLEVE)

ISOBUTYL ACETATE

DOT Number: UN 1213 *DOT Hazard Class:* Flammable liquid *DOT Guide Number:* 26
Synonyms: acetic acid, isobutyl ester
STCC Number: 4909207 *Reportable Qty:* 5000/2270
Mfg Name: Union Carbide *Phone No:* 1-203-794-2000

Physical Description:

Physical Form: Liquid *Color:* Colorless *Odor:* Pleasant, fruity
Other Information: Lighter than and insoluble in water. Vapors are heavier than air. Weighs 6.2 lbs/2.8 kg per gallon/3.8 l.

Chemical Properties:

Specific Gravity: .9 *Vapor Density:* 4 *Boiling Point: 244° F(117.7° C)*
Melting Point: −143° F(−97.2° C) *Vapor Pressure:* 13 mm Hg at 68° F(20° C)
Solubility in water: No *Degree of Solubility: .67%*
Other Information: n/a

Health Hazards:

Inhalation Hazard: Dizziness, nausea, loss of consciousness.
Ingestion Hazard: Will cause nausea.
Absorption Hazard: Irritating to the skin and eyes.
Hazards to Wildlife: Dangerous to aquatic life.
Decontamination Procedures: Wash away any material with copious amounts of soap and water.
First Aid Procedures: Remove victim to fresh air, call emergency medical care. If not breathing give CPR. If breathing is difficult administer oxygen. Treat for shock.

Fire Hazards:

Flashpoint: 64° F(17.7° C) *Ignition temperature:* 790° F(421.1° C)
Flammable Explosive High Range: 10.5 *Low Range:* 1.3
Toxic Products of Combustion: n/a
Other Hazards: n/a
Possible extinguishing agents: Apply water from as far a distance as possible. Use foam, dry chemical or carbon dioxide.

Reactivity Hazards:

Reactive With: Plastics *Other Reactions:* n/a

Corrosivity Hazards:

Corrosive With: n/a *Neutralizing Agent:* n/a

Radioactivity Hazards:

Radiation Emitted: n/a *Other Hazards:* n/a

Recommended Protection for Response Personnel:

Avoid breathing vapors, keep upwind. Structural protective clothing provides limited protection. Wash away any material which may have come into contact with the body with copious amounts of soap and water. Consider appropriate evacuation.

ISOBUTYL ALCOHOL

DOT Number: UN 1212 *DOT Hazard Class:* Flammable liquid *DOT Guide Number:* 26
Synonyms: butyl alcohol, isobutanol
STCC Number: 4909131 *Reportable Qty:* 5000/2270
Mfg Name: Ashland Chemical *Phone No:* 1-614-889-3333

Physical Description:

Physical Form: Liquid *Color:* Colorless *Odor:* Mild alcohol
Other Information: Vapors are heavier than air.

Chemical Properties:

Specific Gravity: .8 *Vapor Density:* 2.6 *Boiling Point: 225° F(107.2° C)*
Melting Point: −143° F(−97.2° C) *Vapor Pressure:* 13 mm Hg at 68° F(20° C)
Solubility in water: Yes *Degree of Solubility: .67%*
Other Information: n/a

Health Hazards:

Inhalation Hazard: Will cause dizziness, nausea, or headache.
Ingestion Hazard: Harmful if swallowed.
Absorption Hazard: Irritating to the eyes.
Hazards to Wildlife: Dangerous to aquatic life.
Decontamination Procedures: Wash away any material with copious amounts of soap and water.
First Aid Procedures: Remove victim to fresh air, call emergency medical care. If not breathing give CPR. If breathing is difficult administer oxygen. Treat for shock.

Fire Hazards:

Flashpoint: 82° F(27.7° C) *Ignition temperature:* 780° F(415.5° C)
Flammable Explosive High Range: 10.6 *Low Range:* 1.7
Toxic Products of Combustion: n/a
Other Hazards: Flashback along vapor trail may occur. May explode if ignited in an enclosed area.
Possible extinguishing agents: Apply water from as far a distance as possible. Use alcohol foam, dry chemical, or carbon dioxide.

Reactivity Hazards:

Reactive With: n/a *Other Reactions:* n/a

Corrosivity Hazards:

Corrosive With: n/a *Neutralizing Agent:* n/a

Radioactivity Hazards:

Radiation Emitted: n/a *Other Hazards:* n/a

Recommended Protection for Response Personnel:

Avoid breathing vapors, keep upwind. Structural protective clothing provides limited protection. Wash away any material which may have come into contact with the body with copious amounts of soap and water. Consider appropriate evacuation.

ISOBUTYLAMINE

DOT Number: UN 1214 *DOT Hazard Class:* Flammable liquid *DOT Guide Number:* 68
Synonyms: iso-butyl amine
STCC Number: 4908186 *Reportable Qty:* 1000/454
Mfg Name: Aldrich Chemical *Phone No:* 1-414-273-3850

Physical Description:

Physical Form: Liquid *Color:* Colorless *Odor:* Fish
Other Information: Weighs 6.1 lbs/2.7 kg per gallon/3.8 l.

Chemical Properties:

Specific Gravity: .7 *Vapor Density:* 2.5 *Boiling Point: 150° F(65.5° C)*
Melting Point: n/a *Vapor Pressure:* 100 mm Hg at 65° F(18.3° C) *Solubility in water:* Yes
Other Information: Lighter than water. Vapors are heavier than air.

Health Hazards:

Inhalation Hazard: Coughing, difficulty in breathing, loss of consciousness.
Ingestion Hazard: Nausea, loss of consciousness.
Absorption Hazard: Will burn the skin and eyes.
Hazards to Wildlife: n/a
Decontamination Procedures: Wash away any material with copious amounts of soap and water.
First Aid Procedures: Remove victim to fresh air, call emergency medical care. If not breathing give CPR. If breathing is difficult administer oxygen. Treat for shock.

Fire Hazards:

Flashpoint: 15° F(−9.4° C) *Ignition temperature:* 712° F(377.7° C)
Flammable Explosive High Range: 9 *Low Range:* 3.4
Toxic Products of Combustion: Toxic oxides of nitrogen may be produced in a fire.
Other Hazards: Flashback along vapor trail may occur. Vapors may explode if ignited in an enclosed area.
Possible extinguishing agents: Apply water from as far a distance as possible. Use alcohol foam, dry chemical, or carbon dioxide.

Reactivity Hazards:

Reactive With: n/a *Other Reactions:* n/a

Corrosivity Hazards:

Corrosive With: n/a *Neutralizing Agent:* n/a

Radioactivity Hazards:

Radiation Emitted: n/a *Other Hazards:* n/a

Recommended Protection for Response Personnel:

Avoid breathing vapors, keep upwind. Wear a sealed chemical suit (butyl rubber). Wash away any material which may have come into contact with the body with copious amounts of soap and water. Consider appropriate evacuation.

ISOBUTYLENE

DOT Number: UN 1055 *DOT Hazard Class:* Liquefied gas *DOT Guide Number:* 22
Synonyms: isobutene
STCC Number: 4905748 *Reportable Qty:* n/a
Mfg Name: Petro-Tex Chemical Corp. *Phone No:* 1-713-477-9211

Physical Description:

Physical Form: Gas *Color:* Colorless *Odor:* Gasolinelike
Other Information: It is used to produce high octane and aviation gas.

Chemical Properties:

Specific Gravity: n/a *Vapor Density:* 1.9 *Boiling Point: 20° F(–6.6° C)*
Melting Point: n/a *Vapor Pressure:* n/a *Solubility in water:* No
Other Information: It is shipped as a liquefied gas under its own pressure.

Health Hazards:

Inhalation Hazard: Dizziness, loss of consciousness.
Ingestion Hazard: n/a
Absorption Hazard: Will cause frostbite.
Hazards to Wildlife: n/a
Decontamination Procedures: Wash away any material with copious amounts of soap and water.
First Aid Procedures: Remove victim to fresh air, call emergency medical care. If not breathing give CPR. If breathing is difficult administer oxygen. Treat for shock.

Fire Hazards:

Flashpoint: Gas *Ignition temperature:* 869° F(465° C)
Flammable Explosive High Range: 9.6 *Low Range:* 1.8
Toxic Products of Combustion: n/a
Other Hazards: Under fire conditions, cylinders or tank cars may violently rupture/rocket!! (BLEVE)
Possible extinguishing agents: Apply water from as far a distance as possible to cool the containers.

Reactivity Hazards:

Reactive With: n/a *Other Reactions:* n/a

Corrosivity Hazards:

Corrosive With: n/a *Neutralizing Agent:* n/a

Radioactivity Hazards:

Radiation Emitted: n/a *Other Hazards:* n/a

Recommended Protection for Response Personnel:

Avoid breathing vapors, keep upwind. Structural protective clothing provides limited protection. Wash away any material which may have come into contact with the body with copious amounts of soap and water. If the fire becomes uncontrollable, consider appropriate evacuation. (BLEVE)

ISOBUTYLNITRILE

DOT Number: UN 2284 *DOT Hazard Class:* Flammable liquid *DOT Guide Number:* 28
Synonyms: IBN, isopropyl cyanide
STCC Number: 4909208 *Reportable Qty:* n/a
Mfg Name: Eastman Chemical *Phone No:* 1-615-229-2000

Physical Description:

Physical Form: Liquid *Color:* Colorless *Odor:* Almond
Other Information: n/a

Chemical Properties:

Specific Gravity: .8 *Vapor Density:* 2.38 *Boiling Point: 215° F(101.6° C)*
Melting Point: n/a *Vapor Pressure:* 100 mm Hg at 130° F(54.4° C) *Solubility in water:* Slight
Other Information: Floats on water. Flammable vapor is produced.

Health Hazards:

Inhalation Hazard: Poisonous if inhaled.
Ingestion Hazard: Poisonous if swallowed.
Absorption Hazard: Poisonous if the skin is exposed.
Hazards to Wildlife: n/a
Decontamination Procedures: Wash away any material with copious amounts of soap and water.
First Aid Procedures: Remove victim to fresh air, call emergency medical care. If not breathing give CPR. If breathing is difficult administer oxygen. Treat for shock.

Fire Hazards:

Flashpoint: 47° F(8.3° C) *Ignition temperature:* 900° F(482.2° C)
Flammable Explosive High Range: n/a *Low Range:* n/a
Toxic Products of Combustion: Toxic oxides of nitrogen are formed in fires.
Other Hazards: n/a
Possible extinguishing agents: Apply water from as far a distance as possible. Use alcohol foam, dry chemical, or carbon dioxide. Do not extinguish the fire unless the flow can be stopped.

Reactivity Hazards:

Reactive With: n/a *Other Reactions:* n/a

Corrosivity Hazards:

Corrosive With: n/a *Neutralizing Agent:* n/a

Radioactivity Hazards:

Radiation Emitted: n/a *Other Hazards:* n/a

Recommended Protection for Response Personnel:

Avoid breathing vapors, keep upwind. Wear appropriate chemical clothing. Wash away any material which may come into contact with the body with copious amounts of soap and water. Consider appropriate evacuation.

ISOBUTYRALDEHYDE

DOT Number: UN 2045 *DOT Hazard Class:* Flammable liquid *DOT Guide Number:* 26
Synonyms: isobutanal, isobutyric aldehyde, isobutylaldehyde
STCC Number: 4908185 *Reportable Qty:* n/a *CHEMTREC Phone No:* 1-800-424-9300

Physical Description:

Physical Form: Liquid *Color:* Colorless *Odor:* Pungent
Other Information: It is used as a solvent, for making flavors, gas additives, perfumes, brake fluid, plasticizers, and in other chemicals.

Chemical Properties:

Specific Gravity: .790 *Vapor Density:* 2.5 *Boiling Point: 140-149° F(60-65° C)*
Melting Point: −86° F(−65.5° C) Vapor Pressure: 130 mm Hg at 68° F(20° C)
Solubility in water: Yes *Degree of Solubility: 6%*
Other Information: Soluble, lighter than water. Weighs 6.6 lbs/2.9 kg per gallon/3.8 l.

Health Hazards:

Inhalation Hazard: May cause difficult breathing, coughing, drowsiness
Ingestion Hazard: Strong irritant
Absorption Hazard: Irritating or burn the skin and eyes.
Hazards to Wildlife: n/a
Decontamination Procedures: Wash away any material with copious amounts of soap and water.
First Aid Procedures: Remove victim to fresh air, call emergency medical care. If not breathing give CPR. If breathing is difficult administer oxygen. Treat for shock.

Fire Hazards:

Flashpoint: −40° F(−40° C) *Ignition temperature:* 490° F(254.4° C)
Flammable Explosive High Range: 10.6 *Low Range:* 1
Toxic Products of Combustion: Acrid fumes and smoke may be toxic.
Other Hazards: Flashback along vapor trail may occur. Vapors may explode if ignited in an enclosed area. The containers may rupture violently in a fire.
Possible extinguishing agents: Apply water from as far a distance as possible. Use foam, dry chemical, or carbon dioxide.

Reactivity Hazards:

Reactive With: Strong oxidizers *Other Reactions:* n/a

Corrosivity Hazards:

Corrosive With: Mild steel *Neutralizing Agent:* n/a

Radioactivity Hazards:

Radiation Emitted: n/a *Other Hazards:* n/a

Recommended Protection for Response Personnel:

Avoid breathing vapors, keep upwind. Wear a sealed chemical suit (butyl rubber, polyethylene). Wash away any material which may have come into contact with the body with copious amounts of soap and water. The containers may rupture violently in a fire. Violent polymerization may occur in the presence of nonoxidizing material acids or organic acids. Consider appropriate evacuation.

ISOBUTYRIC ACID

DOT Number: UN 2529 *DOT Hazard Class:* Corrosive *DOT Guide Number:* 29
Synonyms: dimethyl acetic acid
STCC Number: 4931438 *Reportable Qty:* 5000/2270
Mfg Name: Eastman Organic *Phone No:* 1-800-445-6325

Physical Description:

Physical Form: Liquid *Color:* Colorless *Odor:* Rancid, butter
Other Information: Weighs 7.9 lbs/3.5 kg per gallon/3.8 l.

Chemical Properties:

Specific Gravity: 1 *Vapor Density:* 3 *Boiling Point: 306° F(152.2° C)*
Melting Point: n/a *Vapor Pressure:* 1.5 mm Hg at 68° F(20° C) *Solubility in water:* Yes
Other Information: n/a

Health Hazards:

Inhalation Hazard: Harmful if inhaled.
Ingestion Hazard: Harmful if swallowed.
Absorption Hazard: Will burn the skin and eyes.
Hazards to Wildlife: n/a
Decontamination Procedures: Wash away any material with copious amounts of soap and water.
First Aid Procedures: Remove victim to fresh air, call emergency medical care. If not breathing give CPR. If breathing is difficult administer oxygen. Treat for shock.

Fire Hazards:

Flashpoint: 132° F(55.5° C) *Ignition temperature:* 900° F(482.2° C)
Flammable Explosive High Range: 9.2 *Low Range:* 2
Toxic Products of Combustion: n/a
Other Hazards: Do not breathe vapors. Stay clear of dust and vapor cloud.
Possible extinguishing agents: Apply water from as far a distance as possible. Use alcohol foam, dry chemical, or carbon dioxide.

Reactivity Hazards:

Reactive With: n/a *Other Reactions:* n/a

Corrosivity Hazards:

Corrosive With: Aluminum and other metals
Neutralizing Agent: Crushed limestone, soda ash, or lime

Radioactivity Hazards:

Radiation Emitted: n/a *Other Hazards:* n/a

Recommended Protection for Response Personnel:

Avoid breathing vapors, keep upwind. Structural protective clothing provides limited protection. Wash away any material which may have come into contact with the body with copious amounts of soap and water. Consider appropriate evacuation.

ISODECALDEHYDE

DOT Number: n/a *DOT Hazard Class:* n/a *DOT Guide Number:* n/a
Synonyms: isodecaldeyde mixed isomers, trimethylheptanals
STCC Number: n/a *Reportable Qty:* n/a
Mfg Name: Union Carbide *Phone No:* 1-203-794-2000

Physical Description:

Physical Form: Liquid *Color:* Colorless *Odor:* Fruity
Other Information: n/a

Chemical Properties:

Specific Gravity: .84 *Vapor Density:* 5.4 *Boiling Point: n/a*
Melting Point: n/a *Vapor Pressure:* n/a *Solubility in water:* n/a
Other Information: Floats on water

Health Hazards:

Inhalation Hazard: n/a
Ingestion Hazard: n/a
Absorption Hazard: Irritating to the skin and eyes.
Hazards to Wildlife: n/a
Decontamination Procedures: Wash away any material with copious amounts of soap and water.
First Aid Procedures: Remove victim to fresh air, call emergency medical care. If not breathing give CPR. If breathing is difficult administer oxygen. Treat for shock.

Fire Hazards:

Flashpoint: 185° F(85° C) *Ignition temperature:* n/a
Flammable Explosive High Range: n/a *Low Range:* n/a
Toxic Products of Combustion: n/a
Other Hazards: n/a
Possible extinguishing agents: Cool all affected containers with water. Use foam, dry chemical, carbon dioxide. Water may be ineffective on fires.

Reactivity Hazards:

Reactive With: n/a *Other Reactions:* n/a

Corrosivity Hazards:

Corrosive With: n/a *Neutralizing Agent:* n/a

Radioactivity Hazards:

Radiation Emitted: n/a *Other Hazards:* n/a

Recommended Protection for Response Personnel:

Avoid breathing vapors, keep upwind. Structural protective clothing provides limited protection. Wash away any material which may have come into contact with the body with copious amounts of soap and water. Consider appropriate evacuation.

ISODECYL ACRYLATE

DOT Number: n/a *DOT Hazard Class:* n/a *DOT Guide Number:* n/a
Synonyms: none given
STCC Number: n/a *Reportable Qty:* n/a
Mfg Name: Union Carbide *Phone No:* 1-203-794-2000

Physical Description:

Physical Form: Liquid *Color:* Colorless *Odor:* Weak
Other Information: n/a

Chemical Properties:

Specific Gravity: .88 *Vapor Density:* n/a *Boiling Point: Polymerizes*
Melting Point: −148° F(−100° C) Vapor Pressure: n/a *Solubility in water:* n/a
Other Information: Floats on water

Health Hazards:

Inhalation Hazard: n/a
Ingestion Hazard: n/a
Absorption Hazard: Irritating to the skin and eyes.
Hazards to Wildlife: n/a
Decontamination Procedures: Wash away any material with copious amounts of soap and water.
First Aid Procedures: Remove victim to fresh air, call emergency medical care. If not breathing give CPR. If breathing is difficult administer oxygen. Treat for shock.

Fire Hazards:

Flashpoint: 240° F(115.5° C) *Ignition temperature:* n/a
Flammable Explosive High Range: n/a *Low Range:* n/a
Toxic Products of Combustion: n/a
Other Hazards: May polymerize to a gummy solid in fire.
Possible extinguishing agents: Cool all affected containers with water. Use foam, dry chemical, carbon dioxide. Water may be ineffective on fires.

Reactivity Hazards:

Reactive With: n/a *Other Reactions:* n/a

Corrosivity Hazards:

Corrosive With: n/a *Neutralizing Agent:* n/a

Radioactivity Hazards:

Radiation Emitted: n/a *Other Hazards:* n/a

Recommended Protection for Response Personnel:

Avoid breathing vapors, keep upwind. Wear appropriate sealed chemical suit. Wash away any material which may have come into contact with the body with copious amounts of soap and water. Consider appropriate evacuation.

ISODECYL ALCOHOL

DOT Number: n/a *DOT Hazard Class:* n/a *DOT Guide Number:* n/a
Synonyms: none given
STCC Number: n/a *Reportable Qty:* n/a
Mfg Name: Exxon Chemical Co. *Phone No:* 1-203-655-5200

Physical Description:

Physical Form: Liquid *Color:* Colorless *Odor:* Mild alcohol
Other Information: n/a

Chemical Properties:

Specific Gravity: .84 *Vapor Density:* n/a *Boiling Point: 428° F(220° C)*
Melting Point: 140° F(60° C) *Vapor Pressure:* n/a *Solubility in water:* n/a
Other Information: Floats on water

Health Hazards:

Inhalation Hazard: n/a
Ingestion Hazard: n/a
Absorption Hazard: Will burn the skin and eyes.
Hazards to Wildlife: n/a
Decontamination Procedures: Wash away any material with copious amounts of soap and water.
First Aid Procedures: Remove victim to fresh air, call emergency medical care. If not breathing give CPR. If breathing is difficult administer oxygen. Treat for shock.

Fire Hazards:

Flashpoint: 220° F(104.4° C) *Ignition temperature:* n/a
Flammable Explosive High Range: n/a *Low Range:* n/a
Toxic Products of Combustion: n/a
Other Hazards: n/a
Possible extinguishing agents: Cool all affected containers with water. Use foam, dry chemical, or carbon dioxide. Water may be ineffective on fires.

Reactivity Hazards:

Reactive With: n/a *Other Reactions:* n/a

Corrosivity Hazards:

Corrosive With: n/a *Neutralizing Agent:* n/a

Radioactivity Hazards:

Radiation Emitted: n/a *Other Hazards:* n/a

Recommended Protection for Response Personnel:

Avoid breathing vapors, keep upwind. Structural protective clothing provides limited protection. Wash away any material which may have come into contact with the body with copious amounts of soap and water. Consider appropriate evacuation.

ISOHEXANE

DOT Number: UN 1208 *DOT Hazard Class:* Flammable liquid *DOT Guide Number:* 27
Synonyms: 2-methylpentane
STCC Number: n/a *Reportable Qty:* n/a
Mfg Name: Phillips Petroleum *Phone No:* 1-918-661-6600

Physical Description:

Physical Form: Liquid *Color:* Clear *Odor:* Gasoline
Other Information: n/a

Chemical Properties:

Specific Gravity: .65 *Vapor Density:* 3 *Boiling Point: 140° F(60° C)*
Melting Point: −244° F(−153.3 C) Vapor Pressure: n/a
Solubility in water: n/a
Other Information: Floats on water. Flammable, irritating vapor is produced.

Health Hazards:

Inhalation Hazard: Dizziness, headache, difficulty in breathing, loss of consciousness.
Ingestion Hazard: Nausea or vomiting.
Absorption Hazard: Irritating to the skin and eyes.
Hazards to Wildlife: n/a
Decontamination Procedures: Wash away any material with copious amounts of soap and water.
First Aid Procedures: Remove victim to fresh air, call emergency medical care. If not breathing give CPR. If breathing is difficult administer oxygen. Treat for shock.

Fire Hazards:

Flashpoint: −20° F(−28.8° C) *Ignition temperature:* 585° F(307.2° C)
Flammable Explosive High Range: 7.7 *Low Range:* 1.2
Toxic Products of Combustion: n/a
Other Hazards: n/a
Possible extinguishing agents: Cool all affected containers with water. Use foam, dry chemical, carbon dioxide. Water may be ineffective on fires.

Reactivity Hazards:

Reactive With: n/a *Other Reactions:* n/a

Corrosivity Hazards:

Corrosive With: n/a *Neutralizing Agent:* n/a

Radioactivity Hazards:

Radiation Emitted: n/a *Other Hazards:* n/a

Recommended Protection for Response Personnel:

Avoid breathing vapors, keep upwind. Structural protective clothing provides limited protection. Wash away any material which may have come into contact with the body with copious amounts of soap and water. Consider appropriate evacuation.

ISOOCTALDEHYDE

DOT Number: n/a *DOT Hazard Class:* n/a *DOT Guide Number:* n/a
Synonyms: dimethylhexanal, isooctylaldehyde, 6-methyl-1-heptanal, oxo octaldehyde,
STCC Number: n/a *Reportable Qty:* n/a
Mfg Name: Exxon Chemical Co. *Phone No:* 1-203-655-5200

Physical Description:

Physical Form: Liquid *Color:* Colorless *Odor:* Mild fruity
Other Information: n/a

Chemical Properties:

Specific Gravity: .825 *Vapor Density:* n/a *Boiling Point: 307-352° F(152-177° C)*
Melting Point: −180° F(−117.7° C) *Vapor Pressure:* n/a
Solubility in water: n/a
Other Information: Floats on water

Health Hazards:

Inhalation Hazard: n/a
Ingestion Hazard: n/a
Absorption Hazard: Irritating to the eyes.
Hazards to Wildlife: n/a
Decontamination Procedures: Wash away any material with copious amounts of soap and water.
First Aid Procedures: Remove victim to fresh air, call emergency medical care. If not breathing give CPR. If breathing is difficult administer oxygen. Treat for shock.

Fire Hazards:

Flashpoint: 104° F(40° C) *Ignition temperature:* 320° F(160° C)
Flammable Explosive High Range: n/a *Low Range:* n/a
Toxic Products of Combustion: n/a
Other Hazards: n/a
Possible extinguishing agents: Cool all affected containers with water. Use foam, dry chemical or carbon dioxide. Water may be ineffective on fires.

Reactivity Hazards:

Reactive With: n/a *Other Reactions:* n/a

Corrosivity Hazards:

Corrosive With: n/a *Neutralizing Agent:* n/a

Radioactivity Hazards:

Radiation Emitted: n/a *Other Hazards:* n/a

Recommended Protection for Response Personnel:

Avoid breathing vapors, keep upwind. Wear a sealed chemical suit (butyl rubber). Wash away any material which may have come into contact with the body with copious amounts of soap and water. Consider appropriate evacuation.

ISOOCTYL ALCOHOL

DOT Number: UN 1987 *DOT Hazard Class:* Combustible liquid *DOT Guide Number:* 26
Synonyms: dimethyl-1-hexanol, 6-methyl-1-heptanol, oxo octyl alcohol
STCC Number: 4913128 *Reportable Qty:* n/a
Mfg Name: Exxon Chemical America *Phone No:* 1-713-870-6000

Physical Description:

Physical Form: Liquid *Color:* Colorless *Odor:* Mild
Other Information: It is used as a solvent, in the making of cutting and lubricating oils, in hydraulic fluids, and in the making of other chemicals.

Chemical Properties:

Specific Gravity: .83 *Vapor Density:* 1 *Boiling Point: 367° F(186.1° C)*
Melting Point: n/a *Vapor Pressure:* n/a *Solubility in water:* No
Other Information: Lighter than and insoluble in water. Vapors are heavier than air.

Health Hazards:

Inhalation Hazard: n/a
Ingestion Hazard: Harmful if swallowed.
Absorption Hazard: Irritating to the skin and the eyes.
Hazards to Wildlife: n/a
Decontamination Procedures: Wash away any material with copious amounts of soap and water.
First Aid Procedures: Remove victim to fresh air, call emergency medical care. If not breathing give CPR. If breathing is difficult administer oxygen. Treat for shock.

Fire Hazards:

Flashpoint: 180° F(82.2° C) *Ignition temperature:* 530° F(276.6° C)
Flammable Explosive High Range: 5.7 *Low Range:* .9
Toxic Products of Combustion: n/a
Other Hazards: n/a
Possible extinguishing agents: Apply water from as far a distance as possible to cool containers. Use alcohol foam, dry chemical, or carbon dioxide.

Reactivity Hazards:

Reactive With: n/a *Other Reactions:* n/a

Corrosivity Hazards:

Corrosive With: n/a *Neutralizing Agent:* n/a

Radioactivity Hazards:

Radiation Emitted: n/a *Other Hazards:* n/a

Recommended Protection for Response Personnel:

Avoid breathing vapors, keep upwind. Structural protective clothing provides limited protection. Wash away any material which may have come into contact with the body with copious amounts of soap and water. If the fire becomes uncontrollable, consider appropriate evacuation.

ISOPENTANE

DOT Number: UN 1265 *DOT Hazard Class:* Flammable liquid *DOT Guide Number:* 27
Synonyms: 2-methyl butane
STCC Number: 4908192 *Reportable Qty:* n/a
Mfg Name: Phillips Petroleum *Phone No:* 1-918-661-6600

Physical Description:

Physical Form: Liquid *Color:* Colorless *Odor:* Petroleum
Other Information: Weighs 7.9 lbs/3.5 kg per gallon/3.8 l.

Chemical Properties:

Specific Gravity: .6 *Vapor Density:* 2.5 *Boiling Point: 82° F(27.7° C)*
Melting Point: n/a *Vapor Pressure:* 1.08 mm Hg at 68° F(20° C) *Solubility in water:* No
Other Information: Lighter than water. Vapors are heavier than air.

Health Hazards:

Inhalation Hazard: Coughing, difficulty breathing, loss of consciousness.
Absorption Hazard: Nausea and vomiting.
Absorption Hazard: Irritating to the skin and eyes.
Hazards to Wildlife: n/a
Decontamination Procedures: Wash away any material with copious amounts of soap and water.
First Aid Procedures: Remove victim to fresh air, call emergency medical care. If not breathing give CPR. If breathing is difficult administer oxygen. Treat for shock.

Fire Hazards:

Flashpoint: −70° F(−56.6° C) *Ignition temperature:* 800° F(426.6° C)
Flammable Explosive High Range: 8.3 *Low Range:* 1.4
Toxic Products of Combustion: n/a
Other Hazards: Highly volatile liquid. Vapors may explode when mixed with air.
Possible extinguishing agents: Apply water from as far a distance as possible to cool containers. Use foam, dry chemical, or carbon dioxide.

Reactivity Hazards:

Reactive With: n/a *Other Reactions:* n/a

Corrosivity Hazards:

Corrosive With: n/a *Neutralizing Agent:* n/a

Radioactivity Hazards:

Type Radiation Emitted : n/a *Other Hazards:* n/a

Recommended Protection for Response Personnel:

Avoid breathing vapors, keep upwind. Wear appropriate chemical clothing, Wash away any material which may have come into contact with the body with copious amounts of soap and water. If the fire becomes uncontrollable, consider evacuation.

ISOPHORONE

DOT Number: n/a *DOT Hazard Class:* n/a *DOT Guide Number:* n/a
Synonyms: 3,5,5-trimethyl-2-cyclohexene-1-100
STCC Number: n/a *Reportable Qty:* n/a
Mfg Name: Union Carbide *Phone No:* 1-203-794-2000

Physical Description:

Physical Form: Liquid *Color:* Colorless *Odor:* Camphor
Other Information: n/a

Chemical Properties:

Specific Gravity: .92 *Vapor Density:* 4.77 *Boiling Point: 419° F(215° C)*
Melting Point: 17° F(−8.3° C) *Vapor Pressure:* 1 mm Hg at 100° F(37.7° C) *Solubility in water:* Yes
Other Information: Floats and mixes slowly with water.

Health Hazards:

Inhalation Hazard: n/a
Ingestion Hazard: Harmful if swallowed.
Absorption Hazard: Irritating to the skin and eyes.
Hazards to Wildlife: n/a
Decontamination Procedures: Wash away any material with copious amounts of soap and water.
First Aid Procedures: Remove victim to fresh air, call emergency medical care. If not breathing give CPR. If breathing is difficult administer oxygen. Treat for shock.

Fire Hazards:

Flashpoint: 184° F(84.4° C) *Ignition temperature:* 864° F(462.2° C)
Flammable Explosive High Range: 3.8 *Low Range:* .8
Toxic Products of Combustion: n/a
Other Hazards: n/a
Possible extinguishing agents: Cool all affected containers with water. Use foam, dry chemical, carbon dioxide. Water may be ineffective on fires.

Reactivity Hazards:

Reactive With: n/a *Other Reactions:* n/a

Corrosivity Hazards:

Corrosive With: n/a *Neutralizing Agent:* n/a

Radioactivity Hazards:

Radiation Emitted: n/a *Other Hazards:* n/a

Recommended Protection for Response Personnel:

Avoid breathing vapors, keep upwind. Structural protective clothing provides limited protection. Wash away any material which may have come into contact with the body with copious amounts of soap and water. Consider appropriate evacuation.

ISOPHTHALIC ACID

DOT Number: n/a *DOT Hazard Class:* n/a *DOT Guide Number:* n/a
Synonyms: benzene-1,3-dicarboxylic acid, IPA, m-phthalic acid
STCC Number: n/a *Reportable Qty:* n/a
Mfg Name: Morgan Chemical *Phone No:* 1-716-632-4000

Physical Description:

Physical Form: Solid *Color:* White *Odor:* Slight unpleasant
Other Information: n/a

Chemical Properties:

Specific Gravity: 1.54 *Vapor Density:* n/a *Boiling Point: n/a*
Melting Point: 653° F(345° C) *Vapor Pressure:* n/a *Solubility in water:* n/a
Other Information: Sinks in watcr.

Health Hazards:

Inhalation Hazard: Coughing, difficulty in breathing.
Ingestion Hazard: Nausea.
Absorption Hazard: Irritating to the skin and eyes.
Hazards to Wildlife: n/a
Decontamination Procedures: Wash away any material with copious amounts of soap and water.
First Aid Procedures: Remove victim to fresh air, call emergency medical care. If not breathing give CPR. If breathing is difficult administer oxygen. Treat for shock.

Fire Hazards:

Flashpoint: Combustible solid *Ignition temperature:* n/a
Flammable Explosive High Range: n/a *Low Range:* n/a
Toxic Products of Combustion: n/a
Other Hazards: Dust forms explosive mixtures in air.
Possible extinguishing agents: Extinguish fire using suitable agent for the type of surrounding fire.

Reactivity Hazards:

Reactive With: n/a *Other Reactions:* n/a

Corrosivity Hazards:

Corrosive With: n/a *Neutralizing Agent:* n/a

Radioactivity Hazards:

Radiation Emitted: n/a *Other Hazards:* n/a

Recommended Protection for Response Personnel:

Avoid breathing vapors, keep upwind. Structural protective clothing provides limited protection. Wash away any material which may have come into contact with the body with copious amounts of soap and water. Consider appropriate evacuation.

ISOPRENE

DOT Number: UN 1218 *DOT Hazard Class:* Flammable liquid *DOT Guide Number:* 27
Synonyms: 2-methyl-1,3-butadiene
STCC Number: 4907230 *Reportable Qty:* 1000/45.4
Mfg Name: Exxon Chemical America *Phone No:* 1-713-870-6000

Physical Description:

Physical Form: Liquid *Color:* Clear *Odor:* Petroleum
Other Information: Weighs 5.7 lbs/2.5 kg per gallon/3.8 l.

Chemical Properties:

Specific Gravity: .7 *Vapor Density:* 2.4 *Boiling Point: 93° F(33.8° C)*
Melting Point: n/a *Vapor Pressure:* n/a *Solubility in water:* No
Other Information: Lighter than water. Vapors are heavier than air.

Health Hazards:

Inhalation Hazard: n/a
Ingestion Hazard: n/a
Absorption Hazard: Irritating to the skin and eyes.
Hazards to Wildlife: Dangerous to aquatic life.
Decontamination Procedures: Wash away any material with copious amounts of soap and water.
First Aid Procedures: Remove victim to fresh air, call emergency medical care. If not breathing give CPR. If breathing is difficult administer oxygen. Treat for shock.

Fire Hazards:

Flashpoint: −65° F(−53.8° C) *Ignition temperature:* 743° F(395° C)
Flammable Explosive High Range: 8.9 *Low Range:* 1.5
Toxic Products of Combustion: n/a
Other Hazards: Material may polymerize. Containers may violently rupture.
Possible extinguishing agents: Apply water from as far a distance as possible to cool containers. Use alcohol foam, dry chemical, or carbon dioxide.

Reactivity Hazards:

Reactive With: n/a *Other Reactions:* n/a

Corrosivity Hazards:

Corrosive With: n/a *Neutralizing Agent:* n/a

Radioactivity Hazards:

Radiation Emitted: n/a *Other Hazards:* n/a

Recommended Protection for Response Personnel:

Avoid breathing vapors, keep upwind. Wear appropriate chemical clothing. Wash away any material which may come into contact with the body with copious amounts of soap and water. Consider appropriate evacuation.

ISOPROPANOL

DOT Number: UN 1219 *DOT Hazard Class:* Flammable liquid *DOT Guide Number:* 26
Synonyms: isopropyl alcohol, IPA, 2-propanol, rubbing alcohol
STCC Number: 4909205 *Reportable Qty:* n/a
Mfg Name: Shell Chemical *Phone No:* 1-713-241-6161

Physical Description:

Physical Form: Liquid *Color:* Colorless *Odor:* Rubbing alcohol
Other Information: It is used to make cosmetics, pharmaceuticals, antifreezes, soaps, acetone, and other chemicals and products.

Chemical Properties:

Specific Gravity: .785 *Vapor Density:* 2.5 *Boiling Point: 180° F(82.2° C)*
Melting Point: −127° F(−88.3° C) *Vapor Pressure:* 33 mm Hg at 68° F(20° C)
Solubility in water: Yes
Other Information: Will mix freely with water. Weighs 6.5 lbs/2.9 kg per gallon/3.8 l.

Health Hazards:

Inhalation Hazard: Irritating to the nose and throat.
Ingestion Hazard: Nausea, vomiting, and headache.
Absorption Hazard: Irritating and stinging to the skin and eyes.
Hazards to Wildlife: Dangerous to aquatic life.
Decontamination Procedures: Wash away any material with copious amounts of soap and water.
First Aid Procedures: Remove victim to fresh air, call emergency medical care. If not breathing give CPR. If breathing is difficult administer oxygen. Treat for shock.

Fire Hazards:

Flashpoint: 53° F(11.6° C) *Ignition temperature:* 750° F(398.8° C)
Flammable Explosive High Range: 12.7 *Low Range:* 2.3
Toxic Products of Combustion: n/a
Other Hazards: Flashback along vapor trail may occur. Containers may explode if ignited in an enclosed area.
Possible *Possible extinguishing agents:* Apply water from as far a distance as possible to cool the containers. Use alcohol foam, dry chemical, or carbon dioxide.

Reactivity Hazards:

Reactive With: n/a *Other Reactions:* n/a

Corrosivity Hazards:

Corrosive With: n/a *Neutralizing Agent:* n/a

Radioactivity Hazards:

Radiation Emitted: n/a *Other Hazards:* n/a

Recommended Protection for Response Personnel:

Avoid breathing vapors, keep upwind. Wear a sealed chemical suit (butyl rubber, neoprene, polyethylene, viton). Wash away any material which may have come into contact with the body with copious amounts of soap and water. If the fire becomes uncontrollable, consider appropriate evacuation.

ISOPROPYL ACETATE

DOT Number: UN 1220 *DOT Hazard Class:* Flammable liquid *DOT Guide Number:* 26
Synonyms: acetic acid, isopropyl ester
STCC Number: 4909210 *Reportable Qty:* n/a
Mfg Name: Union Carbide *Phone No:* 1-203-794-2000

Physical Description:

Physical Form: Liquid *Color:* Colorless *Odor:* Fruity
Other Information: n/a

Chemical Properties:

Specific Gravity: .9 *Vapor Density:* 3.5 *Boiling Point: 194° F(90° C)*
Melting Point: −139° F(−95° C) Vapor Pressure: 43 mm Hg at 68° F(20° C)
Solubility in water: Slight *Degree of Solubility:* 2.9%
Other Information: Lighter than water. Vapors are heavier than air.

Health Hazards:

Inhalation Hazard: Irritating to the nose and throat.
Ingestion Hazard: Harmful if swallowed.
Absorption Hazard: Irritating to the skin and eyes.
Hazards to Wildlife: n/a
Decontamination Procedures: Wash away any material with copious amounts of soap and water.
First Aid Procedures: Remove victim to fresh air, call emergency medical care. If not breathing give CPR. If breathing is difficult administer oxygen. Treat for shock.

Fire Hazards:

Flashpoint: 35° F(1.6° C) *Ignition temperature:* 860° F(460° C)
Flammable Explosive High Range: 8 *Low Range:* 1.8
Toxic Products of Combustion: n/a
Other Hazards: Flashback along vapor trail may occur. Containers may explode if ignited in an enclosed area.
Possible extinguishing agents: Apply water from as far a distance as possible to cool containers. Use alcohol foam, dry chemical, or carbon dioxide.

Reactivity Hazards:

Reactive With: n/a *Other Reactions:* n/a

Corrosivity Hazards:

Corrosive With: n/a *Neutralizing Agent:* n/a

Radioactivity Hazards:

Radiation Emitted: n/a *Other Hazards:* n/a

Recommended Protection for Response Personnel:

Avoid breathing vapors, keep upwind. Structural protective clothing provides limited protection. Wash away any material which may have come into contact with the body with copious amounts of soap and water. If the fire becomes uncontrollable, consider appropriate evacuation.

ISOPROPYL ALCOHOL

DOT Number: UN 1219 *DOT Hazard Class:* Flammable liquid *DOT Guide Number:* 26
Synonyms: isopropanol, petrohol
STCC Number: 4909205 *Reportable Qty:* n/a
Mfg Name: Shell Chemical *Phone No:* 1-713-241-6161

Physical Description:

Physical Form: Liquid *Color:* Colorless *Odor:* Rubbing alcohol
Other Information: It is used to make cosmetics, pharmacueticals, antifreezes, soaps, acetone, and other chemicals and products.

Chemical Properties:

Specific Gravity: .8 *Vapor Density:* 2.1 *Boiling Point: 181° F(82.7° C)*
Melting Point: –128° F(–88.8° C) *Vapor Pressure:* 33 mm Hg at 68° F(20° C)
Solubility in water: Yes
Other Information: n/a

Health Hazards:

Inhalation Hazard: Irritating to nose and throat.
Ingestion Hazard: Harmful if swallowed.
Absorption Hazard: Irritating to the eyes.
Hazards to Wildlife: Dangerous to aquatic life.
Decontamination Procedures: Wash away any material with copious amounts of soap and water.
First Aid Procedures: Remove victim to fresh air, call emergency medical care. If not breathing give CPR. If breathing is difficult administer oxygen. Treat for shock.

Fire Hazards:

Flashpoint: 53° F(11.6° C) *Ignition temperature:* 750° F(398.8° C)
Flammable Explosive High Range: 12.7 *Low Range:* 2
Toxic Products of Combustion: n/a
Other Hazards: Flashback along vapor trail may occur. Containers may explode if ignited in an enclosed area.
Possible extinguishing agents: Apply water from as far a distance as possible to cool containers. Use alcohol foam, dry chemical, or carbon dioxide.

Reactivity Hazards:

Reactive With: n/a *Other Reactions:* n/a

Corrosivity Hazards:

Corrosive With: n/a *Neutralizing Agent:* n/a

Radioactivity Hazards:

Radiation Emitted: n/a *Other Hazards:* n/a

Recommended Protection for Response Personnel:

Avoid breathing vapors, keep upwind. Structural protective clothing provides limited protection. Wash away any material which may have come into contact with the body with copious amounts of soap and water. If the fire becomes uncontrollable, consider appropriate evacuation.

ISOPROPYLAMINE

DOT Number: UN 1221 *DOT Hazard Class:* Flammable liquid *DOT Guide Number:* 68
Synonyms: monoisopropylamine
STCC Number: 4908194 *Reportable Qty:* n/a
Mfg Name: Penwald Corporation *Phone No:* 1-215-587-7000

Physical Description:

Physical Form: Liquid *Color:* Colorless *Odor:* Ammonia like
Other Information: It is used to make other chemicals, and as a solvent.

Chemical Properties:

Specific Gravity: .7 *Vapor Density:* 2 *Boiling Point: 89° F(31.6° C)*
Melting Point: −139° F(−95° C) Vapor Pressure: 478 mm Hg at 68° F(20° C) *Solubility in water:* Yes
Other Information: Lighter than and soluble in water. Vapors are heavier than air.

Health Hazards:

Inhalation Hazard: Irritating to the nose and throat.
Ingestion Hazard: n/a
Absorption Hazard: Irritating to the skin and the eyes.
Hazards to Wildlife: n/a
Decontamination Procedures: Wash away any material with copious amounts of soap and water.
First Aid Procedures: Remove victim to fresh air, call emergency medical care. If not breathing give CPR. If breathing is difficult administer oxygen. Treat for shock.

Fire Hazards:

Flashpoint: −35° F(−37.2° C) *Ignition temperature:* 756° F(402.2° C)
Flammable Explosive High Range: 10.4 *Low Range:* 2
Toxic Products of Combustion: Toxic oxides of nitrogen are produced.
Other Hazards: Flashback along vapor trail may occur. Containers may explode if ignited in an enclosed area.
Possible extinguishing agents: Apply water from as far a distance as possible to cool containers. Use alcohol foam, dry chemical, or carbon dioxide.

Reactivity Hazards:

Reactive With: n/a *Other Reactions:* n/a

Corrosivity Hazards:

Corrosive With: n/a *Neutralizing Agent:* n/a

Radioactivity Hazards:

Radiation Emitted: n/a *Other Hazards:* n/a

Recommended Protection for Response Personnel:

Avoid breathing vapors, keep upwind. Wear a sealed chemical suit (butyl rubber). Wash away any material which may have come into contact with the body with copious amounts of soap and water. If the fire becomes uncontrollable, consider appropriate evacuation.

ISOPROPYL ETHER

DOT Number: UN 1159 *DOT Hazard Class:* Flammable liquid *DOT Guide Number:* 26
Synonyms: diisopropyl ether
STCC Number: n/a *Reportable Qty:* n/a
Mfg Name: Shell Chemical *Phone No:* 1-713-241-6161

Physical Description:

Physical Form: Liquid *Color:* Colorless *Odor:* Sweet
Other Information: n/a

Chemical Properties:

Specific Gravity: .7 *Vapor Density:* 3.5 *Boiling Point: 156° F(68.8° C)*
Melting Point: –121° F(–85° C) *Vapor Pressure:* 119 mm Hg at 68° F(20° C)
Solubility in water: Very slight *Degree of Solubility: .2%*
Other Information: n/a

Health Hazards:

Inhalation Hazard: May cause headache and dizziness.
Ingestion Hazard: n/a
Absorption Hazard: Irritating to the skin and eyes.
Hazards to Wildlife: n/a
Decontamination Procedures: Wash away any material with copious amounts of soap and water.
First Aid Procedures: Remove victim to fresh air, call emergency medical care. If not breathing give CPR. If breathing is difficult administer oxygen. Treat for shock.

Fire Hazards:

Flashpoint: –18° F(–27.7° C) *Ignition temperature:* 830° F(443.3° C)
Flammable Explosive High Range: 7.9 *Low Range:* 1.4
Toxic Products of Combustion: An irritating vapor is produced.
Other Hazards: Flashback along vapor trail may occur. Containers may explode if heated or ignited in an enclosed area.
Possible *Possible extinguishing agents:* Apply water from as far a distance as possible to cool containers. Use alcohol foam, dry chemical, or carbon dioxide.

Reactivity Hazards:

Reactive With: n/a *Other Reactions:* n/a

Corrosivity Hazards:

Corrosive With: n/a *Neutralizing Agent:* n/a

Radioactivity Hazards:

Radiation Emitted: n/a *Other Hazards:* n/a

Recommended Protection for Response Personnel:

Avoid breathing vapors, keep upwind. Wear a sealed chemical suit (polycarbonate, butyl rubber, chlorinated polyethylene, PVC). Wash away any material which may have come into contact with the body with copious amounts of soap and water. If the fire becomes uncontrollable, consider appropriate evacuation.

ISOPROPYL MERCAPTAN

DOT Number: UN 2703 *DOT Hazard Class:* Flammable liquid *DOT Guide Number:* 27
Synonyms: 2-propanethiol, propane-2-thiol
STCC Number: 4908205 *Reportable Qty:* n/a
Mfg Name: Phillips Petroleum *Phone No:* 1-918-661-6600

Physical Description:

Physical Form: Liquid *Color:* White *Odor:* Strong skunk
Other Information: n/a

Chemical Properties:

Specific Gravity: .81 *Vapor Density:* 2.6 *Boiling Point: 127° F(52.7° C)*
Melting Point: n/a *Vapor Pressure:* n/a *Solubility in water:* Yes
Other Information: Floats and mixes on water. Flammable irritating vapor is produced.

Health Hazards:

Inhalation Hazard: Difficulty in breathing, loss of consciousness.
Ingestion Hazard: Will cause nausea and vomiting.
Absorption Hazard: Irritating to the skin and eyes.
Hazards to Wildlife: n/a
Decontamination Procedures: Wash away any material with copious amounts of soap and water.
First Aid Procedures: Remove victim to fresh air, call emergency medical care. If not breathing give CPR. If breathing is difficult administer oxygen. Treat for shock.

Fire Hazards:

Flashpoint: −30° F (−34.4° C) *Ignition temperature:* n/a
Flammable Explosive High Range: n/a *Low Range:* n/a
Toxic Products of Combustion: Irritating sulfur dioxide gases are formed in fires.
Other Hazards: Flashback along vapor trail may occur, vapors may explode if ignited in an enclosed area.
Possible extinguishing agents: Apply water from as far a distance as possible. Use alcohol foam, dry chemical, or carbon dioxide. Do not extinguish the fire unless the flow can be stopped.

Reactivity Hazards:

Reactive With: n/a *Other Reactions:* n/a

Corrosivity Hazards:

Corrosive With: n/a *Neutralizing Agent:* n/a

Radioactivity Hazards:

Radiation Emitted: n/a *Other Hazards:* n/a

Recommended Protection for Response Personnel:

Avoid breathing vapors, keep upwind. Wear a sealed chemical suit (polycarbonate, butyl rubber). Wash away any material which may have come into contact with the body with copious amounts of soap and water. If the fire becomes uncontrollable, consider appropriate evacuation.

ISOPROPYL PERCARBONATE (stabilized)

DOT Number: NA 2134 *DOT Hazard Class:* Organic peroxide *DOT Guide Number:* 52
Synonyms: diisopropyl percarbonate, diisopropyl peroxydicarbonate, isopropyl peroxydicarbonate
STCC Number: 4919550 *Reportable Qty:* n/a
Mfg Name: PPG Industries Corp. *Phone No:* 1-412-434-3131

Physical Description:

Physical Form: Solid *Color:* White *Odor:* Sharp, unpleasant
Other Information: n/a

Chemical Properties:

Specific Gravity: 1.08 *Vapor Density:* 4.1 *Boiling Point: n/a*
Melting Point: 48° F(8.8° C) *Vapor Pressure:* n/a *Solubility in water:* n/a
Other Information: Must be shipped in refrigerated containers.

Health Hazards:

Inhalation Hazard: Difficulty in breathing, coughing.
Ingestion Hazard: Harmful if swallowed.
Absorption Hazard: Irritating to the skin and eyes.
Hazards to Wildlife: n/a
Decontamination Procedures: Wash away any material with copious amounts of soap and water.
First Aid Procedures: Remove victim to fresh air, call emergency medical care. If not breathing give CPR. If breathing is difficult administer oxygen. Treat for shock.

Fire Hazards:

Flashpoint: n/a *Ignition temperature:* n/a
Flammable Explosive High Range: n/a *Low Range:* n/a
Toxic Products of Combustion: Toxic and flammable gases include acetone, isopropyl alcohol, acetaldehyde, and ethane.
Other Hazards: Undergoes autoaccelerative decomposition and may self ignite. Fire is very difficult to extinguish because no air is need for combustion.
Possible extinguishing agents: Apply water from as far a distance as possible. Use alcohol foam, dry chemical, or carbon dioxide. dangerously explosive!!

Reactivity Hazards:

Reactive With: Temperature to self ignite. *Other Reactions:* Metals to form oxygen.

Corrosivity Hazards:

Corrosive With: n/a *Neutralizing Agent:* n/a

Radioactivity Hazards:

Radiation Emitted: n/a *Other Hazards:* n/a

Recommended Protection for Response Personnel:

Avoid breathing vapors, keep upwind. Wear a sealed chemical suit (polycarbonate, butyl rubber). Wash away any material which may have come into contact with the body with copious amounts of soap and water. If the fire becomes uncontrollable, consider appropriate evacuation.

ISOPROPYL PERCARBONATE (unstabilized)

DOT Number: NA 2133 *DOT Hazard Class:* Organic peroxide *DOT Guide Number:* 52
Synonyms: diisopropyl percarbonate, diisopropyl peroxydicarbonate, isopropyl peroxydicarbonate
STCC Number: 4919551 *Reportable Qty:* n/a
Mfg Name: PPG Industries Corp. *Phone No:* 1-412-434-3131

Physical Description:

Physical Form: Solid *Color:* White *Odor:* Sharp, unpleasant
Other Information: n/a

Chemical Properties:

Specific Gravity: 1.08 *Vapor Density:* 4.1 *Boiling Point: n/a*
Melting Point: 48° F(8.8° C) *Vapor Pressure:* n/a *Solubility in water:* n/a
Other Information: Must be shipped in refrigerated containers.

Health Hazards:

Inhalation Hazard: Difficulty in breathing, coughing.
Ingestion Hazard: Harmful if swallowed.
Absorption Hazard: Irritating to the skin and eyes.
Hazards to Wildlife: n/a
Decontamination Procedures: Wash away any material with copious amounts of soap and water.
First Aid Procedures: Remove victim to fresh air, call emergency medical care. If not breathing give CPR. If breathing is difficult administer oxygen. Treat for shock.

Fire Hazards:

Flashpoint: n/a *Ignition temperature:* n/a
Flammable Explosive High Range: n/a *Low Range:* n/a
Toxic Products of Combustion: Toxic and flammable gases include acetone, isopropyl alcohol, acetaldehyde, and ethane.
Other Hazards: Undergoes autoaccelerative decomposition and may self ignite. Fire is very difficult to extinguish because no air is need for combustion.
Possible extinguishing agents: Apply water from as far a distance as possible. Use alcohol foam, dry chemical, or carbon dioxide. Dangerously explosive!!

Reactivity Hazards:

Reactive With: Temperature to self ignite. *Other Reactions:* Metals to form oxygen.

Corrosivity Hazards:

Corrosive With: n/a *Neutralizing Agent:* n/a

Radioactivity Hazards:

Radiation Emitted: n/a *Other Hazards:* n/a

Recommended Protection for Response Personnel:

Avoid breathing vapors, keep upwind. Wear a sealed chemical suit (polycarbonate, butyl rubber). Wash away any material which may have come into contact with the body with copious amounts of soap and water. If the fire becomes uncontrollable, consider appropriate evacuation.

ISOVALERALDEHYDE

DOT Number: UN 1989 *DOT Hazard Class:* Flammable liquid *DOT Guide Number:* 26
Synonyms: isovaleral, isovaleric aldehyde, 3-methylbutanal, 3-methylbutyraldehyde
STCC Number: n/a *Reportable Qty:* n/a
Mfg Name: Aldrich Chemical *Phone No:* 1-414-273-3850

Physical Description:

Physical Form: Liquid *Color:* Colorless *Odor:* Weak suffocating
Other Information: n/a

Chemical Properties:

Specific Gravity: .78 *Vapor Density:* 2.96 *Boiling Point: 189° F(87.2° C)*
Melting Point: −60° F(−51.1° C) Vapor Pressure: n/a *Solubility in water:* n/a
Other Information: Floats on water. Flammable, irritating vapor is produced.

Health Hazards:

Inhalation Hazard: Headache, nausea, vomiting, difficulty in breathing.
Ingestion Hazard: Harmful if swallowed.
Absorption Hazard: Irritating to the skin and eyes.
Hazards to Wildlife: n/a
Decontamination Procedures: Wash away any material with copious amounts of soap and water.
First Aid Procedures: Remove victim to fresh air, call emergency medical care. If not breathing give CPR. If breathing is difficult administer oxygen. Treat for shock.

Fire Hazards:

Flashpoint: 55° F(12.7° C) *Ignition temperature:* n/a
Flammable Explosive High Range: n/a *Low Range:* n/a
Toxic Products of Combustion: n/a
Other Hazards: Flashback along vapor trail may occur. Vapors may explode if ignited in an enclosed area.
Possible extinguishing agents: Alcohol foam, dry chemical, carbon dioxide. Cool all affected containers with water. Water may be ineffective on fires.

Reactivity Hazards:

Reactive With: n/a *Other Reactions:* n/a

Corrosivity Hazards:

Corrosive With: n/a *Neutralizing Agent:* n/a

Radioactivity Hazards:

Radiation Emitted: n/a *Other Hazards:* n/a

Recommended Protection for Response Personnel:

Avoid breathing vapors, keep upwind. Wear a sealed chemical suit (butyl rubber). Wash away any material which may have come into contact with the body with copious amounts of soap and water. Consider appropriate evacuation.

JET FUELS, JP-1

DOT Number: UN 1223 *DOT Hazard Class:* Flammable liquid *DOT Guide Number:* 27
Synonyms: fuel oil 1, kerosene, kerosene, range oil
STCC Number: n/a *Reportable Qty:* n/a
Mfg Name: Shell Oil Co. *Phone No:* 1-713-241-6161

Physical Description:

Physical Form: Liquid *Color:* Colorless *Odor:* Fuel oil
Other Information: It is used as a jet fuel.

Chemical Properties:

Specific Gravity: .8 *Vapor Density:* 3 *Boiling Point: 392-500° F(200-260° C)*
Melting Point: −45/−50° F(−42/−45 C) *Vapor Pressure:* n/a
Solubility in water: No
Other Information: Floats on water

Health Hazards:

Inhalation Hazard: May cause headache and nausea.
Ingestion Hazard: Harmful if swallowed.
Absorption Hazard: Irritating to the skin and eyes.
Hazards to Wildlife: Dangerous to aquatic life.
Decontamination Procedures: Wash away any material with copious amounts of soap and water.
First Aid Procedures: Remove victim to fresh air, call emergency medical care. If not breathing give CPR. If breathing is difficult administer oxygen. Treat for shock.

Fire Hazards:

Flashpoint: 100° F(37.7° C) *Ignition temperature:* 444° F(228.8° C)
Flammable Explosive High Range: 5 *Low Range:* .7
Toxic Products of Combustion: n/a
Other Hazards: Flashback along vapor trail may occur. Containers may explode if ignited or heated in an enclosed area.
Possible extinguishing agents: Apply water from as far a distance as possible to cool containers. Use foam, dry chemical, or carbon dioxide.

Reactivity Hazards:

Reactive With: n/a *Other Reactions:* n/a

Corrosivity Hazards:

Corrosive With: n/a *Neutralizing Agent:* n/a

Radioactivity Hazards:

Radiation Emitted: n/a *Other Hazards:* n/a

Recommended Protection for Response Personnel:

Avoid breathing vapors, keep upwind. Structural protective clothing provides limited protection. Wash away any material which may have come into contact with the body with copious amounts of soap and water. If the fire becomes uncontrollable, consider appropriate evacuation.

JET FUELS, JP-3

DOT Number: UN 1223 *DOT Hazard Class:* Flammable liquid *DOT Guide Number:* 27
Synonyms: none given
STCC Number: n/a *Reportable Qty:* n/a
Mfg Name: Shell Oil Co. *Phone No:* 1-713-241-6161

Physical Description:

Physical Form: Liquid *Color:* Colorless *Odor:* Fuel oil
Other Information: It is used as a jet fuel.

Chemical Properties:

Specific Gravity: .8 *Vapor Density:* 3 *Boiling Point: 86 to 500° F(30-260° C)*
Melting Point: n/a *Vapor Pressure:* n/a *Solubility in water:* No
Other Information: Floats on water. Irritating, combustible vapor is produced.

Health Hazards:

Inhalation Hazard: Headache, coughing, difficulty in breathing, dizziness.
Ingestion Hazard: Harmful if swallowed.
Absorption Hazard: Irritating to the skin and eyes.
Hazards to Wildlife: Dangerous to aquatic life.
Decontamination Procedures: Wash away any material with copious amounts of soap and water.
First Aid Procedures: Remove victim to fresh air, call emergency medical care. If not breathing give CPR. If breathing is difficult administer oxygen. Treat for shock.

Fire Hazards:

Flashpoint: 110 to 150° F(43-65° C) *Ignition temperature:* n/a
Flammable Explosive High Range: n/a *Low Range:* n/a
Toxic Products of Combustion: n/a
Other Hazards: Flashback along vapor trail may occur. Containers may explode if ignited or heated in an enclosed area.
Possible extinguishing agents: Apply water from as far a distance as possible to cool containers. Use foam, dry chemical, or carbon dioxide.

Reactivity Hazards:

Reactive With: n/a *Other Reactions:* n/a

Corrosivity Hazards:

Corrosive With: n/a *Neutralizing Agent:* n/a

Radioactivity Hazards:

Radiation Emitted: n/a *Other Hazards:* n/a

Recommended Protection for Response Personnel:

Avoid breathing vapors, keep upwind. Structural protective clothing provides limited protection. Wash away any material which may have come into contact with the body with copious amounts of soap and water. If the fire becomes uncontrollable, consider appropriate evacuation.

JET FUELS, JP-4

DOT Number: UN 1863
DOT Hazard Class: Flammable liquid
DOT Guide Number: n/a
Synonyms: none given
STCC Number: n/a
Reportable Qty: n/a
Mfg Name: Shell Oil Co.
Phone No: 1-713-241-6161

Physical Description:

Physical Form: Liquid
Color: Colorless
Odor: Fuel oil
Other Information: Floats on water.

Chemical Properties:

Specific Gravity: .81
Vapor Density: n/a
Boiling Point: 349-549° F(176-287° C) at 1 atm.
Melting Point: n/a
Vapor Pressure: n/a
Solubility in water: No
Other Information: n/a

Health Hazards:

Inhalation Hazard: May cause headache and nausea.
Ingestion Hazard: Harmful if swallowed.
Absorption Hazard: Irritating to the skin and eyes.
Hazards to Wildlife: Dangerous to aquatic life.
Decontamination Procedures: Wash away any material with copious amounts of soap and water.
First Aid Procedures: Remove victim to fresh air, call emergency medical care. If not breathing give CPR. If breathing is difficult administer oxygen. Treat for shock.

Fire Hazards:

Flashpoint: –10/+30° F(–23/–1° C)
Ignition temperature: 464° F(240° C)
Flammable Explosive High Range: 8 *Low Range:* 1.3
Toxic Products of Combustion: n/a
Other Hazards: Flashback along vapor trail may occur. Containers may explode if ignited or heated in an enclosed area.
Possible extinguishing agents: Apply water from as far a distance as possible to cool containers. Use alcohol foam, dry chemical, or carbon dioxide.

Reactivity Hazards:

Reactive With: n/a
Other Reactions: n/a

Corrosivity Hazards:

Corrosive With: n/a
Neutralizing Agent: n/a

Radioactivity Hazards:

Radiation Emitted: n/a
Other Hazards: n/a

Recommended Protection for Response Personnel:

Avoid breathing vapors, keep upwind. Structural protective clothing provides limited protection. Wash away any material which may have come into contact with the body with copious amounts of soap and water. If the fire becomes uncontrollable, consider appropriate evacuation.

JET FUELS, JP-5

DOT Number: UN 2761 *DOT Hazard Class:* Flammable liquid *DOT Guide Number:* 55
Synonyms: kerosene heavy
STCC Number: n/a *Reportable Qty:* n/a
Mfg Name: Shell Oil Co. *Phone No:* 1-713-241-6161

Physical Description:

Physical Form: Liquid *Color:* Colorless *Odor:* Fuel oil
Other Information: It is used as a jet fuel.

Chemical Properties:

Specific Gravity: .82 *Vapor Density:* 3 *Boiling Point: 349/549° F(176/287° C)*
Melting Point: −54° F(−47.7° C) Vapor Pressure: n/a *Solubility in water:* No
Other Information: Floats on water.

Health Hazards:

Inhalation Hazard: n/a
Ingestion Hazard: Harmful if swallowed.
Absorption Hazard: Irritating to the skin and eyes.
Hazards to Wildlife: Dangerous to aquatic life.
Decontamination Procedures: Wash away any material with copious amounts of soap and water.
First Aid Procedures: Remove victim to fresh air, call emergency medical care. If not breathing give CPR. If breathing is difficult administer oxygen. Treat for shock.

Fire Hazards:

Flashpoint: 140° F(60° C) *Ignition temperature:* 475° F(246.1° C)
Flammable Explosive High Range: 4.6 *Low Range:* .6
Toxic Products of Combustion: n/a
Other Hazards: Flashback along vapor trail may occur. Containers may explode if ignited or heated in an enclosed area.
Possible *Possible extinguishing agents:* Apply water from as far a distance as possible to cool containers. Use foam, dry chemical, or carbon dioxide.

Reactivity Hazards:

Reactive With: n/a *Other Reactions:* n/a

Corrosivity Hazards:

Corrosive With: n/a *Neutralizing Agent:* n/a

Radioactivity Hazards:

Radiation Emitted: n/a *Other Hazards:* n/a

Recommended Protection for Response Personnel:

Avoid breathing vapors, keep upwind. Structural protective clothing provides limited protection. Wash away any material which may have come into contact with the body with copious amounts of soap and water. If the fire becomes uncontrollable, consider appropriate evacuation.

KEPONE

DOT Number: NA 2761 *DOT Hazard Class:* ORM-E *DOT Guide Number:* 55
Synonyms: Chlordecone, GC-1189, ENT-16391, Merx,
STCC Number: 4960140 *Reportable Qty:* 1/.0454
Mfg Name: Allied Corp. *Phone No:* 1-201-455-2000

Physical Description:

Physical Form: Solid *Color:* Colorless *Odor:* Odorless
Other Information: In case of damage to, or leaking from these containers, contact the Pesticide Safety Team Network at 1-800-424-9300.

Chemical Properties:

Specific Gravity: n/a *Vapor Density:* n/a *Boiling Point: Decomposes*
Melting Point: n/a *Vapor Pressure:* n/a *Solubility in water:* Yes
Other Information: Avoid contact with solid or dust, keep people away.

Health Hazards:

Inhalation Hazard: Poisonous if inhaled.
Ingestion Hazard: Poisonous if swallowed.
Absorption Hazard: Poisonous if the skin is exposed.
Hazards to Wildlife: Dangerous to aquatic life.
Decontamination Procedures: Wash away any material with copious amounts of soap and water.
First Aid Procedures: Remove victim to fresh air, call emergency medical care. If not breathing give CPR. If breathing is difficult administer oxygen. Treat for shock.

Fire Hazards:

Flashpoint: n/a *Ignition temperature:* n/a
Flammable Explosive High Range: n/a *Low Range:* n/a
Toxic Products of Combustion: n/a
Other Hazards: n/a
Possible extinguishing agents: Extinguish fire using suitable agent for the type of surrounding fire.

Reactivity Hazards:

Reactive With: n/a *Other Reactions:* n/a

Corrosivity Hazards:

Corrosive With: n/a *Neutralizing Agent:* n/a

Radioactivity Hazards:

Radiation Emitted: n/a *Other Hazards:* n/a

Recommended Protection for Response Personnel:

Avoid breathing vapors, keep upwind. Structural protective clothing provides limited protection. Wash away any material which may have come into contact with the body with copious amounts of soap and water. Consider appropriate evacuation.

KEROSENE

DOT Number: UN 1223 *DOT Hazard Class:* Combustible liquid *DOT Guide Number:* 27
Synonyms: fuel oil number 1, Jet fuel-JP-1, range oil
STCC Number: 4915171 *Reportable Qty:* n/a
Mfg Name: Shell Oil Co. *Phone No:* 1-713-241-6161

Physical Description:

Physical Form: Liquid *Color:* Colorless *Odor:* Fuel oil
Other Information: n/a

Chemical Properties:

Specific Gravity: .8 *Vapor Density:* 4.1 *Boiling Point: 304-574° F(151-301° C)*
Melting Point: n/a *Vapor Pressure:* n/a *Solubility in water:* Insoluble
Other Information: Floats on watcr.

Health Hazards:

Inhalation Hazard: n/a
Ingestion Hazard: Harmful if swallowed.
Absorption Hazard: Irritating to the skin and eyes.
Hazards to Wildlife: Dangerous to aquatic life.
Decontamination Procedures: Wash away any material with copious amounts of soap and water.
First Aid Procedures: Remove victim to fresh air, call emergency medical care. If not breathing give CPR. If breathing is difficult administer oxygen. Treat for shock.

Fire Hazards:

Flashpoint: 100° F(37.7° C) *Ignition temperature:* 444° F(228° C)
Flammable Explosive High Range: 5 *Low Range:* .7
Toxic Products of Combustion: n/a
Other Hazards: n/a
Possible extinguishing agents: Do not extinguish the fire unless the flow can be stopped. Apply water from as far a distance as possible. use foam, dry chemical, or carbon dioxide.

Reactivity Hazards:

Reactive With: n/a *Other Reactions:* n/a

Corrosivity Hazards:

Corrosive With: n/a *Neutralizing Agent:* n/a

Radioactivity Hazards:

Radiation Emitted: n/a *Other Hazards:* n/a

Recommended Protection for Response Personnel:

Avoid breathing vapors, keep upwind. Structural protective clothing provides limited protection. Wash away any material which may have come into contact with the body with copious amounts of soap and water. Consider appropriate evacuation.

LACTIC ACID

DOT Number: NA 1760 *DOT Hazard Class:* Corrosive *DOT Guide Number:* 60
Synonyms: 2-hydroxypropanoic acid, α-hydroxypropionic acid, racemic lactic acid, milk acid
STCC Number: 4931757 *Reportable Qty:* n/a
Mfg Name: Monsanto Company *Phone No:* 1-314-694-1000

Physical Description:

Physical Form: Liquid *Color:* Colorless to yellow *Odor:* Weak, unpleasant
Other Information: It is used to make cultured dairy products, as a food preservative, and to make chemicals.

Chemical Properties:

Specific Gravity: 1.2 *Vapor Density:* 3.1 *Boiling Point: n/a*
Melting Point: n/a *Vapor Pressure:* n/a *Solubility in water:* Yes
Other Information: Sinks and mixes with water.

Health Hazards:

Inhalation Hazard: Causes coughing or difficulty in breathing.
Ingestion Hazard: Causes nausea.
Absorption Hazard: Will burn the skin and eyes.
Hazards to Wildlife: Dangerous to aquatic life.
Decontamination Procedures: Wash away any material with copious amounts of soap and water.
First Aid Procedures: Remove victim to fresh air, call emergency medical care. If not breathing give CPR. If breathing is difficult administer oxygen. Treat for shock.

Fire Hazards:

Flashpoint: n/a *Ignition temperature:* n/a
Flammable Explosive High Range: n/a *Low Range:* n/a
Toxic Products of Combustion: n/a
Other Hazards: Burns with difficulty.
Possible *Possible extinguishing agents:* Apply water from as far a distance as possible to cool containers. Use alcohol foam, dry chemical, or carbon dioxide.

Reactivity Hazards:

Reactive With: n/a *Other Reactions:* n/a

Corrosivity Hazards:

Corrosive With: Metal and tissue. *Neutralizing Agent:* Crushed limestone, soda ash, or lime.

Radioactivity Hazards:

Radiation Emitted: n/a *Other Hazards:* n/a

Recommended Protection for Response Personnel:

Avoid breathing vapors, keep upwind. Structural protective clothing provides limited protection. Wash away any material which may have come into contact with the body with copious amounts of soap and water. If the fire becomes uncontrollable, consider appropriate evacuation.

LATEX, LIQUID SYNTHETIC

DOT Number: n/a *DOT Hazard Class:* n/a *DOT Guide Number:* n/a
Synonyms: plastic latex, poly dimethyl siloxane, synthetic rubber latex,
STCC Number: n/a *Reportable Qty:* n/a
Mfg Name: W.R. Grace & Co. *Phone No:* 1-617-861-6000

Physical Description:

Physical Form: Liquid *Color:* White *Odor:* Characteristic
Other Information: n/a

Chemical Properties:

Specific Gravity: 1.06 *Vapor Density:* n/a *Boiling Point: Very high*
Melting Point: n/a *Vapor Pressure:* n/a *Solubility in water:* Yes
Other Information: Mixcs with water.

Health Hazards:

Inhalation Hazard: n/a
Ingestion Hazard: n/a
Absorption Hazard: Irritating to the eyes.
Hazards to Wildlife: n/a
Decontamination Procedures: Wash away any material with copious amounts of soap and water.
First Aid Procedures: Remove victim to fresh air, call emergency medical care. If not breathing give CPR. If breathing is difficult administer oxygen. Treat for shock.

Fire Hazards:

Flashpoint: n/a *Ignition temperature:* n/a
Flammable Explosive High Range: n/a *Low Range:* n/a
Toxic Products of Combustion: If latex dries out and then burns, hydrochloric acid, hydrogen cyanide and styrene gases may be generated. All are irritating and poisonous.
Other Hazards: Heat may coagulate the latex and form sticky plastic lumps which may burn.
Possible extinguishing agents: Extinguish fire using suitable agent for the type of surrounding fire.

Reactivity Hazards:

Reactive With: Heat and acids to a gummy, flammable material. *Other Reactions:* n/a

Corrosivity Hazards:

Corrosive With: n/a *Neutralizing Agent:* n/a

Radioactivity Hazards:

Radiation Emitted: n/a *Other Hazards:* n/a

Recommended Protection for Response Personnel:

Avoid breathing vapors, keep upwind. Structural protective clothing provides limited protection. Wash away any material which may have come into contact with the body with copious amounts of soap and water. If the fire becomes uncontrollable, consider appropriate evacuation.

LAUROYL PEROXIDE

DOT Number: UN 2124 *DOT Hazard Class:* Organic peroxide *DOT Guide Number:* 48
Synonyms: dilauroyl peroxide
STCC Number: 4919560 *Reportable Qty:* n/a
Mfg Name: Witco Chemical Corp. *Phone No:* 1-212-605-3800

Physical Description:

Physical Form: Solid *Color:* White *Odor:* Faint soapy
Other Information: It is shipped either dry or 30% water wet.

Chemical Properties:

Specific Gravity: .91 *Vapor Density:* n/a *Boiling Point: Decomposes*
Melting Point: n/a *Vapor Pressure:* n/a *Solubility in water:* Yes
Other Information: Wet or dry supports combustion of burning materials.

Health Hazards:

Inhalation Hazard: n/a
Ingestion Hazard: Harmful if swallowed.
Absorption Hazard: Irritating to the skin and eyes.
Hazards to Wildlife: n/a
Decontamination Procedures: Wash away any material with copious amounts of soap and water.
First Aid Procedures: Remove victim to fresh air, call emergency medical care. If not breathing give CPR. If breathing is difficult administer oxygen. Treat for shock.

Fire Hazards:

Flashpoint: n/a *Ignition temperature:* n/a
Flammable Explosive High Range: 8 *Low Range:* 1.3
Toxic Products of Combustion: n/a
Other Hazards: May cause fire upon contact with combustibles, containers may explode in fire.
Possible extinguishing agents: Apply water from as far a distance as possible to cool containers. Use alcohol foam, dry chemical, or carbon dioxide.

Reactivity Hazards:

Reactive With: Combustible materials *Other Reactions:* Heat

Corrosivity Hazards:

Corrosive With: n/a *Neutralizing Agent:* n/a

Radioactivity Hazards:

Radiation Emitted: n/a *Other Hazards:* n/a

Recommended Protection for Response Personnel:

Avoid breathing vapors, keep upwind. Wear a sealed chemical suit (polycarbonate, butyl rubber). Wash away any material which may have come into contact with the body with copious amounts of soap and water. If the fire becomes uncontrollable, consider appropriate evacuation.

LAUROYL PEROXIDE (less than 42%)

DOT Number: UN 2893 *DOT Hazard Class:* Organic peroxide *DOT Guide Number:* 48
Synonyms: dilauroyl peroxide
STCC Number: 4919167 *Reportable Qty:* n/a
Mfg Name: Witco Chemical Corp. *Phone No:* 1-212-605-3800

Physical Description:

Physical Form: Liquid *Color:* Clear *Odor:* Faint soapy
Other Information: It is used in the manufacture of paints, plastics, and rubber.

Chemical Properties:

Specific Gravity: n/a *Vapor Density:* n/a *Boiling Point: n/a*
Melting Point: n/a *Vapor Pressure:* n/a *Solubility in water:* No
Other Information: Wet or dry supports combustion of burning materials.

Health Hazards:

Inhalation Hazard: n/a
Ingestion Hazard: Harmful if swallowed.
Absorption Hazard: Irritating to the skin and eyes.
Hazards to Wildlife: n/a
Decontamination Procedures: Wash away any material with copious amounts of soap and water.
First Aid Procedures: Remove victim to fresh air, call emergency medical care. If not breathing give CPR. If breathing is difficult administer oxygen. Treat for shock.

Fire Hazards:

Flashpoint: n/a *Ignition temperature:* n/a
Flammable Explosive High Range: n/a *Low Range:* n/a
Toxic Products of Combustion: n/a
Other Hazards: May cause fire upon contact with combustibles. Containers may explode in fire.
Possible extinguishing agents: Apply water from as far a distance as possible to cool containers. Use alcohol foam, dry chemical, or carbon dioxide.

Reactivity Hazards:

Reactive With: Combustible materials *Other Reactions:* Heat

Corrosivity Hazards:

Corrosive With: n/a *Neutralizing Agent:* n/a

Radioactivity Hazards:

Radiation Emitted: n/a *Other Hazards:* n/a

Recommended Protection for Response Personnel:

Avoid breathing vapors, keep upwind. Wear a sealed chemical suit (polycarbonate, butyl rubber). Wash away any material which may have come into contact with the body with copious amounts of soap and water. If the fire becomes uncontrollable, consider appropriate evacuation.

LAURYL MERCAPTAN

DOT Number: n/a *DOT Hazard Class:* n/a *DOT Guide Number:* n/a
Synonyms: 1-dodecanethiol, dodecyl mercaptan
STCC Number: n/a *Reportable Qty:* n/a
Mfg Name: Phillips Petroleum *Phone No:* 1-718-661-6600

Physical Description:

Physical Form: Liquid *Color:* Colorless *Odor:* Mild skunk
Other Information: n/a

Chemical Properties:

Specific Gravity: .85 *Vapor Density:* 6.9 *Boiling Point: Very high*
Melting Point: 19° F(7.2° C) *Vapor Pressure:* n/a *Solubility in water:* n/a
Other Information: Floats on water

Health Hazards:

Inhalation Hazard: n/a
Ingestion Hazard: Nausea
Absorption Hazard: Irritating to the skin and eyes.
Hazards to Wildlife: n/a
Decontamination Procedures: Wash away any material with copious amounts of soap and water.
First Aid Procedures: Remove victim to fresh air, call emergency medical care. If not breathing give CPR. If breathing is difficult administer oxygen. Treat for shock.

Fire Hazards:

Flashpoint: 262° F(127.7° C) *Ignition temperature:* n/a
Flammable Explosive High Range: n/a *Low Range:* n/a
Toxic Products of Combustion: Poisonous and irritating gases (e.g. sulfur dioxide) may be formed in fires.
Other Hazards: n/a
Possible *Possible extinguishing agents:* Apply water from as far a distance as possible to cool containers. Use alcohol foam, dry chemical, or carbon dioxide.

Reactivity Hazards:

Reactive With: n/a *Other Reactions:* n/a

Corrosivity Hazards:

Corrosive With: n/a *Neutralizing Agent:* n/a

Radioactivity Hazards:

Radiation Emitted: n/a *Other Hazards:* n/a

Recommended Protection for Response Personnel:

Avoid breathing vapors, keep upwind. Wear a sealed chemical suit (polycarbonate, butyl rubber). Wash away any material which may have come into contact with the body with copious amounts of soap and water. If the fire becomes uncontrollable, consider appropriate evacuation.

LEAD ACETATE

DOT Number: NA 1616 *DOT Hazard Class:* ORM-E *DOT Guide Number:* 53
Synonyms: natural lead acetate, salt of saturn, sugar of lead
STCC Number: 4966640 *Reportable Qty:* 5000/2270
Mfg Name: Mallinckrodt Inc. *Phone No:* 1-314-895-2000

Physical Description:

Physical Form: Solid *Color:* White *Odor:* Odorless
Other Information: It is used to make lead compounds, in chemical analysis, as an insecticide, and for many other uses.

Chemical Properties:

Specific Gravity: 2.55 *Vapor Density:* *Boiling Point: Decomposes*
Melting Point: n/a *Vapor Pressure:* n/a *Solubility in water:* Yes
Other Information: Sinks and mixes with water.

Health Hazards:

Inhalation Hazard: Poisonous if inhaled.
Ingestion Hazard: Nausea, vomiting, loss of consciousness.
Absorption Hazard: Irritating to the skin and the eyes.
Hazards to Wildlife: Dangerous to aquatic life.
Decontamination Procedures: Wash away any material with copious amounts of soap and water.
First Aid Procedures: Remove victim to fresh air, call emergency medical care. If not breathing give CPR. If breathing is difficult administer oxygen. Treat for shock.

Fire Hazards:

Flashpoint: n/a *Ignition temperature:* n/a
Flammable Explosive High Range: n/a *Low Range:* n/a
Toxic Products of Combustion: Irritating acid fumes may be formed in fires.
Other Hazards: n/a
Possible extinguishing agents: Extinguish fire using suitable agent for the type of surrounding fire.

Reactivity Hazards:

Reactive With: n/a *Other Reactions:* n/a

Corrosivity Hazards:

Corrosive With: n/a *Neutralizing Agent:* n/a

Radioactivity Hazards:

Radiation Emitted: n/a *Other Hazards:* n/a

Recommended Protection for Response Personnel:

Avoid breathing vapors, keep upwind. Wear a sealed chemical suit (polycarbonate, butyl rubber, PVC, chlorinated polyethylene). Wash away any material which may have come into contact with the body with copious amounts of soap and water. Consider appropriate evacuation.

LEAD ARSENATE

DOT Number: UN 1617 *DOT Hazard Class:* Poison B *DOT Guide Number:* 53
Synonyms: acid plumbous arsenate
STCC Number: 4923318 *Reportable Qty:* 5000/2270
Mfg Name: Chevron Chemical *Phone No:* 1-415-233-3737

Physical Description:

Physical Form: Solid *Color:* Yellow white to yellow *Odor:* Odorless
Other Information: n/a

Chemical Properties:

Specific Gravity: 5.79 *Vapor Density:* 11 *Boiling Point: n/a*
Melting Point: n/a *Vapor Pressure:* n/a *Solubility in water:* Insoluble
Other Information: n/a

Health Hazards:

Inhalation Hazard: Poisonous if inhaled.
Ingestion Hazard: Poisonous if swallowed.
Absorption Hazard: Poisonous if absorbed.
Hazards to Wildlife: Dangerous to aquatic life.
Decontamination Procedures: Wash away any material with copious amounts of soap and water.
First Aid Procedures: Remove victim to fresh air, call emergency medical care. If not breathing give CPR. If breathing is difficult administer oxygen. Treat for shock.

Fire Hazards:

Flashpoint: n/a *Ignition temperature:* n/a
Flammable Explosive High Range: n/a *Low Range:* n/a
Toxic Products of Combustion: n/a
Other Hazards: n/a
Possible extinguishing agents: Extinguish fire using suitable agent for type of surrounding fire.

Reactivity Hazards:

Reactive With: n/a *Other Reactions:* n/a

Corrosivity Hazards:

Corrosive With: n/a *Neutralizing Agent:* n/a

Radioactivity Hazards:

Radiation Emitted: n/a *Other Hazards:* n/a

Recommended Protection for Response Personnel:

Avoid breathing vapors, keep upwind. Wear a sealed chemical suit (polycarbonate, butyl rubber). Wash away any material which may have come into contact with the body with copious amounts of soap and water. Consider appropriate evacuation.

LEAD CHLORIDE

DOT Number: NA 2291 *DOT Hazard Class:* ORM-B *DOT Guide Number:* 53
Synonyms: lead(II) chloride, lead dichloride
STCC Number: 4944130 *Reportable Qty:* 100/45.4
Mfg Name: J.T. Baker Chemical *Phone No:* 1-210-859-2151

Physical Description:

Physical Form: Solid *Color:* White *Odor:* n/a
Other Information: It is used to make other lead compounds, and in analytic chemistry.

Chemical Properties:

Specific Gravity: 5.85 *Vapor Density:* 9.6 *Boiling Point: 1742° F(950° C)*
Melting Point: n/a *Vapor Pressure:* 1 mm Hg at 1016° F(546.6° C) *Solubility in water:* Yes
Other Information: Insoluble in cold water, soluble in hot water. Sinks in water.

Health Hazards:

Inhalation Hazard: Poisonous if inhaled.
Ingestion Hazard: Causes metallic taste, abdominal pain, vomiting, and diarrhea.
Absorption Hazard: n/a
Hazards to Wildlife: Dangerous to aquatic life.
Decontamination Procedures: Wash away any material with copious amounts of soap and water.
First Aid Procedures: Remove victim to fresh air, call emergency medical care. If not breathing give CPR. If breathing is difficult administer oxygen. Treat for shock.

Fire Hazards:

Flashpoint: n/a *Ignition temperature:* n/a
Flammable Explosive High Range: n/a *Low Range:* n/a
Toxic Products of Combustion: Toxic metal fumes
Other Hazards: n/a
Possible extinguishing agents: Extinguish fire using suitable agent for the type of surrounding fire.

Reactivity Hazards:

Reactive With: n/a *Other Reactions:* n/a

Corrosivity Hazards:

Corrosive With: Aluminum *Neutralizing Agent:* n/a

Radioactivity Hazards:

Radiation Emitted: n/a *Other Hazards:* n/a

Recommended Protection for Response Personnel:

Avoid breathing vapors, keep upwind. Wear appropriate chemical clothing, Wash away any material which may come into contact with the body with copious amounts of soap and water. Consider appropriate evacuation.

LEAD FLUORIDE

DOT Number: NA 2811 *DOT Hazard Class:* ORM-B *DOT Guide Number:* 53
Synonyms: lead difluoride, plumbous fluoride
STCC Number: 4944140 *Reportable Qty:* 1000/454
Mfg Name: A.D. MacKay Inc. *Phone No:* 1-203-655-3000

Physical Description:

Physical Form: Solid *Color:* White *Odor:* Odorless
Other Information: It is used in electronics, and in making optical glasses.

Chemical Properties:

Specific Gravity: 8.24 *Vapor Density:* 8.4 *Boiling Point: n/a*
Melting Point: n/a *Vapor Pressure:* 10 mm Hg at 1659° F(903.8° C) *Solubility in water:* Insoluble
Other Information: Sinks in water.

Health Hazards:

Inhalation Hazard: Poisonous if inhaled.
Ingestion Hazard: Cause nausea, vomiting, loss of consciousness.
Absorption Hazard: Irritating to the skin and eyes.
Hazards to Wildlife: Dangerous to aquatic life.
Decontamination Procedures: Wash away any material with copious amounts of soap and water.
First Aid Procedures: Remove victim to fresh air, call emergency medical care. If not breathing give CPR. If breathing is difficult administer oxygen. Treat for shock.

Fire Hazards:

Flashpoint: n/a *Ignition temperature:* n/a
Flammable Explosive High Range: n/a *Low Range:* n/a
Toxic Products of Combustion: n/a
Other Hazards: n/a
Possible extinguishing agents: Extinguish fire using suitable agent for the type of surrounding fire.

Reactivity Hazards:

Reactive With: n/a *Other Reactions:* n/a

Corrosivity Hazards:

Corrosive With: Aluminum *Neutralizing Agent:* n/a

Radioactivity Hazards:

Radiation Emitted: n/a *Other Hazards:* n/a

Recommended Protection for Response Personnel:

Avoid breathing vapors, keep upwind. Wear a sealed chemical suit (polycarbonate, butyl rubber). Wash away any material which may have come into contact with the body with copious amounts of soap and water. Consider appropriate evacuation.

LEAD FLUOROBORATE

DOT Number: UN 2291 *DOT Hazard Class:* ORM-B *DOT Guide Number:* 53
Synonyms: lead fluoborate, lead fluoroborate solution
STCC Number: n/a *Reportable Qty:* n/a
Mfg Name: Allied Corp. *Phone No:* 1-201-445-2000

Physical Description:

Physical Form: Liquid *Color:* Colorless *Odor:* Odorless
Other Information: n/a

Chemical Properties:

Specific Gravity: 1.75 *Vapor Density:* 13 *Boiling Point: n/a*
Melting Point: n/a *Vapor Pressure:* n/a *Solubility in water:* Yes
Other Information: Sinks and mixes with water.

Health Hazards:

Inhalation Hazard: Poisonous if inhaled.
Ingestion Hazard: Nausea, vomiting, loss of consciousness.
Absorption Hazard: Will burn the skin and eyes.
Hazards to Wildlife: n/a
Decontamination Procedures: Wash away any material with copious amounts of soap and water.
First Aid Procedures: Remove victim to fresh air, call emergency medical care. If not breathing give CPR. If breathing is difficult administer oxygen. Treat for shock.

Fire Hazards:

Flashpoint: n/a *Ignition temperature:* n/a
Flammable Explosive High Range: n/a *Low Range:* n/a
Toxic Products of Combustion: Toxic and irritating hydrogen fluoride gas may be formed in fires.
Other Hazards: n/a
Possible extinguishing agents: Extinguish fire using suitable agent for the type of surrounding fire.

Reactivity Hazards:

Reactive With: n/a *Other Reactions:* n/a

Corrosivity Hazards:

Corrosive With: Most metals. *Neutralizing Agent:* Crushed limestone, soda ash, or lime.

Radioactivity Hazards:

Radiation Emitted: n/a *Other Hazards:* n/a

Recommended Protection for Response Personnel:

Avoid breathing vapors, keep upwind. Wear appropriate sealed chemical suit. Wash away any material which may have come into contact with the body with copious amounts of soap and water. Consider appropriate evacuation.

LEAD IODIDE

DOT Number: NA 2811 *DOT Hazard Class:* ORM-E *DOT Guide Number:* 53
Synonyms: none given
STCC Number: 4966950 *Reportable Qty:* 100/45.4
Mfg Name: A.D. MacKay Inc. *Phone No:* 1-203-655-3000

Physical Description:

Physical Form: Solid *Color:* Bright yellow *Odor:* Odorless
Other Information: It is used in printing, photography, to seed clouds, and in many other uses.

Chemical Properties:

Specific Gravity: 6.16 *Vapor Density:* 8.4 *Boiling Point: n/a*
Melting Point: n/a *Vapor Pressure:* n/a *Solubility in water:* Insoluble
Other Information: Sinks and mixes with water.

Health Hazards:

Inhalation Hazard: Poisonous if inhaled.
Ingestion Hazard: Cause nausea, vomiting, loss of consciousness.
Absorption Hazard: Irritating to the skin and eyes.
Hazards to Wildlife: May be toxic to waterfowl.
Decontamination Procedures: Wash away any material with copious amounts of soap and water.
First Aid Procedures: Remove victim to fresh air, call emergency medical care. If not breathing give CPR. If breathing is difficult administer oxygen. Treat for shock.

Fire Hazards:

Flashpoint: n/a *Ignition temperature:* n/a
Flammable Explosive High Range: n/a *Low Range:* n/a
Toxic Products of Combustion: Toxic oxides of nitrogen are produced in fires.
Other Hazards: n/a
Possible extinguishing agents: Extinguish fire using suitable agent for the type of surrounding fire.

Reactivity Hazards:

Reactive With: n/a *Other Reactions:* n/a

Corrosivity Hazards:

Corrosive With: n/a *Neutralizing Agent:* n/a

Radioactivity Hazards:

Radiation Emitted: n/a *Other Hazards:* n/a

Recommended Protection for Response Personnel:

Avoid breathing vapors, keep upwind. Structural protective clothing provides limited protection. Wash away any material which may have come into contact with the body with copious amounts of soap and water. Consider appropriate evacuation.

LEAD NITRATE

DOT Number: UN 1469 *DOT Hazard Class:* Oxidizer *DOT Guide Number:* 42
Synonyms: nitric acid, lead(II) salt
STCC Number: 4918726 *Reportable Qty:* 100/45.4
Mfg Name: J.T. Baker Chemical *Phone No:* 1-201-859-2151

Physical Description:

Physical Form: Solid *Color:* White *Odor:* Odorless
Other Information: n/a

Chemical Properties:

Specific Gravity: 4.53 *Vapor Density:* 11 *Boiling Point: n/a*
Melting Point: n/a *Vapor Pressure:* n/a *Solubility in water:* Yes
Other Information: Will accelerate thc burning of combustible materials.

Health Hazards:

Inhalation Hazard: Poisonous if inhaled.
Ingestion Hazard: Cause nausea or loss of consciousness.
Absorption Hazard: Irritating to the skin and eyes.
Hazards to Wildlife: Dangerous to aquatic life.
Decontamination Procedures: Wash away any material with copious amounts of soap and water.
First Aid Procedures: Remove victim to fresh air, call emergency medical care. If not breathing give CPR. If breathing is difficult administer oxygen. Treat for shock.

Fire Hazards:

Flashpoint: n/a *Ignition temperature:* n/a
Flammable Explosive High Range: n/a *Low Range:* n/a
Toxic Products of Combustion: Toxic oxides of nitrogen are produced in fires containing this material.
Other Hazards: Prolonged exposure of this material to fire or heat may result in an explosion.
Possible extinguishing agents: Apply water from as far a distance as possible.

Reactivity Hazards:

Reactive With: Wood and paper *Other Reactions:* n/a

Corrosivity Hazards:

Corrosive With: n/a *Neutralizing Agent:* n/a

Radioactivity Hazards:

Radiation Emitted: n/a *Other Hazards:* n/a

Recommended Protection for Response Personnel:

Avoid breathing vapors, keep upwind. Structural protective clothing provides limited protection. Wash away any material which may have come into contact with the body with copious amounts of soap and water. If the fire becomes uncontrollable, consider appropriate evacuation.

LEAD STEARATE

DOT Number: NA 2811 *DOT Hazard Class:* ORM-E *DOT Guide Number:* 53
Synonyms: natural lead stearate; stearic acid, lead salt
STCC Number: 4966960 *Reportable Qty:* 5000/2270
Mfg Name: J.T. Baker Chemical *Phone No:* 1-201-859-2151

Physical Description:

Physical Form: Solid *Color:* White *Odor:* Slight fatty
Other Information: It is used as a corrosion inhibitor, as a drier for paints, and in many other uses.

Chemical Properties:

Specific Gravity: 1.3-1.4 *Vapor Density:* 26 *Boiling Point: n/a*
Melting Point: n/a *Vapor Pressure:* n/a *Solubility in water:* Slight
Other Information: Sinks in water.

Health Hazards:

Inhalation Hazard: Poisonous if inhaled.
Ingestion Hazard: Cause headache, abdominal pain, nausea, vomiting.
Absorption Hazard: n/a
Hazards to Wildlife: Dangerous to aquatic life.
Decontamination Procedures: Wash away any material with copious amounts of soap and water.
First Aid Procedures: Remove victim to fresh air, call emergency medical care. If not breathing give CPR. If breathing is difficult administer oxygen. Treat for shock.

Fire Hazards:

Flashpoint: 450° F(232° C) *Ignition temperature:* n/a
Flammable Explosive High Range: n/a *Low Range:* n/a
Toxic Products of Combustion: Toxic fumes are emitted.
Other Hazards: Dust may explode at high temperatures or source of ignition.
Possible extinguishing agents: Extinguish fire using suitable agent for the type of surrounding fire.

Reactivity Hazards:

Reactive With: n/a *Other Reactions:* n/a

Corrosivity Hazards:

Corrosive With: n/a *Neutralizing Agent:* n/a

Radioactivity Hazards:

Radiation Emitted: n/a *Other Hazards:* n/a

Recommended Protection for Response Personnel:

Avoid breathing vapors, keep upwind. Wear appropriate chemical clothing. Wash away any material which may come into contact with the body with copious amounts of soap and water. Consider appropriate evacuation.

LEAD SULFIDE

DOT Number: NA 2991 *DOT Hazard Class:* ORM-E *DOT Guide Number:* 53
Synonyms: galena, plumbous sulfide
STCC Number: 4966987 *Reportable Qty:* 5000/2270 *CHEMTREC Phone No:* 1-800-424-9300

Physical Description:

Physical Form: Solid *Color:* Black or silver *Odor:* Odorless
Other Information: It is used to make ceramics, in the electronic industry, and in many other uses.

Chemical Properties:

Specific Gravity: 7.5 *Vapor Density:* n/a *Boiling Point: 2337.6° F(1281° C)*
Melting Point: n/a *Vapor Pressure:* n/a *Solubility in water:* No
Other Information: Sinks in water.

Health Hazards:

Inhalation Hazard: Poisonous if inhaled.
Ingestion Hazard: Poisonous if swallowed.
Absorption Hazard: Irritating to the skin and eyes.
Hazards to Wildlife: Dangerous to aquatic life.
Decontamination Procedures: Wash away any material with copious amounts of soap and water.
First Aid Procedures: Remove victim to fresh air, call emergency medical care. If not breathing give CPR. If breathing is difficult administer oxygen. Treat for shock.

Fire Hazards:

Flashpoint: n/a *Ignition temperature:* n/a
Flammable Explosive High Range: n/a *Low Range:* n/a
Toxic Products of Combustion: Toxic sulfur oxide fumes are emitted.
Other Hazards: n/a
Possible extinguishing agents: Extinguish fire using suitable agent for the type of surrounding fire.

Reactivity Hazards:

Reactive With: n/a
Other Reactions: n/a

Corrosivity Hazards:

Corrosive With: n/a *Neutralizing Agent:* n/a

Radioactivity Hazards:

Radiation Emitted: n/a *Other Hazards:* n/a

Recommended Protection for Response Personnel:

Avoid breathing vapors, keep upwind. Wear appropriate chemical clothing. Wash away any material which may come into contact with the body with copious amounts of soap and water. Consider appropriate evacuation.

LEAD TETRAACETATE

DOT Number: n/a *DOT Hazard Class:* ORM-E *DOT Guide Number:* n/a
Synonyms: lead(IV) acetate
STCC Number: n/a *Reportable Qty:* n/a
Mfg Name: Ventron Corp. *Phone No:* 1-617-774-3100

Physical Description:

Physical Form: Wet crystals *Color:* Faintly pink *Odor:* Vinegarlike
Other Information: n/a

Chemical Properties:

Specific Gravity: 2.2 *Vapor Density:* n/a *Boiling Point: n/a*
Melting Point: 347° F(175° C) *Vapor Pressure:* n/a *Solubility in water:* Reacts.
Other Information: Reacts with water.

Health Hazards:

Inhalation Hazard: n/a
Ingestion Hazard: Nausea, vomiting, loss of consciousness.
Absorption Hazard: Irritating to the skin and eyes.
Hazards to Wildlife: n/a
Decontamination Procedures: Wash away any material with copious amounts of soap and water.
First Aid Procedures: Remove victim to fresh air, call emergency medical care. If not breathing give CPR. If breathing is difficult administer oxygen. Treat for shock.

Fire Hazards:

Flashpoint: n/a *Ignition temperature:* n/a
Flammable Explosive High Range: n/a *Low Range:* n/a
Toxic Products of Combustion: n/a
Other Hazards: Will increase the intensity of a fire when in contact with combustible material.
Possible extinguishing agents: Extinguish fire using suitable agent for the type of surrounding fire. Cool all affected containers with flooding quantities of water.

Reactivity Hazards:

Reactive With: Water to form lead dioxide and acetic acid. *Other Reactions:* n/a

Corrosivity Hazards:

Corrosive With: Metals when moist. *Neutralizing Agent:* Crushed limestone, soda ash, or lime.

Radioactivity Hazards:

Radiation Emitted: n/a *Other Hazards:* n/a

Recommended Protection for Response Personnel:

Avoid breathing vapors, keep upwind. Wear a sealed chemical suit (polycarbonate, butyl rubber). Wash away any material which may have come into contact with the body with copious amounts of soap and water. Consider appropriate evacuation.

LEAD THIOCYANATE

DOT Number: NA 2291 *DOT Hazard Class:* ORM-E *DOT Guide Number:* 53
Synonyms: lead sulfocyanate
STCC Number: 4966356 *Reportable Qty:* 100/454
Mfg Name: Atomergic Chemetals Corp. *Phone No:* 1-516-349-8800

Physical Description:

Physical Form: Solid *Color:* White *Odor:* Odorless
Other Information: It is used in making explosives, in safety matches, and in dyeing.

Chemical Properties:

Specific Gravity: 3.82 *Vapor Density:* 11 *Boiling Point: n/a*
Melting Point: n/a *Vapor Pressure:* n/a *Solubility in water:* Slightly
Other Information: Sinks and mixes in water.

Health Hazards:

Inhalation Hazard: Poisonous if inhaled.
Ingestion Hazard: Cause nausea, vomiting, loss of consciousness.
Absorption Hazard: Irritating to the skin and eyes.
Hazards to Wildlife: Dangerous to waterfowl.
Decontamination Procedures: Wash away any material with copious amounts of soap and water.
First Aid Procedures: Remove victim to fresh air, call emergency medical care. If not breathing give CPR. If breathing is difficult administer oxygen. Treat for shock.

Fire Hazards:

Flashpoint: n/a *Ignition temperature:* n/a
Flammable Explosive High Range: n/a *Low Range:* n/a
Toxic Products of Combustion: Irritating sulfur dioxide gases may be formed in fires.
Other Hazards: n/a
Possible extinguishing agents: Extinguish fire using suitable agent for the type of surrounding fire.

Reactivity Hazards:

Reactive With: n/a *Other Reactions:* n/a

Corrosivity Hazards:

Corrosive With: n/a *Neutralizing Agent:* n/a

Radioactivity Hazards:

Radiation Emitted: n/a *Other Hazards:* n/a

Recommended Protection for Response Personnel:

Avoid breathing vapors, keep upwind. Structural protective clothing provides limited protection. Wash away any material which may have come into contact with the body with copious amounts of soap and water. Consider appropriate evacuation.

LEAD THIOSULFATE

DOT Number: n/a *DOT Hazard Class:* n/a *DOT Guide Number:* n/a
Synonyms: lead hyposulfite; thiosulfuric acid, lead salt
STCC Number: n/a *Reportable Qty:* n/a
Mfg Name: City Chemical Corp. *Phone No:* 1-212-929-2723

Physical Description:

Physical Form: Solid *Color:* White *Odor:* n/a
Other Information: n/a

Chemical Properties:

Specific Gravity: 5.18 *Vapor Density:* 10 *Boiling Point: n/a*
Melting Point: Decomposes *Vapor Pressure:* n/a *Solubility in water:* Yes
Other Information: Sinks and mixes slowly with water.

Health Hazards:

Inhalation Hazard: Poisonous if inhaled.
Ingestion Hazard: Abdominal pain, diarrhea, weakness, nausea, vomiting.
Absorption Hazard: n/a
Hazards to Wildlife: Dangerous to aquatic life.
Decontamination Procedures: Wash away any material with copious amounts of soap and water.
First Aid Procedures: Remove victim to fresh air, call emergency medical care. If not breathing give CPR. If breathing is difficult administer oxygen. Treat for shock.

Fire Hazards:

Flashpoint: n/a *Ignition temperature:* n/a
Flammable Explosive High Range: n/a *Low Range:* n/a
Toxic Products of Combustion: Toxic metal fumes and oxides of sulfur.
Other Hazards: n/a
Possible extinguishing agents: Extinguish fire using suitable agent for the type of surrounding fire.

Reactivity Hazards:

Reactive With: n/a *Other Reactions:* n/a

Corrosivity Hazards:

Corrosive With: n/a *Neutralizing Agent:* n/a

Radioactivity Hazards:

Radiation Emitted: n/a *Other Hazards:* n/a

Recommended Protection for Response Personnel:

Avoid breathing vapors, keep upwind. Wear appropriate sealed chemical suit. Wash away any material which may have come into contact with the body with copious amounts of soap and water. Consider appropriate evacuation.

LEAD TUNGSTATE

DOT Number: n/a *DOT Hazard Class:* n/a *DOT Guide Number:* n/a
Synonyms: lead wolframate, raspite, scheelite, stolzite
STCC Number: n/a *Reportable Qty:* n/a
Mfg Name: A.D. MacKay Inc. *Phone No:* 1-203-655-3000

Physical Description:

Physical Form: Solid *Color:* White to pale yellow *Odor:* n/a
Other Information: n/a

Chemical Properties:

Specific Gravity: 8.23 *Vapor Density:* n/a *Boiling Point: n/a*
Melting Point: 2053° F(1123° C) *Vapor Pressure:* n/a *Solubility in water:* n/a
Other Information: Sinks in water.

Health Hazards:

Inhalation Hazard: Poisonous if inhaled.
Ingestion Hazard: Abdominal pain, diarrhea, weakness, nausea, vomiting.
Absorption Hazard: n/a
Hazards to Wildlife: Dangerous to aquatic life.
Decontamination Procedures: Wash away any material with copious amounts of soap and water.
First Aid Procedures: Remove victim to fresh air, call emergency medical care. If not breathing give CPR. If breathing is difficult administer oxygen. Treat for shock.

Fire Hazards:

Flashpoint: n/a *Ignition temperature:* n/a
Flammable Explosive High Range: n/a *Low Range:* n/a
Toxic Products of Combustion: Toxic metal fumes may be produced in fires.
Other Hazards: n/a
Possible extinguishing agents: Extinguish fire using suitable agent for the type of surrounding fire.

Reactivity Hazards:

Reactive With: n/a *Other Reactions:* n/a

Corrosivity Hazards:

Corrosive With: n/a *Neutralizing Agent:* n/a

Radioactivity Hazards:

Radiation Emitted: n/a *Other Hazards:* n/a

Recommended Protection for Response Personnel:

Avoid breathing vapors, keep upwind. Wear appropriate sealed chemical suit. Wash away any material which may have come into contact with the body with copious amounts of soap and water. Consider appropriate evacuation.

LINDANE

DOT Number: NA 2761 *DOT Hazard Class:* ORM-A *DOT Guide Number:* 55
Synonyms: none given
STCC Number: 4941152 *Reportable Qty:* 1/.4544
Mfg Name: Prentiss Drug & Chemical Co. *Phone No:* 1-516-326-1919

Physical Description:

Physical Form: Solid *Color:* White to yellow *Odor:* Odorless
Other Information: In case of damage to, or leaking from these containers, contact the Pesticide Safety Team Network at 1-800-424-9300.

Chemical Properties:

Specific Gravity: n/a *Vapor Density:* n/a *Boiling Point: Decomposes*
Melting Point: 234° F(112° C) *Vapor Pressure:* .0317 mm Hg at 68° F(20° C)
Solubility in water: Slight *Degree of Solubility:* .001%
Other Information: May be dissolved in combustible solvents.

Health Hazards:

Inhalation Hazard: Hazardous if inhaled.
Ingestion Hazard: Hazardous if swallowed.
Absorption Hazard: Irritating to the skin and eyes.
Hazards to Wildlife: n/a
Decontamination Procedures: Wash away any material with copious amounts of soap and water.
First Aid Procedures: Remove victim to fresh air, call emergency medical care. If not breathing give CPR. If breathing is difficult administer oxygen. Treat for shock.

Fire Hazards:

Flashpoint: n/a *Ignition temperature:* n/a
Flammable Explosive High Range: n/a *Low Range:* n/a
Toxic Products of Combustion: n/a
Other Hazards: n/a
Possible extinguishing agents: Extinguish fire using suitable agent for the type of surrounding fire.

Reactivity Hazards:

Reactive With: n/a *Other Reactions:* n/a

Corrosivity Hazards:

Corrosive With: n/a *Neutralizing Agent:* n/a

Radioactivity Hazards:

Radiation Emitted: n/a *Other Hazards:* n/a

Recommended Protection for Response Personnel:

Avoid breathing vapors, keep upwind. Structural protective clothing provides limited protection. Wash away any material which may have come into contact with the body with copious amounts of soap and water. Consider appropriate evacuation.

LINEAR ALCOHOL

DOT Number: n/a *DOT Hazard Class:* n/a *DOT Guide Number:* n/a
Synonyms: dodecanol, pentadecanol, tetradecanol, trideconol
STCC Number: n/a *Reportable Qty:* n/a
Mfg Name: Ethyl Corp., Chemicals Division *Phone No:* 1-804-788-5000

Physical Description:

Physical Form: Solid or liquid *Color:* Colorless *Odor:* Mild alcohol
Other Information: n/a

Chemical Properties:

Specific Gravity: .84 *Vapor Density:* n/a *Boiling Point: 486° F(252° C)*
Melting Point: 66° F(19° C) *Vapor Pressure:* n/a *Solubility in water:* n/a
Other Information: Floats on water.

Health Hazards:

Inhalation Hazard: n/a
Ingestion Hazard: n/a
Absorption Hazard: Irritating to the skin, will burn the eyes.
Hazards to Wildlife: n/a
Decontamination Procedures: Wash away any material with copious amounts of soap and water.
First Aid Procedures: Remove victim to fresh air, call emergency medical care. If not breathing give CPR. If breathing is difficult administer oxygen. Treat for shock.

Fire Hazards:

Flashpoint: 180 to 285° F(82-96° C) *Ignition temperature:* n/a
Flammable Explosive High Range: n/a ILow Range: n/a
Toxic Products of Combustion: n/a
Other Hazards: n/a
Possible extinguishing agents: Alcohol foam, dry chemical or carbon dioxide. Cool all affected containers with water.

Reactivity Hazards:

Reactive With: n/a *Other Reactions:* n/a

Corrosivity Hazards:

Corrosive With: n/a *Neutralizing Agent:* n/a

Radioactivity Hazards:

Radiation Emitted: n/a *Other Hazards:* n/a

Recommended Protection for Response Personnel:

Avoid breathing vapors, keep upwind. Structural protective clothing provides limited protection. Wash away any material which may have come into contact with the body with copious amounts of soap and water. Consider appropriate evacuation.

LIQUEFIED NATURAL GAS (LNG)

DOT Number: UN 1972 *DOT Hazard Class:* Flammable gas *DOT Guide Number:* 22
Synonyms: LNG
STCC Number: n/a *Reportable Qty:* n/a
Mfg Name: Phillips Petroleum *Phone No:* 1-918-661-6600

Physical Description:

Physical Form: Gas *Color:* Colorless *Odor:* Odorless to weak skunk
Other Information: n/a

Chemical Properties:

Specific Gravity: .43 *Vapor Density:* .55 to 1° F(–17.5/–17.2° C) *Boiling Point: –258° F(–161° C)*
Melting Point: –296° F(–182° C) *Vapor Pressure:* High *Solubility in water:* No
Other Information: Floats and boils on water. Flammable, visible vapor cloud is produced.

Health Hazards:

Inhalation Hazard: Dizziness, difficulty in breathing, loss of consciousness.
Ingestion Hazard: n/a
Absorption Hazard: Will cause frostbite.
Hazards to Wildlife: Not harmful.
Decontamination Procedures: Wash away any material with copious amounts of soap and water.
First Aid Procedures: Remove victim to fresh air, call emergency medical care. If not breathing give CPR. If breathing is difficult administer oxygen. Treat for shock.

Fire Hazards:

Flashpoint: Gas *Ignition temperature:* 999° F(537° C)
Flammable Explosive High Range: 14 *Low Range:* 5.3
Toxic Products of Combustion: n/a
Other Hazards: Flashback along vapor trail may occur. Vapors may explode if ignited in an enclosed area.
Possible extinguishing agents: Do not extinguish large spill fires!! allow to burn while cooling adjacent equipment with water. Shut off leak if possible. Extinguish small fires with dry chemical.

Reactivity Hazards:

Reactive With: n/a *Other Reactions:* n/a

Corrosivity Hazards:

Corrosive With: n/a *Neutralizing Agent:* n/a

Radioactivity Hazards:

Radiation Emitted: n/a *Other Hazards:* n/a

Recommended Protection for Response Personnel:

Avoid breathing vapors, keep upwind. Structural protective clothing provides limited protection. Wash away any material which may have come into contact with the body with copious amounts of soap and water. If the fire becomes uncontrollable, consider appropriate evacuation. Cylinders may violently rupture and rocket (BLEVE)

LIQUEFIED PETROLEUM

DOT Number: UN 1075 *DOT Hazard Class:* Flammable gas *DOT Guide Number:* 22
Synonyms: LPG (bottled gas)
STCC Number: 4905752 *Reportable Qty:* n/a
Mfg Name: Phillips Petroleum *Phone No:* 1-918-661-6600

Physical Description:

Physical Form: Gas *Color:* Colorless *Odor:* Weak odor, skunklike
Other Information: It is shipped as a liquid under its own vapor pressure.

Chemical Properties:

Specific Gravity: .51 *Vapor Density:* 1.5 *Boiling Point: −40 to +31° F(−40/−1° C)*
Melting Point: n/a *Vapor Pressure:* 2.1-8.6 atm at 68° F(20° C) *Solubility in water:* Insoluble
Other Information: n/a

Health Hazards:

Inhalation Hazard: Dizziness, difficulty in breathing, loss of consciousness.
Ingestion Hazard: n/a
Absorption Hazard: Will cause frostbite.
Hazards to Wildlife: n/a
Decontamination Procedures: Wash away any material with copious amounts of soap and water.
First Aid Procedures: Remove victim to fresh air, call emergency medical care. If not breathing give CPR. If breathing is difficult administer oxygen. Treat for shock.

Fire Hazards:

Flashpoint: −156° F(−104° C) *Ignition temperature:* 871° F(466° C)
Flammable Explosive High Range: 9.5 *Low Range:* 1.9
Toxic Products of Combustion: n/a
Other Hazards: Flashback along vapor trail may occur. Containers may explode in an enclosed area. Under fire conditions, cylinders may rupture/rocket. (BLEVE)
Possible extinguishing agents: Apply water from as far a distance as possible. Do not extinguish the fire unless the flow can be stopped.

Reactivity Hazards:

Reactive With: n/a *Other Reactions:* n/a

Corrosivity Hazards:

Corrosive With: n/a *Neutralizing Agent:* n/a

Radioactivity Hazards:

Radiation Emitted: n/a *Other Hazards:* n/a

Recommended Protection for Response Personnel:

Avoid breathing vapors, keep upwind. Structural protective clothing provides limited protection. Wash away any material which may have come into contact with the body with copious amounts of soap and water. If the fire becomes uncontrollable, consider appropriate evacuation. (BLEVE)

LITHARGE

DOT Number: n/a *DOT Hazard Class:* n/a *DOT Guide Number:* n/a
Synonyms: lead monoxide, lead oxide, lead protoxide, massicot, yellow plumbous oxide
STCC Number: n/a *Reportable Qty:* n/a
Mfg Name: American Cyanamid Co. *Phone No:* 1-201-831-2000

Physical Description:

Physical Form: Solid *Color:* Gray, yellow, green or red-brown *Odor:* Odorless
Other Information: n/a

Chemical Properties:

Specific Gravity: 9.5 *Vapor Density:* n/a *Boiling Point: Decomposes*
Melting Point: n/a *Vapor Pressure:* n/a *Solubility in water:* n/a
Other Information: Sinks in water

Health Hazards:

Inhalation Hazard: Harmful if inhaled.
Ingestion Hazard: n/a
Absorption Hazard: Irritating to the eyes.
Hazards to Wildlife: n/a
Decontamination Procedures: Wash away any material with copious amounts of soap and water.
First Aid Procedures: Remove victim to fresh air, call emergency medical care. If not breathing give CPR. If breathing is difficult administer oxygen. Treat for shock.

Fire Hazards:

Flashpoint: n/a *Ignition temperature:* n/a
Flammable Explosive High Range: n/a *Low Range:* n/a
Toxic Products of Combustion: n/a
Other Hazards: n/a
Possible extinguishing agents: Extinguish fire using suitable agent for the type of surrounding fire.

Reactivity Hazards:

Reactive With: n/a *Other Reactions:* n/a

Corrosivity Hazards:

Corrosive With: n/a *Neutralizing Agent:* n/a

Radioactivity Hazards:

Radiation Emitted: n/a *Other Hazards:* n/a

Recommended Protection for Response Personnel:

Avoid breathing vapors, keep upwind. Structural protective clothing provides limited protection. Wash away any material which may have come into contact with the body with copious amounts of soap and water. Consider appropriate evacuation.

LITHIUM

DOT Number: UN 1415 *DOT Hazard Class:* Flammable solid *DOT Guide Number:* 40
Synonyms: none given
STCC Number: 4916428 *Reportable Qty:* n/a *CHEMTREC Phone No:* 1-800-424-9300

Physical Description:

Physical Form: Soft solid *Color:* White to light silver *Odor:* Odorless
Other Information: n/a

Chemical Properties:

Specific Gravity: .53 *Vapor Density:* n/a *Boiling Point: n/a*
Melting Point: n/a *Vapor Pressure:* n/a *Solubility in water:* Reacts violently!!
Other Information: Reacts with water to form lithium hydroxide, a corrosive material and hydrogen, a flammable gas.

Health Hazards:

Inhalation Hazard: n/a
Ingestion Hazard: Harmful if swallowed.
Absorption Hazard: Will burn the skin and eyes.
Hazards to Wildlife: n/a
Decontamination Procedures: Wash away any material with copious amounts of soap and water.
First Aid Procedures: Remove victim to fresh air, call emergency medical care, If not breathing give CPR. If breathing is difficult administer oxygen. Treat for shock.

Fire Hazards:

Flashpoint: n/a *Ignition temperature:* n/a
Flammable Explosive High Range: n/a *Low Range:* n/a
Toxic Products of Combustion: Irritating gases may be produced when heated.
Other Hazards: May ignite combustibles if they are damp.
Possible extinguishing agents: Use graphite, soda ash, powdered sodium chloride or a suitable dry powder. Do not use water.

Reactivity Hazards:

Reactive With: Water *Other Reactions:* Damp combustibles

Corrosivity Hazards:

Corrosive With: n/a *Neutralizing Agent:* n/a

Radioactivity Hazards:

Radiation Emitted: n/a *Other Hazards:* n/a

Recommended Protection for Response Personnel:

Avoid breathing vapors, keep upwind. Wear appropriate chemical clothing. Wash away any material which may come into contact with the body with copious amounts of soap and water. Consider appropriate evacuation.

LITHIUM ALUMINUM HYDRIDE

DOT Number: UN 1410 *DOT Hazard Class:* Flammable solid *DOT Guide Number:* 40
Synonyms: HAH
STCC Number: 4916420 *Reportable Qty:* n/a
Mfg Name: Ventron Corp. *Phone No:* 1-617-774-3100

Physical Description:

Physical Form: Solid *Color:* Gray to white *Odor:* Odorless
Other Information: It is used to make other chemicals, as a polymerization catalyst, as a hydrogen source, and as a propellant.

Chemical Properties:

Specific Gravity: .92 *Vapor Density:* 1.3 *Boiling Point: Decomposes*
Melting Point: n/a *Vapor Pressure:* n/a *Solubility in water:* Reacts
Other Information: Reacts violently with water. A flammable gas is produced.

Health Hazards:

Inhalation Hazard: n/a
Ingestion Hazard: Harmful if swallowed.
Absorption Hazard: Will burn the skin and eyes.
Hazards to Wildlife: n/a
Decontamination Procedures: Wash away any material with copious amounts of soap and water.
First Aid Procedures: Remove victim to fresh air, call emergency medical care. If not breathing give CPR. If breathing is difficult administer oxygen. Treat for shock.

Fire Hazards:

Flashpoint: n/a *Ignition temperature:* n/a
Flammable Explosive High Range: n/a *Low Range:* n/a
Toxic Products of Combustion: Decomposes at 257° F(125° C) to form hydrogen gas.
Other Hazards: Flammable gas is released on contact with water, metals, or acids.
Possible extinguishing agents: Do not use water. Use graphite, soda ash, powdered sodium chloride, or a suitable dry powder.

Reactivity Hazards:

Reactive With: Water to form lithium hydroxide, and hydrogen. *Other Reactions:* n/a

Corrosivity Hazards:

Corrosive With: Metal in the presence of moisture. *Neutralizing Agent:* n/a

Radioactivity Hazards:

Radiation Emitted: n/a *Other Hazards:* n/a

Recommended Protection for Response Personnel:

Avoid breathing vapors, keep upwind. Structural protective clothing provides limited protection. Wash away any material which may have come into contact with the body with copious amounts of soap and water. Do not breathe dust and fumes from the burning material. Consider appropriate evacuation.

LITHIUM BICHROMATE

DOT Number: n/a *DOT Hazard Class:* n/a *DOT Guide Number:* n/a
Synonyms: lithium bichromate dihydrate, lithium dichromate
STCC Number: n/a *Reportable Qty:* n/a
Mfg Name: Great Western Inorganics *Phone No:* 1-303-423-9770

Physical Description:

Physical Form: Solid *Color:* Orange red to black brown *Odor:* n/a
Other Information: n/a

Chemical Properties:

Specific Gravity: 2.34 *Vapor Density:* n/a *Boiling Point: 368° F(187° C)*
Melting Point: 230 to 266° F(110-130° C) *Vapor Pressure:* n/a
Solubility in water: Yes
Other Information: Sinks and mixes with water.

Health Hazards:

Inhalation Hazard: Difficulty in breathing.
Ingestion Hazard: Dizziness, nausea, vomiting, coma.
Absorption Hazard: Will burn the skin and eyes.
Hazards to Wildlife: Dangerous to aquatic life.
Decontamination Procedures: Wash away any material with copious amounts of soap and water.
First Aid Procedures: Remove victim to fresh air, call emergency medical care. If not breathing give CPR. If breathing is difficult administer oxygen. Treat for shock.

Fire Hazards:

Flashpoint: n/a *Ignition temperature:* n/a
Flammable Explosive High Range: n/a *Low Range:* n/a
Toxic Products of Combustion: n/a
Other Hazards: May decompose giving off oxygen which supports further combustion.
Possible extinguishing agents: Flood spill area with water.

Reactivity Hazards:

Reactive With: Combustibles *Other Reactions:* n/a

Corrosivity Hazards:

Corrosive With: n/a *Neutralizing Agent:* n/a

Radioactivity Hazards:

Radiation Emitted: n/a *Other Hazards:* n/a

Recommended Protection for Response Personnel:

Avoid breathing vapors, keep upwind. Wear appropriate sealed chemical clothing. Wash away any material which may have come into contact with the body with copious amounts of soap and water. Consider appropriate evacuation.

LITHIUM CHROMATE

DOT Number: NA 9134 *DOT Hazard Class:* ORM-E *DOT Guide Number:* 31
Synonyms: chromic acid, dilithium salt; chromium lithium oxide
STCC Number: 4963720 *Reportable Qty:* 1000/454
Mfg Name: Great Western Inorganics *Phone No:* 1-303-423-9770

Physical Description:

Physical Form: Solid *Color:* Yellow *Odor:* Odorless
Other Information: It is used as a corrosion inhibitor, and in the manufacture of other chemicals.

Chemical Properties:

Specific Gravity: n/a *Vapor Density:* 5.7 *Boiling Point: n/a*
Melting Point: n/a *Vapor Pressure:* n/a *Solubility in water:* Yes
Other Information: Mixes with water.

Health Hazards:

Inhalation Hazard: Will cause difficulty in breathing.
Ingestion Hazard: Will cause nausea, vomiting, dizziness, coma.
Absorption Hazard: Will burn the skin and eyes.
Hazards to Wildlife: Dangerous to aquatic life.
Decontamination Procedures: Wash away any material with copious amounts of soap and water.
First Aid Procedures: Remove victim to fresh air, call emergency medical care. If not breathing give CPR. If breathing is difficult administer oxygen. Treat for shock.

Fire Hazards:

Flashpoint: n/a *Ignition temperature:* n/a
Flammable Explosive High Range: n/a *Low Range:* n/a
Toxic Products of Combustion: n/a
Other Hazards: May cause fire upon contact with combustibles.
Possible extinguishing agents: Extinguish fire using suitable agent for the type of surrounding fire.

Reactivity Hazards:

Reactive With: Oxidizes combustibles *Other Reactions:* n/a

Corrosivity Hazards:

Corrosive With: n/a *Neutralizing Agent:* n/a

Radioactivity Hazards:

Radiation Emitted: n/a *Other Hazards:* n/a

Recommended Protection for Response Personnel:

Avoid breathing vapors, keep upwind. Wear appropriate chemical clothing. Wash away any material which may have come into contact with the body with copious amounts of soap and water. Consider appropriate evacuation.

LITHIUM HYDRIDE

DOT Number: UN 1414 *DOT Hazard Class:* Flammable solid *DOT Guide Number:* 40
Synonyms: none given
STCC Number: 4916424 *Reportable Qty:* n/a
Mfg Name: Ventron Corp. *Phone No:* 1-617-774-3100

Physical Description:

Physical Form: Solid *Color:* Grayish white *Odor:* Odorless
Other Information: n/a

Chemical Properties:

Specific Gravity: .78 *Vapor Density:* .3 *Boiling Point: Decomposes*
Melting Point: 1267° F(686° C) *Vapor Pressure:* 0 mm Hg at 68° F(20° C) *Solubility in water:* Reacts
Other Information: Rcacts violently with water. A flammable gas is produced.

Health Hazards:

Inhalation Hazard: Poisonous if inhaled.
Ingestion Hazard: Cause nausea, loss of consciousness.
Absorption Hazard: Will burn the skin and eyes.
Hazards to Wildlife: n/a
Decontamination Procedures: Wash away any material with copious amounts of soap and water.
First Aid Procedures: Remove victim to fresh air, call emergency medical care. If not breathing give CPR. If breathing is difficult administer oxygen. Treat for shock.

Fire Hazards:

Flashpoint: n/a *Ignition temperature:* 392° F(200° C)
Flammable Explosive High Range: n/a *Low Range:* n/a
Toxic Products of Combustion: Irritating alkali fumes may be formed in fire.
Other Hazards: Reacts violently with water to form hydrogen gas, which may explode in the air.
Possible extinguishing agents: Do not use water. Use graphite, soda ash, or powdered sodium chloride.

Reactivity Hazards:

Reactive With: Water to form flammable hydrogen gas.
Other Reactions: May ignite combustible materials, if they are damp.

Corrosivity Hazards:

Corrosive With: n/a *Neutralizing Agent:* n/a

Radioactivity Hazards:

Radiation Emitted: n/a *Other Hazards:* n/a

Recommended Protection for Response Personnel:

Avoid breathing vapors, keep upwind. Wear appropriate chemical clothing. Wash away any material which may have come into contact with the body with copious amounts of soap and water. Avoid breathing dust and the fumes from the burning material. Consider appropriate evacuation.

MAGNESIUM

DOT Number: NA 1869 *DOT Hazard Class:* Flammable solid *DOT Guide Number:* 76
Synonyms: none given
STCC Number: 4913436 *Reportable Qty:* n/a *CHEMTREC Phone No:* 1-800-424-9300

Physical Description:

Physical Form: Solid *Color:* Silvery *Odor:* Odorless
Other Information: n/a

Chemical Properties:

Specific Gravity: 1.74 *Vapor Density:* n/a *Boiling Point: 2012° F(1100° C)*
Melting Point: 1202° F(650° C) *Vapor Pressure:* 0 mm Hg at 68° F(20° C) *Solubility in water:* Insoluble
Other Information: The more highly divided forms react with water to liberate hydrogen, a flammable gas.

Health Hazards:

Inhalation Hazard: n/a
Ingestion Hazard: Harmful if swallowed.
Absorption Hazard: Irritating to the eyes.
Hazards to Wildlife: n/a
Decontamination Procedures: Wash away any material with copious amounts of soap and water.
First Aid Procedures: Remove victim to fresh air, call emergency medical care. If not breathing give CPR. If breathing is difficult administer oxygen. Treat for shock.

Fire Hazards:

Flashpoint: n/a *Ignition temperature:* 883° F(473° C)
Flammable Explosive High Range: n/a *Low Range:* n/a
Toxic Products of Combustion: n/a
Other Hazards: Reacts with water and acids to release hydrogen gas. The flame is very bright.
Possible extinguishing agents: Use graphite, soda ash, powdered sodium chloride or a suitable dry powder. Do not use water!

Reactivity Hazards:

Reactive With: Water and acids to release flammable hydrogen gas. *Other Reactions:* n/a

Corrosivity Hazards:

Corrosive With: n/a *Neutralizing Agent:* n/a

Radioactivity Hazards:

Radiation Emitted: n/a *Other Hazards:* n/a

Recommended Protection for Response Personnel:

Avoid breathing vapors, keep upwind. Wear appropriate chemical clothing. Wash away any material which may have come into contact with the body with copious amounts of soap and water. Consider appropriate evacuation.

MAGNESIUM PERCHLORATE

DOT Number: NA 1475 *DOT Hazard Class:* Oxidizer *DOT Guide Number:* 35
Synonyms: anhydrone, hehydrite
STCC Number: 4918729 *Reportable Qty:* n/a
Mfg Name: Atomergic Chemetals Corp. *Phone No:* 1-516-349-8800

Physical Description:

Physical Form: Solid *Color:* White *Odor:* Odorless
Other Information: It is used as a drying agent, and as an oxidizing agent.

Chemical Properties:

Specific Gravity: 2.21 *Vapor Density:* 7.7 *Boiling Point: Decomposes*
Melting Point: n/a *Vapor Pressure:* n/a *Solubility in water:* Yes
Other Information: Sinks and mixes with water.

Health Hazards:

Inhalation Hazard: Will cause difficulty in breathing.
Ingestion Hazard: Will cause nausea, vomiting, loss of consciousness.
Absorption Hazard: Irritating to the skin and eyes.
Hazards to Wildlife: n/a
Decontamination Procedures: Wash away any material with copious amounts of soap and water.
First Aid Procedures: Remove victim to fresh air, call emergency medical care. If not breathing give CPR. If breathing is difficult administer oxygen. Treat for shock.

Fire Hazards:

Flashpoint: n/a *Ignition temperature:* n/a
Flammable Explosive High Range: n/a *Low Range:* n/a
Toxic Products of Combustion: n/a
Other Hazards: Can form explosive mixtures with combustibles, or finely powdered metals.
Possible extinguishing agents: Flood with water. Apply water from as far a distance as possible.

Reactivity Hazards:

Reactive With: Water in which it dissolves with liberation of heat. May cause splattering.
Other Reactions: Contact with wood, paper, oils, grease, or finely divided metals may cause an explosion.

Corrosivity Hazards:

Corrosive With: n/a *Neutralizing Agent:* n/a

Radioactivity Hazards:

Radiation Emitted: n/a *Other Hazards:* n/a

Recommended Protection for Response Personnel:

Avoid breathing vapors, keep upwind. Structural protective clothing provides limited protection. Wash away any material which may have come into contact with the body with copious amounts of soap and water. If the fire becomes uncontrollable, consider appropriate evacuation.

MALATHION

DOT Number: NA 2783 *DOT Hazard Class:* ORM-A *DOT Guide Number:* 55
Synonyms: Cythion insecticide
STCC Number: 4941156 *Reportable Qty:* 100/45.4
Mfg Name: Stauffer Chemical Co. *Phone No:* 1-203-222-3000

Physical Description:

Physical Form: Liquid *Color:* Yellow to dark brown *Odor:* Skunklike
Other Information: It is used as an insecticide.

Chemical Properties:

Specific Gravity: 1.23 *Vapor Density:* n/a *Boiling Point: Decomposes*
Melting Point: 37° F(3° C) *Vapor Pressure:* .00004 mm Hg at 68° F(20° C)
Solubility in water: No *Degree of Solubility:* .0145%
Other Information: In case of damage to or leaking from containers, contact the Pesticide Safety Network Team at 1-800-424-9300.

Health Hazards:

Inhalation Hazard: Poisonous if inhaled.
Ingestion Hazard: Poisonous if swallowed.
Absorption Hazard: Poisonous to the skin and the eyes.
Hazards to Wildlife: n/a
Decontamination Procedures: Wash away any material with copious amounts of soap and water.
First Aid Procedures: Remove victim to fresh air, call emergency medical care. If not breathing give CPR. If breathing is difficult administer oxygen. Treat for shock.

Fire Hazards:

Flashpoint: 325° F(163° C) *Ignition temperature:* n/a
Flammable Explosive High Range: n/a *Low Range:* n/a
Toxic Products of Combustion: Poisonous gases are produced in fire and when heated.
Other Hazards: Containers may explode in a fire.
Possible extinguishing agents: Extinguish fire using suitable agent for type of surrounding fire.

Reactivity Hazards:

Reactive With: n/a *Other Reactions:* n/a

Corrosivity Hazards:

Corrosive With: Acids and caustics. *Neutralizing Agent:* Liquid bleach

Radioactivity Hazards:

Radiation Emitted: n/a *Other Hazards:* n/a

Recommended Protection for Response Personnel:

Avoid breathing vapors, keep upwind. Wear a sealed chemical suit (polycarbonate, butyl rubber, PVC, nitrile, neoprene). Wash away any material which may have come into contact with the body with copious amounts of soap and water. Consider appropriate evacuation.

MALEIC ACID

DOT Number: NA 2215 *DOT Hazard Class:* ORM-A *DOT Guide Number:* 60
Synonyms: maleinic acid, malenic acid, toxilic acid
STCC Number: 4941155 *Reportable Qty:* 5000/2270
Mfg Name: Allied Corp. *Phone No:* 1-201-455-2000

Physical Description:

Physical Form: Solid *Color:* White *Odor:* Odorless
Other Information: It is used to make other chemicals, for dyeing and finishing naturally occurring fibers.

Chemical Properties:

Specific Gravity: 1.59 *Vapor Density:* 4 *Boiling Point: Decomposes*
Melting Point: n/a *Vapor Pressure:* n/a *Solubility in water:* Yes
Other Information: Sinks and mixes with water.

Health Hazards:

Inhalation Hazard: Will cause coughing, difficulty in breathing.
Ingestion Hazard: Harmful if swallowed.
Absorption Hazard: Irritating to the skin and the eyes.
Hazards to Wildlife: Dangerous to aquatic life.
Decontamination Procedures: Wash away any material with copious amounts of soap and water.
First Aid Procedures: Remove victim to fresh air, call emergency medical care. If not breathing give CPR. If breathing is difficult administer oxygen. Treat for shock.

Fire Hazards:

Flashpoint: n/a *Ignition temperature:* n/a
Flammable Explosive High Range: n/a *Low Range:* n/a
Toxic Products of Combustion: Irritating smoke of maleic anhydride may be formed in fires.
Other Hazards: n/a
Possible extinguishing agents: Do not extinguish the fire unless the flow can be stopped. Apply water from as far a distance as possible. use alcohol foam, dry chemical, or carbon dioxide.

Reactivity Hazards:

Reactive With: n/a *Other Reactions:* n/a

Corrosivity Hazards:

Corrosive With: Metals when wet.
Neutralizing Agent: Flush with water, rinse with solution of sodium bicarbonate, or soda ash.

Radioactivity Hazards:

Type *Radiation Emitted:* n/a *Other Hazards:* n/a

Recommended Protection for Response Personnel:

Avoid breathing vapors, keep upwind. Wear a sealed chemical suit (polycarbonate, butyl rubber, chlorinated polyethylene, PVC, viton). Wash away any material which may have come into contact with the body with copious amounts of soap and water. Consider appropriate evacuation.

MALEIC ANHYDRIDE

DOT Number: UN 2215 *DOT Hazard Class:* ORM-A *DOT Guide Number:* 60
Synonyms: toxilic anhydride
STCC Number: 4941161 *Reportable Qty:* 5000/2270
Mfg Name: Monsanto Co. *Phone No:* 1-314-694-1000

Physical Description:

Physical Form: Molten or solid crystal tablets *Color:* Colorless *Odor:* Choking
Other Information: It is used to make paints, plastics, and other chemicals.

Chemical Properties:

Specific Gravity: .9 *Vapor Density:* 3.4 *Boiling Point: 396° F(202° C)*
Melting Point: 127° F(53° C) *Vapor Pressure:* .16 mm Hg at 68° F(20° C) *Solubility in water:* Yes
Other Information: Decomposes slowly to form maleic acid, another ORM-A.

Health Hazards:

Inhalation Hazard: Harmful if inhaled.
Ingestion Hazard: Harmful if swallowed.
Absorption Hazard: Will burn the skin and the eyes.
Hazards to Wildlife: Dangerous to aquatic life.
Decontamination Procedures: Wash away any material with copious amounts of soap and water.
First Aid Procedures: Remove victim to fresh air, call emergency medical care. If not breathing give CPR. If breathing is difficult administer oxygen. Treat for shock.

Fire Hazards:

Flashpoint: 215° F(102° C) *Ignition temperature:* 890° F(477° C)
Flammable Explosive High Range: 7.1 *Low Range:* 1.4
Toxic Products of Combustion: n/a
Other Hazards: Dust cloud may be ignited by sparks or flame.
Possible extinguishing agents: Apply water from as far a distance as possible. Use alcohol foam, dry chemical, or carbon dioxide.

Reactivity Hazards:

Reactive With: Hot water *Other Reactions:* n/a

Corrosivity Hazards:

Corrosive With: n/a *Neutralizing Agent:* n/a

Radioactivity Hazards:

Radiation Emitted: n/a *Other Hazards:* n/a

Recommended Protection for Response Personnel:

Avoid breathing vapors, keep upwind. Wear a sealed chemical suit (polycarbonate, butyl rubber, viton, PVC, chlorinated polyethylene). Wash away any material which may have come into contact with the body with copious amounts of soap and water. Consider appropriate evacuation.

MALEIC HYDRAZIDE

DOT Number: n/a *DOT Hazard Class:* n/a *DOT Guide Number:* n/a

Synonyms: 1,2-dihydro-3,6-pyridazinedione, 6-hydroxy-3-(2h)-pyridazinone, maleic acid hydrazide, Regulox

STCC Number: n/a *Reportable Qty:* n/a

Mfg Name: Ansul Co. *Phone No:* 1-715-735-7411

Physical Description:

Physical Form: Solid *Color:* White *Odor:* Odorless

Other Information: n/a

Chemical Properties:

Specific Gravity: 1.60 *Vapor Density:* n/a *Boiling Point: Decomposes*

Melting Point: 558° F(292° C) *Vapor Pressure:* n/a *Solubility in water:* No

Other Information: Sinks in water

Health Hazards:

Inhalation Hazard: n/a

Ingestion Hazard: Harmful if swallowed.

Absorption Hazard: Irritating to the skin and the eyes.

Hazards to Wildlife: n/a

Decontamination Procedures: Wash away any material with copious amounts of soap and water.

First Aid Procedures: Remove victim to fresh air, call emergency medical care. If not breathing give CPR. If breathing is difficult administer oxygen. Treat for shock.

Fire Hazards:

Flashpoint: Combustible solid *Ignition temperature:* n/a

Flammable Explosive High Range: n/a *Low Range:* n/a

Toxic Products of Combustion: Toxic nitrogen oxides are produced in fire.

Other Hazards: n/a

Possible extinguishing agents: Extinguish fire using suitable agent for the type of surrounding fire.

Reactivity Hazards:

Reactive With: n/a *Other Reactions:* n/a

Corrosivity Hazards:

Corrosive With: n/a *Neutralizing Agent:* n/a

Radioactivity Hazards:

Radiation Emitted: n/a *Other Hazards:* n/a

Recommended Protection for Response Personnel:

Avoid breathing vapors, keep upwind. Wear appropriate sealed chemical suit. Wash away any material which may have come into contact with the body with copious amounts of soap and water. Consider appropriate evacuation.

MERCAPTODIMETHUR

DOT Number: NA 2757 *DOT Hazard Class:* ORM-E *DOT Guide Number:* 55
Synonyms: Bay 37344, Mesurol, Methiocarb
STCC Number: 4962145 *Reportable Qty:* 10/4.54 *CHEMTREC Phone No:* 1-800-424-9300

Physical Description:

Physical Form: Solid *Color:* White *Odor:* Mild
Other Information: In case of damage to, or leaking from these containers, contact the Pesticide Safety Team Network at 1-800-424-9300.

Chemical Properties:

Specific Gravity: 1 *Vapor Density:* n/a *Boiling Point: Very high*
Melting Point: n/a *Vapor Pressure:* n/a *Solubility in water:* Yes
Other Information: Sinks in water. Material is used as a pesticide.

Health Hazards:

Inhalation Hazard: Poisonous if inhaled.
Ingestion Hazard: Poisonous if swallowed.
Absorption Hazard: Poisonous if the skin is exposed.
Hazards to Wildlife: Dangerous to both aquatic life and waterfowl.
Decontamination Procedures: Wash away any material with copious amounts of soap and water.
First Aid Procedures: Remove victim to fresh air, call emergency medical care. If not breathing give CPR. If breathing is difficult administer oxygen. Treat for shock.

Fire Hazards:

Flashpoint: n/a *Ignition temperature:* n/a
Flammable Explosive High Range: n/a *Low Range:* n/a
Toxic Products of Combustion: Toxic fumes of sulfur and nitrogen are emitted.
Other Hazards: Dust may be explosive.
Possible extinguishing agents: Use suitable agent for the type of surrounding fire.

Reactivity Hazards:

Reactive With: n/a *Other Reactions:* n/a

Corrosivity Hazards:

Corrosive With: n/a *Neutralizing Agent:* n/a

Radioactivity Hazards:

Radiation Emitted: n/a *Other Hazards:* n/a

Recommended Protection for Response Personnel:

Avoid breathing vapors, keep upwind. Wear appropriate chemical clothing. Wash away any material which may come into contact with the body with copious amounts of soap and water. Consider appropriate evacuation.

MERCURIC ACETATE

DOT Number: UN 1629
DOT Hazard Class: Poison B
DOT Guide Number: 53
Synonyms: none given
STCC Number: 4923241
Reportable Qty: n/a
Mfg Name: Ventron Corp.
Phone No: 1-716-774-3100

Physical Description:

Physical Form: Solid
Color: White
Odor: Vinegar
Other Information: n/a

Chemical Properties:

Specific Gravity: 3.27
Vapor Density: 11
Boiling Point: n/a
Melting Point: n/a
Vapor Pressure: n/a
Solubility in water: Yes
Other Information: n/a

Health Hazards:

Inhalation Hazard: Poisonous if inhaled.

Ingestion Hazard: Poisonous if swallowed.

Absorption Hazard: n/a

Hazards to Wildlife: Dangerous to aquatic life.

Decontamination Procedures: Wash away any material with copious amounts of soap and water.

First Aid Procedures: Remove victim to fresh air, call emergency medical care. If not breathing give CPR. If breathing is difficult administer oxygen. Treat for shock.

Fire Hazards:

Flashpoint: n/a
Ignition temperature: n/a

Flammable Explosive High Range: n/a
Low Range: n/a

Toxic Products of Combustion: Poisonous gases may be produced when heated.

Other Hazards: Avoid contact with solid and dust.

Possible extinguishing agents: Extinguish fire using suitable agent for type of surrounding fire. Use alcohol foam, dry chemical, or carbon dioxide.

Reactivity Hazards:

Reactive With: n/a
Other Reactions: n/a

Corrosivity Hazards:

Corrosive With: n/a
Neutralizing Agent: n/a

Radioactivity Hazards:

Radiation Emitted: n/a
Other Hazards: n/a

Recommended Protection for Response Personnel:

Avoid breathing vapors, keep upwind. Wear a sealed chemical suit (polycarbonate, butyl rubber). Wash away any material which may have come into contact with the body with copious amounts of soap and water. Consider appropriate evacuation.

MERCURIC CHLORIDE

DOT Number: UN 1624 *DOT Hazard Class:* Poison B *DOT Guide Number:* 53
Synonyms: calochlor, mercury bichloride
STCC Number: 4923245 *Reportable Qty:* n/a
Mfg Name: Mallinckrodt Inc. *Phone No:* 1-314-895-2000

Physical Description:

Physical Form: Solid *Color:* White *Odor:* n/a
Other Information: n/a

Chemical Properties:

Specific Gravity: 5.4 *Vapor Density:* 8.7 *Boiling Point: n/a*
Melting Point: n/a *Vapor Pressure:* 1 mm Hg at 277° F(136° C)
Other Information: n/a

Health Hazards:

Inhalation Hazard: Poisonous if inhaled.
Ingestion Hazard: Poisonous if swallowed.
Absorption Hazard: Poisonous if absorbed.
Hazards to Wildlife: Dangerous to aquatic life.
Decontamination Procedures: Wash away any material with copious amounts of soap and water.
First Aid Procedures: Remove victim to fresh air, call emergency medical care. If not breathing give CPR. If breathing is difficult administer oxygen. Treat for shock.

Fire Hazards:

Flashpoint: n/a *Ignition temperature:* n/a
Flammable Explosive High Range: n/a *Low Range:* n/a
Toxic Products of Combustion: Poisonous gases may be produced when heated.
Other Hazards: Avoid contact with solid and dust.
Possible extinguishing agents: Extinguish fire using suitable agent for type of surrounding fire. Use alcohol foam, dry chemical, or carbon dioxide.

Reactivity Hazards:

Reactive With: n/a *Other Reactions:* n/a

Corrosivity Hazards:

Corrosive With: n/a *Neutralizing Agent:* n/a

Radioactivity Hazards:

Radiation Emitted: n/a *Other Hazards:* n/a

Recommended Protection for Response Personnel:

Avoid breathing vapors, keep upwind. Wear a sealed chemical suit (polycarbonate, butyl rubber, chlorinated polyethylene, nitrile, neoprene). Wash away any material which may have come into contact with the body with copious amounts of soap and water. Consider appropriate evacuation.

MERCURIC CYANIDE

DOT Number: UN 1636 *DOT Hazard Class:* Poison B *DOT Guide Number:* 53
Synonyms: cianurina, mercury(II) cyanide
STCC Number: 4923246 *Reportable Qty:* 1/.454
Mfg Name: Mallinckrodt Inc. *Phone No:* 1-314-895-2000

Physical Description:

Physical Form: Solid *Color:* White *Odor:* n/a
Other Information: n/a

Chemical Properties:

Specific Gravity: 4 *Vapor Density:* 8.7 *Boiling Point: n/a*
Melting Point: n/a *Vapor Pressure:* n/a *Solubility in water:* Yes
Other Information: n/a

Health Hazards:

Inhalation Hazard: Poisonous if inhaled.
Ingestion Hazard: Poisonous if swallowed.
Absorption Hazard: Poisonous if absorbed.
Hazards to Wildlife: Dangerous to aquatic life.
Decontamination Procedures: Wash away any material with copious amounts of soap and water.
First Aid Procedures: Remove victim to fresh air, call emergency medical care. If not breathing give CPR. If breathing is difficult administer oxygen. Treat for shock.

Fire Hazards:

Flashpoint: n/a *Ignition temperature:* n/a
Flammable Explosive High Range: n/a *Low Range:* n/a
Toxic Products of Combustion: Poisonous gases may be produced when heated.
Other Hazards: It is gradually decomposed by water and rapidly decomposes by acids to form hydrogen cyanide.
Possible extinguishing agents: Extinguish fire using suitable agent for type of surrounding fire. Use alcohol foam, dry chemical, or carbon dioxide.

Reactivity Hazards:

Reactive With: Any acidic material *Other Reactions:* n/a

Corrosivity Hazards:

Corrosive With: n/a *Neutralizing Agent:* n/a

Radioactivity Hazards:

Radiation Emitted: n/a *Other Hazards:* n/a

Recommended Protection for Response Personnel:

Avoid breathing vapors, keep upwind. Wear a sealed chemical suit (butyl rubber, polycarbonate). Wash away any material which may have come into contact with the body with copious amounts of soap and water. Consider appropriate evacuation.

MERCURIC IODIDE (liquid)

DOT Number: UN 1638 *DOT Hazard Class:* Poison B *DOT Guide Number:* 53
Synonyms: mercuric iodide (red), mercury biniodide
STCC Number: 4923249 *Reportable Qty:* n/a
Mfg Name: Mallinckrodt Inc. *Phone No:* 1-314-895-2000

Physical Description:

Physical Form: Liquid *Color:* Red *Odor:* Odorless
Other Information: n/a

Chemical Properties:

Specific Gravity: 6.3 *Vapor Density:* 16 *Boiling Point: 669° F(354° C)*
Melting Point: 495° F(257° C) *Vapor Pressure:* 1 mm Hg at 315° F(157° C) *Solubility in water:* Yes
Other Information: Sinks and mixes with water.

Health Hazards:

Inhalation Hazard: Poisonous if inhaled.
Ingestion Hazard: Poisonous if swallowed.
Absorption Hazard: Poisonous if absorbed.
Hazards to Wildlife: n/a
Decontamination Procedures: Wash away any material with copious amounts of soap and water.
First Aid Procedures: Remove victim to fresh air, call emergency medical care. If not breathing give CPR. If breathing is difficult administer oxygen. Treat for shock.

Fire Hazards:

Flashpoint: n/a *Ignition temperature:* n/a
Flammable Explosive High Range: n/a *Low Range:* n/a
Toxic Products of Combustion: Fumes may contain toxic mercury vapor.
Other Hazards: n/a
Possible extinguishing agents: Extinguish fire using suitable agent for the type of surrounding fire. Use foam, dry chemical, or carbon dioxide.

Reactivity Hazards:

Reactive With: n/a *Other Reactions:* n/a

Corrosivity Hazards:

Corrosive With: n/a *Neutralizing Agent:* n/a

Radioactivity Hazards:

Radiation Emitted: n/a *Other Hazards:* n/a

Recommended Protection for Response Personnel:

Avoid breathing vapors, keep upwind. Wear a sealed chemical suit (polycarbonate, butyl rubber). Wash away any material which may have come into contact with the body with copious amounts of soap and water. Consider appropriate evacuation.

MERCURIC IODIDE (solid)

DOT Number: UN 1638 *DOT Hazard Class:* Poison B *DOT Guide Number:* 53
Synonyms: mercuric iodide (red), mercury biniodide
STCC Number: 4923247 *Reportable Qty:* n/a
Mfg Name: Mallinckrodt Inc. *Phone No:* 1-314-895-2000

Physical Description:

Physical Form: Solid *Color:* Red *Odor:* Odorless
Other Information: n/a

Chemical Properties:

Specific Gravity: 6.3 *Vapor Density:* 16 *Boiling Point: 669° F(354° C)*
Melting Point: 495° F(257° C) *Vapor Pressure:* 1 mm Hg at 315° F(157° C) *Solubility in water:* Yes
Other Information: Sinks in and mixes with water.

Health Hazards:

Inhalation Hazard: Poisonous if inhaled.
Ingestion Hazard: Poisonous if swallowed.
Absorption Hazard: Poisonous if absorbed.
Hazards to Wildlife: n/a
Decontamination Procedures: Wash away any material with copious amounts of soap and water.
First Aid Procedures: Remove victim to fresh air, call emergency medical care. If not breathing give CPR. If breathing is difficult administer oxygen. Treat for shock.

Fire Hazards:

Flashpoint: n/a *Ignition temperature:* n/a
Flammable Explosive High Range: n/a *Low Range:* n/a
Toxic Products of Combustion: Fumes may contain toxic mercury vapor.
Other Hazards: n/a
Possible extinguishing agents: Extinguish fire using suitable agent for the type of surrounding fire. Use foam, dry chemical, or carbon dioxide.

Reactivity Hazards:

Reactive With: n/a *Other Reactions:* n/a

Corrosivity Hazards:

Corrosive With: n/a *Neutralizing Agent:* n/a

Radioactivity Hazards:

Radiation Emitted: n/a *Other Hazards:* n/a

Recommended Protection for Response Personnel:

Avoid breathing vapors, keep upwind. Wear a sealed chemical suit (polycarbonate, butyl rubber). Wash away any material which may have come into contact with the body with copious amounts of soap and water. Consider appropriate evacuation.

MERCURIC NITRATE

DOT Number: UN 1625 *DOT Hazard Class:* Oxidizer *DOT Guide Number:* 42
Synonyms: mercury binitrate, mercury pernitrate
STCC Number: 4918769 *Reportable Qty:* 10/4.54
Mfg Name: Mallinckrodt Inc. *Phone No:* 1-314-895-2000

Physical Description:

Physical Form: Solid *Color:* White *Odor:* Sharp
Other Information: It is used to make other chemicals, and in medicine.

Chemical Properties:

Specific Gravity: 4.3 *Vapor Density:* 11 *Boiling Point: Decomposes*
Melting Point: n/a *Vapor Pressure:* n/a *Solubility in water:* Yes
Other Information: Sinks in and mixes with water.

Health Hazards:

Inhalation Hazard: Poisonous if inhaled.
Ingestion Hazard: Poisonous if swallowed.
Absorption Hazard: Poisonous if the skin is exposed.
Hazards to Wildlife: n/a
Decontamination Procedures: Wash away any material with copious amounts of soap and water.
First Aid Procedures: Remove victim to fresh air, call emergency medical care. If not breathing give CPR. If breathing is difficult administer oxygen. Treat for shock.

Fire Hazards:

Flashpoint: n/a *Ignition temperature:* n/a
Flammable Explosive High Range: n/a *Low Range:* n/a
Toxic Products of Combustion: Vapor from fire may contain toxic mercury and oxides of nitrogen.
Other Hazards: May increase the intensity of a fire.
Possible extinguishing agents: Flood with water. Apply water from as far a distance as possible.

Reactivity Hazards:

Reactive With: Water to form a cloudy, acid solution. *Other Reactions:* Wood and paper may cause a fire.

Corrosivity Hazards:

Corrosive With: Solution will corrode most metals.
Neutralizing Agent: Flush well with water. Rinse with solution of sodium bicarbonate, or soda ash.

Radioactivity Hazards:

Radiation Emitted: n/a *Other Hazards:* n/a

Recommended Protection for Response Personnel:

Avoid breathing vapors, keep upwind. Wear a sealed chemical suit (polycarbonate, butyl rubber). Wash away any material which may have come into contact with the body with copious amounts of soap and water. If the fire becomes uncontrollable, consider appropriate evacuation.

MERCURIC OXIDE

DOT Number: UN 1641 *DOT Hazard Class:* Poison B *DOT Guide Number:* 53
Synonyms: mercury oxide, mercuric oxide red, mercuric oxide yellow
STCC Number: 4923251 *Reportable Qty:* n/a
Mfg Name: Mallinckrodt Inc. *Phone No:* 1-314-895-2000

Physical Description:

Physical Form: Solid *Color:* Red, orange, or yellow *Odor:* Odorless
Other Information: Solids change color when hot.

Chemical Properties:

Specific Gravity: 11.1 *Vapor Density:* 7.5 *Boiling Point: n/a*
Melting Point: n/a *Vapor Pressure:* n/a *Solubility in water:* No
Other Information: Heavier than and insoluble in water.

Health Hazards:

Inhalation Hazard: Poisonous if inhaled.
Ingestion Hazard: Poisonous if swallowed.
Absorption Hazard: Poisonous if absorbed.
Hazards to Wildlife: n/a
Decontamination Procedures: Wash away any material with copious amounts of soap and water.
First Aid Procedures: Remove victim to fresh air, call emergency medical care. If not breathing give CPR. If breathing is difficult administer oxygen. Treat for shock.

Fire Hazards:

Flashpoint: n/a *Ignition temperature:* n/a
Flammable Explosive High Range: n/a *Low Range:* n/a
Toxic Products of Combustion: Fumes may contain poisonous mercury vapor.
Other Hazards: Decomposes at 933° F(501° C) into mercury and oxygen which can increase the intensity of a fire.
Possible extinguishing agents: Extinguish fire using suitable agent for the type of surrounding fire. Use foam, dry chemical, or carbon dioxide.

Reactivity Hazards:

Reactive With: n/a *Other Reactions:* n/a

Corrosivity Hazards:

Corrosive With: n/a *Neutralizing Agent:* n/a

Radioactivity Hazards:

Radiation Emitted: n/a *Other Hazards:* n/a

Recommended Protection for Response Personnel:

Avoid breathing vapors, keep upwind. Wear a sealed chemical suit (polycarbonate, butyl rubber). Wash away any material which may have come into contact with the body with copious amounts of soap and water. Consider appropriate evacuation.

MERCURIC SULFATE

DOT Number: UN 1645 *DOT Hazard Class:* Poison B *DOT Guide Number:* 53
Synonyms: mercury bisulfate, mercury persulfate
STCC Number: 4923257 *Reportable Qty:* 10/4.54
Mfg Name: Fisher Scientific *Phone No:* 1-412-349-3322

Physical Description:

Physical Form: Solid *Color:* White *Odor:* Odorless
Other Information: It is used in medicine, in gold and silver extraction, and to make other mercury compounds.

Chemical Properties:

Specific Gravity: 6.47 *Vapor Density:* 10 *Boiling Point: Decomposes*
Melting Point: n/a *Vapor Pressure:* n/a *Solubility in water:* n/a
Other Information: Sinks in water.

Health Hazards:

Inhalation Hazard: Causes coughing, pain, difficulty in breathing. Dust is toxic.
Ingestion Hazard: Poisonous if swallowed.
Absorption Hazard: Will burn the skin and the eyes.
Hazards to Wildlife: Dangerous to aquatic life.
Decontamination Procedures: Wash away any material with copious amounts of soap and water.
First Aid Procedures: Remove victim to fresh air, call emergency medical care. If not breathing give CPR. If breathing is difficult administer oxygen. Treat for shock.

Fire Hazards:

Flashpoint: n/a *Ignition temperature:* n/a
Flammable Explosive High Range: n/a *Low Range:* n/a
Toxic Products of Combustion: n/a
Other Hazards: n/a
Possible extinguishing agents: Extinguish fire using suitable agent for the type of surrounding fire. Use foam, dry chemical, or carbon dioxide.

Reactivity Hazards:

Reactive With: Decomposed by water to yellow mercuric subsulfate and free sulfuric acid, a corrosive material.
Other Reactions: n/a

Corrosivity Hazards:

Corrosive With: n/a *Neutralizing Agent:* n/a

Radioactivity Hazards:

Radiation Emitted: n/a *Other Hazards:* n/a

Recommended Protection for Response Personnel:

Avoid breathing vapors, keep upwind. Wear a sealed chemical suit (polycarbonate, PVC). Wash away any material which may have come into contact with the body with copious amounts of soap and water. If the fire becomes uncontrollable, consider appropriate evacuation.

MERCURIC SULFIDE

DOT Number: UN 2025 *DOT Hazard Class:* Poison B *DOT Guide Number:* 53
Synonyms: artificial cinnabar, chinese red, ethiops mineral, mercuric sulfide black, mercuric sulfide red, vermilion
STCC Number: n/a *Reportable Qty:* n/a
Mfg Name: Ventron Corp. *Phone No:* 1-617-774-3100

Physical Description:

Physical Form: Solid *Color:* Red or black *Odor:* Odorless
Other Information: n/a

Chemical Properties:

Specific Gravity: 8 *Vapor Density:* 8 *Boiling Point: n/a*
Melting Point: n/a *Vapor Pressure:* n/a *Solubility in water:* n/a
Other Information: Sinks in water.

Health Hazards:

Inhalation Hazard: Poisonous if inhaled.
Ingestion Hazard: Poisonous if swallowed.
Absorption Hazard: Poisonous to the skin and the eyes.
Hazards to Wildlife: n/a
Decontamination Procedures: Wash away any material with copious amounts of soap and water.
First Aid Procedures: Remove victim to fresh air, call emergency medical care. If not breathing give CPR. If breathing is difficult administer oxygen. Treat for shock.

Fire Hazards:

Flashpoint: Combustible solid *Ignition temperature:* n/a
Flammable Explosive High Range: n/a *Low Range:* n/a
Toxic Products of Combustion: Smoke from fire contains poisonous mercury vapor and irritating sulfur dioxide gas.
Other Hazards: In fire, may soften and molten sulfur may flow and burn.
Possible extinguishing agents: Water, foam and sand. Other agents may be ineffective.

Reactivity Hazards:

Reactive With: n/a *Other Reactions:* n/a

Corrosivity Hazards:

Corrosive With: n/a *Neutralizing Agent:* n/a

Radioactivity Hazards:

Radiation Emitted: n/a *Other Hazards:* n/a

Recommended Protection for Response Personnel:

Avoid breathing vapors, keep upwind. Wear a sealed chemical suit (polycarbonate, butyl rubber). Wash away any material which may have come into contact with the body with copious amounts of soap and water. Consider appropriate evacuation.

MERCURIC THIOCYANATE

DOT Number: UN 1646 *DOT Hazard Class:* Poison B *DOT Guide Number:* 53
Synonyms: mercuric sulfocyanate, mercuric sulfocyanide, mercury rhodanide
STCC Number: 4923258 *Reportable Qty:* 10/4.54
Mfg Name: R.S.A. Corp. *Phone No:* 1-914-693-1818

Physical Description:

Physical Form: Solid *Color:* White to tan *Odor:* Odorless
Other Information: When heated or burned, the material expands to many times in volume.

Chemical Properties:

Specific Gravity: .99 *Vapor Density:* 1.78 *Boiling Point: 329° F(165° C)*
Melting Point: n/a *Vapor Pressure:* 116.7 psia at 68° F(20° C) *Solubility in water:* Yes
Other Information: Sinks and mixes slowly with water.

Health Hazards:

Inhalation Hazard: Poisonous if inhaled.
Ingestion Hazard: Poisonous if swallowed.
Absorption Hazard: Irritating to the skin and the eyes.
Hazards to Wildlife: Harmful to aquatic life.
Decontamination Procedures: Wash away any material with copious amounts of soap and water.
First Aid Procedures: Remove victim to fresh air, call emergency medical care. If not breathing give CPR. If breathing is difficult administer oxygen. Treat for shock.

Fire Hazards:

Flashpoint: 250° F(121° C) *Ignition temperature:* n/a
Flammable Explosive High Range: n/a *Low Range:* n/a
Toxic Products of Combustion: When heated, material decomposes into nitrogen oxides and sulfur compounds.
Other Hazards: When heated, material swells to many times its volume.
Possible extinguishing agents: Extinguish fire using suitable agent for the type of surrounding fire. Use alcohol foam, dry chemical, or carbon dioxide.

Reactivity Hazards:

Reactive With: n/a *Other Reactions:* n/a

Corrosivity Hazards:

Corrosive With: n/a *Neutralizing Agent:* n/a

Radioactivity Hazards:

Radiation Emitted: n/a *Other Hazards:* n/a

Recommended Protection for Response Personnel:

Avoid breathing vapors, keep upwind. Wear a sealed chemical suit. Wash away any material which may have come into contact with the body with copious amounts of soap and water. Consider appropriate evacuation.

MERCUROUS CHLORIDE

DOT Number: UN 2025 *DOT Hazard Class:* Poison B *DOT Guide Number:* 53
Synonyms: calomel, mercury monochloride, mercury protochloride, mercury subchloride, mild mercury chloride
STCC Number: n/a *Reportable Qty:* n/a
Mfg Name: Mallinckrodt Inc. *Phone No:* 1-314-895-2000

Physical Description:

Physical Form: Solid *Color:* White *Odor:* Odorless
Other Information: n/a

Chemical Properties:

Specific Gravity: 7.15 *Vapor Density:* n/a *Boiling Point: n/a*
Melting Point: n/a *Vapor Pressure:* n/a *Solubility in water:* n/a
Other Information: Sinks in water

Health Hazards:

Inhalation Hazard: Poisonous if inhaled.
Ingestion Hazard: Poisonous if swallowed.
Absorption Hazard: Irritating to the skin and the eyes.
Hazards to Wildlife: n/a
Decontamination Procedures: Wash away any material with copious amounts of soap and water.
First Aid Procedures: Remove victim to fresh air, call emergency medical care. If not breathing give CPR. If breathing is difficult administer oxygen. Treat for shock.

Fire Hazards:

Flashpoint: n/a *Ignition temperature:* n/a
Flammable Explosive High Range: n/a *Low Range:* n/a
Toxic Products of Combustion: Fumes from fire may contain toxic vapors of substance.
Other Hazards: Vaporizes and escapes as a sublimate.
Possible extinguishing agents: Water, alcohol foam, dry chemical, and carbon dioxide.

Reactivity Hazards:

Reactive With: n/a *Other Reactions:* n/a

Corrosivity Hazards:

Corrosive With: n/a *Neutralizing Agent:* n/a

Radioactivity Hazards:

Radiation Emitted: n/a *Other Hazards:* n/a

Recommended Protection for Response Personnel:

Avoid breathing vapors, keep upwind. Wear a sealed chemical suit (polycarbonate, butyl rubber).Wash away any material which may have come into contact with the body with copious amounts of soap and water. Consider appropriate evacuation.

MERCUROUS NITRATE

DOT Number: UN 1627 *DOT Hazard Class:* Oxidizer *DOT Guide Number:* 42
Synonyms: mercurous nitrate monohydrate, mercury protonitrate
STCC Number: 4918752 *Reportable Qty:* 10/4.54
Mfg Name: Mallinckrodt Inc. *Phone No:* 1-314-895-2000

Physical Description:

Physical Form: Solid *Color:* White *Odor:* Slight
Other Information: It is used in medicine, and in chemical analysis.

Chemical Properties:

Specific Gravity: 4.78 *Vapor Density:* n/a *Boiling Point: Decomposes*
Melting Point: n/a *Vapor Pressure:* n/a *Solubility in water:* Slight
Other Information: Sinks in water.

Health Hazards:

Inhalation Hazard: Poisonous if inhaled.
Ingestion Hazard: Poisonous if swallowed.
Absorption Hazard: Poisonous if the skin is exposed.
Hazards to Wildlife: n/a
Decontamination Procedures: Wash away any material with copious amounts of soap and water.
First Aid Procedures: Remove victim to fresh air, call emergency medical care. If not breathing give CPR. If breathing is difficult administer oxygen. Treat for shock.

Fire Hazards:

Flashpoint: n/a *Ignition temperature:* n/a
Flammable Explosive High Range: n/a *Low Range:* n/a
Toxic Products of Combustion: Smoke from fire may contain toxic mercury vapor and oxides of nitrogen.
Other Hazards: Will increase the intensity of the fire.
Possible extinguishing agents: Flood with water. Apply water from as far a distance as possible.

Reactivity Hazards:

Reactive With: Water which dissolves then forms a cloudy, acid solution.
Other Reactions: With wood or paper which may cause fire.

Corrosivity Hazards:

Corrosive With: Metal in the presence of moisture.
Neutralizing Agent: Flush with with water, rinse with solution of sodium bicarbonate or soda ash.

Radioactivity Hazards:

Radiation Emitted: n/a *Other Hazards:* n/a

Recommended Protection for Response Personnel:

Avoid breathing vapors, keep upwind. Wear a sealed chemical suit (polycarbonate, butyl rubber).Wash away any material which may have come into contact with the body with copious amounts of soap and water. If the fire becomes uncontrollable, consider appropriate evacuation.

MERCURY

DOT Number: NA 2809 *DOT Hazard Class:* ORM-B *DOT Guide Number:* 60
Synonyms: quicksilver
STCC Number: 4944325 *Reportable Qty:* 1/.454
Mfg Name: Belmont Metals *Phone No:* 1-718-342-4900

Physical Description:

Physical Form: Silvery liquid metal *Color:* Silver *Odor:* Odorless
Other Information: It is used as a catalyst in instruments, boilers, mirror coatings and other uses.

Chemical Properties:

Specific Gravity: n/a *Vapor Density:* 7 *Boiling Point: 674° F(357° C)*
Melting Point: −38° F(−39° C) *Vapor Pressure:* .0012 mm Hg at 68° F(20° C)
Solubility in water: No *Degree of Solubility:* .002%
Other Information: n/a

Health Hazards:

Inhalation Hazard: Poisonous if inhaled.
Ingestion Hazard: Poisonous if swallowed.
Absorption Hazard: Poisonous if absorbed.
Hazards to Wildlife: Dangerous to aquatic life.
Decontamination Procedures: Wash away any material with copious amounts of soap and water.
First Aid Procedures: Remove victim to fresh air, call emergency medical care. If not breathing give CPR. If breathing is difficult administer oxygen. Treat for shock.

Fire Hazards:

Flashpoint: n/a *Ignition temperature:* n/a
Flammable Explosive High Range: n/a *Low Range:* n/a
Toxic Products of Combustion: n/a
Other Hazards: n/a
Possible extinguishing agents: Extinguish fire using suitable agent for type of surrounding fire.

Reactivity Hazards:

Reactive With: n/a *Other Reactions:* n/a

Corrosivity Hazards:

Corrosive With: n/a *Neutralizing Agent:* n/a

Radioactivity Hazards:

Radiation Emitted: n/a *Other Hazards:* n/a

Recommended Protection for Response Personnel:

Avoid breathing vapors, keep upwind. Wear a sealed chemical suit (polycarbonate, butyl rubber, viton, nitrile, PVC, chlorinated polyethylene, neoprene). Wash away any material which may have come into contact with the body with copious amounts of soap and water. Caution, be aware of mercury poisoning. Consider appropriate evacuation.

MERCURY AMMONIUM CHLORIDE

DOT Number: UN 1630 *DOT Hazard Class:* Poison B *DOT Guide Number:* 53
Synonyms: Albus, ammoniated mercury, mercuric chloride (ammoniated), mercury amide chloride, mercury ammonium chloride
STCC Number: 4923242 *Reportable Qty:* n/a
Mfg Name: Ventron Corp. *Phone No:* 1-617-774-3100

Physical Description:

Physical Form: Solid *Color:* White *Odor:* Odorless
Other Information: n/a

Chemical Properties:

Specific Gravity: 5.7 *Vapor Density:* 8.7 *Boiling Point: n/a*
Melting Point: n/a *Vapor Pressure:* n/a *Solubility in water:* Yes
Other Information: Sinks, soluble in water.

Health Hazards:

Inhalation Hazard: Poisonous if inhaled.
Ingestion Hazard: Poisonous if swallowed.
Absorption Hazard: Poisonous if absorbed.
Hazards to Wildlife: n/a
Decontamination Procedures: Wash away any material with copious amounts of soap and water.
First Aid Procedures: Remove victim to fresh air, call emergency medical care. If not breathing give CPR. If breathing is difficult administer oxygen. Treat for shock.

Fire Hazards:

Flashpoint: n/a *Ignition temperature:* n/a
Flammable Explosive High Range: n/a *Low Range:* n/a
Toxic Products of Combustion: Toxic mercury compounds.
Other Hazards: n/a
Possible extinguishing agents: Extinguish fire using suitable agent for the type of surrounding fire. Use alcohol foam, dry chemical, or carbon dioxide.

Reactivity Hazards:

Reactive With: n/a *Other Reactions:* n/a

Corrosivity Hazards:

Corrosive With: n/a *Neutralizing Agent:* n/a

Radioactivity Hazards:

Radiation Emitted: n/a *Other Hazards:* n/a

Recommended Protection for Response Personnel:

Avoid breathing vapors, keep upwind. Wear a sealed chemical suit (polycarbonate, butyl rubber). Wash away any material which may have come into contact with the body with copious amounts of soap and water. Consider appropriate evacuation.

MESITYL OXIDE

DOT Number: UN 1229 *DOT Hazard Class:* Flammable liquid *DOT Guide Number:* 26
Synonyms: methyl isobutenyl ketone
STCC Number: 4909223 *Reportable Qty:* n/a
Mfg Name: Union Carbide *Phone No:* 1-203-794-2000

Physical Description:

Physical Form: Liquid *Color:* Colorless to light yellow *Odor:* Strong peppermint
Other Information: It is used in paint removers, as a solvent for plastics, and as an insect repellent.

Chemical Properties:

Specific Gravity: .9 *Vapor Density:* 3.4 *Boiling Point: 266° F(130° C)*
Melting Point: –51° F(–46° C) *Vapor Pressure:* 8 mm Hg at 68° F(20° C)
Solubility in water: Slight *Degree of Solubility:* 3%
Other Information: Lighter than and slightly soluble in water. Vapors are heavier than air.

Health Hazards:

Inhalation Hazard: Dizziness, difficulty in breathing.
Ingestion Hazard: Harmful if swallowed.
Absorption Hazard: Irritating to the skin and eyes.
Hazards to Wildlife: n/a
Decontamination Procedures: Wash away any material with copious amounts of soap and water.
First Aid Procedures: Remove victim to fresh air, call emergency medical care. If not breathing give CPR. If breathing is difficult administer oxygen. Treat for shock.

Fire Hazards:

Flashpoint: 87° F(31° C) *Ignition temperature:* 652° F(344° C)
Flammable Explosive High Range: 7.2 *Low Range:* 1.4
Toxic Products of Combustion: None
Other Hazards: Containers may explode in fire. Flashback along vapor trail may occur. Vapors may explode if ignited in an enclosed area.
Possible extinguishing agents: Apply water from as far a distance as possible. Use alcohol foam, dry chemical, or carbon dioxide.

Reactivity Hazards:

Reactive With: n/a *Other Reactions:* n/a

Corrosivity Hazards:

Corrosive With: n/a *Neutralizing Agent:* n/a

Radioactivity Hazards:

Radiation Emitted: n/a *Other Hazards:* n/a

Recommended Protection for Response Personnel:

Avoid breathing vapors, keep upwind. Wear a sealed chemical suit (polycarbonate, butyl rubber). Wash away any material which may have come into contact with the body with copious amounts of soap and water. Consider appropriate evacuation.

METHALLYL ALCOHOL

DOT Number: UN 2614 *DOT Hazard Class:* Flammable liquid *DOT Guide Number:* 26
Synonyms: none given
STCC Number: 4910353 *Reportable Qty:* n/a *CHEMTREC Phone No:* 1-800-424-9300

Physical Description:

Physical Form: Liquid *Color:* Colorless *Odor:* Sharp pungent
Other Information: It is used in the chemical processing industry.

Chemical Properties:

Specific Gravity: .9 *Vapor Density:* 2.5 *Boiling Point: 237° F(114° C)*
Melting Point: n/a *Vapor Pressure:* n/a *Solubility in water:* Slight
Other Information: Soluble in water, alcohols, esters, ketones, and hydrocarbons.

Health Hazards:

Inhalation Hazard: Moderately toxic.
Ingestion Hazard: Moderately toxic.
Absorption Hazard: Irritating to the skin and eyes.
Hazards to Wildlife: n/a
Decontamination Procedures: Wash away any material with copious amounts of soap and water.
First Aid Procedures: Remove victim to fresh air, call emergency medical care. If not breathing give CPR. If breathing is difficult administer oxygen. Treat for shock.

Fire Hazards:

Flashpoint: 92° F(33° C) *Ignition temperature:* n/a
Flammable Explosive High Range: n/a *Low Range:* n/a
Toxic Products of Combustion: n/a
Other Hazards: n/a
Possible extinguishing agents: Apply water from as far a distance as possible. Use alcohol foam, dry chemical, or carbon dioxide.

Reactivity Hazards:

Reactive With: n/a *Other Reactions:* n/a

Corrosivity Hazards:

Corrosive With: n/a *Neutralizing Agent:* n/a

Radioactivity Hazards:

Radiation Emitted: n/a *Other Hazards:* n/a

Recommended Protection for Response Personnel:

Avoid breathing vapors, keep upwind. Wear appropriate chemical clothing. Wash away any material which may have come into contact with the body with copious amounts of soap and water. Consider appropriate evacuation.

METHANE

DOT Number: UN 1971 *DOT Hazard Class:* Flammable liquid *DOT Guide Number:* 17
Synonyms: marsh gas, natural gas
STCC Number: 4905755 *Reportable Qty:* n/a
Mfg Name: E.I. Du Pont *Phone No:* 1-800-441-3637

Physical Description:

Physical Form: Liquid *Color:* Colorless *Odor:* Odorless
Other Information: It is used in making other chemicals, and as a constituent of the fuel, natural gas.

Chemical Properties:

Specific Gravity: .42 *Vapor Density:* .6 *Boiling Point: – 259° F(– 162° C)*
Melting Point: n/a *Vapor Pressure:* .55 mm Hg at 68° F(20° C) *Solubility in water:* No
Other Information: Can only be shipped in cylinders. Material is lighter than air.

Health Hazards:

Inhalation Hazard: Dizziness, difficulty in breathing, loss of consciousness.
Ingestion Hazard: Displaces oxygen.
Absorption Hazard: Will cause frostbite.
Hazards to Wildlife: n/a
Decontamination Procedures: Wash away any material with copious amounts of soap and water.
First Aid Procedures: Remove victim to fresh air, call emergency medical care. If not breathing give CPR. If breathing is difficult administer oxygen. Treat for shock.

Fire Hazards:

Flashpoint: n/a *Ignition temperature:* 999° F
Flammable Explosive High Range: 15 *Low Range:* 5
Toxic Products of Combustion: n/a
Other Hazards: Containers may explode in fire. Flashback along vapor trail may occur. Vapors may explode if ignited in an enclosed area.
Possible extinguishing agents: Apply water from as far a distance as possible.

Reactivity Hazards:

Reactive With: n/a *Other Reactions:* n/a

Corrosivity Hazards:

Corrosive With: n/a *Neutralizing Agent:* n/a

Radioactivity Hazards:

Radiation Emitted: n/a *Other Hazards:* n/a

Recommended Protection for Response Personnel:

Avoid breathing vapors, keep upwind. Wear appropriate chemical clothing. Wash away any material which may have come into contact with the body with copious amounts of soap and water. If fire becomes uncontrollable, consider appropriate evacuation.

METHANEARSONIC ACID, SODIUM SALT

DOT Number: UN 1557 *DOT Hazard Class:* Poison B *DOT Guide Number:* 53
Synonyms: disodium methanearsonate, disodium methyl arsenate, DSMA, monosodium methanearsonate, monosodium methyl arsenate, MSMA
STCC Number: n/a *Reportable Qty:* n/a
Mfg Name: Ansul Co. *Phone No:* 1-715-735-7411

Physical Description:

Physical Form: Solid *Color:* Colorless *Odor:* Odorless
Other Information: n/a

Chemical Properties:

Specific Gravity: 1 *Vapor Density:* n/a *Boiling Point: Decomposes*
Melting Point: n/a *Vapor Pressure:* n/a *Solubility in water:* Yes
Other Information: May float or sink in water.

Health Hazards:

Inhalation Hazard: n/a
Ingestion Hazard: Nausea, vomiting, loss of consciousness.
Absorption Hazard: Irritating to the skin and the eyes.
Hazards to Wildlife: Dangerous to aquatic life.
Decontamination Procedures: Wash away any material with copious amounts of soap and water.
First Aid Procedures: Remove victim to fresh air, call emergency medical care. If not breathing give CPR. If breathing is difficult administer oxygen. Treat for shock.

Fire Hazards:

Flashpoint: n/a *Ignition temperature:* n/a
Flammable Explosive High Range: n/a *Low Range:* n/a
Toxic Products of Combustion: Toxic gases may be generated in fires.
Other Hazards: n/a
Possible extinguishing agents: Extinguish fire using suitable agent for the type of surrounding fire.

Reactivity Hazards:

Reactive With: n/a *Other Reactions:* n/a

Corrosivity Hazards:

Corrosive With: n/a *Neutralizing Agent:* n/a

Radioactivity Hazards:

Radiation Emitted: n/a *Other Hazards:* n/a

Recommended Protection for Response Personnel:

Avoid breathing vapors, keep upwind. Structural protective clothing provides limited protection. Wash away any material which may have come into contact with the body with copious amounts of soap and water. Consider appropriate evacuation.

METHANOL

DOT Number: UN 1230 *DOT Hazard Class:* Flammable liquid *DOT Guide Number:* 28
Synonyms: methyl alcohol, methyl hydroxide, wood alcohol, wood spirit
STCC Number: 4909230 *Reportable Qty:* n/a
Mfg Name: Allied Corp. *Phone No:* 1-201-455-2000

Physical Description:

Physical Form: Liquid *Color:* Colorless *Odor:* Sweet pungent
Other Information: It is used as a solvent for inks, resins, adhesives and dyes.

Chemical Properties:

Specific Gravity: .792 *Vapor Density:* 1.1 *Boiling Point: 148° F(64° C)*
Melting Point: −144° F(−98° C) *Vapor Pressure:* 97 mm Hg at 68° F(20° C) *Solubility in water:* Yes
Other Information: Fully soluble in water. Weighs 6.6 lbs/3 kg per gallon/3.8 l.

Health Hazards:

Inhalation Hazard: Headache, weakness, and drowsiness.
Ingestion Hazard: Poisonous if swallowed.
Absorption Hazard: Irritating to the skin and eyes.
Hazards to Wildlife: Dangerous to aquatic life.
Decontamination Procedures: Wash away any material with copious amounts of soap and water.
First Aid Procedures: Remove victim to fresh air, call emergency medical care. If not breathing give CPR. If breathing is difficult administer oxygen. Treat for shock.

Fire Hazards:

Flashpoint: 54° F(12° C) *Ignition temperature:* 725° F(385° C)
Flammable Explosive High Range: 36.5 *Low Range:* 6
Toxic Products of Combustion: May include toxic formaldehyde, carbon monoxide, and possibly unburned methanol.
Other Hazards: Flashback along vapor trail may occur. The containers may explode in a fire.
Possible extinguishing agents: Apply water from as far a distance as possible. Use alcohol foam, dry chemical, or carbon dioxide.

Reactivity Hazards:

Reactive With: n/a *Other Reactions:* n/a

Corrosivity Hazards:

Corrosive With: n/a *Neutralizing Agent:* n/a

Radioactivity Hazards:

Radiation Emitted: n/a *Other Hazards:* n/a

Recommended Protection for Response Personnel:

Avoid breathing vapors, keep upwind. Structural protective clothing provides limited protection. Wash away any material which may have come into contact with the body with copious amounts of soap and water. If the fire becomes uncontrollable, consider appropriate evacuation.

METHOXYCHLOR

DOT Number: NA 2761 *DOT Hazard Class:* ORM-E *DOT Guide Number:* 55
Synonyms: DMDT, Marlate 50, methoxy DDT
STCC Number: 4960646 *Reportable Qty:* 1/.454
Mfg Name: E.I. Du Pont *Phone No:* 1-800-441-3637

Physical Description:

Physical Form: Solid *Color:* White to light yellow *Odor:* Fruity
Other Information: It is used as a pesticide.

Chemical Properties:

Specific Gravity: 1.4 *Vapor Density:* 12 *Boiling Point: Decomposes*
Melting Point: 180° F(82° C) *Vapor Pressure:* Very low
Solubility in water: No *Degree of Solubility:* .01%
Other Information: In case of damage to or leaking from containers, contact the Pesticide Safety Network at 1-800-424-9300.

Health Hazards:

Inhalation Hazard: Poisonous if inhaled.
Ingestion Hazard: Poisonous if swallowed.
Absorption Hazard: Irritating to the skin and eyes.
Hazards to Wildlife: Dangerous to waterfowl and aquatic life.
Decontamination Procedures: Wash away any material with copious amounts of soap and water.
First Aid Procedures: Remove victim to fresh air, call emergency medical care. If not breathing give CPR. If breathing is difficult administer oxygen. Treat for shock.

Fire Hazards:

Flashpoint: n/a *Ignition temperature:* n/a
Flammable Explosive High Range: n/a *Low Range:* n/a
Toxic Products of Combustion: Irritating and toxic hydrogen chloride gas may be formed in a fire.
Other Hazards: Avoid contact with solid and dust. Keep people away!
Possible extinguishing agents: Extinguish fire using suitable agent for type of surrounding fire.

Reactivity Hazards:

Reactive With: n/a *Other Reactions:* n/a

Corrosivity Hazards:

Corrosive With: n/a *Neutralizing Agent:* n/a

Radioactivity Hazards:

Radiation Emitted: n/a *Other Hazards:* n/a

Recommended Protection for Response Personnel:

Avoid breathing vapors, keep upwind. Structural protective clothing provides limited protection. Wash away any material which may have come into contact with the body with copious amounts of soap and water. Consider appropriate evacuation.

METHYL ACETATE

DOT Number: UN 1231 *DOT Hazard Class:* Flammable liquid *DOT Guide Number:* 26
Synonyms: acetic acid, methyl ester
STCC Number: 4908220 *Reportable Qty:* n/a
Mfg Name: Monsanto Co. *Phone No:* 1-314-694-1000

Physical Description:

Physical Form: Liquid *Color:* Colorless *Odor:* Mild sweet, fragrant
Other Information: It is lighter than and soluble in water. Vapors are heavier than air.

Chemical Properties:

Specific Gravity: .9 *Vapor Density:* 2.8 *Boiling Point: 140° F(60° C)*
Melting Point: −144° F(−98° C) *Vapor Pressure:* 173 mm Hg at 68° F(20° C)
Solubility in water: Yes *Degree of Solubility:* 24.5%
Other Information: n/a

Health Hazards:

Inhalation Hazard: Headaches and dizziness.
Ingestion Hazard: n/a
Absorption Hazard: Irritating to the skin and the eyes.
Hazards to Wildlife: n/a
Decontamination Procedures: Wash away any material with copious amounts of soap and water.
First Aid Procedures: Remove victim to fresh air, call emergency medical care. If not breathing give CPR. If breathing is difficult administer oxygen. Treat for shock.

Fire Hazards:

Flashpoint: 14° F(−10° C) *Ignition temperature:* 850° F(454° C)
Flammable Explosive High Range: 16 *Low Range:* 3.1
Toxic Products of Combustion: Irritating vapors are produced in a fire.
Other Hazards: Containers may explode in a fire. Flashback along vapor trail may occur. Vapors may explode if ignited in an enclosed area.
Possible extinguishing agents: Apply water from as far a distance as possible. Use alcohol foam, dry chemical, or carbon dioxide.

Reactivity Hazards:

Reactive With: Water to form acetic acid and methyl alcohol. *Other Reactions:* n/a

Corrosivity Hazards:

Corrosive With: n/a *Neutralizing Agent:* n/a

Radioactivity Hazards:

Radiation Emitted: n/a *Other Hazards:* n/a

Recommended Protection for Response Personnel:

Avoid breathing vapors, keep upwind. Structural protective clothing provides limited protection. Wash away any material which may have come into contact with the body with copious amounts of soap and water. If the fire becomes uncontrollable, consider appropriate evacuation.

METHYL ACETYLENE

DOT Number: UN 1060 *DOT Hazard Class:* Flammable gas *DOT Guide Number:* 17
Synonyms: MAPP gas
STCC Number: n/a *Reportable Qty:* n/a
Mfg Name: Dow Chemical *Phone No:* 1-517-636-4400

Physical Description:

Physical Form: Gas *Color:* Colorless *Odor:* Garlic
Other Information: Floats and boils on water, flammable, visible vapor cloud is produced.

Chemical Properties:

Specific Gravity: .567 *Vapor Density:* 1.48 *Boiling Point: – 10° F(–23° C)*
Melting Point: –153° F(–103° C) *Vapor Pressure:* 3800 mm Hg at 68° F(20° C) *Solubility in water:* Insoluble
Other Information: n/a

Health Hazards:

Inhalation Hazard: Difficulty in breathing.
Ingestion Hazard: n/a
Absorption Hazard: Will cause frostbite.
Hazards to Wildlife: n/a
Decontamination Procedures: Wash away any material with copious amounts of soap and water.
First Aid Procedures: Remove victim to fresh air, call emergency medical care. If not breathing give CPR. If breathing is difficult administer oxygen. Treat for shock.

Fire Hazards:

Flashpoint: –101° F(–74° C) *Ignition temperature:* 850° F(454° C)
Flammable Explosive High Range: 11.7 *Low Range:* 1.7
Toxic Products of Combustion: n/a
Other Hazards: Containers may explode in a fire. Flashback along vapor trail may occur. Vapors may explode if ignited in an enclosed area.
Possible extinguishing agents: Apply water from as far a distance as possible to cool the containers.

Reactivity Hazards:

Reactive With: Copper to form explosive compounds. *Other Reactions:* n/a

Corrosivity Hazards:

Corrosive With: n/a *Neutralizing Agent:* n/a

Radioactivity Hazards:

Radiation Emitted: n/a *Other Hazards:* n/a

Recommended Protection for Response Personnel:

Avoid breathing vapors, keep upwind. Wear appropriate chemical clothing. Wash away any material which may come into contact with the body with copious amounts of soap and water. If the fire becomes uncontrollable, consider appropriate evacuation.

METHYL ACRYLATE

DOT Number: UN 1919 *DOT Hazard Class:* Flammable liquid *DOT Guide Number:* 26
Synonyms: acrylic acid, acrylic ester
STCC Number: 4907245 *Reportable Qty:* n/a
Mfg Name: H. Muehlstein & Co. *Phone No:* 1-203-622-6500

Physical Description:

Physical Form: Liquid *Color:* Colorless *Odor:* Sweet sharp
Other Information: n/a

Chemical Properties:

Specific Gravity: 1 *Vapor Density:* 3 *Boiling Point: 176° F(80° C)*
Melting Point: −130° F(54.4° C) *Vapor Pressure:* 68.2 mm Hg at 68° F(20° C)
Solubility in water: Slight *Degree of Solubility: 5.5%*
Other Information: Lighter than and slightly soluble in water. Vapors are heavier than air.

Health Hazards:

Inhalation Hazard: Dizziness, difficulty in breathing.
Ingestion Hazard: Harmful if swallowed.
Absorption Hazard: Will burn the skin and the eyes.
Hazards to Wildlife: n/a
Decontamination Procedures: Wash away any material with copious amounts of soap and water.
First Aid Procedures: Remove victim to fresh air, call emergency medical care. If not breathing give CPR. If breathing is difficult administer oxygen. Treat for shock.

Fire Hazards:

Flashpoint: 13° F(−10.6° C) *Ignition temperature:* 875° F(468.3° C)
Flammable Explosive High Range: 25 *Low Range:* 2.8
Toxic Products of Combustion: n/a
Other Hazards: Subject to polymerization with the evolution of heat. If this takes place, the container may violently rupture.
Possible extinguishing agents: Apply water from as far a distance as possible. Use foam, dry chemical, or carbon dioxide.

Reactivity Hazards:

Reactive With: n/a *Other Reactions:* n/a

Corrosivity Hazards:

Corrosive With: n/a *Neutralizing Agent:* n/a

Radioactivity Hazards:

Radiation Emitted: n/a *Other Hazards:* n/a

Recommended Protection for Response Personnel:

Avoid breathing vapors, keep upwind. Wear a sealed chemical suit (neoprene, butyl rubber). Wash away any material which may have come into contact with the body with copious amounts of soap and water. If fire becomes uncontrollable, consider appropriate evacuation.

METHYL ALCOHOL

DOT Number: UN 1230 *DOT Hazard Class:* Flammable liquid *DOT Guide Number:* 28
Synonyms: methanol, wood alcohol
STCC Number: 4909230 *Reportable Qty:* 5000/2270
Mfg Name: Allied Corp. *Phone No:* 1-201-455-2000

Physical Description:

Physical Form: Liquid *Color:* Colorless *Odor:* Sweet pungent
Other Information: It is used as a raw material to make chemicals, and to remove water from automobile fuels.

Chemical Properties:

Specific Gravity: .8 *Vapor Density:* 1.1 *Boiling Point: 147° F(63.9° C)*
Melting Point: −144° F(−97.8° C) *Vapor Pressure:* 97 mm Hg at 68° F(20° C)
Solubility in water: Yes
Other Information: Fully soluble in water. Weighs 6.6 lbs/3.0 kg per gallon/3.8 l.

Health Hazards:

Inhalation Hazard: Dizziness, difficulty in breathing.
Ingestion Hazard: Poisonous if swallowed.
Absorption Hazard: Irritating to the skin and eyes.
Hazards to Wildlife: Dangerous to aquatic life.
Decontamination Procedures: Wash away any material with copious amounts of soap and water.
First Aid Procedures: Remove victim to fresh air, call emergency medical care. If not breathing give CPR. If breathing is difficult administer oxygen. Treat for shock.

Fire Hazards:

Flashpoint: 52° F(11.1° C) *Ignition temperature:* 867° F(463.9° C)
Flammable Explosive High Range: 36 *Low Range:* 6
Toxic Products of Combustion: n/a
Other Hazards: Flashback along vapor trail may occur. Containers may explode in fire.
Possible extinguishing agents: Apply water from as far a distance as possible. Use alcohol foam, dry chemical, or carbon dioxide.

Reactivity Hazards:

Reactive With: n/a *Other Reactions:* n/a

Corrosivity Hazards:

Corrosive With: n/a *Neutralizing Agent:* n/a

Radioactivity Hazards:

Radiation Emitted: n/a *Other Hazards:* n/a

Recommended Protection for Response Personnel:

Avoid breathing vapors, keep upwind. Structural protective clothing provides limited protection. Wash away any material which may have come into contact with the body with copious amounts of soap and water. If the fire becomes uncontrollable, consider appropriate evacuation.

METHYL ALLYL CHLORIDE

DOT Number: UN 2554 *DOT Hazard Class:* Flammable liquid *DOT Guide Number:* 26
Synonyms: β-methallyl chloride
STCC Number: 4908223 *Reportable Qty:* n/a
Mfg Name: Aldrich Chemical *Phone No:* 1-414-273-3850

Physical Description:

Physical Form: Liquid *Color:* Straw colored *Odor:* Sharp penetrating
Other Information: n/a

Chemical Properties:

Specific Gravity: .9 *Vapor Density:* 3.1 *Boiling Point: 162° F(72° C)*
Melting Point: n/a *Vapor Pressure:* n/a *Solubility in water:* No
Other Information: Lighter than and insoluble in water. Vapors are heavier than air.

Health Hazards:

Inhalation Hazard: Harmful if inhaled.
Ingestion Hazard: Harmful if swallowed.
Absorption Hazard: Irritating to the skin and eyes.
Hazards to Wildlife: n/a
Decontamination Procedures: Wash away any material with copious amounts of soap and water.
First Aid Procedures: Remove victim to fresh air, call emergency medical care. If not breathing give CPR. If breathing is difficult administer oxygen. Treat for shock.

Fire Hazards:

Flashpoint: 11° F(−12° C) *Ignition temperature:* n/a
Flammable Explosive High Range: 8.1 *Low Range:* 3.2
Toxic Products of Combustion: Irritating gases may be produced when heated.
Other Hazards: Containers may explode in fire. Flashback along vapor trail may occur. Vapors may explode if ignited in an enclosed area.
Possible extinguishing agents: Apply water from as far a distance as possible. Use foam, dry chemical, or carbon dioxide.

Reactivity Hazards:

Reactive With: n/a *Other Reactions:* n/a

Corrosivity Hazards:

Corrosive With: n/a *Neutralizing Agent:* n/a

Radioactivity Hazards:

Radiation Emitted: n/a *Other Hazards:* n/a

Recommended Protection for Response Personnel:

Avoid breathing vapors, keep upwind. Wear appropriate chemical clothing. Wash away any material which may have come into contact with the body with copious amounts of soap and water. If fire becomes uncontrollable, consider appropriate evacuation.

METHYLAMINE

DOT Number: UN 1061 *DOT Hazard Class:* Flammable gas *DOT Guide Number:* 19
Synonyms: anhydrous methylamine
STCC Number: n/a *Reportable Qty:* n/a
Mfg Name: E.I. Du Pont *Phone No:* 1-800-441-3637

Physical Description:

Physical Form: Gas *Color:* Colorless *Odor:* Ammonialike
Other Information: n/a

Chemical Properties:

Specific Gravity: .66 *Vapor Density:* 1 *Boiling Point: 21° F(−6.1° C)*
Melting Point: −136° F(−93.3° C) *Vapor Pressure:* n/a
Solubility in water: Yes *Degree of Solubility:* 959%
Other Information: n/a

Health Hazards:

Inhalation Hazard: Poisonous if inhaled.
Ingestion Hazard: n/a
Absorption Hazard: Irritating to the skin and eyes.
Hazards to Wildlife: n/a
Decontamination Procedures: Wash away any material with copious amounts of soap and water.
First Aid Procedures: Remove victim to fresh air, call emergency medical care. If not breathing give CPR. If breathing is difficult administer oxygen. Treat for shock.

Fire Hazards:

Flashpoint: 32° F(0° C) *Ignition temperature:* 806° F(430° C)
Flammable Explosive High Range: 21 *Low Range:* 5
Toxic Products of Combustion: n/a
Other Hazards: Flashback along vapor trail may occur. Containers may explode in a fire.
Possible extinguishing agents: Apply water from as far a distance as possible.

Reactivity Hazards:

Reactive With: n/a *Other Reactions:* n/a

Corrosivity Hazards:

Corrosive With: n/a *Neutralizing Agent:* n/a

Radioactivity Hazards:

Radiation Emitted: n/a *Other Hazards:* n/a

Recommended Protection for Response Personnel:

Avoid breathing vapors, keep upwind. Wear a sealed chemical suit (butyl rubber). Wash away any material which may have come into contact with the body with copious amounts of soap and water. If the fire becomes uncontrollable, consider appropriate evacuation.

METHYL AMYL ACETATE

DOT Number: UN 1233 *DOT Hazard Class:* Flammable liquid *DOT Guide Number:* 26
Synonyms: MAAC, hexyl acetate, 4-methyl-2-pentanol acetate, 4-methyl-2-pentyl acetate
STCC Number: 4909235 *Reportable Qty:* n/a
Mfg Name: Shulton Inc. *Phone No:* 1-201-831-1234

Physical Description:

Physical Form: Liquid *Color:* Colorless *Odor:* Fruity
Other Information: n/a

Chemical Properties:

Specific Gravity: .86 *Vapor Density:* 4.97 *Boiling Point: 295° F(146.1° C)*
Melting Point: −82° F(−63.3° C) *Vapor Pressure:* 3.8 mm Hg at 68° F(20° C) *Solubility in water:* No
Other Information: Lighter than and insoluble in water. Vapors are heavier than air.

Health Hazards:

Inhalation Hazard: n/a
Ingestion Hazard: n/a
Absorption Hazard: Irritating to the skin and the eyes.
Hazards to Wildlife: n/a
Decontamination Procedures: Wash away any material with copious amounts of soap and water.
First Aid Procedures: Remove victim to fresh air, call emergency medical care. If not breathing give CPR. If breathing is difficult administer oxygen. Treat for shock.

Fire Hazards:

Flashpoint: 95° F(35° C) *Ignition temperature:* 510° F(265.6° C)
Flammable Explosive High Range: 5.7 *Low Range:* .9
Toxic Products of Combustion: n/a
Other Hazards: Containers may explode in a fire. Flashback along vapor trail may occur. Vapors may explode if ignited in an enclosed area.
Possible extinguishing agents: Apply water from as far a distance as possible. Use foam, dry chemical, or carbon dioxide.

Reactivity Hazards:

Reactive With: n/a *Other Reactions:* n/a

Corrosivity Hazards:

Corrosive With: n/a *Neutralizing Agent:* n/a

Radioactivity Hazards:

Radiation Emitted: n/a *Other Hazards:* n/a

Recommended Protection for Response Personnel:

Avoid breathing vapors, keep upwind. Structural protective clothing provides limited protection. Wash away any material which may have come into contact with the body with copious amounts of soap and water. If the fire becomes uncontrollable, consider appropriate evacuation.

METHYL AMYL ALCOHOL

DOT Number: UN 2053 *DOT Hazard Class:* Combustible liquid *DOT Guide Number:* 26
Synonyms: MAOH, MIBC, MIC
STCC Number: n/a *Reportable Qty:* n/a
Mfg Name: Ashland Chemical *Phone No:* 1-614-889-3333

Physical Description:

Physical Form: Liquid *Color:* Colorless *Odor:* Mild alcohol
Other Information: Lighter than and insoluble in water. Vapors are heavier than air.

Chemical Properties:

Specific Gravity: .8 *Vapor Density:* 3.5 *Boiling Point: 269° F(131.7° C)*
Melting Point: –130° F(–90° C) *Vapor Pressure:* 2.8 mm Hg at 68° F(20° C)
Solubility in water: Slight *Degree of Solubility:* 1.5%
Other Information: n/a

Health Hazards:

Inhalation Hazard: Causes dizziness, difficulty in breathing.
Ingestion Hazard: Harmful if swallowed.
Absorption Hazard: Irritating to the skin and the eyes.
Hazards to Wildlife: n/a
Decontamination Procedures: Wash away any material with copious amounts of soap and water.
First Aid Procedures: Remove victim to fresh air, call emergency medical care. If not breathing give CPR. If breathing is difficult administer oxygen. Treat for shock.

Fire Hazards:

Flashpoint: 106° F(41.1° C) *Ignition temperature:* 583° F(306.1° C)
Flammable Explosive High Range: 5.5 *Low Range:* 1
Toxic Products of Combustion: n/a
Other Hazards: n/a
Possible extinguishing agents: Apply water from as far a distance as possible. Use alcohol foam, dry chemical, or carbon dioxide.

Reactivity Hazards:

Reactive With: n/a *Other Reactions:* n/a

Corrosivity Hazards:

Corrosive With: n/a *Neutralizing Agent:* n/a

Radioactivity Hazards:

Radiation Emitted: n/a *Other Hazards:* n/a

Recommended Protection for Response Personnel:

Avoid breathing vapors, keep upwind. Wear a sealed chemical suit (polycarbonate, butyl rubber). Wash away any material which may have come into contact with the body with copious amounts of soap and water. Consider appropriate evacuation.

METHYL AMYL KETONE

DOT Number: UN 1110 *DOT Hazard Class:* Combustible liquid *DOT Guide Number:* 26
Synonyms: 2-heptanone
STCC Number: 4913132 *Reportable Qty:* n/a *CHEMTREC Phone No:* 1-800-424-9300

Physical Description:

Physical Form: Liquid *Color:* Colorless *Odor:* Bananalike
Other Information: Used as a synthetic flavoring, and in perfumes.

Chemical Properties:

Specific Gravity: .8 *Vapor Density:* 3.9 *Boiling Point: 302° F(150° C)*
Melting Point: −31° F(−35° C) *Vapor Pressure:* 2 mm Hg at 68° F(20° C)
Solubility in water: Slight *Degree of Solubility: .43%*
Other Information: Lighter than and slightly soluble in water. Its vapors are heavier than air. Weighs 6.8 lbs/3.1 kg per gallon/3.8 l.

Health Hazards:

Inhalation Hazard: Irritating to mucous membranes.
Ingestion Hazard: n/a
Absorption Hazard: Irritating to the skin and eyes.
Hazards to Wildlife: n/a
Decontamination Procedures: Wash away any material with copious amounts of soap and water.
First Aid Procedures: Remove victim to fresh air, call emergency medical care. If not breathing give CPR. If breathing is difficult administer oxygen. Treat for shock.

Fire Hazards:

Flashpoint: 102° F(38.9° C) *Ignition temperature:* 740° F(393.3° C)
Flammable Explosive High Range: 7.9 *Low Range:* 1.1
Toxic Products of Combustion: n/a
Other Hazards: n/a
Possible extinguishing agents: Apply water from as far a distance as possible. Use alcohol foam, dry chemical, or carbon dioxide.

Reactivity Hazards:

Reactive With: n/a *Other Reactions:* n/a

Corrosivity Hazards:

Corrosive With: n/a *Neutralizing Agent:* n/a

Radioactivity Hazards:

Radiation Emitted: n/a *Other Hazards:* n/a

Recommended Protection for Response Personnel:

Avoid breathing vapors, keep upwind. Wear appropriate chemical clothing, Wash away any material which may have come into contact with the body with copious amounts of soap and water. Consider appropriate evacuation.

METHYLANILINE

DOT Number: UN 2294 *DOT Hazard Class:* n/a *DOT Guide Number:* 55
Synonyms: anilinomethane, n-methylaminobenzene, methylaniline(mono), methyl phenylamine
STCC Number: n/a *Reportable Qty:* n/a
Mfg Name: Eastman Organic *Phone No:* 1-800-445-6325

Physical Description:

Physical Form: Liquid *Color:* Yellow to light brown *Odor:* Chemical
Other Information: n/a

Chemical Properties:

Specific Gravity: .99 *Vapor Density:* 3.9 *Boiling Point: 384° F(195.5° C)*
Melting Point: −71° F(−57.2° C) *Vapor Pressure:* n/a *Solubility in water:* n/a
Other Information: May float or sink in water.

Health Hazards:

Inhalation Hazard: n/a
Ingestion Hazard: Harmful if swallowed.
Absorption Hazard: Irritating to the skin and the eyes.
Hazards to Wildlife: n/a
Decontamination Procedures: Wash away any material with copious amounts of soap and water.
First Aid Procedures: Remove victim to fresh air, call emergency medical care. If not breathing give CPR. If breathing is difficult administer oxygen. Treat for shock.

Fire Hazards:

Flashpoint: 175° F(79.4° C) *Ignition temperature:* n/a
Flammable Explosive High Range: n/a *Low Range:* n/a
Toxic Products of Combustion: Toxic vapors are generated when heated.
Other Hazards: n/a
Possible extinguishing agents: Foam, dry chemical, carbon dioxide. Water may be ineffective on fires. Cool all affected containers with water. Apply water from as far a distance as possible.

Reactivity Hazards:

Reactive With: Oxidizing materials *Other Reactions:* n/a

Corrosivity Hazards:

Corrosive With: May attack some forms of plastics. *Neutralizing Agent:* n/a

Radioactivity Hazards:

Radiation Emitted: n/a *Other Hazards:* n/a

Recommended Protection for Response Personnel:

Avoid breathing vapors, keep upwind. Wear a sealed chemical suit (polycarbonate, butyl rubber). Wash away any material which may have come into contact with the body with copious amounts of soap and water. If the fire becomes uncontrollable, consider appropriate evacuation.

METHYL BROMIDE

DOT Number: NA 1062 *DOT Hazard Class:* Poison B *DOT Guide Number:* 55
Synonyms: Embafume, M-B-C fumigant
STCC Number: 4921440 *Reportable Qty:* 1000/454
Mfg Name: Dow Chemical *Phone No:* 1-517-636-4400

Physical Description:

Physical Form: Liquid *Color:* Colorless *Odor:* Chloroform
Other Information: n/a

Chemical Properties:

Specific Gravity: 1.7 *Vapor Density:* 3.3 *Boiling Point: 38.4° F(3.6° C)*
Melting Point: −137° F(−93.9° C) Vapor Pressure: 1.8 atm at 68° F(20° C)
Solubility in water: No *Degree of Solubility:* .1%
Other Information: n/a

Health Hazards:

Inhalation Hazard: Poisonous if inhaled.
Ingestion Hazard: Poisonous if swallowed.
Absorption Hazard: Will burn the skin and eyes.
Hazards to Wildlife: n/a
Decontamination Procedures: Wash away any material with copious amounts of soap and water.
First Aid Procedures: Remove victim to fresh air, call emergency medical care. If not breathing give CPR. If breathing is difficult administer oxygen. Treat for shock.

Fire Hazards:

Flashpoint: n/a *Ignition temperature:* 999° F(537.2° C)
Flammable Explosive High Range: 16 *Low Range:* 10
Toxic Products of Combustion: Poisonous and irritating gases are produced in a fire.
Other Hazards: Under most circumstances, the material is noncombustible.
Possible extinguishing agents: Apply water from as far a distance as possible. Use foam, dry chemical, or carbon dioxide.

Reactivity Hazards:

Reactive With: n/a *Other Reactions:* n/a

Corrosivity Hazards:

Corrosive With: n/a *Neutralizing Agent:* n/a

Radioactivity Hazards:

Radiation Emitted: n/a *Other Hazards:* n/a

Recommended Protection for Response Personnel:

Avoid breathing vapors, keep upwind. Wear a sealed chemical suit (viton, nitrile). Wash away any material which may have come into contact with the body with copious amounts of soap and water. Consider appropriate evacuation.

METHYL BUTENE

DOT Number: UN 2460 *DOT Hazard Class:* Flammable liquid *DOT Guide Number:* 26
Synonyms: 2-methylbutane
STCC Number: 4909225 *Reportable Qty:* n/a *CHEMTREC Phone No:* 1-800-424-9300

Physical Description:

Physical Form: Liquid *Color:* Colorless *Odor:* Petroleumlike
Other Information: n/a

Chemical Properties:

Specific Gravity: .6 *Vapor Density:* n/a *Boiling Point: 82° F(27.8° C)*
Melting Point: n/a *Vapor Pressure:* n/a *Solubility in water:* No
Other Information: Lighter than and insoluble in water. Vapors are heavier than air.

Health Hazards:

Inhalation Hazard: n/a
Ingestion Hazard: n/a
Absorption Hazard: n/a
Hazards to Wildlife: n/a
Decontamination Procedures: Wash away any material with copious amounts of soap and water.
First Aid Procedures: Remove victim to fresh air, call emergency medical care. If not breathing give CPR. If breathing is difficult administer oxygen. Treat for shock.

Fire Hazards:

Flashpoint: −60° F(−51.1° C) *Ignition temperature:* 788° F(420° C)
Flammable Explosive High Range: 7.6 *Low Range:* 1.4
Toxic Products of Combustion: n/a
Other Hazards: n/a
Possible extinguishing agents: Apply water from as far a distance as possible. Use alcohol foam, dry chemical, or carbon dioxide.

Reactivity Hazards:

Reactive With: n/a *Other Reactions:* n/a

Corrosivity Hazards:

Corrosive With: n/a *Neutralizing Agent:* n/a

Radioactivity Hazards:

Radiation Emitted: n/a *Other Hazards:* n/a

Recommended Protection for Response Personnel:

Avoid breathing vapors, keep upwind. Wear appropriate chemical clothing. Wash away any material which may have come into contact with the body with copious amounts of soap and water. Consider appropriate evacuation.

METHYL CHLORIDE

DOT Number: UN 1063 *DOT Hazard Class:* Flammable gas *DOT Guide Number:* 18
Synonyms: artic, chloromethane
STCC Number: 4905761 *Reportable Qty:* 1/.454
Mfg Name: Dow Chemical *Phone No:* 1-517-636-4400

Physical Description:

Physical Form: Gas *Color:* Colorless *Odor:* Sweet
Other Information: Used to make other chemicals, and as a herbicide.

Chemical Properties:

Specific Gravity: 1 *Vapor Density:* 1.8 *Boiling Point: −11° F(−23.9° C)*
Melting Point: −144° F(−97.8° C) *Vapor Pressure:* 4.8 atm at 68° F(20° C)
Solubility in water: Slight
Other Information: It is shipped as a liquefied gas under its own vapor pressure.

Health Hazards:

Inhalation Hazard: Causes nausea and difficulty in breathing.
Ingestion Hazard: n/a
Absorption Hazard: Will cause frostbite.
Hazards to Wildlife: No
Decontamination Procedures: Wash away any material with copious amounts of soap and water.
First Aid Procedures: Remove victim to fresh air, call emergency medical care. If not breathing give CPR. If breathing is difficult administer oxygen. Treat for shock.

Fire Hazards:

Flashpoint: −50° F(−45.6° C) *Ignition temperature:* 1170° F(632.2° C)
Flammable Explosive High Range: 17.4 *Low Range:* 8.1
Toxic Products of Combustion: Poisonous gases are produced in a fire.
Other Hazards: Flashback on vapor trail may occur. Under fire conditions, the cylinders may violently rupture and rocket. (BLEVE)
Possible extinguishing agents: Apply water from as far a distance as possible.

Reactivity Hazards:

Reactive With: Zinc, aluminum, magnesium and their alloys. Note: the reaction is not violent.
Other Reactions: n/a

Corrosivity Hazards:

Corrosive With: n/a *Neutralizing Agent:* n/a

Radioactivity Hazards:

Radiation Emitted: n/a *Other Hazards:* n/a

Recommended Protection for Response Personnel:

Avoid breathing vapors, keep upwind. Structural protective clothing provides limited protection. Wash away any material which may have come into contact with the body with copious amounts of soap and water. If the fire becomes uncontrollable, consider appropriate evacuation. (BLEVE)

METHYL CHLOROFORMATE

DOT Number: UN 1238 *DOT Hazard Class:* Flammable liquid *DOT Guide Number:* 57
Synonyms: chlorformic acid, methyl ester; methyl chlorocarbonate
STCC Number: 4907429 *Reportable Qty:* 100/454
Mfg Name: Aldrich Chemical *Phone No:* 1-414-273-3850

Physical Description:

Physical Form: Liquid *Color:* Colorless to light yellow *Odor:* Unpleasant
Other Information: It is used to make other chemicals, and in insecticides.

Chemical Properties:

Specific Gravity: 1.22 *Vapor Density:* 3.3 *Boiling Point: 160° F(71.1° C)*
Melting Point: n/a *Vapor Pressure:* n/a *Solubility in water:* Reacts
Other Information: Sinks and reacts with water. A flammable, irritating vapor is produced.

Health Hazards:

Inhalation Hazard: Causes difficulty in breathing.
Ingestion Hazard: Poisonous if swallowed.
Absorption Hazard: Will burn the skin and eyes.
Hazards to Wildlife: n/a
Decontamination Procedures: Wash away any material with copious amounts of soap and water.
First Aid Procedures: Remove victim to fresh air, call emergency medical care. If not breathing give CPR. If breathing is difficult administer oxygen. Treat for shock.

Fire Hazards:

Flashpoint: 54° F(12.2° C) *Ignition temperature:* n/a
Flammable Explosive High Range: n/a *Low Range:* 6.7
Toxic Products of Combustion: Irritating and toxic hydrogen chloride and phosgene gases may be formed.
Other Hazards: Containers may explode in fire. Flashback along vapor trail may occur. Vapors may explode if ignited in an enclosed area.
Possible extinguishing agents: Do not extinguish the fire unless the flow can be stopped. Use alcohol foam, dry chemical, or carbon dioxide. Apply water from as far a distance as possible.

Reactivity Hazards:

Reactive With: Water to form hydrochloric acid. *Other Reactions:* n/a

Corrosivity Hazards:

Corrosive With: Metals, tissue, and rubber.
Neutralizing Agent: Crushed limestone, soda ash, or lime.

Radioactivity Hazards:

Radiation Emitted: n/a *Other Hazards:* n/a

Recommended Protection for Response Personnel:

Avoid breathing vapors, keep upwind. Wear appropriate chemical clothing. Wash away any material which may have come into contact with the body with copious amounts of soap and water. If the fire becomes uncontrollable, consider appropriate evacuation.

METHYL CYANIDE

DOT Number: UN 1648 *DOT Hazard Class:* Flammable liquid *DOT Guide Number:* 28
Synonyms: acetonitrile
STCC Number: 4907433 *Reportable Qty:* n/a *CHEMTREC Phone No:* 1-800-424-9300

Physical Description:

Physical Form: Liquid *Color:* Colorless *Odor:* Ethereal
Other Information: n/a

Chemical Properties:

Specific Gravity: .8 *Vapor Density:* 1.4 *Boiling Point: 179° F(81.7° C)*
Melting Point: −50° F(−45.6° C) *Vapor Pressure:* 73 mm Hg at 68° F(20° C) *Solubility in water:* Yes
Other Information: Lighter than and soluble in water. Vapors are heavier than air.

Health Hazards:

Inhalation Hazard: Hazardous
Ingestion Hazard: Hazardous
Absorption Hazard: Toxic
Hazards to Wildlife: Dangerous to aquatic life.
Decontamination Procedures: Wash away any material with copious amounts of soap and water.
First Aid Procedures: Remove victim to fresh air, call emergency medical care. If not breathing give CPR. If breathing is difficult administer oxygen. Treat for shock.

Fire Hazards:

Flashpoint: 42° F(5.6° C) *Ignition temperature:* 975° F(523.9° C)
Flammable Explosive High Range: 16 *Low Range:* 3
Toxic Products of Combustion: Oxides of nitrogen are produced during combustion of this material.
Other Hazards: Do not attempt to extinguish the fire unless the flow can be stopped.
Possible extinguishing agents: Apply water from as far a distance as possible. Use alcohol foam, dry chemical, or carbon dioxide.

Reactivity Hazards:

Reactive With: n/a *Other Reactions:* n/a

Corrosivity Hazards:

Corrosive With: n/a *Neutralizing Agent:* n/a

Radioactivity Hazards:

Radiation Emitted: n/a *Other Hazards:* n/a

Recommended Protection for Response Personnel:

Avoid breathing vapors, keep upwind. Wear appropriate chemical clothing. Wash away any material which may have come into contact with the body with copious amounts of soap and water. If container is exposed to direct flame or if fire becomes uncontrollable, consider appropriate evacuation.

METHYL CYCLOPENTANE

DOT Number: UN 2298 *DOT Hazard Class:* Flammable liquid *DOT Guide Number:* 26
Synonyms: cyclopentane methyl
STCC Number: n/a *Reportable Qty:* n/a
Mfg Name: Phillips Petroleum *Phone No:* 1-918-661-6600

Physical Description:

Physical Form: Liquid *Color:* Colorless *Odor:* Gasolinelike
Other Information: Floats on water.

Chemical Properties:

Specific Gravity: .8 *Vapor Density:* 2.9 *Boiling Point: 116° F(46.7° C)*
Melting Point: n/a *Vapor Pressure:* 2.9 mm Hg at 68° F(20° C) *Solubility in water:* n/a
Other Information: n/a

Health Hazards:

Inhalation Hazard: Causes dizziness, difficulty in breathing.
Ingestion Hazard: n/a
Absorption Hazard: Irritating to the skin and eyes.
Hazards to Wildlife: n/a
Decontamination Procedures: Wash away any material with copious amounts of soap and water.
First Aid Procedures: Remove victim to fresh air, call emergency medical care. If not breathing give CPR. If breathing is difficult administer oxygen. Treat for shock.

Fire Hazards:

Flashpoint: 0° F(−17.7° C) *Ignition temperature:* 624° F(328.9° C)
Flammable Explosive High Range: 8.7 *Low Range:* 1.1
Toxic Products of Combustion: n/a
Other Hazards: Containers may explode in fire. Flashback along vapor trail may occur. Vapors may explode if ignited in an enclosed area.
Possible extinguishing agents: Apply water from as far a distance as possible. Use foam, dry chemical, or carbon dioxide.

Reactivity Hazards:

Reactive With: n/a *Other Reactions:* n/a

Corrosivity Hazards:

Corrosive With: n/a *Neutralizing Agent:* n/a

Radioactivity Hazards:

Radiation Emitted: n/a *Other Hazards:* n/a

Recommended Protection for Response Personnel:

Avoid breathing vapors, keep upwind. Wear appropriate chemical clothing. Wash away any material which may come into contact with the body with copious amounts of soap and water. If container is exposed to direct flame or if fire becomes uncontrollable, consider appropriate evacuation.

METHYL DICHLOROSILANE

DOT Number: UN 1242 *DOT Hazard Class:* Flammable liquid *DOT Guide Number:* 29
Synonyms: none given
STCC Number: 4907625 *Reportable Qty:* n/a
Mfg Name: Union Carbide *Phone No:* 1-203-794-2000

Physical Description:

Physical Form: Liquid *Color:* Colorless *Odor:* Sharp, irritating
Other Information: n/a

Chemical Properties:

Specific Gravity: 1.1 *Vapor Density:* 4 *Boiling Point: 106° F(41.1° C)*
Melting Point: −135° F(−92.7° C) *Vapor Pressure:* n/a
Solubility in water: No
Other Information: Reacts violently with water. Irritating gas is produced upon contact with water.

Health Hazards:

Inhalation Hazard: Causes difficulty in breathing.
Ingestion Hazard: Harmful if swallowed.
Absorption Hazard: Will burn the skin and the eyes.
Hazards to Wildlife: n/a
Decontamination Procedures: Wash away any material with copious amounts of soap and water.
First Aid Procedures: Remove victim to fresh air, call emergency medical care. If not breathing give CPR. If breathing is difficult administer oxygen. Treat for shock.

Fire Hazards:

Flashpoint: −14° F(−25.5° C) *Ignition temperature:* 600° F(315.5° C)
Flammable Explosive High Range: 55 *Low Range:* 6
Toxic Products of Combustion: Toxic hydrogen chloride and phosgene gases may be formed.
Other Hazards: Containers may explode in a fire. Flashback along vapor trail may occur. Vapors may explode if ignited in an enclosed area.
Possible extinguishing agents: Apply water from as far a distance as possible. Use alcohol foam, dry chemical, or carbon dioxide. Do not extinguish the fire unless the flow can be stopped. Cool all affected containers. Do not use water on the material itself!

Reactivity Hazards:

Reactive With: Water to form hydrochloric acid. *Other Reactions:* n/a

Corrosivity Hazards:

Corrosive With: Common metals and surface moisture to release hydrogen chloride.
Neutralizing Agent: Crushed limestone, soda ash, or lime.

Radioactivity Hazards:

Radiation Emitted: n/a *Other Hazards:* n/a

Recommended Protection for Response Personnel:

Avoid breathing vapors, keep upwind. Wear a sealed chemical suit (butyl rubber). Wash away any material which may have come into contact with the body with copious amounts of soap and water. Consider appropriate evacuation.

METHYLENE CHLORIDE

DOT Number: UN 1593 *DOT Hazard Class:* ORM-A *DOT Guide Number:* 74
Synonyms: dichloromethane
STCC Number: 4941132 *Reportable Qty:* 1000/454 *CHEMTREC Phone No:* 1-800-424-9300

Physical Description:

Physical Form: Liquid *Color:* Colorless *Odor:* Chloroformlike
Other Information: It is used as a solvent and a paint remover.

Chemical Properties:

Specific Gravity: 1.3 *Vapor Density:* 2.9 *Boiling Point: 104° F(40° C)*
Melting Point: −142° F(−96.6° C) *Vapor Pressure:* 350 mm Hg at 68° F(20° C)
Solubility in water: Slight
Other Information: Vapors are narcotic in high concentrations.

Health Hazards:

Inhalation Hazard: Fatigue, weak, light headed.
Ingestion Hazard: Harmful if swallowed.
Absorption Hazard: Irritating to the skin and eyes.
Hazards to Wildlife: n/a
Decontamination Procedures: Wash away any material with copious amounts of soap and water.
First Aid Procedures: Remove victim to fresh air, call emergency medical care. If not breathing give CPR. If breathing is difficult administer oxygen. Treat for shock.

Fire Hazards:

Flashpoint: None *Ignition temperature:* 1033° F(556.1° C)
Flammable Explosive High Range: 19 *Low Range:* 12
Toxic Products of Combustion: If exposed to high temperatures, material may emit toxic chloride fumes.
Other Hazards: n/a
Possible extinguishing agents: Extinguish fire using suitable agent for type of surrounding fire.

Reactivity Hazards:

Reactive With: n/a *Other Reactions:* n/a

Corrosivity Hazards:

Corrosive With: n/a *Neutralizing Agent:* n/a

Radioactivity Hazards:

Radiation Emitted: n/a *Other Hazards:* n/a

Recommended Protection for Response Personnel:

Avoid breathing vapors, keep upwind. Wear appropriate chemical clothing. Wash away any material which may have come into contact with the body with copious amounts of soap and water. Consider appropriate evacuation.

METHYL ETHYL KETONE

DOT Number: UN 1193 *DOT Hazard Class:* Flammable liquid *DOT Guide Number:* 26
Synonyms: 2-butanone, MEK
STCC Number: 4909243 *Reportable Qty:* 5000/2270
Mfg Name: Shell Chemical *Phone No:* 1-713-241-6161

Physical Description:

Physical Form: Liquid *Color:* Colorless *Odor:* Pleasant
Other Information: It is used as a solvent and for making other chemicals.

Chemical Properties:

Specific Gravity: .8 *Vapor Density:* 2.5 *Boiling Point: 176° F(80° C)*
Melting Point: n/a *Vapor Pressure:* 3.5 psia at 68° F(20° C) *Solubility in water:* Yes
Other Information: Weighs 6.7 lbs/3.0 kg per gallon/3.8 l.

Health Hazards:

Inhalation Hazard: Nausea, headaches, difficulty in breathing.
Ingestion Hazard: Harmful if swallowed.
Absorption Hazard: Will burn the eyes.
Hazards to Wildlife: Dangerous to aquatic life.
Decontamination Procedures: Wash away any material with copious amounts of soap and water.
First Aid Procedures: Remove victim to fresh air, call emergency medical care. If not breathing give CPR. If breathing is difficult administer oxygen. Treat for shock.

Fire Hazards:

Flashpoint: 16° F(−8.8° C) *Ignition temperature:* 759° F(403.8° C)
Flammable Explosive High Range: 11.4 *Low Range:* 1.4
Toxic Products of Combustion: Toxic constituents (elements)
Other Hazards: Flashback along vapor trail may occur. Vapors may explode if ignited in an enclosed area.
Possible extinguishing agents: Apply water from as far a distance as possible. Use alcohol foam, dry chemical, or carbon dioxide.

Reactivity Hazards:

Reactive With: Strong oxidizing materials.
Other Reactions: Will dissolve or soften some plastics.

Corrosivity Hazards:

Corrosive With: n/a *Neutralizing Agent:* n/a

Radioactivity Hazards:

Radiation Emitted: n/a *Other Hazards:* n/a

Recommended Protection for Response Personnel:

Avoid breathing vapors, keep upwind. Structural protective clothing provides limited protection. Wash away any material which may have come into contact with the body with copious amounts of soap and water. Consider appropriate evacuation.

METHYL ETHYL PYRIDINE

DOT Number: UN 2300 *DOT Hazard Class:* Corrosive *DOT Guide Number:* 60
Synonyms: 5-ethyl-2-methylpyridine, 5-ethyl-2-picoline, MEP
STCC Number: 4935660 *Reportable Qty:* n/a
Mfg Name: Union Carbide *Phone No:* 1-203-794-2000

Physical Description:

Physical Form: Liquid *Color:* Colorless *Odor:* Sharp
Other Information: n/a

Chemical Properties:

Specific Gravity: .92 *Vapor Density:* 4.2 *Boiling Point: 352° F(162.7° C)*
Melting Point: n/a *Vapor Pressure:* n/a *Solubility in water:* Slightly
Other Information: Floats on water.

Health Hazards:

Inhalation Hazard: n/a
Ingestion Hazard: Harmful if swallowed.
Absorption Hazard: Will burn the skin and the eyes.
Hazards to Wildlife: n/a
Decontamination Procedures: Wash away any material with copious amounts of soap and water.
First Aid Procedures: Remove victim to fresh air, call emergency medical care. If not breathing give CPR. If breathing is difficult administer oxygen. Treat for shock.

Fire Hazards:

Flashpoint: 155° F(68.3° C) *Ignition temperature:* 939° F(503.8° C)
Flammable Explosive High Range: 6.6 *Low Range:* 1.1
Toxic Products of Combustion: Toxic oxides of nitrogen are produced during combustion of this material.
Other Hazards: n/a
Possible extinguishing agents: Do not extinguish the fire unless the flow can be stopped use alcohol foam, dry chemical, or carbon dioxide. Apply water from as far a distance as possible.

Reactivity Hazards:

Reactive With: n/a *Other Reactions:* n/a

Corrosivity Hazards:

Corrosive With: Metals and tissue. *Neutralizing Agent:* Crushed limestone, soda ash, or lime.

Radioactivity Hazards:

Radiation Emitted: n/a *Other Hazards:* n/a

Recommended Protection for Response Personnel:

Avoid breathing vapors, keep upwind. Structural protective clothing provides limited protection. Wash away any material which may have come into contact with the body with copious amounts of soap and water. If the fire becomes uncontrollable, consider appropriate evacuation.

METHYL FORMAL

DOT Number: UN 1234 *DOT Hazard Class:* Flammable liquid *DOT Guide Number:* 26
Synonyms: dimethoxymethane, dimethylformal, formaldehyde dimethylacetal, methylal, methylene dimethyl ether
STCC Number: n/a *Reportable Qty:* n/a
Mfg Name: Aldrich Chemical *Phone No:* 1-414-273-3850

Physical Description:

Physical Form: Liquid *Color:* Colorless *Odor:* Mild sweet
Other Information: n/a

Chemical Properties:

Specific Gravity: .86 *Vapor Density:* 2.6 *Boiling Point: 108° F(42.2° C)*
Melting Point: −157° F(−105° C) *Vapor Pressure:* n/a *Solubility in water:* Yes
Other Information: Mixes with water. Flammable, irritating vapor is produced.

Health Hazards:

Inhalation Hazard: Harmful if inhaled.
Ingestion Hazard: Harmful if swallowed.
Absorption Hazard: Irritating to the skin and the eyes.
Hazards to Wildlife: n/a
Decontamination Procedures: Wash away any material with copious amounts of soap and water.
First Aid Procedures: Remove victim to fresh air, call emergency medical care. If not breathing give CPR. If breathing is difficult administer oxygen. Treat for shock.

Fire Hazards:

Flashpoint: 0° F(−17.7° C) *Ignition temperature:* 459° F(237.2° C)
Flammable Explosive High Range: 17.6 *Low Range:* 1.6
Toxic Products of Combustion: Irritating formaldehyde gas may be present in smoke.
Other Hazards: Containers may explode in fire. Flashback along vapor trail may occur. Vapors may explode if ignited in an enclosed area.
Possible extinguishing agents: Alcohol foam, dry chemical, carbon dioxide. Water may be ineffective on fire.

Reactivity Hazards:

Reactive With: n/a *Other Reactions:* n/a

Corrosivity Hazards:

Corrosive With: n/a *Neutralizing Agent:* n/a

Radioactivity Hazards:

Radiation Emitted: n/a *Other Hazards:* n/a

Recommended Protection for Response Personnel:

Avoid breathing vapors, keep upwind. Structural protective clothing provides limited protection. Wash away any material which may have come into contact with the body with copious amounts of soap and water. Consider appropriate evacuation.

METHYL FORMATE

DOT Number: UN 1243 *DOT Hazard Class:* Flammable liquid *DOT Guide Number:* 26
Synonyms: formic acid, methyl ester
STCC Number: 4908225 *Reportable Qty:* n/a
Mfg Name: Eastman Organic *Phone No:* 1-800-445-6325

Physical Description:

Physical Form: Liquid *Color:* Colorless *Odor:* Agreeable
Other Information: Lighter than water. Heavier than air.

Chemical Properties:

Specific Gravity: 1 *Vapor Density:* 2.1 *Boiling Point: 90° F(32.2° C)*
Melting Point: −148° F(−100° C) *Vapor Pressure:* 476 mm Hg at 68° F(20° C)
Solubility in water: Yes *Degree of Solubility:* 30%
Other Information: n/a

Health Hazards:

Inhalation Hazard: Harmful if inhaled.
Ingestion Hazard: Harmful if swallowed.
Absorption Hazard: Irritating to the skin and eyes.
Hazards to Wildlife: n/a
Decontamination Procedures: Wash away any material with copious amounts of soap and water.
First Aid Procedures: Remove victim to fresh air, call emergency medical care. If not breathing give CPR. If breathing is difficult administer oxygen. Treat for shock.

Fire Hazards:

Flashpoint: −2° F(−18.8° C) *Ignition temperature:* 840° F(448.8° C)
Flammable Explosive High Range: 23 *Low Range:* 4.5
Toxic Products of Combustion: none
Other Hazards: Flashback along vapor trail may occur. Vapors may explode in an enclosed area. Containers may explode in a fire.
Possible extinguishing agents: Apply water from as far a distance as possible. Use alcohol foam, dry chemical, or carbon dioxide.

Reactivity Hazards:

Reactive With: Water to form formic acid and methyl alcohol. The reaction is not hazardous.
Other Reactions: n/a

Corrosivity Hazards:

Corrosive With: n/a *Neutralizing Agent:* n/a

Radioactivity Hazards:

Radiation Emitted: n/a *Other Hazards:* n/a

Recommended Protection for Response Personnel:

Avoid breathing vapors, keep upwind. Structural protective clothing provides limited protection. Wash away any material which may have come into contact with the body with copious amounts of soap and water. If the fire becomes uncontrollable, consider appropriate evacuation.

METHYLHYDRAZINE

DOT Number: UN 1244 *DOT Hazard Class:* Flammable liquid *DOT Guide Number:* 57
Synonyms: MMH, monomethylhydrazine
STCC Number: 4906230 *Reportable Qty:* 1/.454
Mfg Name: Eastman Organic *Phone No:* 1-800-445-6325

Physical Description:

Physical Form: Liquid *Color:* Colorless *Odor:* Ammonialike
Other Information: It is used to make other chemicals, as a solvent, and is also used as a rocket propellant.

Chemical Properties:

Specific Gravity: .9 *Vapor Density:* 1.6 *Boiling Point: 190° F(87.7° C)*
Melting Point: n/a *Vapor Pressure:* n/a *Solubility in water:* Slight
Other Information: Lighter than and slightly soluble in water. Vapors are heavier than air.

Health Hazards:

Inhalation Hazard: Poisonous if inhaled.
Ingestion Hazard: Poisonous if swallowed.
Absorption Hazard: Poisonous if absorbed.
Hazards to Wildlife: n/a
Decontamination Procedures: Wash away any material with copious amounts of soap and water.
First Aid Procedures: Remove victim to fresh air, call emergency medical care. If not breathing give CPR. If breathing is difficult administer oxygen. Treat for shock.

Fire Hazards:

Flashpoint: 17° F(−8.3° C) *Ignition temperature:* 382° F(194.4° C)
Flammable Explosive High Range: 92 *Low Range:* 2.5
Toxic Products of Combustion: Toxic products of nitrogen are produced during combustion of this material.
Other Hazards: Containers may explode in a fire. Flashback along vapor trail may occur.
Possible extinguishing agents: Apply water from as far a distance as possible. Use alcohol foam, dry chemical, or carbon dioxide.

Reactivity Hazards:

Reactive With: Air, but heat may cause ignition of rags, rust, or other combustibles. *Other Reactions:* n/a

Corrosivity Hazards:

Corrosive With: n/a *Neutralizing Agent:* n/a

Radioactivity Hazards:

Radiation Emitted: n/a *Other Hazards:* n/a

Recommended Protection for Response Personnel:

Avoid breathing vapors, keep upwind. Wear appropriate chemical clothing. Wash away any material which may have come into contact with the body with copious amounts of soap and water. If fire becomes uncontrollable, or if containers are exposed to direct flame, consider appropriate evacuation.

METHYL ISOBUTYL CARBINOL

DOT Number: UN 2053 *DOT Hazard Class:* Combustible liquid *DOT Guide Number:* 26
Synonyms: MAOH, MIBC, MIC
STCC Number: 4913143 *Reportable Qty:* n/a
Mfg Name: Shell Chemical *Phone No:* 1-713-241-6161

Physical Description:

Physical Form: Liquid *Color:* Colorless *Odor:* Mild alcohol
Other Information: Lighter than and insoluble in water. Vapors are heavier than air.

Chemical Properties:

Specific Gravity: .8 *Vapor Density:* 3.5 *Boiling Point: 269° F(131.6° C)*
Melting Point: −130° F(−90° C) *Vapor Pressure:* 2.8 mm Hg at 68° F(20° C)
Solubility in water: Slight *Degree of Solubility:* 1.5%
Other Information: n/a

Health Hazards:

Inhalation Hazard: Causes dizziness, difficulty in breathing.
Ingestion Hazard: Harmful if swallowed.
Absorption Hazard: Irritating to the skin and eyes.
Hazards to Wildlife: Dangerous to aquatic life.
Decontamination Procedures: Wash away any material with copious amounts of soap and water.
First Aid Procedures: Remove victim to fresh air, call emergency medical care. If not breathing give CPR. If breathing is difficult administer oxygen. Treat for shock.

Fire Hazards:

Flashpoint: 106° F(41.1° C) *Ignition temperature:* n/a
Flammable Explosive High Range: 5.5 *Low Range:* 1
Toxic Products of Combustion: n/a
Other Hazards: n/a
Possible extinguishing agents: Apply water from as far a distance as possible. Use alcohol foam, dry chemical, or carbon dioxide.

Reactivity Hazards:

Reactive With: n/a *Other Reactions:* n/a

Corrosivity Hazards:

Corrosive With: n/a *Neutralizing Agent:* n/a

Radioactivity Hazards:

Radiation Emitted: n/a *Other Hazards:* n/a

Recommended Protection for Response Personnel:

Avoid breathing vapors, keep upwind. Wear a sealed chemical suit (polycarbonate, butyl rubber). Wash away any material which may have come into contact with the body with copious amounts of soap and water. Consider appropriate evacuation.

METHYL ISOBUTYL KETONE

DOT Number: UN 1245 *DOT Hazard Class:* Flammable liquid *DOT Guide Number:* 26
Synonyms: hexone, isopropylacetone, MIBK, MIK
STCC Number: 4909245 *Reportable Qty:* 5000/2270
Mfg Name: Shell Chemical *Phone No:* 1-713-241-6161

Physical Description:

Physical Form: Liquid *Color:* Colorless *Odor:* Mild, pleasant
Other Information: n/a

Chemical Properties:

Specific Gravity: .80 *Vapor Density:* 3.5 *Boiling Point: 244° F(117.7° C)*
Melting Point: n/a *Vapor Pressure:* 15.7 mm Hg at 68° F(20° C) *Solubility in water:* Slight
Other Information: Floats and mixes slowly with water. A flammable, irritating vapor is produced.

Health Hazards:

Inhalation Hazard: Causes dizziness, loss of consciousness.
Ingestion Hazard: Harmful if swallowed.
Absorption Hazard: Irritating to the skin and eyes.
Hazards to Wildlife: n/a
Decontamination Procedures: Wash away any material with copious amounts of soap and water.
First Aid Procedures: Remove victim to fresh air, call emergency medical care. If not breathing give CPR. If breathing is difficult administer oxygen. Treat for shock.

Fire Hazards:

Flashpoint: 64° F(17.7° C) *Ignition temperature:* 840° F(448.8° C)
Flammable Explosive High Range: 8 *Low Range:* 1.2
Toxic Products of Combustion: Irritating vapors are generated when heated.
Other Hazards: Flashback along vapor trail may occur. Vapors may explode if ignited in an enclosed area.
Possible extinguishing agents: Do not extinguish the fire unless the flow can be stopped. Use alcohol foam, dry chemical, or carbon dioxide. Apply water from as far a distance as possible.

Reactivity Hazards:

Reactive With: n/a *Other Reactions:* n/a

Corrosivity Hazards:

Corrosive With: n/a *Neutralizing Agent:* n/a

Radioactivity Hazards:

Radiation Emitted: n/a *Other Hazards:* n/a

Recommended Protection for Response Personnel:

Avoid breathing vapors, keep upwind. Structural protective clothing provides limited protection. Wash away any material which may have come into contact with the body with copious amounts of soap and water. Consider appropriate evacuation.

METHYL ISOCYANATE

DOT Number: UN 2480 *DOT Hazard Class:* Flammable liquid *DOT Guide Number:* 30
Synonyms: isocyanatomethane, isocyanic acid, MIC
STCC Number: 4907448 *Reportable Qty:* 1/.454 *CHEMTREC Phone No:* 1-800-424-9300

Physical Description:

Physical Form: Liquid *Color:* Colorless *Odor:* Sharp
Other Information: n/a

Chemical Properties:

Specific Gravity: .96 *Vapor Density:* 2 *Boiling Point: 102.4° F(39.1° C)*
Melting Point: −112° F(−80° C) *Vapor Pressure:* 338 mm Hg at 68° F(20° C)
Solubility in water: Yes *Degree of Solubility:* 6.7%
Other Information: It is soluble in water. Weighs 8.1 lbs/3.6 kg per gallon/3.8 l. Manufacturer suggests prompt evacuation of 10,000 foot radius for significant discharges.

Health Hazards:

Inhalation Hazard: MIC vapors can cause death!!
Ingestion Hazard: Poisonous if swallowed.
Absorption Hazard: Severe burns and tissue destruction. *Hazards to Wildlife:* n/a
Decontamination Procedures: Wash away any material with copious amounts of soap and water.
First Aid Procedures: Remove victim to fresh air, call emergency medical care. If not breathing give CPR. If breathing is difficult administer oxygen. Treat for shock.

Fire Hazards:

Flashpoint: 0° F(−17.7° C) *Ignition temperature:* 995° F(535° C)
Flammable Explosive High Range: 26 *Low Range:* 5.3
Toxic Products of Combustion: Poisonous vapors are generated when heated.
Other Hazards: Flashback along vapor trail may occur. Containers may violently rupture in a fire and possibly explode. Unburned vapors may pose a toxic hazard over considerable downwind distances.
Possible extinguishing agents: Do not extinguish the fire unless the flow can be stopped. Use alcohol foam, dry chemical, or carbon dioxide. Apply water from as far a distance as possible.

Reactivity Hazards:

Reactive With: Water to form carbon dioxide, methylamine, dimethylurea and or trimethylbiuret.
Other Reactions: n/a

Corrosivity Hazards:

Corrosive With: Steel or iron, aluminum, zinc, galvanized iron, copper or tin, or their alloys as these may initiate polymerization actions. *Neutralizing Agent:* n/a

Radioactivity Hazards:

Radiation Emitted: n/a *Other Hazards:* n/a

Recommended Protection for Response Personnel:

Avoid breathing vapors, keep upwind. Wear a sealed chemical suit (manufacturer suggests butyl type). Wash away any material which may have come into contact with the body with copious amounts of soap and water. Consider appropriate evacuation.

METHYL ISOPROPENYL KETONE

DOT Number: UN 1246 *DOT Hazard Class:* Flammable liquid *DOT Guide Number:* 26
Synonyms: methyl-1-butene-3-one; isopropenyl methylketone
STCC Number: 4907240 *Reportable Qty:* n/a
Mfg Name: Eastern Chemical Div. *Phone No:* 1-516-273-0900

Physical Description:

Physical Form: Liquid *Color:* Colorless *Odor:* Sweet, pleasant
Other Information: If contaminated or subject to heat, material may polymerize.

Chemical Properties:

Specific Gravity: .85 *Vapor Density:* 2.9 *Boiling Point: 208° F(97.7° C)*
Melting Point: n/a *Vapor Pressure:* n/a *Solubility in water:* No
Other Information: Floats on water. A flammable, irritating vapor is produced.

Health Hazards:

Inhalation Hazard: n/a
Ingestion Hazard: Harmful if swallowed.
Absorption Hazard: Irritating to the skin and the eyes.
Hazards to Wildlife: n/a
Decontamination Procedures: Wash away any material with copious amounts of soap and water.
First Aid Procedures: Remove victim to fresh air, call emergency medical care. If not breathing give CPR. If breathing is difficult administer oxygen. Treat for shock.

Fire Hazards:

Flashpoint: 73° F(22.7° C) *Ignition temperature:* n/a
Flammable Explosive High Range: 9 *Low Range:* 1.8
Toxic Products of Combustion: May polymerize and explode.
Other Hazards: If polymerization takes place, the container may violently rupture.
Possible extinguishing agents: Do not extinguish the fire unless the flow can be stopped use foam, dry chemical, or carbon dioxide. Apply water from as far a distance as possible.

Reactivity Hazards:

Reactive With: n/a *Other Reactions:* n/a

Corrosivity Hazards:

Corrosive With: n/a *Neutralizing Agent:* n/a

Radioactivity Hazards:

Radiation Emitted: n/a *Other Hazards:* n/a

Recommended Protection for Response Personnel:

Avoid breathing vapors, keep upwind. Wear appropriate chemical clothing. Wash away any material which may have come into contact with the body with copious amounts of soap and water. If the fire becomes uncontrollable, or if the container is exposed to direct flame, consider appropriate evacuation.

METHYL MERCAPTAN

DOT Number: UN 1064 *DOT Hazard Class:* Flammable gas *DOT Guide Number:* 13
Synonyms: mercaptomethane, methanethiol, methyl sulfhydrate, thiomethyl alcohol
STCC Number: 4905520 *Reportable Qty:* 100/45.4
Mfg Name: Pennwald Corp. *Phone No:* 1-215-587-7000

Physical Description:

Physical Form: Gas *Color:* Colorless *Odor:* Rotten cabbage
Other Information: It is shipped as a liquefied gas under its own vapor pressure.

Chemical Properties:

Specific Gravity: .9 *Vapor Density:* 1.7 *Boiling Point: 42.4° F(5.7° C)*
Melting Point: −186° F(−121.1° C) *Vapor Pressure:* 1 atm at 68° F(20° C)
Solubility in water: Yes *Degree of Solubility:* 2.4%
Other Information: Weighs 7.3 lbs/3.3 kg per gallon/3.8 l.

Health Hazards:

Inhalation Hazard: Poisonous if inhaled.
Ingestion Hazard: Poisonous if swallowed.
Absorption Hazard: Irritating to the skin and eyes.
Hazards to Wildlife: Dangerous to aquatic life.
Decontamination Procedures: Wash away any material with copious amounts of soap and water.
First Aid Procedures: Remove victim to fresh air, call emergency medical care. If not breathing give CPR. If breathing is difficult administer oxygen. Treat for shock.

Fire Hazards:

Flashpoint: 0° F(−17.7° C) *Ignition temperature:* n/a
Flammable Explosive High Range: 12.8 *Low Range:* 3.9
Toxic Products of Combustion: Poisonous gases are produced in a fire.
Other Hazards: Flashback along vapor trail may occur. Vapors may explode if ignited in an enclosed area. Containers may explode in a fire. (BLEVE)
Possible extinguishing agents: Apply water from as far a distance as possible.

Reactivity Hazards:

Reactive With: n/a *Other Reactions:* n/a

Corrosivity Hazards:

Corrosive With: n/a *Neutralizing Agent:* n/a

Radioactivity Hazards:

Radiation Emitted: n/a *Other Hazards:* n/a

Recommended Protection for Response Personnel:

Avoid breathing vapors, keep upwind. Structural protective clothing provides limited protection. Wash away any material which may have come into contact with the body with copious amounts of soap and water. If the fire becomes uncontrollable, or if containers are exposed to direct flame, consider appropriate evacuation. (BLEVE)

METHYL METHACRYLATE

DOT Number: UN 1247 *DOT Hazard Class:* Flammable liquid *DOT Guide Number:* 26
Synonyms: methacrylic acid, methyl ester
STCC Number: 4907250 *Reportable Qty:* 1000/454
Mfg Name: E.I. Du Pont *Phone No:* 1-800-414-3637

Physical Description:

Physical Form: Liquid *Color:* Colorless *Odor:* Pleasant
Other Information: It is used to make plastics. Weighs 7.8lbs/3.5 kg per gallon/3.8 l.

Chemical Properties:

Specific Gravity: .9 *Vapor Density:* 3.6 *Boiling Point: 212° F(100° C)*
Melting Point: −54° F(−47.7° C) *Vapor Pressure:* 35 mm Hg at 68° F(20° C)
Solubility in water: Slight *Degree of Solubility:* 1.5%
Other Information: Lighter than water. Vapors are heavier than air.

Health Hazards:

Inhalation Hazard: Dizziness, difficulty in breathing.
Ingestion Hazard: Harmful if swallowed.
Absorption Hazard: Will burn the skin and eyes.
Hazards to Wildlife: Dangerous to aquatic life.
Decontamination Procedures: Wash away any material with copious amounts of soap and water.
First Aid Procedures: Remove victim to fresh air, call emergency medical care. If not breathing give CPR. If breathing is difficult administer oxygen. Treat for shock.

Fire Hazards:

Flashpoint: 50° F(10° C) *Ignition temperature:* 790° F(421.1° C)
Flammable Explosive High Range: 8.28 *Low Range:* 1.7
Toxic Products of Combustion: Polymerization of material may occur when heated.
Other Hazards: Containers may violently rupture when heated.
Possible extinguishing agents: Apply water from as far a distance as possible. Use foam, dry chemical, or carbon dioxide.

Reactivity Hazards:

Reactive With: n/a *Other Reactions:* n/a

Corrosivity Hazards:

Corrosive With: n/a *Neutralizing Agent:* n/a

Radioactivity Hazards:

Radiation Emitted: n/a *Other Hazards:* n/a

Recommended Protection for Response Personnel:

Avoid breathing vapors, keep upwind. Structural protective clothing provides limited protection. Wash away any material which may have come into contact with the body with copious amounts of soap and water. If the fire becomes uncontrollable, or if the containers are exposed to direct flame, consider appropriate evacuation.

METHYL n-BUTYL KETONE

DOT Number: UN 1993 *DOT Hazard Class:* Flammable liquid *DOT Guide Number:* 27
Synonyms: n-butyl methyl ketone, 2-hexanone
STCC Number: 4909242 *Reportable Qty:* n/a
Mfg Name: Eastman Chemical *Phone No:* 1-615-229-2000

Physical Description:

Physical Form: Liquid *Color:* Clear *Odor:* Disagreeable
Other Information: n/a

Chemical Properties:

Specific Gravity: .81 *Vapor Density:* 3.4 *Boiling Point: 262° F(127.7° C)*
Melting Point: −70° F(−56.6° C) *Vapor Pressure:* 10 mm Hg at 68° F(20° C) *Solubility in water:* Yes
Other Information: Floats and mixes with water. Vapors are heavier than air.

Health Hazards:

Inhalation Hazard: Harmful if inhaled.
Ingestion Hazard: Harmful if swallowed.
Absorption Hazard: Irritating to the skin and the eyes.
Hazards to Wildlife: n/a
Decontamination Procedures: Wash away any material with copious amounts of soap and water.
First Aid Procedures: Remove victim to fresh air, call emergency medical care. If not breathing give CPR. If breathing is difficult administer oxygen. Treat for shock.

Fire Hazards:

Flashpoint: 77° F(25° C) *Ignition temperature:* 795° F(423.8° C)
Flammable Explosive High Range: 8 *Low Range:* 1.3
Toxic Products of Combustion: n/a
Other Hazards: Containers may explode in fire. Flashback along vapor trail may occur. Vapors may explode if ignited in an enclosed area.
Possible extinguishing agents: Do not extinguish the fire unless the flow can be stopped. Cool all affected containers with water from as far a distance as possible. Use alcohol foam, dry chemical, or carbon dioxide.

Reactivity Hazards:

Reactive With: n/a *Other Reactions:* n/a

Corrosivity Hazards:

Corrosive With: n/a *Neutralizing Agent:* n/a

Radioactivity Hazards:

Radiation Emitted: n/a *Other Hazards:* n/a

Recommended Protection for Response Personnel:

Avoid breathing vapors, keep upwind. Structural protective clothing provides limited protection. Wash away any material which may have come into contact with the body with copious amounts of soap and water. Consider appropriate evacuation.

METHYL PARATHION

DOT Number: NA 2783 *DOT Hazard Class:* Poison B *DOT Guide Number:* 55
Synonyms: Metron, MPT, Paridol
STCC Number: 4921447 *Reportable Qty:* 100/45.4
Mfg Name: Monsanto Co. *Phone No:* 1-314-694-1000

Physical Description:

Physical Form: Liquid *Color:* Brown *Odor:* Rotten eggs
Other Information: n/a

Chemical Properties:

Specific Gravity: 1.36 *Vapor Density:* 9.1 *Boiling Point: n/a*
Melting Point: n/a *Vapor Pressure:* n/a *Solubility in water:* Yes
Other Information: In case of damage to, or leaking from containers, contact the Pesticide Safety Team Network at 1-800-424-9300.

Health Hazards:

Inhalation Hazard: Poisonous if inhaled.
Ingestion Hazard: Poisonous if swallowed.
Absorption Hazard: Poisonous if absorbed.
Hazards to Wildlife: Dangerous to waterfowl and aquatic life.
Decontamination Procedures: Wash away any material with copious amounts of soap and water.
First Aid Procedures: Remove victim to fresh air, call emergency medical care. If not breathing give CPR. If breathing is difficult administer oxygen. Treat for shock.

Fire Hazards:

Flashpoint: 115° F(46.1° C) *Ignition temperature:* n/a
Flammable Explosive High Range: n/a *Low Range:* n/a
Toxic Products of Combustion: Poisonous gases are produced when heated, and when involved in a fire.
Other Hazards: Containers may explode in a fire.
Possible extinguishing agents: Apply water from as far a distance as possible. Use foam, dry chemical, or carbon dioxide.

Reactivity Hazards:

Reactive With: Temperatures above 122° F(50° C) with possible explosive force. *Other Reactions:* n/a

Corrosivity Hazards:

Corrosive With: n/a *Neutralizing Agent:* n/a

Radioactivity Hazards:

Radiation Emitted: n/a *Other Hazards:* n/a

Recommended Protection for Response Personnel:

Avoid breathing vapors, keep upwind. Wear a sealed chemical suit (polycarbonate, butyl rubber). Wash away any material which may have come into contact with the body with copious amounts of soap and water. If the fire becomes uncontrollable, or if the containers are exposed to direct flame, consider appropriate evacuation.

METHYL PHOSPHONOTHIOIC DICHLORIDE

DOT Number: NA 1760 *DOT Hazard Class:* Corrosive *DOT Guide Number:* 60
Synonyms: MPTD
STCC Number: 4933336 *Reportable Qty:* n/a
Mfg Name: Ethyl Chemical Corp. *Phone No:* 1-804-788-5000

Physical Description:

Physical Form: Liquid *Color:* Colorless *Odor:* Sharp, unpleasant
Other Information: n/a

Chemical Properties:

Specific Gravity: 1.42 *Vapor Density:* 5.1 *Boiling Point: n/a*
Melting Point: n/a *Vapor Pressure:* n/a *Solubility in water:* Yes
Other Information: Sinks and mixes violently with water.

Health Hazards:

Inhalation Hazard: Causes coughing or difficulty in breathing.

Ingestion Hazard: Causes nausea and vomiting.

Absorption Hazard: Will burn the skin and eyes.

Hazards to Wildlife: n/a

Decontamination Procedures: Wash away any material with copious amounts of soap and water.

First Aid Procedures: Remove victim to fresh air, call emergency medical care. If not breathing give CPR. If breathing is difficult administer oxygen. Treat for shock.

Fire Hazards:

Flashpoint: 122° F(50° C) *Ignition temperature:* n/a

Flammable Explosive High Range: n/a *Low Range:* n/a

Toxic Products of Combustion: Irritating hydrogen chloride and sulfur dioxide and other fumes may be formed in fires.

Other Hazards: n/a

Possible extinguishing agents: Apply water from as far a distance as possible. Use alcohol foam, dry chemical, or carbon dioxide.

Reactivity Hazards:

Reactive With: Water to from hydrochloric acid. *Other Reactions:* n/a

Corrosivity Hazards:

Corrosive With: Metals and tissue. *Neutralizing Agent:* Crushed limestone, soda ash, or lime.

Radioactivity Hazards:

Radiation Emitted: n/a *Other Hazards:* n/a

Recommended Protection for Response Personnel:

Avoid breathing vapors, keep upwind. Wear appropriate chemical clothing. Wash away any material which may have come into contact with the body with copious amounts of soap and water. Consider appropriate evacuation.

METHYLPYRROLIDONE

DOT Number: n/a *DOT Hazard Class:* n/a *DOT Guide Number:* n/a
Synonyms: n-methylpyrrolidone, 1-methyl-2-pyrrolidinone, n-methylpyrrolidinone, n-methyl-α-pyrrolidone
STCC Number: n/a *Reportable Qty:* n/a
Mfg Name: Aldrich Chemical *Phone No:* 1-414-273-3850

Physical Description:

Physical Form: Liquid *Color:* White *Odor:* Fishy
Other Information: n/a

Chemical Properties:

Specific Gravity: 1.03 *Vapor Density:* 3.4 *Boiling Point: 396° F(202.2° C)*
Melting Point: 1° F(−17.2° C) *Vapor Pressure:* n/a *Solubility in water:* n/a
Other Information: May float or sink in water.

Health Hazards:

Inhalation Hazard: Nausea and vomiting.
Ingestion Hazard: n/a
Absorption Hazard: Irritating to the skin and the eyes.
Hazards to Wildlife: n/a
Decontamination Procedures: Wash away any material with copious amounts of soap and water.
First Aid Procedures: Remove victim to fresh air, call emergency medical care. If not breathing give CPR. If breathing is difficult administer oxygen. Treat for shock.

Fire Hazards:

Flashpoint: 204° F(95.5° C) *Ignition temperature:* n/a
Flammable Explosive High Range: n/a *Low Range:* n/a
Toxic Products of Combustion: Toxic oxides of nitrogen may be formed in fires.
Other Hazards: n/a
Possible extinguishing agents: Alcohol foam, dry chemical, carbon dioxide. Water may be ineffective on fire.

Reactivity Hazards:

Reactive With: n/a *Other Reactions:* n/a

Corrosivity Hazards:

Corrosive With: n/a *Neutralizing Agent:* n/a

Radioactivity Hazards:

Radiation Emitted: n/a *Other Hazards:* n/a

Recommended Protection for Response Personnel:

Avoid breathing vapors, keep upwind. Wear appropriate sealed chemical clothing. Wash away any material which may have come into contact with the body with copious amounts of soap and water. If the fire becomes uncontrollable, consider appropriate evacuation.

METHYLSTRYENE

DOT Number: n/a *DOT Hazard Class:* n/a *DOT Guide Number:* n/a
Synonyms: isopropenyl benzene, 1-methyl-1-phenylethylene, phenylpropylene
STCC Number: n/a *Reportable Qty:* n/a
Mfg Name: Dow Chemical *Phone No:* 1-517-636-4400

Physical Description:

Physical Form: Liquid *Color:* Colorless *Odor:* n/a
Other Information: n/a

Chemical Properties:

Specific Gravity: .91 *Vapor Density:* n/a *Boiling Point: 329° F(165° C)*
Melting Point: –10° F(–23.3° C) *Vapor Pressure:* n/a *Solubility in water:* n/a
Other Information: Floats on water.

Health Hazards:

Inhalation Hazard: n/a
Ingestion Hazard: Nausea and vomiting
Absorption Hazard: Will burn the skin and the eyes.
Hazards to Wildlife: Dangerous to aquatic life.
Decontamination Procedures: Wash away any material with copious amounts of soap and water.
First Aid Procedures: Remove victim to fresh air, call emergency medical care. If not breathing give CPR. If breathing is difficult administer oxygen. Treat for shock.

Fire Hazards:

Flashpoint: 137° F(58.3° C) *Ignition temperature:* 1066° F(574.4° C)
Flammable Explosive High Range: 6.1 *Low Range:* 1.9
Toxic Products of Combustion: n/a
Other Hazards: n/a
Possible extinguishing agents: Use foam, dry chemical, carbon dioxide. Water may be ineffective on fire. Cool all affected containers with water. Apply water from as far a distance as possible.

Reactivity Hazards:

Reactive With: n/a *Other Reactions:* n/a

Corrosivity Hazards:

Corrosive With: May attack some forms of plastic. *Neutralizing Agent:* n/a

Radioactivity Hazards:

Radiation Emitted: n/a *Other Hazards:* n/a

Recommended Protection for Response Personnel:

Avoid breathing vapors, keep upwind. Structural protective clothing provides limited protection. Wash away any material which may have come into contact with the body with copious amounts of soap and water. If the fire becomes uncontrollable, consider appropriate evacuation.

METHYL TRICHLOROSILANE

DOT Number: UN 1250 *DOT Hazard Class:* Flammable liquid *DOT Guide Number:* 29
Synonyms: trichloromethylsilane
STCC Number: 4907630 *Reportable Qty:* n/a
Mfg Name: Union Carbide *Phone No:* 1-203-794-2000

Physical Description:

Physical Form: Liquid *Color:* Colorless *Odor:* Sharp, irritating
Other Information: n/a

Chemical Properties:

Specific Gravity: 1.29 *Vapor Density:* 5.16 *Boiling Point: 151° F(66.1° C)*
Melting Point: n/a *Vapor Pressure:* n/a *Solubility in water:* n/a
Other Information: A fuming liquid decomposed by water. Its vapors are heavier than air.

Health Hazards:

Inhalation Hazard: Difficulty in breathing.
Ingestion Hazard: Poisonous if swallowed.
Absorption Hazard: Will burn the skin and eyes.
Hazards to Wildlife: n/a
Decontamination Procedures: Wash away any material with copious amounts of soap and water.
First Aid Procedures: Remove victim to fresh air, call emergency medical care. If not breathing give CPR. If breathing is difficult administer oxygen. Treat for shock.

Fire Hazards:

Flashpoint: 15° F(−9.4° C) *Ignition temperature:* 760° F(404.4° C)
Flammable Explosive High Range: 20 *Low Range:* 7.6
Toxic Products of Combustion: Poisonous gases are produced when heated, or when involved in a fire.
Other Hazards: Containers may explode in a fire. Flashback along vapor trail may occur.
Possible extinguishing agents: Dry chemical or carbon dioxide.

Reactivity Hazards:

Reactive With: Water to form hydrogen chloride.
Other Reactions: Surface moisture to involve hydrogen chloride which is corrosive to metals.

Corrosivity Hazards:

Corrosive With: Surface moisture
Neutralizing Agent: Flood with water, rinse with sodium bicarbonate or lime solution.

Radioactivity Hazards:

Radiation Emitted: n/a *Other Hazards:* n/a

Recommended Protection for Response Personnel:

Avoid breathing vapors, keep upwind. Wear a sealed chemical suit (butyl rubber). Wash away any material which may have come into contact with the body with copious amounts of soap and water. If the fire becomes uncontrollable, or if the containers are exposed to direct flame, consider appropriate evacuation.

METHYL VINYL KETONE

DOT Number: UN 1251 *DOT Hazard Class:* Flammable liquid *DOT Guide Number:* 28
Synonyms: 3-buten-2-one
STCC Number: n/a *Reportable Qty:* n/a
Mfg Name: Aldrich Chemical *Phone No:* 1-414-273-3850

Physical Description:

Physical Form: Liquid *Color:* Colorless to light yellow *Odor:* Strong irritating
Other Information: n/a

Chemical Properties:

Specific Gravity: .86 *Vapor Density:* 2.4 *Boiling Point: 177° F(80.5° C)*
Melting Point: n/a *Vapor Pressure:* n/a *Solubility in water:* Yes
Other Information: When mixed with water, it produces an irritating vapor which is heavier than air.

Health Hazards:

Inhalation Hazard: Coughing, difficulty in breathing.
Ingestion Hazard: Poisonous if swallowed.
Absorption Hazard: Will burn the skin and eyes.
Hazards to Wildlife: n/a
Decontamination Procedures: Wash away any material with copious amounts of soap and water.
First Aid Procedures: Remove victim to fresh air, call emergency medical care. If not breathing give CPR. If breathing is difficult administer oxygen. Treat for shock.

Fire Hazards:

Flashpoint: 20° F(−6.6° C) *Ignition temperature:* 915° F(490.5° C)
Flammable Explosive High Range: 15.6 *Low Range:* 2.1
Toxic Products of Combustion: Material will polymerize when heated or exposed to sunlight causing the containers to violently rupture.
Other Hazards: Containers may explode in a fire. Flashback along vapor trail may occur.
Possible extinguishing agents: Apply water from as far a distance as possible. Use alcohol foam, dry chemical, or carbon dioxide.

Reactivity Hazards:

Reactive With: Heat and sunlight. *Other Reactions:* n/a

Corrosivity Hazards:

Corrosive With: n/a *Neutralizing Agent:* n/a

Radioactivity Hazards:

Radiation Emitted: n/a *Other Hazards:* n/a

Recommended Protection for Response Personnel:

Avoid breathing vapors, keep upwind. Wear appropriate chemical clothing. Wash away any material which may have come into contact with the body with copious amounts of soap and water. If fire becomes uncontrollable, or if containers are exposed to direct flame, consider appropriate evacuation.

MINERAL SPIRITS

DOT Number: UN 1300 *DOT Hazard Class:* Combustible liquid *DOT Guide Number:* 27
Synonyms: naphtha, petroleum benzine, petroleum naphtha, petroleum ether, petroleum spirits
STCC Number: n/a *Reportable Qty:* n/a
Mfg Name: Phillips Petroleum *Phone No:* 1-918-661-6600

Physical Description:

Physical Form: Liquid *Color:* Colorless *Odor:* Gasolinelike
Other Information: n/a

Chemical Properties:

Specific Gravity: .78 *Vapor Density:* 4.3 *Boiling Point: 310 to 395° F(154-201° C)*
Melting Point: n/a *Vapor Pressure:* .3 psia *Solubility in water:* n/a
Other Information: Floats on water

Health Hazards:

Inhalation Hazard: n/a
Ingestion Hazard: Harmful if swallowed.
Absorption Hazard: Irritating to the skin and the eyes.
Hazards to Wildlife: n/a
Decontamination Procedures: Wash away any material with copious amounts of soap and water.
First Aid Procedures: Remove victim to fresh air, call emergency medical care. If not breathing give CPR. If breathing is difficult administer oxygen. Treat for shock.

Fire Hazards:

Flashpoint: 105 to 140° F(40-60° C) *Ignition temperature:* 540° F(282.2° C)
Flammable Explosive High Range: 5 *Low Range:* .8
Toxic Products of Combustion: n/a
Other Hazards: n/a
Possible extinguishing agents: Use foam, dry chemical, carbon dioxide. Water may be ineffective on fires. Cool all affected containers with water. Apply water from as far a distance as possible.

Reactivity Hazards:

Reactive With: n/a *Other Reactions:* n/a

Corrosivity Hazards:

Corrosive With: n/a *Neutralizing Agent:* n/a

Radioactivity Hazards:

Radiation Emitted: n/a *Other Hazards:* n/a

Recommended Protection for Response Personnel:

Avoid breathing vapors, keep upwind. Structural protective clothing provides limited protection. Wash away any material which may have come into contact with the body with copious amounts of soap and water. If the fire becomes uncontrollable, consider appropriate evacuation.

MIREX

DOT Number: UN 1615 *DOT Hazard Class:* Poison B *DOT Guide Number:* n/a
Synonyms: dechlorane, ENT 25,719, perchlorodihomocubane
STCC Number: n/a *Reportable Qty:* n/a
Mfg Name: Allied Corp. *Phone No:* 1-201-455-2000

Physical Description:

Physical Form: Solid *Color:* White *Odor:* Odorless
Other Information: It is used as an insecticide.

Chemical Properties:

Specific Gravity: n/a *Vapor Density:* n/a *Boiling Point: 905° F(485° C)*
Melting Point: n/a *Vapor Pressure:* n/a *Solubility in water:* n/a
Other Information: Avoid contact with solid. Keep people away!!

Health Hazards:

Inhalation Hazard: Poisonous if inhaled.
Ingestion Hazard: Poisonous if swallowed.
Absorption Hazard: Poisonous to the skin and the eyes.
Hazards to Wildlife: Dangerous to aquatic life.
Decontamination Procedures: Wash away any material with copious amounts of soap and water.
First Aid Procedures: Remove victim to fresh air, call emergency medical care. If not breathing give CPR. If breathing is difficult administer oxygen. Treat for shock.

Fire Hazards:

Flashpoint: n/a *Ignition temperature:* n/a
Flammable Explosive High Range: n/a *Low Range:* n/a
Toxic Products of Combustion: n/a
Other Hazards: Supports combustion
Possible extinguishing agents: Water, alcohol foam, dry chemical and carbon dioxide.

Reactivity Hazards:

Reactive With: n/a *Other Reactions:* n/a

Corrosivity Hazards:

Corrosive With: n/a *Neutralizing Agent:* n/a

Radioactivity Hazards:

Radiation Emitted: n/a *Other Hazards:* n/a

Recommended Protection for Response Personnel:

Avoid breathing vapors, keep upwind. Wear appropriate sealed chemical clothing. Wash away any material which may have come into contact with the body with copious amounts of soap and water. Consider appropriate evacuation.

MOLYBDIC TRIOXIDE

DOT Number: n/a *DOT Hazard Class:* n/a *DOT Guide Number:* n/a
Synonyms: molybdenum trioxide, molybdic anhydride
STCC Number: n/a *Reportable Qty:* n/a
Mfg Name: Allied Corp. *Phone No:* 1-201-455-2000

Physical Description:

Physical Form: Solid *Color:* Colorless to white or yellow *Odor:* Odorless
Other Information: n/a

Chemical Properties:

Specific Gravity: 4.69 *Vapor Density:* n/a *Boiling Point: n/a*
Melting Point: n/a *Vapor Pressure:* n/a *Solubility in water:* n/a
Other Information: Sinks in water.

Health Hazards:

Inhalation Hazard: n/a
Ingestion Hazard: Harmful if swallowed.
Absorption Hazard: Irritating to the skin and the eyes.
Hazards to Wildlife: Dangerous to aquatic life.
Decontamination Procedures: Wash away any material with copious amounts of soap and water.
First Aid Procedures: Remove victim to fresh air, call emergency medical care. If not breathing give CPR. If breathing is difficult administer oxygen. Treat for shock.

Fire Hazards:

Flashpoint: n/a *Ignition temperature:* n/a
Flammable Explosive High Range: n/a *Low Range:* n/a
Toxic Products of Combustion: n/a
Other Hazards: n/a
Possible extinguishing agents: Extinguish fire using suitable agent for the type of surrounding fire.

Reactivity Hazards:

Reactive With: n/a *Other Reactions:* n/a

Corrosivity Hazards:

Corrosive With: n/a *Neutralizing Agent:* n/a

Radioactivity Hazards:

Radiation Emitted: n/a *Other Hazards:* n/a

Recommended Protection for Response Personnel:

Avoid breathing vapors, keep upwind. Structural protective clothing provides limited protection. Wash away any material which may have come into contact with the body with copious amounts of soap and water. Consider appropriate evacuation.

MONOCHLOROACETIC ACID (liquid)

DOT Number: UN 1750 *DOT Hazard Class:* Corrosive *DOT Guide Number:* 59
Synonyms: chloracetic acid, chloroacetic acid, chloroethanoic acid
STCC Number: 4931444 *Reportable Qty:* n/a
Mfg Name: Mallinckrodt Inc. *Phone No:* 1-314-895-2000

Physical Description:

Physical Form: Liquid *Color:* Colorless *Odor:* n/a
Other Information: It is used as a herbicide and a bacteriostat.

Chemical Properties:

Specific Gravity: 1.6 *Vapor Density:* 3.3 *Boiling Point: n/a*
Melting Point: n/a *Vapor Pressure:* n/a *Solubility in water:* n/a
Other Information: n/a

Health Hazards:

Inhalation Hazard: Hazardous if inhaled.
Ingestion Hazard: Hazardous if swallowed.
Absorption Hazard: Hazardous to the skin and the eyes.
Hazards to Wildlife: n/a
Decontamination Procedures: Wash away any material with copious amounts of soap and water.
First Aid Procedures: Remove victim to fresh air, call emergency medical care. If not breathing give CPR. If breathing is difficult administer oxygen. Treat for shock.

Fire Hazards:

Flashpoint: 259° F(126.1° C) *Ignition temperature:* 932° F(500° C)
Flammable Explosive High Range: n/a *Low Range:* n/a
Toxic Products of Combustion: n/a
Other Hazards: n/a
Possible extinguishing agents: Apply water from as far a distance as possible. Cool all affected containers with large quantities of water. Extinguish fire using suitable agent for the type of surrounding fire.

Reactivity Hazards:

Reactive With: n/a *Other Reactions:* n/a

Corrosivity Hazards:

Corrosive With: Metals and tissue. *Neutralizing Agent:* Crushed limestone, soda ash, or lime.

Radioactivity Hazards:

Radiation Emitted: n/a *Other Hazards:* n/a

Recommended Protection for Response Personnel:

Avoid breathing vapors, keep upwind. Structural protective clothing provides limited protection. Wash away any material which may have come into contact with the body with copious amounts of soap and water. Consider appropriate evacuation.

MONOCHLOROACETIC ACID (solid)

DOT Number: UN 1750 *DOT Hazard Class:* Corrosive *DOT Guide Number:* 59
Synonyms: chloracetic acid, chloroacetic acid, chloroethanoic acid
STCC Number: 4931444 *Reportable Qty:* n/a
Mfg Name: Mallinckrodt Inc. *Phone No:* 1-314-895-2000

Physical Description:

Physical Form: Solid *Color:* Colorless *Odor:* n/a
Other Information: It is used as a herbicide and a bacteriostat.

Chemical Properties:

Specific Gravity: 1.6 *Vapor Density:* 3.3 *Boiling Point: n/a*
Melting Point: n/a *Vapor Pressure:* n/a *Solubility in water:* n/a
Other Information: n/a

Health Hazards:

Inhalation Hazard: Hazardous if inhaled.
Ingestion Hazard: Hazardous if swallowed.
Absorption Hazard: Hazardous to the skin and the eyes.
Hazards to Wildlife: n/a
Decontamination Procedures: Wash away any material with copious amounts of soap and water.
First Aid Procedures: Remove victim to fresh air, call emergency medical care. If not breathing give CPR. If breathing is difficult administer oxygen. Treat for shock.

Fire Hazards:

Flashpoint: 259° F(126.1° C) *Ignition temperature:* 932° F(500° C)
Flammable Explosive High Range: n/a *Low Range:* n/a
Toxic Products of Combustion: n/a
Other Hazards: n/a
Possible extinguishing agents: Apply water from as far a distance as possible. Cool all affected containers with large quantities of water. extinguish fire using suitable agent for the type of surrounding fire.

Reactivity Hazards:

Reactive With: n/a *Other Reactions:* n/a

Corrosivity Hazards:

Corrosive With: Metals and tissue. *Neutralizing Agent:* Crushed limestone, soda ash, or lime.

Radioactivity Hazards:

Radiation Emitted: n/a *Other Hazards:* n/a

Recommended Protection for Response Personnel:

Avoid breathing vapors, keep upwind. Structural protective clothing provides limited protection. Wash away any material which may have come into contact with the body with copious amounts of soap and water. Consider appropriate evacuation.

MONOCHLORODIFLUOROMETHANE

DOT Number: UN 1018 *DOT Hazard Class:* Nonflammable gas *DOT Guide Number:* 12
Synonyms: Algeon 22, Aroton #4, F22, Freon 22, R22, Yukon 22
STCC Number: 4904552 *Reportable Qty:* n/a
Mfg Name: E.I. Du Pont *Phone No:* 1-800-441-3637

Physical Description:

Physical Form: Gas *Color:* Colorless *Odor:* Sweet
Other Information: It is used for refrigerant, low temperature solvent, and in the manufacture of teflon.

Chemical Properties:

Specific Gravity: 1.41 *Vapor Density:* 4.7 *Boiling Point: −40.9° F(−40.5° C)*
Melting Point: −230.8° F(−146° C) *Vapor Pressure:* 10 atm at 75.2° F(24° C)
Solubility in water: Yes
Other Information: It is slightly soluble and heavier than water. Weighs 11.8 lbs/5.3 kg per gallon/3.8 l.

Health Hazards:

Inhalation Hazard: Lightheadiness, giddiness, shortness of breath.
Ingestion Hazard: Will cause frostbite.
Absorption Hazard: Will cause frostbite.
Hazards to Wildlife: n/a
Decontamination Procedures: Wash away any material with copious amounts of soap and water.
First Aid Procedures: Remove victim to fresh air, call emergency medical care. If not breathing give CPR. If breathing is difficult administer oxygen. Treat for shock.

Fire Hazards:

Flashpoint: n/a *Ignition temperature:* n/a
Flammable Explosive High Range: n/a *Low Range:* n/a
Toxic Products of Combustion: Highly toxic fluorine and phosgene gases.
Other Hazards: Containers may rupture violently in a fire due to overpressurization.
Possible extinguishing agents: Extinguish fire using suitable agent for the type of surrounding fire.

Reactivity Hazards:

Reactive With: Some metals *Other Reactions:* n/a

Corrosivity Hazards:

Corrosive With: n/a *Neutralizing Agent:* n/a

Radioactivity Hazards:

Radiation Emitted: n/a *Other Hazards:* n/a

Recommended Protection for Response Personnel:

Avoid breathing vapors, keep upwind. Structural protective clothing provides limited protection. Wash away any material which may have come into contact with the body with copious amounts of soap and water. Consider appropriate evacuation.

MONOETHANOLAMINE

DOT Number: UN 2491 *DOT Hazard Class:* Corrosive *DOT Guide Number:* 60
Synonyms: β-amineoethyl alcohol
STCC Number: 4935665 *Reportable Qty:* n/a
Mfg Name: Dow Chemical *Phone No:* 1-517-636-4400

Physical Description:

Physical Form: Liquid *Color:* Colorless *Odor:* Ammonialike
Other Information: It is soluble in water, with releases of heat.

Chemical Properties:

Specific Gravity: 1.01 *Vapor Density:* 2.11 *Boiling Point: 338° F(170° C)*
Melting Point: n/a *Vapor Pressure:* n/a *Solubility in water:* Yes
Other Information: Sinks and mixes with water.

Health Hazards:

Inhalation Hazard: n/a
Ingestion Hazard: Harmful if swallowed.
Absorption Hazard: Irritating to the skin and eyes.
Hazards to Wildlife: Dangerous to aquatic life.
Decontamination Procedures: Wash away any material with copious amounts of soap and water.
First Aid Procedures: Remove victim to fresh air, call emergency medical care. If not breathing give CPR. If breathing is difficult administer oxygen. Treat for shock.

Fire Hazards:

Flashpoint: 185° F(85° C) *Ignition temperature:* n/a
Flammable Explosive High Range: n/a *Low Range:* n/a
Toxic Products of Combustion: Toxic oxides of nitrogen are produced during combustion of this material.
Other Hazards: Vapors may explode in an enclosed area. Flashback along vapor trail may occur.
Possible extinguishing agents: Apply water from as far a distance as possible. Use alcohol foam, dry chemical, or carbon dioxide.

Reactivity Hazards:

Reactive With: Water *Other Reactions:* n/a

Corrosivity Hazards:

Corrosive With: Tissue, may attack copper, brass, and rubber. *Neutralizing Agent:* n/a

Radioactivity Hazards:

Radiation Emitted: n/a *Other Hazards:* n/a

Recommended Protection for Response Personnel:

Avoid breathing vapors, keep upwind. Structural protective clothing provides limited protection. Wash away any material which may have come into contact with the body with copious amounts of soap and water. Consider appropriate evacuation.

MONOISOPROPANOLIMINE

DOT Number: n/a *DOT Hazard Class:* n/a *DOT Guide Number:* n/a
Synonyms: 1-amino-2-propanol, 2-hydroxypropylamine, isopropanolamine, MIPA
STCC Number: n/a *Reportable Qty:* n/a
Mfg Name: Dow Chemical *Phone No:* 1-517-636-4400

Physical Description:

Physical Form: Liquid *Color:* Colorless *Odor:* Slight ammonia
Other Information: n/a

Chemical Properties:

Specific Gravity: n/a *Vapor Density:* 2.6 *Boiling Point: 320° F(160° C)*
Melting Point: 35° F(1.6° C) *Vapor Pressure:* n/a *Solubility in water:* Yes
Other Information: Floats and mixes with water.

Health Hazards:

Inhalation Hazard: n/a
Ingestion Hazard: Harmful if swallowed.
Absorption Hazard: Irritating to the skin and the eyes.
Hazards to Wildlife: n/a
Decontamination Procedures: Wash away any material with copious amounts of soap and water.
First Aid Procedures: Remove victim to fresh air, call emergency medical care. If not breathing give CPR. If breathing is difficult administer oxygen. Treat for shock.

Fire Hazards:

Flashpoint: 165° F(73.8° C) *Ignition temperature:* 706° F(374.4° C)
Flammable Explosive High Range: 12 *Low Range:* 2.2
Toxic Products of Combustion: Irritating vapor is generated when heated.
Other Hazards: n/a
Possible extinguishing agents: Extinguish fire using suitable agent for the type of surrounding fire. Apply water from as far a distance as possible. Cool all affected containers with water.

Reactivity Hazards:

Reactive With: Powerful oxidizers *Other Reactions:* n/a

Corrosivity Hazards:

Corrosive With: n/a *Neutralizing Agent:* n/a

Radioactivity Hazards:

Radiation Emitted: n/a *Other Hazards:* n/a

Recommended Protection for Response Personnel:

Avoid breathing vapors, keep upwind. Structural protective clothing provides limited protection. Wash away any material which may have come into contact with the body with copious amounts of soap and water. If the fire becomes uncontrollable, consider appropriate evacuation.

MONOMETHYLAMINE

DOT Number: UN 1061 *DOT Hazard Class:* Flammable gas *DOT Guide Number:* 19
Synonyms: carbinamine, monomethyl anhydrous
STCC Number: 4905530 *Reportable Qty:* 1000/454 *CHEMTREC Phone No:* 1-800-424-9300

Physical Description:

Physical Form: Gas *Color:* Colorless *Odor:* Fishy
Other Information: It is used in the manufacture of pharmaceuticals, insecticides, rocket fuels, explosives, dyes and textiles.

Chemical Properties:

Specific Gravity: .69 *Vapor Density:* 1.1 *Boiling Point: 20.3° F(−6.5° C)*
Melting Point: −136° F(−93.3° C) *Vapor Pressure:* 2 atm at 50.2° F(10.1° C)
Solubility in water: Yes
Other Information: Vapors are heavier than air. Weighs 5.8 lbs/2.6 kg per gallon/3.8 l.

Health Hazards:

Inhalation Hazard: Vapors are severely irritating.
Ingestion Hazard: Burns to the mouth.
Absorption Hazard: Chemical burns to the skin. Serious eye injury and loss of sight.
Hazards to Wildlife: n/a
Decontamination Procedures: Wash away any material with copious amounts of soap and water.
First Aid Procedures: Remove victim to fresh air, call emergency medical care. If not breathing give CPR. If breathing is difficult administer oxygen. Treat for shock.

Fire Hazards:

Flashpoint: 14° F(−10° C) *Ignition temperature:* 806° F(430° C)
Flammable Explosive High Range: 21 *Low Range:* 4.3
Toxic Products of Combustion: Toxic oxides of nitrogen, carbon monoxide, and hydrogen cyanide.
Other Hazards: May travel a considerable distance to a source of ignition and flashback. Containers may rupture violently in a fire.
Possible extinguishing agents: Do not extinguish the fire unless the flow can be stopped. Apply water from as far a distance as possible. Use alcohol foam, dry chemical, or carbon dioxide.

Reactivity Hazards:

Reactive With: n/a *Other Reactions:* n/a

Corrosivity Hazards:

Corrosive With: Copper, copper alloys, zinc alloys, aluminum, and galvanized surfaces. May attack some forms of plastics, rubber, and coatings. *Neutralizing Agent:* n/a

Radioactivity Hazards:

Radiation Emitted: n/a *Other Hazards:* n/a

Recommended Protection for Response Personnel:

Avoid breathing vapors, keep upwind. Wear a sealed chemical suit (butyl rubber, neoprene, nitrile, polyethylene). Wash away any material which may have come into contact with the body with copious amounts of soap and water. If the fire becomes uncontrollable, or if the containers are exposed to direct flame, consider appropriate evacuation.

MORPHOLINE

DOT Number: UN 2054 *DOT Hazard Class:* Flammable liquid *DOT Guide Number:* 29
Synonyms: tetrahydro-2H-1,4-oxazine
STCC Number: 4907846 *Reportable Qty:* n/a
Mfg Name: Dow Chemical *Phone No:* 1-517-636-4400

Physical Description:

Physical Form: Liquid *Color:* Colorless *Odor:* Fishlike
Other Information: It is used to make other chemicals, as a corrosion inhibitor, and in detergents.

Chemical Properties:

Specific Gravity: 1 *Vapor Density:* 3 *Boiling Point: 262° F(127.7° C)*
Melting Point: 23° F(–5° C) *Vapor Pressure:* 7 mm Hg at 68° F(20° C) *Solubility in water:* Yes
Other Information: Lighter than and soluble in water. Vapors are heavier than air.

Health Hazards:

Inhalation Hazard: Headache, difficulty in breathing.
Ingestion Hazard: Nausea
Absorption Hazard: Irritating to the skin and the eyes.
Hazards to Wildlife: n/a
Decontamination Procedures: Wash away any material with copious amounts of soap and water.
First Aid Procedures: Remove victim to fresh air, call emergency medical care. If not breathing give CPR. If breathing is difficult administer oxygen. Treat for shock.

Fire Hazards:

Flashpoint: 98° F(36.6° C) *Ignition temperature:* 555° F(290.5° C)
Flammable Explosive High Range: 11.2 *Low Range:* 1.4
Toxic Products of Combustion: n/a
Other Hazards: Vapors may explode in an enclosed area. Flashback along vapor trail may occur.
Possible extinguishing agents: Apply water from as far a distance as possible. Use alcohol foam, dry chemical, or carbon dioxide.

Reactivity Hazards:

Reactive With: n/a *Other Reactions:* n/a

Corrosivity Hazards:

Corrosive With: Tissues *Neutralizing Agent:* n/a

Radioactivity Hazards:

Radiation Emitted: n/a *Other Hazards:* n/a

Recommended Protection for Response Personnel:

Avoid breathing vapors, keep upwind. Structural protective clothing provides limited protection. Wash away any material which may have come into contact with the body with copious amounts os soap and water. If the fire becomes uncontrollable, or if the containers are exposed to direct flame, consider appropriate evacuation.

MOTOR FUEL ANTIKNOCK COMPOUND

DOT Number: UN 1649 *DOT Hazard Class:* Poison B *DOT Guide Number:* 56
Synonyms: none given
STCC Number: 4921445 *Reportable Qty:* Varies
Mfg Name: E.I. Du Pont *Phone No:* 1-800-441-3637

Physical Description:

Physical Form: Liquid *Color:* Red, orange, or blue *Odor:* Sweet, fruity
Other Information: It is used to increase the octane of gasolines.

Chemical Properties:

Specific Gravity: 1.5-1.7 *Vapor Density:* n/a *Boiling Point: 200° F(93.3° C)*
Melting Point: n/a *Vapor Pressure:* n/a *Solubility in water:* Barely
Other Information: Sinks in water.

Health Hazards:

Inhalation Hazard: n/a
Ingestion Hazard: Poisonous if swallowed.
Absorption Hazard: Poisonous if the skin is exposed.
Hazards to Wildlife: n/a
Decontamination Procedures: Wash away any material with copious amounts of soap and water.
First Aid Procedures: Remove victim to fresh air, call emergency medical care. If not breathing give CPR. If breathing is difficult administer oxygen. Treat for shock.

Fire Hazards:

Flashpoint: 30-264° F(−1.1-128.8° C) *Ignition temperature:* 212° F(100° C)
Flammable Explosive High Range: n/a *Low Range:* n/a
Toxic Products of Combustion: Toxic lead containing gases. Containers may explode if heated.
Other Hazards: n/a
Possible extinguishing agents: Apply water from as far a distance as possible. Use foam, dry chemical, or carbon dioxide. Do not extinguish the fire unless the flow can be stopped.

Reactivity Hazards:

Reactive With: Oxidizing materials, active metals, and rust. *Other Reactions:* n/a

Corrosivity Hazards:

Corrosive With: n/a *Neutralizing Agent:* n/a

Radioactivity Hazards:

Radiation Emitted: n/a *Other Hazards:* n/a

Recommended Protection for Response Personnel:

Avoid breathing vapors, keep upwind. Wear a sealed chemical suit (PVC, butyl rubber). Wash away any material which may have come into contact with the body with copious amounts of soap and water. If the fire becomes uncontrollable, or if the containers are exposed to direct flame, consider appropriate evacuation.

NABAM

DOT Number: UN 1609 *DOT Hazard Class:* Poison B *DOT Guide Number:* n/a
Synonyms: Chem Bam; disodium ethylenebis; Dithane; EBDC, sodium salt
STCC Number: n/a *Reportable Qty:* n/a
Mfg Name: Rohm & Haas Co. *Phone No:* 1-215-592-3000

Physical Description:

Physical Form: Solid or solution *Color:* Colorless to light amber *Odor:* Slight sulfide
Other Information: n/a

Chemical Properties:

Specific Gravity: 1.14 *Vapor Density:* 8.8 *Boiling Point: Decomposes*
Melting Point: n/a *Vapor Pressure:* n/a *Solubility in water:* Yes
Other Information: Mixes with water.

Health Hazards:

Inhalation Hazard: Poisonous if inhaled.
Ingestion Hazard: Poisonous if swallowed.
Absorption Hazard: Irritating to the skin and eyes.
Hazards to Wildlife: Dangerous to aquatic life.
Decontamination Procedures: Wash away any material with copious amounts of soap and water.
First Aid Procedures: Remove victim to fresh air, call emergency medical care. If not breathing give CPR. If breathing is difficult administer oxygen. Treat for shock.

Fire Hazards:

Flashpoint: n/a *Ignition temperature:* n/a
Flammable Explosive High Range: n/a *Low Range:* n/a
Toxic Products of Combustion: Poisonous and flammable gases are produced if the solution boils.
Other Hazards: If material boils, poisonous hydrogen sulfide, and highly flammable carbon disulfide vapors form.
Possible extinguishing agents: Apply water from as far a distance as possible. Keep containers cool, extinguish fire using suitable agent for the type of surrounding fire.

Reactivity Hazards:

Reactive With: Boiling water *Other Reactions:* n/a

Corrosivity Hazards:

Corrosive With: n/a *Neutralizing Agent:* n/a

Radioactivity Hazards:

Radiation Emitted: n/a *Other Hazards:* n/a

Recommended Protection for Response Personnel:

Avoid breathing vapors, keep upwind. Wear a sealed chemical suit (polycarbonate, butyl rubber). Wash away any material which may have come into contact with the body with copious amounts of soap and water. Consider appropriate evacuation.

NALED (liquid)

DOT Number: NA 2783 *DOT Hazard Class:* ORM-E *DOT Guide Number:* 55
Synonyms: arthodibrom, Bronex, Dibrom
STCC Number: 4961656 *Reportable Qty:* 10/4.54
Mfg Name: Chevron Chemical *Phone No:* 1-415-233-3737

Physical Description:

Physical Form: Liquid *Color:* Light straw *Odor:* Slightly pungent
Other Information: In case of damage to, or leaking from these containers, contact the Pesticide Safety Network Team at 1-800-424-9300.

Chemical Properties:

Specific Gravity: 1.97 *Vapor Density:* 13.1 *Boiling Point: 392° F(200° C)*
Melting Point: n/a *Vapor Pressure:* .002 mm Hg at 68° F(20° C) *Solubility in water:* Yes
Other Information: It is used as a pesticide. Sinks and mixes slowly with water.

Health Hazards:

Inhalation Hazard: Poisonous if inhaled.
Ingestion Hazard: Poisonous if swallowed.
Absorption Hazard: Irritating to the skin and eyes.
Hazards to Wildlife: Dangerous to both aquatic life and waterfowl.
Decontamination Procedures: Wash away any material with copious amounts of soap and water.
First Aid Procedures: Remove victim to fresh air, call emergency medical care. If not breathing give CPR. If breathing is difficult administer oxygen. Treat for shock.

Fire Hazards:

Flashpoint: n/a *Ignition temperature:* n/a
Flammable Explosive High Range: n/a *Low Range:* n/a
Toxic Products of Combustion: n/a
Other Hazards: n/a
Possible extinguishing agents: Extinguish fire using suitable agent for the type of surrounding fire.

Reactivity Hazards:

Reactive With: Unstable in alkali conditions. Degraded by sunlight. *Other Reactions:* n/a

Corrosivity Hazards:

Corrosive With: n/a *Neutralizing Agent:* n/a

Radioactivity Hazards:

Radiation Emitted: n/a *Other Hazards:* n/a

Recommended Protection for Response Personnel:

Avoid breathing vapors, keep upwind. Wear appropriate chemical clothing. Wash away any material which may have come into contact with the body with copious amounts of soap and water. Consider appropriate evacuation.

NALED (solid)

DOT Number: NA 2783 *DOT Hazard Class:* ORM-E *DOT Guide Number:* 55
Synonyms: arthodibrom, Bronex, Dibrom
STCC Number: 4961657 *Reportable Qty:* 10/4.54
Mfg Name: Chevron Chemical *Phone No:* 1-415-233-3737

Physical Description:

Physical Form: Solid *Color:* White *Odor:* Slightly pungent
Other Information: In case of damage to, or leaking from these containers, contact the Pesticide Safety Network Team at 1-800-424-9300.

Chemical Properties:

Specific Gravity: 1.97 *Vapor Density:* 13.1 *Boiling Point: 392° F(200° C)*
Melting Point: n/a *Vapor Pressure:* .002 mm Hg at 68° F(20° C) *Solubility in water:* Yes
Other Information: It is used as a pesticide. Sinks and mixes slowly with water.

Health Hazards:

Inhalation Hazard: Poisonous if inhaled.
Ingestion Hazard: Poisonous if swallowed.
Absorption Hazard: Irritating to the skin and eyes.
Hazards to Wildlife: Dangerous to both aquatic life and waterfowl.
Decontamination Procedures: Wash away any material with copious amounts of soap and water.
First Aid Procedures: Remove victim to fresh air, call emergency medical care. If not breathing give CPR. If breathing is difficult administer oxygen. Treat for shock.

Fire Hazards:

Flashpoint: n/a *Ignition temperature:* n/a
Flammable Explosive High Range: n/a *Low Range:* n/a
Toxic Products of Combustion: n/a
Other Hazards: n/a
Possible extinguishing agents: Extinguish fire using suitable agent for the type of surrounding fire.

Reactivity Hazards:

Reactive With: Unstable in the presence of iron. *Other Reactions:* n/a

Corrosivity Hazards:

Corrosive With: n/a *Neutralizing Agent:* n/a

Radioactivity Hazards:

Radiation Emitted: n/a *Other Hazards:* n/a

Recommended Protection for Response Personnel:

Avoid breathing vapors, keep upwind. Wear appropriate chemical clothing. Wash away any material which may come into contact with the body with copious amounts of soap and water. Consider appropriate evacuation.

NAPHTHA

DOT Number: NA 2553 *DOT Hazard Class:* n/a *DOT Guide Number:* 27
Synonyms: cold tar; mixture of benzene, toluene, and xylenes
STCC Number: n/a *Reportable Qty:* n/a
Mfg Name: Crowley Tar Products Inc. *Phone No:* 1-212-682-1200

Physical Description:

Physical Form: Liquid *Color:* Colorless to pale yellow *Odor:* Gasolinelike
Other Information: n/a

Chemical Properties:

Specific Gravity: .86 *Vapor Density:* 1 *Boiling Point: 230-375° F(110-190.5° C)*
Melting Point: n/a *Vapor Pressure:* 5 mm Hg at 68° F(20° C) *Solubility in water:* No
Other Information: Floats on water. An irritating vapor is produced.

Health Hazards:

Inhalation Hazard: Headache, dizziness, difficulty breathing, loss of consciousness.
Ingestion Hazard: Causes nausea and vomiting.
Absorption Hazard: Irritating to the skin and eyes.
Hazards to Wildlife: n/a
Decontamination Procedures: Wash away any material with copious amounts of soap and water.
First Aid Procedures: Remove victim to fresh air, call emergency medical care. If not breathing give CPR. If breathing is difficult administer oxygen. Treat for shock.

Fire Hazards:

Flashpoint: 107° F(41.6° C) *Ignition temperature:* 531° F(277.2° C)
Flammable Explosive High Range: n/a *Low Range:* n/a
Toxic Products of Combustion: n/a
Other Hazards: n/a
Possible extinguishing agents: Apply water from as far a distance as possible. Use foam, dry chemical, or carbon dioxide.

Reactivity Hazards:

Reactive With: n/a *Other Reactions:* n/a

Corrosivity Hazards:

Corrosive With: n/a *Neutralizing Agent:* n/a

Radioactivity Hazards:

Radiation Emitted: n/a *Other Hazards:* n/a

Recommended Protection for Response Personnel:

Avoid breathing vapors, keep upwind. Wear a sealed chemical suit (nitrile, neoprene, polycarbonate). Wash away any material which may have come into contact with the body with copious amounts of soap and water. Consider appropriate evacuation.

NAPHTHA, SOLVENT

DOT Number: UN 1256 *DOT Hazard Class:* Flammable liquid *DOT Guide Number:* 27
Synonyms: light naphtha, petroleum solvent
STCC Number: n/a *Reportable Qty:* n/a
Mfg Name: Union Oil Co. *Phone No:* 1-213-977-7600

Physical Description:

Physical Form: Liquid *Color:* Colorless *Odor:* Gasolinelike
Other Information: n/a

Chemical Properties:

Specific Gravity: .86 *Vapor Density:* 1 *Boiling Point: 266° F(130° C)*
Melting Point: n/a *Vapor Pressure:* n/a *Solubility in water:* No
Other Information: Floats on water. Vapor is produced.

Health Hazards:

Inhalation Hazard: Dizziness, loss of consciousness.
Ingestion Hazard: Harmful if swallowed.
Absorption Hazard: Irritating to the skin and eyes.
Hazards to Wildlife: n/a
Decontamination Procedures: Wash away any material with copious amounts of soap and water.
First Aid Procedures: Remove victim to fresh air, call emergency medical care. If not breathing give CPR. If breathing is difficult administer oxygen. Treat for shock.

Fire Hazards:

Flashpoint: 100° F(37.7° C) *Ignition temperature:* 444° F(228.8° C)
Flammable Explosive High Range: 5 *Low Range:* .8
Toxic Products of Combustion: n/a
Other Hazards: n/a
Possible extinguishing agents: Apply water from as far a distance as possible. Use foam, dry chemical, or carbon dioxide.

Reactivity Hazards:

Reactive With: n/a *Other Reactions:* n/a

Corrosivity Hazards:

Corrosive With: n/a *Neutralizing Agent:* n/a

Radioactivity Hazards:

Radiation Emitted: n/a *Other Hazards:* n/a

Recommended Protection for Response Personnel:

Avoid breathing vapors, keep upwind. Wear a sealed chemical suit (PVC). Wash away any material which may have come into contact with the body with copious amounts of soap and water. Consider appropriate evacuation.

NAPHTHA (STODDARD SOLVENT)

DOT Number: UN 1268 *DOT Hazard Class:* Flammable liquid *DOT Guide Number:* 27
Synonyms: dry cleaners naptha, petroleum solvent, spotting naptha
STCC Number: n/a *Reportable Qty:* n/a
Mfg Name: Union Oil Co. *Phone No:* 1-213-977-7600

Physical Description:

Physical Form: Liquid *Color:* Colorless *Odor:* Gasolinelike
Other Information: n/a

Chemical Properties:

Specific Gravity: .78 *Vapor Density:* 1 *Boiling Point: 320-390° F(160-198° C)*
Melting Point: n/a *Vapor Pressure:* n/a *Solubility in water:* No
Other Information: Floats on water

Health Hazards:

Inhalation Hazard: n/a
Ingestion Hazard: Harmful if swallowed.
Absorption Hazard: Irritating to the skin and eyes.
Hazards to Wildlife: n/a
Decontamination Procedures: Wash away any material with copious amounts of soap and water.
First Aid Procedures: Remove victim to fresh air, call emergency medical care. If not breathing give CPR. If breathing is difficult administer oxygen. Treat for shock.

Fire Hazards:

Flashpoint: 110° F(43.3° C) *Ignition temperature:* 540° F(282.2° C)
Flammable Explosive High Range: 5 *Low Range:* .8
Toxic Products of Combustion: n/a
Other Hazards: n/a
Possible extinguishing agents: Apply water from as far a distance as possible. Use foam, dry chemical, or carbon dioxide.

Reactivity Hazards:

Reactive With: n/a *Other Reactions:* n/a

Corrosivity Hazards:

Corrosive With: n/a *Neutralizing Agent:* n/a

Radioactivity Hazards:

Radiation Emitted: n/a *Other Hazards:* n/a

Recommended Protection for Response Personnel:

Avoid breathing vapors, keep upwind. Wear a sealed chemical suit (PVC). Wash away any material which may have come into contact with the body with copious amounts of soap and water. Consider appropriate evacuation.

NAPHTHA (WMP)

DOT Number: UN 1255 *DOT Hazard Class:* Flammable liquid *DOT Guide Number:* 27
Synonyms: light naptha, painters naptha, petroleum solvent
STCC Number: n/a *Reportable Qty:* n/a
Mfg Name: Union Oil Co. *Phone No:* 1-213-977-7600

Physical Description:

Physical Form: Liquid *Color:* Colorless *Odor:* Gasolinelike
Other Information: n/a

Chemical Properties:

Specific Gravity: .75 *Vapor Density:* 4.2 *Boiling Point: 200-300° F(93-148° C)*
Melting Point: n/a *Vapor Pressure:* n/a *Solubility in water:* No
Other Information: Floats on water, flammable, irritating vapor is produced.

Health Hazards:

Inhalation Hazard: Dizziness, loss of consciousness.
Ingestion Hazard: Nausea and vomiting.
Absorption Hazard: Irritating to the skin and eyes.
Hazards to Wildlife: n/a
Decontamination Procedures: Wash away any material with copious amounts of soap and water.
First Aid Procedures: Remove victim to fresh air, call emergency medical care. If not breathing give CPR. If breathing is difficult administer oxygen. Treat for shock.

Fire Hazards:

Flashpoint: 20-55° F(−6.6-12.7° C) *Ignition temperature:* 450° F(232.2° C)
Flammable Explosive High Range: 6.7 *Low Range:* .9
Toxic Products of Combustion: n/a
Other Hazards: Flashback along vapor trail may occur. Vapors may explode if ignited in an enclosed area.
Possible extinguishing agents: Apply water from as far a distance as possible. Use foam, dry chemical, or carbon dioxide.

Reactivity Hazards:

Reactive With: n/a *Other Reactions:* n/a

Corrosivity Hazards:

Corrosive With: n/a *Neutralizing Agent:* n/a

Radioactivity Hazards:

Radiation Emitted: n/a *Other Hazards:* n/a

Recommended Protection for Response Personnel:

Avoid breathing vapors, keep upwind. Wear appropriate sealed chemical suit. Wash away any material which may have come into contact with the body with copious amounts of soap and water. Consider appropriate evacuation.

NAPHTHALENE

DOT Number: NA 2304 *DOT Hazard Class:* n/a *DOT Guide Number:* 32
Synonyms: tar camphor, white tar
STCC Number: n/a *Reportable Qty:* n/a
Mfg Name: Allied Corp. *Phone No:* 1-201-455-2000

Physical Description:

Physical Form: Solid *Color:* Colorless *Odor:* Mothball
Other Information: n/a

Chemical Properties:

Specific Gravity: 1.1 *Vapor Density:* 4.4 *Boiling Point: 424° F(217.7° C)*
Melting Point: 165-176° F(73.8-80° C) *Vapor Pressure:* .05 mm Hg at 68° F(20° C)
Solubility in water: No
Other Information: Floats, solidifies, and sinks in water.

Health Hazards:

Inhalation Hazard: n/a
Ingestion Hazard: n/a
Absorption Hazard: Irritating to the skin and eyes.
Hazards to Wildlife: Dangerous to aquatic life.
Decontamination Procedures: Wash away any material with copious amounts of soap and water.
First Aid Procedures: Remove victim to fresh air, call emergency medical care. If not breathing give CPR. If breathing is difficult administer oxygen. Treat for shock.

Fire Hazards:

Flashpoint: 174° F(78.8° C) *Ignition temperature:* 979° F(526.1° C)
Flammable Explosive High Range: 5.9 *Low Range:* .9
Toxic Products of Combustion: Toxic vapor is given off in a fire.
Other Hazards: n/a
Possible extinguishing agents: Apply water from as far a distance as possible. Use foam, dry chemical, carbon dioxide, or water fog.

Reactivity Hazards:

Reactive With: n/a *Other Reactions:* n/a

Corrosivity Hazards:

Corrosive With: n/a *Neutralizing Agent:* n/a

Radioactivity Hazards:

Radiation Emitted: n/a *Other Hazards:* n/a

Recommended Protection for Response Personnel:

Avoid breathing vapors, keep upwind. Wear appropriate chemical clothing. Wash away any material which may have come into contact with the body with copious amounts of soap and water. Consider appropriate evacuation.

NAPHTHENIC ACID

DOT Number: NA 9137 *DOT Hazard Class:* ORM-E *DOT Guide Number:* 31
Synonyms: none given
STCC Number: 4962356 *Reportable Qty:* 100/45.4
Mfg Name: Texaco Inc. *Phone No:* 1-914-253-4000

Physical Description:

Physical Form: Liquid *Color:* Black to gold *Odor:* None listed
Other Information: It is used to make paint driers, detergents, and solvents.

Chemical Properties:

Specific Gravity: .98 *Vapor Density:* 4.4 *Boiling Point: 270-470° F(115-243° C)*
Melting Point: n/a *Vapor Pressure:* n/a *Solubility in water:* No
Other Information: May float or sink in water.

Health Hazards:

Inhalation Hazard: Causes coughing, difficulty in breathing.
Ingestion Hazard: Causes nausea.
Absorption Hazard: Irritating to the skin and eyes.
Hazards to Wildlife: Dangerous to aquatic life.
Decontamination Procedures: Wash away any material with copious amounts of soap and water.
First Aid Procedures: Remove victim to fresh air, call emergency medical care. If not breathing give CPR. If breathing is difficult administer oxygen. Treat for shock.

Fire Hazards:

Flashpoint: 300° F(148.8° C) *Ignition temperature:* n/a
Flammable Explosive High Range: n/a *Low Range:* 1
Toxic Products of Combustion: n/a
Other Hazards: n/a
Possible extinguishing agents: Apply water from as far a distance as possible. Use foam, dry chemical, or carbon dioxide. Do not extinguish the fire unless the flow can be stopped.

Reactivity Hazards:

Reactive With: n/a *Other Reactions:* n/a

Corrosivity Hazards:

Corrosive With: n/a *Neutralizing Agent:* n/a

Radioactivity Hazards:

Radiation Emitted: n/a *Other Hazards:* n/a

Recommended Protection for Response Personnel:

Avoid breathing vapors, keep upwind. Wear appropriate chemical clothing. Wash away any material which may come into contact with the body with copious amounts of soap and water. Consider appropriate evacuation.

NAPHTHYLAMINE

DOT Number: UN 2077 | *DOT Hazard Class:* n/a | *DOT Guide Number:* 55
Synonyms: α-naphthylamine
STCC Number: n/a | *Reportable Qty:* n/a
Mfg Name: Aldrich Chemical | *Phone No:* 1-414-273-3850

Physical Description:

Physical Form: Solid | *Color:* Light to dark brown | *Odor:* Weak ammonia
Other Information: n/a

Chemical Properties:

Specific Gravity: 1.12 | *Vapor Density:* 4.4 | *Boiling Point: 572° F(300° C)*
Melting Point: n/a | *Vapor Pressure:* n/a | *Solubility in water:* n/a
Other Information: Sinks in water.

Health Hazards:

Inhalation Hazard: Poisonous if inhaled.
Ingestion Hazard: Poisonous if swallowed.
Absorption Hazard: Poisonous if the skin is exposed.
Hazards to Wildlife: n/a
Decontamination Procedures: Wash away any material with copious amounts of soap and water.
First Aid Procedures: Remove victim to fresh air, call emergency medical care. If not breathing give CPR. If breathing is difficult administer oxygen. Treat for shock.

Fire Hazards:

Flashpoint: 315° F(157.2° C) | *Ignition temperature:* n/a
Flammable Explosive High Range: n/a | *Low Range:* n/a
Toxic Products of Combustion: Toxic nitrogen oxides are produced in fires.
Other Hazards: n/a
Possible extinguishing agents: Apply water from as far a distance as possible. Use foam, dry chemical, or carbon dioxide.

Reactivity Hazards:

Reactive With: n/a | *Other Reactions:* n/a

Corrosivity Hazards:

Corrosive With: n/a | *Neutralizing Agent:* n/a

Radioactivity Hazards:

Radiation Emitted: n/a | *Other Hazards:* n/a

Recommended Protection for Response Personnel:

Avoid breathing vapors, keep upwind. Wear a sealed chemical suit (polycarbonate, butyl rubber). Wash away any material which may have come into contact with the body with copious amounts of soap and water. Consider appropriate evacuation.

NATURAL GAS

DOT Number: UN 1972 *DOT Hazard Class:* Flammable gas *DOT Guide Number:* 22
Synonyms: none given
STCC Number: 4905770 *Reportable Qty:* n/a *CHEMTREC Phone No:* 1-800-424-9300

Physical Description:

Physical Form: Gas *Color:* Colorless *Odor:* Odorless
Other Information: It is used to make other chemicals.

Chemical Properties:

Specific Gravity: .42 *Vapor Density:* .6 *Boiling Point: – 259° F(– 161.6° C)*
Melting Point: n/a *Vapor Pressure:* .55 mm Hg at 68° F(20° C) *Solubility in water:* No
Other Information: Lighter than air, material can only be shipped in cylinders.

Health Hazards:

Inhalation Hazard: Dizziness, difficulty in breathing, loss of consciousness.
Ingestion Hazard: Displaces oxygen.
Absorption Hazard: Will cause frostbite.
Hazards to Wildlife: n/a
Decontamination Procedures: Wash away any material with copious amounts of soap and water.
First Aid Procedures: Remove victim to fresh air, call emergency medical care. If not breathing give CPR. If breathing is difficult administer oxygen. Treat for shock.

Fire Hazards:

Flashpoint: n/a *Ignition temperature:* 900-1170° F (482-632° C)
Flammable Explosive High Range: 15 *Low Range:* 5
Toxic Products of Combustion: n/a
Other Hazards: Vapors may explode in an enclosed area. Flashback along vapor trail may occur. Containers may explode in a fire.
Possible extinguishing agents: Apply water from as far a distance as possible.

Reactivity Hazards:

Reactive With: n/a *Other Reactions:* n/a

Corrosivity Hazards:

Corrosive With: n/a *Neutralizing Agent:* n/a

Radioactivity Hazards:

Radiation Emitted: n/a *Other Hazards:* n/a

Recommended Protection for Response Personnel:

Avoid breathing vapors, keep upwind. Wear appropriate chemical clothing. Wash away any material which may have come into contact with the body with copious amounts of soap and water. If the fire becomes uncontrollable, consider appropriate evacuation. (BLEVE)

NEOHEXANE

DOT Number: UN 1208 *DOT Hazard Class:* Flammable liquid *DOT Guide Number:* 27
Synonyms: 2,2-dimethylbutane
STCC Number: 4908250 *Reportable Qty:* n/a
Mfg Name: Phillips Petroleum *Phone No:* 1-918-661-6600

Physical Description:

Physical Form: Liquid *Color:* Colorless *Odor:* Petroleum like
Other Information: Lighter than and insoluble in water. Vapors are heavier than air.

Chemical Properties:

Specific Gravity: .6 *Vapor Density:* 3 *Boiling Point: 122° F(50° C)*
Melting Point: n/a *Vapor Pressure:* 400 mm Hg at 88° F(31.1° C) *Solubility in water:* No
Other Information: Floats on water. Irritating vapors are produced.

Health Hazards:

Inhalation Hazard: Dizziness, difficulty in breathing.
Ingestion Hazard: Nausea and vomiting.
Absorption Hazard: Irritating to the skin and eyes.
Hazards to Wildlife: n/a
Decontamination Procedures: Wash away any material with copious amounts of soap and water.
First Aid Procedures: Remove victim to fresh air, call emergency medical care. If not breathing give CPR. If breathing is difficult administer oxygen. Treat for shock.

Fire Hazards:

Flashpoint: −54° F(−47.7° C) *Ignition temperature:* 761° F(405° C)
Flammable Explosive High Range: 7 *Low Range:* 1.2
Toxic Products of Combustion: n/a
Other Hazards: Vapors may explode in an enclosed area. Flashback along vapor trail may occur. Containers may explode in a fire.
Possible extinguishing agents: Apply water from as far a distance as possible. Use foam, dry chemical, or carbon dioxide.

Reactivity Hazards:

Reactive With: n/a *Other Reactions:* n/a

Corrosivity Hazards:

Corrosive With: n/a *Neutralizing Agent:* n/a

Radioactivity Hazards:

Radiation Emitted: n/a *Other Hazards:* n/a

Recommended Protection for Response Personnel:

Avoid breathing vapors, keep upwind. Wear appropriate chemical clothing. Wash away any material which may come into contact with the body with copious amounts of soap and water. If the fire becomes uncontrollable, consider appropriate evacuation.

NICKEL ACETATE

DOT Number: n/a *DOT Hazard Class:* n/a *DOT Guide Number:* n/a
Synonyms: acetic acid, nickel(II) salt; nickel acetate tetrahydrate, nickelous acetate
STCC Number: n/a *Reportable Qty:* n/a
Mfg Name: J.T. Baker Chemical *Phone No:* 1-201-859-2151

Physical Description:

Physical Form: Solid *Color:* Dull green *Odor:* Odorless
Other Information: n/a

Chemical Properties:

Specific Gravity: 1.74 *Vapor Density:* n/a *Boiling Point: Decomposes*
Melting Point: n/a *Vapor Pressure:* n/a *Solubility in water:* Yes
Other Information: Sinks and mixes slowly with water.

Health Hazards:

Inhalation Hazard: Coughing, difficulty in breathing.
Ingestion Hazard: Nausea and vomiting.
Absorption Hazard: Irritating to the skin and eyes.
Hazards to Wildlife: n/a
Decontamination Procedures: Wash away any material with copious amounts of soap and water.
First Aid Procedures: Remove victim to fresh air, call emergency medical care. If not breathing give CPR. If breathing is difficult administer oxygen. Treat for shock.

Fire Hazards:

Flashpoint: n/a *Ignition temperature:* n/a
Flammable Explosive High Range: n/a *Low Range:* n/a
Toxic Products of Combustion: n/a
Other Hazards: n/a
Possible extinguishing agents: Extinguish fire using suitable agent for the type of surrounding fire.

Reactivity Hazards:

Reactive With: n/a *Other Reactions:* n/a

Corrosivity Hazards:

Corrosive With: n/a *Neutralizing Agent:* n/a

Radioactivity Hazards:

Radiation Emitted: n/a *Other Hazards:* n/a

Recommended Protection for Response Personnel:

Avoid breathing vapors, keep upwind. Structural protective clothing provides limited protection. Wash away any material which may have come into contact with the body with copious amounts of soap and water. Consider appropriate evacuation.

NICKEL AMMONIUM SULFATE

DOT Number: NA 9138 *DOT Hazard Class:* ORM-E *DOT Guide Number:* 31
Synonyms: ammonium disulfatonickelate(II), ammonium nickel sulfate
STCC Number: 4966360 *Reportable Qty:* 5000/2270
Mfg Name: Mallinckrodt Inc. *Phone No:* 1-314-895-2000

Physical Description:

Physical Form: Solid *Color:* Dark green blue *Odor:* Odorless
Other Information: It is used in electroplating nickel.

Chemical Properties:

Specific Gravity: 1.19 *Vapor Density:* n/a *Boiling Point: n/a*
Melting Point: n/a *Vapor Pressure:* n/a *Solubility in water:* Yes
Other Information: Sinks and mixes slowly in water.

Health Hazards:

Inhalation Hazard: Causes coughing, difficulty in breathing.
Ingestion Hazard: Causes nausea and vomiting.
Absorption Hazard: Irritating to the skin and the eyes.
Hazards to Wildlife: Dangerous to aquatic life.
Decontamination Procedures: Wash away any material with copious amounts of soap and water.
First Aid Procedures: Remove victim to fresh air, call emergency medical care. If not breathing give CPR. If breathing is difficult administer oxygen. Treat for shock.

Fire Hazards:

Flashpoint: n/a *Ignition temperature:* n/a
Flammable Explosive High Range: n/a *Low Range:* n/a
Toxic Products of Combustion: Toxic nitrogen oxides are formed in fires.
Other Hazards: n/a
Possible extinguishing agents: Extinguish fire using suitable agent for the type of surrounding fire.

Reactivity Hazards:

Reactive With: n/a *Other Reactions:* n/a

Corrosivity Hazards:

Corrosive With: n/a *Neutralizing Agent:* n/a

Radioactivity Hazards:

Radiation Emitted: n/a *Other Hazards:* n/a

Recommended Protection for Response Personnel:

Avoid breathing vapors, keep upwind. Structural protective clothing provides limited protection. Wash away any material which may have come into contact with the body with copious amounts of soap and water. COnsider appropriate evacuation.

NICKEL BROMIDE

DOT Number: n/a *DOT Hazard Class:* n/a *DOT Guide Number:* n/a
Synonyms: nickel bromide trihydrate
STCC Number: n/a *Reportable Qty:* n/a
Mfg Name: Great Western Inorganic *Phone No:* 1-303-423-9790

Physical Description:

Physical Form: Solid *Color:* Yellowish green *Odor:* Odorless
Other Information: n/a

Chemical Properties:

Specific Gravity: 4 *Vapor Density:* n/a *Boiling Point: Decomposes*
Melting Point: n/a *Vapor Pressure:* n/a *Solubility in water:* Yes
Other Information: Sinks and mixes with water.

Health Hazards:

Inhalation Hazard: Coughing, difficulty in breathing.
Ingestion Hazard: Nausea and vomiting.
Absorption Hazard: Irritating to the skin and eyes.
Hazards to Wildlife: n/a
Decontamination Procedures: Wash away any material with copious amounts of soap and water.
First Aid Procedures: Remove victim to fresh air, call emergency medical care. If not breathing give CPR. If breathing is difficult administer oxygen. Treat for shock.

Fire Hazards:

Flashpoint: n/a *Ignition temperature:* n/a
Flammable Explosive High Range: n/a *Low Range:* n/a
Toxic Products of Combustion: Irritating hydrogen bromide vapors may be formed in fires.
Other Hazards: n/a
Possible extinguishing agents: Extinguish fire using suitable agent for the type of surrounding fire.

Reactivity Hazards:

Reactive With: n/a *Other Reactions:* n/a

Corrosivity Hazards:

Corrosive With: n/a *Neutralizing Agent:* n/a

Radioactivity Hazards:

Radiation Emitted: n/a *Other Hazards:* n/a

Recommended Protection for Response Personnel:

Avoid breathing vapors, keep upwind. Structural protective clothing provides limited protection. Wash away any material which may have come into contact with the body with copious amounts of soap and water. Consider appropriate evacuation.

NICKEL CARBONYL

DOT Number: UN 1259 *DOT Hazard Class:* Flammable liquid *DOT Guide Number:* 57
Synonyms: nickel tetracarbonyl
STCC Number: 4906050 *Reportable Qty:* 1/.454
Mfg Name: A.D. McKay Inc. *Phone No:* 1-203-655-7410

Physical Description:

Physical Form: Liquid *Color:* Colorless to yellow *Odor:* Musty, stale
Other Information: It is used to nickel coat steel and other metals, and to make very pure nickel powder.

Chemical Properties:

Specific Gravity: 1.32 *Vapor Density:* 5.89 *Boiling Point: 110° F(43.3° C)*
Melting Point: −13° F(−25° C) *Vapor Pressure:* 321 mm Hg at 68° F(20° C)
Solubility in water: Insoluble *Degree of Solubility:* .018%
Other Information: It may only be shipped in cylinders. Heavier than and is insoluble in water.

Health Hazards:

Inhalation Hazard: Poisonous if inhaled.
Ingestion Hazard: Poisonous if swallowed.
Absorption Hazard: Poisonous if absorbed.
Hazards to Wildlife: n/a
Decontamination Procedures: Wash away any material with copious amounts of soap and water.
First Aid Procedures: Remove victim to fresh air, call emergency medical care. If not breathing give CPR. If breathing is difficult administer oxygen. Treat for shock.

Fire Hazards:

Flashpoint: −4° F(−20° C) *Ignition temperature:* 200° F(93.3° C)
Flammable Explosive High Range: None *Low Range:* 2
Toxic Products of Combustion: Poisonous gases may be produced in a fire.
Other Hazards: Vapors may explode in an enclosed area. Flashback along vapor trail may occur. Containers may explode in a fire.
Possible extinguishing agents: Apply water from as far a distance as possible. Use foam, dry chemical, or carbon dioxide.

Reactivity Hazards:

Reactive With: n/a *Other Reactions:* n/a

Corrosivity Hazards:

Corrosive With: n/a *Neutralizing Agent:* n/a

Radioactivity Hazards:

Radiation Emitted: n/a *Other Hazards:* n/a

Recommended Protection for Response Personnel:

Avoid breathing vapors, keep upwind. Wear a sealed chemical suit (polycarbonate, butyl rubber). Wash away any material which may have come into contact with the body with copious amounts of soap and water. If fire becomes uncontrollable, consider appropriate evacuation.

NICKEL CHLORIDE

DOT Number: NA 9139 *DOT Hazard Class:* ORM-E *DOT Guide Number:* 31
Synonyms: nickel chloride hexahydrate
STCC Number: 4966360 *Reportable Qty:* 5000/2270
Mfg Name: Allied Chemical *Phone No:* 1-201-455-2000

Physical Description:

Physical Form: Solid *Color:* Green *Odor:* Odorless
Other Information: It is used for electroplating nickel and chemical analysis.

Chemical Properties:

Specific Gravity: 3.55 *Vapor Density:* 1 *Boiling Point: n/a*
Melting Point: n/a *Vapor Pressure:* n/a *Solubility in water:* Yes
Other Information: n/a

Health Hazards:

Inhalation Hazard: Coughing, difficulty in breathing.
Ingestion Hazard: Nausea and vomiting.
Absorption Hazard: Irritating to the skin and eyes.
Hazards to Wildlife: Dangerous to aquatic life.
Decontamination Procedures: Wash away any material with copious amounts of soap and water.
First Aid Procedures: Remove victim to fresh air, call emergency medical care. If not breathing give CPR. If breathing is difficult administer oxygen. Treat for shock.

Fire Hazards:

Flashpoint: n/a *Ignition temperature:* n/a
Flammable Explosive High Range: n/a *Low Range:* n/a
Toxic Products of Combustion: n/a
Other Hazards: n/a
Possible extinguishing agents: Use suitable agent for type of surrounding fire.

Reactivity Hazards:

Reactive With: n/a *Other Reactions:* n/a

Corrosivity Hazards:

Corrosive With: n/a *Neutralizing Agent:* n/a

Radioactivity Hazards:

Radiation Emitted: n/a *Other Hazards:* n/a

Recommended Protection for Response Personnel:

Avoid breathing vapors, keep upwind. Structural protective clothing provides limited protection. Wash away any material which may have come into contact with the body with copious amounts of soap and water. Consider appropriate evacuation.

NICKEL CYANIDE

DOT Number: UN 1653 *DOT Hazard Class:* Poison B *DOT Guide Number:* 53
Synonyms: none given
STCC Number: 4923275 *Reportable Qty:* 1/.454
Mfg Name: Varlac-oid Chemical Co. *Phone No:* 1-201-387-0038

Physical Description:

Physical Form: Solid *Color:* Apple green *Odor:* Weak almond
Other Information: n/a

Chemical Properties:

Specific Gravity: 2.4 *Vapor Density:* 3.8 *Boiling Point: Decomposes*
Melting Point: n/a *Vapor Pressure:* n/a *Solubility in water:* No
Other Information: n/a

Health Hazards:

Inhalation Hazard: Poisonous if inhaled.
Ingestion Hazard: Poisonous if swallowed.
Absorption Hazard: Irritating to the skin and eyes.
Hazards to Wildlife: Dangerous to aquatic life.
Decontamination Procedures: Wash away any material with copious amounts of soap and water.
First Aid Procedures: Remove victim to fresh air, call emergency medical care. If not breathing give CPR. If breathing is difficult administer oxygen. Treat for shock.

Fire Hazards:

Flashpoint: n/a *Ignition temperature:* n/a
Flammable Explosive High Range: n/a *Low Range:* n/a
Toxic Products of Combustion: Toxic oxides of nitrogen are produced when material is involved in a fire.
Other Hazards: n/a
Possible extinguishing agents: Use suitable agent for type of surrounding fire.

Reactivity Hazards:

Reactive With: n/a *Other Reactions:* n/a

Corrosivity Hazards:

Corrosive With: n/a *Neutralizing Agent:* n/a

Radioactivity Hazards:

Radiation Emitted: n/a *Other Hazards:* n/a

Recommended Protection for Response Personnel:

Avoid breathing vapors, keep upwind. Wear a sealed chemical suit (polycarbonate, butyl rubber). Wash away any material which may have come into contact with the body with copious amounts of soap and water. Consider appropriate evacuation.

NICKEL FLUOROBORATE

DOT Number: n/a *DOT Hazard Class:* n/a *DOT Guide Number:* n/a
Synonyms: nickel(2)fluoroborate, nickel fluoroborate solution
STCC Number: n/a *Reportable Qty:* n/a
Mfg Name: Allied Corp. *Phone No:* 1-201-455-2000

Physical Description:

Physical Form: Liquid *Color:* Green *Odor:* n/a
Other Information: n/a

Chemical Properties:

Specific Gravity: 1.5 *Vapor Density:* n/a *Boiling Point: n/a*
Vapor Pressure: n/a *Solubility in water:* Yes
Other Information: Sinks and mixes with water.

Health Hazards:

Inhalation Hazard: Harmful if inhaled.
Ingestion Hazard: Nausea and vomiting.
Absorption Hazard: Irritating to the skin and eyes.
Hazards to Wildlife: n/a
Decontamination Procedures: Wash away any material with copious amounts of soap and water.
First Aid Procedures: Remove victim to fresh air, call emergency medical care. If not breathing give CPR. If breathing is difficult administer oxygen. Treat for shock.

Fire Hazards:

Flashpoint: n/a *Ignition temperature:* n/a
Flammable Explosive High Range: n/a *Low Range:* n/a
Toxic Products of Combustion: n/a
Other Hazards: n/a
Possible extinguishing agents: Extinguish fire using suitable agent for the type of surrounding fire.

Reactivity Hazards:

Reactive With: n/a *Other Reactions:* n/a

Corrosivity Hazards:

Corrosive With: n/a *Neutralizing Agent:* n/a

Radioactivity Hazards:

Radiation Emitted: n/a *Other Hazards:* n/a

Recommended Protection for Response Personnel:

Avoid breathing vapors, keep upwind. Wear appropriate sealed chemical suit. Wash away any material which may have come into contact with the body with copious amounts of soap and water. Consider appropriate evacuation.

NICKEL FORMATE

DOT Number: n/a *DOT Hazard Class:* n/a *DOT Guide Number:* n/a
Synonyms: nickel formate dihydrate
STCC Number: n/a *Reportable Qty:* n/a
Mfg Name: Hall Chemical Co. *Phone No:* 1-216-944-8500

Physical Description:

Physical Form: Liquid *Color:* Green *Odor:* Odorless
Other Information: n/a

Chemical Properties:

Specific Gravity: 2.15 *Vapor Density:* n/a *Boiling Point: Decomposes*
Vapor Pressure: n/a *Solubility in water:* Yes
Other Information: Sinks and mixes with water.

Health Hazards:

Inhalation Hazard: Coughing, difficulty in breathing.
Ingestion Hazard: Nausea and vomiting.
Absorption Hazard: Irritating to the skin and eyes.
Hazards to Wildlife: n/a
Decontamination Procedures: Wash away any material with copious amounts of soap and water.
First Aid Procedures: Remove victim to fresh air, call emergency medical care. If not breathing give CPR. If breathing is difficult administer oxygen. Treat for shock.

Fire Hazards:

Flashpoint: n/a *Ignition temperature:* n/a
Flammable Explosive High Range: n/a *Low Range:* n/a
Toxic Products of Combustion: Irritating formic acid vapors may be formed in fires.
Other Hazards: n/a
Possible extinguishing agents: Extinguish fire using suitable agent for the type of surrounding fire.

Reactivity Hazards:

Reactive With: n/a *Other Reactions:* n/a

Corrosivity Hazards:

Corrosive With: n/a *Neutralizing Agent:* n/a

Radioactivity Hazards:

Radiation Emitted: n/a *Other Hazards:* n/a

Recommended Protection for Response Personnel:

Avoid breathing vapors, keep upwind. Wear a sealed chemical suit (polycarbonate, butyl rubber). Wash away any material which may have come into contact with the body with copious amounts of soap and water. Consider appropriate evacuation.

NICKEL HYDROXIDE

DOT Number: NA 9140 *DOT Hazard Class:* ORM-E *DOT Guide Number:* 31
Synonyms: green nickel oxide, nickel hydroxide, nickelous hydroxide
STCC Number: 4963863 *Reportable Qty:* 1000/454
Mfg Name: Hall Chemical Co. *Phone No:* 1-216-944-8500

Physical Description:

Physical Form: Solid *Color:* Black to green *Odor:* n/a
Other Information: It is used to make other nickel compounds.

Chemical Properties:

Specific Gravity: 4.1 *Vapor Density:* 3.2 *Boiling Point: n/a*
Melting Point: 466° F(241.1° C) *Vapor Pressure:* n/a *Solubility in water:* No
Other Information: Sinks and is insoluble in water.

Health Hazards:

Inhalation Hazard: Harmful if inhaled.
Ingestion Hazard: Harmful if swallowed.
Absorption Hazard: Irritating to the eyes.
Hazards to Wildlife: Dangerous to aquatic life.
Decontamination Procedures: Wash away any material with copious amounts of soap and water.
First Aid Procedures: Remove victim to fresh air, call emergency medical care. If not breathing give CPR. If breathing is difficult administer oxygen. Treat for shock.

Fire Hazards:

Flashpoint: 752° F(400° C) *Ignition temperature:* 752° F(400° C)
Flammable Explosive High Range: n/a *Low Range:* n/a
Toxic Products of Combustion: n/a
Other Hazards: Converts to black nickelic oxide in fire.
Possible extinguishing agents: Extinguish fire using suitable agent for the type of surrounding fire.

Reactivity Hazards:

Reactive With: n/a *Other Reactions:* n/a

Corrosivity Hazards:

Corrosive With: n/a *Neutralizing Agent:* n/a

Radioactivity Hazards:

Radiation Emitted: n/a *Other Hazards:* n/a

Recommended Protection for Response Personnel:

Avoid breathing vapors, keep upwind. Wear a sealed chemical suit (polycarbonate, butyl rubber). Wash away any material which may have come into contact with the body with copious amounts of soap and water. Consider appropriate evacuation.

NICKEL NITRATE

DOT Number: NA 2725 *DOT Hazard Class:* Oxidizer *DOT Guide Number:* 35
Synonyms: nickel nitrate hexahydride
STCC Number: 4966364 *Reportable Qty:* 5000/2270
Mfg Name: Mallinckrodt Inc. *Phone No:* 1-314-895-2000

Physical Description:

Physical Form: Solid *Color:* Green *Odor:* Odorless
Other Information: It is used in nickel plating and to make nickel catalysts in chemical manufacturing.

Chemical Properties:

Specific Gravity: 2.5 *Vapor Density:* 10 *Boiling Point: Decomposes*
Melting Point: n/a *Vapor Pressure:* n/a *Solubility in water:* Yes
Other Information: Sinks and mixes in water.

Health Hazards:

Inhalation Hazard: Causes coughing, difficulty in breathing.
Ingestion Hazard: Causes nausea and vomiting.
Absorption Hazard: Irritating to the skin and eyes.
Hazards to Wildlife: Dangerous to aquatic life.
Decontamination Procedures: Wash away any material with copious amounts of soap and water.
First Aid Procedures: Remove victim to fresh air, call emergency medical care. If not breathing give CPR. If breathing is difficult administer oxygen. Treat for shock.

Fire Hazards:

Flashpoint: n/a *Ignition temperature:* n/a
Flammable Explosive High Range: n/a *Low Range:* n/a
Toxic Products of Combustion: Toxic nitrogen oxides are formed in fires.
Other Hazards: Will increase the intensity of a fire.
Possible extinguishing agents: Extinguish fire using suitable agent for the type of surrounding fire. Flood with water. Apply water from as far a distance as possible.

Reactivity Hazards:

Reactive With: Wood or paper may cause fire. *Other Reactions:* n/a

Corrosivity Hazards:

Corrosive With: n/a *Neutralizing Agent:* n/a

Radioactivity Hazards:

Radiation Emitted: n/a *Other Hazards:* n/a

Recommended Protection for Response Personnel:

Avoid breathing vapors, keep upwind. Wear a sealed chemical suit (polycarbonate, butyl rubber). Wash away any material which may have come into contact with the body with copious amounts of soap and water. If the fire becomes uncontrollable, consider appropriate evacuation.

NICKEL SULFATE

DOT Number: NA 9141 *DOT Hazard Class:* ORM-E *DOT Guide Number:* 31
Synonyms: nickelous sulfate
STCC Number: 4966363 *Reportable Qty:* 5000/2270
Mfg Name: Mallinckrodt Inc. *Phone No:* 1-314-895-2000

Physical Description:

Physical Form: Solid *Color:* Pale green *Odor:* Odorless
Other Information: It is used in making other nickel compounds.

Chemical Properties:

Specific Gravity: 3.68 *Vapor Density:* 5.3 *Boiling Point: Decomposes*
Melting Point: n/a *Vapor Pressure:* n/a *Solubility in water:* Yes
Other Information: Sinks and mixes slowly with water.

Health Hazards:

Inhalation Hazard: n/a
Ingestion Hazard: n/a
Absorption Hazard: Irritating to the skin and eyes.
Hazards to Wildlife: Dangerous to aquatic life.
Decontamination Procedures: Wash away any material with copious amounts of soap and water.
First Aid Procedures: Remove victim to fresh air, call emergency medical care. If not breathing give CPR. If breathing is difficult administer oxygen. Treat for shock.

Fire Hazards:

Flashpoint: n/a *Ignition temperature:* n/a
Flammable Explosive High Range: n/a *Low Range:* n/a
Toxic Products of Combustion: n/a
Other Hazards: n/a
Possible extinguishing agents: Extinguish fire using suitable agent for the type of surrounding fire.

Reactivity Hazards:

Reactive With: n/a *Other Reactions:* n/a

Corrosivity Hazards:

Corrosive With: n/a *Neutralizing Agent:* n/a

Radioactivity Hazards:

Radiation Emitted: n/a *Other Hazards:* n/a

Recommended Protection for Response Personnel:

Avoid breathing vapors, keep upwind. Wear a sealed chemical suit (polycarbonate, butyl rubber, chlorinated polyethylene, viton, PVC, nitrile, neoprene). Wash away any material which may have come into contact with the body with copious amounts of soap and water. Consider appropriate evacuation.

NICOTINE

DOT Number: UN 1654 *DOT Hazard Class:* Poison B *DOT Guide Number:* 55
Synonyms: 1-methyl-2-(3)-(pyridyl)pyrrolidine
STCC Number: 4921449 *Reportable Qty:* 100/45.4
Mfg Name: Atomergic Chemetals Corp. *Phone No:* 1-516-349-8800

Physical Description:

Physical Form: Liquid *Color:* Colorless to brown *Odor:* Fishy
Other Information: n/a

Chemical Properties:

Specific Gravity: 1 *Vapor Density:* 5.6 *Boiling Point: 475° F(246.1° C)*
Melting Point: –110° F(–78.8° C) *Vapor Pressure:* .425 mm Hg at 68° F(20° C)
Solubility in water: Yes
Other Information: n/a

Health Hazards:

Inhalation Hazard: Poisonous if inhaled.
Ingestion Hazard: Poisonous if swallowed.
Absorption Hazard: Poisonous to the skin and the eyes.
Hazards to Wildlife: Dangerous to waterfowl and aquatic life.
Decontamination Procedures: Wash away any material with copious amounts of soap and water.
First Aid Procedures: Remove victim to fresh air, call emergency medical care. If not breathing give CPR. If breathing is difficult administer oxygen. Treat for shock.

Fire Hazards:

Flashpoint: n/a *Ignition temperature:* 471° F(243.8° C)
Flammable Explosive High Range: 4 *Low Range:* .7
Toxic Products of Combustion: Toxic oxides of nitrogen are produced during combustion of this material.
Other Hazards: Avoid contact with liquid.
Possible extinguishing agents: Apply water from as far a distance as possible. Use alcohol foam, dry chemical, or carbon dioxide.

Reactivity Hazards:

Reactive With: n/a *Other Reactions:* n/a

Corrosivity Hazards:

Corrosive With: n/a *Neutralizing Agent:* n/a

Radioactivity Hazards:

Radiation Emitted: n/a *Other Hazards:* n/a

Recommended Protection for Response Personnel:

Avoid breathing vapors, keep upwind. Wear a sealed chemical suit (polycarbonate, butyl rubber). Wash away any material which may have come into contact with the body with copious amounts of soap and water. Consider appropriate evacuation.

NICOTINE SULFATE (liquid)

DOT Number: UN 1658 *DOT Hazard Class:* Poison B *DOT Guide Number:* 55
Synonyms: Black Leaf 40 (40% water solution), natural nicotine sulfate
STCC Number: 4921451 *Reportable Qty:* 100/45.4
Mfg Name: Prentiss Drug & Chemical Co. *Phone No:* 1-516-326-1919

Physical Description:

Physical Form: Liquid *Color:* Colorless *Odor:* Tobacco
Other Information: n/a

Chemical Properties:

Specific Gravity: 1.15 *Vapor Density:* 1 *Boiling Point: n/a*
Melting Point: n/a *Vapor Pressure:* 1 mm Hg at 143° F(61.6° C) *Solubility in water:* Yes
Other Information: Mixes with water.

Health Hazards:

Inhalation Hazard: Poisonous if inhaled.
Ingestion Hazard: Poisonous if swallowed.
Absorption Hazard: Poisonous if the skin is exposed.
Hazards to Wildlife: Dangerous to both aquatic life and waterfowl.
Decontamination Procedures: Wash away any material with copious amounts of soap and water.
First Aid Procedures: Remove victim to fresh air, call emergency medical care. If not breathing give CPR. If breathing is difficult administer oxygen. Treat for shock.

Fire Hazards:

Flashpoint: n/a *Ignition temperature:* n/a
Flammable Explosive High Range: n/a *Low Range:* n/a
Toxic Products of Combustion: Toxic decomposition products are released in a fire.
Other Hazards: n/a
Possible extinguishing agents: Extinguish fire using suitable agent for the type of surrounding fire. Use alcohol foam, dry chemical, or carbon dioxide.

Reactivity Hazards:

Reactive With: n/a *Other Reactions:* n/a

Corrosivity Hazards:

Corrosive With: n/a *Neutralizing Agent:* n/a

Radioactivity Hazards:

Radiation Emitted: n/a *Other Hazards:* n/a

Recommended Protection for Response Personnel:

Avoid breathing vapors, keep upwind. Wear a sealed chemical suit (polycarbonate, butyl rubber). Wash away any material which may have come into contact with the body with copious amounts of soap and water. Consider appropriate evacuation.

NICOTINE SULFATE (solid)

DOT Number: UN 1658 *DOT Hazard Class:* Poison B *DOT Guide Number:* 55
Synonyms: black leaf 40 (40% water solution), natural nicotine sulfate
STCC Number: 4921452 *Reportable Qty:* 100/45.4
Mfg Name: Prentiss Drug & Chemical Co. *Phone No:* 1-516-326-1919

Physical Description:

Physical Form: Solid *Color:* White to light brown *Odor:* Odorless
Other Information: n/a

Chemical Properties:

Specific Gravity: 1.15 *Vapor Density:* 1 *Boiling Point: n/a*
Melting Point: n/a *Vapor Pressure:* 1 mm Hg at 143° F(61.6° C) *Solubility in water:* Yes
Other Information: Mixes with water.

Health Hazards:

Inhalation Hazard: Poisonous if inhaled.
Ingestion Hazard: Poisonous if swallowed.
Absorption Hazard: Poisonous if the skin is exposed.
Hazards to Wildlife: Dangerous to both aquatic life and waterfowl.
Decontamination Procedures: Wash away any material with copious amounts of soap and water.
First Aid Procedures: Remove victim to fresh air, call emergency medical care. If not breathing give CPR. If breathing is difficult administer oxygen. Treat for shock.

Fire Hazards:

Flashpoint: n/a *Ignition temperature:* n/a
Flammable Explosive High Range: n/a *Low Range:* n/a
Toxic Products of Combustion: Toxic oxides of nitrogen are produced during combustion of this material.
Other Hazards: n/a
Possible extinguishing agents: Extinguish fire using suitable agent for the type of surrounding fire. Use alcohol foam, dry chemical, or carbon dioxide.

Reactivity Hazards:

Reactive With: n/a *Other Reactions:* n/a

Corrosivity Hazards:

Corrosive With: n/a *Neutralizing Agent:* n/a

Radioactivity Hazards:

Radiation Emitted: n/a *Other Hazards:* n/a

Recommended Protection for Response Personnel:

Avoid breathing vapors, keep upwind. Wear a sealed chemical suit (polycarbonate, butyl rubber). Wash away any material which may have come into contact with the body with copious amounts of soap and water. Consider appropriate evacuation.

NITRALIN

DOT Number: UN 1609 *DOT Hazard Class:* Poison A *DOT Guide Number:* n/a
Synonyms: 4-(methylsulfonyl)-2,6-dinitro-N,N-dipropylaniline, planavin
STCC Number: n/a *Reportable Qty:* n/a
Mfg Name: Shell Chemical *Phone No:* 1-713-241-6161

Physical Description:

Physical Form: Solid *Color:* Light yellow to orange *Odor:* Mild
Other Information: n/a

Chemical Properties:

Specific Gravity: 1 *Vapor Density:* n/a *Boiling Point: 437° F(225° C)*
Melting Point: 304° F(151.1° C) *Vapor Pressure:* n/a *Solubility in water:* n/a
Other Information: Sinks in water.

Health Hazards:

Inhalation Hazard: Poisonous if inhaled.
Ingestion Hazard: Poisonous if swallowed.
Absorption Hazard: Irritating to the skin and eyes.
Hazards to Wildlife: n/a
Decontamination Procedures: Wash away any material with copious amounts of soap and water.
First Aid Procedures: Remove victim to fresh air, call emergency medical care. If not breathing give CPR. If breathing is difficult administer oxygen. Treat for shock.

Fire Hazards:

Flashpoint: Combustible solid *Ignition temperature:* 435° F(223.8° C)
Flammable Explosive High Range: n/a *Low Range:* n/a
Toxic Products of Combustion: Irritating oxides of sulfur and nitrogen are formed in fires.
Other Hazards: Decomposes vigorously in a self sustaining reaction at or above 437° F(225° C).
Possible extinguishing agents: Water.

Reactivity Hazards:

Reactive With: n/a *Other Reactions:* n/a

Corrosivity Hazards:

Corrosive With: n/a *Neutralizing Agent:* n/a

Radioactivity Hazards:

Radiation Emitted: n/a *Other Hazards:* n/a

Recommended Protection for Response Personnel:

Avoid breathing vapors, keep upwind. Structural protective clothing provides limited protection. Wash away any material which may have come into contact with the body with copious amounts of soap and water. Consider appropriate evacuation.

NITRATING ACID

DOT Number: UN 1796 *DOT Hazard Class:* Oxidizer *DOT Guide Number:* 73
Synonyms: acid mixtures, nitric sulfuric acid mixtures, mixed acid
STCC Number: 4918525 *Reportable Qty:* 1000/454 *CHEMTREC Phone No:* 1-800-424-9300

Physical Description:

Physical Form: Liquid *Odor:* Acrid *Color:* Colorless to light yellow, gray to light brown
Other Information: It is used for industrial oxidizing and nitrating acid.

Chemical Properties:

Specific Gravity: 1.53 to 1.79 *Vapor Density:* n/a *Boiling Point: 179.6-206.6° F(82-97° C)*
Melting Point: −49° F(−45° C) *Vapor Pressure:* 1 to 45 mm Hg at 68° F(20° C) *Solubility in water:* Yes
Other Information: Reacts violently with water. Weighs 12.7 to 15.9 lbs/5.7-7.2 kg per gallon/3.8 l.

Health Hazards:

Inhalation Hazard: Irritation to nose, mouth, and throat.
Ingestion Hazard: Immediate pain and burns of the mouth.
Absorption Hazard: Immediate and deep severe burns to the skin, eyes may be permanently damaged.
Hazards to Wildlife: n/a
Decontamination Procedures: Wash away any material with copious amounts of soap and water.
First Aid Procedures: Remove victim to fresh air, call emergency medical care. If not breathing give CPR. If breathing is difficult administer oxygen. Treat for shock.

Fire Hazards:

Flashpoint: n/a *Ignition temperature:* n/a
Flammable Explosive High Range: n/a *Low Range:* n/a
Toxic Products of Combustion: Toxic oxides of nitrogen are produced.
Other Hazards: Avoid breathing vapors, dust, and fumes from this material.
Possible extinguishing agents: Cool all affected containers with flooding quantities of water. Apply water from as far a distance as possible.

Reactivity Hazards:

Reactive With: Water producing heat, fumes and splattering.
Other Reactions: Combustible or oxidizable materials.

Corrosivity Hazards:

Corrosive With: Metals and tissue *Neutralizing Agent:* Crushed limestone, soda ash, or lime.

Radioactivity Hazards:

Radiation Emitted: n/a *Other Hazards:* n/a

Recommended Protection for Response Personnel:

Avoid breathing vapors, keep upwind. Wear a sealed chemical suit (neoprene, nitrile, chlorinated polyethylene, PVC, viton). Wash away any material which may have come into contact with the body with copious amounts of soap and water. If the fire becomes uncontrollable, or if the containers are exposed to direct flame, consider appropriate evacuation.

NITRIC ACID

DOT Number: UN 2031 *DOT Hazard Class:* Oxidizer *DOT Guide Number:* 44
Synonyms: none given
STCC Number: 4918528 *Reportable Qty:* 1000/454
Mfg Name: Allied Corp. *Phone No:* 1-201-455-2000

Physical Description:

Physical Form: Liquid *Color:* Colorless to light brown *Odor:* Choking
Other Information: It is used to make other chemicals. As a regent in chemical analysis for ore flotation, and many other uses.

Chemical Properties:

Specific Gravity: 1.49 *Vapor Density:* 1 *Boiling Point: 192° F(88.8° C)*
Melting Point: −42/−61° F(−41/−51° C) *Vapor Pressure:* 2.6 to 103 mm Hg at 68° F(20° C)
Solubility in water: Yes
Other Information: Weighs 12.6 lbs/5.7 kg per gallon/3.8 l.

Health Hazards:

Inhalation Hazard: Difficulty in breathing, loss of consciousness.
Ingestion Hazard: Harmful if swallowed.
Absorption Hazard: Will burn the skin and eyes.
Hazards to Wildlife: Dangerous to aquatic life.
Decontamination Procedures: Wash away any material with copious amounts of soap and water.
First Aid Procedures: Remove victim to fresh air, call emergency medical care. If not breathing give CPR. If breathing is difficult administer oxygen. Treat for shock.

Fire Hazards:

Flashpoint: n/a *Ignition temperature:* n/a
Flammable Explosive High Range: n/a *Low Range:* n/a
Toxic Products of Combustion: Poisonous gases are produced when heated.
Other Hazards: May cause fire upon contact with combustibles.
Possible extinguishing agents: Use water from as far a distance as possible.

Reactivity Hazards:

Reactive With: Water, wood, paper, cloth, and most metals.
Other Reactions: Toxic oxides of nitrogen are formed.

Corrosivity Hazards:

Corrosive With: Metal and tissue *Neutralizing Agent:* Crushed limestone, soda ash, or lime.

Radioactivity Hazards:

Radiation Emitted: n/a *Other Hazards:* n/a

Recommended Protection for Response Personnel:

Avoid breathing vapors, keep upwind. Wear a sealed chemical suit (PVC). Wash away any material which may come into contact with the body with copious amounts of soap and water. Consider appropriate evacuation.

NITRIC OXIDE

DOT Number: UN 1660 *DOT Hazard Class:* Poison A *DOT Guide Number:* 20
Synonyms: nitrogen monoxide
STCC Number: 4920330 *Reportable Qty:* 10/4.54
Mfg Name: Union Carbide *Phone No:* 1-203-794-2000

Physical Description:

Physical Form: Gas *Color:* Colorless *Odor:* Sharp, unpleasant
Other Information: n/a

Chemical Properties:

Specific Gravity: 1.6 *Vapor Density:* 1.4 *Boiling Point: −241° F(−151.6° C)*
Melting Point: 263° F(128.3° C) *Vapor Pressure:* 26,000 mm Hg at 68° F(20° C) *Solubility in water:* Reacts
Other Information: It may only be shipped in cylinders not equipped with safety release devices.

Health Hazards:

Inhalation Hazard: Poisonous if inhaled.
Ingestion Hazard: Poisonous if swallowed.
Absorption Hazard: Poisonous if absorbed.
Hazards to Wildlife: n/a
Decontamination Procedures: Wash away any material with copious amounts of soap and water.
First Aid Procedures: Remove victim to fresh air, call emergency medical care. If not breathing give CPR. If breathing is difficult administer oxygen. Treat for shock.

Fire Hazards:

Flashpoint: n/a *Ignition temperature:* n/a
Flammable Explosive High Range: n/a *Low Range:* n/a
Toxic Products of Combustion: n/a
Other Hazards: Will accelerate the combustion of burning materials. Prolonged exposure to fire or heat will result in the violent rupture and rocketing of cylinders.
Possible extinguishing agents: Use suitable agent for type of surrounding fire. Apply water from as far a distance as possible. Use foam, dry chemical, or carbon dioxide.

Reactivity Hazards:

Reactive With: Water to form nitric acid. Air to form nitrogen tetroxide. *Other Reactions:* n/a

Corrosivity Hazards:

Corrosive With: n/a *Neutralizing Agent:* n/a

Radioactivity Hazards:

Radiation Emitted: n/a *Other Hazards:* n/a

Recommended Protection for Response Personnel:

Avoid breathing vapors, keep upwind. Wear a sealed chemical suit (polycarbonate, butyl rubber). Wash away any material which may have come into contact with the body with copious amounts of soap and water. Consider appropriate evacuation.

NITRILOTRIACETIC ACID AND SALTS

DOT Number: n/a *DOT Hazard Class:* n/a *DOT Guide Number:* n/a
Synonyms: disodium nitrilotriacetate, NTA, triglycine, trisodium nitrilotriacetate
STCC Number: n/a *Reportable Qty:* n/a
Mfg Name: W.R. Grace & Co. *Phone No:* 1-212-819-5500

Physical Description:

Physical Form: Solid *Color:* White *Odor:* Mild
Other Information: n/a

Chemical Properties:

Specific Gravity: 1 *Vapor Density:* n/a *Boiling Point: Decomposes*
Melting Point: n/a *Vapor Pressure:* n/a *Solubility in water:* Yes
Other Information: Sinks and mixes with water.

Health Hazards:

Inhalation Hazard: n/a
Ingestion Hazard: n/a
Absorption Hazard: Irritating to the skin and eyes.
Hazards to Wildlife: Dangerous to aquatic life.
Decontamination Procedures: Wash away any material with copious amounts of soap and water.
First Aid Procedures: Remove victim to fresh air, call emergency medical care. If not breathing give CPR. If breathing is difficult administer oxygen. Treat for shock.

Fire Hazards:

Flashpoint: n/a *Ignition temperature:* n/a
Flammable Explosive High Range: n/a *Low Range:* n/a
Toxic Products of Combustion: n/a
Other Hazards: n/a
Possible extinguishing agents: Extinguish fire using suitable agent for the type of surrounding fire. Apply water from as far a distance as possible.

Reactivity Hazards:

Reactive With: n/a *Other Reactions:* n/a

Corrosivity Hazards:

Corrosive With: n/a *Neutralizing Agent:* n/a

Radioactivity Hazards:

Radiation Emitted: n/a *Other Hazards:* n/a

Recommended Protection for Response Personnel:

Avoid breathing vapors, keep upwind. Wear appropriate sealed chemical suit. Wash away any material which may have come into contact with the body with copious amounts of soap and water. Consider appropriate evacuation.

NITROANILINE

DOT Number: UN 1661 *DOT Hazard Class:* Poison B *DOT Guide Number:* 55
Synonyms: 2- or p-nitroaniline
STCC Number: 4921554 *Reportable Qty:* 5000/2270
Mfg Name: Monsanto Chemical *Phone No:* 1-314-964-1000

Physical Description:

Physical Form: Solid *Color:* Orange *Odor:* Musty
Other Information: It is used in dye, and is a corrosion inhibitor.

Chemical Properties:

Specific Gravity: 1.44 *Vapor Density:* 4.77 *Boiling Point: 637° F(336° C)*
Melting Point: 293° F(145° C) *Vapor Pressure:* 1 mm Hg at 68° F(20° C)
Solubility in water: n/a *Degree of Solubility:* .08%
Other Information: n/a

Health Hazards:

Inhalation Hazard: Headaches, dizziness, loss of consciousness.
Ingestion Hazard: Same as inhalation.
Absorption Hazard: Irritating to the skin and eyes.
Hazards to Wildlife: Dangerous to aquatic life.
Decontamination Procedures: Wash away any material with copious amounts of soap and water.
First Aid Procedures: Remove victim to fresh air, call emergency medical care. If not breathing give CPR. If breathing is difficult administer oxygen. Treat for shock.

Fire Hazards:

Flashpoint: 390° F(198.8° C) *Ignition temperature:* 970° F(521° C)
Flammable Explosive High Range: n/a *Low Range:* n/a
Toxic Products of Combustion: Toxic and explosive nitric oxide vapors when heated to high temperatures.
Other Hazards: n/a
Possible extinguishing agents: Apply water from as far a distance as possible. Use foam, dry chemical, or carbon dioxide.

Reactivity Hazards:

Reactive With: n/a *Other Reactions:* n/a

Corrosivity Hazards:

Corrosive With: n/a *Neutralizing Agent:* n/a

Radioactivity Hazards:

Radiation Emitted: n/a *Other Hazards:* n/a

Recommended Protection for Response Personnel:

Avoid breathing vapors, keep upwind. Wear a sealed chemical suit (polycarbonate, butyl rubber). Wash away any material which may have come into contact with the body with copious amounts of soap and water. Consider appropriate evacuation.

NITROBENZENE

DOT Number: UN 1662 *DOT Hazard Class:* Poison B *DOT Guide Number:* 55
Synonyms: nitrobenzol, oil of mirbane
STCC Number: 4921455 *Reportable Qty:* 1000/454
Mfg Name: E.I. Du Pont *Phone No:* 1-800-441-3637

Physical Description:

Physical Form: Liquid *Color:* Pale yellow to dark brown *Odor:* Shoe polish
Other Information: Weigh 10 lbs/4.5 kg per gallon/3.8 l.

Chemical Properties:

Specific Gravity: 1.2 *Vapor Density:* 3.4 *Boiling Point: 412° F(211° C)*
Melting Point: 41° F(5° C) *Vapor Pressure:* 1 mm Hg at 68° F(20° C)
Solubility in water: No *Degree of Solubility: .2%*
Other Information: Avoid contact with liquid.

Health Hazards:

Inhalation Hazard: n/a
Ingestion Hazard: Poisonous if swallowed.
Absorption Hazard: Poisonous if absorbed.
Hazards to Wildlife: Dangerous to aquatic life.
Decontamination Procedures: Wash away any material with copious amounts of soap and water.
First Aid Procedures: Remove victim to fresh air, call emergency medical care. If not breathing give CPR. If breathing is difficult administer oxygen. Treat for shock.

Fire Hazards:

Flashpoint: 190° F(87.7° C) *Ignition temperature:* 900° F(482.2° C)
Flammable Explosive High Range: n/a *Low Range:* 1.8
Toxic Products of Combustion: Toxic oxides of nitrogen are produced during combustion.
Other Hazards: n/a
Possible extinguishing agents: Apply water from as far a distance as possible. Use alcohol foam, dry chemical, or carbon dioxide.

Reactivity Hazards:

Reactive With: n/a *Other Reactions:* n/a

Corrosivity Hazards:

Corrosive With: n/a *Neutralizing Agent:* n/a

Radioactivity Hazards:

Radiation Emitted: n/a *Other Hazards:* n/a

Recommended Protection for Response Personnel:

Avoid breathing vapors, keep upwind. Wear a sealed chemical suit (viton). Wash away any material which may have come into contact with the body with copious amounts of soap and water. Consider appropriate evacuation.

NITROCHLOROBENZENE-meta

DOT Number: UN 1578 *DOT Hazard Class:* Poison B *DOT Guide Number:* 55
Synonyms: 1-chloro-3-nitrobenzene, chloro-m-nitrobenzene, 3-chloronitrobenzene
STCC Number: 4921458 *Reportable Qty:* n/a *CHEMTREC Phone No:* 1-800-424-9300

Physical Description:

Physical Form: Solid *Color:* Yellow *Odor:* Ammonialike
Other Information: It is used as a chemical intermediate especially for dyes.

Chemical Properties:

Specific Gravity: 1.34 *Vapor Density:* 5.4 *Boiling Point: 456° F(235.5° C)*
Melting Point: 112° F(44.4° C) *Vapor Pressure:* Very low *Solubility in water:* No
Other Information: It is shipped as a molten liquid that may solidify in route. Weighs 11.2 lbs/5.0 kg per gallon/3.8 l, or 95.7 lbs per cubic foot as a solid.

Health Hazards:

Inhalation Hazard: Poisonous if inhaled.
Ingestion Hazard: Poisonous if swallowed.
Absorption Hazard: Irritating to the skin and eyes.
Hazards to Wildlife: n/a
Decontamination Procedures: Wash away any material with copious amounts of soap and water.
First Aid Procedures: Remove victim to fresh air, call emergency medical care. If not breathing give CPR. If breathing is difficult administer oxygen. Treat for shock.

Fire Hazards:

Flashpoint: 261° F(127.2° C) *Ignition temperature:* n/a
Flammable Explosive High Range: n/a *Low Range:* n/a
Toxic Products of Combustion: Toxic oxides of nitrogen, hydrogen chloride, carbon monoxide, chlorides, and possibly chlorine.
Other Hazards: Will burn, but is difficult to ignite.
Possible extinguishing agents: Extinguish fire using suitable agent for the type of surrounding fire.

Reactivity Hazards:

Reactive With: n/a *Other Reactions:* n/a

Corrosivity Hazards:

Corrosive With: n/a *Neutralizing Agent:* n/a

Radioactivity Hazards:

Radiation Emitted: n/a *Other Hazards:* n/a

Recommended Protection for Response Personnel:

Avoid breathing vapors, keep upwind. Wear a sealed chemical suit (butyl rubber). Wash away any material which may have come into contact with the body with copious amounts of soap and water. Consider appropriate evacuation.

NITROCHLOROBENZENE-ortho

DOT Number: UN 1578 *DOT Hazard Class:* Poison B *DOT Guide Number:* 55
Synonyms: 1-chloro-2-nitrobenzene, chloro-1-nitrobenzene, ONCB
STCC Number: 4921457 *Reportable Qty:* n/a *CHEMTREC Phone No:* 1-800-424-9300

Physical Description:

Physical Form: Liquid *Color:* Yellow *Odor:* Ammonialike
Other Information: It is used as a chemical intermediate for dyes, pigments, and in other chemicals.

Chemical Properties:

Specific Gravity: 1.73 *Vapor Density:* 5.4 *Boiling Point: 474° F(245.5° C)*
Melting Point: 95° F(35° C) *Vapor Pressure:* 8 mm Hg at 246° F(118.8° C)
Solubility in water: Yes *Degree of Solubility:* .05%
Other Information: It is shipped as a molten liquid that may solidify in route. Weighs 11.4 lbs/5.1 kg per gallon/3.8 l.

Health Hazards:

Inhalation Hazard: Poisonous if inhaled.
Ingestion Hazard: Poisonous if swallowed.
Absorption Hazard: Irritating to the skin and eyes.
Hazards to Wildlife: n/a
Decontamination Procedures: Wash away any material with copious amounts of soap and water.
First Aid Procedures: Remove victim to fresh air, call emergency medical care. If not breathing give CPR. If breathing is difficult administer oxygen. Treat for shock.

Fire Hazards:

Flashpoint: 261° F(127.2° C) *Ignition temperature:* n/a
Flammable Explosive High Range: n/a *Low Range:* n/a
Toxic Products of Combustion: Toxic oxides of nitrogen, hydrogen chloride, carbon monoxide, chlorides, and possibly chlorine.
Other Hazards: Will burn, but is difficult to ignite.
Possible extinguishing agents: Do not extinguish the fire unless the flow can be stopped. Apply water from as far a distance as possible. Use foam, dry chemical, or carbon dioxide.

Reactivity Hazards:

Reactive With: n/a *Other Reactions:* n/a

Corrosivity Hazards:

Corrosive With: Steel, may attack plastics, rubber, and coatings. *Neutralizing Agent:* Crushed limestone, soda ash, or lime.

Radioactivity Hazards:

Radiation Emitted: n/a *Other Hazards:* n/a

Recommended Protection for Response Personnel:

Avoid breathing vapors, keep upwind. Wear a sealed chemical suit (butyl rubber). Wash away any material which may have come into contact with the body with copious amounts of soap and water. Consider appropriate evacuation.

NITROCHLOROBENZENE-para

DOT Number: UN 1578 *DOT Hazard Class:* Poison B *DOT Guide Number:* 55
Synonyms: 4-chloro-1-nitrobenzene, PNCB, PCNB
STCC Number: 4921459 *Reportable Qty:* n/a *CHEMTREC Phone No:* 1-800-424-9300

Physical Description:

Physical Form: Solid *Color:* Yellow *Odor:* Sweet, pleasant
Other Information: It is used in the manufacture of chemicals, dyes, pharmaceuticals, gasoline gum inhibitor, and corrosion inhibitor.

Chemical Properties:

Specific Gravity: 1.3 *Vapor Density:* 5.4 *Boiling Point: 465° F(240.5° C)*
Melting Point: 182° F(83.3° C) *Vapor Pressure:* .09 mm Hg at 75° F(23.8° C) *Solubility in water:* No
Other Information: Practically insoluble in water and heavier. Weighs 10.8 lbs/4.8 kg per gallon/3.8 l.

Health Hazards:

Inhalation Hazard: Poisonous if inhaled.
Ingestion Hazard: Poisonous if swallowed.
Absorption Hazard: Irritating to the skin and eyes.
Hazards to Wildlife: n/a
Decontamination Procedures: Wash away any material with copious amounts of soap and water.
First Aid Procedures: Remove victim to fresh air, call emergency medical care. If not breathing give CPR. If breathing is difficult administer oxygen. Treat for shock.

Fire Hazards:

Flashpoint: 261° F(127.2° C) *Ignition temperature:* n/a
Flammable Explosive High Range: n/a *Low Range:* n/a
Toxic Products of Combustion: Toxic oxides of nitrogen, hydrogen chloride, carbon monoxide, chlorides, and possibly chlorine.
Other Hazards: Will burn, but is difficult to ignite.
Possible extinguishing agents: Extinguish fire using suitable agent for the type of surrounding fire.

Reactivity Hazards:

Reactive With: n/a *Other Reactions:* n/a

Corrosivity Hazards:

Corrosive With: n/a *Neutralizing Agent:* n/a

Radioactivity Hazards:

Radiation Emitted: n/a *Other Hazards:* n/a

Recommended Protection for Response Personnel:

Avoid breathing vapors, keep upwind. Wear a sealed chemical suit (butyl rubber). Wash away any material which may have come into contact with the body with copious amounts of soap and water. Consider appropriate evacuation.

NITROETHANE

DOT Number: UN 2842 *DOT Hazard Class:* Flammable liquid *DOT Guide Number:* 26
Synonyms: none given
STCC Number: 4909192 *Reportable Qty:* n/a
Mfg Name: Aldrich Chemical *Phone No:* 1-414-273-3850

Physical Description:

Physical Form: Liquid *Color:* Colorless *Odor:* Fruity
Other Information: It is used as a propellant, and as a solvent.

Chemical Properties:

Specific Gravity: 1.1 *Vapor Density:* 2.6 *Boiling Point: 237° F(113.8° C)*
Melting Point: −130° F(−90° C) *Vapor Pressure:* 15.6 mm Hg at 68° F(20° C)
Solubility in water: Slight *Degree of Solubility:* 4.5%
Other Information: May float or sink in water. Vapors are much heavier than air.

Health Hazards:

Inhalation Hazard: Causes coughing, difficulty in breathing.
Ingestion Hazard: Causes nausea and vomiting.
Absorption Hazard: Irritating to the skin and eyes.
Hazards to Wildlife: n/a
Decontamination Procedures: Wash away any material with copious amounts of soap and water.
First Aid Procedures: Remove victim to fresh air, call emergency medical care. If not breathing give CPR. If breathing is difficult administer oxygen. Treat for shock.

Fire Hazards:

Flashpoint: 82° F(27.7° C) *Ignition temperature:* 778° F(414.4° C)
Flammable Explosive High Range: n/a *Low Range:* 3.4
Toxic Products of Combustion: Toxic oxides of nitrogen may be formed in fires.
Other Hazards: Containers may explode in a fire. Flashback along vapor trail may occur, vapors may explode if ignited in an enclosed area.
Possible extinguishing agents: Do not extinguish the fire unless the flow can be stopped. Use alcohol foam, dry chemical, or carbon dioxide. Apply water from as far a distance as possible.

Reactivity Hazards:

Reactive With: May attack some forms of plastic. *Other Reactions:* n/a

Corrosivity Hazards:

Corrosive With: n/a *Neutralizing Agent:* n/a

Radioactivity Hazards:

Radiation Emitted: n/a *Other Hazards:* n/a

Recommended Protection for Response Personnel:

Avoid breathing vapors, keep upwind. Structural protective clothing provides limited protection. Wash away any material which may have come into contact with the body with copious amounts of soap and water. Consider appropriate evacuation.

NITROGEN

DOT Number: UN 1977 *DOT Hazard Class:* Nonflammable Gas *DOT Guide Number:* 21
Synonyms: liquid nitrogen
STCC Number: 4904565 *Reportable Qty:* n/a
Mfg Name: Airco Industrial Gases *Phone No:* 1-201-464-8100

Physical Description:

Physical Form: Gas *Color:* Colorless *Odor:* Odorless
Other Information: It is used in food processing, purging air conditioning and refrigerant systems, and in pressurizing aircraft tires.

Chemical Properties:

Specific Gravity: .807 *Vapor Density:* .97 *Boiling Point:* $-320°F(-195.5°C)$
Melting Point: n/a *Vapor Pressure:* n/a *Solubility in water:* No
Other Information: Floats and boils on water.

Health Hazards:

Inhalation Hazard: Asphyxiation by the displacement of air.
Ingestion Hazard: n/a
Absorption Hazard: Will cause frostbite.
Hazards to Wildlife: No
Decontamination Procedures: Wash away any material with copious amounts of soap and water.
First Aid Procedures: Remove victim to fresh air, call emergency medical care. If not breathing give CPR. If breathing is difficult administer oxygen. Treat for shock.

Fire Hazards:

Flashpoint: n/a *Ignition temperature:* n/a
Flammable Explosive High Range: n/a *Low Range:* n/a
Toxic Products of Combustion: n/a
Other Hazards: Containers may explode when heated.
Possible extinguishing agents: Extinguish fire using suitable agent for the type of surrounding fire. Apply water from as far a distance as possible.

Reactivity Hazards:

Reactive With: Water will vigorously vaporize nitrogen.
Other Reactions: Low temperature may cause brittleness in rubber or plastics.

Corrosivity Hazards:

Corrosive With: n/a *Neutralizing Agent:* n/a

Radioactivity Hazards:

Radiation Emitted: n/a *Other Hazards:* n/a

Recommended Protection for Response Personnel:

Avoid breathing vapors, keep upwind. Wear appropriate chemical clothing. Wash away any material which may have come into contact with the body with copious amounts of soap and water. Consider appropriate evacuation.

NITROGEN TETROXIDE

DOT Number: UN 1067 *DOT Hazard Class:* Poison A *DOT Guide Number:* 20
Synonyms: nitrogen dioxide, nitrogen peroxide, red nitrogen
STCC Number: 4920360 *Reportable Qty:* 1000/454
Mfg Name: Hercules Inc. *Phone No:* 1-302-594-5000

Physical Description:

Physical Form: Liquefied compressed gas *Color:* Red brown
Odor: Sharp unpleasant chemical
Other Information: It is used to produce nitric acid, as an oxidizer in rocket fuels, and as a polymerization inhibitor in acrylics.

Chemical Properties:

Specific Gravity: 1.45 *Vapor Density:* 1.6 *Boiling Point: 70.1° F(21.1° C)*
Melting Point: n/a *Vapor Pressure:* 400 mm Hg at 167° F(75° C) *Solubility in water:* Reacts
Other Information: Sinks and reacts with water. A poisonous, brown vapor is formed.

Health Hazards:

Inhalation Hazard: Poisonous if inhaled.
Ingestion Hazard: Poisonous if swallowed.
Absorption Hazard: Will burn the skin and the eyes.
Hazards to Wildlife: Dangerous to aquatic life.
Decontamination Procedures: Wash away any material with copious amounts of soap and water.
First Aid Procedures: Remove victim to fresh air, call emergency medical care. If not breathing give CPR. If breathing is difficult administer oxygen. Treat for shock.

Fire Hazards:

Flashpoint: n/a *Ignition temperature:* n/a
Flammable Explosive High Range: n/a *Low Range:* n/a
Toxic Products of Combustion: Produces toxic gases when heated.
Other Hazards: May cause fire and explode upon contact with combustibles.
Possible extinguishing agents: Extinguish fire using suitable agent for the type of surrounding fire. Apply water from as far a distance as possible. Use foam, dry chemical, or carbon dioxide.

Reactivity Hazards:

Reactive With: Water to from nitric acid and nitric oxide.
Other Reactions: Combustible materials such as wood or paper.

Corrosivity Hazards:

Corrosive With: Metals when wet. *Neutralizing Agent:* Crushed limestone, soda ash, or lime.

Radioactivity Hazards:

Radiation Emitted: n/a *Other Hazards:* n/a

Recommended Protection for Response Personnel:

Avoid breathing vapors, keep upwind. Wear appropriate chemical clothing. Wash away any material which may have come into contact with the body with copious amounts of soap and water. Prolonged exposure to heat or flame may cause the cylinders to violently rupture/rocket. Consider appropriate evacuation depending upon the amount spilled.

NITROMETHANE

DOT Number: UN 1261 *DOT Hazard Class:* Flammable liquid *DOT Guide Number:* 26
Synonyms: nitrocarbol
STCC Number: 4907030 *Reportable Qty:* n/a
Mfg Name: Mallinckrodt Inc. *Phone No:* 1-314-895-2000

Physical Description:

Physical Form: Liquid *Color:* Clear *Odor:* Strong
Other Information: May only be shipped in 110 gallon or less containers.

Chemical Properties:

Specific Gravity: 1.1 *Vapor Density:* 2.1 *Boiling Point: 214° F(101.1° C)*
Melting Point: −20° F(−28.8° C) *Vapor Pressure:* 27.8 mm Hg at 68° F(20° C)
Solubility in water: Slight *Degree of Solubility:* 9.5%
Other Information: Decomposes violently under fire or heat conditions.

Health Hazards:

Inhalation Hazard: Harmful if inhaled.
Ingestion Hazard: Harmful if swallowed.
Absorption Hazard: No hazard.
Hazards to Wildlife: n/a
Decontamination Procedures: Wash away any material with copious amounts of soap and water.
First Aid Procedures: Remove victim to fresh air, call emergency medical care. If not breathing give CPR. If breathing is difficult administer oxygen. Treat for shock.

Fire Hazards:

Flashpoint: 95° F(35° C) *Ignition temperature:* 875° F(468.3° C)
Flammable Explosive High Range: n/a *Low Range:* 7.3
Toxic Products of Combustion: Toxic oxides of nitrogen are produced during combustion.
Other Hazards: Flashback along vapor trail may occur. Vapors may explode if ignited in an enclosed area.
Possible extinguishing agents: Apply water from as far a distance as possible. Use alcohol foam, dry chemical, or carbon dioxide.

Reactivity Hazards:

Reactive With: Wet material corrodes steel and copper. *Other Reactions:* None

Corrosivity Hazards:

Corrosive With: n/a *Neutralizing Agent:* n/a

Radioactivity Hazards:

Radiation Emitted: n/a *Other Hazards:* n/a

Recommended Protection for Response Personnel:

Avoid breathing vapors, keep upwind. Structural protective clothing provides limited protection. Wash away any material which may have come into contact with the body with copious amounts of soap and water. Consider appropriate evacuation.

NITROPHENOL

DOT Number: NA 1663 *DOT Hazard Class:* ORM-E *DOT Guide Number:* 55
Synonyms: 2-hydroxynitrobenzene, o-nitrophenol, ONP
STCC Number: 4963394 *Reportable Qty:* 100/45.4
Mfg Name: Monsanto Chemical *Phone No:* 1-314-694-1000

Physical Description:

Physical Form: Solid *Color:* Yellow *Odor:* n/a
Other Information: It is used to manufacture other chemicals.

Chemical Properties:

Specific Gravity: 1.49 *Vapor Density:* 4.8 *Boiling Point: 417° F(213.8° C)*
Melting Point: n/a *Vapor Pressure:* 1 mm Hg at 121° F(49.4° C) *Solubility in water:* Yes
Other Information: Mixes with water.

Health Hazards:

Inhalation Hazard: Headache or loss of consciousness.
Ingestion Hazard: Headache, nausea, or loss of consciousness.
Absorption Hazard: Irritating to the skin and eyes.
Hazards to Wildlife: Dangerous to aquatic life.
Decontamination Procedures: Wash away any material with copious amounts of soap and water.
First Aid Procedures: Remove victim to fresh air, call emergency medical care. If not breathing give CPR. If breathing is difficult administer oxygen. Treat for shock.

Fire Hazards:

Flashpoint: n/a *Ignition temperature:* n/a
Flammable Explosive High Range: n/a *Low Range:* n/a
Toxic Products of Combustion: Toxic and irritating fumes of unburned material and oxides of nitrogen.
Other Hazards: n/a
Possible extinguishing agents: Apply water from as far a distance as possible. Use foam, dry chemical, or carbon dioxide.

Reactivity Hazards:

Reactive With: n/a *Other Reactions:* n/a

Corrosivity Hazards:

Corrosive With: n/a *Neutralizing Agent:* n/a

Radioactivity Hazards:

Radiation Emitted: n/a *Other Hazards:* n/a

Recommended Protection for Response Personnel:

Avoid breathing vapors, keep upwind. Structural protective clothing provides limited protection. Wash away any material which may have come into contact with the body with copious amounts of soap and water. Consider appropriate evacuation.

NITROPROPANE

DOT Number: UN 2608 *DOT Hazard Class:* Flammable liquid *DOT Guide Number:* 26
Synonyms: sec-nitropropane, 2-NP
STCC Number: 4909193 *Reportable Qty:* n/a
Mfg Name: Aldrich Chemical *Phone No:* 1-414-273-3850

Physical Description:

Physical Form: Liquid *Color:* Colorless *Odor:* Mild, fruity
Other Information: Liquid weighs the same as water. It is soluble in water. vapors are much heavier than air.

Chemical Properties:

Specific Gravity: 1 *Vapor Density:* 3.1 *Boiling Point: 248° F(120° C)*
Melting Point: −135° F(−92.7° C) *Vapor Pressure:* 12.9 mm Hg at 68° F(20° C)
Solubility in water: Slight *Degree of Solubility:* 1.7%
Other Information: n/a

Health Hazards:

Inhalation Hazard: Headache, dizziness, coughing, difficulty in breathing.
Ingestion Hazard: Nausea and vomiting.
Absorption Hazard: Irritating to the skin and eyes.
Hazards to Wildlife: n/a
Decontamination Procedures: Wash away any material with copious amounts of soap and water.
First Aid Procedures: Remove victim to fresh air, call emergency medical care. If not breathing give CPR. If breathing is difficult administer oxygen. Treat for shock.

Fire Hazards:

Flashpoint: 75° F(23.8° C) *Ignition temperature:* 802° F(427.7° C)
Flammable Explosive High Range: 11 *Low Range:* 2.6
Toxic Products of Combustion: Toxic oxides of nitrogen are released during combustion.
Other Hazards: Flashback along vapor trail may occur. Vapors may explode if ignited in an enclosed area. Containers may explode in a fire.
Possible extinguishing agents: Apply water from as far a distance as possible. Use alcohol foam, dry chemical, or carbon dioxide.

Reactivity Hazards:

Reactive With: Some forms of plastic. *Other Reactions:* None

Corrosivity Hazards:

Corrosive With: n/a *Neutralizing Agent:* n/a

Radioactivity Hazards:

Radiation Emitted: n/a *Other Hazards:* n/a

Recommended Protection for Response Personnel:

Avoid breathing vapors, keep upwind. Wear a sealed chemical suit (polycarbonate, butyl rubber). Wash away any material which may have come into contact with the body with copious amounts of soap and water. Consider appropriate evacuation.

NITROSYL CHLORIDE

DOT Number: UN 1069 *DOT Hazard Class:* Nonflammable gas *DOT Guide Number:* 16
Synonyms: none given
STCC Number: 4904330 *Reportable Qty:* n/a
Mfg Name: Ideal Gas Products Inc. *Phone No:* 1-201-287-8766

Physical Description:

Physical Form: Gas *Color:* Yellow to yellowish red *Odor:* Choking
Other Information: Vapors are heavier than air.

Chemical Properties:

Specific Gravity: 1.36 *Vapor Density:* 2.3 *Boiling Point: 21.6° F(−5.7° C)*
Melting Point: n/a *Vapor Pressure:* 76 mm Hg at 122° F(50° C)
Solubility in water: Decomposed by water.
Other Information: Avoid contact with liquid and vapor.

Health Hazards:

Inhalation Hazard: Poisonous if inhaled.
Ingestion Hazard: Poisonous if swallowed.
Absorption Hazard: Poisonous if absorbed.
Hazards to Wildlife: n/a
Decontamination Procedures: Wash away any material with copious amounts of soap and water.
First Aid Procedures: Remove victim to fresh air, call emergency medical care. If not breathing give CPR. If breathing is difficult administer oxygen. Treat for shock.

Fire Hazards:

Flashpoint: n/a *Ignition temperature:* n/a
Flammable Explosive High Range: n/a *Low Range:* n/a
Toxic Products of Combustion: Poisonous gases are produced when heated.
Other Hazards: n/a
Possible extinguishing agents: Extinguish fire using suitable agent for type of surrounding fire.

Reactivity Hazards:

Reactive With: Water to form toxic red oxides of nitrogen. *Other Reactions:* Most metals.

Corrosivity Hazards:

Corrosive With: Most metals. *Neutralizing Agent:* Crushed limestone, soda ash, or lime.

Radioactivity Hazards:

Radiation Emitted: n/a *Other Hazards:* n/a

Recommended Protection for Response Personnel:

Avoid breathing vapors, keep upwind. Wear a sealed chemical suit (polycarbonate, butyl rubber). Wash away any material which may have come into contact with the body with copious amounts of soap and water. Consider appropriate evacuation.

NITROTOLUENE

DOT Number: NA 1664 *DOT Hazard Class:* ORM-E *DOT Guide Number:* 55
Synonyms: m-nitrotoluene
STCC Number: 4963131 *Reportable Qty:* 1000/454
Mfg Name: E.I. Du Pont *Phone No:* 1-800-441-3637

Physical Description:

Physical Form: Liquid *Color:* Yellow *Odor:* Weak, aromatic odor
Other Information: It is used to make other chemicals.

Chemical Properties:

Specific Gravity: 1.16 *Vapor Density:* 4.73 *Boiling Point: 450° F(232.2° C)*
Melting Point: 25° F(−3.8° C) *Vapor Pressure:* .12 to .15 mm Hg at 68° F(20° C)
Solubility in water: No *Degree of Solubility:* .05 to .005%
Other Information: Weighs 9.7 lbs/4.3 kg to the gallon/3.8 l.

Health Hazards:

Inhalation Hazard: Headaches, dizziness, difficulty breathing, loss consciousness.
Ingestion Hazard: Nausea, vomiting, loss of consciousness, difficulty in breathing.
Absorption Hazard: Same as above
Hazards to Wildlife: Dangerous to aquatic life.
Decontamination Procedures: Wash away any material with copious amounts of soap and water.
First Aid Procedures: Remove victim to fresh air, call emergency medical care. If not breathing give CPR. If breathing is difficult administer oxygen. Treat for shock.

Fire Hazards:

Flashpoint: 223° F(106.1° C) *Ignition temperature:* n/a
Flammable Explosive High Range: None *Low Range:* 2.2 to 1.6
Toxic Products of Combustion: Poisonous gases may be produced in fire.
Other Hazards: n/a
Possible extinguishing agents: Apply water from as far a distance as possible. Use foam, dry chemical, or carbon dioxide.

Reactivity Hazards:

Reactive With: Strong oxidizers or sulfuric acid. *Other Reactions:* None

Corrosivity Hazards:

Corrosive With: n/a *Neutralizing Agent:* n/a

Radioactivity Hazards:

Radiation Emitted: n/a *Other Hazards:* n/a

Recommended Protection for Response Personnel:

Avoid breathing vapors, keep upwind. Wear appropriate chemical clothing, Wash away any material which may have come into contact with the body with copious amounts of soap and water. Consider appropriate evacuation.

NITROUS OXIDE

DOT Number: UN 1070 *DOT Hazard Class:* Nonflammable gas *DOT Guide Number:* 14
Synonyms: laughing gas
STCC Number: 4904340 *Reportable Qty:* n/a
Mfg Name: Matheson Gas Products *Phone No:* 1-201-867-4100

Physical Description:

Physical Form: Gas *Color:* Colorless *Odor:* Odorless to sweet tasting
Other Information: It is used as an anesthetic, in pressure packing, and to manufacture other chemicals.

Chemical Properties:

Specific Gravity: 1.26 *Vapor Density:* 1.53 *Boiling Point: – 129° F(– 89.4° C)*
Melting Point: n/a *Vapor Pressure:* n/a *Solubility in water:* Yes
Other Information: Vapors may impair decision making process.

Health Hazards:

Inhalation Hazard: Dizziness, difficulty in breathing, loss of consciousness.
Ingestion Hazard: n/a
Absorption Hazard: Will cause frostbite.
Hazards to Wildlife: No
Decontamination Procedures: Wash away any material with copious amounts of soap and water.
First Aid Procedures: Remove victim to fresh air, call emergency medical care. If not breathing give CPR. If breathing is difficult administer oxygen. Treat for shock.

Fire Hazards:

Flashpoint: n/a *Ignition temperature:* n/a
Flammable Explosive High Range: None *Low Range:* None
Toxic Products of Combustion: None
Other Hazards: Will accelerate the combustion of combustible materials. May only be shipped in cylinders.
Possible extinguishing agents: Apply water from as far a distance as possible.

Reactivity Hazards:

Reactive With: Combustibles *Other Reactions:* None

Corrosivity Hazards:

Corrosive With: n/a *Neutralizing Agent:* n/a

Radioactivity Hazards:

Radiation Emitted: n/a *Other Hazards:* n/a

Recommended Protection for Response Personnel:

Avoid breathing vapors, keep upwind. Wear appropriate chemical clothing. Wash away any material which may have come into contact with the body with copious amounts of soap and water. If the fire becomes uncontrollable, consider appropriate evacuation. Cylinders may rupture and rocket when exposed to heat or fire.

NONANE

DOT Number: UN 1920 *DOT Hazard Class:* Flammable liquid *DOT Guide Number:* 27
Synonyms: n-nonane
STCC Number: n/a *Reportable Qty:* n/a
Mfg Name: Phillips Petroleum *Phone No:* 1-918-661-6600

Physical Description:

Physical Form: Liquid *Color:* Colorless *Odor:* Gasolinelike
Other Information: n/a

Chemical Properties:

Specific Gravity: .72 *Vapor Density:* 4.4 *Boiling Point: 304° F(151.1° C)*
Melting Point: −64° F(−53.3° C) *Vapor Pressure:* 10 mm Hg at 100° F(37.7° C) *Solubility in water:* n/a
Other Information: Floats on water.

Health Hazards:

Inhalation Hazard: n/a
Ingestion Hazard: Nausea and vomiting.
Absorption Hazard: Irritating to the skin and eyes.
Hazards to Wildlife: n/a
Decontamination Procedures: Wash away any material with copious amounts of soap and water.
First Aid Procedures: Remove victim to fresh air, call emergency medical care. If not breathing give CPR. If breathing is difficult administer oxygen. Treat for shock.

Fire Hazards:

Flashpoint: 88° F(31.1° C) *Ignition temperature:* 401° F(205° C)
Flammable Explosive High Range: 2.9 *Low Range:* .87
Toxic Products of Combustion: n/a
Other Hazards: n/a
Possible extinguishing agents: Water, foam, dry chemical, carbon dioxide. Water may be ineffective on fires. Cool all affected containers with water.

Reactivity Hazards:

Reactive With: n/a *Other Reactions:* n/a

Corrosivity Hazards:

Corrosive With: n/a *Neutralizing Agent:* n/a

Radioactivity Hazards:

Radiation Emitted: n/a *Other Hazards:* n/a

Recommended Protection for Response Personnel:

Avoid breathing vapors, keep upwind. Structural protective clothing provides limited protection. Wash away any material which may have come into contact with the body with copious amounts of soap and water. Consider appropriate evacuation.

NONANOL

DOT Number: n/a *DOT Hazard Class:* n/a *DOT Guide Number:* n/a
Synonyms: 1-nonanol, nonyl alcohol, octylcarbinol, pelargonic alcohol
STCC Number: n/a *Reportable Qty:* n/a
Mfg Name: Givaudan Corp. *Phone No:* 1-201-365-8000

Physical Description:

Physical Form: Liquid *Color:* Colorless *Odor:* Rose or fruity
Other Information: n/a

Chemical Properties:

Specific Gravity: .83 *Vapor Density:* 1.04 *Boiling Point: 415° F(212.7° C)*
Melting Point: 23° F(−5° C) *Vapor Pressure:* n/a *Solubility in water:* No
Other Information: Floats on water.

Health Hazards:

Inhalation Hazard: n/a
Ingestion Hazard: Harmful if swallowed.
Absorption Hazard: n/a
Hazards to Wildlife: n/a
Decontamination Procedures: Wash away any material with copious amounts of soap and water.
First Aid Procedures: Remove victim to fresh air, call emergency medical care. If not breathing give CPR. If breathing is difficult administer oxygen. Treat for shock.

Fire Hazards:

Flashpoint: 165° F(73.8° C) *Ignition temperature:* n/a
Flammable Explosive High Range: 6.1 *Low Range:* .8
Toxic Products of Combustion: n/a
Other Hazards: n/a
Possible extinguishing agents: Alcohol foam, dry chemical, carbon dioxide. water may be ineffective on fires. Cool all affected containers with water.

Reactivity Hazards:

Reactive With: n/a *Other Reactions:* n/a

Corrosivity Hazards:

Corrosive With: n/a *Neutralizing Agent:* n/a

Radioactivity Hazards:

Radiation Emitted: n/a *Other Hazards:* n/a

Recommended Protection for Response Personnel:

Avoid breathing vapors, keep upwind. Structural protective clothing provides limited protection. Wash away any material which may have come into contact with the body with copious amounts of soap and water. Consider appropriate evacuation.

NONYLPHENOL

DOT Number: n/a *DOT Hazard Class:* n/a *DOT Guide Number:* n/a
Synonyms: none given
STCC Number: n/a *Reportable Qty:* n/a
Mfg Name: Monsanto Co. *Phone No:* 1-314-694-1000

Physical Description:

Physical Form: Liquid *Color:* Light yellow *Odor:* Medicinal
Other Information: n/a

Chemical Properties:

Specific Gravity: .95 *Vapor Density:* 7.6 *Boiling Point: 579° F(303.8° C)*
Melting Point: n/a *Vapor Pressure:* n/a *Solubility in water:* n/a
Other Information: Floats on water.

Health Hazards:

Inhalation Hazard: n/a
Ingestion Hazard: Harmful if swallowed.
Absorption Hazard: Will burn the skin and eyes.
Hazards to Wildlife: n/a
Decontamination Procedures: Wash away any material with copious amounts of soap and water.
First Aid Procedures: Remove victim to fresh air, call emergency medical care. If not breathing give CPR. If breathing is difficult administer oxygen. Treat for shock.

Fire Hazards:

Flashpoint: 285° F(140.5° C) *Ignition temperature:* n/a
Flammable Explosive High Range: n/a *Low Range:* 1
Toxic Products of Combustion: n/a
Other Hazards: n/a
Possible extinguishing agents: Alcohol foam, dry chemical, carbon dioxide. Water may be ineffective on fires. Cool all affected containers with water.

Reactivity Hazards:

Reactive With: n/a *Other Reactions:* n/a

Corrosivity Hazards:

Corrosive With: n/a *Neutralizing Agent:* n/a

Radioactivity Hazards:

Radiation Emitted: n/a *Other Hazards:* n/a

Recommended Protection for Response Personnel:

Avoid breathing vapors, keep upwind. Wear a sealed chemical suit (butyl rubber). Wash away any material which may have come into contact with the body with copious amounts of soap and water. Consider appropriate evacuation.

NUCLEAR REACTOR FUEL

DOT Number: UN 2918 *DOT Hazard Class:* Radioactive materal *DOT Guide Number:* n/a
Synonyms: fuel, spent fuel
STCC Number: 4929115 *Reportable Qty:* n/a *CHEMTREC Phone No:* 1-800-424-9300

Physical Description:

Physical Form: Solid *Color:* n/a *Odor:* n/a
Other Information: Radioactive material emits certain rays that are hazardous. These can only be detected with special equipment.

Chemical Properties:

Specific Gravity: n/a *Vapor Density:* n/a *Boiling Point: n/a*
Melting Point: n/a *Vapor Pressure:* n/a *Solubility in water:* n/a
Other Information: It is packaged in a massive steel encased cask. Generally weighs 20 to 125 tons each (18,143 to 113,398 kg each).

Health Hazards:

Inhalation Hazard: Radioactive poisoning.
Ingestion Hazard: Radioactive poisoning.
Absorption Hazard: Radioactive poisoning.
Hazards to Wildlife: Radioactive poisoning.
Decontamination Procedures: None available.
First Aid Procedures: Remove victim to fresh air, call emergency medical care. If not breathing give CPR. If breathing is difficult administer oxygen. Treat for shock.

Fire Hazards:

Flashpoint: n/a *Ignition temperature:* n/a
Flammable Explosive High Range: None *Low Range:* None
Toxic Products of Combustion: Radiation fallout.
Other Hazards: High emittance of radioactive energy,
Possible extinguishing agents: Apply water from as far a distance as possible. extinguish fire using suitable agent for type of surrounding fire.

Reactivity Hazards:

Reactive With: Itself *Other Reactions:* None

Corrosivity Hazards:

Corrosive With: n/a *Neutralizing Agent:* n/a

Radioactivity Hazards:

Radiation Emitted: Alpha, beta and gamma.
Other Hazards: In case of trouble with shipment of these containers, all unauthorized personnel should be kept as far away as possible until qualified personnel and equipment can be obtained.

Recommended Protection for Response Personnel:

Avoid breathing vapors, keep upwind. Wear appropriate chemical clothing, wash away any material which may have come into contact with the body with copious amounts of soap and water. Consider appropriate evacuation.

OCTANE

DOT Number: UN 1262 *DOT Hazard Class:* Flammable liquid *DOT Guide Number:* 27
Synonyms: normal octane
STCC Number: 4909250 *Reportable Qty:* n/a
Mfg Name: Phillips Petroleum *Phone No:* 1-918-661-6600

Physical Description:

Physical Form: Liquid *Color:* Colorless *Odor:* Petroleum
Other Information: n/a

Chemical Properties:

Specific Gravity: .7 *Vapor Density:* 3.9 *Boiling Point: 258° F(125.5° C)*
Melting Point: −70° F(−56.6° C) *Vapor Pressure:* 11 mm Hg at 68° F(20° C)
Solubility in water: No *Degree of Solubility:* .04%
Other Information: Lighter than and insoluble in water. its vapors are heavier than air.

Health Hazards:

Inhalation Hazard: Headaches, dizziness, difficulty in breathing, loss of consciousness.
Ingestion Hazard: Nausea and vomiting.
Absorption Hazard: Irritating to the skin and eyes.
Hazards to Wildlife: n/a
Decontamination Procedures: Wash away any material with copious amounts of soap and water.
First Aid Procedures: Remove victim to fresh air, call emergency medical care. If not breathing give CPR. If breathing is difficult administer oxygen. Treat for shock.

Fire Hazards:

Flashpoint: 56° F(13.3° C) *Ignition temperature:* 403° F(206.1° C)
Flammable Explosive High Range: 5.6 *Low Range:* 1
Toxic Products of Combustion: n/a
Other Hazards: Containers may explode in fire. Flashback along vapor trail may occur. Vapors may ignite if in an enclosed area.
Possible extinguishing agents: Apply water from as far a distance as possible. Use foam, dry chemical, or carbon dioxide.

Reactivity Hazards:

Reactive With: n/a *Other Reactions:* n/a

Corrosivity Hazards:

Corrosive With: n/a *Neutralizing Agent:* n/a

Radioactivity Hazards:

Radiation Emitted: n/a *Other Hazards:* n/a

Recommended Protection for Response Personnel:

Avoid breathing vapors, keep upwind. Wear appropriate chemical clothing. Wash away any material which may have come into contact with the body with copious amounts of soap and water. Containers involved in a fire can rupture or rocket. Consider appropriate evacuation.

OCTANOL

DOT Number: n/a *DOT Hazard Class:* n/a *DOT Guide Number:* n/a
Synonyms: C-8, heptylcarbinol, 1-octanol, octyl alcohol
STCC Number: n/a *Reportable Qty:* n/a
Mfg Name: Fisher Scientific *Phone No:* 1-412-349-3322

Physical Description:

Physical Form: Liquid *Color:* Colorless *Odor:* Sweet
Other Information: n/a

Chemical Properties:

Specific Gravity: .38 *Vapor Density:* 4.5 *Boiling Point: 383° F(195° C)*
Melting Point: 5° F(−15° C) *Vapor Pressure:* n/a *Solubility in water:* n/a
Other Information: Floats on water.

Health Hazards:

Inhalation Hazard: n/a
Ingestion Hazard: n/a
Absorption Hazard: Irritating to the skin, will burn the eyes.
Hazards to Wildlife: n/a
Decontamination Procedures: Wash away any material with copious amounts of soap and water.
First Aid Procedures: Remove victim to fresh air, call emergency medical care. If not breathing give CPR. If breathing is difficult administer oxygen. Treat for shock.

Fire Hazards:

Flashpoint: 178° F(81.1° C) *Ignition temperature:* n/a
Flammable Explosive High Range: n/a *Low Range:* n/a
Toxic Products of Combustion: n/a
Other Hazards: n/a
Possible extinguishing agents: Foam, dry chemical, carbon dioxide. Cool all affected containers with water. Water may be ineffective on fires.

Reactivity Hazards:

Reactive With: n/a *Other Reactions:* n/a

Corrosivity Hazards:

Corrosive With: n/a *Neutralizing Agent:* n/a

Radioactivity Hazards:

Radiation Emitted: n/a *Other Hazards:* n/a

Recommended Protection for Response Personnel:

Avoid breathing vapors, keep upwind. Structural protective clothing provides limited protection. Wash away any material which may have come into contact with the body with copious amounts of soap and water. Consider appropriate evacuation.

OCTENE

DOT Number: n/a *DOT Hazard Class:* Flammable liquid *DOT Guide Number:* n/a
Synonyms: caprylene, α-octylene
STCC Number: n/a *Reportable Qty:* n/a
Mfg Name: Phillips Petroleum *Phone No:* 1-918-661-6600

Physical Description:

Physical Form: Liquid *Color:* Colorless *Odor:* Gasolinelike
Other Information: n/a

Chemical Properties:

Specific Gravity: .71 *Vapor Density:* 3.9 *Boiling Point: 250° F(121.1° C)*
Melting Point: −151° F(−101.6° C) *Vapor Pressure:* 36 mm Hg at 100° F(37.7° C)
Solubility in water: n/a
Other Information: Floats on water. Flammable, harmful vapor is produced.

Health Hazards:

Inhalation Hazard: Dizziness
Ingestion Hazard: Nausea and vomiting.
Absorption Hazard: Irritating to the skin and eyes.
Hazards to Wildlife: n/a
Decontamination Procedures: Wash away any material with copious amounts of soap and water.
First Aid Procedures: Remove victim to fresh air, call emergency medical care. If not breathing give CPR. If breathing is difficult administer oxygen. Treat for shock.

Fire Hazards:

Flashpoint: 70° F(21.1° C) *Ignition temperature:* 493° F(256.1° C)
Flammable Explosive High Range: n/a *Low Range: .9*
Toxic Products of Combustion: n/a
Other Hazards: Flashback along vapor trail may occur. Vapors may explode if ignited in an enclosed area.
Possible extinguishing agents: Foam, dry chemical, carbon dioxide. Water may be ineffective on fires. Apply water from as far a distance as possible. Cool all affected containers with water.

Reactivity Hazards:

Reactive With: n/a *Other Reactions:* n/a

Corrosivity Hazards:

Corrosive With: n/a *Neutralizing Agent:* n/a

Radioactivity Hazards:

Radiation Emitted: n/a *Other Hazards:* n/a

Recommended Protection for Response Personnel:

Avoid breathing vapors, keep upwind. Structural protective clothing provides limited protection. Wash away any material which may have come into contact with the body with copious amounts of soap and water. If the fire becomes uncontrollable, consider appropriate evacuation.

OCTYL EPOXY TALLATE

DOT Number: n/a *DOT Hazard Class:* Flammable liquid *DOT Guide Number:* n/a
Synonyms: epoxidized tall oil, octyl ester
STCC Number: n/a *Reportable Qty:* n/a
Mfg Name: U.S.S. Chemicals *Phone No:* 1-412-433-1121

Physical Description:

Physical Form: Liquid *Color:* Pale yellow *Odor:* Mild
Other Information: n/a

Chemical Properties:

Specific Gravity: 1 *Vapor Density:* n/a *Boiling Point: Decomposes*
Melting Point: n/a *Vapor Pressure:* n/a *Solubility in water:* n/a
Other Information: Floats on water.

Health Hazards:

Inhalation Hazard: n/a
Ingestion Hazard: Harmful if swallowed.
Absorption Hazard: Irritating to the skin and eyes.
Hazards to Wildlife: n/a
Decontamination Procedures: Wash away any material with copious amounts of soap and water.
First Aid Procedures: Remove victim to fresh air, call emergency medical care. If not breathing give CPR. If breathing is difficult administer oxygen. Treat for shock.

Fire Hazards:

Flashpoint: 450° F(232.2° C) *Ignition temperature:* n/a
Flammable Explosive High Range: n/a *Low Range:* n/a
Toxic Products of Combustion: n/a
Other Hazards: Flashback along vapor trail may occur. Vapors may explode if ignited in an enclosed area.
Possible extinguishing agents: Foam, dry chemical, carbon dioxide. Water may be ineffective on fires. Apply water from as far a distance as possible. Cool all affected containers with water.

Reactivity Hazards:

Reactive With: n/a *Other Reactions:* n/a

Corrosivity Hazards:

Corrosive With: May attack some forms of plastic. *Neutralizing Agent:* n/a

Radioactivity Hazards:

Radiation Emitted: n/a *Other Hazards:* n/a

Recommended Protection for Response Personnel:

Avoid breathing vapors, keep upwind. Structural protective clothing provides limited protection. Wash away any material which may have come into contact with the body with copious amounts of soap and water. If the fire becomes uncontrollable, consider appropriate evacuation.

OIL

DOT Number: NA 1270 *DOT Hazard Class:* Flammable liquid *DOT Guide Number:* 27
Synonyms: petroleum oil, N.O.S.
STCC Number: 4910245 *Reportable Qty:* n/a
Mfg Name: Shell Oil Co. *Phone No:* 1-713-241-6161

Physical Description:

Physical Form: Liquid *Color:* Yellow to black *Odor:* Lube or fuel oil odor
Other Information: n/a

Chemical Properties:

Specific Gravity: .81 *Vapor Density:* n/a *Boiling Point: 550° F(287.7° C)*
Melting Point: Very low *Vapor Pressure:* Very low *Solubility in water:* No
Other Information: Lighter than and insoluble in water. Vapors are heavier than air.

Health Hazards:

Inhalation Hazard: No hazard
Ingestion Hazard: Harmful if swallowed.
Absorption Hazard: Irritating to the skin and eyes.
Hazards to Wildlife: Dangerous to both waterfowl and aquatic life.
Decontamination Procedures: Wash away any material with copious amounts of soap and water.
First Aid Procedures: Remove victim to fresh air, call emergency medical care. If not breathing give CPR. If breathing is difficult administer oxygen. Treat for shock.

Fire Hazards:

Flashpoint: 275° F(135° C) *Ignition temperature:* n/a
Flammable Explosive High Range: n/a *Low Range:* n/a
Toxic Products of Combustion: n/a
Other Hazards: n/a
Possible extinguishing agents: Apply water from as far a distance as possible. Use foam, dry chemical, or carbon dioxide.

Reactivity Hazards:

Reactive With: n/a *Other Reactions:* n/a

Corrosivity Hazards:

Corrosive With: n/a *Neutralizing Agent:* n/a

Radioactivity Hazards:

Radiation Emitted: n/a *Other Hazards:* n/a

Recommended Protection for Response Personnel:

Avoid breathing vapors, keep upwind. Structural protective clothing provides limited protection. Wash away any material which may have come into contact with the body with copious amounts of soap and water. Consider appropriate evacuation.

OLEIC ACID

DOT Number: n/a *DOT Hazard Class:* n/a *DOT Guide Number:* n/a
Synonyms: cis-8-heptadecylene carboxylic acid, cis-9-octadecenoic acid, red oil
STCC Number: n/a *Reportable Qty:* n/a
Mfg Name: Witco Chemical Co. *Phone No:* 1-212-605-3800

Physical Description:

Physical Form: Liquid *Color:* Colorless to pale yellow *Odor:* Mild
Other Information: n/a

Chemical Properties:

Specific Gravity: .89 *Vapor Density:* n/a *Boiling Point: 432° F(222.2° C)*
Melting Point: 57° F(13.8° C) *Vapor Pressure:* 1 mm Hg at 350° F(176.6° C) *Solubility in water:* n/a
Other Information: Floats on water.

Health Hazards:

Inhalation Hazard: n/a
Ingestion Hazard: Causes nausea
Absorption Hazard: Irritating to the skin and eyes.
Hazards to Wildlife: n/a
Decontamination Procedures: Wash away any material with copious amounts of soap and water.
First Aid Procedures: Remove victim to fresh air, call emergency medical care. If not breathing give CPR. If breathing is difficult administer oxygen. Treat for shock.

Fire Hazards:

Flashpoint: 390-425° F(198-218° C) *Ignition temperature:* 685° F(362.7° C)
Flammable Explosive High Range: n/a *Low Range:* n/a
Toxic Products of Combustion: n/a
Other Hazards: n/a
Possible extinguishing agents: Dry chemical, carbon dioxide. Water or foam may be ineffective on fires. Water or foam may cause frothing. Cool all exposed containers with water. Apply water from as far a distance as possible.

Reactivity Hazards:

Reactive With: n/a *Other Reactions:* n/a

Corrosivity Hazards:

Corrosive With: n/a *Neutralizing Agent:* n/a

Radioactivity Hazards:

Radiation Emitted: n/a *Other Hazards:* n/a

Recommended Protection for Response Personnel:

Avoid breathing vapors, keep upwind. Structural protective clothing provides limited protection. Wash away any material which may have come into contact with the body with copious amounts of soap and water. If the fire becomes uncontrollable, consider appropriate evacuation.

OLEIC ACID, POTASSIUM SALT

DOT Number: n/a *DOT Hazard Class:* n/a *DOT Guide Number:* n/a
Synonyms: potassium oleate
STCC Number: n/a *Reportable Qty:* n/a
Mfg Name: Diamond Shamrock Chemical Corp. *Phone No:* 1-216-946-2064

Physical Description:

Physical Form: Solid or liquid *Color:* Brown *Odor:* Soapy
Other Information: n/a

Chemical Properties:

Specific Gravity: 1.1 *Vapor Density:* n/a *Boiling Point: Decomposes*
Melting Point: 455-464° F(235-240° C) *Vapor Pressure:* n/a
Solubility in water: Yes
Other Information: Sinks and mixes slowly with water.

Health Hazards:

Inhalation Hazard: n/a
Ingestion Hazard: Nausea and vomiting.
Absorption Hazard: Irritating to the skin and eyes.
Hazards to Wildlife: n/a
Decontamination Procedures: Wash away any material with copious amounts of soap and water.
First Aid Procedures: Remove victim to fresh air, call emergency medical care. If not breathing give CPR. If breathing is difficult administer oxygen. Treat for shock.

Fire Hazards:

Flashpoint: 140° F(60° C) *Ignition temperature:* n/a
Flammable Explosive High Range: n/a *Low Range:* n/a
Toxic Products of Combustion: n/a
Other Hazards: n/a
Possible extinguishing agents: Alcohol foam, dry chemical, carbon dioxide. Water may be ineffective on fires.

Reactivity Hazards:

Reactive With: n/a *Other Reactions:* n/a

Corrosivity Hazards:

Corrosive With: n/a *Neutralizing Agent:* n/a

Radioactivity Hazards:

Radiation Emitted: n/a *Other Hazards:* n/a

Recommended Protection for Response Personnel:

Avoid breathing vapors, keep upwind. Structural protective clothing provides limited protection. Wash away any material which may have come into contact with the body with copious amounts of soap and water. If the fire becomes uncontrollable, consider appropriate evacuation.

OLEIC ACID, SODIUM SALT

DOT Number: n/a *DOT Hazard Class:* n/a *DOT Guide Number:* n/a
Synonyms: Eunatrol, sodium oleate
STCC Number: n/a *Reportable Qty:* n/a
Mfg Name: Diamond Shamrock Chemical Corp. *Phone No:* 1-216-946-2064

Physical Description:

Physical Form: Solid *Color:* Light tan *Odor:* Slight tallowlike
Other Information: n/a

Chemical Properties:

Specific Gravity: 1.1 *Vapor Density:* n/a *Boiling Point: Decomposes*
Melting Point: 450-455° F(232-235° C) *Vapor Pressure:* n/a
Solubility in water: Yes
Other Information: Sinks and mixes slowly with water.

Health Hazards:

Inhalation Hazard: n/a
Ingestion Hazard: Nausea and vomiting.
Absorption Hazard: Irritating to the skin and eyes.
Hazards to Wildlife: n/a
Decontamination Procedures: Wash away any material with copious amounts of soap and water.
First Aid Procedures: Remove victim to fresh air, call emergency medical care. If not breathing give CPR. If breathing is difficult administer oxygen. Treat for shock.

Fire Hazards:

Flashpoint: Combustible solid *Ignition temperature:* n/a
Flammable Explosive High Range: n/a *Low Range:* n/a
Toxic Products of Combustion: n/a
Other Hazards: n/a
Possible extinguishing agents: Extinguish fire using suitable agent for the type of surrounding fire.

Reactivity Hazards:

Reactive With: n/a *Other Reactions:* n/a

Corrosivity Hazards:

Corrosive With: n/a *Neutralizing Agent:* n/a

Radioactivity Hazards:

Radiation Emitted: n/a *Other Hazards:* n/a

Recommended Protection for Response Personnel:

Avoid breathing vapors, keep upwind. Structural protective clothing provides limited protection. Wash away any material which may have come into contact with the body with copious amounts of soap and water. If the fire becomes uncontrollable, consider appropriate evacuation.

OLEUM

DOT Number: NA 1831 *DOT Hazard Class:* Corrosive *DOT Guide Number:* 39
Synonyms: fuming sulfuric acid
STCC Number: 4930030 *Reportable Qty:* 1000/454
Mfg Name: E.I. Du Pont *Phone No:* 1-800-441-3637

Physical Description:

Physical Form: Liquid *Color:* Colorless to black *Odor:* Choking
Other Information: It is used in the manufacturing of chemicals, dyes, explosives, and in petroleum refining.

Chemical Properties:

Specific Gravity: 1.91 *Vapor Density:* 3.3 *Boiling Point: n/a*
Melting Point: n/a *Vapor Pressure:* n/a *Solubility in water:* Yes
Other Information: Weighs 16.5 lbs/7.4 kg per gallon/3.8 l.

Health Hazards:

Inhalation Hazard: Coughing, difficulty in breathing.
Ingestion Hazard: Harmful if swallowed.
Absorption Hazard: Will burn the skin and the eyes.
Hazards to Wildlife: Dangerous to aquatic life.
Decontamination Procedures: Wash away any material with copious amounts of soap and water.
First Aid Procedures: Remove victim to fresh air, call emergency medical care. If not breathing give CPR. If breathing is difficult administer oxygen. Treat for shock.

Fire Hazards:

Flashpoint: n/a *Ignition temperature:* n/a
Flammable Explosive High Range: n/a *Low Range:* n/a
Toxic Products of Combustion: n/a
Other Hazards: n/a
Possible extinguishing agents: Apply water from as far a distance as possible. Use suitable agent for type of surrounding fire. Material itself will not burn.

Reactivity Hazards:

Reactive With: Vigorous reaction with water. *Other Reactions:* Combustibles on contact.

Corrosivity Hazards:

Corrosive With: Metals and tissue
Neutralizing Agent: Crushed limestone, soda ash, or lime.

Radioactivity Hazards:

Radiation Emitted: n/a *Other Hazards:* n/a

Recommended Protection for Response Personnel:

Avoid breathing vapors, keep upwind. Wear a sealed chemical suit (chlorinated polyethylene, viton, PVC). Wash away any material which may have come into contact with the body with copious amounts of soap and water. Consider appropriate evacuation.

OXALIC ACID

DOT Number: n/a *DOT Hazard Class:* n/a *DOT Guide Number:* n/a
Synonyms: dicarboxylic acid, ethane-di-acid, ethanedioic acid, oxalic acid dihydrate
STCC Number: n/a *Reportable Qty:* n/a
Mfg Name: Allied Corp. *Phone No:* 1-201-455-2000

Physical Description:

Physical Form: Solid *Color:* White *Odor:* Odorless
Other Information: n/a

Chemical Properties:

Specific Gravity: 1.9 *Vapor Density:* 4.3 *Boiling Point: Decomposes*
Melting Point: 214° F(101.1° C) *Vapor Pressure:* n/a *Solubility in water:* Yes
Other Information: Sinks and mixes with water.

Health Hazards:

Inhalation Hazard: Difficulty in breathing.
Ingestion Hazard: Nausea or loss of consciousness.
Absorption Hazard: Will burn the skin and eyes.
Hazards to Wildlife: Dangerous to aquatic life.
Decontamination Procedures: Wash away any material with copious amounts of soap and water.
First Aid Procedures: Remove victim to fresh air, call emergency medical care. If not breathing give CPR. If breathing is difficult administer oxygen. Treat for shock.

Fire Hazards:

Flashpoint: n/a *Ignition temperature:* n/a
Flammable Explosive High Range: n/a *Low Range:* n/a
Toxic Products of Combustion: Poisonous gases are produced in fire.
Other Hazards: n/a
Possible extinguishing agents: Extinguish fire using suitable agent for the type of surrounding fire.

Reactivity Hazards:

Reactive With: n/a *Other Reactions:* n/a

Corrosivity Hazards:

Corrosive With: n/a *Neutralizing Agent:* Crushed limestone, soda ash or lime.

Radioactivity Hazards:

Radiation Emitted: n/a *Other Hazards:* n/a

Recommended Protection for Response Personnel:

Avoid breathing vapors, keep upwind. Wear a sealed chemical suit (polycarbonate, butyl rubber, PVC, viton, chlorinated polyethylene). Wash away any material which may have come into contact with the body with copious amounts of soap and water. Consider appropriate evacuation.

OXYGEN

DOT Number: UN 1073 *DOT Hazard Class:* Nonflammable gas *DOT Guide Number:* 23
Synonyms: liquid oxygen, LOX
STCC Number: 4904360 *Reportable Qty:* n/a
Mfg Name: Arco Industrial Gases *Phone No:* 1-201-464-8100

Physical Description:

Physical Form: Liquid *Color:* Light blue *Odor:* Odorless
Other Information: It is shipped as a refrigerated liquid at pressures below 200 psig.

Chemical Properties:

Specific Gravity: 1.14 *Vapor Density:* 1.1 *Boiling Point:* $-297.3°F(-182.9°C)$
Melting Point: n/a *Vapor Pressure:* n/a *Solubility in water:* No
Other Information: n/a

Health Hazards:

Inhalation Hazard: Cause dizziness or difficulty in breathing.
Ingestion Hazard: n/a
Absorption Hazard: Will cause frostbite.
Hazards to Wildlife: No.
Decontamination Procedures: Wash away any material with copious amounts of soap and water.
First Aid Procedures: Remove victim to fresh air, call emergency medical care. If not breathing give CPR. If breathing is difficult administer oxygen. Treat for shock.

Fire Hazards:

Flashpoint: n/a *Ignition temperature:* n/a
Flammable Explosive High Range: n/a *Low Range:* n/a
Toxic Products of Combustion: n/a
Other Hazards: Containers may explode in a fire.
Possible extinguishing agents: Apply water from as far a distance as possible.

Reactivity Hazards:

Reactive With: Combustible and oxidizable materials.
Other Reactions: Leakage or spilled material will readily evaporate to a gaseous state.

Corrosivity Hazards:

Corrosive With: n/a *Neutralizing Agent:* n/a

Radioactivity Hazards:

Radiation Emitted: n/a *Other Hazards:* n/a

Recommended Protection for Response Personnel:

Avoid breathing vapors, keep upwind. Structural protective clothing provides limited protection. Wash away any material which may have come into contact with the body with copious amounts of soap and water. Consider appropriate evacuation.

PARAFORMALDEHYDE

DOT Number: UN 2213 *DOT Hazard Class:* ORM-A *DOT Guide Number:* 32
Synonyms: polyformaldehyde, polyoxymethylene
STCC Number: 4941143 *Reportable Qty:* 100/45.4
Mfg Name: Mallinckrodt Inc. *Phone No:* 1-314-895-2000

Physical Description:

Physical Form: Solid *Color:* White *Odor:* Irritating
Other Information: It is used for fungicides, bactericides, in the manufacture of adhesives, and in many other uses.

Chemical Properties:

Specific Graity: 1.46 *Vapor Density:* 1 *Boiling Point: Decomposes*
Melting Point: n/a *Vapor Pressure:* n/a *Solubility in water:* Yes
Other Information: Sinks and mixes with water.

Health Hazards:

Inhalation Hazard: Harmful if inhaled.
Ingestion Hazard: Vomiting, nausea, loss of consciousness.
Absorption Hazard: Irritating to the skin and eyes.
Hazards to Wildlife: Dangerous to aquatic life.
Decontamination Procedures: Wash away any material with copious amounts of soap and water.
First Aid Procedures: Remove victim to fresh air, call emergency medical care. If not breathing give CPR. If breathing is difficult administer oxygen. Treat for shock.

Fire Hazards:

Flashpoint: 185° F(85° C) *Ignition temperature:* 572° F(300° C)
Flammable Explosive High Range: 73 *Low Range:* 7
Toxic Products of Combustion: n/a
Other Hazards: Changes to formaldehyde gas which is highly flammable.
Possible extinguishing agents: Apply water from as far a distance as possible. Use alcohol foam, dry chemical, or carbon dioxide.

Reactivity Hazards:

Reactive With: Water to form solution of formaldehyde. *Other Reactions:* n/a

Corrosivity Hazards:

Corrosive With: n/a *Neutralizing Agent:* n/a

Radioactivity Hazards:

Radiation Emitted: n/a *Other Hazards:* n/a

Recommended Protection for Response Personnel:

Avoid breathing vapors, keep upwind. Structural protective clothing provides limited protection. Wash away any material which may have come into contact with the body with copious amounts of soap and water. Consider appropriate evacuation.

PARALDEHYDE

DOT Number: UN 1264 *DOT Hazard Class:* Flammable liquid *DOT Guide Number:* 26
Synonyms: paracetaldehyde, 1,3,5-trioxane
STCC Number: 4909260 *Reportable Qty:* 1000/454
Mfg Name: Lonza Inc. *Phone No:* 1-201-794-2400

Physical Description:

Physical Form: Liquid *Color:* Colorless *Odor:* Pleasant
Other Information: n/a

Chemical Properties:

Specific Gravity: 1 *Vapor Density:* 4.5 *Boiling Point: 255° F(123.8° C)*
Melting Point: n/a *Vapor Pressure:* n/a *Solubility in water:* Slight
Other Information: Lighter than and slightly soluble in water. Vapors are heavier than air.

Health Hazards:

Inhalation Hazard: Harmful if inhaled.
Ingestion Hazard: Headaches, incoordination, drowsiness or coma.
Absorption Hazard: Irritating to the skin and eyes.
Hazards to Wildlife: n/a
Decontamination Procedures: Wash away any material with copious amounts of soap and water.
First Aid Procedures: Remove victim to fresh air, call emergency medical care. If not breathing give CPR. If breathing is difficult administer oxygen. Treat for shock.

Fire Hazards:

Flashpoint: 96° F(35.5° C) *Ignition temperature:* 460° F(237.7° C)
Flammable Explosive High Range: n/a *Low Range:* 1.3
Toxic Products of Combustion: Poisonous gases are produced in a fire.
Other Hazards: Flashback along vapor trail may occur. Vapors may explode if ignited in an enclosed area.
Possible *Possible extinguishing agents:* Apply water from as far a distance as possible. use alcohol foam, dry chemical, or carbon dioxide.

Reactivity Hazards:

Reactive With: Heat or flame. *Other Reactions:* n/a

Corrosivity Hazards:

Corrosive With: n/a *Neutralizing Agent:* n/a

Radioactivity Hazards:

Radiation Emitted: n/a *Other Hazards:* n/a

Recommended Protection for Response Personnel:

Avoid breathing vapors, keep upwind. Wear appropriate chemical clothing. Wash away any material which may come into contact with the body with copious amounts of soap and water. Consider appropriate evacuation.

PARATHION

DOT Number: NA 2783 *DOT Hazard Class:* Poison B *DOT Guide Number:* 55
Synonyms: ethyl parathion
STCC Number: 4921469 *Reportable Qty:* 1/.454
Mfg Name: Monsanto Chemical Co. *Phone No:* 1-314-694-1000

Physical Description:

Physical Form: Liquid *Color:* Yellow *Odor:* n/a
Other Information: It weighs 10.5 lbs/4.7 kg per gallon/3.8 l.

Chemical Properties:

Specific Gravity: 1.27 *Vapor Density:* 10 *Boiling Point: 707° F(375° C)*
Melting Point: 43° F(6.1° C) *Vapor Pressure:* .0004 mm Hg at 68° F(20° C)
Solubility in water: Slight *Degree of Solubility:* .00002%
Other Information: In case of damage to or leaking from this container, contact the Pesticide Safety Team Network at 1-800-424-9300.

Health Hazards:

Inhalation Hazard: Poisonous if inhaled.
Ingestion Hazard: Poisonous if swallowed.
Absorption Hazard: Poisonous if absorbed.
Hazards to Wildlife: Dangerous to both waterfowl and aquatic life.
Decontamination Procedures: Wash away any material with copious amounts of soap and water.
First Aid Procedures: Remove victim to fresh air, call emergency medical care. If not breathing give CPR. If breathing is difficult administer oxygen. Treat for shock.

Fire Hazards:

Flashpoint: n/a *Ignition temperature:* n/a
Flammable Explosive High Range: n/a *Low Range:* n/a
Toxic Products of Combustion: Poisonous gases are produced when heated.
Other Hazards: n/a
Possible extinguishing agents: Apply water from as far a distance as possible. Use alcohol foam, dry chemical, or carbon dioxide.

Reactivity Hazards:

Reactive With: Water *Other Reactions:* n/a

Corrosivity Hazards:

Corrosive With: n/a *Neutralizing Agent:* n/a

Radioactivity Hazards:

Radiation Emitted: n/a *Other Hazards:* n/a

Recommended Protection for Response Personnel:

Avoid breathing vapors, keep upwind. Wear a sealed chemical suit (polycarbonate, butyl rubber). Wash away any material which may have come into contact with the body with copious amounts of soap and water. Consider appropriate evacuation.

PENTABORANE

DOT Number: UN 1380 *DOT Hazard Class:* Flammable liquid *DOT Guide Number:* 75
Synonyms: (9)-pentaboron nonahydride
STCC Number: 4906060 *Reportable Qty:* n/a
Mfg Name: Callery Chemical Co. *Phone No:* 1-412-538-3510

Physical Description:

Physical Form: Liquid *Color:* Colorless *Odor:* Strong sour milk
Other Information: It may only be shipped in cylinders.

Chemical Properties:

Specific Gravity: .6 *Vapor Density:* 2.2 *Boiling Point: 140° F(60° C)*
Melting Point: −52° F(−46.6° C) *Vapor Pressure:* 171 mm Hg at 68° F(20° C)
Solubility in water: Reacts
Other Information: Material is heavier than water. Slowly decomposed by water.

Health Hazards:

Inhalation Hazard: Poisonous if inhaled.
Ingestion Hazard: Poisonous if swallowed.
Absorption Hazard: Poisonous if absorbed.
Hazards to Wildlife: n/a
Decontamination Procedures: Wash away any material with copious amounts of soap and water.
First Aid Procedures: Remove victim to fresh air, call emergency medical care. If not breathing give CPR. If breathing is difficult administer oxygen. Treat for shock.

Fire Hazards:

Flashpoint: 86° F(30° C) *Ignition temperature:* n/a
Flammable Explosive High Range: 98 *Low Range:* .42
Toxic Products of Combustion: Poisonous products of combustion.
Other Hazards: Pyrophoric, ignites when exposed to air.
Possible *Possible extinguishing agents:* Do not use water. Use dry chemical or carbon dioxide.

Reactivity Hazards:

Reactive With: Water to form hydrogen gas. *Other Reactions:* Air

Corrosivity Hazards:

Corrosive With: Natural, synthetic rubbers, greases, and some lubricants. *Neutralizing Agent:* n/a

Radioactivity Hazards:

Radiation Emitted: n/a *Other Hazards:* n/a

Recommended Protection for Response Personnel:

Avoid breathing vapors, keep upwind. Wear appropriate chemical clothing. Wash away any material which may come into contact with the body with copious amounts of soap and water. If fire becomes uncontrollable, consider appropriate evacuation. Under fire conditions, the cylinders may violently rupture/rocket!! (BLEVE!!)

PENTACHLOROPHENOL

DOT Number: NA 2020 *DOT Hazard Class:* ORM-E *DOT Guide Number:* 53
Synonyms: Dowicide 7, penta, Santophen 20
STCC Number: 4961380 *Reportable Qty:* 10/4.54
Mfg Name: Dow Chemical Co. *Phone No:* 1-519-636-4400

Physical Description:

Physical Form: Solid *Color:* White to light brown *Odor:* n/a
Other Information: It is used as a fungicide, wood preservative, and for many other uses.

Chemical Properties:

Specific Gravity: 1.98 *Vapor Density:* n/a *Boiling Point: 590° F(310° C)*
Melting Point: 360-374° F(182-190° C) *Vapor Pressure:* .0002 mm Hg at 68° F(20° C)
Solubility in water: Yes *Degree of Solubility:* .002%
Other Information: Sinks in water.

Health Hazards:

Inhalation Hazard: Causes coughing, difficulty in breathing.
Ingestion Hazard: Poisonous if swallowed.
Absorption Hazard: Will burn the skin and the eyes.
Hazards to Wildlife: Dangerous to both aquatic life and waterfowl.
Decontamination Procedures: Wash away any material with copious amounts of soap and water.
First Aid Procedures: Remove victim to fresh air, call emergency medical care. If not breathing give CPR. If breathing is difficult administer oxygen. Treat for shock.

Fire Hazards:

Flashpoint: n/a *Ignition temperature:* n/a
Flammable Explosive High Range: n/a *Low Range:* n/a
Toxic Products of Combustion: Generates toxic and irritating vapors.
Other Hazards: n/a
Possible extinguishing agents: Extinguish fire using suitable agent for the type of surrounding fire.

Reactivity Hazards:

Reactive With: n/a *Other Reactions:* n/a

Corrosivity Hazards:

Corrosive With: n/a *Neutralizing Agent:* n/a

Radioactivity Hazards:

Radiation Emitted: n/a *Other Hazards:* n/a

Recommended Protection for Response Personnel:

Avoid breathing vapors, keep upwind. Wear a sealed chemical suit (polycarbonate, butyl rubber). Wash away any material which may have come into contact with the body with copious amounts of soap and water. Consider appropriate evacuation.

PENTADECANOL

DOT Number: n/a *DOT Hazard Class:* n/a *DOT Guide Number:* n/a
Synonyms: 1-pentadecanol, pentadecyl alcohol
STCC Number: n/a *Reportable Qty:* n/a *CHEMTREC Phone No:* 1-800-424-9300

Physical Description:

Physical Form: Liquid *Color:* Colorless *Odor:* Faint alcohol
Other Information: n/a

Chemical Properties:

Specific Gravity: .83 *Vapor Density:* n/a *Boiling Point: 521° F(271.6° C)*
Melting Point: 111° F(43.8° C) *Vapor Pressure:* n/a *Solubility in water:* n/a
Other Information: Floats on water.

Health Hazards:

Inhalation Hazard: n/a
Ingestion Hazard: n/a
Absorption Hazard: Irritating to the skin and eyes.
Hazards to Wildlife: Dangerous to both aquatic life and waterfowl.
Decontamination Procedures: Wash away any material with copious amounts of soap and water.
First Aid Procedures: Remove victim to fresh air, call emergency medical care. If not breathing give CPR. If breathing is difficult administer oxygen. Treat for shock.

Fire Hazards:

Flashpoint: n/a *Ignition temperature:* n/a
Flammable Explosive High Range: n/a *Low Range:* n/a
Toxic Products of Combustion: n/a
Other Hazards: n/a
Possible extinguishing agents: Foam, dry chemical, carbon dioxide. Water may be ineffective on fires. Cool all exposed containers with water.

Reactivity Hazards:

Reactive With: n/a *Other Reactions:* n/a

Corrosivity Hazards:

Corrosive With: n/a *Neutralizing Agent:* n/a

Radioactivity Hazards:

Radiation Emitted: n/a *Other Hazards:* n/a

Recommended Protection for Response Personnel:

Avoid breathing vapors, keep upwind. Wear a sealed chemical suit (polycarbonate, butyl rubber, viton, PVC, nitrile, neoprene, chlorinated polyethylene). Wash away any material which may have come into contact with the body with copious amounts of soap and water. If the fire becomes uncontrollable, consider appropriate evacuation.

PENTAERYTHRITOL

DOT Number: n/a *DOT Hazard Class:* n/a *DOT Guide Number:* n/a
Synonyms: Monope, PE, pentaerythrite, Penetek, Pentek
STCC Number: n/a *Reportable Qty:* n/a
Mfg Name: IMC Chemical *Phone No:* 1-312-296-0600

Physical Description:

Physical Form: Solid *Color:* White *Odor:* Odorless
Other Information: n/a

Chemical Properties:

Specific Gravity: 1.39 *Vapor Density:* n/a *Boiling Point: 520° F(271.1° C)*
Melting Point: n/a *Vapor Pressure:* n/a *Solubility in water:* Yes
Other Information: Sinks and mixes slowly with water.

Health Hazards:

Inhalation Hazard: n/a
Ingestion Hazard: n/a
Absorption Hazard: n/a
Hazards to Wildlife: n/a
Decontamination Procedures: Wash away any material with copious amounts of soap and water.
First Aid Procedures: Remove victim to fresh air, call emergency medical care. If not breathing give CPR. If breathing is difficult administer oxygen. Treat for shock.

Fire Hazards:

Flashpoint: n/a *Ignition temperature:* 842° F(450° C)
Flammable Explosive High Range: n/a *Low Range:* n/a
Toxic Products of Combustion: n/a
Other Hazards: n/a
Possible extinguishing agents: Extinguish fire using suitable agent for the type of surrounding fire.

Reactivity Hazards:

Reactive With: n/a *Other Reactions:* n/a

Corrosivity Hazards:

Corrosive With: n/a *Neutralizing Agent:* n/a

Radioactivity Hazards:

Radiation Emitted: n/a *Other Hazards:* n/a

Recommended Protection for Response Personnel:

Avoid breathing vapors, keep upwind. Structural protective clothing provides limited protection. Wash away any material which may have come into contact with the body with copious amounts of soap and water. Consider appropriate evacuation.

PENTANE

DOT Number: UN 1265 *DOT Hazard Class:* Flammable liquid *DOT Guide Number:* 27
Synonyms: normal pentane
STCC Number: 4908255 *Reportable Qty:* n/a
Mfg Name: Ashland Oil Inc. *Phone No:* 1-614-889-3333

Physical Description:

Physical Form: Liquid *Color:* Colorless *Odor:* Petroleum
Other Information: n/a

Chemical Properties:

Specific Gravity: .6 *Vapor Density:* 2.5 *Boiling Point: 97° F(36.1° C)*
Melting Point: −200° F(−128.8° C) *Vapor Pressure:* 426 mm Hg at 68° F(20° C)
Solubility in water: No *Degree of Solubility:* .04%
Other Information: Lighter than and insoluble in water. Vapors are heavier than air.

Health Hazards:

Inhalation Hazard: Dizziness, difficulty in breathing.
Ingestion Hazard: Harmful if swallowed.
Absorption Hazard: n/a
Hazards to Wildlife: Dangerous to aquatic life.
Decontamination Procedures: Wash away any material with copious amounts of soap and water.
First Aid Procedures: Remove victim to fresh air, call emergency medical care. If not breathing give CPR. If breathing is difficult administer oxygen. Treat for shock.

Fire Hazards:

Flashpoint: −40° F(−40° C) *Ignition temperature:* 500° F(260° C)
Flammable Explosive High Range: 7.8 *Low Range:* 1.5
Toxic Products of Combustion: n/a
Other Hazards: Flashback along vapor trail may occur. Containers may explode when heated. Vapors may explode if ignited in an enclosed area.
Possible extinguishing agents: Apply water from as far a distance as possible. Use foam, dry chemical, or carbon dioxide.

Reactivity Hazards:

Reactive With: n/a *Other Reactions:* n/a

Corrosivity Hazards:

Corrosive With: n/a *Neutralizing Agent:* n/a

Radioactivity Hazards:

Radiation Emitted: n/a *Other Hazards:* n/a

Recommended Protection for Response Personnel:

Avoid breathing vapors, keep upwind. Structural protective clothing provides limited protection. Wash away any material which may have come into contact with the body with copious amounts of soap and water. If the fire becomes uncontrollable, consider appropriate evacuation. Under fire conditions, the cylinders may violently rupture/rocket.

PENTENE

DOT Number: UN 1108 *DOT Hazard Class:* Flammable liquid DOT Guide Number: 26
Synonyms: α-n-amylene, 2-methyl butene-2, propylethylene
STCC Number: n/a *Reportable Qty:* n/a
Mfg Name: Phillips Petroleum *Phone No:* 1-918-661-6600

Physical Description:

Physical Form: Liquid *Color:* Colorless *Odor:* Gasoline
Other Information: n/a

Chemical Properties:

Specific Gravity: .64 *Vapor Density:* 2.4 *Boiling Point: 85° F(29.4° C)*
Melting Point: −265° F(−165° C) *Vapor Pressure:* n/a
Solubility in water: n/a
Other Information: Floats on water. Flammable vapor is produced.

Health Hazards:

Inhalation Hazard: Dizziness.
Ingestion Hazard: Harmful if swallowed.
Absorption Hazard: n/a
Hazards to Wildlife: n/a
Decontamination Procedures: Wash away any material with copious amounts of soap and water.
First Aid Procedures: Remove victim to fresh air, call emergency medical care. If not breathing give CPR. If breathing is difficult administer oxygen. Treat for shock.

Fire Hazards:

Flashpoint: −60° F(−51.1° C) *Ignition temperature:* 527° F(275° C)
Flammable Explosive High Range: 8.7 *Low Range:* 1.4
Toxic Products of Combustion: n/a
Other Hazards: Containers may explode in fire. Flashback along vapor trail may occur.
Possible extinguishing agents: Foam, dry chemical, carbon dioxide. Stop flow of vapor!! Water may be ineffective on fire. Cool all exposed containers with water. Apply water from as far a distance as possible.

Reactivity Hazards:

Reactive With: n/a *Other Reactions:* n/a

Corrosivity Hazards:

Corrosive With: n/a *Neutralizing Agent:* n/a

Radioactivity Hazards:

Radiation Emitted: n/a *Other Hazards:* n/a

Recommended Protection for Response Personnel:

Avoid breathing vapors, keep upwind. Structural protective clothing provides limited protection. Wash away any material which may have come into contact with the body with copious amounts of soap and water. If the fire becomes uncontrollable, consider appropriate evacuation.

PERACETIC ACID

DOT Number: NA 2131
DOT Hazard Class: Organic peroxide
DOT Guide Number: 51
Synonyms: peroxyacetic acid
STCC Number: 4919570
Reportable Qty: n/a
Mfg Name: FMC Corp.
Phone No: 1-312-861-5900

Physical Description:

Physical Form: Liquid *Color:* Colorless *Odor:* Acrid
Other Information: Used usually as a solvent.

Chemical Properties:

Specific Gravity: 1.5 *Vapor Density:* 2.6 *Boiling Point: 221° F(105° C)*
Melting Point: n/a *Vapor Pressure:* n/a *Solubility in water:* Yes
Other Information: It must be shipped in a solution not more than 40% in strength.

Health Hazards:

Inhalation Hazard: n/a
Ingestion Hazard: Harmful if swallowed.
Absorption Hazard: Irritating to the skin and eyes.
Hazards to Wildlife: n/a
Decontamination Procedures: Wash away any material with copious amounts of soap and water.
First Aid Procedures: Remove victim to fresh air, call emergency medical care. If not breathing give CPR. If breathing is difficult administer oxygen. Treat for shock.

Fire Hazards:

Flashpoint: 104° F(40° C) *Ignition temperature:* 392° F(200° C)
Flammable Explosive High Range: n/a *Low Range:* n/a
Toxic Products of Combustion: n/a
Other Hazards: May cause fire on contact with combustibles. Containers may explode in a fire.
Possible extinguishing agents: Apply water from as far a distance as possible. Use alcohol foam, dry chemical, or carbon dioxide.

Reactivity Hazards:

Reactive With: Organic materials such as wood, cotton, or straw. *Other Reactions:* n/a

Corrosivity Hazards:

Corrosive With: Metals including aluminum. *Neutralizing Agent:* n/a

Radioactivity Hazards:

Radiation Emitted: n/a *Other Hazards:* n/a

Recommended Protection for Response Personnel:

Avoid breathing vapors, keep upwind. Wear a sealed chemical suit (polycarbonate, butyl rubber). Wash away any material which may have come into contact with the body with copious amounts of soap and water. If the fire becomes uncontrollable, consider appropriate evacuation. under fire conditions, the cylinders may violently rupture/rocket.

PERCHLORETHYLENE

DOT Number: UN 1897 *DOT Hazard Class:* ORM-A *DOT Guide Number:* 74
Synonyms: perclene, perk, tetracap, tetrachloroethylene
STCC Number: 4940355 *Reportable Qty:* 1/.454
Mfg Name: Dow Chemical Co. *Phone No:* 1-513-636-4400

Physical Description:

Physical Form: Liquid *Color:* Colorless *Odor:* Sweet
Other Information: It is used in dry cleaning solvent, vapor degreasing solvent, drying agent for metals, and for the manufacture of other chemicals.

Chemical Properties:

Specific Gravity: 1.11 *Vapor Density:* 5.8 *Boiling Point: 250° F(121.1° C)*
Melting Point: −8° F(−22.2° C) *Vapor Pressure:* 14 mm Hg at 68° F(20° C) *Solubility in water:* No
Other Information: Insoluble in water. Vapors are heavier than air. Weighs 13.5 lbs/6.1 kg per gallon/3.8 l.

Health Hazards:

Inhalation Hazard: Difficulty in breathing, loss of consciousness.
Ingestion Hazard: Harmful if swallowed.
Absorption Hazard: Irritating to the skin and eyes.
Hazards to Wildlife: n/a
Decontamination Procedures: Wash away any material with copious amounts of soap and water.
First Aid Procedures: Remove victim to fresh air, call emergency medical care. If not breathing give CPR. If breathing is difficult administer oxygen. Treat for shock.

Fire Hazards:

Flashpoint: n/a *Ignition temperature:* n/a
Flammable Explosive High Range: n/a *Low Range:* n/a
Toxic Products of Combustion: Toxic, irritating gases may be generated in a fire.
Other Hazards: n/a
Possible extinguishing agents: Extinguish fire using suitable agent for the type of surrounding fire.

Reactivity Hazards:

Reactive With: n/a *Other Reactions:* n/a

Corrosivity Hazards:

Corrosive With: n/a *Neutralizing Agent:* n/a

Radioactivity Hazards:

Radiation Emitted: n/a *Other Hazards:* n/a

Recommended Protection for Response Personnel:

Avoid breathing vapors, keep upwind. Structural protective clothing provides limited protection. Wash away any material which may have come into contact with the body with copious amounts of soap and water. Consider appropriate evacuation.

PERCHLORIC ACID, less than 50%

DOT Number: UN 1802 *DOT Hazard Class:* Oxidizer *DOT Guide Number:* 47
Synonyms: dioxonium perchlorate solution, perchlorate acid solution
STCC Number: 4918522 *Reportable Qty:* n/a
Mfg Name: Mallinckrodt Inc. *Phone No:* 1-314-895-2000

Physical Description:

Physical Form: Liquid *Color:* Colorless *Odor:* Odorless
Other Information: n/a

Chemical Properties:

Specific Gravity: 1.6 *Vapor Density:* n/a *Boiling Point: Decomposes*
Melting Point: n/a *Vapor Pressure:* n/a *Solubility in water:* Yes
Other Information: Sinks and mixes in water.

Health Hazards:

Inhalation Hazard: Causes coughing, difficulty in breathing.
Ingestion Hazard: Causes nausea and vomiting.
Absorption Hazard: Will burn the skin and eyes.
Hazards to Wildlife: n/a
Decontamination Procedures: Wash away any material with copious amounts of soap and water.
First Aid Procedures: Remove victim to fresh air, call emergency medical care. If not breathing give CPR. If breathing is difficult administer oxygen. Treat for shock.

Fire Hazards:

Flashpoint: n/a *Ignition temperature:* n/a
Flammable Explosive High Range: n/a *Low Range:* n/a
Toxic Products of Combustion: Poisonous gases may be produced in a fire.
Other Hazards: Will increase the intensity of the fire. May cause fire on contact with combustibles. Containers may explode in a fire.
Possible extinguishing agents: Use water in flooding quantities as a fog. Apply water from as far a distance as possible.

Reactivity Hazards:

Reactive With: Combustible materials *Other Reactions:* n/a

Corrosivity Hazards:

Corrosive With: Metals and tissue *Neutralizing Agent:* Crushed limestone, soda ash, or lime.

Radioactivity Hazards:

Radiation Emitted: n/a *Other Hazards:* n/a

Recommended Protection for Response Personnel:

Avoid breathing vapors, keep upwind. Wear a sealed chemical suit (polycarbonate, butyl rubber, viton, neoprene). Wash away any material which may have come into contact with the body with copious amounts of soap and water. If the fire becomes uncontrollable, consider appropriate evacuation.

PERCHLORIC ACID, over 50%

DOT Number: UN 1873 *DOT Hazard Class:* Oxidizer *DOT Guide Number:* 47
Synonyms: dioxonium perchlorate solution, perchlorate acid solution
STCC Number: 4918523 *Reportable Qty:* n/a
Mfg Name: Mallinckrodt Inc. *Phone No:* 1-314-895-2000

Physical Description:

Physical Form: Liquid *Color:* Colorless *Odor:* Odorless
Other Information: n/a

Chemical Properties:

Specific Gravity: 1.6 *Vapor Density:* 3.5 *Boiling Point: Decomposes*
Melting Point: n/a *Vapor Pressure:* n/a *Solubility in water:* Yes
Other Information: Sinks and mixes in water.

Health Hazards:

Inhalation Hazard: Causes coughing, difficulty in breathing.
Ingestion Hazard: Causes nausea and vomiting.
Absorption Hazard: Will burn the skin and the eyes.
Hazards to Wildlife: n/a
Decontamination Procedures: Wash away any material with copious amounts of soap and water.
First Aid Procedures: Remove victim to fresh air, call emergency medical care. If not breathing give CPR. If breathing is difficult administer oxygen. Treat for shock.

Fire Hazards:

Flashpoint: n/a *Ignition temperature:* n/a
Flammable Explosive High Range: n/a *Low Range:* n/a
Toxic Products of Combustion: Poisonous gases may be produced in a fire.
Other Hazards: Will increase the intensity of the fire. May cause fire on contact with combustibles. Containers may explode in a fire.
Possible extinguishing agents: Use water in flooding quantities as a fog. Apply water from as far a distance as possible.

Reactivity Hazards:

Reactive With: Combustible materials *Other Reactions:* n/a

Corrosivity Hazards:

Corrosive With: Metals and tissue *Neutralizing Agent:* Crushed limestone, soda ash, or lime.

Radioactivity Hazards:

Radiation Emitted: n/a *Other Hazards:* n/a

Recommended Protection for Response Personnel:

Avoid breathing vapors, keep upwind. Wear a sealed chemical suit (polycarbonate, butyl rubber, viton, neoprene). Wash away any material which may have come into contact with the body with copious amounts of soap and water. If the fire becomes uncontrollable, consider appropriate evacuation.

PERCHLOROMETHYL MERCAPTAN

DOT Number: UN 1670 *DOT Hazard Class:* Poison B *DOT Guide Number:* 55
Synonyms: trichloromethyl sulfur chloride
STCC Number: 4921473 *Reportable Qty:* 100/454
Mfg Name: Stauffer Chemical Co. *Phone No:* 1-203-222-3000

Physical Description:

Physical Form: Liquid *Color:* Yellow to orange red *Odor:* Strong unpleasant
Other Information: n/a

Chemical Properties:

Specific Gravity: 1.7 *Vapor Density:* 6.4 *Boiling Point: 300° F(148.8° C)*
Melting Point: n/a *Vapor Pressure:* 65 mm Hg at 68° F(20° C) *Solubility in water:* Insoluble
Other Information: Insoluble and sinks in water.

Health Hazards:

Inhalation Hazard: Poisonous if inhaled.
Ingestion Hazard: Poisonous if swallowed.
Absorption Hazard: Poisonous if absorbed.
Hazards to Wildlife: n/a
Decontamination Procedures: Wash away any material with copious amounts of soap and water.
First Aid Procedures: Remove victim to fresh air, call emergency medical care. If not breathing give CPR. If breathing is difficult administer oxygen. Treat for shock.

Fire Hazards:

Flashpoint: n/a *Ignition temperature:* n/a
Flammable Explosive High Range: n/a *Low Range:* n/a
Toxic Products of Combustion: Poisonous gases may be produced when heated.
Other Hazards: n/a
Possible extinguishing agents: Apply water from as far a distance as possible. Use foam, dry chemical, or carbon dioxide.

Reactivity Hazards:

Reactive With: Water to give off carbon dioxide, hydrochloric acid, and sulfur.
Other Reactions: Steel and iron to give off carbon tetrachloride.

Corrosivity Hazards:

Corrosive With: Most metals *Neutralizing Agent:* Crushed limestone, soda ash, or lime.

Radioactivity Hazards:

Radiation Emitted: n/a *Other Hazards:* n/a

Recommended Protection for Response Personnel:

Avoid breathing vapors, keep upwind. Wear a sealed chemical suit (butyl rubber). Wash away any material which may have come into contact with the body with copious amounts of soap and water. Consider appropriate evacuation.

PETROLATUM

DOT Number: n/a *DOT Hazard Class:* n/a *DOT Guide Number:* n/a
Synonyms: petrolatum jelly, petroleum jelly, Vasoline, yellow petrolatum
STCC Number: n/a *Reportable Qty:* n/a
Mfg Name: Standard Oil Co. *Phone No:* 1-216-575-4141

Physical Description:

Physical Form: Liquid *Color:* Dark brown, green, amber or white *Odor:* n/a
Other Information: n/a

Chemical Properties:

Specific Gravity: .86 *Vapor Density:* n/a *Boiling Point: Very high*
Melting Point: 100-135° F(37.7-57.2° C) *Vapor Pressure:* n/a
Solubility in water: No
Other Information: Floats on water.

Health Hazards:

Inhalation Hazard: n/a
Ingestion Hazard: n/a
Absorption Hazard: Irritating to the eyes.
Hazards to Wildlife: n/a
Decontamination Procedures: Wash away any material with copious amounts of soap and water.
First Aid Procedures: Remove victim to fresh air, call emergency medical care. If not breathing give CPR. If breathing is difficult administer oxygen. Treat for shock.

Fire Hazards:

Flashpoint: 360-430° F(182-221° C) *Ignition temperature:* n/a
Flammable Explosive High Range: n/a *Low Range:* n/a
Toxic Products of Combustion: n/a
Other Hazards: n/a
Possible extinguishing agents: Extinguish fire using suitable agent for the type of surrounding fire. Cool all exposed containers with water.

Reactivity Hazards:

Reactive With: n/a *Other Reactions:* n/a

Corrosivity Hazards:

Corrosive With: n/a *Neutralizing Agent:* n/a

Radioactivity Hazards:

Radiation Emitted: n/a *Other Hazards:* n/a

Recommended Protection for Response Personnel:

Avoid breathing vapors, keep upwind. Wear a sealed chemical suit (polycarbonate, butyl rubber). Wash away any material which may have come into contact with the body with copious amounts of soap and water. Consider appropriate evacuation.

PETROLEUM NAPHTHA

DOT Number: UN 1255 *DOT Hazard Class:* Flammable liquid *DOT Guide Number:* 27
Synonyms: petroleum solvent
STCC Number: 4910259 *Reportable Qty:* n/a
Mfg Name: Crowley Chemical Co. *Phone No:* 1-212-682-1200

Physical Description:

Physical Form: Liquid *Color:* Colorless *Odor:* Gasoline
Other Information: It is used in making cleaning preparations, in solvents, and as a raw material for making other chemicals.

Chemical Properties:

Specific Gravity: .6 *Vapor Density:* 2.5 *Boiling Point: 95-140° F(35-60° C)*
Melting Point: n/a *Vapor Pressure:* 40 mm Hg at 68° F(20° C)
Solubility in water: No *Degree of Solubility:* .04%
Other Information: Floats on water. Flammable vapor is produced.

Health Hazards:

Inhalation Hazard: n/a
Ingestion Hazard: Harmful if swallowed.
Absorption Hazard: n/a
Hazards to Wildlife: n/a
Decontamination Procedures: Wash away any material with copious amounts of soap and water.
First Aid Procedures: Remove victim to fresh air, call emergency medical care. If not breathing give CPR. If breathing is difficult administer oxygen. Treat for shock.

Fire Hazards:

Flashpoint: 0° F(−17.7° C) *Ignition temperature:* 550° F(287.7° C)
Flammable Explosive High Range: 5.9 *Low Range:* 1.1
Toxic Products of Combustion: n/a
Other Hazards: Flashback along vapor trail may occur. Vapors may explode if ignited in an enclosed area.
Possible extinguishing agents: Do not extinguish the fire unless the flow can be stopped. Apply water from as far a distance as possible. Use foam, dry chemical, or carbon dioxide.

Reactivity Hazards:

Reactive With: n/a *Other Reactions:* n/a

Corrosivity Hazards:

Corrosive With: n/a *Neutralizing Agent:* n/a

Radioactivity Hazards:

Radiation Emitted: n/a *Other Hazards:* n/a

Recommended Protection for Response Personnel:

Avoid breathing vapors, keep upwind. Structural protective clothing provides limited protection. Wash away any material which may have come into contact with the body with copious amounts of soap and water. If the fire becomes uncontrollable, consider appropriate evacuation.

PHENOL

DOT Number: UN 1671 *DOT Hazard Class:* Poison B *DOT Guide Number:* 55
Synonyms: carbolic acid, phenic acid
STCC Number: 4921220 *Reportable Qty:* 1000/454
Mfg Name: Allied Corp. *Phone No:* 1-212-455-2000

Physical Description:

Physical Form: Solid or liquid *Color:* Colorless to pink or red *Odor:* Sweet tarry
Other Information: It is used to make plastics, adhesives, and other chemicals. Weighs 9.9 lbs/4.4 kg per gallon/3.8 l.

Chemical Properties:

Specific Gravity: 1.1 *Vapor Density:* 3.2 *Boiling Point: 358° F(181.1° C)*
Melting Point: 106° F(41.1° C) *Vapor Pressure:* .36 mm Hg at 68° F(20° C)
Solubility in water: Yes *Degree of Solubility:* 8.4%
Other Information: Soluble in water, vapors are heavier than air.

Health Hazards:

Inhalation Hazard: n/a
Ingestion Hazard: Poisonous if swallowed.
Absorption Hazard: Will burn or numb skin and the eyes.
Hazards to Wildlife: Dangerous to aquatic life.
Decontamination Procedures: Wash away any material with copious amounts of soap and water.
First Aid Procedures: Remove victim to fresh air, call emergency medical care. If not breathing give CPR. If breathing is difficult administer oxygen. Treat for shock.

Fire Hazards:

Flashpoint: 175° F(79.4° C) *Ignition temperature:* 1319° F(715° C)
Flammable Explosive High Range: 8.6 *Low Range:* 1.8
Toxic Products of Combustion: Poisonous gases are produced in fire.
Other Hazards: Avoid contact with liquid and solid.
Possible extinguishing agents: Apply water from as far a distance as possible. Use foam, dry chemical, or carbon dioxide.

Reactivity Hazards:

Reactive With: n/a *Other Reactions:* n/a

Corrosivity Hazards:

Corrosive With: Lead and alloys, certain plastics and rubbers.
Neutralizing Agent: Crushed limestone, soda ash, or lime.

Radioactivity Hazards:

Radiation Emitted: n/a *Other Hazards:* n/a

Recommended Protection for Response Personnel:

Avoid breathing vapors, keep upwind. Structural protective clothing provides limited protection. Wash away any material which may have come into contact with the body with copious amounts of soap and water. Consider appropriate evacuation.

PHENYLDICHLOROARSINE

DOT Number: NA 1556 *DOT Hazard Class:* Poison B *DOT Guide Number:* 55
Synonyms: phenylarsenic dichloride
STCC Number: 4924474 *Reportable Qty:* 1/.454
Mfg Name: Ventron Corp. *Phone No:* 1-617-774-3100

Physical Description:

Physical Form: Liquid *Color:* Colorless to yellow *Odor:* Weak unpleasant
Other Information: n/a

Chemical Properties:

Specific Gravity: 1.66 *Vapor Density:* 7.7 *Boiling Point: 495° F(257.2° C)*
Melting Point: n/a *Vapor Pressure:* .021 mm Hg at 68° F(20° C) *Solubility in water:* n/a
Other Information: Sinks in water.

Health Hazards:

Inhalation Hazard: n/a
Ingestion Hazard: Poisonous if swallowed.
Absorption Hazard: Will burn the skin and the eyes.
Hazards to Wildlife: n/a
Decontamination Procedures: Wash away any material with copious amounts of soap and water.
First Aid Procedures: Remove victim to fresh air, call emergency medical care. If not breathing give CPR. If breathing is difficult administer oxygen. Treat for shock.

Fire Hazards:

Flashpoint: n/a *Ignition temperature:* n/a
Flammable Explosive High Range: n/a *Low Range:* n/a
Toxic Products of Combustion: Highly toxic arsenic fumes are formed when hot.
Other Hazards: n/a
Possible extinguishing agents: Extinguish fire using suitable agent for the type of surrounding fire. Use alcohol foam, dry chemical, or carbon dioxide.

Reactivity Hazards:

Reactive With: Water to from hydrochloric acid. *Other Reactions:* n/a

Corrosivity Hazards:

Corrosive With: Metals in the presence of moisture.
Neutralizing Agent: Crushed limestone, soda ash, or lime.

Radioactivity Hazards:

Radiation Emitted: n/a *Other Hazards:* n/a

Recommended Protection for Response Personnel:

Avoid breathing vapors, keep upwind. Wear a sealed chemical suit (butyl rubber). Wash away any material which may have come into contact with the body with copious amounts of soap and water. Consider appropriate evacuation dependent upon the amount of material spilled.

PHENYLHYDRAZINE HYDROCHLORIDE

DOT Number: UN 2572 *DOT Hazard Class:* Poison B *DOT Guide Number:* 53
Synonyms: phenylhydrazinium chloride
STCC Number: n/a *Reportable Qty:* n/a
Mfg Name: Eastern Chemical *Phone No:* 1-516-273-0900

Physical Description:

Physical Form: Solid *Color:* White to tan *Odor:* Weak
Other Information: n/a

Chemical Properties:

Specific Gravity: 1 *Vapor Density:* 3.7 *Boiling Point: Decomposes*
Melting Point: 469° F(242.7° C) *Vapor Pressure:* n/a *Solubility in water:* Yes
Other Information: Sinks and mixes with water.

Health Hazards:

Inhalation Hazard: Coughing, difficulty in breathing.
Ingestion Hazard: Poisonous if swallowed.
Absorption Hazard: Irritating to the eyes.
Hazards to Wildlife: n/a
Decontamination Procedures: Wash away any material with copious amounts of soap and water.
First Aid Procedures: Remove victim to fresh air, call emergency medical care. If not breathing give CPR. If breathing is difficult administer oxygen. Treat for shock.

Fire Hazards:

Flashpoint: Combustible solid *Ignition temperature:* n/a
Flammable Explosive High Range: n/a *Low Range:* n/a
Toxic Products of Combustion: Toxic and irritating oxides of nitrogen and hydrogen chloride may be formed in fire.
Other Hazards: The solid may sublime without melting and deposit on cool surfaces.
Possible extinguishing agents: Extinguish fire using suitable agent for the type of surrounding fire.

Reactivity Hazards:

Reactive With: n/a *Other Reactions:* n/a

Corrosivity Hazards:

Corrosive With: May be corrosive to metals.
Neutralizing Agent: Crushed limestone, soda ash, or lime.

Radioactivity Hazards:

Radiation Emitted: n/a *Other Hazards:* n/a

Recommended Protection for Response Personnel:

Avoid breathing vapors, keep upwind. Wear a sealed chemical suit (butyl rubber). Wash away any material which may have come into contact with the body with copious amounts of soap and water. Consider appropriate evacuation.

PHOSDRIN

DOT Number: UN 2783 *DOT Hazard Class:* Poison B *DOT Guide Number:* 55
Synonyms: Menite, Pevinphos, Phosfene
STCC Number: n/a *Reportable Qty:* n/a
Mfg Name: Shell Chemical Co. *Phone No:* 1-713-241-6161

Physical Description:

Physical Form: Liquid *Color:* Yellow to orange *Odor:* Mild to none
Other Information: n/a

Chemical Properties:

Specific Gravity: 1.25 *Vapor Density:* 7.7 *Boiling Point: 617° F(325° C)*
Melting Point: −89° F(−67.2° C) *Vapor Pressure:* n/a *Solubility in water:* Yes
Other Information: Sinks and mixes with water.

Health Hazards:

Inhalation Hazard: Poisonous if inhaled.
Ingestion Hazard: Poisonous if swallowed.
Absorption Hazard: Poisonous to the skin and eyes.
Hazards to Wildlife: Dangerous to aquatic life.
Decontamination Procedures: Wash away any material with copious amounts of soap and water.
First Aid Procedures: Remove victim to fresh air, call emergency medical care. If not breathing give CPR. If breathing is difficult administer oxygen. Treat for shock.

Fire Hazards:

Flashpoint: 175° F(79.4° C) *Ignition temperature:* n/a
Flammable Explosive High Range: n/a *Low Range:* n/a
Toxic Products of Combustion: Poisonous gases are produced in fire and when heated.
Other Hazards: Emits highly toxic fumes!!
Possible extinguishing agents: Extinguish fire using suitable agent for the type of surrounding fire. Cool all affected containers with flooding quantities of water. Apply water from as far a distance as possible.

Reactivity Hazards:

Reactive With: Hydrolyzes with water rapidly. *Other Reactions:* n/a

Corrosivity Hazards:

Corrosive With: Metals. *Neutralizing Agent:* Crushed limestone, soda ash, or lime.

Radioactivity Hazards:

Radiation Emitted: n/a *Other Hazards:* n/a

Recommended Protection for Response Personnel:

Avoid breathing vapors, keep upwind. Wear a sealed chemical suit (polycarbonate, butyl rubber). Wash away any material which may have come into contact with the body with copious amounts of soap and water. Consider appropriate evacuation.

PHOSGENE

DOT Number: UN 1076 *DOT Hazard Class:* Poison A *DOT Guide Number:* 15
Synonyms: carbonyl chloride
STCC Number: 4920540 *Reportable Qty:* 10/454
Mfg Name: Scientific Gas Products *Phone No:* 1-201-754-7700

Physical Description:

Physical Form: Gas *Color:* Colorless gas to light yellow liquid *Odor:* Sweet or sharp
Other Information: It is shipped as a liquid or liquefied gas in cylinders or in tontype containers (tank cars). These cylinders or tank cars may not have relief valves.

Chemical Properties:

Specific Gravity: .57 *Vapor Density:* 1.17 *Boiling Point: − 126° F(−87.7° C)*
Melting Point: n/a *Vapor Pressure:* 1180 mm Hg at 68° F(20° C) *Solubility in water:* Reacts
Other Information: Weighs 11.6 lbs/5.2 kg per gallon/3.8 l.

Health Hazards:

Inhalation Hazard: Poisonous if inhaled.
Ingestion Hazard: n/a
Absorption Hazard: n/a
Hazards to Wildlife: n/a
Decontamination Procedures: Wash away any material with copious amounts of soap and water.
First Aid Procedures: Remove victim to fresh air, call emergency medical care. If not breathing give CPR. If breathing is difficult administer oxygen. Treat for shock.

Fire Hazards:

Flashpoint: n/a *Ignition temperature:* 212° F(100° C)
Flammable Explosive High Range: n/a *Low Range:* n/a
Toxic Products of Combustion: Poisonous gases are produced when heated.
Other Hazards: Prolonged exposure to heat will cause the cylinders or tank cars to violently rupture/rocket. (BLEVE)
Possible extinguishing agents: Apply water from as far a distance as possible. Use alcohol foam, dry chemical, or carbon dioxide.

Reactivity Hazards:

Reactive With: n/a *Other Reactions:* n/a

Corrosivity Hazards:

Corrosive With: n/a *Neutralizing Agent:* n/a

Radioactivity Hazards:

Radiation Emitted: n/a *Other Hazards:* n/a

Recommended Protection for Response Personnel:

Avoid breathing vapors, keep upwind. Wear a sealed chemical suit (polycarbonate, butyl rubber). Wash away any material which may have come into contact with the body with copious amounts of soap and water. Consider appropriate evacuation.

PHOSPHORIC ACID

DOT Number: UN 1805 *DOT Hazard Class:* Corrosive *DOT Guide Number:* 60
Synonyms: ortho phosphoric
STCC Number: 4930248 *Reportable Qty:* 5000/2270
Mfg Name: Allied Corp. *Phone No:* 1-201-455-2000

Physical Description:

Physical Form: Liquid *Color:* Colorless *Odor:* Odorless
Other Information: It is used in making fertilizers, detergents, food processing, and many other uses.

Chemical Properties:

Specific Gravity: 1.89 *Vapor Density:* 3.4 *Boiling Point: 500° F(260° C)*
Melting Point: 70° F(21.1° C) *Vapor Pressure:* .03 mm Hg at 68° F(20° C) *Solubility in water:* Slight
Other Information: Soluble in water with releases of heat. Weighs 15.6 lbs/7.0 kg per gallon/3.8 l.

Health Hazards:

Inhalation Hazard: n/a
Ingestion Hazard: Nausea, vomiting, loss of consciousness.
Absorption Hazard: Will burn the skin and the eyes.
Hazards to Wildlife: Dangerous to aquatic life.
Decontamination Procedures: Wash away any material with copious amounts of soap and water.
First Aid Procedures: Remove victim to fresh air, call emergency medical care. If not breathing give CPR. If breathing is difficult administer oxygen. Treat for shock.

Fire Hazards:

Flashpoint: n/a *Ignition temperature:* 212° F(100° C)
Flammable Explosive High Range: n/a *Low Range:* n/a
Toxic Products of Combustion: n/a
Other Hazards: Avoid contact with liquid.
Possible extinguishing agents: Apply water from as far a distance as possible. extinguish fire using suitable agent for type of surrounding fire.

Reactivity Hazards:

Reactive With: Water with liberation of heat.
Other Reactions: Metals to form flammable hydrogen gas.

Corrosivity Hazards:

Corrosive With: Metals and tissue *Neutralizing Agent:* Crushed limestone, soda ash, or lime.

Radioactivity Hazards:

Radiation Emitted: n/a *Other Hazards:* n/a

Recommended Protection for Response Personnel:

Avoid breathing vapors, keep upwind. Wear a sealed chemical suit (butyl rubber, PVC, chlorinated polyethylene, neoprene, viton). Wash away any material which may have come into contact with the body with copious amounts of soap and water. Consider appropriate evacuation.

PHOSPHORIC ACID TRIETHYLENEIMINE

DOT Number: UN 2501 *DOT Hazard Class:* Corrosive *DOT Guide Number:* 55
Synonyms: APO, triethylenephosphoramide
STCC Number: 4999953 *Reportable Qty:* n/a *CHEMTREC Phone No:* 1-800-424-9300

Physical Description:

Physical Form: Solid *Color:* White *Odor:* n/a
Other Information: n/a

Chemical Properties:

Specific Gravity: 1 *Vapor Density:* 6 *Boiling Point: Decomposes*
Melting Point: n/a *Vapor Pressure:* n/a *Solubility in water:* Yes
Other Information: n/a

Health Hazards:

Inhalation Hazard: n/a
Ingestion Hazard: Harmful if swallowed.
Absorption Hazard: Irritating to the skin and eyes.
Hazards to Wildlife: Dangerous to waterfowl.
Decontamination Procedures: Wash away any material with copious amounts of soap and water.
First Aid Procedures: Remove victim to fresh air, call emergency medical care. If not breathing give CPR. If breathing is difficult administer oxygen. Treat for shock.

Fire Hazards:

Flashpoint: n/a *Ignition temperature:* n/a
Flammable Explosive High Range: n/a *Low Range:* n/a
Toxic Products of Combustion: Toxic oxides of nitrogen are produced.
Other Hazards: Avoid breathing dust or vapors.
Possible extinguishing agents: Apply water from as far a distance as possible. Use alcohol foam, dry chemical, or carbon dioxide.

Reactivity Hazards:

Reactive With: Acids and strong caustics. *Other Reactions:* n/a

Corrosivity Hazards:

Corrosive With: Metals and tissue. *Neutralizing Agent:* n/a

Radioactivity Hazards:

Radiation Emitted: n/a *Other Hazards:* n/a

Recommended Protection for Response Personnel:

Avoid breathing vapors, keep upwind. Wear appropriate chemical clothing. Wash away any material which may have come into contact with the body with copious amounts of soap and water. Consider appropriate evacuation.

PHOSPHORUS-BLACK

DOT Number: n/a *DOT Hazard Class:* n/a *DOT Guide Number:* n/a
Synonyms: menite, pevinphos, phosfene
STCC Number: n/a *Reportable Qty:* n/a
Mfg Name: Atomergic Chemetals Corp. *Phone No:* 1-516-349-8800

Physical Description:

Physical Form: Solid *Color:* Black *Odor:* n/a
Other Information: n/a

Chemical Properties:

Specific Gravity: 2.7 *Vapor Density:* 4.27 *Boiling Point: 889° F(476.1° C)*
Melting Point: n/a *Vapor Pressure:* n/a *Solubility in water:* No
Other Information: Sinks in water.

Health Hazards:

No medical data available, contact the Poison Control Center Team at 1-800-732-2200
Absorption Hazard: n/a
Hazards to Wildlife: n/a
Decontamination Procedures: Wash away any material with copious amounts of soap and water.
First Aid Procedures: Remove victim to fresh air, call emergency medical care. If not breathing give CPR. If breathing is difficult administer oxygen. Treat for shock.

Fire Hazards:

Flashpoint: n/a *Ignition temperature:* 752° F(400° C)
Flammable Explosive High Range: n/a *Low Range:* n/a
Toxic Products of Combustion: n/a
Other Hazards: n/a
Possible *Possible extinguishing agents:* n/a

Reactivity Hazards:

Reactive With: n/a *Other Reactions:* n/a

Corrosivity Hazards:

Corrosive With: n/a *Neutralizing Agent:* n/a

Radioactivity Hazards:

Radiation Emitted: n/a *Other Hazards:* n/a

Recommended Protection for Response Personnel:

Avoid breathing vapors, keep upwind. Wear appropriate sealed chemical suit. Consider appropriate evacuation. Use extreme care and caution with this material!!

PHOSPHORUS OXYCHLORIDE

DOT Number: UN 1810 *DOT Hazard Class:* Corrosive *DOT Guide Number:* 39
Synonyms: phosphoryl chloride
STCC Number: 4932352 *Reportable Qty:* 1000/454
Mfg Name: Monsanto Chemical Co. *Phone No:* 1-314-694-1000

Physical Description:

Physical Form: Liquid *Color:* Colorless to light yellow *Odor:* Musty
Other Information: It is used in gasoline additives and hydraulic fluids.

Chemical Properties:

Specific Gravity: 1.67 *Vapor Density:* 5.3 *Boiling Point: 225° F(107.2° C)*
Melting Point: n/a *Vapor Pressure:* 40 mm Hg at 81° F(27.2° C) *Solubility in water:* Reacts.
Other Information: Fumes in air, sinks and reacts with water. Poisonous gas is produced.

Health Hazards:

Inhalation Hazard: Harmful if inhaled.
Ingestion Hazard: Harmful if swallowed.
Absorption Hazard: Will burn the skin and the eyes.
Hazards to Wildlife: n/a
Decontamination Procedures: Wash away any material with copious amounts of soap and water.
First Aid Procedures: Remove victim to fresh air, call emergency medical care. If not breathing give CPR. If breathing is difficult administer oxygen. Treat for shock.

Fire Hazards:

Flashpoint: n/a *Ignition temperature:* n/a
Flammable Explosive High Range: n/a *Low Range:* n/a
Toxic Products of Combustion: Poisonous, irritating, corrosive gases are generated when heated or in contact with water.
Other Hazards: n/a
Possible extinguishing agents: Do not use water. Use dry chemical, dry sand, or carbon dioxide.

Reactivity Hazards:

Reactive With: Water with vigorous reaction with evolution of hydrochloric fumes. *Other Reactions:* n/a

Corrosivity Hazards:

Corrosive With: Metals except nickel and lead.
Neutralizing Agent: Crushed limestone, soda ash, or lime.

Radioactivity Hazards:

Radiation Emitted: n/a *Other Hazards:* n/a

Recommended Protection for Response Personnel:

Avoid breathing vapors, keep upwind. Wear a sealed chemical suit (polycarbonate, butyl rubber). Wash away any material which may have come into contact with the body with copious amounts of soap and water. Consider appropriate evacuation.

PHOSPHORUS PENTASULFIDE

DOT Number: UN 1340 *DOT Hazard Class:* Flammable solid *DOT Guide Number:* 41
Synonyms: phosphoric sulfide
STCC Number: 4916320 *Reportable Qty:* 100/45.4
Mfg Name: Monsanto Chemical Co. *Phone No:* 1-314-694-1000

Physical Description:

Physical Form: Solid *Color:* Yellow to green *Odor:* Odorless or rotten egg
Other Information: It is used for making lube oil additives, insecticides, flotation agents, safety matches, blown asphalt, and other products and chemicals. Weighs 127 lbs. per cubic foot.

Chemical Properties:

Specific Gravity: 2.03 *Vapor Density:* n/a *Boiling Point: 955° F(512.7° C)*
Melting Point: 527° F(275° C) *Vapor Pressure:* 1 mm Hg at 572° F(300° C) *Solubility in water:* Reacts
Other Information: Heavier than water. Will slowly react with water or moist air to form toxic hydrogen sulfide. Gas and corrosive phosphoric acid.

Health Hazards:

Inhalation Hazard: n/a
Ingestion Hazard: Harmful if swallowed.
Absorption Hazard: Irritating to the skin and eyes.
Hazards to Wildlife: Dangerous to aquatic life.
Decontamination Procedures: Wash away any material with copious amounts of soap and water.
First Aid Procedures: Remove victim to fresh air, call emergency medical care. If not breathing give CPR. If breathing is difficult administer oxygen. Treat for shock.

Fire Hazards:

Flashpoint: n/a *Ignition temperature:* 527° F(275° C)
Flammable Explosive High Range: n/a *Low Range:* n/a
Toxic Products of Combustion: Sulfur dioxide, phosphorus pentoxide which vapors are toxic, irritating, and corrosive.
Other Hazards: Placarded with "Apply No Water" symbol.
Possible extinguishing agents: Use sand and carbon dioxide. Do not use water.

Reactivity Hazards:

Reactive With: Water or atmospheric moisture to liberate toxic hydrogen sulfide gas.
Other Reactions: n/a

Corrosivity Hazards:

Corrosive With: When wet, will attack most metals to form flammable hydrogen gas.
Neutralizing Agent: Crushed limestone, soda ash, or lime.

Radioactivity Hazards:

Radiation Emitted: n/a *Other Hazards:* n/a

Recommended Protection for Response Personnel:

Avoid breathing vapors, keep upwind. Wear a sealed chemical suit (polycarbonate, butyl rubber). Wash away any material which may have come into contact with the body with copious amounts of soap and water. Consider appropriate evacuation.

PHOSPHORUS-RED

DOT Number: UN 1338 *DOT Hazard Class:* Flammable solid *DOT Guide Number:* 32
Synonyms: amorphous phosphorus
STCC Number: n/a *Reportable Qty:* n/a
Mfg Name: Monsanto Chemical Co. *Phone No:* 1-314-694-1000

Physical Description:

Physical Form: Solid *Color:* Reddish brown *Odor:* Odorless
Other Information: n/a

Chemical Properties:

Specific Gravity: 2.2 *Vapor Density:* 4.8 *Boiling Point: Catches fire*
Melting Point: n/a *Vapor Pressure:* n/a *Solubility in water:* n/a
Other Information: Sinks in water.

Health Hazards:

Inhalation Hazard: n/a
Ingestion Hazard: Harmful if swallowed.
Absorption Hazard: Will burn the eyes.
Hazards to Wildlife: Dangerous to aquatic life.
Decontamination Procedures: Wash away any material with copious amounts of soap and water.
First Aid Procedures: Remove victim to fresh air, call emergency medical care. If not breathing give CPR. If breathing is difficult administer oxygen. Treat for shock.

Fire Hazards:

Flashpoint: Flammable solid *Ignition temperature:* 395° F(201.6° C)
Flammable Explosive High Range: n/a *Low Range:* n/a
Toxic Products of Combustion: Heat may cause reversion to yellow phosphorus which is toxic and spontaneously flammable upon contact with air. Burning yields toxic oxides of phosphorus.
Other Hazards: n/a
Possible extinguishing agents: Flood area with water. Cool exposed containers with water. Continue cooling after fire has been extinguished.

Reactivity Hazards:

Reactive With: Oxidizing agents *Other Reactions:* Strong alkaline hydroxides

Corrosivity Hazards:

Corrosive With: n/a *Neutralizing Agent:* n/a

Radioactivity Hazards:

Radiation Emitted: n/a *Other Hazards:* n/a

Recommended Protection for Response Personnel:

Avoid breathing vapors, keep upwind. Wear appropriate sealed chemical suit. Wash away any material which may have come into contact with the body with copious amounts of soap and water. If the fire becomes uncontrollable, consider appropriate evacuation.

PHOSPHORUS TRIBROMIDE

DOT Number: UN 1808 *DOT Hazard Class:* Corrosive *DOT Guide Number:* 39
Synonyms: phosphorus bromide
STCC Number: 4932358 *Reportable Qty:* n/a
Mfg Name: A.D. MacKay Inc. *Phone No:* 1-203-655-3000

Physical Description:

Physical Form: Liquid *Color:* Colorless to pale yellow *Odor:* Sharp penetrating
Other Information: n/a

Chemical Properties:

Specific Gravity: 2.86 *Vapor Density:* 9.3 *Boiling Point: 343° F(172.7° C)*
Melting Point: n/a *Vapor Pressure:* 10 mm Hg at 118° F(47.7° C) *Solubility in water:* Yes
Other Information: Sinks and mixes violently in water.

Health Hazards:

Inhalation Hazard: n/a
Ingestion Hazard: Causes nausea.
Absorption Hazard: Will burn the skin and the eyes.
Hazards to Wildlife: n/a
Decontamination Procedures: Wash away any material with copious amounts of soap and water.
First Aid Procedures: Remove victim to fresh air, call emergency medical care. If not breathing give CPR. If breathing is difficult administer oxygen. Treat for shock.

Fire Hazards:

Flashpoint: n/a *Ignition temperature:* n/a
Flammable Explosive High Range: n/a *Low Range:* n/a
Toxic Products of Combustion: Irritating hydrogen bromide and phosphoric acid gases may be formed in a fire.
Other Hazards: n/a
Possible extinguishing agents: Do not use water. Use dry chemical, dry sand, or carbon dioxide.

Reactivity Hazards:

Reactive With: Violently reacts with water evolving hydrogen bromide, an irritating and corrosive gas.
Other Reactions: n/a

Corrosivity Hazards:

Corrosive With: Metals and tissue *Neutralizing Agent:* Crushed limestone, soda ash, or lime.

Radioactivity Hazards:

Radiation Emitted: n/a *Other Hazards:* n/a

Recommended Protection for Response Personnel:

Avoid breathing vapors, keep upwind. Wear appropriate chemical clothing. Wash away any material which may have come into contact with the body with copious amounts of soap and water. Consider appropriate evacuation.

PHOSPHORUS TRICHLORIDE

DOT Number: UN 1809 *DOT Hazard Class:* Corrosive solid *DOT Guide Number:* 39
Synonyms: phosphorus chloride
STCC Number: 4932359 *Reportable Qty:* 1000/454
Mfg Name: Monsanto Chemical Co. *Phone No:* 1-314-694-1000

Physical Description:

Physical Form: Liquid *Color:* Colorless to slight yellow *Odor:* Sharp irritating
Other Information: It is used during electro-deposition of metal on rubber, and for making pesticides, gasoline additives, germicides, medical products, and other chemicals.

Chemical Properties:

Specific Gravity: 1.57 *Vapor Density:* 4.7 *Boiling Point: 169° F(76.1° C)*
Melting Point: –169° F(–111.6° C) *Vapor Pressure:* 100 mm Hg at 68° F(20° C) *Solubility in water:* Reacts
Other Information: Avoid contact with liquid and vapor.

Health Hazards:

Inhalation Hazard: Harmful if inhaled.
Ingestion Hazard: Poisonous if swallowed.
Absorption Hazard: Will burn the skin and the eyes.
Hazards to Wildlife: n/a
Decontamination Procedures: Wash away any material with copious amounts of soap and water.
First Aid Procedures: Remove victim to fresh air, call emergency medical care. If not breathing give CPR. If breathing is difficult administer oxygen. Treat for shock.

Fire Hazards:

Flashpoint: n/a *Ignition temperature:* n/a
Flammable Explosive High Range: n/a *Low Range:* n/a
Toxic Products of Combustion: n/a
Other Hazards: It is a strong oxidizer which may ignite other combustible materials upon contact.
Possible extinguishing agents: Dry chemical, dry sand or carbon dioxide. Do not use water.

Reactivity Hazards:

Reactive With: Water which may cause flashes of fire and hydrochloric acid fumes. *Other Reactions:* n/a

Corrosivity Hazards:

Corrosive With: Common construction materials and water to form hydrochloric acid which reacts with metals to form flammable hydrogen gas. *Neutralizing Agent:* Crushed limestone, soda ash, or lime.

Radioactivity Hazards:

Radiation Emitted: n/a *Other Hazards:* n/a

Recommended Protection for Response Personnel:

Avoid breathing vapors, keep upwind. Wear a sealed chemical suit (polycarbonate, butyl rubber, viton, PVC, chlorinated polyethylene), Wash away any material which may have come into contact with the body with copious amounts of soap and water. Consider appropriate evacuation.

PHOSPHORUS-WHITE

DOT Number: UN 1381 *DOT Hazard Class:* Flammable solid *DOT Guide Number:* n/a
Synonyms: yellow phosphorus, Willie P
STCC Number: 4916140 *Reportable Qty:* 1/.454
Mfg Name: Monsanto Chemical Co. *Phone No:* 1-314-694-1000

Physical Description:

Physical Form: Solid *Color:* White or yellow *Odor:* Sharp, garlic
Other Information: It is used in munitions, pyrotechnics, explosives, smoke bombs, artificial fertilizers, rat poisons, and in other products and chemicals.

Chemical Properties:

Specific Gravity: 1.82 *Vapor Density:* 4.4 *Boiling Point: 535.5° F(279.4° C)*
Melting Point: n/a *Vapor Pressure:* 1 mm Hg at 76° F(24.4° C) *Solubility in water:* n/a
Other Information: Fumes and burns in air. Sinks in water.

Health Hazards:

Inhalation Hazard: n/a
Ingestion Hazard: Causes nausea, vomiting, loss of consciousness.
Absorption Hazard: Will burn the skin and the eyes.
Hazards to Wildlife: Dangerous to aquatic life.
Decontamination Procedures: Wash away any material with copious amounts of soap and water.
First Aid Procedures: Remove victim to fresh air, call emergency medical care. If not breathing give CPR. If breathing is difficult administer oxygen. Treat for shock.

Fire Hazards:

Flashpoint: Spontaneously ignites *Ignition temperature:* 86° F(30° C)
Flammable Explosive High Range: n/a *Low Range:* n/a
Toxic Products of Combustion: Fumes from burning phosphorus are highly irritating.
Other Hazards: May ignite when in contact with air. Dense white smoke is formed.
Possible extinguishing agents: Flood with water. When the fire is out, cover all suspected material with wet sand until material can be permanently disposed of.

Reactivity Hazards:

Reactive With: Air *Other Reactions:* n/a

Corrosivity Hazards:

Corrosive With: n/a *Neutralizing Agent:* n/a

Radioactivity Hazards:

Radiation Emitted: n/a *Other Hazards:* n/a

Recommended Protection for Response Personnel:

Avoid breathing vapors, keep upwind. Wear appropriate chemical clothing. Wash away any material which may come into contact with the body with copious amounts of soap and water. Consider appropriate evacuation.

PHTHALIC ANHYDRIDE

DOT Number: NA 2214 *DOT Hazard Class:* Corrosive *DOT Guide Number:* 60
Synonyms: pan, phthalic acid anhydride
STCC Number: 4934223 *Reportable Qty:* 5000/2270
Mfg Name: Monsanto Chemical Co. *Phone No:* 1-314-694-1000

Physical Description:

Physical Form: Solid flakes or liquid heated *Color:* Colorless to pale yellow *Odor:* Choking
Other Information: It is used in the manufacture of other chemical compounds such as artificial resins.

Chemical Properties:

Specific Gravity: 1.5 *Vapor Density:* 5.1 *Boiling Point: 543° F(283.8° C)*
Melting Point: 268° F(131.1° C) *Vapor Pressure:* .05 mm Hg at 68° F(20° C) *Solubility in water:* No
Degree of Solubility: .62%
Other Information: Solids sink in water. Liquid solidifies and sinks in water.

Health Hazards:

Inhalation Hazard: Will cause coughing.
Ingestion Hazard: Harmful if swallowed.
Absorption Hazard: Will burn the skin and the eyes.
Hazards to Wildlife: Dangerous to aquatic life.
Decontamination Procedures: Wash away any material with copious amounts of soap and water.
First Aid Procedures: Remove victim to fresh air, call emergency medical care. If not breathing give CPR. If breathing is difficult administer oxygen. Treat for shock.

Fire Hazards:

Flashpoint: 305° F(151.6° C) *Ignition temperature:* 1058° F(570° C)
Flammable Explosive High Range: 10.5 *Low Range:* 1.7
Toxic Products of Combustion: n/a
Other Hazards: n/a
Possible extinguishing agents: Apply water from as far a distance as possible. Use alcohol foam, dry chemical, or carbon dioxide.

Reactivity Hazards:

Reactive With: Liquid splatters upon contact with water. *Other Reactions:* n/a

Corrosivity Hazards:

Corrosive With: When mixed with water forms a corrosive acidic solution.
Neutralizing Agent: Crushed limestone, soda ash, or lime.

Radioactivity Hazards:

Radiation Emitted: n/a *Other Hazards:* n/a

Recommended Protection for Response Personnel:

Avoid breathing vapors, keep upwind. Structural protective clothing provides limited protection. Wash away any material which may have come into contact with the body with copious amounts of soap and water. Consider appropriate evacuation.

PINE OIL

DOT Number: UN 1272 *DOT Hazard Class:* Combustible liquid *DOT Guide Number:* 26
Synonyms: oil of pine, oil of 5-Siberian, synthetic pine oil, Yarmar
STCC Number: 4915170 *Reportable Qty:* n/a *CHEMTREC Phone No:* 1-800-424-9300

Physical Description:

Physical Form: Liquid *Color:* Colorless to light amber *Odor:* Pinelike
Other Information: It is used as a wetting and dispersing agent for processing various textiles and paper, for making soap, insecticides, and disinfectants.

Chemical Properties:

Specific Gravity: .86 *Vapor Density:* 1 *Boiling Point: 400° F(204.4° C)*
Melting Point: n/a *Vapor Pressure:* n/a *Solubility in water:* No
Other Information: Lighter than and practically insoluble in water. Weighs 7 lbs/3.1 kg per gallon/3.8 l.

Health Hazards:

Inhalation Hazard: Irritating.
Ingestion Hazard: Irritating.
Absorption Hazard: Irritating to the skin and eyes.
Hazards to Wildlife: n/a
Decontamination Procedures: Wash away any material with copious amounts of soap and water.
First Aid Procedures: Remove victim to fresh air, call emergency medical care. If not breathing give CPR. If breathing is difficult administer oxygen. Treat for shock.

Fire Hazards:

Flashpoint: 172° F(77.7° C) *Ignition temperature:* 671° F(355° C)
Flammable Explosive High Range: n/a *Low Range:* n/a
Toxic Products of Combustion: n/a
Other Hazards: Will burn, but is difficult to ignite. There is a possibility that containers may violently rupture in a fire.
Possible extinguishing agents: Do not extinguish the fire unless the flow can be stopped. Use alcohol foam, dry chemical, carbon dioxide, water spray.

Reactivity Hazards:

Reactive With: n/a *Other Reactions:* n/a

Corrosivity Hazards:

Corrosive With: n/a *Neutralizing Agent:* n/a

Radioactivity Hazards:

Radiation Emitted: n/a *Other Hazards:* n/a

Recommended Protection for Response Personnel:

Avoid breathing vapors, keep upwind. Structural protective clothing provides limited protection. Wash away any material which may have come into contact with the body with copious amounts of soap and water. Consider appropriate evacuation.

PIPERAZINE

DOT Number: UN 2579 *DOT Hazard Class:* Corrosive *DOT Guide Number:* 60
Synonyms: Lumerical, piperazidine
STCC Number: 4936524 *Reportable Qty:* n/a
Mfg Name: Dow Chemical Co. *Phone No:* 1-517-636-4400

Physical Description:

Physical Form: Solid *Color:* White *Odor:* Mild, fishy
Other Information: n/a

Chemical Properties:

Specific Gravity: 1.1 *Vapor Density:* 3 *Boiling Point: 299° F(148.3° C)*
Melting Point: n/a *Vapor Pressure:* n/a *Solubility in water:* Yes
Other Information: Sinks and mixes with water.

Health Hazards:

Inhalation Hazard: Causes coughing, difficulty in breathing.
Ingestion Hazard: Causes nausea and vomiting.
Absorption Hazard: Will burn the eyes, irritating to the skin.
Hazards to Wildlife: n/a
Decontamination Procedures: Wash away any material with copious amounts of soap and water.
First Aid Procedures: Remove victim to fresh air, call emergency medical care. If not breathing give CPR. If breathing is difficult administer oxygen. Treat for shock.

Fire Hazards:

Flashpoint: 225° F(107.2° C) *Ignition temperature:* 851° F(455° C)
Flammable Explosive High Range: n/a *Low Range:* n/a
Toxic Products of Combustion: Toxic oxides of nitrogen are produced in a fire.
Other Hazards: n/a
Possible extinguishing agents: Extinguish fire using suitable agent for the type of surrounding fire.

Reactivity Hazards:

Reactive With: n/a *Other Reactions:* n/a

Corrosivity Hazards:

Corrosive With: Aluminum, magnesium, and zinc.
Neutralizing Agent: Crushed limestone, soda ash, or lime.

Radioactivity Hazards:

Radiation Emitted: n/a *Other Hazards:* n/a

Recommended Protection for Response Personnel:

Avoid breathing vapors, keep upwind. Structural protective clothing provides limited protection. Wash away any material which may have come into contact with the body with copious amounts of soap and water. Consider appropriate evacuation.

POLYBUTENE

DOT Number: n/a *DOT Hazard Class:* n/a *DOT Guide Number:* n/a
Synonyms: butene resins, polyisobutylene plastics, polyisobutylene resins, polyisobutylene waxes
STCC Number: n/a *Reportable Qty:* n/a
Mfg Name: George A. Rowley *Phone No:* 1-215-537-1000

Physical Description:

Physical Form: Liquid *Color:* Colorless *Odor:* Odorless
Other Information: n/a

Chemical Properties:

Specific Gravity: .81 to .91 *Vapor Density:* n/a *Boiling Point: Very high*
Melting Point: n/a *Vapor Pressure:* n/a *Solubility in water:* n/a
Other Information: Floats on water.

Health Hazards:

Inhalation Hazard: n/a
Ingestion Hazard: n/a
Absorption Hazard: n/a
Hazards to Wildlife: n/a
Decontamination Procedures: Wash away any material with copious amounts of soap and water.
First Aid Procedures: Remove victim to fresh air, call emergency medical care. If not breathing give CPR. If breathing is difficult administer oxygen. Treat for shock.

Fire Hazards:

Flashpoint: 215-470° F(101-243° C) *Ignition temperature:* n/a
Flammable Explosive High Range: n/a *Low Range:* n/a
Toxic Products of Combustion: n/a
Other Hazards: n/a
Possible extinguishing agents: Foam, dry chemical, carbon dioxide. Water may be ineffective on fires. Cool all affected containers with water.

Reactivity Hazards:

Reactive With: n/a *Other Reactions:* n/a

Corrosivity Hazards:

Corrosive With: n/a *Neutralizing Agent:* n/a

Radioactivity Hazards:

Radiation Emitted: n/a *Other Hazards:* n/a

Recommended Protection for Response Personnel:

Avoid breathing vapors, keep upwind. Structural protective clothing provides limited protection. Wash away any material which may have come into contact with the body with copious amounts of soap and water. Consider appropriate evacuation.

POLYCHLORINATED BIPHENYL

DOT Number: UN 2315 *DOT Hazard Class:* ORM-E *DOT Guide Number:* 31
Synonyms: Arochlor, halogenated waxes, PCB
STCC Number: 4961666 *Reportable Qty:* 10/4.54
Mfg Name: Monsanto Chemical Company *Phone No:* 1-314-694-1000

Physical Description:

Physical Form: Liquid *Color:* Light yellow *Odor:* Weak
Other Information: It is used in coolants for transformers, and in electrical capacitors.

Chemical Properties:

Specific Gravity: 1.3-1.8 *Vapor Density:* 1 *Boiling Point: Very high*
Melting Point: n/a *Vapor Pressure:* n/a *Solubility in water:* No
Other Information: Sinks in water. Material is extremely persistent in the environment.

Health Hazards:

Inhalation Hazard: n/a
Ingestion Hazard: n/a
Absorption Hazard: Irritating to the skin and eyes.
Hazards to Wildlife: Dangerous to both aquatic life and waterfowl.
Decontamination Procedures: Wash away any material with copious amounts of soap and water.
First Aid Procedures: Remove victim to fresh air, call emergency medical care. If not breathing give CPR. If breathing is difficult administer oxygen. Treat for shock.

Fire Hazards:

Flashpoint: 286° F(141.1° C) *Ignition temperature:* n/a
Flammable Explosive High Range: n/a *Low Range:* n/a
Toxic Products of Combustion: Irritating gases are generated in fires.
Other Hazards: n/a
Possible extinguishing agents: Extinguish fire using suitable agent for the type of surrounding fire.

Reactivity Hazards:

Reactive With: n/a *Other Reactions:* n/a

Corrosivity Hazards:

Corrosive With: n/a *Neutralizing Agent:* n/a

Radioactivity Hazards:

Radiation Emitted: n/a *Other Hazards:* n/a

Recommended Protection for Response Personnel:

Avoid breathing vapors, keep upwind. Wear a sealed chemical suit (polycarbonate, butyl rubber). Wash away any material which may have come into contact with the body with copious amounts of soap and water. Consider appropriate evacuation.

POLYMETHYLENE POLYPHENYL ISOCYANATE

DOT Number: n/a *DOT Hazard Class:* n/a *DOT Guide Number:* n/a
Synonyms: PAPI
STCC Number: n/a *Reportable Qty:* n/a
Mfg Name: Upjohn Company *Phone No:* 1-713-479-1541

Physical Description:

Physical Form: Liquid *Color:* Dark brown *Odor:* Weak
Other Information: n/a

Chemical Properties:

Specific Gravity: 1.2 *Vapor Density:* n/a *Boiling Point: 392° F(200° C)*
Melting Point: n/a *Vapor Pressure:* n/a *Solubility in water:* n/a
Other Information: Sinks in water.

Health Hazards:

Inhalation Hazard: n/a
Ingestion Hazard: Poisonous if swallowed.
Absorption Hazard: Irritating to the skin and eyes.
Hazards to Wildlife: n/a
Decontamination Procedures: Wash away any material with copious amounts of soap and water.
First Aid Procedures: Remove victim to fresh air, call emergency medical care. If not breathing give CPR. If breathing is difficult administer oxygen. Treat for shock.

Fire Hazards:

Flashpoint: 425° F(218.3° C) *Ignition temperature:* n/a
Flammable Explosive High Range: n/a *Low Range:* n/a
Toxic Products of Combustion: n/a
Other Hazards: Containers may explode in fire.
Possible extinguishing agents: Dry chemical or carbon dioxide. Cool exposed containers with water.

Reactivity Hazards:

Reactive With: Water, slowly forming a heavy scum and liberating carbon dioxide gas. Dangerous pressure can build up if the container is sealed. *Other Reactions:* n/a

Corrosivity Hazards:

Corrosive With: n/a *Neutralizing Agent:* n/a

Radioactivity Hazards:

Radiation Emitted: n/a *Other Hazards:* n/a

Recommended Protection for Response Personnel:

Avoid breathing vapors, keep upwind. Structural protective clothing provides limited protection. Wash away any material which may have come into contact with the body with copious amounts of soap and water. Consider appropriate evacuation.

POLYPHOSPHORIC ACID

DOT Number: n/a *DOT Hazard Class:* Corrosive *DOT Guide Number:* n/a
Synonyms: condensed phosphoric acid
STCC Number: n/a *Reportable Qty:* n/a
Mfg Name: Allied Corp. *Phone No:* 1-201-455-2000

Physical Description:

Physical Form: Liquid *Color:* Colorless *Odor:* Odorless
Other Information: n/a

Chemical Properties:

Specific Gravity: 2.05 *Vapor Density:* 1 *Boiling Point: 1022° F(550° C)*
Melting Point: 100° F(37.7° C) *Vapor Pressure:* n/a *Solubility in water:* Yes
Other Information: Sinks and mixes with water.

Health Hazards:

Inhalation Hazard: n/a
Ingestion Hazard: Harmful if swallowed.
Absorption Hazard: Will burn the skin and eyes.
Hazards to Wildlife: Dangerous to aquatic life.
Decontamination Procedures: Wash away any material with copious amounts of soap and water.
First Aid Procedures: Remove victim to fresh air, call emergency medical care. If not breathing give CPR. If breathing is difficult administer oxygen. Treat for shock.

Fire Hazards:

Flashpoint: n/a *Ignition temperature:* n/a
Flammable Explosive High Range: n/a *Low Range:* n/a
Toxic Products of Combustion: n/a
Other Hazards: n/a
Possible *Possible extinguishing agents:* Water, alcohol foam, dry chemical, carbon dioxide.

Reactivity Hazards:

Reactive With: Water generating heat to form phosphoric acid. *Other Reactions:* n/a

Corrosivity Hazards:

Corrosive With: Metals to liberate flammable hydrogen gas. *Neutralizing Agent:* n/a

Radioactivity Hazards:

Radiation Emitted: n/a *Other Hazards:* n/a

Recommended Protection for Response Personnel:

Avoid breathing vapors, keep upwind. Structural protective clothing provides limited protection. Wash away any material which may have come into contact with the body with copious amounts of soap and water. Consider appropriate evacuation.

POLYPROPYLENE

DOT Number: n/a *DOT Hazard Class:* n/a *DOT Guide Number:* n/a
Synonyms: propylene polymer
STCC Number: n/a *Reportable Qty:* n/a
Mfg Name: Exxon Chemical Co. *Phone No:* 1-203-655-5200

Physical Description:

Physical Form: Solid *Color:* Tan to white *Odor:* Odorless
Other Information: n/a

Chemical Properties:

Specific Gravity: .9 *Vapor Density:* n/a *Boiling Point: Decomposes*
Melting Point: n/a *Vapor Pressure:* n/a *Solubility in water:* n/a
Other Information: Floats on water.

Health Hazards:

Inhalation Hazard: n/a
Ingestion Hazard: n/a
Absorption Hazard: n/a
Hazards to Wildlife: n/a
Decontamination Procedures: Wash away any material with copious amounts of soap and water.
First Aid Procedures: Remove victim to fresh air, call emergency medical care. If not breathing give CPR. If breathing is difficult administer oxygen. Treat for shock.

Fire Hazards:

Flashpoint: Combustible solid *Ignition temperature:* n/a
Flammable Explosive High Range: n/a *Low Range:* n/a
Toxic Products of Combustion: n/a
Other Hazards: n/a
Possible extinguishing agents: Foam, dry chemical, carbon dioxide. Water may be ineffective on fires. Cool all affected containers with water.

Reactivity Hazards:

Reactive With: n/a *Other Reactions:* n/a

Corrosivity Hazards:

Corrosive With: Metals to liberate flammable hydrogen gas. *Neutralizing Agent:* n/a

Radioactivity Hazards:

Radiation Emitted: n/a *Other Hazards:* n/a

Recommended Protection for Response Personnel:

Avoid breathing vapors, keep upwind. Structural protective clothing provides limited protection. Wash away any material which may have come into contact with the body with copious amounts of soap and water. Consider appropriate evacuation.

POTASSIUM

DOT Number: UN 2257 *DOT Hazard Class:* Flammable solid *DOT Guide Number:* 40
Synonyms: none given
STCC Number: n/a *Reportable Qty:* n/a
Mfg Name: A.D. MacKay Inc. *Phone No:* 1-203-655-7401

Physical Description:

Physical Form: Solid *Color:* Silver white *Odor:* Odorless
Other Information: n/a

Chemical Properties:

Specific Gravity: .86 *Vapor Density:* 1 *Boiling Point: 1425° F(773.8° C)*
Melting Point: 145° F(62.7° C) *Vapor Pressure:* n/a *Solubility in water:* Reacts
Other Information: Reacts violently with water. Flammable vapor is produced.

Health Hazards:

Inhalation Hazard: n/a
Ingestion Hazard: n/a
Absorption Hazard: Will burn the skin and eyes.
Hazards to Wildlife: Dangerous to aquatic life.
Decontamination Procedures: Wash away any material with copious amounts of soap and water.
First Aid Procedures: Remove victim to fresh air, call emergency medical care. If not breathing give CPR. If breathing is difficult administer oxygen. Treat for shock.

Fire Hazards:

Flashpoint: Combustible solid *Ignition temperature:* n/a
Flammable Explosive High Range: n/a *Low Range:* n/a
Toxic Products of Combustion: n/a
Other Hazards: Reacts violently with water forming flammable and explosive hydrogen gas. May ignite spontaneously in air!
Possible extinguishing agents: Use graphite, sand, sodium chloride. Do not use water!!

Reactivity Hazards:

Reactive With: Water to form flammable hydrogen gas, and a strong caustic solution.
Other Reactions: May ignite combustible materials if they are damp.

Corrosivity Hazards:

Corrosive With: n/a *Neutralizing Agent:* n/a

Radioactivity Hazards:

Radiation Emitted: n/a *Other Hazards:* n/a

Recommended Protection for Response Personnel:

Avoid breathing vapors, keep upwind. Structural protective clothing provides limited protection. Wash away any material which may have come into contact with the body with copious amounts of soap and water. If the fire becomes uncontrollable, consider appropriate evacuation.

POTASSIUM ARSENATE

DOT Number: UN 1677 *DOT Hazard Class:* Poison B *DOT Guide Number:* 53
Synonyms: Macquer's salt, potassium acid arsenate
STCC Number: 4923277 *Reportable Qty:* 100/454
Mfg Name: Cerac Inc. *Phone No:* 1-414-289-9800

Physical Description:

Physical Form: Solid *Color:* White *Odor:* Odorless
Other Information: n/a

Chemical Properties:

Specific Gravity: 2.8 *Vapor Density:* 6.2 *Boiling Point: Decomposes*
Melting Point: n/a *Vapor Pressure:* n/a *Solubility in water:* Yes
Other Information: Mixes with water.

Health Hazards:

Inhalation Hazard: Poisonous if inhaled.
Ingestion Hazard: Poisonous if swallowed.
Absorption Hazard: Irritating to the skin and eyes.
Hazards to Wildlife: Dangerous to aquatic life.
Decontamination Procedures: Wash away any material with copious amounts of soap and water.
First Aid Procedures: Remove victim to fresh air, call emergency medical care. If not breathing give CPR. If breathing is difficult administer oxygen. Treat for shock.

Fire Hazards:

Flashpoint: n/a *Ignition temperature:* n/a
Flammable Explosive High Range: n/a *Low Range:* n/a
Toxic Products of Combustion: n/a
Other Hazards: n/a
Possible extinguishing agents: Extinguish fire using suitable agent for the type of surrounding fire. Use alcohol foam, dry chemical, or carbon dioxide.

Reactivity Hazards:

Reactive With: n/a *Other Reactions:* n/a

Corrosivity Hazards:

Corrosive With: n/a *Neutralizing Agent:* n/a

Radioactivity Hazards:

Radiation Emitted: n/a *Other Hazards:* n/a

Recommended Protection for Response Personnel:

Avoid breathing vapors, keep upwind. Structural protective clothing provides limited protection. Wash away any material which may have come into contact with the body with copious amounts of soap and water. Consider appropriate evacuation.

POTASSIUM ARSENITE

DOT Number: UN 1678 *DOT Hazard Class:* Poison B *DOT Guide Number:* 54
Synonyms: arsenious acid, potassium salt; Fowler's solution
STCC Number: 4923278 *Reportable Qty:* 1000/454
Mfg Name: Gallard-Schlesinger *Phone No:* 1-516-333-5600

Physical Description:

Physical Form: Solid *Color:* White *Odor:* Odorless
Other Information: n/a

Chemical Properties:

Specific Gravity: n/a *Vapor Density:* 8.8 *Boiling Point: n/a*
Melting Point: n/a *Vapor Pressure:* n/a *Solubility in water:* Yes
Other Information: Mixes with water.

Health Hazards:

Inhalation Hazard: n/a
Ingestion Hazard: Poisonous if swallowed.
Absorption Hazard: Irritating to the skin and eyes.
Hazards to Wildlife: Dangerous to aquatic life.
Decontamination Procedures: Wash away any material with copious amounts of soap and water.
First Aid Procedures: Remove victim to fresh air, call emergency medical care. If not breathing give CPR. If breathing is difficult administer oxygen. Treat for shock.

Fire Hazards:

Flashpoint: n/a *Ignition temperature:* n/a
Flammable Explosive High Range: n/a *Low Range:* n/a
Toxic Products of Combustion: n/a
Other Hazards: n/a
Possible extinguishing agents: Extinguish fire using suitable agent for the type of surrounding fire. Use alcohol foam, dry chemical, or carbon dioxide.

Reactivity Hazards:

Reactive With: n/a *Other Reactions:* n/a

Corrosivity Hazards:

Corrosive With: n/a *Neutralizing Agent:* n/a

Radioactivity Hazards:

Radiation Emitted: n/a *Other Hazards:* n/a

Recommended Protection for Response Personnel:

Avoid breathing vapors, keep upwind. Structural protective clothing provides limited protection. Wash away any material which may have come into contact with the body with copious amounts of soap and water. Consider appropriate evacuation.

POTASSIUM CHLORATE

DOT Number: UN 1485 *DOT Hazard Class:* Oxidizer *DOT Guide Number:* 35
Synonyms: chlorate of potash, chlorate of potassium, Potcrate
STCC Number: 4918722 *Reportable Qty:* n/a
Mfg Name: Pennwald Corp. *Phone No:* 1-201-455-2000

Physical Description:

Physical Form: Solid *Color:* White *Odor:* Odorless
Other Information: It is used to make matches, paper, explosives, and for many other uses.

Chemical Properties:

Specific Gravity: 2.34 *Vapor Density:* 4.2 *Boiling Point: Decomposes*
Melting Point: n/a *Vapor Pressure:* n/a *Solubility in water:* Yes
Other Information: Mixes with water.

Health Hazards:

Inhalation Hazard: n/a
Ingestion Hazard: Nausea, vomiting, loss of consciousness.
Absorption Hazard: Irritating to the skin and the eyes.
Hazards to Wildlife: n/a
Decontamination Procedures: Wash away any material with copious amounts of soap and water.
First Aid Procedures: Remove victim to fresh air, call emergency medical care. If not breathing give CPR. If breathing is difficult administer oxygen. Treat for shock.

Fire Hazards:

Flashpoint: n/a *Ignition temperature:* n/a
Flammable Explosive High Range: n/a *Low Range:* n/a
Toxic Products of Combustion: Toxic fumes are formed in fires.
Other Hazards: May cause fire upon contact with combustibles.
Possible *Possible extinguishing agents:* Flood with water. Apply water from as far a distance as possible.

Reactivity Hazards:

Reactive With: Combustible materials may cause fire. *Other Reactions:* n/a

Corrosivity Hazards:

Corrosive With: n/a *Neutralizing Agent:* n/a

Radioactivity Hazards:

Radiation Emitted: n/a *Other Hazards:* n/a

Recommended Protection for Response Personnel:

Avoid breathing vapors, keep upwind. Structural protective clothing provides limited protection. Wash away any material which may have come into contact with the body with copious amounts of soap and water. If the fire becomes uncontrollable, consider appropriate evacuation.

POTASSIUM CHROMATE

DOT Number: NA 9142 *DOT Hazard Class:* ORM-E *DOT Guide Number:* 31
Synonyms: natural potassium chromate, potassium chromate(VI)
STCC Number: 4963364 *Reportable Qty:* 1000/454
Mfg Name: Allied Corp. *Phone No:* 1-201-455-2000

Physical Description:

Physical Form: Solid *Color:* Bright yellow *Odor:* Odorless
Other Information: It is used in chemical analysis, as a pigment in paints and inks, as a fungicide, and to make other chromium compounds.

Chemical Properties:

Specific Gravity: 2.73 *Vapor Density:* 6.7 *Boiling Point: Decomposes*
Melting Point: n/a *Vapor Pressure:* n/a *Solubility in water:* Yes
Other Information: Sinks and mixes with water.

Health Hazards:

Inhalation Hazard: Causes coughing or difficulty in breathing.
Ingestion Hazard: Poisonous if swallowed.
Absorption Hazard: Irritating to the skin and eyes.
Hazards to Wildlife: Dangerous to aquatic life.
Decontamination Procedures: Wash away any material with copious amounts of soap and water.
First Aid Procedures: Remove victim to fresh air, call emergency medical care. If not breathing give CPR. If breathing is difficult administer oxygen. Treat for shock.

Fire Hazards:

Flashpoint: n/a *Ignition temperature:* n/a
Flammable Explosive High Range: n/a *Low Range:* n/a
Toxic Products of Combustion: n/a
Other Hazards: Will increase the intensity of a fire. May cause fire upon contact with combustibles.
Possible *Possible extinguishing agents:* Extinguish fire using suitable agent for the type of surrounding fire.

Reactivity Hazards:

Reactive With: Combustible materials may cause fire. *Other Reactions:* n/a

Corrosivity Hazards:

Corrosive With: n/a *Neutralizing Agent:* n/a

Radioactivity Hazards:

Radiation Emitted: n/a *Other Hazards:* n/a

Recommended Protection for Response Personnel:

Avoid breathing vapors, keep upwind. Wear a sealed chemical suit (polycarbonate, butyl rubber). Wash away any material which may have come into contact with the body with copious amounts of soap and water. Consider appropriate evacuation.

POTASSIUM CYANIDE (liquid)

DOT Number: UN 1680 *DOT Hazard Class:* Poison B *DOT Guide Number:* 55
Synonyms: cyanide
STCC Number: 4923225 *Reportable Qty:* 10/4.54
Mfg Name: Mallinckrodt Inc. *Phone No:* 1-314-895-2000

Physical Description:

Physical Form: Liquid *Color:* n/a *Odor:* Almond
Other Information: It is used for gold and silver extraction, in chemical analysis, to make other chemicals, and in insecticides.

Chemical Properties:

Specific Gravity: 1.52 *Vapor Density:* n/a *Boiling Point: 1174° F(634.4° C)*
Melting Point: n/a *Vapor Pressure:* n/a *Solubility in water:* Yes
Other Information: Sinks and mixes with water.

Health Hazards:

Inhalation Hazard: Poisonous if inhaled.
Ingestion Hazard: Poisonous if swallowed.
Absorption Hazard: Poisonous if the skin is exposed.
Hazards to Wildlife: Dangerous to aquatic life.
Decontamination Procedures: Wash away any material with copious amounts of soap and water.
First Aid Procedures: Remove victim to fresh air, call emergency medical care. If not breathing give CPR. If breathing is difficult administer oxygen. Treat for shock.

Fire Hazards:

Flashpoint: n/a *Ignition temperature:* n/a
Flammable Explosive High Range n/a *Low Range:* n/a
Toxic Products of Combustion: n/a
Other Hazards: n/a
Possible extinguishing agents: Extinguish fire using suitable agent for the type of surrounding fire. Use alcohol foam, dry chemical, or carbon dioxide.

Reactivity Hazards:

Reactive With: Water to form poisonous hydrogen cyanide gas.
Other Reactions: Contact with weak acids cause deadly hydrogen cyanide gas.

Corrosivity Hazards:

Corrosive With: n/a *Neutralizing Agent:* n/a

Radioactivity Hazards:

Radiation Emitted: n/a *Other Hazards:* n/a

Recommended Protection for Response Personnel:

Avoid breathing vapors, keep upwind. Wear a sealed chemical suit (polycarbonate, butyl rubber, viton, PVC, chlorinated polyethylene, nitrile, neoprene). Wash away any material which may have come into contact with the body with copious amounts of soap and water. Consider appropriate evacuation.

POTASSIUM CYANIDE (solid)

DOT Number: UN 1680 *DOT Hazard Class:* Poison B *DOT Guide Number:* 55
Synonyms: cyanide
STCC Number: 4923226 *Reportable Qty:* 10/4.54
Mfg Name: Mallinckrodt Inc. *Phone No:* 1-314-895-2000

Physical Description:

Physical Form: Solid *Color:* White *Odor:* Almond
Other Information: It is used for gold and silver extraction, in chemical analysis, to make other chemicals, and in insecticides.

Chemical Properties:

Specific Gravity: 1.52 *Vapor Density:* 2.2 *Boiling Point: 1174° F(634.4° C)*
Melting Point: n/a *Vapor Pressure:* n/a *Solubility in water:* Yes
Other Information: Sinks and mixes with water.

Health Hazards:

Inhalation Hazard: Poisonous if inhaled.
Ingestion Hazard: Poisonous if swallowed.
Absorption Hazard: Poisonous if the skin is exposed.
Hazards to Wildlife: Dangerous to aquatic life.
Decontamination Procedures: Wash away any material with copious amounts of soap and water.
First Aid Procedures: Remove victim to fresh air, call emergency medical care. If not breathing give CPR. If breathing is difficult administer oxygen. Treat for shock.

Fire Hazards:

Flashpoint: n/a *Ignition temperature:* n/a
Flammable Explosive High Range: n/a *Low Range:* n/a
Toxic Products of Combustion: n/a
Other Hazards: n/a
Possible extinguishing agents: Extinguish fire using suitable agent for the type of surrounding fire. Use alcohol foam, dry chemical, or carbon dioxide.

Reactivity Hazards:

Reactive With: Absorbs water from air to form a syrup.
Other Reactions: Contact with weak acids cause deadly hydrogen cyanide gas.

Corrosivity Hazards:

Corrosive With: n/a *Neutralizing Agent:* n/a

Radioactivity Hazards:

Radiation Emitted: n/a *Other Hazards:* n/a

Recommended Protection for Response Personnel:

Avoid breathing vapors, keep upwind. Wear a sealed chemical suit (polycarbonate, butyl rubber, viton, PVC, chlorinated polyethylene, nitrile, neoprene). Wash away any material which may have come into contact with the body with copious amounts of soap and water. Consider appropriate evacuation.

POTASSIUM DICHLOROISOCYANURATE

DOT Number: NA 2465 *DOT Hazard Class:* Oxidizer *DOT Guide Number:* 42
Synonyms: potassium dichloro-s-triazinetrione
STCC Number: 4918430 *Reportable Qty:* n/a *CHEMTREC Phone No:* 1-800-424-9300

Physical Description:

Physical Form: Solid *Color:* White *Odor:* Chlorinelike
Other Information: It is used as a dry bleach in household cleaning compounds, and in swimming pool disinfectants.

Chemical Properties:

Specific Gravity: .96 *Vapor Density:* n/a *Boiling Point: Decomposes*
Melting Point: n/a *Vapor Pressure:* n/a *Solubility in water:* Yes
Other Information: Mixes with water.

Health Hazards:

Inhalation Hazard: Coughing, difficulty in breathing.
Ingestion Hazard: Harmful if swallowed.
Absorption Hazard: Irritating to the skin and eyes.
Hazards to Wildlife: n/a
Decontamination Procedures: Wash away any material with copious amounts of soap and water.
First Aid Procedures: Remove victim to fresh air, call emergency medical care. If not breathing give CPR. If breathing is difficult administer oxygen. Treat for shock.

Fire Hazards:

Flashpoint: n/a *Ignition temperature:* n/a
Flammable Explosive High Range: n/a *Low Range:* n/a
Toxic Products of Combustion: May form chlorine and other gases in fire.
Other Hazards: May cause fire on contact with combustibles.
Possible *Possible extinguishing agents:* Flood with water.

Reactivity Hazards:

Reactive With: Water to form a bleach solution.
Other Reactions: Organic matter or easily chlorinated or oxidized materials may result in a fire.

Corrosivity Hazards:

Corrosive With: n/a *Neutralizing Agent:* n/a

Radioactivity Hazards:

Radiation Emitted: n/a *Other Hazards:* n/a

Recommended Protection for Response Personnel:

Avoid breathing vapors, keep upwind. Wear appropriate chemical clothing. Wash away any material which may have come into contact with the body with copious amounts of soap and water. If the fire becomes uncontrollable, consider appropriate evacuation.

POTASSIUM DICHROMATE

DOT Number: NA 1479 *DOT Hazard Class:* ORM-A *DOT Guide Number:* 35
Synonyms: bichrome, potassium bichromate
STCC Number: 4941160 *Reportable Qty:* 1000/454
Mfg Name: Allied Corp. *Phone No:* 1-201-455-2000

Physical Description:

Physical Form: Solid *Color:* Red to orange *Odor:* Odorless
Other Information: It is used in the production of other chemicals, in pyrotechnics, explosives, dyeing, printing, tanning leather, and in many other uses.

Chemical Properties:

Specific Gravity: 2.68 *Vapor Density:* 10 *Boiling Point: Decomposes*
Melting Point: n/a *Vapor Pressure:* n/a *Solubility in water:* Yes
Other Information: Mixes with water.

Health Hazards:

Inhalation Hazard: Difficulty in breathing.

Ingestion Hazard: Nausea, vomiting, loss of consciousness.

Absorption Hazard: Will burn the skin and the eyes.

Hazards to Wildlife: Dangerous to aquatic life.

Decontamination Procedures: Wash away any material with copious amounts of soap and water.

First Aid Procedures: Remove victim to fresh air, call emergency medical care. If not breathing give CPR. If breathing is difficult administer oxygen. Treat for shock.

Fire Hazards:

Flashpoint: n/a *Ignition temperature:* n/a
Flammable Explosive High Range: n/a *Low Range:* n/a
Toxic Products of Combustion: May decompose forming oxygen which supports combustion of materials.
Other Hazards: May cause fire on contact with combustibles.
Possible extinguishing agents: Extinguish fire using suitable agent for the type of surrounding fire.

Reactivity Hazards:

Reactive With: Finely divided combustibles. *Other Reactions:* n/a

Corrosivity Hazards:

Corrosive With: n/a *Neutralizing Agent:* n/a

Radioactivity Hazards:

Radiation Emitted: n/a *Other Hazards:* n/a

Recommended Protection for Response Personnel:

Avoid breathing vapors, keep upwind. Wear a sealed chemical suit (polycarbonate, butyl rubber). Wash away any material which may have come into contact with the body with copious amounts of soap and water. Consider appropriate evacuation.

POTASSIUM HYDROXIDE (liquid)

DOT Number: UN 1814 *DOT Hazard Class:* Corrosive *DOT Guide Number:* 60
Synonyms: caustic potash, lye
STCC Number: 4935230 *Reportable Qty:* 1000/454
Mfg Name: Monsanto Chemical Co. *Phone No:* 1-314-694-1000

Physical Description:

Physical Form: Liquid *Color:* Colorless *Odor:* Odorless
Other Information: It is used to make soaps, other potassium compounds, in liquid fertilizers, and for many other uses.

Chemical Properties:

Specific Gravity: 2.04 *Vapor Density:* n/a *Boiling Point: Very high*
Melting Point: n/a *Vapor Pressure:* n/a *Solubility in water:* Yes
Other Information: Mixes with water. Weighs 12.8 lbs/5.8 kg per gallon/3.8 l.

Health Hazards:

Inhalation Hazard: Harmful if inhaled.
Ingestion Hazard: Harmful if swallowed.
Absorption Hazard: Will burn the skin and the eyes.
Hazards to Wildlife: Dangerous to aquatic life.
Decontamination Procedures: Wash away any material with copious amounts of soap and water.
First Aid Procedures: Remove victim to fresh air, call emergency medical care. If not breathing give CPR. If breathing is difficult administer oxygen. Treat for shock.

Fire Hazards:

Flashpoint: n/a *Ignition temperature:* n/a
Flammable Explosive High Range: n/a *Low Range:* n/a
Toxic Products of Combustion: n/a
Other Hazards: Flammable gas may be produced on contact with other metals. May cause fire on contact with moisture and other combustibles.
Possible extinguishing agents: Extinguish fire using suitable agent for the type of surrounding fire.

Reactivity Hazards:

Reactive With: Water. *Other Reactions:* Metals when wet producing flammable hydrogen gas.

Corrosivity Hazards:

Corrosive With: Metals and tissue. *Neutralizing Agent:* Vinegar or other dilute acid.

Radioactivity Hazards:

Radiation Emitted: n/a *Other Hazards:* n/a

Recommended Protection for Response Personnel:

Avoid breathing vapors, keep upwind. Wear a sealed chemical suit (butyl rubber, viton, PVC, chlorinated polyethylene, nitrile). Wash away any material which may have come into contact with the body with copious amounts of soap and water. Consider appropriate evacuation.

POTASSIUM HYDROXIDE (solid)

DOT Number: UN 1813 *DOT Hazard Class:* Corrosive *DOT Guide Number:* 60
Synonyms: caustic potash, lye
STCC Number: 4935225 *Reportable Qty:* 1000/454
Mfg Name: Monsanto Chemical Co. *Phone No:* 1-314-694-1000

Physical Description:

Physical Form: Solid *Color:* White *Odor:* Odorless
Other Information: It is used in soap manufacturing, bleach, as an electrolyte in alkaline batteries, and as a food additive.

Chemical Properties:

Specific Gravity: 2.04 *Vapor Density:* 1.9 *Boiling Point: Very high*
Melting Point: n/a *Vapor Pressure:* n/a *Solubility in water:* Yes
Other Information: Sinks and mixes slowly with water.

Health Hazards:

Inhalation Hazard: Harmful if inhaled.
Ingestion Hazard: Harmful if swallowed.
Absorption Hazard: Will burn the skin and the eyes.
Hazards to Wildlife: Dangerous to aquatic life.
Decontamination Procedures: Wash away any material with copious amounts of soap and water.
First Aid Procedures: Remove victim to fresh air, call emergency medical care. If not breathing give CPR. If breathing is difficult administer oxygen. Treat for shock.

Fire Hazards:

Flashpoint: n/a *Ignition temperature:* n/a
Flammable Explosive High Range: n/a *Low Range:* n/a
Toxic Products of Combustion: n/a
Other Hazards: Flammable gas may be produced on contact with other metals. May cause fire on contact with moisture and other combustibles.
Possible extinguishing agents: Extinguish fire using suitable agent for the type of surrounding fire.

Reactivity Hazards:

Reactive With: Water. *Other Reactions:* Metals when wet producing flammable hydrogen gas.

Corrosivity Hazards:

Corrosive With: Metals and tissue. *Neutralizing Agent:* Vinegar or other dilute acid.

Radioactivity Hazards:

Radiation Emitted: n/a *Other Hazards:* n/a

Recommended Protection for Response Personnel:

Avoid breathing vapors, keep upwind. Wear a sealed chemical suit (butyl rubber, viton, PVC, chlorinated polyethylene, nitrile). Wash away any material which may have come into contact with the body with copious amounts of soap and water. Consider appropriate evacuation.

POTASSIUM NITRATE

DOT Number: UN 1486 *DOT Hazard Class:* Oxidizer *DOT Guide Number:* 35
Synonyms: niter; nitre; nitric acid, potassium salt; saltpeter
STCC Number: 4918736 *Reportable Qty:* n/a *CHEMTREC Phone No:* 1-800-424-9300

Physical Description:

Physical Form: Solid *Color:* White *Odor:* Odorless
Other Information: It is used for solid rocket propellants, explosives, pharmaceuticals, and meat processing.

Chemical Properties:

Specific Gravity: 2.1 *Vapor Density:* 3 *Boiling Point: Decomposes*
Melting Point: 633° F(333.8° C) *Vapor Pressure:* n/a *Solubility in water:* Yes
Other Information: Will sink and dissolve fairly rapidly. Weighs 131.6 lbs per cubic foot.

Health Hazards:

Inhalation Hazard: Irritating.
Ingestion Hazard: Nausea, dizziness.
Absorption Hazard: Irritating to the skin and eyes.
Hazards to Wildlife: n/a
Decontamination Procedures: Wash away any material with copious amounts of soap and water.
First Aid Procedures: Remove victim to fresh air, call emergency medical care. If not breathing give CPR. If breathing is difficult administer oxygen. Treat for shock.

Fire Hazards:

Flashpoint: n/a *Ignition temperature:* n/a
Flammable Explosive High Range: n/a *Low Range:* n/a
Toxic Products of Combustion: Toxic oxides of nitrogen are produced.
Other Hazards: May accelerate the burning of combustible materials. Containers may explode in a fire.
Possible extinguishing agents: Flood with water. Cool all affected containers. Apply water from as far a distance as possible.

Reactivity Hazards:

Reactive With: Oxidizable materials. *Other Reactions:* n/a

Corrosivity Hazards:

Corrosive With: n/a *Neutralizing Agent:* n/a

Radioactivity Hazards:

Radiation Emitted: n/a *Other Hazards:* n/a

Recommended Protection for Response Personnel:

Avoid breathing vapors, keep upwind. Structural protective clothing provides limited protection. Wash away any material which may have come into contact with the body with copious amounts of soap and water. Consider appropriate evacuation.

POTASSIUM OXALATE

DOT Number: n/a *DOT Hazard Class:* Poison B *DOT Guide Number:* n/a
Synonyms: potassium oxalate monohydrate
STCC Number: n/a *Reportable Qty:* n/a
Mfg Name: Pfizer Chemical *Phone No:* 1-201-546-7721

Physical Description:

Physical Form: Solid *Color:* White *Odor:* Odorless
Other Information: n/a

Chemical Properties:

Specific Gravity: 2.13 *Vapor Density:* n/a *Boiling Point: Decomposes*
Melting Point: n/a *Vapor Pressure:* n/a *Solubility in water:* Yes
Other Information: Sinks and mixes with water.

Health Hazards:

Inhalation Hazard: Coughing or difficulty in breathing.
Ingestion Hazard: Poisonous if swallowed.
Absorption Hazard: Irritating to the skin and eyes.
Hazards to Wildlife: n/a
Decontamination Procedures: Wash away any material with copious amounts of soap and water.
First Aid Procedures: Remove victim to fresh air, call emergency medical care. If not breathing give CPR. If breathing is difficult administer oxygen. Treat for shock.

Fire Hazards:

Flashpoint: n/a *Ignition temperature:* n/a
Flammable Explosive High Range: n/a *Low Range:* n/a
Toxic Products of Combustion: n/a
Other Hazards: n/a
Possible extinguishing agents: Extinguish fire using suitable agent for the type of surrounding fire.

Reactivity Hazards:

Reactive With: n/a *Other Reactions:* n/a

Corrosivity Hazards:

Corrosive With: n/a *Neutralizing Agent:* n/a

Radioactivity Hazards:

Radiation Emitted: n/a *Other Hazards:* n/a

Recommended Protection for Response Personnel:

Avoid breathing vapors, keep upwind. Wear a sealed chemical suit (polycarbonate, butyl rubber). Wash away any material which may have come into contact with the body with copious amounts of soap and water. Consider appropriate evacuation.

POTASSIUM PERMANGANATE

DOT Number: UN 1490 *DOT Hazard Class:* Oxidizer *DOT Guide Number:* 35
Synonyms: none given
STCC Number: 4918740 *Reportable Qty:* 100/45.4
Mfg Name: Mallinckrodt Inc. *Phone No:* 1-314-895-2000

Physical Description:

Physical Form: Solid *Color:* Purple *Odor:* Odorless
Other Information: It is used to make other chemicals, a disinfectant, and for many other uses.

Chemical Properties:

Specific Gravity: 2.70 *Vapor Density:* 5.4 *Boiling Point: Decomposes*
Melting Point: n/a *Vapor Pressure:* n/a *Solubility in water:* Yes
Other Information: Sinks and mixes slowly with water.

Health Hazards:

Inhalation Hazard: n/a
Ingestion Hazard: Nausea, vomiting, loss of consciousness.
Absorption Hazard: Irritating to the skin and eyes.
Hazards to Wildlife: Dangerous to aquatic life.
Decontamination Procedures: Wash away any material with copious amounts of soap and water.
First Aid Procedures: Remove victim to fresh air, call emergency medical care. If not breathing give CPR. If breathing is difficult administer oxygen. Treat for shock.

Fire Hazards:

Flashpoint: n/a *Ignition temperature:* n/a
Flammable Explosive High Range: n/a *Low Range:* n/a
Toxic Products of Combustion: May cause fire on contact with combustibles.
Other Hazards: Containers may explode in a fire.
Possible *Possible extinguishing agents:* Flood with water.

Reactivity Hazards:

Reactive With: Rubbers and most fibers
Other Reactions: May cause ignition of wood. Some acids such as sulfuric acid may cause an explosion.

Corrosivity Hazards:

Corrosive With: n/a *Neutralizing Agent:* n/a

Radioactivity Hazards:

Radiation Emitted: n/a *Other Hazards:* n/a

Recommended Protection for Response Personnel:

Avoid breathing vapors, keep upwind. Wear a sealed chemical suit (polycarbonate, butyl rubber). Wash away any material which may have come into contact with the body with copious amounts of soap and water. Consider appropriate evacuation.

POTASSIUM PEROXIDE

DOT Number: UN 1491 *DOT Hazard Class:* Oxidizer *DOT Guide Number:* 47
Synonyms: potassium superoxide
STCC Number: 4918530 *Reportable Qty:* n/a
Mfg Name: Ventron Inc. *Phone No:* 1-617-774-3100

Physical Description:

Physical Form: Solid *Color:* Yellow *Odor:* Odorless
Other Information: It is used as a bleach, bleaching and oxidizing agent.

Chemical Properties:

Specific Gravity: 1 *Vapor Density:* 3.8 *Boiling Point: Decomposes*
Melting Point: n/a *Vapor Pressure:* n/a *Solubility in water:* Yes
Other Information: Sinks and mixes violently with water.

Health Hazards:

Inhalation Hazard: Coughing or difficulty in breathing.
Ingestion Hazard: Causes nausea.
Absorption Hazard: Will burn the skin and the eyes.
Hazards to Wildlife: Dangerous to aquatic life.
Decontamination Procedures: Wash away any material with copious amounts of soap and water.
First Aid Procedures: Remove victim to fresh air, call emergency medical care. If not breathing give CPR. If breathing is difficult administer oxygen. Treat for shock.

Fire Hazards:

Flashpoint: n/a *Ignition temperature:* n/a
Flammable Explosive High Range: n/a *Low Range:* n/a
Toxic Products of Combustion: n/a
Other Hazards: Increases intensity and can start fire when in contact with organic combustibles.
Possible extinguishing agents: Dry chemical, graphite, or dry earth.

Reactivity Hazards:

Reactive With: Water with liberation of heat and oxygen and the formation of caustic solution.
Other Reactions: Can form explosive and selfigniting mixtures with wood and other combustible materials.

Corrosivity Hazards:

Corrosive With: n/a *Neutralizing Agent:* n/a

Radioactivity Hazards:

Radiation Emitted: n/a *Other Hazards:* n/a

Recommended Protection for Response Personnel:

Avoid breathing vapors, keep upwind. Wear a sealed chemical suit (polycarbonate, butyl rubber). Wash away any material which may have come into contact with the body with copious amounts of soap and water. If the fire becomes uncontrollable, consider appropriate evacuation.

PROPANE

DOT Number: UN 1978 *DOT Hazard Class:* Flammable gas *DOT Guide Number:* 22
Synonyms: dimethyl methane
STCC Number: 4905781 *Reportable Qty:* n/a
Mfg Name: Shell Oil Co. *Phone No:* 1-713-241-6161

Physical Description:

Physical Form: Gas *Color:* Colorless *Odor:* Odorless, may have skunk odor
Other Information: It is shipped as a liquefied gas under its own vapor pressure. For transportation, it may be stenched.

Chemical Properties:

Specific Gravity: .59 *Vapor Density:* 1.6 *Boiling Point: −44° F(−42.2° C)*
Melting Point: −360° F(−217.7° C) Vapor Pressure: 8.6 atm at 68° F(20° C)
Solubility in water: Insoluble
Other Information: Liquid floats and boils on water. A flammable, visible vapor cloud is produced.

Health Hazards:

Inhalation Hazard: Dizziness, difficulty in breathing, loss of consciousness.
Ingestion Hazard: n/a
Absorption Hazard: May cause frostbite.
Hazards to Wildlife: No
Decontamination Procedures: Wash away any material with copious amounts of soap and water.
First Aid Procedures: Remove victim to fresh air, call emergency medical care. If not breathing give CPR. If breathing is difficult administer oxygen. Treat for shock.

Fire Hazards:

Flashpoint: −156° F(−104.4° C) *Ignition temperature:* 842° F(450° C)
Flammable Explosive High Range: 9.5 *Low Range:* 2.1
Toxic Products of Combustion: n/a
Other Hazards: Containers may explode in fire. Flashback along vapor trail may occur. Vapors may explode if ignited in an enclosed area.
Possible extinguishing agents: Apply water from as far a distance as possible.

Reactivity Hazards:

Reactive With: n/a *Other Reactions:* n/a

Corrosivity Hazards:

Corrosive With: n/a *Neutralizing Agent:* n/a

Radioactivity Hazards:

Radiation Emitted: n/a *Other Hazards:* n/a

Recommended Protection for Response Personnel:

Avoid breathing vapors, keep upwind. Structural protective clothing provides limited protection. Wash away any material which may have come into contact with the body with copious amounts of soap and water. If the fire becomes uncontrollable, consider appropriate evacuation. Under fire conditions, cylinders or tank cars may violently rupture/rocket. (BLEVE)

PROPARGITE

DOT Number: NA 2765 *DOT Hazard Class:* ORM-E *DOT Guide Number:* 55
Synonyms: Do 14, Naugatuck-do 14, Omite, propargil liquid
STCC Number: 4961165 *Reportable Qty:* 10/4.54 *CHEMTREC Phone No:* 1-800-424-9300

Physical Description:

Physical Form: Liquid *Color:* Dark amber *Odor:* Odorless
Other Information: In case of damage to, or leaking from these containers, contact the Pesticide Safety Network Team at 1-880-424-9300.

Chemical Properties:

Specific Gravity: 1.08 *Vapor Density:* n/a *Boiling Point: Decomposes*
Melting Point: n/a *Vapor Pressure:* n/a *Solubility in water:* Yes
Other Information: Sinks and mixes slowly with water. Material is used as a pesticide.

Health Hazards:

Inhalation Hazard: Poisonous if inhaled.
Ingestion Hazard: Harmful if swallowed.
Absorption Hazard: Irritating to the skin and eyes.
Hazards to Wildlife: Dangerous to both aquatic life and waterfowl.
Decontamination Procedures: Wash away any material with copious amounts of soap and water.
First Aid Procedures: Remove victim to fresh air, call emergency medical care. If not breathing give CPR. If breathing is difficult administer oxygen. Treat for shock.

Fire Hazards:

Flashpoint: 82° F(27.7° C) *Ignition temperature:* n/a
Flammable Explosive High Range: n/a *Low Range:* n/a
Toxic Products of Combustion: Irritating fumes of sulfur oxide are produced.
Other Hazards: Containers may rupture in fire conditions.
Possible extinguishing agents: Extinguish fire using suitable agent for the type of surrounding fire.

Reactivity Hazards:

Reactive With: n/a *Other Reactions:* n/a

Corrosivity Hazards:

Corrosive With: n/a *Neutralizing Agent:* n/a

Radioactivity Hazards:

Radiation Emitted: n/a *Other Hazards:* n/a

Recommended Protection for Response Personnel:

Avoid breathing vapors, keep upwind. Wear appropriate chemical clothing. Wash away any material which may have come into contact with the body with copious amounts of soap and water. Consider appropriate evacuation.

PROPIOLACTONE-β

DOT Number: n/a *DOT Hazard Class:* n/a *DOT Guide Number:* n/a
Synonyms: β-propiolactone, β-propionolactone
STCC Number: n/a *Reportable Qty:* n/a
Mfg Name: Eastman Organic *Phone No:* 1-800-445-6325

Physical Description:

Physical Form: Liquid *Color:* Colorless *Odor:* Irritating
Other Information: n/a

Chemical Properties:

Specific Gravity: 1.1 *Vapor Density:* 2.5 *Boiling Point: 311° F(155° C)*
Melting Point: −28° F(−33.3° C) *Vapor Pressure:* 3.4 mm Hg at 77° F(25° C)
Solubility in water: Yes *Degree of Solubility:* 37 v/v
Other Information: Mixes with water.

Health Hazards:

Inhalation Hazard: n/a
Ingestion Hazard: Poisonous if swallowed.
Absorption Hazard: Poisonous if the skin is exposed.
Hazards to Wildlife: n/a
Decontamination Procedures: Wash away any material with copious amounts of soap and water.
First Aid Procedures: Remove victim to fresh air, call emergency medical care. If not breathing give CPR. If breathing is difficult administer oxygen. Treat for shock.

Fire Hazards:

Flashpoint: 165° F(73.8° C) *Ignition temperature:* n/a
Flammable Explosive High Range: n/a *Low Range:* 2.9
Toxic Products of Combustion: Vapors of unburned material are very toxic.
Other Hazards: Containers may explode in fire.
Possible extinguishing agents: Use water, foam, dry chemical or carbon dioxide.

Reactivity Hazards:

Reactive With: Water to form propiolactone acid. *Other Reactions:* n/a

Corrosivity Hazards:

Corrosive With: n/a *Neutralizing Agent:* n/a

Radioactivity Hazards:

Radiation Emitted: n/a *Other Hazards:* n/a

Recommended Protection for Response Personnel:

Avoid breathing vapors, keep upwind. Wear a sealed chemical suit (polycarbonate, butyl rubber). Wash away any material which may have come into contact with the body with copious amounts of soap and water. Consider appropriate evacuation.

PROPIONALDEHYDE

DOT Number: UN 1275 *DOT Hazard Class:* Flammable liquid *DOT Guide Number:* 26
Synonyms: propanal, propionic aldehyde, propyl aldehyde
STCC Number: 4908270 *Reportable Qty:* n/a
Mfg Name: Union Carbide *Phone No:* 1-203-794-2000

Physical Description:

Physical Form: Liquid *Color:* Colorless *Odor:* Suffocating, unpleasant
Other Information: n/a

Chemical Properties:

Specific Gravity: .8 *Vapor Density:* 2 *Boiling Point: 120° F(48.8° C)*
Melting Point: n/a *Vapor Pressure:* n/a *Solubility in water:* Yes
Other Information: Floats and mixes slowly with water. Flammable irritating vapor is produced.

Health Hazards:

Inhalation Hazard: Causes nausea and vomiting.
Ingestion Hazard: Harmful if swallowed.
Absorption Hazard: Irritating to the skin and eyes.
Hazards to Wildlife: n/a
Decontamination Procedures: Wash away any material with copious amounts of soap and water.
First Aid Procedures: Remove victim to fresh air, call emergency medical care. If not breathing give CPR. If breathing is difficult administer oxygen. Treat for shock.

Fire Hazards:

Flashpoint: −22° F(−30° C) *Ignition temperature:* 405° F(207.2° C)
Flammable Explosive High Range: 17 *Low Range:* 2.6
Toxic Products of Combustion: n/a
Other Hazards: Flashback along vapor trail may occur. Vapors may explode if ignited in an enclosed area.
Possible extinguishing agents: Do not extinguish the fire unless the flow can be stopped. Apply water from as far a distance as possible. Use alcohol foam, dry chemical, or carbon dioxide.

Reactivity Hazards:

Reactive With: n/a *Other Reactions:* n/a

Corrosivity Hazards:

Corrosive With: n/a *Neutralizing Agent:* n/a

Radioactivity Hazards:

Radiation Emitted: n/a *Other Hazards:* n/a

Recommended Protection for Response Personnel:

Avoid breathing vapors, keep upwind. Structural protective clothing provides limited protection. Wash away any material which may have come into contact with the body with copious amounts of soap and water. If the fire becomes uncontrollable, consider appropriate evacuation.

PROPIONIC ACID

DOT Number: UN 1848 *DOT Hazard Class:* Corrosive DOT Guide Number: 60
Synonyms: methanecarboxylic acid, propanoic acid
STCC Number: 4931448 *Reportable Qty:* 5000/2270
Mfg Name: Mallinckrodt Inc. *Phone No:* 1-314-895-2000

Physical Description:

Physical Form: Liquid *Color:* Colorless *Odor:* Sharp, rancid
Other Information: n/a

Chemical Properties:

Specific Gravity: .99 *Vapor Density:* 2.5 *Boiling Point: 285° F(140.5° C)*
Melting Point: n/a *Vapor Pressure:* 10 mm Hg at 103° F(39.4° C) *Solubility in water:* Yes
Other Information: Mixes with water. Irritating vapor is produced. Weighs 8.3 lbs/3.7 kg per gallon/3.8 l.

Health Hazards:

Inhalation Hazard: n/a
Ingestion Hazard: Harmful if swallowed.
Absorption Hazard: Will burn the skin and the eyes.
Hazards to Wildlife: Dangerous to aquatic life.
Decontamination Procedures: Wash away any material with copious amounts of soap and water.
First Aid Procedures: Remove victim to fresh air, call emergency medical care. If not breathing give CPR. If breathing is difficult administer oxygen. Treat for shock.

Fire Hazards:

Flashpoint: 126° F(52.2° C) *Ignition temperature:* 870° F(465.5° C)
Flammable Explosive High Range: 12.1 *Low Range:* 2.9
Toxic Products of Combustion: n/a
Other Hazards: n/a
Possible extinguishing agents: Apply water from as far a distance as possible. Use alcohol foam, dry chemical, or carbon dioxide.

Reactivity Hazards:

Reactive With: n/a *Other Reactions:* n/a

Corrosivity Hazards:

Corrosive With: Metals and tissue *Neutralizing Agent:* Crushed limestone, soda ash, or lime.

Radioactivity Hazards:

Radiation Emitted: n/a *Other Hazards:* n/a

Recommended Protection for Response Personnel:

Avoid breathing vapors, keep upwind. Structural protective clothing provides limited protection. Wash away any material which may have come into contact with the body with copious amounts of soap and water. Consider appropriate evacuation.

PROPIONIC ANHYDRIDE

DOT Number: UN 2496 *DOT Hazard Class:* Corrosive *DOT Guide Number:* 29
Synonyms: methylacetic anhydride, propanoic anhydride
STCC Number: 4931449 *Reportable Qty:* 5000/2270
Mfg Name: Union Carbide *Phone No:* 1-203-794-2000

Physical Description:

Physical Form: Liquid *Color:* Colorless *Odor:* Sharp
Other Information: n/a

Chemical Properties:

Specific Gravity: 1 *Vapor Density:* 4.5 *Boiling Point: 336° F(168.8° C)*
Melting Point: n/a *Vapor Pressure:* 1 mm Hg at 68° F(20° C) *Solubility in water:* Decomposes
Other Information: Sinks and mixes slowly with water. Weighs 8.4 lbs/3.8 kg per gallon/3.8 l.

Health Hazards:

Inhalation Hazard: n/a
Ingestion Hazard: Harmful if swallowed.
Absorption Hazard: Will burn the skin and the eyes.
Hazards to Wildlife: Dangerous to aquatic life.
Decontamination Procedures: Wash away any material with copious amounts of soap and water.
First Aid Procedures: Remove victim to fresh air, call emergency medical care. If not breathing give CPR. If breathing is difficult administer oxygen. Treat for shock.

Fire Hazards:

Flashpoint: 145° F(62.7° C) *Ignition temperature:* 545° F(285° C)
Flammable Explosive High Range: 9.5 *Low Range:* 1.3
Toxic Products of Combustion: n/a
Other Hazards: n/a
Possible extinguishing agents: Apply water from as far a distance as possible. Use alcohol foam, dry chemical, or carbon dioxide.

Reactivity Hazards:

Reactive With: Water to form a weak propionic acid. *Other Reactions:* n/a

Corrosivity Hazards:

Corrosive With: Metals and tissue. *Neutralizing Agent:* Crushed limestone, soda ash, or lime.

Radioactivity Hazards:

Radiation Emitted: n/a *Other Hazards:* n/a

Recommended Protection for Response Personnel:

Avoid breathing vapors, keep upwind. Structural protective clothing provides limited protection. Wash away any material which may have come into contact with the body with copious amounts of soap and water. Consider appropriate evacuation.

PROPYL ACETATE

DOT Number: UN 1276 *DOT Hazard Class:* Flammable liquid *DOT Guide Number:* 26
Synonyms: acetic acid, n-propyl ester
STCC Number: 4909268 *Reportable Qty:* n/a
Mfg Name: Ashland Chemical *Phone No:* 1-614-889-3333

Physical Description:

Physical Form: Liquid *Color:* Colorless *Odor:* Mild
Other Information: n/a

Chemical Properties:

Specific Gravity: .9 *Vapor Density:* 3.5 *Boiling Point: 215° F(101.6° C)*
Melting Point: −140° F(−95.5° C) *Vapor Pressure:* 25 mm Hg at 68° F(20° C)
Solubility in water: Slight *Degree of Solubility:* 2%
Other Information: Floats. Lighter than and slightly soluble in water. Vapors are heavier than air.

Health Hazards:

Inhalation Hazard: Nausea, vomiting, dizziness, loss of consciousness.
Ingestion Hazard: Harmful if swallowed.
Absorption Hazard: Irritating to the skin and eyes.
Hazards to Wildlife: Dangerous to aquatic life.
Decontamination Procedures: Wash away any material with copious amounts of soap and water.
First Aid Procedures: Remove victim to fresh air, call emergency medical care. If not breathing give CPR. If breathing is difficult administer oxygen. Treat for shock.

Fire Hazards:

Flashpoint: 55° F(12.7° C) *Ignition temperature:* 842° F(450° C)
Flammable Explosive High Range: 8 *Low Range:* 1.7
Toxic Products of Combustion: n/a
Other Hazards: Flashback along vapor trail may occur. Vapors may explode if ignited in an enclosed area.
Possible extinguishing agents: Apply water from as far a distance as possible. Use alcohol foam, dry chemical, or carbon dioxide.

Reactivity Hazards:

Reactive With: Water to form a flammable irritating vapor. *Other Reactions:* None

Corrosivity Hazards:

Corrosive With: n/a *Neutralizing Agent:* n/a

Radioactivity Hazards:

Radiation Emitted: n/a *Other Hazards:* n/a

Recommended Protection for Response Personnel:

Avoid breathing vapors, keep upwind. Wear a sealed chemical suit (butyl rubber, chlorinated polyethylene, PVC). Wash away any material which may have come into contact with the body with copious amounts of soap and water. Consider appropriate evacuation.

PROPYL ALCOHOL

DOT Number: UN 1274 *DOT Hazard Class:* Flammable liquid *DOT Guide Number:* 26
Synonyms: 1-propanol, n-propyl alcohol
STCC Number: n/a *Reportable Qty:* n/a
Mfg Name: Mallinckrodt Inc. *Phone No:* 1-314-895-2000

Physical Description:

Physical Form: Liquid *Color:* Colorless *Odor:* Alcohol
Other Information: n/a

Chemical Properties:

Specific Gravity: .8 *Vapor Density:* 2.1 *Boiling Point: 207° F(97.2° C)*
Melting Point: −195° F(−126.1° C) *Vapor Pressure:* 15 mm Hg at 68° F(20° C)
Solubility in water: Yes
Other Information: n/a

Health Hazards:

Inhalation Hazard: Nausea, dizziness, headache.
Ingestion Hazard: Harmful if swallowed.
Absorption Hazard: Will burn the eyes.
Hazards to Wildlife: Dangerous to aquatic life.
Decontamination Procedures: Wash away any material with copious amounts of soap and water.
First Aid Procedures: Remove victim to fresh air, call emergency medical care. If not breathing give CPR. If breathing is difficult administer oxygen. Treat for shock.

Fire Hazards:

Flashpoint: 74° F(23.3° C) *Ignition temperature:* 775° F(412.7° C)
Flammable Explosive High Range: 13.7 *Low Range:* 2.2
Toxic Products of Combustion: n/a
Other Hazards: Flashback along vapor trail may occur. Vapors may explode if ignited in an enclosed area.
Possible extinguishing agents: Apply water from as far a distance as possible. Use alcohol foam, dry chemical, or carbon dioxide.

Reactivity Hazards:

Reactive With: n/a *Other Reactions:* n/a

Corrosivity Hazards:

Corrosive With: n/a *Neutralizing Agent:* n/a

Radioactivity Hazards:

Radiation Emitted: n/a *Other Hazards:* n/a

Recommended Protection for Response Personnel:

Avoid breathing vapors, keep upwind. Structural protective clothing provides limited protection. Wash away any material which may have come into contact with the body with copious amounts of soap and water. Consider appropriate evacuation.

PROPYLAMINE

DOT Number: UN 1277 *DOT Hazard Class:* Flammable liquid *DOT Guide Number:* 68
Synonyms: 1-aminopropane
STCC Number: 4908269 *Reportable Qty:* 5000/2270
Mfg Name: Pennwald Corp. *Phone No:* 1-205-587-7000

Physical Description:

Physical Form: Liquid *Color:* Colorless *Odor:* Ammonia
Other Information: It is used in chemical analysis and in making other chemicals.

Chemical Properties:

Specific Gravity: .7 *Vapor Density:* 2 *Boiling Point: 120° F(48.8° C)*
Melting Point: n/a *Vapor Pressure:* 248 mm Hg at 68° F(20° C) *Solubility in water:* Yes
Other Information: Lighter than and soluble in water. Vapors are heavier than air.

Health Hazards:

Inhalation Hazard: Harmful if inhaled.
Ingestion Hazard: Harmful if swallowed.
Absorption Hazard: Will burn the eyes.
Hazards to Wildlife: Dangerous to aquatic life.
Decontamination Procedures: Wash away any material with copious amounts of soap and water.
First Aid Procedures: Remove victim to fresh air, call emergency medical care. If not breathing give CPR. If breathing is difficult administer oxygen. Treat for shock.

Fire Hazards:

Flashpoint: −35° F(−37.2° C) *Ignition temperature:* 604° F(317.7° C)
Flammable Explosive High Range: 10.4 *Low Range:* 2
Toxic Products of Combustion: Toxic oxides of nitrogen are produced.
Other Hazards: Flashback along vapor trail may occur. Vapors may explode if ignited in an enclosed area.
Possible extinguishing agents: Apply water from as far a distance as possible. Use alcohol foam, dry chemical, or carbon dioxide.

Reactivity Hazards:

Reactive With: n/a *Other Reactions:* n/a

Corrosivity Hazards:

Corrosive With: n/a *Neutralizing Agent:* n/a

Radioactivity Hazards:

Radiation Emitted: n/a *Other Hazards:* n/a

Recommended Protection for Response Personnel:

Avoid breathing vapors, keep upwind. Wear appropriate chemical clothing. Wash away any material which may have come into contact with the body with copious amounts of soap and water. If fire becomes uncontrollable, consider appropriate evacuation.

PROPYLENE

DOT Number: UN 1077 *DOT Hazard Class:* Flammable gas *DOT Guide Number:* 22
Synonyms: propene
STCC Number: 4905782 *Reportable Qty:* n/a
Mfg Name: Exxon Chemical America *Phone No:* 1-713-870-6000

Physical Description:

Physical Form: Gas *Color:* Colorless *Odor:* Mild odor
Other Information: It is used in making other chemicals. It is shipped as a liquefied gas under its own vapor pressure. For transportation, material may be stenched!

Chemical Properties:

Specific Gravity: .1 *Vapor Density:* 1.5 *Boiling Point: −53° F(−47.2° C)*
Melting Point: n/a *Vapor Pressure:* 10 mm Hg at 68° F(20° C) *Solubility in water:* Yes
Other Information: Floats, boils on water. A flammable, visible vapor cloud is produced.

Health Hazards:

Inhalation Hazard: Dizziness, loss of consciousness.
Ingestion Hazard: n/a
Absorption Hazard: Will cause frostbite.
Hazards to Wildlife: No.
Decontamination Procedures: Wash away any material with copious amounts of soap and water.
First Aid Procedures: Remove victim to fresh air, call emergency medical care. If not breathing give CPR. If breathing is difficult administer oxygen. Treat for shock.

Fire Hazards:

Flashpoint: −162° F(−107.7° C) *Ignition temperature:* 851° F(455° C)
Flammable Explosive High Range: 11.1 *Low Range:* 2
Toxic Products of Combustion: n/a
Other Hazards: Flashback along vapor trail may occur. Vapors may explode if ignited in an enclosed area. Containers may explode in a fire.
Possible extinguishing agents: Apply water from as far a distance as possible.

Reactivity Hazards:

Reactive With: n/a *Other Reactions:* n/a

Corrosivity Hazards:

Corrosive With: n/a *Neutralizing Agent:* n/a

Radioactivity Hazards:

Radiation Emitted: n/a *Other Hazards:* n/a

Recommended Protection for Response Personnel:

Avoid breathing vapors, keep upwind. Wear a sealed chemical suit (viton, chlorinated polyethylene). Wash away any material which may have come into contact with the body with copious amounts of soap and water. If the fire becomes uncontrollable, consider appropriate evacuation. Under fire conditions, cylinders may violently rupture/rocket. (BLEVE)

PROPYLENE GLYCOL

DOT Number: n/a *DOT Hazard Class:* n/a *DOT Guide Number:* n/a
Synonyms: 1,2-dihydroxpropane, methylene glycol, 1,2-propanediol
STCC Number: n/a *Reportable Qty:* n/a
Mfg Name: Dow Chemical Co. *Phone No:* 1-517-636-4400

Physical Description:

Physical Form: Liquid *Color:* Colorless *Odor:* Odorless
Other Information: n/a

Chemical Properties:

Specific Gravity: 1.04 *Vapor Density:* 2.6 *Boiling Point: 369° F(187.2° C)*
Melting Point: −76° F(−60° C) *Vapor Pressure:* .08 mm Hg at 68° F(20° C) *Solubility in water:* Yes
Other Information: Mixes with water

Health Hazards:

Inhalation Hazard: n/a
Ingestion Hazard: n/a
Absorption Hazard: n/a
Hazards to Wildlife: n/a
Decontamination Procedures: Wash away any material with copious amounts of soap and water.
First Aid Procedures: Remove victim to fresh air, call emergency medical care. If not breathing give CPR. If breathing is difficult administer oxygen. Treat for shock.

Fire Hazards:

Flashpoint: 210° F(98.8° C) *Ignition temperature:* 790° F(421.1° C)
Flammable Explosive High Range: 12.5 *Low Range:* 2.6
Toxic Products of Combustion: n/a
Other Hazards: n/a
Possible extinguishing agents: Water, alcohol foam, dry chemical, carbon dioxide.

Reactivity Hazards:

Reactive With: n/a *Other Reactions:* n/a

Corrosivity Hazards:

Corrosive With: n/a *Neutralizing Agent:* n/a

Radioactivity Hazards:

Radiation Emitted: n/a *Other Hazards:* n/a

Recommended Protection for Response Personnel:

Avoid breathing vapors, keep upwind. Structural protective clothing provides limited protection. Wash away any material which may have come into contact with the body with copious amounts of soap and water. If the fire becomes uncontrollable, consider appropriate evacuation.

PROPYLENE OXIDE

DOT Number: UN 1280 *DOT Hazard Class:* Flammable liquid *DOT Guide Number:* 26
Synonyms: 1,2-epoxypropane
STCC Number: 4906620 *Reportable Qty:* 100/45.4
Mfg Name: Dow Chemical Co. *Phone No:* 1-517-636-4400

Physical Description:

Physical Form: Liquid *Color:* Colorless *Odor:* Sweet alcohol
Other Information: It is used as a fumigant in making detergents and lubricants, and to make other chemicals.

Chemical Properties:

Specific Gravity: .83 *Vapor Density:* 2 *Boiling Point: 94° F(34.4° C)*
Melting Point: −170° F(−112.2° C) *Vapor Pressure:* 442 mm Hg at 68° F(20° C)
Solubility in water: Yes *Degree of Solubility:* 41%
Other Information: Lighter than and soluble in water. Vapors are heavier than air.

Health Hazards:

Inhalation Hazard: Headache, nausea, vomiting, loss of consciousness.
Ingestion Hazard: Harmful if swallowed.
Absorption Hazard: Will burn the skin and the eyes. *Hazards to Wildlife:* n/a
Decontamination Procedures: Wash away any material with copious amounts of soap and water.
First Aid Procedures: Remove victim to fresh air, call emergency medical care. If not breathing give CPR. If breathing is difficult administer oxygen. Treat for shock.

Fire Hazards:

Flashpoint: −35° F(−37.2° C) *Ignition temperature:* 840° F(448.8° C)
Flammable Explosive High Range: 36 *Low Range:* 2.3
Toxic Products of Combustion: n/a
Other Hazards: Flashback along vapor trail may occur. Vapors may explode if ignited in an enclosed area. Containers may explode in a fire. If contaminated, polymerization occurs with evolution of heat, possible rupture of containers may occur.
Possible extinguishing agents: Apply water from as far a distance as possible. Use alcohol foam, dry chemical, or carbon dioxide.

Reactivity Hazards:

Reactive With: n/a *Other Reactions:* n/a

Corrosivity Hazards:

Corrosive With: n/a *Neutralizing Agent:* n/a

Radioactivity Hazards:

Radiation Emitted: n/a *Other Hazards:* n/a

Recommended Protection for Response Personnel:

Avoid breathing vapors, keep upwind. Structural protective clothing provides limited protection. Wash away any material which may have come into contact with the body with copious amounts of soap and water. If the fire becomes uncontrollable, consider appropriate evacuation. Under fire conditions, the cylinders may rupture/rocket. (BLEVE)

PROPYLENE TETRAMER

DOT Number: UN 2850 *DOT Hazard Class:* n/a *DOT Guide Number:* 27
Synonyms: dodecene(nonlinear), tetrapropylene
STCC Number: n/a *Reportable Qty:* n/a
Mfg Name: Atlantic Richfield Co. *Phone No:* 1-213-486-3511

Physical Description:

Physical Form: Liquid *Color:* Colorless *Odor:* n/a
Other Information: n/a

Chemical Properties:

Specific Gravity: .29 *Vapor Density:* 1 *Boiling Point: 365-385° F(185-196° C)*
Melting Point: n/a *Vapor Pressure:* n/a *Solubility in water:* n/a
Other Information: Floats on water.

Health Hazards:

Inhalation Hazard: n/a
Ingestion Hazard: Harmful if swallowed.
Absorption Hazard: Irritating to the skin and eyes.
Hazards to Wildlife: n/a
Decontamination Procedures: Wash away any material with copious amounts of soap and water.
First Aid Procedures: Remove victim to fresh air, call emergency medical care. If not breathing give CPR. If breathing is difficult administer oxygen. Treat for shock.

Fire Hazards:

Flashpoint: 120° F(48.8° C) *Ignition temperature:* 400° F(204.4° C)
Flammable Explosive High Range: n/a *Low Range:* n/a
Toxic Products of Combustion: n/a
Other Hazards: n/a
Possible extinguishing agents: Water, dry chemical, carbon dioxide. Cool all exposed containers with water.

Reactivity Hazards:

Reactive With: n/a *Other Reactions:* n/a

Corrosivity Hazards:

Corrosive With: n/a *Neutralizing Agent:* n/a

Radioactivity Hazards:

Radiation Emitted: n/a *Other Hazards:* n/a

Recommended Protection for Response Personnel:

Avoid breathing vapors, keep upwind. Structural protective clothing provides limited protection. Wash away any material which may have come into contact with the body with copious amounts of soap and water. If the fire becomes uncontrollable, consider appropriate evacuation.

PROPYLENEIMINE

DOT Number: UN 1921 *DOT Hazard Class:* Flammable liquid *DOT Guide Number:* 30
Synonyms: 2-methylazaridine, propyleneimine
STCC Number: 4907040 *Reportable Qty:* 1/.454
Mfg Name: Polysciences Inc. *Phone No:* 1-215-343-6484

Physical Description:

Physical Form: Liquid *Color:* Colorless *Odor:* Strong ammonialike
Other Information: It is used as an organic intermediate.

Chemical Properties:

Specific Gravity: .8 *Vapor Density:* 2 *Boiling Point: 151° F(66.1° C)*
Melting Point: −85° F(−65° C) *Vapor Pressure:* n/a *Solubility in water:* Yes
Other Information: Mixes with water. Flammable, irritating vapor is produced.

Health Hazards:

Inhalation Hazard: Nausea, vomiting, difficulty in breathing.
Ingestion Hazard: Poisonous if swallowed.
Absorption Hazard: Will burn the skin and eyes.
Hazards to Wildlife: n/a
Decontamination Procedures: Wash away any material with copious amounts of soap and water.
First Aid Procedures: Remove victim to fresh air, call emergency medical care. If not breathing give CPR. If breathing is difficult administer oxygen. Treat for shock.

Fire Hazards:

Flashpoint: 25° F(−3.8° C) *Ignition temperature:* n/a
Flammable Explosive High Range: n/a *Low Range:* n/a
Toxic Products of Combustion: Irritating nitrogen oxides are produced.
Other Hazards: Containers may explode in a fire. Flashback along vapor trail may occur. Vapors may explode if ignited in an enclosed area.
Possible extinguishing agents: Apply water from as far a distance as possible. Use alcohol foam, dry chemical, or carbon dioxide.

Reactivity Hazards:

Reactive With: Reacts slowly with water to form propanolamine. *Other Reactions:* n/a

Corrosivity Hazards:

Corrosive With: n/a *Neutralizing Agent:* n/a

Radioactivity Hazards:

Radiation Emitted: n/a *Other Hazards:* n/a

Recommended Protection for Response Personnel:

Avoid breathing vapors, keep upwind, wear a sealed chem suit, (butyl rubber). Wash away any material which may have come into contact with the body with copious amounts of soap and water. If the fire becomes uncontrollable consider evacuation. Prolonged exposure of the containers to heat or flame may cause the cylinders to violently rupture/rocket. (BLEVE!!)

PROPYL MERCAPTAN

DOT Number: UN 2402 *DOT Hazard Class:* Flammable liquid *DOT Guide Number:* 27
Synonyms: 1-propanethiol, propane-1-thiol
STCC Number: n/a *Reportable Qty:* n/a
Mfg Name: Phillips Petroleum *Phone No:* 1-918-661-6600

Physical Description:

Physical Form: Liquid *Color:* Colorless *Odor:* Skunklike
Other Information: n/a

Chemical Properties:

Specific Gravity: .84 *Vapor Density:* 2.6 *Boiling Point: 153° F(67.2° C)*
Melting Point: −171° F(−112.7° C) *Vapor Pressure:* n/a
Solubility in water: n/a
Other Information: Floats on water. Flammable, irritating vapor is produced.

Health Hazards:

Inhalation Hazard: Difficulty in breathing.
Ingestion Hazard: Harmful if swallowed.
Absorption Hazard: Irritating to the skin and eyes.
Hazards to Wildlife: n/a
Decontamination Procedures: Wash away any material with copious amounts of soap and water.
First Aid Procedures: Remove victim to fresh air, call emergency medical care. If not breathing give CPR. If breathing is difficult administer oxygen. Treat for shock.

Fire Hazards:

Flashpoint: 5° F(−15° C) *Ignition temperature:* n/a
Flammable Explosive High Range: n/a *Low Range:* n/a
Toxic Products of Combustion: Toxic sulfur dioxide is generated.
Other Hazards: Containers may explode in fire. Flashback along vapor trail may occur. Vapors may explode if ignited in an enclosed area.
Possible extinguishing agents: Water, foam, dry chemical, carbon dioxide. Water may be ineffective on fires. Apply water from as far a distance as possible.

Reactivity Hazards:

Reactive With: n/a *Other Reactions:* n/a

Corrosivity Hazards:

Corrosive With: n/a *Neutralizing Agent:* n/a

Radioactivity Hazards:

Radiation Emitted: n/a *Other Hazards:* n/a

Recommended Protection for Response Personnel:

Avoid breathing vapors, keep upwind. Wear a sealed chemical suit (polycarbonate, butyl rubber). Wash away any material which may have come into contact with the body with copious amounts of soap and water. If the fire becomes uncontrollable, consider appropriate evacuation.

PYRETHRINS ACID (liquid)

DOT Number: NA 9184 *DOT Hazard Class:* ORM-E *DOT Guide Number:* 31
Synonyms: Persian insect powder, pyrethrum flowers
STCC Number: 4963872 *Reportable Qty:* 1/.454
Mfg Name: FMC Corp. *Phone No:* 1-312-861-5900

Physical Description:

Physical Form: Liquid *Color:* Yellow to brown *Odor:* Characteristic of carrier
Other Information: In case of damage to, or leaking from these containers, contact the Pesticide Safety Network Team at 1-800-424-9300.

Chemical Properties:

Specific Gravity: n/a *Vapor Density:* 11 *Boiling Point: 338° F(170° C)*
Melting Point: n/a *Vapor Pressure:* n/a *Solubility in water:* Yes
Other Information: Sinks in water.

Health Hazards:

Inhalation Hazard: Sneezing, nasal discharge, nasal stuffiness.
Ingestion Hazard: Nausea, vomiting, headache.
Absorption Hazard: Irritating to the skin and eyes.
Hazards to Wildlife: Dangerous to both aquatic life and waterfowl.
Decontamination Procedures: Wash away any material with copious amounts of soap and water.
First Aid Procedures: Remove victim to fresh air, call emergency medical care. If not breathing give CPR. If breathing is difficult administer oxygen. Treat for shock.

Fire Hazards:

Flashpoint: n/a *Ignition temperature:* n/a
Flammable Explosive High Range: n/a *Low Range:* n/a
Toxic Products of Combustion: Highly toxic fumes are emitted.
Other Hazards: n/a
Possible extinguishing agents: Extinguish fire using suitable agent for the type of surrounding fire.

Reactivity Hazards:

Reactive With: n/a *Other Reactions:* n/a

Corrosivity Hazards:

Corrosive With: n/a *Neutralizing Agent:* n/a

Radioactivity Hazards:

Radiation Emitted: n/a *Other Hazards:* n/a

Recommended Protection for Response Personnel:

Avoid breathing vapors, keep upwind. Structural protective clothing provides limited protection. Wash away any material which may have come into contact with the body with copious amounts of soap and water. Consider appropriate evacuation.

PYRETHRINS ACID (solid)

DOT Number: NA 9184 *DOT Hazard Class:* ORM-E *DOT Guide Number:* 31
Synonyms: Persian insect powder, pyrethrum flowers
STCC Number: 4963877 *Reportable Qty:* 1/.454
Mfg Name: FMC Corp. *Phone No:* 1-312-861-5900

Physical Description:

Physical Form: Solid *Color:* Colorless to white *Odor:* Characteristic of carrier
Other Information: In case of damage to, or leaking from these containers, contact the Pesticide Safety Network Team at 1-800-424-9300.

Chemical Properties:

Specific Gravity: n/a *Vapor Density:* n/a *Boiling Point: 338° F(170° C)*
Melting Point: n/a *Vapor Pressure:* n/a *Solubility in water:* Yes
Other Information: Sinks in water.

Health Hazards:

Inhalation Hazard: Sneezing, nasal discharge, nasal stuffiness.
Ingestion Hazard: Nausea, vomiting, headache.
Absorption Hazard: Irritating to the skin and eyes.
Hazards to Wildlife: Dangerous to both aquatic life and waterfowl.
Decontamination Procedures: Wash away any material with copious amounts of soap and water.
First Aid Procedures: Remove victim to fresh air, call emergency medical care. If not breathing give CPR. If breathing is difficult administer oxygen. Treat for shock.

Fire Hazards:

Flashpoint: n/a *Ignition temperature:* n/a
Flammable Explosive High Range: n/a *Low Range:* n/a
Toxic Products of Combustion: Highly toxic fumes are emitted.
Other Hazards: n/a
Possible *Possible extinguishing agents:* Extinguish fire using suitable agent for the type of surrounding fire.

Reactivity Hazards:

Reactive With: n/a *Other Reactions:* n/a

Corrosivity Hazards:

Corrosive With: n/a *Neutralizing Agent:* n/a

Radioactivity Hazards:

Radiation Emitted: n/a *Other Hazards:* n/a

Recommended Protection for Response Personnel:

Avoid breathing vapors, keep upwind. Structural protective clothing provides limited protection. Wash away any material which may have come into contact with the body with copious amounts of soap and water. Consider appropriate evacuation.

PYRIDINE

DOT Number: UN 1282 *DOT Hazard Class:* Flammable liquid *DOT Guide Number:* 26
Synonyms: none given
STCC Number: 4909277 *Reportable Qty:* 1000/454
Mfg Name: Mallinckrodt Inc. *Phone No:* 1-314-694-1000

Physical Description:

Physical Form: Liquid *Color:* Colorless to yellow *Odor:* Sharp nauseating
Other Information: n/a

Chemical Properties:

Specific Gravity: 1 *Vapor Density:* 2.7 *Boiling Point: 239° F(115° C)*
Melting Point: −44° F(−42.2° C) *Vapor Pressure:* 18 mm Hg at 68° F(20° C)
Solubility in water: Yes
Other Information: Lighter than and soluble in water. Vapors are heavier than air.

Health Hazards:

Inhalation Hazard: Poisonous if inhaled.
Ingestion Hazard: Poisonous if swallowed.
Absorption Hazard: Poisonous if absorbed.
Hazards to Wildlife: Dangerous to aquatic life.
Decontamination Procedures: Wash away any material with copious amounts of soap and water.
First Aid Procedures: Remove victim to fresh air, call emergency medical care. If not breathing give CPR. If breathing is difficult administer oxygen. Treat for shock.

Fire Hazards:

Flashpoint: 68° F(20° C) *Ignition temperature:* 900° F(482.2° C)
Flammable Explosive High Range: 12.4 *Low Range:* 1.8
Toxic Products of Combustion: Toxic oxides of nitrogen are produced.
Other Hazards: Flashback along vapor trail may occur. Vapors may explode if ignited in an enclosed area.
Possible extinguishing agents: Apply water from as far a distance as possible. Use alcohol foam, dry chemical, or carbon dioxide.

Reactivity Hazards:

Reactive With: n/a *Other Reactions:* n/a

Corrosivity Hazards:

Corrosive With: n/a *Neutralizing Agent:* n/a

Radioactivity Hazards:

Radiation Emitted: n/a *Other Hazards:* n/a

Recommended Protection for Response Personnel:

Avoid breathing vapors, keep upwind. Wear a sealed chemical suit (butyl rubber). Wash away any material which may have come into contact with the body with copious amounts of soap and water. Consider appropriate evacuation.

PYROGALLIC ACID

DOT Number: n/a *DOT Hazard Class:* n/a *DOT Guide Number:* n/a
Synonyms: 1,2,3-benzenetriol, pyrogallol, 1,2,3-trihydroxybenzene
STCC Number: n/a *Reportable Qty:* n/a
Mfg Name: Aldrich Chemical *Phone No:* 1-414-273-3850

Physical Description:

Physical Form: Solid *Color:* White to gray *Odor:* Odorless
Other Information: n/a

Chemical Properties:

Specific Gravity: 1.45 *Vapor Density:* 4.3 *Boiling Point: 588° F(308.8° C)*
Melting Point: 268° F(131.1° C) *Vapor Pressure:* 10 mm Hg at 334° F(167.7° C) *Solubility in water:* Yes
Other Information: Sinks and mixes with water.

Health Hazards:

Inhalation Hazard: Coughing or difficulty in breathing.
Ingestion Hazard: Nausea, vomiting, loss of consciousness.
Absorption Hazard: Irritating to the skin and eyes.
Hazards to Wildlife: Dangerous to aquatic life.
Decontamination Procedures: Wash away any material with copious amounts of soap and water.
First Aid Procedures: Remove victim to fresh air, call emergency medical care. If not breathing give CPR. If breathing is difficult administer oxygen. Treat for shock.

Fire Hazards:

Flashpoint: Combustible solid *Ignition temperature:* n/a
Flammable Explosive High Range: n/a *Low Range:* n/a
Toxic Products of Combustion: n/a
Other Hazards: n/a
Possible extinguishing agents: Extinguish fire using suitable agent for the type of surrounding fire.

Reactivity Hazards:

Reactive With: n/a *Other Reactions:* n/a

Corrosivity Hazards:

Corrosive With: n/a *Neutralizing Agent:* n/a

Radioactivity Hazards:

Radiation Emitted: n/a *Other Hazards:* n/a

Recommended Protection for Response Personnel:

Avoid breathing vapors, keep upwind. Wear a sealed chemical suit (polycarbonate, butyl rubber). Wash away any material which may have come into contact with the body with copious amounts of soap and water. if the fire becomes uncontrollable, consider appropriate evacuation.

QUINOLINE

DOT Number: NA 2656 *DOT Hazard Class:* ORM-E *DOT Guide Number:* 29
Synonyms: 1-benzazine, Leucol
STCC Number: 4963367 *Reportable Qty:* 5000/2270
Mfg Name: Atomergic Chemetals Corp. *Phone No:* 1-516-349-8800

Physical Description:

Physical Form: Liquid *Color:* Colorless to brown *Odor:* Strong, unpleasant
Other Information: It is used in the manufacture of pharmaceuticals, dyes, paints, and other chemicals.

Chemical Properties:

Specific Gravity: 1.1 *Vapor Density:* 4.5 *Boiling Point: 460° F(237.7° C)*
Melting Point: n/a *Vapor Pressure:* 1 mm Hg at 139° F(59.4° C) *Solubility in water:* No
Other Information: Slightly soluble in cold water. Absorbs moisture from the air. Weighs 9.7 lbs/4.3 kg per gallon/3.8 l.

Health Hazards:

Inhalation Hazard: n/a
Ingestion Hazard: Nausea and vomiting.
Absorption Hazard: Irritating to the skin and eyes.
Hazards to Wildlife: Dangerous to aquatic life.
Decontamination Procedures: Wash away any material with copious amounts of soap and water.
First Aid Procedures: Remove victim to fresh air, call emergency medical care. If not breathing give CPR. If breathing is difficult administer oxygen. Treat for shock.

Fire Hazards:

Flashpoint: 225° F(107.2° C) *Ignition temperature:* 896° F(480° C)
Flammable Explosive High Range: n/a *Low Range:* n/a
Toxic Products of Combustion: Toxic oxides of nitrogen are produced.
Other Hazards: n/a
Possible extinguishing agents: Apply water from as far a distance as possible. Use foam, dry chemical, or carbon dioxide.

Reactivity Hazards:

Reactive With: n/a *Other Reactions:* n/a

Corrosivity Hazards:

Corrosive With: n/a *Neutralizing Agent:* n/a

Radioactivity Hazards:

Radiation Emitted: n/a *Other Hazards:* n/a

Recommended Protection for Response Personnel:

Avoid breathing vapors, keep upwind. Structural protective clothing provides limited protection. Wash away any material which may have come into contact with the body with copious amounts of soap and water. Consider appropriate evacuation.

RADIOACTIVE MATERIAL (SPECIAL FORM) N.O.S.

DOT Number: NA 2974 *DOT Hazard Class:* Radioactive Mateiral *DOT Guide Number:* 63
Synonyms: none given
STCC Number: 4929950 *Reportable Qty:* n/a *CHEMTREC Phone No:* 1-800-424-9300

Physical Description:

Physical Form: Solid *Color:* n/a *Odor:* n/a
Other Information: Radioactive material emits certain rays that are hazardous. These can only be detected with special equipment.

Chemical Properties:

Specific Gravity: n/a *Vapor Density:* n/a *Boiling Point: n/a*
Melting Point: n/a *Vapor Pressure:* n/a *Solubility in water:* n/a
Other Information: Radioactive material (special form) is various devices that contain radioactive material.

Health Hazards:

Inhalation Hazard: Radioactive poisoning
Ingestion Hazard: Radioactive poisoning
Absorption Hazard: Radioactive poisoning
Hazards to Wildlife: Radioactive poisoning
Decontamination Procedures: None available!!
First Aid Procedures: Remove victim to fresh air, call emergency medical care. If not breathing give CPR. If breathing is difficult administer oxygen. Treat for shock.

Fire Hazards:

Flashpoint: n/a *Ignition temperature:* n/a
Flammable Explosive High Range: None *Low Range:* None
Toxic Products of Combustion: Radiation fallout.
Other Hazards: High emitance of radioactive energy.
Possible *Possible extinguishing agents:* Apply water from as far a distance as possible. Extinguish fire using suitable agent for type of surrounding fire.

Reactivity Hazards:

Reactive With: Itself! *Other Reactions:* None.

Corrosivity Hazards:

Corrosive With: n/a *Neutralizing Agent:* n/a

Radioactivity Hazards:

Radiation Emitted: Alpha, beta, and gamma.
Other Hazards: In case of trouble with shipment of these containers, all unauthorized personnel should be kept as far away as possible until qualified personnel and equipment can be obtained.

Recommended Protection for Response Personnel:

Avoid breathing vapors, keep upwind. Wear appropriate chemical clothing. Wash away any material which may come into contact with the body with copious amounts of soap and water. Consider appropriate evacuation.

RESORCINOL

DOT Number: NA 2876 *DOT Hazard Class:* ORM-E *DOT Guide Number:* 55
Synonyms: 1,3 benzenediol, Resorcin
STCC Number: 4966774 *Reportable Qty:* 5000/2270
Mfg Name: Ashland Chemical *Phone No:* 1-614-889-3333

Physical Description:

Physical Form: Solid *Color:* White *Odor:* Faint
Other Information: It is used to make plastics, pharmecueticals, and many other uses.

Chemical Properties:

Specific Gravity: 1.28 *Vapor Density:* 3.80 *Boiling Point: 531° F(277.2° C)*
Melting Point: n/a *Vapor Pressure:* n/a *Solubility in water:* Yes
Other Information: Soluble in water. Weighs 10.5 lbs/4.7 kg per gallon/3.8 l.

Health Hazards:

Inhalation Hazard: Coughing, difficulty in breathing.
Ingestion Hazard: Nausea or loss of consciousness.
Absorption Hazard: Irritating to the skin and eyes.
Hazards to Wildlife: Dangerous to aquatic life.
Decontamination Procedures: Wash away any material with copious amounts of soap and water.
First Aid Procedures: Remove victim to fresh air, call emergency medical care. If not breathing give CPR. If breathing is difficult administer oxygen. Treat for shock.

Fire Hazards:

Flashpoint: 261° F(127.2° C) *Ignition temperature:* 1126° F(607.7° C)
Flammable Explosive High Range: n/a *Low Range:* 1.4
Toxic Products of Combustion: n/a
Other Hazards: Containers may explode in a fire.
Possible extinguishing agents: Extinguish fire using suitable agent for type of surrounding fire.

Reactivity Hazards:

Reactive With: n/a *Other Reactions:* n/a

Corrosivity Hazards:

Corrosive With: n/a *Neutralizing Agent:* n/a

Radioactivity Hazards:

Radiation Emitted: n/a *Other Hazards:* n/a

Recommended Protection for Response Personnel:

Avoid breathing vapors, keep upwind. Structural protective clothing provides limited protection. Wash away any material which may have come into contact with the body with copious amounts of soap and water. Consider appropriate evacuation.

ROSIN SOLUTION

DOT Number: NA 1993 *DOT Hazard Class:* Combustible liquid *DOT Guide Number:* 27
Synonyms: colophony solution
STCC Number: 4915344 *Reportable Qty:* n/a *CHEMTREC Phone No:* 1-800-424-9300

Physical Description:

Physical Form: Liquid *Color:* Colorless to amber, red or black *Odor:* Pine, pitch
Other Information: It is used for paper size, solder flux, yellow laundry soaps, core oil, modified alkalis, and paint driers.

Chemical Properties:

Specific Gravity: 1 *Vapor Density:* n/a *Boiling Point: n/a*
Melting Point: n/a *Vapor Pressure:* n/a *Solubility in water:* n/a
Other Information: Product is secretion of pine and fir trees, and is principally comprised of abietic acid and its derivatives.

Health Hazards:

Inhalation Hazard: Irritating.
Ingestion Hazard: Vomiting, stomach pains.
Absorption Hazard: Irritating to the skin and eyes.
Hazards to Wildlife: n/a
Decontamination Procedures: Wash away any material with copious amounts of soap and water.
First Aid Procedures: Remove victim to fresh air, call emergency medical care. If not breathing give CPR. If breathing is difficult administer oxygen. Treat for shock.

Fire Hazards:

Flashpoint: 100-199° F(37.7-92.7° C) *Ignition temperature:* n/a
Flammable Explosive High Range: n/a *Low Range:* n/a
Toxic Products of Combustion: n/a
Other Hazards: There is some potential that containers may rupture in a fire.
Possible extinguishing agents: Apply water from as far a distance as possible. Use foam, dry chemical or carbon dioxide.

Reactivity Hazards:

Reactive With: n/a *Other Reactions:* n/a

Corrosivity Hazards:

Corrosive With: May attack some forms of plastics, rubbers, and coatings. *Neutralizing Agent:* n/a

Radioactivity Hazards:

Radiation Emitted: n/a *Other Hazards:* n/a

Recommended Protection for Response Personnel:

Avoid breathing vapors, keep upwind. Structural protective clothing provides limited protection. Wash away any material which may have come into contact with the body with copious amounts of soap and water. Consider appropriate evacuation.

SALICYLIC ACID

DOT Number: n/a *DOT Hazard Class:* n/a *DOT Guide Number:* n/a
Synonyms: o-hydroxybenzoic acid, Retarder W
STCC Number: n/a *Reportable Qty:* n/a
Mfg Name: Monsanto Co. *Phone No:* 1-314-694-1000

Physical Description:

Physical Form: Solid *Color:* White to light tan *Odor:* Odorless
Other Information: n/a

Chemical Properties:

Specific Gravity: 1.17 *Vapor Density:* 4.8 *Boiling Point: Decomposes*
Melting Point: 315° F(157.2° C) *Vapor Pressure:* n/a *Solubility in water:* Yes
Other Information: Sinks and mixes slowly with water.

Health Hazards:

Inhalation Hazard: Coughing or difficulty in breathing.
Ingestion Hazard: Vomiting.
Absorption Hazard: Irritating to the skin and eyes.
Hazards to Wildlife: Dangerous to aquatic life.
Decontamination Procedures: Wash away any material with copious amounts of soap and water.
First Aid Procedures: Remove victim to fresh air, call emergency medical care. If not breathing give CPR. If breathing is difficult administer oxygen. Treat for shock.

Fire Hazards:

Flashpoint: Combustible solid *Ignition temperature:* n/a
Flammable Explosive High Range: n/a *Low Range:* n/a
Toxic Products of Combustion: Irritating vapors of unburned material and phenol may be formed in fires.
Other Hazards: Sublimes and forms vapor or dust that may explode.
Possible extinguishing agents: Water, foam, dry chemical, carbon dioxide. Water and foam may cause frothing. Water and foam may be ineffective on fire. Cool exposed containers with water.

Reactivity Hazards:

Reactive With: n/a *Other Reactions:* n/a

Corrosivity Hazards:

Corrosive With: n/a *Neutralizing Agent:* n/a

Radioactivity Hazards:

Radiation Emitted: n/a *Other Hazards:* n/a

Recommended Protection for Response Personnel:

Avoid breathing vapors, keep upwind. Wear a sealed chemical suit (polycarbonate, butyl rubber, viton, PVC, chlorinated polyethylene, nitrile, and neoprene). Wash away any material which may have come into contact with the body with copious amounts of soap and water. If the fire becomes uncontrollable, consider appropriate evacuation.

SELENIUM DIOXIDE

DOT Number: n/a *DOT Hazard Class:* Poison B *DOT Guide Number:* n/a
Synonyms: selenious anhydride, selenium oxide
STCC Number: n/a *Reportable Qty:* n/a
Mfg Name: A.D. MacKay Inc. *Phone No:* 1-203-655-3000

Physical Description:

Physical Form: Solid *Color:* White *Odor:* Sour
Other Information: n/a

Chemical Properties:

Specific Gravity: 3.95 *Vapor Density:* 3.8 *Boiling Point: 599° F(315° C)*
Melting Point: n/a *Vapor Pressure:* n/a *Solubility in water:* Yes
Other Information: Sinks and mixes with water.

Health Hazards:

Inhalation Hazard: Poisonous if inhaled.
Ingestion Hazard: Coughing, nausea or vomiting.
Absorption Hazard: Poisonous to the skin and eyes.
Hazards to Wildlife: Dangerous to aquatic life.
Decontamination Procedures: Wash away any material with copious amounts of soap and water.
First Aid Procedures: Remove victim to fresh air, call emergency medical care. If not breathing give CPR. If breathing is difficult administer oxygen. Treat for shock.

Fire Hazards:

Flashpoint: n/a *Ignition temperature:* n/a
Flammable Explosive High Range: n/a *Low Range:* n/a
Toxic Products of Combustion: Sublimes and forms toxic vapors when heated.
Other Hazards: n/a
Possible extinguishing agents: Water, alcohol foam, dry chemical, carbon dioxide.

Reactivity Hazards:

Reactive With: n/a *Other Reactions:* n/a

Corrosivity Hazards:

Corrosive With: Most metals in the presence of moisture. *Neutralizing Agent:* n/a

Radioactivity Hazards:

Radiation Emitted: n/a *Other Hazards:* n/a

Recommended Protection for Response Personnel:

Avoid breathing vapors, keep upwind. Structural protective clothing provides limited protection. Wash away any material which may have come into contact with the body with copious amounts of soap and water. If the fire becomes uncontrollable, consider appropriate evacuation.

SELENIUM TRIOXIDE

DOT Number: n/a *DOT Hazard Class:* n/a *DOT Guide Number:* n/a
Synonyms: selenic anhydride
STCC Number: n/a *Reportable Qty:* n/a
Mfg Name: A.D. MacKay Inc. *Phone No:* 1-203-655-3000

Physical Description:

Physical Form: Solid *Color:* White *Odor:* n/a
Other Information: n/a

Chemical Properties:

Specific Gravity: 3.6 *Vapor Density:* n/a *Boiling Point: Decomposes*
Melting Point: 244° F(117.7° C) *Vapor Pressure:* n/a *Solubility in water:* n/a
Other Information: Avoid contact with solid or dust. Keep people away!!

Health Hazards:

Inhalation Hazard: Poisonous if inhaled.
Ingestion Hazard: Coughing, nausea and vomiting.
Absorption Hazard: Poisonous to the skin and eyes.
Hazards to Wildlife: n/a
Decontamination Procedures: Wash away any material with copious amounts of soap and water.
First Aid Procedures: Remove victim to fresh air, call emergency medical care. If not breathing give CPR. If breathing is difficult administer oxygen. Treat for shock.

Fire Hazards:

Flashpoint: n/a *Ignition temperature:* n/a
Flammable Explosive High Range: n/a *Low Range:* n/a
Toxic Products of Combustion: n/a
Other Hazards: n/a
Possible extinguishing agents: Extinguish fire using suitable agent for the type of surrounding fire.

Reactivity Hazards:

Reactive With: Reacts vigorously with water to form selenic acid solution. *Other Reactions:* n/a

Corrosivity Hazards:

Corrosive With: Corrodes all metals when moisture is present.
Neutralizing Agent: Crushed limestone, soda ash or lime.

Radioactivity Hazards:

Radiation Emitted: n/a *Other Hazards:* n/a

Recommended Protection for Response Personnel:

Avoid breathing vapors, keep upwind. Wear a sealed chemical suit (polycarbonate, butyl rubber). Wash away any material which may have come into contact with the body with copious amounts of soap and water. If the fire becomes uncontrollable, consider appropriate evacuation.

SILICON CHLORIDE

DOT Number: UN 1818 *DOT Hazard Class:* Corrosive *DOT Guide Number:* 39
Synonyms: silicon tetrachloride, tetrachlorosilane
STCC Number: 4932370 *Reportable Qty:* n/a
Mfg Name: A.D. MacKay Inc. *Phone No:* 1-203-655-3000

Physical Description:

Physical Form: Liquid *Color:* Colorless to light yellow *Odor:* Suffocating
Other Information: It is used in smoke screens, and to make various silicone-containing chemicals and in chemical analysis.

Chemical Properties:

Specific Gravity: 1.48 *Vapor Density:* 5.8 *Boiling Point: 135.7° F(57.6° C)*
Melting Point: −89.9/−94° F(−67/−70° C) *Vapor Pressure:* 200 mm Hg at 69.8° F(21° C)
Solubility in water: Reacts
Other Information: Reacts vigorously with water.

Health Hazards:

Inhalation Hazard: Will cause difficulty in breathing.
Ingestion Hazard: Poisonous if swallowed.
Absorption Hazard: Will burn the skin and the eyes.
Hazards to Wildlife: n/a
Decontamination Procedures: Wash away any material with copious amounts of soap and water.
First Aid Procedures: Remove victim to fresh air, call emergency medical care. If not breathing give CPR. If breathing is difficult administer oxygen. Treat for shock.

Fire Hazards:

Flashpoint: n/a *Ignition temperature:* n/a
Flammable Explosive High Range: n/a *Low Range:* n/a
Toxic Products of Combustion: n/a
Other Hazards: Avoid contact with smoke or dust.
Possible extinguishing agents: Do not use water. Use dry sand, dry chemical, or carbon dioxide.

Reactivity Hazards:

Reactive With: Water to give off hydrochloric acid with evolution of heat. *Other Reactions:* n/a

Corrosivity Hazards:

Corrosive With: Metal and tissue in the presence of moisture.
Neutralizing Agent: Crushed limestone, soda ash, or lime.

Radioactivity Hazards:

Radiation Emitted: n/a *Other Hazards:* n/a

Recommended Protection for Response Personnel:

Avoid breathing vapors, keep upwind. Wear a sealed chemical suit (polycarbonate, butyl rubber). Wash away any material which may have come into contact with the body with copious amounts of soap and water. Avoid contact with the fumes and vapors. Consider appropriate evacuation.

SILVER ACETATE

DOT Number: n/a *DOT Hazard Class:* n/a *DOT Guide Number:* n/a
Synonyms: none given
STCC Number: n/a *Reportable Qty:* n/a
Mfg Name: Mallinckrodt Inc. *Phone No:* 1-314-895-2000

Physical Description:

Physical Form: Solid *Color:* White to gray *Odor:* Odorless
Other Information: n/a

Chemical Properties:

Specific Gravity: 3.26 *Vapor Density:* n/a *Boiling Point: Decomposes*
Melting Point: n/a *Vapor Pressure:* n/a *Solubility in water:* n/a
Other Information: Sinks in water.

Health Hazards:

Inhalation Hazard: Coughing or difficulty in breathing.
Ingestion Hazard: Harmful if swallowed.
Absorption Hazard: Irritating to the skin and eyes.
Hazards to Wildlife: n/a
Decontamination Procedures: Wash away any material with copious amounts of soap and water.
First Aid Procedures: Remove victim to fresh air, call emergency medical care. If not breathing give CPR. If breathing is difficult administer oxygen. Treat for shock.

Fire Hazards:

Flashpoint: n/a *Ignition temperature:* n/a
Flammable Explosive High Range: n/a *Low Range:* n/a
Toxic Products of Combustion: n/a
Other Hazards: n/a
Possible extinguishing agents: Extinguish fire using suitable agent for the type of surrounding fire.

Reactivity Hazards:

Reactive With: n/a *Other Reactions:* n/a

Corrosivity Hazards:

Corrosive With: n/a *Neutralizing Agent:* n/a

Radioactivity Hazards:

Radiation Emitted: n/a *Other Hazards:* n/a

Recommended Protection for Response Personnel:

Avoid breathing vapors, keep upwind. Structural protective clothing provides limited protection. Wash away any material which may have come into contact with the body with copious amounts of soap and water. Consider appropriate evacuation.

SILVER CARBONATE

DOT Number: n/a *DOT Hazard Class:* n/a *DOT Guide Number:* n/a
Synonyms: none given
STCC Number: n/a *Reportable Qty:* n/a
Mfg Name: Mallinckrodt Inc. *Phone No:* 1-314-895-2000

Physical Description:

Physical Form: Solid *Color:* Yellow to brown *Odor:* Odorless
Other Information: n/a

Chemical Properties:

Specific Gravity: 6.1 *Vapor Density:* n/a *Boiling Point: Decomposes*
Melting Point: n/a *Vapor Pressure:* n/a *Solubility in water:* n/a
Other Information: Sinks in water.

Health Hazards:

Inhalation Hazard: Coughing or difficulty in breathing.
Ingestion Hazard: Harmful if swallowed.
Absorption Hazard: Irritating to the skin and eyes.
Hazards to Wildlife: n/a
Decontamination Procedures: Wash away any material with copious amounts of soap and water.
First Aid Procedures: Remove victim to fresh air, call emergency medical care. If not breathing give CPR. If breathing is difficult administer oxygen. Treat for shock.

Fire Hazards:

Flashpoint: n/a *Ignition temperature:* n/a
Flammable Explosive High Range: n/a *Low Range:* n/a
Toxic Products of Combustion: n/a
Other Hazards: Decomposes to silver oxide, silver, and carbon dioxide. the reaction is not hazardous!
Possible extinguishing agents: Extinguish fire using suitable agent for the type of surrounding fire.

Reactivity Hazards:

Reactive With: n/a *Other Reactions:* n/a

Corrosivity Hazards:

Corrosive With: n/a *Neutralizing Agent:* n/a

Radioactivity Hazards:

Radiation Emitted: n/a *Other Hazards:* n/a

Recommended Protection for Response Personnel:

Avoid breathing vapors, keep upwind. Structural protective clothing provides limited protection. Wash away any material which may have come into contact with the body with copious amounts of soap and water. Consider appropriate evacuation.

SILVER FLUORIDE

DOT Number: n/a *DOT Hazard Class:* n/a *DOT Guide Number:* n/a
Synonyms: argentous fluoride, silver monofluoride
STCC Number: n/a *Reportable Qty:* n/a
Mfg Name: Atomergic Chemical Metals Inc. *Phone No:* 1-516-349-8800

Physical Description:

Physical Form: Solid *Color:* Yellow to gray *Odor:* Odorless
Other Information: n/a

Chemical Properties:

Specific Gravity: 5.82 *Vapor Density:* n/a *Boiling Point: 2118° F(1158.8° C)*
Melting Point: n/a *Vapor Pressure:* n/a *Solubility in water:* Yes
Other Information: Sinks and mixes with water.

Health Hazards:

Inhalation Hazard: Coughing or difficulty in breathing.
Ingestion Hazard: Nausea, vomiting, loss of consciousness.
Absorption Hazard: Irritating to the skin and eyes.
Hazards to Wildlife: n/a
Decontamination Procedures: Wash away any material with copious amounts of soap and water.
First Aid Procedures: Remove victim to fresh air, call emergency medical care. If not breathing give CPR. If breathing is difficult administer oxygen. Treat for shock.

Fire Hazards:

Flashpoint: n/a *Ignition temperature:* n/a
Flammable Explosive High Range: n/a *Low Range:* n/a
Toxic Products of Combustion: n/a
Other Hazards: n/a
Possible extinguishing agents: Extinguish fire using suitable agent for the type of surrounding fire.

Reactivity Hazards:

Reactive With: n/a *Other Reactions:* n/a

Corrosivity Hazards:

Corrosive With: n/a *Neutralizing Agent:* n/a

Radioactivity Hazards:

Radiation Emitted: n/a *Other Hazards:* n/a

Recommended Protection for Response Personnel:

Avoid breathing vapors, keep upwind. Wear a sealed chemical suit (polycarbonate, butyl rubber). Wash away any material which may have come into contact with the body with copious amounts of soap and water. Consider appropriate evacuation.

SILVER IODATE

DOT Number: n/a *DOT Hazard Class:* n/a *DOT Guide Number:* n/a
Synonyms: none given
STCC Number: n/a *Reportable Qty:* n/a
Mfg Name: Atomergic Chemical Metals Inc. *Phone No:* 1-516-349-8800

Physical Description:

Physical Form: Solid *Color:* White *Odor:* Odorless
Other Information: n/a

Chemical Properties:

Specific Gravity: 5.53 *Vapor Density:* n/a *Boiling Point: Decomposes*
Melting Point: n/a *Vapor Pressure:* n/a *Solubility in water:* n/a
Other Information: Sinks in water.

Health Hazards:

Inhalation Hazard: Coughing or difficulty in breathing.
Ingestion Hazard: Harmful if swallowed.
Absorption Hazard: Irritating to the skin and eyes.
Hazards to Wildlife: n/a
Decontamination Procedures: Wash away any material with copious amounts of soap and water.
First Aid Procedures: Remove victim to fresh air, call emergency medical care. If not breathing give CPR. If breathing is difficult administer oxygen. Treat for shock.

Fire Hazards:

Flashpoint: n/a *Ignition temperature:* n/a
Flammable Explosive High Range: n/a *Low Range:* n/a
Toxic Products of Combustion: n/a
Other Hazards: n/a
Possible extinguishing agents: Extinguish fire using suitable agent for the type of surrounding fire.

Reactivity Hazards:

Reactive With: n/a *Other Reactions:* n/a

Corrosivity Hazards:

Corrosive With: n/a *Neutralizing Agent:* n/a

Radioactivity Hazards:

Radiation Emitted: n/a *Other Hazards:* n/a

Recommended Protection for Response Personnel:

Avoid breathing vapors, keep upwind. Structural protective clothing provides limited protection). Wash away any material which may have come into contact with the body with copious amounts of soap and water. Consider appropriate evacuation.

SILVER NITRATE

DOT Number: UN 1493 *DOT Hazard Class:* Oxidizer *DOT Guide Number:* 45
Synonyms: lunar caustic
STCC Number: 4918742 *Reportable Qty:* 1/.454
Mfg Name: Gallard-Schlesinger *Phone No:* 1-516-333-5600

Physical Description:

Physical Form: Solid *Color:* Colorless to black *Odor:* Odorless
Other Information: Turns black on exposure to light or organic material.

Chemical Properties:

Specific Gravity: 4.35 *Vapor Density:* 5.8 *Boiling Point: Decomposes*
Melting Point: n/a *Vapor Pressure:* n/a *Solubility in water:* Yes
Other Information: n/a

Health Hazards:

Inhalation Hazard: n/a
Ingestion Hazard: Harmful if swallowed.
Absorption Hazard: Irritating to the skin and eyes.
Hazards to Wildlife: Dangerous to aquatic life.
Decontamination Procedures: Wash away any material with copious amounts of soap and water.
First Aid Procedures: Remove victim to fresh air, call emergency medical care. If not breathing give CPR. If breathing is difficult administer oxygen. Treat for shock.

Fire Hazards:

Flashpoint: n/a *Ignition temperature:* n/a
Flammable Explosive High Range: n/a *Low Range:* n/a
Toxic Products of Combustion: Toxic oxides of nitrogen are produced in a fire.
Other Hazards: Prolonged exposure of this material to heat or fire may result in an explosion. This material will accelerate the burning of other combustible materials.
Possible extinguishing agents: Apply water from as far a distance as possible.

Reactivity Hazards:

Reactive With: n/a *Other Reactions:* n/a

Corrosivity Hazards:

Corrosive With: n/a *Neutralizing Agent:* n/a

Radioactivity Hazards:

Radiation Emitted: n/a *Other Hazards:* n/a

Recommended Protection for Response Personnel:

Avoid breathing vapors, keep upwind. Structural protective clothing provides Limited protection. Wash away any material which may have come into contact with the body with copious amounts of soap and water. Consider appropriate evacuation.

SILVER OXIDE

DOT Number: n/a *DOT Hazard Class:* n/a *DOT Guide Number:* n/a
Synonyms: argentous oxide
STCC Number: n/a *Reportable Qty:* n/a
Mfg Name: Gallard-Schlesinger *Phone No:* 1-516-333-5600

Physical Description:

Physical Form: Solid *Color:* Black to brown *Odor:* Odorless
Other Information: n/a

Chemical Properties:

Specific Gravity: 7.14 *Vapor Density:* 8 *Boiling Point: Decomposes*
Melting Point: n/a *Vapor Pressure:* n/a *Solubility in water:* n/a
Other Information: Sinks in water.

Health Hazards:

Inhalation Hazard: Coughing or difficulty in breathing.
Ingestion Hazard: Harmful if swallowed.
Absorption Hazard: Irritating to the skin and eyes.
Hazards to Wildlife: n/a
Decontamination Procedures: Wash away any material with copious amounts of soap and water.
First Aid Procedures: Remove victim to fresh air, call emergency medical care. If not breathing give CPR. If breathing is difficult administer oxygen. Treat for shock.

Fire Hazards:

Flashpoint: n/a *Ignition temperature:* n/a
Flammable Explosive High Range: n/a *Low Range:* n/a
Toxic Products of Combustion: n/a
Other Hazards: Decomposes to metallic silver and oxygen. If large quantities are involved, the oxygen will intensify the fire.
Possible extinguishing agents: Extinguish fire using suitable agent for the type of surrounding fire.

Reactivity Hazards:

Reactive With: n/a *Other Reactions:* n/a

Corrosivity Hazards:

Corrosive With: n/a *Neutralizing Agent:* n/a

Radioactivity Hazards:

Radiation Emitted: n/a *Other Hazards:* n/a

Recommended Protection for Response Personnel:

Avoid breathing vapors, keep upwind. Structural protective clothing provides limited protection). Wash away any material which may have come into contact with the body with copious amounts of soap and water. If the fire becomes uncontrollable, consider appropriate evacuation.

SILVER SULFATE

DOT Number: n/a *DOT Hazard Class:* n/a *DOT Guide Number:* n/a
Synonyms: none given
STCC Number: n/a *Reportable Qty:* n/a
Mfg Name: Gallard-Schlesinger *Phone No:* 1-516-333-5600

Physical Description:

Physical Form: Solid *Color:* White to gray *Odor:* Odorless
Other Information: n/a

Chemical Properties:

Specific Gravity: 5.45 *Vapor Density:* n/a *Boiling Point: n/a*
Melting Point: n/a *Vapor Pressure:* n/a *Solubility in water:* Yes
Other Information: Sinks and mixes with water.

Health Hazards:

Inhalation Hazard: Coughing or difficulty in breathing.
Ingestion Hazard: Harmful if swallowed.
Absorption Hazard: Irritating to the skin and eyes.
Hazards to Wildlife: Dangerous to aquatic life.
Decontamination Procedures: Wash away any material with copious amounts of soap and water.
First Aid Procedures: Remove victim to fresh air, call emergency medical care. If not breathing give CPR. If breathing is difficult administer oxygen. Treat for shock.

Fire Hazards:

Flashpoint: n/a *Ignition temperature:* n/a
Flammable Explosive High Range: n/a *Low Range:* n/a
Toxic Products of Combustion: n/a
Other Hazards: n/a
Possible extinguishing agents: Extinguish fire using suitable agent for the type of surrounding fire.

Reactivity Hazards:

Reactive With: n/a *Other Reactions:* n/a

Corrosivity Hazards:

Corrosive With: n/a *Neutralizing Agent:* n/a

Radioactivity Hazards:

Radiation Emitted: n/a *Other Hazards:* n/a

Recommended Protection for Response Personnel:

Avoid breathing vapors, keep upwind. Structural protective clothing provides limited protection). Wash away any material which may have come into contact with the body with copious amounts of soap and water. Consider appropriate evacuation.

SODIUM

DOT Number: UN 1428 *DOT Hazard Class:* Flammable solid *DOT Guide Number:* 40
Synonyms: metallic sodium, natrium, sodium atom
STCC Number: 4916456 *Reportable Qty:* 1000/454
Mfg Name: E.I. Du Pont *Phone No:* 1-800-441-3637

Physical Description:

Physical Form: Solid *Color:* Silver white, grayish white *Odor:* Odorless
Other Information: It is used for titanium reduction, nuclear reaction coolant, electric power cable, automobile engine exhaust valve coolant, and radioactive medicine.

Chemical Properties:

Specific Gravity: .97 *Vapor Density:* .8 *Boiling Point: 1121° F(605° C)*
Melting Point: 207.5° F(97.5° C) *Vapor Pressure:* 1 mm Hg at 822° F(438.8° C) *Solubility in water:* Reacts
Other Information: May be shipped as a solid or molten liquid. Weighs 60.6 lbs per cubic foot, and is slightly lighter than water.

Health Hazards:

Inhalation Hazard: Poisonous if inhaled. *Ingestion Hazard:* Poisonous if swallowed.
Absorption Hazard: Severe thermal and chemical burns.
Hazards to Wildlife: Dangerous to aquatic life.
Decontamination Procedures: Wash away any material with copious amounts of soap and water.
First Aid Procedures: Remove victim to fresh air, call emergency medical care. If not breathing give CPR. If breathing is difficult administer oxygen. Treat for shock.

Fire Hazards:

Flashpoint: n/a *Ignition temperature:* 250° F(121.1° C)
Flammable Explosive High Range: n/a *Low Range:* n/a
Toxic Products of Combustion: Highly irritating and toxic sodium oxides and sodium hydroxides.
Other Hazards: May generate flammable gases upon release. Burns violently with explosions that may splatter the material.
Possible extinguishing agents: Do not use water. Use graphite, soda ash, powdered sodium chloride or a suitable dry powder.

Reactivity Hazards:

Reactive With: Water and air. *Other Reactions:* n/a

Corrosivity Hazards:

Corrosive With: Attacks clothing, leather, and some metals. Contact with metals may evolve hydrogen gas.
Neutralizing Agent: n/a

Radioactivity Hazards:

Radiation Emitted: n/a *Other Hazards:* n/a

Recommended Protection for Response Personnel:

Avoid breathing vapors, keep upwind. Wear a sealed chemical suit (butyl rubber). Wash away any material which may have come into contact with the body with copious amounts of soap and water. Avoid contact with the fumes and vapors. Consider appropriate evacuation.

SODIUM ALKYLBENZENESULFONATES

DOT Number: n/a *DOT Hazard Class:* n/a *DOT Guide Number:* n/a
Synonyms: alkylbenzenesulfonic acid sodium salt, solfonated alkybenzene sodium salt
STCC Number: n/a *Reportable Qty:* n/a
Mfg Name: Witco Chemical Co. *Phone No:* 1-202-605-3800

Physical Description:

Physical Form: Thick liquid or solid *Color:* Pale yellow *Odor:* Faint detergent
Other Information: n/a

Chemical Properties:

Specific Gravity: 1 *Vapor Density:* n/a *Boiling Point: Decomposes*
Melting Point: n/a *Vapor Pressure:* n/a *Solubility in water:* Yes
Other Information: Sinks and mixes with water. Soap bubbles may be produced.

Health Hazards:

Inhalation Hazard: n/a
Ingestion Hazard: Nausea or vomiting.
Absorption Hazard: Irritating to the skin and eyes.
Hazards to Wildlife: Dangerous to aquatic life.
Decontamination Procedures: Wash away any material with copious amounts of soap and water.
First Aid Procedures: Remove victim to fresh air, call emergency medical care. If not breathing give CPR. If breathing is difficult administer oxygen. Treat for shock.

Fire Hazards:

Flashpoint: n/a *Ignition temperature:* n/a
Flammable Explosive High Range: n/a *Low Range:* n/a
Toxic Products of Combustion: Irritating vapors may be generated.
Other Hazards: n/a
Possible extinguishing agents: Extinguish fire using suitable agent for the type of surrounding fire.

Reactivity Hazards:

Reactive With: n/a *Other Reactions:* n/a

Corrosivity Hazards:

Corrosive With: n/a *Neutralizing Agent:* n/a

Radioactivity Hazards:

Radiation Emitted: n/a *Other Hazards:* n/a

Recommended Protection for Response Personnel:

Avoid breathing vapors, keep upwind. Structural protective clothing provides limited protection). Wash away any material which may have come into contact with the body with copious amounts of soap and water. Consider appropriate evacuation.

SODIUM ALKYLSULFATES

DOT Number: n/a *DOT Hazard Class:* n/a *DOT Guide Number:* n/a
Synonyms: sodium hydrogen alkyl sulfate
STCC Number: n/a *Reportable Qty:* n/a
Mfg Name: E.I. Du Pont *Phone No:* 1-800-441-3637

Physical Description:

Physical Form: Thick liquid or solid *Color:* Pale yellow *Odor:* Faint detergent
Other Information: n/a

Chemical Properties:

Specific Gravity: n/a *Vapor Density:* n/a *Boiling Point: Decomposes*
Melting Point: n/a *Vapor Pressure:* n/a *Solubility in water:* Yes
Other Information: Sinks and mixes with water. Soap bubbles may be produced.

Health Hazards:

Inhalation Hazard: n/a
Ingestion Hazard: Nausea or vomiting.
Absorption Hazard: Irritating to the skin and eyes.
Hazards to Wildlife: Dangerous to aquatic life.
Decontamination Procedures: Wash away any material with copious amounts of soap and water.
First Aid Procedures: Remove victim to fresh air, call emergency medical care. If not breathing give CPR. If breathing is difficult administer oxygen. Treat for shock.

Fire Hazards:

Flashpoint: n/a *Ignition temperature:* n/a
Flammable Explosive High Range: n/a *Low Range:* n/a
Toxic Products of Combustion: Irritating vapors may be generated.
Other Hazards: n/a
Possible extinguishing agents: Extinguish fire using suitable agent for the type of surrounding fire.

Reactivity Hazards:

Reactive With: n/a *Other Reactions:* n/a

Corrosivity Hazards:

Corrosive With: n/a *Neutralizing Agent:* n/a

Radioactivity Hazards:

Radiation Emitted: n/a *Other Hazards:* n/a

Recommended Protection for Response Personnel:

Avoid breathing vapors, keep upwind. Structural protective clothing provides limited protection). Wash away any material which may have come into contact with the body with copious amounts of soap and water. Consider appropriate evacuation.

SODIUM AMIDE

DOT Number: UN 1425 *DOT Hazard Class:* Flammable solid *DOT Guide Number:* 76
Synonyms: sodamide
STCC Number: 4916453 *Reportable Qty:* n/a
Mfg Name: Ventron Corp. *Phone No:* 1-617-774-3100

Physical Description:

Physical Form: Solid *Color:* Grayish to white *Odor:* Ammonia
Other Information: It is used to make other chemicals, and in chemical analysis.

Chemical Properties:

Specific Gravity: 1.39 *Vapor Density:* 1.3 *Boiling Point: 752° F(400° C)*
Melting Point: n/a *Vapor Pressure:* n/a *Solubility in water:* Reacts
Other Information: n/a

Health Hazards:

Inhalation Hazard: n/a
Ingestion Hazard: n/a
Absorption Hazard: Will burn the skin and the eyes.
Hazards to Wildlife: n/a
Decontamination Procedures: Wash away any material with copious amounts of soap and water.
First Aid Procedures: Remove victim to fresh air, call emergency medical care. If not breathing give CPR. If breathing is difficult administer oxygen. Treat for shock.

Fire Hazards:

Flashpoint: n/a *Ignition temperature:* n/a
Flammable Explosive High Range: n/a *Low Range:* n/a
Toxic Products of Combustion: Toxic and irritating ammonia gas may be produced in a fire.
Other Hazards: Violently reacts with water.
Possible extinguishing agents: Graphite, soda ash, or powdered sodium chloride. Do not use water.

Reactivity Hazards:

Reactive With: Water to form a caustic soda solution. *Other Reactions:* n/a

Corrosivity Hazards:

Corrosive With: n/a *Neutralizing Agent:* n/a

Radioactivity Hazards:

Radiation Emitted: n/a *Other Hazards:* n/a

Recommended Protection for Response Personnel:

Avoid breathing vapors, keep upwind. Wear a sealed chemical suit (polycarbonate, butyl rubber). Wash away any material which may have come into contact with the body with copious amounts of soap and water. Consider appropriate evacuation.

SODIUM ARSENATE

DOT Number: UN 1685 *DOT Hazard Class:* Poison B *DOT Guide Number:* 53
Synonyms: disodium arsenate heptahydrate
STCC Number: 4923290 *Reportable Qty:* 1000/454
Mfg Name: Mallinckrodt Inc. *Phone No:* 1-314-895-2000

Physical Description:

Physical Form: Solid *Color:* White *Odor:* Odorless
Other Information: n/a

Chemical Properties:

Specific Gravity: 1.87 *Vapor Density:* 4.5 *Boiling Point: 356° F(180° C)*
Melting Point: n/a *Vapor Pressure:* n/a *Solubility in water:* Yes
Other Information: n/a

Health Hazards:

Inhalation Hazard: Difficulty in breathing.
Ingestion Hazard: Headache, vomiting, loss of consciousness.
Absorption Hazard: Irritating to the skin and eyes.
Hazards to Wildlife: Dangerous to aquatic life.
Decontamination Procedures: Wash away any material with copious amounts of soap and water.
First Aid Procedures: Remove victim to fresh air, call emergency medical care. If not breathing give CPR. If breathing is difficult administer oxygen. Treat for shock.

Fire Hazards:

Flashpoint: n/a *Ignition temperature:* n/a
Flammable Explosive High Range: n/a *Low Range:* n/a
Toxic Products of Combustion: n/a
Other Hazards: Avoid contact with solid and dust. Keep people away!!
Possible extinguishing agents: Extinguish fire using suitable agent for type of surrounding fire. Use foam, dry chemical, or carbon dioxide.

Reactivity Hazards:

Reactive With: n/a *Other Reactions:* n/a

Corrosivity Hazards:

Corrosive With: n/a *Neutralizing Agent:* n/a

Radioactivity Hazards:

Radiation Emitted: n/a *Other Hazards:* n/a

Recommended Protection for Response Personnel:

Avoid breathing vapors, keep upwind. Wear a sealed chemical suit (polycarbonate, butyl rubber). Wash away any material which may have come into contact with the body with copious amounts of soap and water. Consider appropriate evacuation.

SODIUM ARSENITE (liquid)

DOT Number: UN 1686 *DOT Hazard Class:* Poison B *DOT Guide Number:* 54
Synonyms: sodium meta arsenite
STCC Number: 4923291 *Reportable Qty:* 1000/454
Mfg Name: Mallinckrodt Inc. *Phone No:* 1-314-895-2000

Physical Description:

Physical Form: Liquid *Color:* White to gray *Odor:* Odorless
Other Information: It is used as an antiseptic, in insecticides, in herbicides, and to preserve.

Chemical Properties:

Specific Gravity: 1.87 *Vapor Density:* 4.5 *Boiling Point: Decomposes*
Melting Point: 1139° F(615° C) *Vapor Pressure:* n/a *Solubility in water:* Yes
Other Information: n/a

Health Hazards:

Inhalation Hazard: Poisonous if inhaled.
Ingestion Hazard: Poisonous if swallowed.
Absorption Hazard: Irritating to the skin and eyes.
Hazards to Wildlife: Dangerous to both waterfowl and aquatic life.
Decontamination Procedures: Wash away any material with copious amounts of soap and water.
First Aid Procedures: Remove victim to fresh air, call emergency medical care. If not breathing give CPR. If breathing is difficult administer oxygen. Treat for shock.

Fire Hazards:

Flashpoint: n/a *Ignition temperature:* n/a
Flammable Explosive High Range: n/a *Low Range:* n/a
Toxic Products of Combustion: Toxic arsenic gases may be formed in fires.
Other Hazards: Avoid contact with liquid or vapors. Keep people away!!
Possible extinguishing agents: Extinguish fire using suitable agent for type of surrounding fire. Use alcohol foam, dry chemical or carbon dioxide.

Reactivity Hazards:

Reactive With: n/a *Other Reactions:* n/a

Corrosivity Hazards:

Corrosive With: n/a *Neutralizing Agent:* n/a

Radioactivity Hazards:

Radiation Emitted: n/a *Other Hazards:* n/a

Recommended Protection for Response Personnel:

Avoid breathing vapors, keep upwind. Wear a sealed chemical suit (polycarbonate, butyl rubber). Wash away any material which may have come into contact with the body with copious amounts of soap and water. Consider appropriate evacuation.

SODIUM ARSENITE (solid)

DOT Number: UN 2027 *DOT Hazard Class:* Poison B *DOT Guide Number:* 53
Synonyms: sodium meta arsenite
STCC Number: 4923291 *Reportable Qty:* 1000/454
Mfg Name: Mallinckrodt Inc. *Phone No:* 1-314-895-2000

Physical Description:

Physical Form: Solid *Color:* White to gray *Odor:* Odorless
Other Information: It is used as an antiseptic, in insecticides, in herbicides, and to preserve.

Chemical Properties:

Specific Gravity: 1.87 *Vapor Density:* 4.5 *Boiling Point: Decomposes*
Melting Point: 1139° F(615° C) *Vapor Pressure:* n/a *Solubility in water:* Yes
Other Information: n/a

Health Hazards:

Inhalation Hazard: Poisonous if inhaled.
Ingestion Hazard: Poisonous if swallowed.
Absorption Hazard: Irritating to the skin and eyes.
Hazards to Wildlife: Dangerous to both waterfowl and aquatic life.
Decontamination Procedures: Wash away any material with copious amounts of soap and water.
First Aid Procedures: Remove victim to fresh air, call emergency medical care. If not breathing give CPR. If breathing is difficult administer oxygen. Treat for shock.

Fire Hazards:

Flashpoint: n/a *Ignition temperature:* n/a
Flammable Explosive High Range: n/a *Low Range:* n/a
Toxic Products of Combustion: Poisonous gases may be produced when heated.
Other Hazards: Avoid contact with solid and dust. Keep people away!!
Possible extinguishing agents: Extinguish fire using suitable agent for type of surrounding fire. Use alcohol foam, dry chemical or carbon dioxide.

Reactivity Hazards:

Reactive With: n/a *Other Reactions:* n/a

Corrosivity Hazards:

Corrosive With: n/a *Neutralizing Agent:* n/a

Radioactivity Hazards:

Radiation Emitted: n/a *Other Hazards:* n/a

Recommended Protection for Response Personnel:

Avoid breathing vapors, keep upwind. Wear a sealed chemical suit (polycarbonate, butyl rubber). Wash away any material which may have come into contact with the body with copious amounts of soap and water. Consider appropriate evacuation.

SODIUM AZIDE

DOT Number: UN 1687 *DOT Hazard Class:* Poison B *DOT Guide Number:* 56
Synonyms: hydrazoic acid, sodium salt
STCC Number: 4923465 *Reportable Qty:* 100/45.4
Mfg Name: Mallinckrodt Inc. *Phone No:* 1-314-895-2000

Physical Description:

Physical Form: Solid *Color:* White *Odor:* Odorless
Other Information: n/a

Chemical Properties:

Specific Gravity: 1.85 *Vapor Density:* 2.2 *Boiling Point: Decomposes*
Melting Point: n/a *Vapor Pressure:* n/a *Solubility in water:* Yes
Other Information: n/a

Health Hazards:

Inhalation Hazard: Poisonous if inhaled.
Ingestion Hazard: Poisonous if swallowed.
Absorption Hazard: n/a
Hazards to Wildlife: Dangerous to aquatic life.
Decontamination Procedures: Wash away any material with copious amounts of soap and water.
First Aid Procedures: Remove victim to fresh air, call emergency medical care. If not breathing give CPR. If breathing is difficult administer oxygen. Treat for shock.

Fire Hazards:

Flashpoint: n/a *Ignition temperature:* n/a
Flammable Explosive High Range: n/a *Low Range:* n/a
Toxic Products of Combustion: Toxic oxides of nitrogen are produced.
Other Hazards: May explode if subject to shock. Containers may explode in fire.
Possible extinguishing agents: Apply water from as far a distance as possible. Use alcohol foam, dry chemical, or carbon dioxide.

Reactivity Hazards:

Reactive With: Water to form an alkaline solution.
Other Reactions: Lead, silver, mercury, and copper to form explosive sensitive materials.

Corrosivity Hazards:

Corrosive With: n/a *Neutralizing Agent:* n/a

Radioactivity Hazards:

Radiation Emitted: n/a *Other Hazards:* n/a

Recommended Protection for Response Personnel:

Avoid breathing vapors, keep upwind. Wear a sealed chemical suit (polycarbonate, butyl rubber). Wash away any material which may have come into contact with the body with copious amounts of soap and water. Consider appropriate evacuation.

SODIUM BIFLUORIDE

DOT Number: UN 2439 *DOT Hazard Class:* Corrosive *DOT Guide Number:* 60
Synonyms: sodium difluoride
STCC Number: 4932355 *Reportable Qty:* 100/45.4
Mfg Name: Instel Corp. *Phone No:* 1-212-758-5880

Physical Description:

Physical Form: Solid *Color:* Colorless to white *Odor:* Odorless
Other Information: It is used as a preservative for anatomical and zoological specimens, in metal plating, and many other uses.

Chemical Properties:

Specific Gravity: 2.08 *Vapor Density:* 2.1 *Boiling Point: Decomposes*
Melting Point: n/a *Vapor Pressure:* n/a *Solubility in water:* Yes
Other Information: n/a

Health Hazards:

Inhalation Hazard: Irritating to the eyes, ears, and the throat.
Ingestion Hazard: Nausea, vomiting, diarrhea, abdominal pains.
Absorption Hazard: n/a
Hazards to Wildlife: Dangerous to aquatic life.
Decontamination Procedures: Wash away any material with copious amounts of soap and water.
First Aid Procedures: Remove victim to fresh air, call emergency medical care. If not breathing give CPR. If breathing is difficult administer oxygen. Treat for shock.

Fire Hazards:

Flashpoint: n/a *Ignition temperature:* n/a
Flammable Explosive High Range: n/a *Low Range:* n/a
Toxic Products of Combustion: None
Other Hazards: Keep people away.
Possible extinguishing agents: Apply water from as far a distance as possible. Use suitable agent for type of surrounding fire.

Reactivity Hazards:

Reactive With: Water liberating heat to form a corrosive solution. *Other Reactions:* n/a

Corrosivity Hazards:

Corrosive With: Will attack natural rubber, leather, and many organic materials. May generate hydrogen gas upon contact with some metals.
Neutralizing Agent: Crushed limestone, soda ash, or lime.

Radioactivity Hazards:

Radiation Emitted: n/a *Other Hazards:* n/a

Recommended Protection for Response Personnel:

Avoid breathing vapors, keep upwind. Wear appropriate chemical clothing. Wash away any material which may have come into contact with the body with copious amounts of soap and water. Consider appropriate evacuation.

SODIUM BISULFIDE (liquid)

DOT Number: NA 2693 *DOT Hazard Class:* Corrosive *DOT Guide Number:* 60
Synonyms: sodium acid sulfite, sodium pyrosulfide
STCC Number: 4932376 *Reportable Qty:* 5000/2270 *CHEMTREC Phone No:* 1-800-424-9300

Physical Description:

Physical Form: Liquid *Color:* Colorless *Odor:* Burning sulfur
Other Information: It is used as a food preservative, in photography, in paper making, to make other chemicals, and for many other uses.

Chemical Properties:

Specific Gravity: 1.48 *Vapor Density:* n/a *Boiling Point: Decomposes*
Melting Point: n/a *Vapor Pressure:* n/a *Solubility in water:* Yes
Other Information: Sinks and mixes with water.

Health Hazards:

Inhalation Hazard: Harmful if inhaled.
Ingestion Hazard: Harmful if swallowed.
Absorption Hazard: Irritating to the skin and eyes.
Hazards to Wildlife: Dangerous to aquatic life.
Decontamination Procedures: Wash away any material with copious amounts of soap and water.
First Aid Procedures: Remove victim to fresh air, call emergency medical care. If not breathing give CPR. If breathing is difficult administer oxygen. Treat for shock.

Fire Hazards:

Flashpoint: n/a *Ignition temperature:* n/a
Flammable Explosive High Range: n/a *Low Range:* n/a
Toxic Products of Combustion: n/a
Other Hazards: n/a
Possible extinguishing agents: Extinguish fire using suitable agent for the type of surrounding fire.

Reactivity Hazards:

Reactive With: n/a *Other Reactions:* n/a

Corrosivity Hazards:

Corrosive With: Metals and tissue. *Neutralizing Agent:* Crushed limestone, soda ash, or lime.

Radioactivity Hazards:

Radiation Emitted: n/a *Other Hazards:* n/a

Recommended Protection for Response Personnel:

Avoid breathing vapors, keep upwind. Wear appropriate chemical clothing. Wash away any material which may have come into contact with the body with copious amounts of soap and water. Consider appropriate evacuation.

SODIUM BISULFIDE (solid)

DOT Number: UN 1821 *DOT Hazard Class:* ORM-B *DOT Guide Number:* 60
Synonyms: sodium acid sulfite, sodium pyrosulfide
STCC Number: 4944155 *Reportable Qty:* 5000/2270 *CHEMTREC Phone No:* 1-800-424-9300

Physical Description:

Physical Form: Solid *Color:* White *Odor:* Odorless to irritating
Other Information: It is used in the manufacture of other chemicals, in food processing, photography, and in many other uses.

Chemical Properties:

Specific Gravity: 1.48 *Vapor Density:* n/a *Boiling Point: Decomposes*
Melting Point: n/a *Vapor Pressure:* n/a *Solubility in water:* Yes
Other Information: Sinks and mixes with water.

Health Hazards:

Inhalation Hazard: Harmful if inhaled.
Ingestion Hazard: Harmful if swallowed.
Absorption Hazard: Irritating to the skin and eyes.
Hazards to Wildlife: Dangerous to aquatic life.
Decontamination Procedures: Wash away any material with copious amounts of soap and water.
First Aid Procedures: Remove victim to fresh air, call emergency medical care. If not breathing give CPR. If breathing is difficult administer oxygen. Treat for shock.

Fire Hazards:

Flashpoint: n/a *Ignition temperature:* n/a
Flammable Explosive High Range: n/a *Low Range:* n/a
Toxic Products of Combustion: n/a
Other Hazards: n/a
Possible extinguishing agents: Extinguish fire using suitable agent for the type of surrounding fire.

Reactivity Hazards:

Reactive With: n/a *Other Reactions:* n/a

Corrosivity Hazards:

Corrosive With: n/a *Neutralizing Agent:* n/a

Radioactivity Hazards:

Radiation Emitted: n/a *Other Hazards:* n/a

Recommended Protection for Response Personnel:

Avoid breathing vapors, keep upwind. Wear appropriate chemical clothing. Wash away any material which may have come into contact with the body with copious amounts of soap and water. Consider appropriate evacuation.

SODIUM BORATE

DOT Number: n/a *DOT Hazard Class:* n/a *DOT Guide Number:* n/a
Synonyms: borax anhydrous, sodium biborate, sodium pyroborate, sodium tetraborate anhydrous
STCC Number: n/a *Reportable Qty:* n/a
Mfg Name: Mallinckrodt Inc. *Phone No:* 1-314-895-2000

Physical Description:

Physical Form: Solid *Color:* White *Odor:* Odorless
Other Information: n/a

Chemical Properties:

Specific Gravity: 2.37 *Vapor Density:* n/a *Boiling Point: n/a*
Melting Point: n/a *Vapor Pressure:* n/a *Solubility in water:* Yes
Other Information: Sinks and mixes slowly with water.

Health Hazards:

Inhalation Hazard: n/a
Ingestion Hazard: Headache, dizziness, nausea or vomiting.
Absorption Hazard: Irritating to the skin and eyes.
Hazards to Wildlife: Dangerous to aquatic life.
Decontamination Procedures: Wash away any material with copious amounts of soap and water.
First Aid Procedures: Remove victim to fresh air, call emergency medical care. If not breathing give CPR. If breathing is difficult administer oxygen. Treat for shock.

Fire Hazards:

Flashpoint: n/a *Ignition temperature:* n/a
Flammable Explosive High Range: n/a *Low Range:* n/a
Toxic Products of Combustion: n/a
Other Hazards: Compound melts to a glassy material that may flow in large quantities and ignite combustibles elsewhere.
Possible extinguishing agents: Extinguish fire using suitable agent for the type of surrounding fire.

Reactivity Hazards:

Reactive With: n/a *Other Reactions:* n/a

Corrosivity Hazards:

Corrosive With: n/a *Neutralizing Agent:* n/a

Radioactivity Hazards:

Radiation Emitted: n/a *Other Hazards:* n/a

Recommended Protection for Response Personnel:

Avoid breathing vapors, keep upwind. Structural protective clothing provides limited protection. Wash away any material which may have come into contact with the body with copious amounts of soap and water. Consider appropriate evacuation.

SODIUM BOROHYDRIDE

DOT Number: UN 1426 *DOT Hazard Class:* Flammable solid *DOT Guide Number:* 32
Synonyms: borohydride
STCC Number: 4916437 *Reportable Qty:* n/a
Mfg Name: A.D. MacKay Inc. *Phone No:* 1-203-655-3000

Physical Description:

Physical Form: Solid *Color:* White *Odor:* Odorless
Other Information: It is used to make other chemicals, treat waste water, and for many other uses.

Chemical Properties:

Specific Gravity: 1.07 *Vapor Density:* 1.3 *Boiling Point: n/a*
Melting Point: n/a *Vapor Pressure:* n/a *Solubility in water:* Yes
Other Information: Sinks and mixes with water. A flammable gas is produced.

Health Hazards:

Inhalation Hazard: Harmful if inhaled.
Ingestion Hazard: Harmful if swallowed.
Absorption Hazard: Harmful to the skin and eyes.
Hazards to Wildlife: n/a
Decontamination Procedures: Wash away any material with copious amounts of soap and water.
First Aid Procedures: Remove victim to fresh air, call emergency medical care. If not breathing give CPR. If breathing is difficult administer oxygen. Treat for shock.

Fire Hazards:

Flashpoint: n/a *Ignition temperature:* n/a
Flammable Explosive High Range: n/a *Low Range:* n/a
Toxic Products of Combustion: Hydrogen gas is formed in fires.
Other Hazards: Material itself is easily ignited and burns vigorously when ignited.
Possible extinguishing agents: Use graphite, soda ash, or powdered sodium chloride, or a suitable dry powder. Do not use water!!!

Reactivity Hazards:

Reactive With: Water to form flammable hydrogen gas.
Other Reactions: Decomposed by water to form sodium hydroxide.

Corrosivity Hazards:

Corrosive With: Glass *Neutralizing Agent:* Dilute with acetic acid.

Radioactivity Hazards:

Radiation Emitted: n/a *Other Hazards:* n/a

Recommended Protection for Response Personnel:

Avoid breathing vapors, keep upwind. Structural protective clothing provides limited protection. Wash away any material which may have come into contact with the body with copious amounts of soap and water. Consider appropriate evacuation.

SODIUM CACODYLATE

DOT Number: UN 1688 *DOT Hazard Class:* Poison B *DOT Guide Number:* 53
Synonyms: Arsycodile, Arsicodile, Phytar
STCC Number: n/a *Reportable Qty:* n/a
Mfg Name: Atomergic Chemetals Corp. *Phone No:* 1-516-349-8800

Physical Description:

Physical Form: Solid or solution *Color:* White (solid) colorless to yellow (solution) *Odor:* Odorless
Other Information: n/a

Chemical Properties:

Specific Gravity: 1 *Vapor Density:* 7.4 *Boiling Point: Decomposes*
Melting Point: n/a *Vapor Pressure:* n/a *Solubility in water:* Yes
Other Information: n/a

Health Hazards:

Inhalation Hazard: Poisonous if inhaled.
Ingestion Hazard: Poisonous if swallowed.
Absorption Hazard: Irritating to the skin and eyes.
Hazards to Wildlife: n/a
Decontamination Procedures: Wash away any material with copious amounts of soap and water.
First Aid Procedures: Remove victim to fresh air, call emergency medical care. If not breathing give CPR. If breathing is difficult administer oxygen. Treat for shock.

Fire Hazards:

Flashpoint: n/a *Ignition temperature:* n/a
Flammable Explosive High Range: n/a *Low Range:* n/a
Toxic Products of Combustion: Arsenic producing fumes are emitted.
Other Hazards: Avoid contact with solid or dust. Keep people away!!
Possible extinguishing agents: Apply water from as far a distance as possible. Use foam, dry chemical, or carbon dioxide.

Reactivity Hazards:

Reactive With: n/a *Other Reactions:* n/a

Corrosivity Hazards:

Corrosive With: Common metals *Neutralizing Agent:* Crushed limestone, soda ash, or lime.

Radioactivity Hazards:

Radiation Emitted: n/a *Other Hazards:* n/a

Recommended Protection for Response Personnel:

Avoid breathing vapors, keep upwind. Structural protective clothing provides limited protection. Wash away any material which may have come into contact with the body with copious amounts of soap and water. Consider appropriate evacuation.

SODIUM CHLORATE

DOT Number: UN 1495 *DOT Hazard Class:* Oxidizer *DOT Guide Number:* 35
Synonyms: chlorate of soda
STCC Number: 4918723 *Reportable Qty:* n/a
Mfg Name: Instel Corp. *Phone No:* 1-212-758-5880

Physical Description:

Physical Form: Solid *Color:* Colorless to pale yellow *Odor:* Odorless
Other Information: It is used to herbicides, explosives, dyes, matches, etc.

Chemical Properties:

Specific Gravity: 2.49 *Vapor Density:* 3.6 *Boiling Point: Decomposes*
Melting Point: n/a *Vapor Pressure:* n/a *Solubility in water:* Yes
Other Information: Weighs 155.4 lbs/70.4 kg per cubic foot.

Health Hazards:

Inhalation Hazard: n/a
Ingestion Hazard: Poisonous if swallowed.
Absorption Hazard: Irritating to the skin and eyes.
Hazards to Wildlife: Dangerous to aquatic life.
Decontamination Procedures: Wash away any material with copious amounts of soap and water.
First Aid Procedures: Remove victim to fresh air, call emergency medical care. If not breathing give CPR. If breathing is difficult administer oxygen. Treat for shock.

Fire Hazards:

Flashpoint: n/a *Ignition temperature:* n/a
Flammable Explosive High Range: n/a *Low Range:* n/a
Toxic Products of Combustion: n/a
Other Hazards: Containers may explode in fire. May cause fire upon contact with other combustible materials.
Possible extinguishing agents: Apply water from as far a distance as possible.

Reactivity Hazards:

Reactive With: Organic materials. *Other Reactions:* n/a

Corrosivity Hazards:

Corrosive With: n/a *Neutralizing Agent:* n/a

Radioactivity Hazards:

Radiation Emitted: n/a *Other Hazards:* n/a

Recommended Protection for Response Personnel:

Avoid breathing vapors, keep upwind. Structural protective clothing provides limited protection. Wash away any material which may have come into contact with the body with copious amounts of soap and water. Consider appropriate evacuation.

SODIUM CHROMATE

DOT Number: NA 9145 *DOT Hazard Class:* ORM-E *DOT Guide Number:* 31
Synonyms: sodium chromate anhydrous; sodium chromate(VI), natural
STCC Number: 4936639 *Reportable Qty:* 1000/454
Mfg Name: Allied Corp. *Phone No:* 1-201-455-2000

Physical Description:

Physical Form: Solid *Color:* Yellow *Odor:* Odorless
Other Information: It is used to make pigments for paints, inks, other chemicals, and as a wood preservative.

Chemical Properties:

Specific Gravity: 2.72 *Vapor Density:* 5.6 *Boiling Point: Decomposes*
Melting Point: n/a *Vapor Pressure:* n/a *Solubility in water:* Yes
Other Information: Sinks and mixes with water.

Health Hazards:

Inhalation Hazard: n/a
Ingestion Hazard: Poisonous if swallowed.
Absorption Hazard: Irritating to the skin and eyes.
Hazards to Wildlife: Dangerous to aquatic life.
Decontamination Procedures: Wash away any material with copious amounts of soap and water.
First Aid Procedures: Remove victim to fresh air, call emergency medical care. If not breathing give CPR. If breathing is difficult administer oxygen. Treat for shock.

Fire Hazards:

Flashpoint: n/a *Ignition temperature:* n/a
Flammable Explosive High Range: n/a *Low Range:* n/a
Toxic Products of Combustion: Toxic chromium oxide fumes may be formed in fires.
Other Hazards: Will increase the intensity of a fire. May cause fire on contact with combustibles.
Possible extinguishing agents: Extinguish fire using suitable agent for the type of surrounding fire.

Reactivity Hazards:

Reactive With: Combustible materials *Other Reactions:* n/a

Corrosivity Hazards:

Corrosive With: n/a *Neutralizing Agent:* n/a

Radioactivity Hazards:

Radiation Emitted: n/a *Other Hazards:* n/a

Recommended Protection for Response Personnel:

Avoid breathing vapors, keep upwind. Wear a sealed chemical suit (polycarbonate, butyl rubber). Wash away any material which may have come into contact with the body with copious amounts of soap and water. Consider appropriate evacuation.

SODIUM CYANIDE (liquid)

DOT Number: UN 1689 *DOT Hazard Class:* Poison B *DOT Guide Number:* 55
Synonyms: hydrocyanic acid, sodium salt
STCC Number: 4923227 *Reportable Qty:* 10/4.54
Mfg Name: E.I. Du Pont *Phone No:* 1-800-441-3637

Physical Description:

Physical Form: Liquid solution *Color:* White *Odor:* Almond
Other Information: n/a

Chemical Properties:

Specific Gravity: 1.6 *Vapor Density:* 1.7 *Boiling Point: Very high*
Melting Point: n/a *Vapor Pressure:* 1 mm Hg at 1053° F(567.2° C) *Solubility in water:* Yes
Other Information: n/a

Health Hazards:

Inhalation Hazard: Poisonous if inhaled.
Ingestion Hazard: Poisonous if swallowed.
Absorption Hazard: Poisonous if absorbed.
Hazards to Wildlife: Dangerous to aquatic life.
Decontamination Procedures: Wash away any material with copious amounts of soap and water.
First Aid Procedures: Remove victim to fresh air, call emergency medical care. If not breathing give CPR. If breathing is difficult administer oxygen. Treat for shock.

Fire Hazards:

Flashpoint: n/a *Ignition temperature:* n/a
Flammable Explosive High Range: n/a *Low Range:* n/a
Toxic Products of Combustion: Toxic oxides of nitrogen are produced.
Other Hazards: Avoid contact with liquid or vapors. Keep people away!!
Possible extinguishing agents: Apply water from as far a distance as possible. Use foam, dry chemical, or carbon dioxide.

Reactivity Hazards:

Reactive With: Water to form hydrogen cyanide gas. *Other Reactions:* n/a

Corrosivity Hazards:

Corrosive With: n/a *Neutralizing Agent:* n/a

Radioactivity Hazards:

Radiation Emitted: n/a *Other Hazards:* n/a

Recommended Protection for Response Personnel:

Avoid breathing vapors, keep upwind. Wear a sealed chemical suit (polycarbonate, butyl rubber, viton, PVC, nitrile, neoprene, chlorinated polyethylene). Wash away any material which may have come into contact with the body with copious amounts of soap and water. Consider appropriate evacuation.

SODIUM CYANIDE (solid)

DOT Number: UN 1689 *DOT Hazard Class:* Poison B *DOT Guide Number:* 55
Synonyms: hydrocyanic acid, sodium salt
STCC Number: 4923228 *Reportable Qty:* 10/4.54
Mfg Name: E.I. Du Pont *Phone No:* 1-800-441-3637

Physical Description:

Physical Form: Solid *Color:* White *Odor:* Almond
Other Information: n/a

Chemical Properties:

Specific Gravity: 1.6 *Vapor Density:* 1.7 *Boiling Point: Very high*
Melting Point: n/a *Vapor Pressure:* 1 mm Hg at 1053° F(567.2° C) *Solubility in water:* Yes
Other Information: n/a

Health Hazards:

Inhalation Hazard: Poisonous if inhaled.
Ingestion Hazard: Poisonous if swallowed.
Absorption Hazard: Poisonous if absorbed.
Hazards to Wildlife: Dangerous to aquatic life.
Decontamination Procedures: Wash away any material with copious amounts of soap and water.
First Aid Procedures: Remove victim to fresh air, call emergency medical care. If not breathing give CPR. If breathing is difficult administer oxygen. Treat for shock.

Fire Hazards:

Flashpoint: n/a *Ignition temperature:* n/a
Flammable Explosive High Range: n/a *Low Range:* n/a
Toxic Products of Combustion: Toxic oxides of nitrogen are produced.
Other Hazards: Avoid contact with solid or dust. Keep people away!!
Possible extinguishing agents: Apply water from as far a distance as possible. Use foam, dry chemical, or carbon dioxide.

Reactivity Hazards:

Reactive With: Water to form hydrogen cyanide gas. *Other Reactions:* n/a

Corrosivity Hazards:

Corrosive With: n/a *Neutralizing Agent:* n/a

Radioactivity Hazards:

Radiation Emitted: n/a *Other Hazards:* n/a

Recommended Protection for Response Personnel:

Avoid breathing vapors, keep upwind. Wear a sealed chemical suit (polycarbonate, butyl rubber, viton, PVC, nitrile, neoprene, chlorinated polyethylene). Wash away any material which may have come into contact with the body with copious amounts of soap and water. Consider appropriate evacuation.

SODIUM DICHLORO-s-TRIAZINETRIONE

DOT Number: UN 2465 *DOT Hazard Class:* Oxidizer *DOT Guide Number:* 42
Synonyms: sodium dichloroisocyanurate
STCC Number: 4918435 *Reportable Qty:* n/a
Mfg Name: FMC Corp. *Phone No:* 1-312-861-5900

Physical Description:

Physical Form: Solid *Color:* White *Odor:* Bleach like
Other Information: It is used in dry bleach household compounds, and swimming pool disinfectants.

Chemical Properties:

Specific Gravity: .96 *Vapor Density:* 7.6 *Boiling Point: Decomposes*
Melting Point: n/a *Vapor Pressure:* n/a *Solubility in water:* Yes
Other Information: Reacts with small amounts of water to form chlorine gas.

Health Hazards:

Inhalation Hazard: Coughing, difficulty in breathing.
Ingestion Hazard: Harmful if swallowed.
Absorption Hazard: Irritating to the skin and eyes.
Hazards to Wildlife: n/a
Decontamination Procedures: Wash away any material with copious amounts of soap and water.
First Aid Procedures: Remove victim to fresh air, call emergency medical care. If not breathing give CPR. If breathing is difficult administer oxygen. Treat for shock.

Fire Hazards:

Flashpoint: n/a *Ignition temperature:* n/a
Flammable Explosive High Range: n/a *Low Range:* n/a
Toxic Products of Combustion: Poisonous gases may be produced in fire.
Other Hazards: May cause fire upon contact with combustible materials. containers may explode when heated.
Possible extinguishing agents: Apply water from as far a distance as possible.

Reactivity Hazards:

Reactive With: Water to form bleach.
Other Reactions: Contact with most organic matter or easily chlorinated or oxidized materials may result in fire.

Corrosivity Hazards:

Corrosive With: n/a *Neutralizing Agent:* n/a

Radioactivity Hazards:

Radiation Emitted: n/a *Other Hazards:* n/a

Recommended Protection for Response Personnel:

Avoid breathing vapors, keep upwind. Structural protective clothing provides limited protection. Wash away any material which may have come into contact with the body with copious amounts of soap and water. Consider appropriate evacuation.

SODIUM DICHROMATE

DOT Number: NA 1479 *DOT Hazard Class:* ORM-A *DOT Guide Number:* 31
Synonyms: sodium bichromate
STCC Number: 4941170 *Reportable Qty:* 1000/454
Mfg Name: Allied Corp. *Phone No:* 1-201-455-2000

Physical Description:

Physical Form: Solid *Color:* Red to orange *Odor:* Odorless
Other Information: It is used to manufacture other chemicals, as a corrosion inhibitor, and for many other uses.

Chemical Properties:

Specific Gravity: 2.35 *Vapor Density:* 10 *Boiling Point: Decomposes*
Melting Point: n/a *Vapor Pressure:* n/a *Solubility in water:* Yes
Other Information: Sinks and mixes with water.

Health Hazards:

Inhalation Hazard: Causes difficulty in breathing.
Ingestion Hazard: Causes nausea and vomiting.
Absorption Hazard: Will burn the skin and eyes.
Hazards to Wildlife: Dangerous to aquatic life.
Decontamination Procedures: Wash away any material with copious amounts of soap and water.
First Aid Procedures: Remove victim to fresh air, call emergency medical care. If not breathing give CPR. If breathing is difficult administer oxygen. Treat for shock.

Fire Hazards:

Flashpoint: n/a *Ignition temperature:* n/a
Flammable Explosive High Range: n/a *Low Range:* n/a
Toxic Products of Combustion: n/a
Other Hazards: May cause fire on contact with combustibles.
Possible extinguishing agents: Extinguish fire using suitable agent for the type of surrounding fire.

Reactivity Hazards:

Reactive With: Combustible materials *Other Reactions:* n/a

Corrosivity Hazards:

Corrosive With: n/a *Neutralizing Agent:* n/a

Radioactivity Hazards:

Radiation Emitted: n/a *Other Hazards:* n/a

Recommended Protection for Response Personnel:

Avoid breathing vapors, keep upwind. Wear a sealed chemical suit (polycarbonate, butyl rubber). Wash away any material which may have come into contact with the body with copious amounts of soap and water. Consider appropriate evacuation.

SODIUM FERROCYANIDE

DOT Number: n/a *DOT Hazard Class:* n/a *DOT Guide Number:* n/a
Synonyms: None given
STCC Number: n/a *Reportable Qty:* n/a
Mfg Name: American Cyanamide Co. *Phone No:* 1-201-831-2000

Physical Description:

Physical Form: Solid *Color:* Yellow *Odor:* Odorless
Other Information: n/a

Chemical Properties:

Specific Gravity: 1.46 *Vapor Density:* n/a *Boiling Point: Decomposes*
Melting Point: n/a *Vapor Pressure:* n/a *Solubility in water:* Yes
Other Information: Sinks and mixes with water.

Health Hazards:

Inhalation Hazard: n/a
Ingestion Hazard: Harmful if swallowed.
Absorption Hazard: n/a
Hazards to Wildlife: n/a
Decontamination Procedures: Wash away any material with copious amounts of soap and water.
First Aid Procedures: Remove victim to fresh air, call emergency medical care. If not breathing give CPR. If breathing is difficult administer oxygen. Treat for shock.

Fire Hazards:

Flashpoint: n/a *Ignition temperature:* n/a
Flammable Explosive High Range: n/a *Low Range:* n/a
Toxic Products of Combustion: n/a
Other Hazards: n/a
Possible extinguishing agents: Extinguish fire using suitable agent for the type of surrounding fire.

Reactivity Hazards:

Reactive With: n/a *Other Reactions:* n/a

Corrosivity Hazards:

Corrosive With: n/a *Neutralizing Agent:* n/a

Radioactivity Hazards:

Radiation Emitted: n/a *Other Hazards:* n/a

Recommended Protection for Response Personnel:

Avoid breathing vapors, keep upwind. Structural protective clothing provides limited protection). Wash away any material which may have come into contact with the body with copious amounts of soap and water. Consider appropriate evacuation.

SODIUM FLUORIDE

DOT Number: UN 1690 *DOT Hazard Class:* ORM-B *DOT Guide Number:* 54
Synonyms: none given
STCC Number: 4944150 *Reportable Qty:* 1000/454
Mfg Name: Allied Corp. *Phone No:* 1-201-455-2000

Physical Description:

Physical Form: Solid *Color:* White or tinted blue *Odor:* Odorless
Other Information: It is used as an insecticide, to fluorinate water supplies, as a wood preservative, in cleaning compounds, manufacture of glass, and for many other uses.

Chemical Properties:

Specific Gravity: 2.79 *Vapor Density:* n/a *Boiling Point: Very high*
Melting Point: n/a *Vapor Pressure:* 1 mm Hg at 1970° F(1076° C) *Solubility in water:* Yes
Other Information: Sinks and mixes in water.

Health Hazards:

Inhalation Hazard: n/a
Ingestion Hazard: Poisonous if swallowed.
Absorption Hazard: n/a
Hazards to Wildlife: Dangerous to aquatic life.
Decontamination Procedures: Wash away any material with copious amounts of soap and water.
First Aid Procedures: Remove victim to fresh air, call emergency medical care. If not breathing give CPR. If breathing is difficult administer oxygen. Treat for shock.

Fire Hazards:

Flashpoint: n/a *Ignition temperature:* n/a
Flammable Explosive High Range: n/a *Low Range:* n/a
Toxic Products of Combustion: n/a
Other Hazards: n/a
Possible extinguishing agents: Extinguish fire using suitable agent for the type of surrounding fire.

Reactivity Hazards:

Reactive With: n/a *Other Reactions:* n/a

Corrosivity Hazards:

Corrosive With: Aluminum *Neutralizing Agent:* Crushed limestone, soda ash, or lime.

Radioactivity Hazards:

Radiation Emitted: n/a *Other Hazards:* n/a

Recommended Protection for Response Personnel:

Avoid breathing vapors, keep upwind. Wear a sealed chemical suit (polycarbonate, butyl rubber). Wash away any material which may have come into contact with the body with copious amounts of soap and water. Consider appropriate evacuation.

SODIUM HYDRIDE

DOT Number: UN 1427 *DOT Hazard Class:* Flammable solid *DOT Guide Number:* 40
Synonyms: none given
STCC Number: 4916454 *Reportable Qty:* n/a
Mfg Name: A.D. MacKay Inc. *Phone No:* 1-203-655-3000

Physical Description:

Physical Form: Powder in oil *Color:* Gray *Odor:* Kerosene
Other Information: It is used to make other chemicals.

Chemical Properties:

Specific Gravity: n/a *Vapor Density:* .8 *Boiling Point: Very high*
Melting Point: n/a *Vapor Pressure:* n/a *Solubility in water:* Reacts
Other Information: May react with heat or oxidizing materials.

Health Hazards:

Inhalation Hazard: n/a
Ingestion Hazard: Harmful if swallowed.
Absorption Hazard: Will burn the skin and eyes.
Hazards to Wildlife: Dangerous to aquatic life.
Decontamination Procedures: Wash away any material with copious amounts of soap and water.
First Aid Procedures: Remove victim to fresh air, call emergency medical care. If not breathing give CPR. If breathing is difficult administer oxygen. Treat for shock.

Fire Hazards:

Flashpoint: n/a *Ignition temperature:* n/a
Flammable Explosive High Range: n/a *Low Range:* n/a
Toxic Products of Combustion: n/a
Other Hazards: May explode upon contact with water. Flammable gases are produced.
Possible extinguishing agents: Graphite, soda ash, or powdered limestone. Do not use water!

Reactivity Hazards:

Reactive With: Water to form flammable hydrogen gas.
Other Reactions: May react violently with heat or other oxidizing materials.

Corrosivity Hazards:

Corrosive With: n/a *Neutralizing Agent:* n/a

Radioactivity Hazards:

Radiation Emitted: n/a *Other Hazards:* n/a

Recommended Protection for Response Personnel:

Avoid breathing vapors, keep upwind. Wear a sealed chemical suit (polycarbonate, butyl rubber). Wash away any material which may have come into contact with the body with copious amounts of soap and water. Consider appropriate evacuation.

SODIUM HYDROSULFIDE SOLUTION

DOT Number: NA 2922 *DOT Hazard Class:* Corrosive *DOT Guide Number:* 59
Synonyms: sodium bisulfide, sodium sulfhydrate
STCC Number: 4935268 *Reportable Qty:* 5000/2270
Mfg Name: U.S. Chemical Corp. *Phone No:* 1-617-237-4877

Physical Description:

Physical Form: Liquid *Color:* Light yellow to red *Odor:* Rotten egg
Other Information: It is used in paper pulping, manufacturing of dyes, and dehairing hides.

Chemical Properties:

Specific Gravity: 1.3 *Vapor Density:* 4 *Boiling Point: 212° F(100° C)*
Melting Point: n/a *Vapor Pressure:* n/a *Solubility in water:* Yes
Other Information: n/a

Health Hazards:

Inhalation Hazard: n/a
Ingestion Hazard: Nausea, vomiting, loss of consciousness.
Absorption Hazard: Irritating to the skin and eyes.
Hazards to Wildlife: Dangerous to aquatic life.
Decontamination Procedures: Wash away any material with copious amounts of soap and water.
First Aid Procedures: Remove victim to fresh air, call emergency medical care. If not breathing give CPR. If breathing is difficult administer oxygen. Treat for shock.

Fire Hazards:

Flashpoint: n/a *Ignition temperature:* n/a
Flammable Explosive High Range: n/a *Low Range:* n/a
Toxic Products of Combustion: None
Other Hazards: Keep people away!
Possible extinguishing agents: Apply water from as far a distance as possible. Use suitable agent for type of surrounding fire.

Reactivity Hazards:

Reactive With: n/a *Other Reactions:* n/a

Corrosivity Hazards:

Corrosive With: Metals and tissue. *Neutralizing Agent:* Crushed limestone, soda ash, or lime.

Radioactivity Hazards:

Radiation Emitted: n/a *Other Hazards:* n/a

Recommended Protection for Response Personnel:

Avoid breathing vapors, keep upwind. Wear a sealed chemical suit (polycarbonate, butyl rubber). Wash away any material which may have come into contact with the body with copious amounts of soap and water. Consider appropriate evacuation.

SODIUM HYDROSULFITE

DOT Number: UN 1384 *DOT Hazard Class:* Flammable solid *DOT Guide Number:* 37
Synonyms: hydrosulfite of soda; sodium hydrosulphite; sodium sulfoxylate, disodium salt
STCC Number: 4916737 *Reportable Qty:* n/a *CHEMTREC Phone No:* 1-800-424-9300

Physical Description:

Physical Form: Solid *Color:* White to gray *Odor:* Sulfurous
Other Information: It is used in vat dyeing, bleaching sugar, soap, oils, and in groundwood.

Chemical Properties:

Specific Gravity: n/a *Vapor Density:* 3.6 *Boiling Point: Decomposes*
Melting Point: 125.6° F(52° C) *Vapor Pressure:* n/a *Solubility in water:* Yes
Other Information: Will dissolve rapidly in water.

Health Hazards:

Inhalation Hazard: Harmful if inhaled.

Ingestion Hazard: Nausea, vomiting, and diarrhea.

Absorption Hazard: Irritating to the skin and eyes.

Hazards to Wildlife: n/a

Decontamination Procedures: Wash away any material with copious amounts of soap and water.

First Aid Procedures: Remove victim to fresh air, call emergency medical care. If not breathing give CPR. If breathing is difficult administer oxygen. Treat for shock.

Fire Hazards:

Flashpoint: n/a *Ignition temperature:* n/a
Flammable Explosive High Range: n/a *Low Range:* n/a
Toxic Products of Combustion: Toxic sulfur dioxide gas.
Other Hazards: Containers may rupture violently in a fire. Dust may form explosive mixture with air.
Possible extinguishing agents: If material is on fire, flood with water. If material is not on fire, do not use water.

Reactivity Hazards:

Reactive With: Water to form sodium bisulfite and later to form sodium bisulfate with evolution of heat.
Other Reactions: Oxidizing agents

Corrosivity Hazards:

Corrosive With: n/a *Neutralizing Agent:* n/a

Radioactivity Hazards:

Radiation Emitted: n/a *Other Hazards:* n/a

Recommended Protection for Response Personnel:

Avoid breathing vapors, keep upwind. Structural protective clothing provides limited protection. Wash away any material which may have come into contact with the body with copious amounts of soap and water. Consider appropriate evacuation.

SODIUM HYDROXIDE

DOT Number: UN 1823 *DOT Hazard Class:* Corrosive *DOT Guide Number:* 60
Synonyms: caustic soda, lye, soda lye
STCC Number: 4935235 *Reportable Qty:* 1000/454
Mfg Name: Dow Chemical *Phone No:* 1-517-636-4400

Physical Description:

Physical Form: Solid *Color:* White *Odor:* Odorless
Other Information: It is used in chemical manufacturing, petroleum refining, cleaning compounds, home drain openers, and many other uses

Chemical Properties:

Specific Gravity: 2.13 *Vapor Density:* 1.4 *Boiling Point: 2534° F(1390° C)*
Melting Point: 590° F(310° C) *Vapor Pressure:* 0 mm Hg at 68° F(20° C)
Solubility in water: Yes *Degree of Solubility:* 50%
Other Information: n/a

Health Hazards:

Inhalation Hazard: Irritating to the eyes, nose, and throat. *Ingestion Hazard:* Harmful if swallowed.
Absorption Hazard: Will burn the skin and eyes. *Hazards to Wildlife:* Dangerous to aquatic life.
Decontamination Procedures: Wash away any material with copious amounts of soap and water.
First Aid Procedures: Remove victim to fresh air, call emergency medical care. If not breathing give CPR. If breathing is difficult administer oxygen. Treat for shock.

Fire Hazards:

Flashpoint: n/a *Ignition temperature:* n/a
Flammable Explosive High Range: n/a *Low Range:* n/a
Toxic Products of Combustion: n/a
Other Hazards: May cause fire upon contact with combustibles. Flammable gas may be produced upon contact with metals.
Possible extinguishing agents: Apply water from as far a distance as possible. Use suitable agent for type of surrounding fire.

Reactivity Hazards:

Reactive With: Water with liberation of much heat which may steam and splatter. *Other Reactions:* Air to absorb moisture and dissolves in it.

Corrosivity Hazards:

Corrosive With: When wet, material will attack aluminum, tin, lead, and zinc to produce flammable hydrogen gas.
Neutralizing Agent: Flush with water. Rinse with diluted acetic acid.

Radioactivity Hazards:

Radiation Emitted: n/a *Other Hazards:* n/a

Recommended Protection for Response Personnel:

Avoid breathing vapors, keep upwind. Wear a sealed chemical suit (butyl rubber, viton, PVC, nitrile, neoprene, chlorinated polyethylene). Wash away any material which may have come into contact with the body with copious amounts of soap and water. Consider appropriate evacuation.

SODIUM HYPOCHLORITE

DOT Number: NA 1791 *DOT Hazard Class:* Corrosive *DOT Guide Number:* 60
Synonyms: Clorox, liquid bleach
STCC Number: 4944143 *Reportable Qty:* 100/45.4
Mfg Name: Clorox Co. *Phone No:* 1-415-271-7000

Physical Description:

Physical Form: Liquid *Color:* Green to yellow *Odor:* Bleach
Other Information: It is used as a bleach for textiles and in paper pulp, for water purification, and to make other chemicals.

Chemical Properties:

Specific Gravity: 1.6 *Vapor Density:* 2.5 *Boiling Point: Decomposes*
Melting Point: n/a *Vapor Pressure:* n/a *Solubility in water:* Yes
Other Information: n/a

Health Hazards:

Inhalation Hazard: n/a
Ingestion Hazard: Harmful if swallowed.
Absorption Hazard: Irritating to the skin and eyes.
Hazards to Wildlife: n/a
Decontamination Procedures: Wash away any material with copious amounts of soap and water.
First Aid Procedures: Remove victim to fresh air, call emergency medical care. If not breathing give CPR. If breathing is difficult administer oxygen. Treat for shock.

Fire Hazards:

Flashpoint: n/a *Ignition temperature:* n/a
Flammable Explosive High Range: n/a *Low Range:* n/a
Toxic Products of Combustion: n/a
Other Hazards: n/a
Possible extinguishing agents: Apply water from as far a distance as possible. Use suitable agent for type of surrounding fire. Cool exposed containers with water.

Reactivity Hazards:

Reactive With: n/a *Other Reactions:* n/a

Corrosivity Hazards:

Corrosive With: Aluminum *Neutralizing Agent:* Use water for dilution, neutralize with soda ash.

Radioactivity Hazards:

Radiation Emitted: n/a *Other Hazards:* n/a

Recommended Protection for Response Personnel:

Avoid breathing vapors, keep upwind. Structural protective clothing provides limited protection. Wash away any material which may have come into contact with the body with copious amounts of soap and water. Consider appropriate evacuation.

SODIUM METHYLATE

DOT Number: UN 1431 *DOT Hazard Class:* Flammable solid *DOT Guide Number:* 40
Synonyms: sodium methoxide
STCC Number: 4916461 *Reportable Qty:* 1000/454
Mfg Name: Olin Corp. *Phone No:* 1-203-356-2000

Physical Description:

Physical Form: Solid *Color:* White *Odor:* Odorless
Other Information: It is used to process edible fats and oils, and to make other chemicals.

Chemical Properties:

Specific Gravity: 1 *Vapor Density:* 1 *Boiling Point: Decomposes*
Melting Point: n/a *Vapor Pressure:* n/a *Solubility in water:* Yes
Other Information: n/a

Health Hazards:

Inhalation Hazard: Irritating to the eyes, nose, and throat.
Ingestion Hazard: Harmful if swallowed.
Absorption Hazard: Irritating to the skin and eyes.
Hazards to Wildlife: n/a
Decontamination Procedures: Wash away any material with copious amounts of soap and water.
First Aid Procedures: Remove victim to fresh air, call emergency medical care. If not breathing give CPR. If breathing is difficult administer oxygen. Treat for shock.

Fire Hazards:

Flashpoint: n/a *Ignition temperature:* n/a
Flammable Explosive High Range: n/a *Low Range:* n/a
Toxic Products of Combustion: n/a
Other Hazards: n/a
Possible extinguishing agents: Dry chemical or carbon dioxide. Do not use water.

Reactivity Hazards:

Reactive With: Water to form caustic soda solution and methyl alcohol. *Other Reactions:* n/a

Corrosivity Hazards:

Corrosive With: Certain plastics, nylons, and polyesters.
Neutralizing Agent: Dilute with water. Neutralize with acetic acid vinegar.

Radioactivity Hazards:

Radiation Emitted: n/a *Other Hazards:* n/a

Recommended Protection for Response Personnel:

Avoid breathing vapors, keep upwind. Wear a sealed chemical suit (polycarbonate, butyl rubber). Wash away any material which may have come into contact with the body with copious amounts of soap and water. Consider appropriate evacuation.

SODIUM NITRATE

DOT Number: UN 1498 *DOT Hazard Class:* Oxidizer *DOT Guide Number:* 35
Synonyms: chile saltpeter, soda niter
STCC Number: 4918746 *Reportable Qty:* n/a
Mfg Name: Mallinckrodt Inc. *Phone No:* 1-314-895-2000

Physical Description:

Physical Form: Solid *Color:* Colorless to white *Odor:* Odorless
Other Information: It is used as solid propellents, explosives, and fertilizers.

Chemical Properties:

Specific Gravity: 2.26 *Vapor Density:* 2.9 *Boiling Point: Decomposes*
Melting Point: n/a *Vapor Pressure:* n/a *Solubility in water:* Yes
Other Information: n/a

Health Hazards:

Inhalation Hazard: n/a
Ingestion Hazard: Dizziness, vomiting, convulsions, collapse.
Absorption Hazard: n/a
Hazards to Wildlife: Dangerous to aquatic life.
Decontamination Procedures: Wash away any material with copious amounts of soap and water.
First Aid Procedures: Remove victim to fresh air, call emergency medical care. If not breathing give CPR. If breathing is difficult administer oxygen. Treat for shock.

Fire Hazards:

Flashpoint: n/a *Ignition temperature:* n/a
Flammable Explosive High Range: n/a *Low Range:* n/a
Toxic Products of Combustion: Toxic oxides of nitrogen are produced in fires involving this material.
Other Hazards: May cause fire and explode on contact with combustibles.
Possible extinguishing agents: Apply water from as far a distance as possible.

Reactivity Hazards:

Reactive With: Oxidizable substances and organic materials. *Other Reactions:* n/a

Corrosivity Hazards:

Corrosive With: n/a *Neutralizing Agent:* n/a

Radioactivity Hazards:

Radiation Emitted: n/a *Other Hazards:* n/a

Recommended Protection for Response Personnel:

Avoid breathing vapors, keep upwind. Structural protective clothing provides limited protection. Wash away any material which may have come into contact with the body with copious amounts of soap and water. Consider appropriate evacuation.

SODIUM NITRITE

DOT Number: UN 1500 *DOT Hazard Class:* Oxidizer *DOT Guide Number:* 35
Synonyms: Ecrinitrit, Filmerine
STCC Number: 4918747 *Reportable Qty:* 100/45.4
Mfg Name: Ashland Chemical *Phone No:* 1-614-889-3333

Physical Description:

Physical Form: Solid *Color:* White *Odor:* Odorless
Other Information: It is used as a food preservative, and to make other chemicals.

Chemical Properties:

Specific Gravity: 2.17 *Vapor Density:* 2.4 *Boiling Point: Decomposes*
Melting Point: n/a *Vapor Pressure:* n/a *Solubility in water:* Yes
Other Information: n/a

Health Hazards:

Inhalation Hazard: Headache, difficulty in breathing, loss of consciousness.
Ingestion Hazard: Headache, vomiting, loss of consciousness.
Absorption Hazard: Irritating to the skin and eyes.
Hazards to Wildlife: Dangerous to aquatic life.
Decontamination Procedures: Wash away any material with copious amounts of soap and water.
First Aid Procedures: Remove victim to fresh air, call emergency medical care. If not breathing give CPR. If breathing is difficult administer oxygen. Treat for shock.

Fire Hazards:

Flashpoint: n/a *Ignition temperature:* n/a
Flammable Explosive High Range: n/a *Low Range:* n/a
Toxic Products of Combustion: Toxic oxides of nitrogen are produced in fires involving this material.
Other Hazards: May cause fire upon contact with combustibles. Will increase the intensity of a fire.
Possible extinguishing agents: Apply water from as far a distance as possible.

Reactivity Hazards:

Reactive With: n/a *Other Reactions:* n/a

Corrosivity Hazards:

Corrosive With: n/a *Neutralizing Agent:* n/a

Radioactivity Hazards:

Radiation Emitted: n/a *Other Hazards:* n/a

Recommended Protection for Response Personnel:

Avoid breathing vapors, keep upwind. Structural protective clothing provides limited protection. Wash away any material which may have come into contact with the body with copious amounts of soap and water. Consider appropriate evacuation.

SODIUM OXALATE

DOT Number: n/a *DOT Hazard Class:* n/a *DOT Guide Number:* n/a
Synonyms: ethanedioic acid, disodium salt
STCC Number: n/a *Reportable Qty:* n/a
Mfg Name: J.T. Baker Chemical Co. *Phone No:* 1-201-859-2151

Physical Description:

Physical Form: Solid *Color:* White *Odor:* Odorless
Other Information: n/a

Chemical Properties:

Specific Gravity: 2.27 *Vapor Density:* 2.4 *Boiling Point: Decomposes*
Melting Point: n/a *Vapor Pressure:* n/a *Solubility in water:* Yes
Other Information: Sinks and mixes slowly with water.

Health Hazards:

Inhalation Hazard: Difficulty in breathing, loss of consciousness.
Ingestion Hazard: Poisonous if swallowed.
Absorption Hazard: Irritating to the skin and eyes.
Hazards to Wildlife: Dangerous to aquatic life.
Decontamination Procedures: Wash away any material with copious amounts of soap and water.
First Aid Procedures: Remove victim to fresh air, call emergency medical care. If not breathing give CPR. If breathing is difficult administer oxygen. Treat for shock.

Fire Hazards:

Flashpoint: n/a *Ignition temperature:* n/a
Flammable Explosive High Range: n/a *Low Range:* n/a
Toxic Products of Combustion: n/a
Other Hazards: n/a
Possible extinguishing agents: Extinguish fire using suitable agent for the type of surrounding fire.

Reactivity Hazards:

Reactive With: n/a *Other Reactions:* n/a

Corrosivity Hazards:

Corrosive With: n/a *Neutralizing Agent:* n/a

Radioactivity Hazards:

Radiation Emitted: n/a *Other Hazards:* n/a

Recommended Protection for Response Personnel:

Avoid breathing vapors, keep upwind. Wear a sealed chemical suit (polycarbonate, butyl rubber). Wash away any material which may have come into contact with the body with copious amounts of soap and water. Consider appropriate evacuation.

SODIUM PHOSPHATE

DOT Number: NA 9148 *DOT Hazard Class:* ORM-E *DOT Guide Number:* 31
Synonyms: sodium phosphate dibasic, sodium phosphate monobasic, sodium phosphate tribasic
STCC Number: 4966383 *Reportable Qty:* 5000/2270
Mfg Name: Stauffer Chemical *Phone No:* 1-203-222-3000

Physical Description:

Physical Form: Solid *Color:* White or colorless *Odor:* Odorless
Other Information: It is used as a cleaning compound, in photography, as a paint remover, and for many other uses.

Chemical Properties:

Specific Gravity: 1.62 *Vapor Density:* 4.9 *Boiling Point: 212° F(100° C)*
Melting Point: n/a *Vapor Pressure:* n/a *Solubility in water:* Yes
Other Information: Sinks and mixes in water.

Health Hazards:

Inhalation Hazard: n/a
Ingestion Hazard: Harmful if swallowed.
Absorption Hazard: Will burn the skin and eyes.
Hazards to Wildlife: Dangerous to aquatic life.
Decontamination Procedures: Wash away any material with copious amounts of soap and water.
First Aid Procedures: Remove victim to fresh air, call emergency medical care. If not breathing give CPR. If breathing is difficult administer oxygen. Treat for shock.

Fire Hazards:

Flashpoint: n/a *Ignition temperature:* n/a
Flammable Explosive High Range: n/a *Low Range:* n/a
Toxic Products of Combustion: Irritating phosphorus oxides.
Other Hazards: n/a
Possible *Possible extinguishing agents:* Extinguish fire using suitable agent for the type of surrounding fire.

Reactivity Hazards:

Reactive With: n/a *Other Reactions:* n/a

Corrosivity Hazards:

Corrosive With: n/a *Neutralizing Agent:* n/a

Radioactivity Hazards:

Radiation Emitted: n/a *Other Hazards:* n/a

Recommended Protection for Response Personnel:

Avoid breathing vapors, keep upwind. Structural protective clothing provides limited protection. Wash away any material which may have come into contact with the body with copious amounts of soap and water. Consider appropriate evacuation.

SODIUM SELENITE

DOT Number: UN 2630 *DOT Hazard Class:* Poison B *DOT Guide Number:* 53
Synonyms: disodiumselenite, disodium salt
STCC Number: 4923350 *Reportable Qty:* 100/45.4
Mfg Name: A.D. MacKay Inc. *Phone No:* 1-203-655-3300

Physical Description:

Physical Form: Solid *Color:* White to pink *Odor:* Odorless
Other Information: n/a

Chemical Properties:

Specific Gravity: n/a *Vapor Density:* 6.5 *Boiling Point: Decomposes*
Melting Point: n/a *Vapor Pressure:* n/a *Solubility in water:* Yes
Other Information: n/a

Health Hazards:

Inhalation Hazard: Poisonous if inhaled.
Ingestion Hazard: Poisonous if swallowed.
Absorption Hazard: Irritating to the skin and eyes.
Hazards to Wildlife: Dangerous to aquatic life.
Decontamination Procedures: Wash away any material with copious amounts of soap and water.
First Aid Procedures: Remove victim to fresh air, call emergency medical care. If not breathing give CPR. If breathing is difficult administer oxygen. Treat for shock.

Fire Hazards:

Flashpoint: n/a *Ignition temperature:* n/a
Flammable Explosive High Range: n/a *Low Range:* n/a
Toxic Products of Combustion: Poisonous gases are produced in fire.
Other Hazards: Avoid contact with solid or dust. Keep people away.
Possible extinguishing agents: Apply water from as far a distance as possible. Use foam, dry chemical, or carbon dioxide.

Reactivity Hazards:

Reactive With: n/a *Other Reactions:* n/a

Corrosivity Hazards:

Corrosive With: n/a *Neutralizing Agent:* n/a

Radioactivity Hazards:

Radiation Emitted: n/a *Other Hazards:* n/a

Recommended Protection for Response Personnel:

Avoid breathing vapors, keep upwind. Wear appropriate chemical clothing. Wash away any material which may come into contact with the body with copious amounts of soap and water. Consider appropriate evacuation.

SODIUM SILICATE

DOT Number: n/a *DOT Hazard Class:* n/a *DOT Guide Number:* n/a
Synonyms: water glass, soluble glass
STCC Number: n/a *Reportable Qty:* n/a
Mfg Name: E.I. Du Pont *Phone No:* 1-800-441-3637

Physical Description:

Physical Form: Liquid *Color:* Colorless *Odor:* Odorless
Other Information: n/a

Chemical Properties:

Specific Gravity: 1.1 to 1.7 *Vapor Density:* n/a *Boiling Point: Decomposes*
Melting Point: n/a *Vapor Pressure:* n/a *Solubility in water:* Yes
Other Information: Sinks and mixes slowly with water.

Health Hazards:

Inhalation Hazard: n/a
Ingestion Hazard: Harmful if swallowed.
Absorption Hazard: n/a
Hazards to Wildlife: Dangerous to aquatic life.
Decontamination Procedures: Wash away any material with copious amounts of soap and water.
First Aid Procedures: Remove victim to fresh air, call emergency medical care. If not breathing give CPR. If breathing is difficult administer oxygen. Treat for shock.

Fire Hazards:

Flashpoint: n/a *Ignition temperature:* n/a
Flammable Explosive High Range: n/a *Low Range:* n/a
Toxic Products of Combustion: n/a
Other Hazards: n/a
Possible extinguishing agents: Extinguish fire using suitable agent for the type of surrounding fire.

Reactivity Hazards:

Reactive With: n/a *Other Reactions:* n/a

Corrosivity Hazards:

Corrosive With: n/a *Neutralizing Agent:* n/a

Radioactivity Hazards:

Radiation Emitted: n/a *Other Hazards:* n/a

Recommended Protection for Response Personnel:

Avoid breathing vapors, keep upwind. Structural protective clothing provides limited protection. Wash away any material which may have come into contact with the body with copious amounts of soap and water. Consider appropriate evacuation.

SODIUM SILICOFLUORIDE

DOT Number: UN 2674 *DOT Hazard Class:* n/a *DOT Guide Number:* 53
Synonyms: Salufer, sodium fluosilicate, sodium hexafluorosilicate
STCC Number: n/a *Reportable Qty:* n/a
Mfg Name: Ashland Chemical *Phone No:* 1-614-889-3333

Physical Description:

Physical Form: Solid *Color:* White *Odor:* Odorless
Other Information: n/a

Chemical Properties:

Specific Gravity: 2.68 *Vapor Density:* 6.5 *Boiling Point: Decomposes*
Melting Point: n/a *Vapor Pressure:* n/a *Solubility in water:* n/a
Other Information: Sinks in water.

Health Hazards:

Inhalation Hazard: Coughing or difficulty in breathing.
Ingestion Hazard: Poisonous if swallowed.
Absorption Hazard: Will burn the skin and eyes.
Hazards to Wildlife: n/a
Decontamination Procedures: Wash away any material with copious amounts of soap and water.
First Aid Procedures: Remove victim to fresh air, call emergency medical care. If not breathing give CPR. If breathing is difficult administer oxygen. Treat for shock.

Fire Hazards:

Flashpoint: n/a *Ignition temperature:* n/a
Flammable Explosive High Range: n/a *Low Range:* n/a
Toxic Products of Combustion: n/a
Other Hazards: Decomposes at red heat.
Possible extinguishing agents: Extinguish fire using suitable agent for the type of surrounding fire.

Reactivity Hazards:

Reactive With: n/a *Other Reactions:* n/a

Corrosivity Hazards:

Corrosive With: n/a *Neutralizing Agent:* n/a

Radioactivity Hazards:

Radiation Emitted: n/a *Other Hazards:* n/a

Recommended Protection for Response Personnel:

Avoid breathing vapors, keep upwind. Wear a sealed chemical suit (polycarbonate, butyl rubber). Wash away any material which may have come into contact with the body with copious amounts of soap and water. Consider appropriate evacuation.

SODIUM SULFIDE

DOT Number: UN 1385 *DOT Hazard Class:* Flammable solid *DOT Guide Number:* 34
Synonyms: none given
STCC Number: 4916740 *Reportable Qty:* n/a
Mfg Name: PPG Industries Inc. *Phone No:* 1-412-434-3131

Physical Description:

Physical Form: Solid *Color:* Yellow to red *Odor:* Rotten egg
Other Information: n/a

Chemical Properties:

Specific Gravity: 1.856 *Vapor Density:* 2.7 *Boiling Point: Very high*
Melting Point: n/a *Vapor Pressure:* n/a *Solubility in water:* Yes
Other Information: Sinks and mixes with water.

Health Hazards:

Inhalation Hazard: Irritating to the eyes, nose, and throat.
Ingestion Hazard: Harmful if swallowed.
Absorption Hazard: n/a
Hazards to Wildlife: Dangerous to aquatic life.
Decontamination Procedures: Wash away any material with copious amounts of soap and water.
First Aid Procedures: Remove victim to fresh air, call emergency medical care. If not breathing give CPR. If breathing is difficult administer oxygen. Treat for shock.

Fire Hazards:

Flashpoint: n/a *Ignition temperature:* n/a
Flammable Explosive High Range: n/a *Low Range:* n/a
Toxic Products of Combustion: Poisonous gases may be produced in a fire.
Other Hazards: Avoid contact with solid or dust.
Possible extinguishing agents: Flood with water.

Reactivity Hazards:

Reactive With: Acids to form flammable poisonous gases upon contact with methyl alcohol.
Other Reactions: n/a

Corrosivity Hazards:

Corrosive With: n/a *Neutralizing Agent:* n/a

Radioactivity Hazards:

Radiation Emitted: n/a *Other Hazards:* n/a

Recommended Protection for Response Personnel:

Avoid breathing vapors, keep upwind. Wear a sealed chemical suit (polycarbonate, butyl rubber, viton, PVC, nitrile, neoprene, chlorinated polyethylene). Wash away any material which may have come into contact with the body with copious amounts of soap and water. Consider appropriate evacuation.

SODIUM THIOCYANATE

DOT Number: n/a *DOT Hazard Class:* n/a *DOT Guide Number:* n/a
Synonyms: sodium rhodanate, sodium rhodanide, sodium sulfocyanate
STCC Number: n/a *Reportable Qty:* n/a
Mfg Name: J.T. Baker Chemical Co. *Phone No:* 1-201-859-2151

Physical Description:

Physical Form: Solid *Color:* White *Odor:* Odorless
Other Information: n/a

Chemical Properties:

Specific Gravity: 1 *Vapor Density:* n/a *Boiling Point: Decomposes*
Melting Point: n/a *Vapor Pressure:* n/a *Solubility in water:* Yes
Other Information: Sinks and mixes in water.

Health Hazards:

Inhalation Hazard: Coughing or difficulty in breathing.
Ingestion Hazard: Nausea, vomiting, loss of consciousness.
Absorption Hazard: Irritating to the skin and eyes.
Hazards to Wildlife: Dangerous to aquatic life.
Decontamination Procedures: Wash away any material with copious amounts of soap and water.
First Aid Procedures: Remove victim to fresh air, call emergency medical care. If not breathing give CPR. If breathing is difficult administer oxygen. Treat for shock.

Fire Hazards:

Flashpoint: n/a *Ignition temperature:* n/a
Flammable Explosive High Range: n/a *Low Range:* n/a
Toxic Products of Combustion: Irritating oxides of sulfur and nitrogen may be formed in fires.
Other Hazards: n/a
Possible extinguishing agents: Extinguish fire using suitable agent for the type of surrounding fire.

Reactivity Hazards:

Reactive With: n/a *Other Reactions:* n/a

Corrosivity Hazards:

Corrosive With: n/a *Neutralizing Agent:* n/a

Radioactivity Hazards:

Radiation Emitted: n/a *Other Hazards:* n/a

Recommended Protection for Response Personnel:

Avoid breathing vapors, keep upwind. Structural protective clothing provides limited protection. Wash away any material which may have come into contact with the body with copious amounts of soap and water. Consider appropriate evacuation.

SORBITOL

DOT Number: n/a *DOT Hazard Class:* n/a *DOT Guide Number:* n/a
Synonyms: hexahydric alcohol, 1,2,3,4,5,6-hexannehexol, d-glucitol, sorbit, sorbol
STCC Number: n/a *Reportable Qty:* n/a
Mfg Name: Durkee Foods Division *Phone No:* 1-216-344-8000

Physical Description:

Physical Form: Liquid *Color:* Colorless *Odor:* Odorless
Other Information: n/a

Chemical Properties:

Specific Gravity: 1.49 *Vapor Density:* n/a *Boiling Point: Very high*
Melting Point: 230° F(110° C) *Vapor Pressure:* n/a *Solubility in water:* Yes
Other Information: Sinks and mixes with water.

Health Hazards:

Inhalation Hazard: n/a
Ingestion Hazard: n/a
Absorption Hazard: Will burn the skin and eyes.
Hazards to Wildlife: Dangerous to aquatic life.
Decontamination Procedures: Wash away any material with copious amounts of soap and water.
First Aid Procedures: Remove victim to fresh air, call emergency medical care. If not breathing give CPR. If breathing is difficult administer oxygen. Treat for shock.

Fire Hazards:

Flashpoint: 542° F(283.3° C) *Ignition temperature:* n/a
Flammable Explosive High Range: n/a *Low Range:* n/a
Toxic Products of Combustion: n/a
Other Hazards: n/a
Possible extinguishing agents: Extinguish fire using suitable agent for the type of surrounding fire.

Reactivity Hazards:

Reactive With: n/a *Other Reactions:* n/a

Corrosivity Hazards:

Corrosive With: n/a *Neutralizing Agent:* n/a

Radioactivity Hazards:

Radiation Emitted: n/a *Other Hazards:* n/a

Recommended Protection for Response Personnel:

Avoid breathing vapors, keep upwind. Structural protective clothing provides limited protection. Wash away any material which may have come into contact with the body with copious amounts of soap and water. If the fire becomes uncontrollable, consider appropriate evacuation.

STANNOUS FLUORIDE

DOT Number: n/a *DOT Hazard Class:* n/a *DOT Guide Number:* n/a
Synonyms: fluoristan, tin difluoride, tin bifluoride
STCC Number: n/a *Reportable Qty:* n/a
Mfg Name: Ozark Mahoney *Phone No:* 1-918-585-2661

Physical Description:

Physical Form: Solid *Color:* White *Odor:* n/a
Other Information: n/a

Chemical Properties:

Specific Gravity: 2.79 *Vapor Density:* n/a *Boiling Point: 1052° F(566.6° C)*
Melting Point: 419° F(215° C) *Vapor Pressure:* n/a *Solubility in water:* Yes
Other Information: Sinks and mixes with water.

Health Hazards:

Inhalation Hazard: n/a
Ingestion Hazard: Harmful if swallowed.
Absorption Hazard: Will burn the eyes.
Hazards to Wildlife: Dangerous to aquatic life.
Decontamination Procedures: Wash away any material with copious amounts of soap and water.
First Aid Procedures: Remove victim to fresh air, call emergency medical care. If not breathing give CPR. If breathing is difficult administer oxygen. Treat for shock.

Fire Hazards:

Flashpoint: n/a *Ignition temperature:* n/a
Flammable Explosive High Range: n/a *Low Range:* n/a
Toxic Products of Combustion: n/a
Other Hazards: n/a
Possible extinguishing agents: Extinguish fire using suitable agent for the type of surrounding fire.

Reactivity Hazards:

Reactive With: n/a *Other Reactions:* n/a

Corrosivity Hazards:

Corrosive With: n/a *Neutralizing Agent:* n/a

Radioactivity Hazards:

Radiation Emitted: n/a *Other Hazards:* n/a

Recommended Protection for Response Personnel:

Avoid breathing vapors, keep upwind. Wear appropriate sealed chemical clothing. Wash away any material which may have come into contact with the body with copious amounts of soap and water. Consider appropriate evacuation.

STEARIC ACID

DOT Number: n/a *DOT Hazard Class:* n/a *DOT Guide Number:* n/a
Synonyms: 1-heptadecanecarboxylic acid, octadecanoic acid, n-octadecylic acid, stearophanic acid
STCC Number: n/a *Reportable Qty:* n/a
Mfg Name: Ashland Chemical *Phone No:* 1-614-889-3333

Physical Description:

Physical Form: Solid *Color:* White *Odor:* Mild
Other Information: n/a

Chemical Properties:

Specific Gravity: .86 *Vapor Density:* 9.8 *Boiling Point: Decomposes*
Melting Point: 157° F(69.4° C) *Vapor Pressure:* 1 mm Hg at 345° F(173.8° C) *Solubility in water:* n/a
Other Information: Floats on water

Health Hazards:

Inhalation Hazard: Coughing or difficulty in breathing.
Ingestion Hazard: Harmful if swallowed.
Absorption Hazard: Irritating to the skin and eyes.
Hazards to Wildlife: n/a
Decontamination Procedures: Wash away any material with copious amounts of soap and water.
First Aid Procedures: Remove victim to fresh air, call emergency medical care. If not breathing give CPR. If breathing is difficult administer oxygen. Treat for shock.

Fire Hazards:

Flashpoint: 365° F(185° C) *Ignition temperature:* 743° F(395° C)
Flammable Explosive High Range: n/a *Low Range:* n/a
Toxic Products of Combustion: n/a
Other Hazards: n/a
Possible extinguishing agents: Foam, dry chemical, carbon dioxide. Water or foam may cause frothing. Water may be ineffective on fires. Cool all exposed containers with water.

Reactivity Hazards:

Reactive With: n/a *Other Reactions:* n/a

Corrosivity Hazards:

Corrosive With: n/a *Neutralizing Agent:* n/a

Radioactivity Hazards:

Radiation Emitted: n/a *Other Hazards:* n/a

Recommended Protection for Response Personnel:

Avoid breathing vapors, keep upwind. Structural protective clothing provides limited protection. Wash away any material which may have come into contact with the body with copious amounts of soap and water. If the fire becomes uncontrollable, consider appropriate evacuation.

STRONTIUM CHROMATE

DOT Number: NA 9149 *DOT Hazard Class:* ORM-E *DOT Guide Number:* 31
Synonyms: chromic acid, strontium salt (1:1); deep lemon yellow; strontium yellow
STCC Number: 4963377 *Reportable Qty:* 1000/454
Mfg Name: Barium & Chemical Inc. *Phone No:* 1-614-387-9776

Physical Description:

Physical Form: Solid *Color:* Yellow *Odor:* Odorless
Other Information: It is used as a pigment, a protective coating against corrosion, and in pyrotechnics.

Chemical Properties:

Specific Gravity: 3.89 *Vapor Density:* 7 *Boiling Point: n/a*
Melting Point: n/a *Vapor Pressure:* n/a *Solubility in water:* Yes
Other Information: Sinks and mixes with water.

Health Hazards:

Inhalation Hazard: n/a
Ingestion Hazard: Harmful if swallowed.
Absorption Hazard: n/a
Hazards to Wildlife: Dangerous to aquatic life.
Decontamination Procedures: Wash away any material with copious amounts of soap and water.
First Aid Procedures: Remove victim to fresh air, call emergency medical care. If not breathing give CPR. If breathing is difficult administer oxygen. Treat for shock.

Fire Hazards:

Flashpoint: n/a *Ignition temperature:* n/a
Flammable Explosive High Range: n/a *Low Range:* n/a
Toxic Products of Combustion: n/a
Other Hazards: n/a
Possible extinguishing agents: Extinguish fire using suitable agent for the type of surrounding fire.

Reactivity Hazards:

Reactive With: Water to form a hazardous solution. *Other Reactions:* Acids and bases (avoid contact).

Corrosivity Hazards:

Corrosive With: n/a *Neutralizing Agent:* n/a

Radioactivity Hazards:

Radiation Emitted: n/a *Other Hazards:* n/a

Recommended Protection for Response Personnel:

Avoid breathing vapors, keep upwind. Wear appropriate chemical clothing. Wash away any material which may come into contact with the body with copious amounts of soap and water. Consider appropriate evacuation.

STRYCHNINE

DOT Number: UN 1692 *DOT Hazard Class:* Poison B *DOT Guide Number:* 53
Synonyms: Kwik-Kill, Mouse-tox, Nux-vomica
STCC Number: 4921477 *Reportable Qty:* 10/4.54
Mfg Name: Gallard-Schlesinger *Phone No:* 1-516-333-5600

Physical Description:

Physical Form: Solid *Color:* Colorless to white *Odor:* Odorless
Other Information: n/a

Chemical Properties:

Specific Gravity: 1.36 *Vapor Density:* 11 *Boiling Point: Decomposes*
Melting Point: 547° F(286.1° C) *Vapor Pressure:* 0 mm Hg at 68° F(20° C)
Solubility in water: Very slight *Degree of Solubility:* .02%
Other Information: Sinks and mixes slowly in water.

Health Hazards:

Inhalation Hazard: Harmful if inhaled.
Ingestion Hazard: Poisonous if swallowed.
Absorption Hazard: Harmful if absorbed.
Hazards to Wildlife: Dangerous to both waterfowl and aquatic life.
Decontamination Procedures: Wash away any material with copious amounts of soap and water.
First Aid Procedures: Remove victim to fresh air, call emergency medical care. If not breathing give CPR. If breathing is difficult administer oxygen. Treat for shock.

Fire Hazards:

Flashpoint: n/a *Ignition temperature:* n/a
Flammable Explosive High Range: n/a *Low Range:* n/a
Toxic Products of Combustion: Toxic oxides of nitrogen are produced during combustion of this material.
Other Hazards: Avoid breathing vapor or dust.
Possible extinguishing agents: Material itself does not burn or with great difficulty. use foam, dry chemical or carbon dioxide.

Reactivity Hazards:

Reactive With: n/a *Other Reactions:* n/a

Corrosivity Hazards:

Corrosive With: n/a *Neutralizing Agent:* n/a

Radioactivity Hazards:

Radiation Emitted: n/a *Other Hazards:* n/a

Recommended Protection for Response Personnel:

Avoid breathing vapors, keep upwind. Wear appropriate chemical clothing. Wash away any material which may have come into contact with the body with copious amounts of soap and water. Consider appropriate evacuation.

STYRENE

DOT Number: UN 2055 *DOT Hazard Class:* Flammable liquid *DOT Guide Number:* 27
Synonyms: styrol, styrolene, vinylbenzine
STCC Number: 4907265 *Reportable Qty:* 1000/454
Mfg Name: Dow Chemical *Phone No:* 1-517-636-4400

Physical Description:

Physical Form: Liquid *Color:* Colorless to light yellow *Odor:* Sweet
Other Information: It is used to make plastics, paints, synthetic rubber, and to make other chemicals.

Chemical Properties:

Specific Gravity: .9 *Vapor Density:* 3.6 *Boiling Point: 295° F(146.1° C)*
Melting Point: −23° F(−30.5° C) *Vapor Pressure:* 4.5 mm Hg at 68° F(20° C)
Solubility in water: No *Degree of Solubility:* .03%
Other Information: Lighter than and insoluble in water. Vapors are heavier than air. Weighs 7.6 lbs/3.4 kg per gallon/3.8 l.

Health Hazards:

Inhalation Hazard: Dizziness, loss of consciousness.
Ingestion Hazard: Harmful if swallowed.
Absorption Hazard: Will burn the skin and the eyes.
Hazards to Wildlife: Dangerous to aquatic life.
Decontamination Procedures: Wash away any material with copious amounts of soap and water.
First Aid Procedures: Remove victim to fresh air, call emergency medical care. If not breathing give CPR. If breathing is difficult administer oxygen. Treat for shock.

Fire Hazards:

Flashpoint: 88° F(31.1° C) *Ignition temperature:* 914° F(490° C)
Flammable Explosive High Range: 7 *Low Range:* 1.1
Toxic Products of Combustion: n/a
Other Hazards: If contaminated, or subject to heat, material may polymerize. if polymerization takes place inside a container, material may violently explode. Containers may explode in fire. Flashback along vapor trail may occur.
Possible extinguishing agents: Apply water from as far a distance as possible. Use foam, dry chemical, or carbon dioxide.

Reactivity Hazards:

Reactive With: Metal salts, peroxides, and strong acids. *Other Reactions:* n/a

Corrosivity Hazards:

Corrosive With: n/a *Neutralizing Agent:* n/a

Radioactivity Hazards:

Radiation Emitted: n/a *Other Hazards:* n/a

Recommended Protection for Response Personnel:

Avoid breathing vapors, keep upwind. Structural protective clothing provides limited protection. Wash away any material which may have come into contact with the body with copious amounts of soap and water. If the fire becomes uncontrollable, consider appropriate evacuation.

SUCROSE

DOT Number: n/a *DOT Hazard Class:* n/a *DOT Guide Number:* n/a
Synonyms: sugar, cane sugar, saccharose, saccharum, sugar
STCC Number: n/a *Reportable Qty:* n/a
Mfg Name: Amstar Corp. *Phone No:* 1-212-489-9000

Physical Description:

Physical Form: Solid *Color:* White *Odor:* Odorless
Other Information: n/a

Chemical Properties:

Specific Gravity: 1.59 *Vapor Density:* n/a *Boiling Point: Decomposes*
Melting Point: 320-367° F(160-186° C) *Vapor Pressure:* n/a
Solubility in water: n/a
Other Information: Sinks in water

Health Hazards:

Inhalation Hazard: Not harmful.
Ingestion Hazard: Not harmful.
Absorption Hazard: Not harmful.
Hazards to Wildlife: n/a
Decontamination Procedures: Wash away any material with copious amounts of soap and water.
First Aid Procedures: Remove victim to fresh air, call emergency medical care. If not breathing give CPR. If breathing is difficult administer oxygen. Treat for shock.

Fire Hazards:

Flashpoint: Combustible solid *Ignition temperature:* n/a
Flammable Explosive High Range: n/a *Low Range:* n/a
Toxic Products of Combustion: Irritating fumes may be formed in fires.
Other Hazards: Melts and chars in fire.
Possible extinguishing agents: Extinguish fire using suitable agent for the type of surrounding fire.

Reactivity Hazards:

Reactive With: n/a *Other Reactions:* n/a

Corrosivity Hazards:

Corrosive With: n/a *Neutralizing Agent:* n/a

Radioactivity Hazards:

Radiation Emitted: n/a *Other Hazards:* n/a

Recommended Protection for Response Personnel:

Avoid breathing vapors, keep upwind. Structural protective clothing provides limited protection. Wash away any material which may have come into contact with the body with copious amounts of soap and water. Consider appropriate evacuation.

SULFOLANE

DOT Number: n/a *DOT Hazard Class:* n/a *DOT Guide Number:* n/a
Synonyms: sulfolane-w, tetrahydrothiophene-1-dioxide, cyclic tetramethylene sulfone
STCC Number: n/a *Reportable Qty:* n/a
Mfg Name: Phillips Petroleum *Phone No:* 1-918-661-6600

Physical Description:

Physical Form: Liquid *Color:* Colorless *Odor:* Weak oily
Other Information: n/a

Chemical Properties:

Specific Gravity: 1.26 *Vapor Density:* n/a *Boiling Point: 545° F(285° C)*
Melting Point: 79° F(26.1° C) *Vapor Pressure:* n/a *Solubility in water:* Yes
Other Information: Solidifies and sinks and mixes with water.

Health Hazards:

Inhalation Hazard: n/a
Ingestion Hazard: Harmful if swallowed.
Absorption Hazard: Irritating to the eyes. Not irritating to the skin.
Hazards to Wildlife: n/a
Decontamination Procedures: Wash away any material with copious amounts of soap and water.
First Aid Procedures: Remove victim to fresh air, call emergency medical care. If not breathing give CPR. If breathing is difficult administer oxygen. Treat for shock.

Fire Hazards:

Flashpoint: 330° F(165.5° C) *Ignition temperature:* n/a
Flammable Explosive High Range: n/a *Low Range:* n/a
Toxic Products of Combustion: Toxic and irritating gases may be generated in fire.
Other Hazards: n/a
Possible extinguishing agents: Extinguish fire using suitable agent for the type of surrounding fire.

Reactivity Hazards:

Reactive With: n/a *Other Reactions:* n/a

Corrosivity Hazards:

Corrosive With: n/a *Neutralizing Agent:* n/a

Radioactivity Hazards:

Radiation Emitted: n/a *Other Hazards:* n/a

Recommended Protection for Response Personnel:

Avoid breathing vapors, keep upwind. Structural protective clothing provides limited protection. Wash away any material which may have come into contact with the body with copious amounts of soap and water. If the fire becomes uncontrollable, consider appropriate evacuation.

SULFUR

DOT Number: UN 1350 *DOT Hazard Class:* ORM-C *DOT Guide Number:* 32
Synonyms: brimstone, flowers of sulfur, sulphur, sulfur of flower
STCC Number: 4945356 *Reportable Qty:* n/a
Mfg Name: Freeport Sulfur Corp. *Phone No:* 1-504-568-439

Physical Description:

Physical Form: Liquid *Color:* Yellow, orange, tan, brown, or gray *Odor:* Rotten egg
Other Information: It is used in sulfuric acid production, petroleum refining, and in pulp and paper manufacturing.

Chemical Properties:

Specific Gravity: 1.80 *Vapor Density:* n/a *Boiling Point: 832.3° F(444.6° C)*
Melting Point: n/a *Vapor Pressure:* n/a *Solubility in water:* Thickens
Other Information: Sinks and thickens in water.

Health Hazards:

Inhalation Hazard: n/a
Ingestion Hazard: Harmful if swallowed.
Absorption Hazard: Will burn the skin and the eyes.
Hazards to Wildlife: Dangerous to aquatic life.
Decontamination Procedures: Wash away any material with copious amounts of soap and water.
First Aid Procedures: Remove victim to fresh air, call emergency medical care. If not breathing give CPR. If breathing is difficult administer oxygen. Treat for shock.

Fire Hazards:

Flashpoint: 405° F(207.2° C) *Ignition temperature:* 450° F(232.2° C)
Flammable Explosive High Range: n/a *Low Range:* n/a
Toxic Products of Combustion: Produces toxic sulfur dioxide gas.
Other Hazards: Burns with a pale blue flame.
Possible extinguishing agents: Apply water from as far a distance as possible.

Reactivity Hazards:

Reactive With: n/a *Other Reactions:* n/a

Corrosivity Hazards:

Corrosive With: n/a *Neutralizing Agent:* n/a

Radioactivity Hazards:

Radiation Emitted: n/a *Other Hazards:* n/a

Recommended Protection for Response Personnel:

Avoid breathing vapors, keep upwind. Wear appropriate chemical clothing. Wash away any material which may come into contact with the body with copious amounts of soap and water. Consider appropriate evacuation.

SULFUR CHLORIDE

DOT Number: UN 1828 *DOT Hazard Class:* Corrosive *DOT Guide Number:* 39
Synonyms: disulfur dichloride, sulfur monochloride, sulfur subchloride, thiosulfurous dichloride
STCC Number: 4932380 *Reportable Qty:* 1000/454 *CHEMTREC Phone No:* 1-800-424-9300

Physical Description:

Physical Form: Liquid *Color:* Amber to yellowish red *Odor:* Nauseating
Other Information: It is used in rubber and oil vulcanizing, purifying sugared juices, military poison gas, and sulfur solvent.

Chemical Properties:

Specific Gravity: 1.69 *Vapor Density:* 4.7 *Boiling Point: 280° F(137.7° C)*
Melting Point: −110° F(−78.8° C) *Vapor Pressure:* 6.8 mm Hg at 68° F(20° C) *Solubility in water:* Reacts
Other Information: Weighs 14.1 lbs/6.3 kg per gallon/3.8 l. Vapors and fumes may be heavier than air.

Health Hazards:

Inhalation Hazard: Severe irritation.
Ingestion Hazard: Severe damage to the mouth and stomach.
Absorption Hazard: Severe burns to the skin and eyes.
Hazards to Wildlife: n/a
Decontamination Procedures: Wash away any material with copious amounts of soap and water.
First Aid Procedures: Remove victim to fresh air, call emergency medical care. If not breathing give CPR. If breathing is difficult administer oxygen. Treat for shock.

Fire Hazards:

Flashpoint: 245° F(118.3° C) *Ignition temperature:* 453° F(233.8° C)
Flammable Explosive High Range: n/a *Low Range:* n/a
Toxic Products of Combustion: Toxic sulfur dioxide, chlorine and or hydrogen chloride gases.
Other Hazards: Will burn, but is difficult to ignite. Containers may violently rupture in a fire.
Possible extinguishing agents: Use dry chemical, dry sand, or carbon dioxide. Do not use water on the material itself.

Reactivity Hazards:

Reactive With: Water which decomposes violently to evolve heat and hydrochloric acid, sulfur, hydrogen sulfide., etc.
Other Reactions: May ignite organic material.

Corrosivity Hazards:

Corrosive With: Metals and tissue. *Neutralizing Agent:* Crushed limestone, soda ash, or lime.

Radioactivity Hazards:

Radiation Emitted: n/a *Other Hazards:* n/a

Recommended Protection for Response Personnel:

Avoid breathing vapors, keep upwind. Wear a sealed chemical suit (butyl rubber). Wash away any material which may have come into contact with the body with copious amounts of soap and water. Consider appropriate evacuation.

SULFUR DIOXIDE

DOT Number: UN 1079 *DOT Hazard Class:* Nonflammable gas *DOT Guide Number:* 16
Synonyms: none given
STCC Number: 4904290 *Reportable Qty:* n/a
Mfg Name: Airco Industrial Gases *Phone No:* 1-201-464-8100

Physical Description:

Physical Form: Gas *Color:* Colorless *Odor:* Sharp irritating
Other Information: It is used in manufacturing chemicals, paper pulping, in metal and food processing, and in many other uses.

Chemical Properties:

Specific Gravity: 1.4 *Vapor Density:* 2.3 *Boiling Point: 14° F(–10° C)*
Melting Point: –104° F(–75.5° C)*Vapor Pressure:* 2538 mm Hg at 70° F(21.1° C)
Solubility in water: Yes *Degree of Solubility:* 10%
Other Information: Liquid sinks and boils in water. Poisonous, visible vapor cloud is produced.

Health Hazards:

Inhalation Hazard: Poisonous if inhaled.
Ingestion Hazard: n/a
Absorption Hazard: Will cause frostbite.
Hazards to Wildlife: Dangerous to aquatic life.
Decontamination Procedures: Wash away any material with copious amounts of soap and water.
First Aid Procedures: Remove victim to fresh air, call emergency medical care. If not breathing give CPR. If breathing is difficult administer oxygen. Treat for shock.

Fire Hazards:

Flashpoint: n/a *Ignition temperature:* n/a
Flammable Explosive High Range: n/a *Low Range:* n/a
Toxic Products of Combustion: n/a
Other Hazards: Avoid breathing vapors!!
Possible *Possible extinguishing agents:* Apply water from as far a distance as possible. Use suitable agent for type of surrounding fire.

Reactivity Hazards:

Reactive With: Water to form corrosive acid. *Other Reactions:* n/a

Corrosivity Hazards:

Corrosive With: Aluminum. *Neutralizing Agent:* Crushed limestone, soda ash, or lime.

Radioactivity Hazards:

Radiation Emitted: n/a *Other Hazards:* n/a

Recommended Protection for Response Personnel:

Avoid breathing vapors, keep upwind. Structural protective clothing provides limited protection. Wash away any material which may have come into contact with the body with copious amounts of soap and water. Consider appropriate evacuation.

SULFUR MONOCHLORIDE

DOT Number: UN 1828 *DOT Hazard Class:* Corrosive *DOT Guide Number:* 39
Synonyms: sulfur chloride
STCC Number: n/a *Reportable Qty:* n/a
Mfg Name: A.D. MacKay Inc. *Phone No:* 1-203-655-3000

Physical Description:

Physical Form: Liquid *Color:* Yellow to red *Odor:* Sharp irritating
Other Information: n/a

Chemical Properties:

Specific Gravity: 1.68 *Vapor Density:* 4.7 *Boiling Point: 280° F(137.7° C)*
Melting Point: −112° F(−80° C) *Vapor Pressure:* 10 mm Hg at 82° F(27.7° C) *Solubility in water:* Yes
Other Information: Mixes and reacts with water. Poisonous vapor is produced.

Health Hazards:

Inhalation Hazard: Poisonous if inhaled.
Ingestion Hazard: Poisonous if swallowed.
Absorption Hazard: Will burn the skin and eyes.
Hazards to Wildlife: Dangerous to aquatic life.
Decontamination Procedures: Wash away any material with copious amounts of soap and water.
First Aid Procedures: Remove victim to fresh air, call emergency medical care. If not breathing give CPR. If breathing is difficult administer oxygen. Treat for shock.

Fire Hazards:

Flashpoint: 245° F(118.3° C) *Ignition temperature:* 453° F(233.8° C)
Flammable Explosive High Range: n/a *Low Range:* n/a
Toxic Products of Combustion: Toxic and corrosive fumes are evolved when heated.
Other Hazards: n/a
Possible extinguishing agents: Dry chemical or carbon dioxide. Cool all exposed containers with water. Water reacts violently with the compound.

Reactivity Hazards:

Reactive With: Violently with water to produce heat and hydrogen chloride fumes. The solution is strongly acidic.
Other Reactions: n/a

Corrosivity Hazards:

Corrosive With: Liquid dissolves rubber and plastic. After reaction with water, the strong acid formed attacks metals generating flammable hydrogen gas.
Neutralizing Agent: Crushed limestone, soda ash, or lime.

Radioactivity Hazards:

Radiation Emitted: n/a *Other Hazards:* n/a

Recommended Protection for Response Personnel:

Avoid breathing vapors, keep upwind. Wear a sealed chemical suit (butyl rubber, viton, PVC, chlorinated polyethylene). Wash away any material which may have come into contact with the body with copious amounts of soap and water. If the fire becomes uncontrollable, consider appropriate evacuation.

SULFURIC ACID

DOT Number: NA 1831 *DOT Hazard Class:* Corrosive *DOT Guide Number:* 39
Synonyms: battery acid, fertilizer acid
STCC Number: 4930030 *Reportable Qty:* 1000/454
Mfg Name: Allied Corp. *Phone No:* 1-201-455-2000

Physical Description:

Physical Form: Liquid *Color:* Colorless *Odor:* Odorless
Other Information: It is used in manufacturing chemicals, dyes, explosives, and in petroleum refining.

Chemical Properties:

Specific Gravity: 1.8 *Vapor Density:* 1 *Boiling Point: 815° F(435° C)*
Melting Point: 37° F(2.7° C) *Vapor Pressure:* .001 mm Hg at 68° F(20° C) *Solubility in water:* Yes
Other Information: Weighs 16.5 lbs/7.4 kg per gallon/3.8 l.

Health Hazards:

Inhalation Hazard: Coughing, difficulty in breathing, loss of consciousness.
Ingestion Hazard: Harmful if swallowed.
Absorption Hazard: Will burn the skin and the eyes.
Hazards to Wildlife: Dangerous to aquatic life.
Decontamination Procedures: Wash away any material with copious amounts of soap and water.
First Aid Procedures: Remove victim to fresh air, call emergency medical care. If not breathing give CPR. If breathing is difficult administer oxygen. Treat for shock.

Fire Hazards:

Flashpoint: 245° F(118.3° C) *Ignition temperature:* n/a
Flammable Explosive High Range: n/a *Low Range:* n/a
Toxic Products of Combustion: Poisonous gases may be produced in a fire.
Other Hazards: May cause fire upon contact with combustibles.
Possible extinguishing agents: Apply water from as far a distance as possible. Use suitable agent for type of surrounding fire.

Reactivity Hazards:

Reactive With: Water with violent reaction with evolution of heat. *Other Reactions:* n/a

Corrosivity Hazards:

Corrosive With: Metals and tissue. *Neutralizing Agent:* Crushed limestone, soda ash, or lime.

Radioactivity Hazards:

Radiation Emitted: n/a *Other Hazards:* n/a

Recommended Protection for Response Personnel:

Avoid breathing vapors, keep upwind. Wear a sealed chemical suit (chlorinated polyethylene, viton, PVC). Wash away any material which may have come into contact with the body with copious amounts of soap and water. Consider appropriate evacuation.

SULFURIC ACID, SPENT

DOT Number: UN 1832 | *DOT Hazard Class:* Corrosive | *DOT Guide Number:* 39
Synonyms: diluted sulfuric acid
STCC Number: 4930042 | *Reportable Qty:* 1000/454 | *CHEMTREC Phone No:* 1-800-424-9300

Physical Description:

Physical Form: Liquid | *Color:* Black | *Odor:* Odorless
Other Information: n/a

Chemical Properties:

Specific Gravity: 1.3 | *Vapor Density:* 1 | *Boiling Point: 212° F(100° C)*
Melting Point: n/a | *Vapor Pressure:* n/a | *Solubility in water:* Yes
Other Information: Weighs 15 lbs/6.8 kg per gallon/3.8 l.

Health Hazards:

Inhalation Hazard: Coughing, difficulty in breathing, loss of consciousness.
Ingestion Hazard: Harmful if swallowed.
Absorption Hazard: Will burn the skin and the eyes.
Hazards to Wildlife: Dangerous to aquatic life.
Decontamination Procedures: Wash away any material with copious amounts of soap and water.
First Aid Procedures: Remove victim to fresh air, call emergency medical care. If not breathing give CPR. If breathing is difficult administer oxygen. Treat for shock.

Fire Hazards:

Flashpoint: n/a | *Ignition temperature:* n/a
Flammable Explosive High Range: n/a | *Low Range:* n/a
Toxic Products of Combustion: Poisonous gases may be produced in a fire.
Other Hazards: May cause fire upon contact with combustibles.
Possible extinguishing agents: Apply water from as far a distance as possible. Use suitable agent for type of surrounding fire.

Reactivity Hazards:

Reactive With: Water in strength above 80% to 90%.
Other Reactions: Flammable gases may be produced upon contact with other metals.

Corrosivity Hazards:

Corrosive With: Metals and tissue. | *Neutralizing Agent:* Crushed limestone, soda ash, or lime.

Radioactivity Hazards:

Radiation Emitted: n/a | *Other Hazards:* n/a

Recommended Protection for Response Personnel:

Avoid breathing vapors, keep upwind. Wear a sealed chemical suit (butyl rubber, viton, PVC, chlorinated polyethylene, neoprene). Wash away any material which may have come into contact with the body with copious amounts of soap and water. Consider appropriate evacuation.

SULFURYL CHLORIDE

DOT Number: UN 1834 *DOT Hazard Class:* Corrosive *DOT Guide Number:* 39
Synonyms: none given
STCC Number: 4930260 *Reportable Qty:* n/a
Mfg Name: A.D. MacKay Inc. *Phone No:* 1-203-655-3000

Physical Description:

Physical Form: Liquid *Color:* Colorless to light yellow *Odor:* Pungent
Other Information: n/a

Chemical Properties:

Specific Gravity: 1.76 *Vapor Density:* 4.6 *Boiling Point: 156° F(68.8° C)*
Melting Point: n/a *Vapor Pressure:* 100 mm Hg at 64° F(17.7° C)
Solubility in water: Mixes and reacts violently.
Other Information: n/a

Health Hazards:

Inhalation Hazard: Coughing, difficulty in breathing, loss of consciousness.
Ingestion Hazard: Harmful if swallowed.
Absorption Hazard: Will burn the skin and the eyes.
Hazards to Wildlife: n/a
Decontamination Procedures: Wash away any material with copious amounts of soap and water.
First Aid Procedures: Remove victim to fresh air, call emergency medical care. If not breathing give CPR. If breathing is difficult administer oxygen. Treat for shock.

Fire Hazards:

Flashpoint: n/a *Ignition temperature:* n/a
Flammable Explosive High Range: n/a *Low Range:* n/a
Toxic Products of Combustion: Toxic and irritating gases are generated in a fire.
Other Hazards: Flammable gases may be produced on contact with metal.
Possible extinguishing agents: Apply water from as far a distance as possible. Use suitable agent for type of surrounding fire.

Reactivity Hazards:

Reactive With: Water to release hydrogen chloride fumes and sulfuric acid. *Other Reactions:* n/a

Corrosivity Hazards:

Corrosive With: Metals and tissue. *Neutralizing Agent:* Crushed limestone, soda ash, or lime.

Radioactivity Hazards:

Radiation Emitted: n/a *Other Hazards:* n/a

Recommended Protection for Response Personnel:

Avoid breathing vapors, keep upwind. Wear a sealed chemical suit (butyl rubber). Wash away any material which may have come into contact with the body with copious amounts of soap and water. Consider appropriate evacuation.

TANNIC ACID

DOT Number: n/a *DOT Hazard Class:* n/a *DOT Guide Number:* n/a
Synonyms: Chinese tannin, gallotannic acid, galltannin, gylcerite, tannin
STCC Number: n/a *Reportable Qty:* n/a
Mfg Name: Mallinckrodt Inc. *Phone No:* 1-314-895-2000

Physical Description:

Physical Form: Solid *Color:* Light yellow to tan *Odor:* Faint
Other Information: n/a

Chemical Properties:

Specific Gravity: 1 *Vapor Density:* 1 *Boiling Point: Decomposes*
Melting Point: n/a *Vapor Pressure:* n/a *Solubility in water:* n/a
Other Information: Sinks and mixes with water.

Health Hazards:

Inhalation Hazard: Coughing.
Ingestion Hazard: Nausea and vomiting.
Absorption Hazard: Irritating to the skin and eyes.
Hazards to Wildlife: Dangerous to aquatic life.
Decontamination Procedures: Wash away any material with copious amounts of soap and water.
First Aid Procedures: Remove victim to fresh air, call emergency medical care. If not breathing give CPR. If breathing is difficult administer oxygen. Treat for shock.

Fire Hazards:

Flashpoint: 390° F(198.8° C) *Ignition temperature:* 980° F(526.6° C)
Flammable Explosive High Range: n/a *Low Range:* n/a
Toxic Products of Combustion: Decomposes at 210° F(98.8° C)to carbon dioxide and pyrogallol which can form irritating vapors.
Other Hazards: n/a
Possible *Possible extinguishing agents:* Extinguish fire using suitable agent for the type of surrounding fire. Cool all affected containers with water. Water and foam may be ineffective on fire.

Reactivity Hazards:

Reactive With: n/a *Other Reactions:* n/a

Corrosivity Hazards:

Corrosive With: n/a *Neutralizing Agent:* n/a

Radioactivity Hazards:

Radiation Emitted: n/a *Other Hazards:* n/a

Recommended Protection for Response Personnel:

Avoid breathing vapors, keep upwind. Structural protective clothing provides limited protection. Wash away any material which may have come into contact with the body with copious amounts of soap and water. If the fire becomes uncontrollable, consider appropriate evacuation.

TETRABUTYL TITANATE

DOT Number: n/a *DOT Hazard Class:* n/a *DOT Guide Number:* n/a
Synonyms: butyl titanate, butyltitanate monomer, orthotitanic acid, tetrabutyl ester, titanium butoxide, titanium tetrabutoxide
STCC Number: n/a *Reportable Qty:* n/a
Mfg Name: Stauffer Chemical *Phone No:* 1-203-222-3000

Physical Description:

Physical Form: Liquid *Color:* Colorless to light yellow *Odor:* Weak alcohol
Other Information: n/a

Chemical Properties:

Specific Gravity: 1 *Vapor Density:* 11.7 *Boiling Point: 593° F(311.6° C)*
Melting Point: −67° F(−55° C) *Vapor Pressure:* n/a *Solubility in water:* n/a
Other Information: May float or sink in water. Reacts with water.

Health Hazards:

Inhalation Hazard: n/a
Ingestion Hazard: Nausea and vomiting.
Absorption Hazard: Irritating to the skin and eyes.
Hazards to Wildlife: n/a
Decontamination Procedures: Wash away any material with copious amounts of soap and water.
First Aid Procedures: Remove victim to fresh air, call emergency medical care. If not breathing give CPR. If breathing is difficult administer oxygen. Treat for shock.

Fire Hazards:

Flashpoint: 170° F(76.6° C) *Ignition temperature:* n/a
Flammable Explosive High Range: 12 *Low Range:* 2
Toxic Products of Combustion: n/a
Other Hazards: May give off dense white smoke. Containers may explode in fire.
Possible extinguishing agents: Use dry chemical or carbon dioxide. Do not use water. cool all exposed containers with water.

Reactivity Hazards:

Reactive With: Water to form butanol and titanium dioxide. *Other Reactions:* n/a

Corrosivity Hazards:

Corrosive With: n/a *Neutralizing Agent:* n/a

Radioactivity Hazards:

Radiation Emitted: n/a *Other Hazards:* n/a

Recommended Protection for Response Personnel:

Avoid breathing vapors, keep upwind. Structural protective clothing provides limited protection. Wash away any material which may have come into contact with the body with copious amounts of soap and water. If the fire becomes uncontrollable, consider appropriate evacuation.

TETRACHLOROETHANE

DOT Number: UN 1702 *DOT Hazard Class:* ORM-A *DOT Guide Number:* 55
Synonyms: acetylene tetrachloride, 1,1,2,2-tetrachloroethane
STCC Number: 4940354 *Reportable Qty:* 1/.454
Mfg Name: Aldrich Chemical *Phone No:* 1-414-273-3850

Physical Description:

Physical Form: Liquid *Color:* Colorless to pale yellow *Odor:* Sweet
Other Information: It is used as a solvent, insecticide, and in paint removers.

Chemical Properties:

Specific Gravity: 1.59 *Vapor Density:* 5.8 *Boiling Point: 295° F(146.1° C)*
Melting Point: −47° F(−43.8° C) *Vapor Pressure:* 8 mm Hg at 68° F(20° C) *Solubility in water:* n/a
Other Information: Sinks in water

Health Hazards:

Inhalation Hazard: Harmful if inhaled.

Ingestion Hazard: Poisonous if swallowed.

Absorption Hazard: Poisonous to the skin and eyes.

Hazards to Wildlife: n/a

Decontamination Procedures: Wash away any material with copious amounts of soap and water.

First Aid Procedures: Remove victim to fresh air, call emergency medical care. If not breathing give CPR. If breathing is difficult administer oxygen. Treat for shock.

Fire Hazards:

Flashpoint: n/a *Ignition temperature:* n/a

Flammable Explosive High Range: n/a *Low Range:* n/a

Toxic Products of Combustion: Irritating hydrogen chloride vapor may be formed in a fire.

Other Hazards: n/a

Possible extinguishing agents: Apply water from as far a distance as possible. Use foam, dry chemical, or carbon dioxide. Extinguish fire using suitable agent for the type of surrounding fire.

Reactivity Hazards:

Reactive With: Common materials *Other Reactions:* May attack some forms of plastic.

Corrosivity Hazards:

Corrosive With: n/a *Neutralizing Agent:* n/a

Radioactivity Hazards:

Radiation Emitted: n/a *Other Hazards:* n/a

Recommended Protection for Response Personnel:

Avoid breathing vapors, keep upwind. Wear a sealed chemical suit (butyl rubber). Wash away any material which may have come into contact with the body with copious amounts of soap and water. Consider appropriate evacuation.

TETRACHLOROETHYLENE

DOT Number: UN 1897 *DOT Hazard Class:* ORM-A *DOT Guide Number:* 74
Synonyms: perchloroethylene, perclene, perk, tetracap
STCC Number: 4940355 *Reportable Qty:* 1/.454
Mfg Name: Dow Chemical *Phone No:* 1-513-636-4400

Physical Description:

Physical Form: Liquid *Color:* Colorless *Odor:* Sweet
Other Information: It is used in dry cleaning solvent, vapor degreasing solvent, drying agent for metals, and for the manufacture of other chemicals.

Chemical Properties:

Specific Gravity: 1.11 *Vapor Density:* 5.8 *Boiling Point: 250° F(121.1° C)*
Melting Point: −8° F(−22.2° C) *Vapor Pressure:* 14 mm Hg at 68° F(20° C) *Solubility in water:* No
Other Information: Insoluble in water. Vapors are heavier than air. Weighs 13.5 lbs/6.1 kg per gallon/3.8 l.

Health Hazards:

Inhalation Hazard: Difficulty in breathing, loss of consciousness.
Ingestion Hazard: Harmful if swallowed.
Absorption Hazard: Irritating to the skin and eyes.
Hazards to Wildlife: n/a
Decontamination Procedures: Wash away any material with copious amounts of soap and water.
First Aid Procedures: Remove victim to fresh air, call emergency medical care. If not breathing give CPR. If breathing is difficult administer oxygen. Treat for shock.

Fire Hazards:

Flashpoint: n/a *Ignition temperature:* n/a
Flammable Explosive High Range: n/a *Low Range:* n/a
Toxic Products of Combustion: Toxic, irritating gases may be generated in a fire.
Other Hazards: n/a
Possible extinguishing agents: Extinguish fire using suitable agent for the type of surrounding fire.

Reactivity Hazards:

Reactive With: n/a *Other Reactions:* n/a

Corrosivity Hazards:

Corrosive With: n/a *Neutralizing Agent:* n/a

Radioactivity Hazards:

Radiation Emitted: n/a *Other Hazards:* n/a

Recommended Protection for Response Personnel:

Avoid breathing vapors, keep upwind. Structural protective clothing provides limited protection. Wash away any material which may have come into contact with the body with copious amounts of soap and water. Consider appropriate evacuation.

TETRADECANOL

DOT Number: n/a *DOT Hazard Class:* n/a *DOT Guide Number:* n/a
Synonyms: myristic alcohol, myristyl alcohol, 1-tetradecanol, tetradecyl alcohol, n-tetradecyl alcohol
STCC Number: n/a *Reportable Qty:* n/a
Mfg Name: Proctor and Gamble *Phone No:* 1-513-562-1100

Physical Description:

Physical Form: Liquid *Color:* Colorless *Odor:* Faint alcohol
Other Information: n/a

Chemical Properties:

Specific Gravity: .82 *Vapor Density:* 7.39 *Boiling Point: 506° F(263.3° C)*
Melting Point: 100° F(37.7° C) *Vapor Pressure:* .01 mm Hg at 68° F(20° C) *Solubility in water:* n/a
Other Information: Solidifies and floats on water.

Health Hazards:

Inhalation Hazard: n/a
Ingestion Hazard: n/a
Absorption Hazard: Irritating to the skin and eyes.
Hazards to Wildlife: n/a
Decontamination Procedures: Wash away any material with copious amounts of soap and water.
First Aid Procedures: Remove victim to fresh air, call emergency medical care. If not breathing give CPR. If breathing is difficult administer oxygen. Treat for shock.

Fire Hazards:

Flashpoint: 285° F(140.5° C) *Ignition temperature:* n/a
Flammable Explosive High Range: n/a *Low Range:* n/a
Toxic Products of Combustion: n/a
Other Hazards: n/a
Possible extinguishing agents: Foam, dry chemical or carbon dioxide. Water or foam may cause frothing. Water may be ineffective on fire. Cool all affected containers with water.

Reactivity Hazards:

Reactive With: n/a *Other Reactions:* n/a

Corrosivity Hazards:

Corrosive With: n/a *Neutralizing Agent:* n/a

Radioactivity Hazards:

Radiation Emitted: n/a *Other Hazards:* n/a

Recommended Protection for Response Personnel:

Avoid breathing vapors, keep upwind. Structural protective clothing provides limited protection. Wash away any material which may have come into contact with the body with copious amounts of soap and water. If the fire becomes uncontrollable, consider appropriate evacuation.

TETRADECENE

DOT Number: n/a *DOT Hazard Class:* n/a *DOT Guide Number:* n/a
Synonyms: dodecylethylene, tetradecylene
STCC Number: n/a *Reportable Qty:* n/a
Mfg Name: Phillips Petroleum *Phone No:* 1-918-661-6600

Physical Description:

Physical Form: Liquid *Color:* Colorless *Odor:* Mild pleasant
Other Information: n/a

Chemical Properties:

Specific Gravity: .77 *Vapor Density:* 6.8 *Boiling Point: 484° F(251.1° C)*
Melting Point: 9° F(−12.7° C) *Vapor Pressure:* n/a *Solubility in water:* No
Other Information: Floats on water.

Health Hazards:

Inhalation Hazard: n/a
Ingestion Hazard: n/a
Absorption Hazard: Irritating to the eyes.
Hazards to Wildlife: n/a
Decontamination Procedures: Wash away any material with copious amounts of soap and water.
First Aid Procedures: Remove victim to fresh air, call emergency medical care. If not breathing give CPR. If breathing is difficult administer oxygen. Treat for shock.

Fire Hazards:

Flashpoint: 230° F(110° C) *Ignition temperature:* 455° F(235° C)
Flammable Explosive High Range: n/a *Low Range:* n/a
Toxic Products of Combustion: n/a
Other Hazards: n/a
Possible extinguishing agents: Foam, dry chemical or carbon dioxide. Water or foam may cause frothing. Water may be ineffective on fire. Cool all exposed containers with water.

Reactivity Hazards:

Reactive With: n/a *Other Reactions:* n/a

Corrosivity Hazards:

Corrosive With: n/a *Neutralizing Agent:* n/a

Radioactivity Hazards:

Radiation Emitted: n/a *Other Hazards:* n/a

Recommended Protection for Response Personnel:

Avoid breathing vapors, keep upwind. Structural protective clothing provides limited protection. Wash away any material which may have come into contact with the body with copious amounts of soap and water. If the fire becomes uncontrollable, consider appropriate evacuation.

TETRAETHYL DITHIOPYROPHOSPHATE (liquid)

DOT Number: UN 1704 *DOT Hazard Class:* Poison B *DOT Guide Number:* 55
STCC Number: 4921482
Synonyms: dithiopyrophosphoric acid, TEDP
Reportable Qty: 100/45.4 *CHEMTREC Phone No:* 1-800-424-9300

Physical Description:

Physical Form: Liquid *Color:* Colorless *Odor:* n/a
Other Information: n/a

Chemical Properties:

Specific Gravity: 1.19 *Vapor Density:* 11 *Boiling Point: Decomposes*
Melting Point: n/a *Vapor Pressure:* n/a
Solubility in water: n/a *Degree of Solubility:* .0025%
Other Information: In case of damage to or leaking from containers, contact the Pesticide Safety Network Team at 1-800-424-9300.

Health Hazards:

Inhalation Hazard: Poisonous if inhaled.
Ingestion Hazard: Poisonous if swallowed.
Absorption Hazard: Poisonous if absorbed.
Hazards to Wildlife: n/a
Decontamination Procedures: Wash away any material with copious amounts of soap and water.
First Aid Procedures: Remove victim to fresh air, call emergency medical care. If not breathing give CPR. If breathing is difficult administer oxygen. Treat for shock.

Fire Hazards:

Flashpoint: n/a *Ignition temperature:* n/a
Flammable Explosive High Range: n/a *Low Range:* n/a
Toxic Products of Combustion: n/a
Other Hazards: Avoid contact with liquid. Keep people away!!
Possible extinguishing agents: Extinguish fire using suitable agent for type of surrounding fire. Use foam, dry chemical or carbon dioxide.

Reactivity Hazards:

Reactive With: Water to form nonhazardous products. *Other Reactions:* None

Corrosivity Hazards:

Corrosive With: Most metals *Neutralizing Agent:* n/a

Radioactivity Hazards:

Radiation Emitted: n/a *Other Hazards:* n/a

Recommended Protection for Response Personnel:

Avoid breathing vapors, keep upwind. Wear a sealed chemical suit (polycarbonate, butyl rubber). Wash away any material which may have come into contact with the body with copious amounts of soap and water. Consider appropriate evacuation.

TETRAETHYL DITHIOPYROPHOSPHATE (solid)

DOT Number: UN 1704 *DOT Hazard Class:* Poison B *DOT Guide Number:* 55
STCC Number: 4921481
Synonyms: dithiopyrophosphoric acid, TEDP
Reportable Qty: 100/45.4 *CHEMTREC Phone No:* 1-800-424-9300

Physical Description:

Physical Form: Solid *Color:* n/a *Odor:* n/a
Other Information: n/a

Chemical Properties:

Specific Gravity: 1.19 *Vapor Density:* n/a *Boiling Point: Decomposes*
Melting Point: n/a *Vapor Pressure:* n/a
Solubility in water: n/a *Degree of Solubility:* .0025%
Other Information: In case of damage to or leaking from containers, contact the Pesticide Safety Network Team at 1-800-424-9300.

Health Hazards:

Inhalation Hazard: Poisonous if inhaled.
Ingestion Hazard: Poisonous if swallowed.
Absorption Hazard: Poisonous if absorbed.
Hazards to Wildlife: n/a
Decontamination Procedures: Wash away any material with copious amounts of soap and water.
First Aid Procedures: Remove victim to fresh air, call emergency medical care. If not breathing give CPR. If breathing is difficult administer oxygen. Treat for shock.

Fire Hazards:

Flashpoint: n/a *Ignition temperature:* n/a
Flammable Explosive High Range: n/a *Low Range:* n/a
Toxic Products of Combustion: n/a
Other Hazards: Avoid contact with liquid. Keep people away!!
Possible extinguishing agents: Extinguish fire using suitable agent for type of surrounding fire. Use foam, dry chemical or carbon dioxide.

Reactivity Hazards:

Reactive With: Water to form nonhazardous products. *Other Reactions:* None

Corrosivity Hazards:

Corrosive With: Most metals in the presence of moisture. *Neutralizing Agent:* n/a

Radioactivity Hazards:

Radiation Emitted: n/a *Other Hazards:* n/a

Recommended Protection for Response Personnel:

Avoid breathing vapors, keep upwind. Wear appropriate chemical clothing. Wash away any material which may come into contact with the body with copious amounts of soap and water. Consider appropriate evacuation.

TETRAETHYL LEAD

DOT Number: NA 1649 *DOT Hazard Class:* Poison B *DOT Guide Number:* 56
Synonyms: lead tetrathyl, TEL
STCC Number: 4921484 *Reportable Qty:* 100/45.4
Mfg Name: E.I. Du Pont *Phone No:* 1-800-441-3637

Physical Description:

Physical Form: Liquid *Color:* Colorless but generally dyed red *Odor:* Fruity
Other Information: It is insoluble in water. Weighs 14 lbs/6.35 kg per gallon/3.8 l.

Chemical Properties:

Specific Gravity: 1.63 *Vapor Density:* 8.6 *Boiling Point: Decomposes at 212° F(100c)*
Melting Point: 216° F(102° C) *Vapor Pressure:* .2 mm Hg at 68° F(20° C) *Solubility in water:* No
Other Information: n/a

Health Hazards:

Inhalation Hazard: Poisonous if inhaled.
Ingestion Hazard: Poisonous if swallowed.
Absorption Hazard: Poisonous if absorbed.
Hazards to Wildlife: Dangerous to aquatic life.
Decontamination Procedures: Wash away any material with copious amounts of soap and water.
First Aid Procedures: Remove victim to fresh air, call emergency medical care. If not breathing give CPR, If breathing is difficult administer oxygen. Treat for shock.

Fire Hazards:

Flashpoint: 185° F(85° C) *Ignition temperature:* n/a
Flammable Explosive High Range: n/a *Low Range:* 1.8
Toxic Products of Combustion: Poisonous gases are produced in a fire.
Other Hazards: Containers may violently explode, rupture, rocket in a fire.
Possible extinguishing agents: Apply water from as far a distance as possible. Use foam, dry chemical, or carbon dioxide.

Reactivity Hazards:

Reactive With: Temperatures above 230° F(110° C). May cause the cylinders to detonate or explode.
Other Reactions: n/a

Corrosivity Hazards:

Corrosive With: n/a *Neutralizing Agent:* n/a

Radioactivity Hazards:

Radiation Emitted: n/a *Other Hazards:* n/a

Recommended Protection for Response Personnel:

Avoid breathing vapors, keep upwind. Wear appropriate chemical clothing. Wash away any material which may come into contact with the body with copious amounts of soap and water. If the fire becomes uncontrollable, consider appropriate evacuation.

TETRAETHYL PYROPHOSPHATE

DOT Number: UN 1705 *DOT Hazard Class:* Poison A *DOT Guide Number:* 15
Synonyms: Killax, Nitos, T.E.P., T.E.P.P.
STCC Number: 4902545 *Reportable Qty:* 10/4.54
Mfg Name: Chevron Chemical *Phone No:* 1-415-233-3737

Physical Description:

Physical Form: Gas/liquid *Color:* Colorless to yellow *Odor:* Faint, fruity
Other Information: Material may only be shipped in cylinders, and these must be packed in wooden or fiberboard boxes.

Chemical Properties:

Specific Gravity: 1.8 *Vapor Density:* 10 *Boiling Point: n/a*
Melting Point: n/a *Vapor Pressure:* n/a *Solubility in water:* Yes
Other Information: In case of damage to, or leaking from containers, contact the Pesticide Safety Network Team at 1-800-424-9300.

Health Hazards:

Inhalation Hazard: Poisonous if inhaled. *Ingestion Hazard:* Poisonous if swallowed.
Absorption Hazard: Poisonous if absorbed.
Hazards to Wildlife: Dangerous to both aquatic life and waterfowl.
Decontamination Procedures: Wash away any material with copious amounts of soap and water.
First Aid Procedures: Remove victim to fresh air, call emergency medical care. If not breathing give CPR. If breathing is difficult administer oxygen. Treat for shock.

Fire Hazards:

Flashpoint: n/a *Ignition temperature:* n/a
Flammable Explosive High Range: n/a *Low Range:* n/a
Toxic Products of Combustion: Highly toxic gases and vapors of unburned material and phosphoric acid are formed in fires.
Other Hazards: Avoid contact with material. Keep people away!!
Possible extinguishing agents: Apply water from as far a distance as possible. Use alcohol foam, dry chemical, or carb-on dioxide.

Reactivity Hazards:

Reactive With: Water to form phosphoric acid. *Other Reactions:* n/a

Corrosivity Hazards:

Corrosive With: Aluminum, copper, zinc, brass, and tin.
Neutralizing Agent: Crushed limestone, soda ash, or lime.

Radioactivity Hazards:

Radiation Emitted: n/a *Other Hazards:* n/a

Recommended Protection for Response Personnel:

Avoid breathing vapors, keep upwind. Wear a sealed chemical suit (polycarbonate, butyl rubber). Wash away any material which may have come into contact with the body with copious amounts of soap and water. Consider appropriate evacuation.

TETRAETHYLENE GLYCOL

DOT Number: n/a *DOT Hazard Class:* n/a *DOT Guide Number:* n/a
Synonyms: bis-(2-(2-hydroxyethoxy)ethyl ether), Hi-dry, 3,6,9-trioxaundecan-1,11-diol
STCC Number: n/a *Reportable Qty:* n/a
Mfg Name: Union Carbide *Phone No:* 1-203-974-2000

Physical Description:

Physical Form: Liquid *Color:* Colorless to straw *Odor:* Mild
Other Information: n/a

Chemical Properties:

Specific Gravity: 1.2 *Vapor Density:* 6.7 *Boiling Point: 621° F(327.2° C)*
Melting Point: 25° F(−3.8° C) *Vapor Pressure:* 1 mm Hg at 210° F(98.8° C) *Solubility in water:* Yes
Other Information: Sinks and mixes with water.

Health Hazards:

Inhalation Hazard: Not harmful.
Ingestion Hazard: Not harmful.
Absorption Hazard: Not harmful.
Hazards to Wildlife: Not harmful.
Decontamination Procedures: Wash away any material with copious amounts of soap and water.
First Aid Procedures: Remove victim to fresh air, call emergency medical care. If not breathing give CPR. If breathing is difficult administer oxygen. Treat for shock.

Fire Hazards:

Flashpoint: 360° F(182.2° C) *Ignition temperature:* n/a
Flammable Explosive High Range: n/a *Low Range:* n/a
Toxic Products of Combustion: n/a
Other Hazards: n/a
Possible extinguishing agents: Alcohol foam, dry chemical, carbon dioxide. Water or foam may cause frothing. Water may be ineffective on fire. cool all exposed containers with water.

Reactivity Hazards:

Reactive With: n/a *Other Reactions:* n/a

Corrosivity Hazards:

Corrosive With: Some forms of plastic. *Neutralizing Agent:* n/a

Radioactivity Hazards:

Radiation Emitted: n/a *Other Hazards:* n/a

Recommended Protection for Response Personnel:

Avoid breathing vapors, keep upwind. Structural protective clothing provides limited protection. Wash away any material which may have come into contact with the body with copious amounts of soap and water. If the fire becomes uncontrollable, consider appropriate evacuation.

TETRAETHYLENEPENTAMINE

DOT Number: NA 2320 *DOT Hazard Class:* Corrosive *DOT Guide Number:* 60
Synonyms: 1,11-diamino-3,6,9-triazaundecane
STCC Number: 4935216 *Reportable Qty:* n/a
Mfg Name: Dow Chemical *Phone No:* 1-513-636-4400

Physical Description:

Physical Form: Liquid *Color:* Yellow *Odor:* Ammonia
Other Information: It is used as a solvent, and in the manufacture of synthetic rubber.

Chemical Properties:

Specific Gravity: .99 *Vapor Density:* 6.5 *Boiling Point: 644° F(340° C)*
Melting Point: −22° F(−30° C) *Vapor Pressure:* 1 mm Hg at 68° F(20° C) *Solubility in water:* Yes
Other Information: May float or sink in water.

Health Hazards:

Inhalation Hazard: n/a
Ingestion Hazard: Causes nausea.
Absorption Hazard: Will burn the skin and the eyes.
Hazards to Wildlife: n/a
Decontamination Procedures: Wash away any material with copious amounts of soap and water.
First Aid Procedures: Remove victim to fresh air, call emergency medical care. If not breathing give CPR. If breathing is difficult administer oxygen. Treat for shock.

Fire Hazards:

Flashpoint: 340° F(171.1° C) *Ignition temperature:* 572° F(300° C)
Flammable Explosive High Range: 4.6 *Low Range:* .8
Toxic Products of Combustion: Ammonia and toxic oxides of nitrogen may be formed in fires.
Other Hazards: n/a
Possible extinguishing agents: Use alcohol foam, dry chemical, or carbon dioxide. Cool all affected containers with flooding quantities of water. Apply water from as far a distance as possible.

Reactivity Hazards:

Reactive With: May attack some forms of plastic. *Other Reactions:* n/a

Corrosivity Hazards:

Corrosive With: n/a *Neutralizing Agent:* n/a

Radioactivity Hazards:

Radiation Emitted: n/a *Other Hazards:* n/a

Recommended Protection for Response Personnel:

Avoid breathing vapors, keep upwind. Wear appropriate sealed chemical suit. Wash away any material which may have come into contact with the body with copious amounts of soap and water. Consider appropriate evacuation.

TETRAFLUOROETHYLENE

DOT Number: UN 1081 *DOT Hazard Class:* Flammable gas *DOT Guide Number:* 17
Synonyms: Teflon, monomer
STCC Number: 4905783 *Reportable Qty:* n/a
Mfg Name: E.I. Du Pont *Phone No:* 1-800-441-3637

Physical Description:

Physical Form: Gas *Color:* Colorless *Odor:* Odorless
Other Information: Material may only be shipped in cylinders.

Chemical Properties:

Specific Gravity: 1.5 *Vapor Density:* 3.87 *Boiling Point: – 105° F(– 76.1° C)*
Melting Point: n/a *Vapor Pressure:* n/a *Solubility in water:* No
Other Information: Vapors are heavier than air.

Health Hazards:

Inhalation Hazard: Irritating to the eyes, nose, and throat.
Ingestion Hazard: n/a
Absorption Hazard: n/a
Hazards to Wildlife: n/a
Decontamination Procedures: Wash away any material with copious amounts of soap and water.
First Aid Procedures: Remove victim to fresh air, call emergency medical care. If not breathing give CPR. If breathing is difficult administer oxygen. Treat for shock.

Fire Hazards:

Flashpoint: n/a *Ignition temperature:* 370° F(187.7° C)
Flammable Explosive High Range: 50 *Low Range:* 10
Toxic Products of Combustion: Poisonous gases are produced in a fire.
Other Hazards: If exposed to heat for long periods, materials may polymerize. if polymerization takes place, cylinders may rupture. Under fire conditions, cylinders may rupture and rocket. (BLEVE)
Possible extinguishing agents: Apply water from as far a distance as possible. Do not extinguish the fire unless the flow can be stopped.

Reactivity Hazards:

Reactive With: n/a *Other Reactions:* n/a

Corrosivity Hazards:

Corrosive With: n/a *Neutralizing Agent:* n/a

Radioactivity Hazards:

Radiation Emitted: n/a *Other Hazards:* n/a

Recommended Protection for Response Personnel:

Avoid breathing vapors, keep upwind. Structural protective clothing provides limited protection. Wash away any material which may have come into contact with the body with copious amounts of soap and water. If the fire becomes uncontrollable, consider appropriate evacuation. (BLEVE)

TETRAHYDROFURAN

DOT Number: UN 2056 *DOT Hazard Class:* Flammable liquid *DOT Guide Number:* 26
Synonyms: diethylene oxide, THF
STCC Number: 4908290 *Reportable Qty:* 1000/454
Mfg Name: E.I. Du Pont *Phone No:* 1-800-441-3637

Physical Description:

Physical Form: Liquid *Color:* Colorless *Odor:* Fruity
Other Information: n/a

Chemical Properties:

Specific Gravity: .88 *Vapor Density:* 2.5 *Boiling Point: 151° F(66.1° C)*
Melting Point: n/a *Vapor Pressure:* n/a *Solubility in water:* Yes
Other Information: Vapors are heavier than air, lighter than and soluble in water.

Health Hazards:

Inhalation Hazard: Nausea, headache, loss of consciousness.
Ingestion Hazard: Harmful if swallowed.
Absorption Hazard: Irritating to the skin and eyes.
Hazards to Wildlife: n/a
Decontamination Procedures: Wash away any material with copious amounts of soap and water.
First Aid Procedures: Remove victim to fresh air, call emergency medical care. If not breathing give CPR. If breathing is difficult administer oxygen. Treat for shock.

Fire Hazards:

Flashpoint: 6° F(−14.4° C) *Ignition temperature:* 610° F(321.1° C)
Flammable Explosive High Range: 11.8 *Low Range:* 2
Toxic Products of Combustion: n/a
Other Hazards: Flashback along vapor trail may occur. Vapors may explode if ignited in an enclosed area.
Possible extinguishing agents: Apply water from as far a distance as possible. Use alcohol foam, dry chemical, or carbon dioxide.

Reactivity Hazards:

Reactive With: n/a *Other Reactions:* n/a

Corrosivity Hazards:

Corrosive With: n/a *Neutralizing Agent:* n/a

Radioactivity Hazards:

Radiation Emitted: n/a *Other Hazards:* n/a

Recommended Protection for Response Personnel:

Avoid breathing vapors, keep upwind. Wear a sealed chemical suit (butyl rubber). Wash away any material which may have come into contact with the body with copious amounts of soap and water. If the fire becomes uncontrollable, consider appropriate evacuation.

TETRAHYDRONAPHTHALENE

DOT Number: n/a *DOT Hazard Class:* n/a *DOT Guide Number:* n/a
Synonyms: 1,2,3,4,tetrahydronaphthalene, Tetralin, Tetranap, Tetramp
STCC Number: n/a *Reportable Qty:* n/a
Mfg Name: E.I. Du Pont *Phone No:* 1-800-441-3637

Physical Description:

Physical Form: Liquid *Color:* Colorless *Odor:* Moldy turpentine
Other Information: n/a

Chemical Properties:

Specific Gravity: .97 *Vapor Density:* 4.6 *Boiling Point: 406° F(207.7° C)*
Melting Point: −23° F(−30.5° C) *Vapor Pressure:* 1 mm Hg at 100° F(37.7° C) *Solubility in water:* n/a
Other Information: Floats on water.

Health Hazards:

Inhalation Hazard: n/a
Ingestion Hazard: Harmful if swallowed.
Absorption Hazard: Irritating to the skin and eyes.
Hazards to Wildlife: n/a
Decontamination Procedures: Wash away any material with copious amounts of soap and water.
First Aid Procedures: Remove victim to fresh air, call emergency medical care. If not breathing give CPR. If breathing is difficult administer oxygen. Treat for shock.

Fire Hazards:

Flashpoint: 176° F(80° C) *Ignition temperature:* 725° F(385° C)
Flammable Explosive High Range: 5 *Low Range:* .8
Toxic Products of Combustion: n/a
Other Hazards: n/a
Possible extinguishing agents: Foam, dry chemical, carbon dioxide. Water may be ineffective on fire. Cool all exposed containers with water. Avoid the use of water!!

Reactivity Hazards:

Reactive With: n/a *Other Reactions:* n/a

Corrosivity Hazards:

Corrosive With: n/a *Neutralizing Agent:* n/a

Radioactivity Hazards:

Radiation Emitted: n/a *Other Hazards:* n/a

Recommended Protection for Response Personnel:

Avoid breathing vapors, keep upwind. Structural protective clothing provides limited protection. Wash away any material which may have come into contact with the body with copious amounts of soap and water. If the fire becomes uncontrollable, consider appropriate evacuation.

THALLIUM SULFATE

DOT Number: NA 1707 *DOT Hazard Class:* Poison B *DOT Guide Number:* 26
Synonyms: sulfuric acid, thallium salt
STCC Number: 4923297 *Reportable Qty:* 10/45.4 *CHEMTREC Phone No:* 1-800-424-9300

Physical Description:

Physical Form: Solid *Color:* Colorless to white *Odor:* Odorless
Other Information: n/a

Chemical Properties:

Specific Gravity: 6.7 *Vapor Density:* n/a *Boiling Point: n/a*
Melting Point: n/a *Vapor Pressure:* n/a *Solubility in water:* Yes
Other Information: Sinks and mixes with water.

Health Hazards:

Inhalation Hazard: n/a
Ingestion Hazard: Poisonous if swallowed.
Absorption Hazard: Poisonous if the skin is exposed.
Hazards to Wildlife: Dangerous to aquatic life.
Decontamination Procedures: Wash away any material with copious amounts of soap and water.
First Aid Procedures: Remove victim to fresh air, call emergency medical care. If not breathing give CPR. If breathing is difficult administer oxygen. Treat for shock.

Fire Hazards:

Flashpoint: n/a *Ignition temperature:* n/a
Flammable Explosive High Range: n/a *Low Range:* n/a
Toxic Products of Combustion: n/a
Other Hazards: Avoid contact with skin. Keep people away!!
Possible extinguishing agents: Use suitable agent for type of surrounding fire. Use alcohol foam, dry chemical, or carbon dioxide.

Reactivity Hazards:

Reactive With: n/a *Other Reactions:* n/a

Corrosivity Hazards:

Corrosive With: n/a *Neutralizing Agent:* n/a

Radioactivity Hazards:

Radiation Emitted: n/a *Other Hazards:* n/a

Recommended Protection for Response Personnel:

Avoid breathing vapors, keep upwind. Wear appropriate chemical clothing. Wash away any material which may have come into contact with the body with copious amounts of soap and water. Consider appropriate evacuation.

THIOPHOSGENE

DOT Number: UN 2474 *DOT Hazard Class:* Poison B *DOT Guide Number:* 55
Synonyms: thiocarbonyl chloride
STCC Number: 4923298 *Reportable Qty:* n/a
Mfg Name: Aldrich Chemical *Phone No:* 1-414-273-3850

Physical Description:

Physical Form: Liquid *Color:* Red *Odor:* Sharp choking
Other Information: n/a

Chemical Properties:

Specific Gravity: 1.5 *Vapor Density:* 4 *Boiling Point: 163° F(72.7° C)*
Melting Point: n/a *Vapor Pressure:* n/a *Solubility in water:* n/a
Other Information: Sinks and reacts slowly with water to produce a poisonous vapor.

Health Hazards:

Inhalation Hazard: Poisonous if inhaled.
Ingestion Hazard: Poisonous if swallowed.
Absorption Hazard: Irritating to the skin and eyes.
Hazards to Wildlife: n/a
Decontamination Procedures: Wash away any material with copious amounts of soap and water.
First Aid Procedures: Remove victim to fresh air, call emergency medical care. If not breathing give CPR. If breathing is difficult administer oxygen. Treat for shock.

Fire Hazards:

Flashpoint: n/a *Ignition temperature:* n/a
Flammable Explosive High Range: n/a *Low Range:* n/a
Toxic Products of Combustion: Poisonous gases are produced in a fire.
Other Hazards: Avoid contact with liquid and vapor. Keep people away!!
Possible extinguishing agents: Use suitable agent for type of surrounding fire. Use foam, dry chemical, or carbon dioxide.

Reactivity Hazards:

Reactive With: Water to form hydrogen chloride, carbon disulfide, and carbon dioxide.
Other Reactions: n/a

Corrosivity Hazards:

Corrosive With: Metals in the presence of moisture.
Neutralizing Agent: Crushed limestone, soda ash, or lime.

Radioactivity Hazards:

Radiation Emitted: n/a *Other Hazards:* n/a

Recommended Protection for Response Personnel:

Avoid breathing vapors, keep upwind. Wear a sealed chemical suit (polycarbonate, butyl rubber). Wash away any material which may have come into contact with the body with copious amounts of soap and water. Consider appropriate evacuation.

THIRAM

DOT Number: NA 2771 *DOT Hazard Class:* ORM-A *DOT Guide Number:* 55
Synonyms: methyl thiram, methyl tuads, thiuram
STCC Number: 4941187 *Reportable Qty:* 10/4.54 *CHEMTREC Phone No:* 1-800-424-9300

Physical Description:

Physical Form: Solid *Color:* White to light yellow *Odor:* n/a
Other Information: It is used in antiseptic sprays, as a catalyst, and in fungicides.

Chemical Properties:

Specific Gravity: 1.34 *Vapor Density:* 10 *Boiling Point: Decomposes*
Melting Point: n/a *Vapor Pressure:* n/a *Solubility in water:* n/a
Other Information: Sinks in water.

Health Hazards:

Inhalation Hazard: Poisonous if inhaled.
Ingestion Hazard: Poisonous if swallowed.
Absorption Hazard: Irritating to the skin and eyes.
Hazards to Wildlife: Dangerous to aquatic life.
Decontamination Procedures: Wash away any material with copious amounts of soap and water.
First Aid Procedures: Remove victim to fresh air, call emergency medical care. If not breathing give CPR. If breathing is difficult administer oxygen. Treat for shock.

Fire Hazards:

Flashpoint: n/a *Ignition temperature:* n/a
Flammable Explosive High Range: n/a *Low Range:* n/a
Toxic Products of Combustion: Toxic and irritating oxides of sulfur are formed. Carbon disulfide may be formed.
Other Hazards: n/a
Possible extinguishing agents: Extinguish fire using suitable agent for the type of surrounding fire. Use foam, dry chemical, or carbon dioxide.

Reactivity Hazards:

Reactive With: n/a *Other Reactions:* n/a

Corrosivity Hazards:

Corrosive With: n/a *Neutralizing Agent:* n/a

Radioactivity Hazards:

Radiation Emitted: n/a *Other Hazards:* n/a

Recommended Protection for Response Personnel:

Avoid breathing vapors, keep upwind. Structural protective clothing provides limited protection. Wash away any material which may have come into contact with the body with copious amounts of soap and water. Consider appropriate evacuation.

THORIUM NITRATE

DOT Number: UN 2976 *DOT Hazard Class:* Radioactive material *DOT Guide Number:* 64
Synonyms: thorium nitrate tetrahydride
STCC Number: 4906310 *Reportable Qty:* n/a
Mfg Name: A.D. Mackay Inc. *Phone No:* 1-203-655-3000

Physical Description:

Physical Form: Solid *Color:* White *Odor:* Odorless
Other Information: Radioactive material emits certain rays that are hazardous. These can only be detected with special equipment.

Chemical Properties:

Specific Gravity: Estimated at 1 *Vapor Density:* 19 *Boiling Point: Decomposes*
Melting Point: n/a *Vapor Pressure:* n/a *Solubility in water:* n/a
Other Information: Mildly chemically toxic. No special packaging in transport is required.

Health Hazards:

Inhalation Hazard: Harmful if inhaled.
Ingestion Hazard: Harmful if swallowed.
Absorption Hazard: Irritating to the skin and eyes.
Hazards to Wildlife: Dangerous to aquatic life.
Decontamination Procedures: Wash away any material with copious amounts of soap and water.
First Aid Procedures: Remove victim to fresh air, call emergency medical care. If not breathing give CPR. If breathing is difficult administer oxygen. Treat for shock.

Fire Hazards:

Flashpoint: n/a *Ignition temperature:* n/a
Flammable Explosive High Range: n/a *Low Range:* n/a
Toxic Products of Combustion: Toxic oxides of nitrogen are produced in a fire.
Other Hazards: May cause fire upon contact with combustibles.
Possible extinguishing agents: Apply water from as far a distance as possible.

Reactivity Hazards:

Reactive With: Oxidizable substances *Other Reactions:* Itself

Corrosivity Hazards:

Corrosive With: n/a *Neutralizing Agent:* n/a

Radioactivity Hazards:

Radiation Emitted: Alpha, beta, and gamma. Beta and gamma emission is small.
Other Hazards: In case of trouble with shipment of these containers, all unauthorized personnel should be kept as far away as possible until qualified personnel and equipment can be obtained.

Recommended Protection for Response Personnel:

Avoid breathing vapors, keep upwind. Structural protective clothing provides limited protection. Wash away any material which may have come into contact with the body with copious amounts of soap and water. Consider appropriate evacuation.

THORIUM ORE

DOT Number: UN 2912 *DOT Hazard Class:* Radioactive Material *DOT Guide Number:* 62
Synonyms: monazite sand
STCC Number: 4926439 *Reportable Qty:* n/a *CHEMTREC Phone No:* 1-800-424-9300

Physical Description:

Physical Form: Solid *Color:* Yellowish to reddish brown *Odor:* n/a
Other Information: It is used in the manufacture of thorium, uranium, and cerium. Radioactive material emits certain rays that are only detected by special equipment.

Chemical Properties:

Specific Gravity: 4.9 to 5.3 *Vapor Density:* n/a *Boiling Point: n/a*
Melting Point: n/a *Vapor Pressure:* n/a *Solubility in water:* Insoluble
Other Information: Weighs 306 to 331 lbs/138-150 kg per cubic foot. Note: thorium ore is only mildly radioactive, and the radiation hazard in transportation is relatively minor.

Health Hazards:

Inhalation Hazard: Radioactive poisoning.
Ingestion Hazard: Radioactive poisoning.
Absorption Hazard: Radioactive poisoning.
Hazards to Wildlife: Radioactive poisoning.
Decontamination Procedures: No procedures available!!
First Aid Procedures: Remove victim to fresh air, call emergency medical care. If not breathing give CPR. If breathing is difficult administer oxygen. Treat for shock.

Fire Hazards:

Flashpoint: n/a *Ignition temperature:* n/a
Flammable Explosive High Range: n/a *Low Range:* n/a
Toxic Products of Combustion: Radiation fallout.
Other Hazards: High emittance of radioactive energy.
Possible extinguishing agents: Apply water from as far a distance as possible. extinguish fire using suitable agent for the type of surrounding fire.

Reactivity Hazards:

Reactive With: Itself *Other Reactions:* n/a

Corrosivity Hazards:

Corrosive With: n/a *Neutralizing Agent:* n/a

Radioactivity Hazards:

Radiation Emitted: Alpha, beta, and gamma!!
Other Hazards: In case of trouble with shipment of this material, all unauthorized personnel should be kept as far away as possible until qualified personnel and equipment can be obtained.

Recommended Protection for Response Personnel:

Avoid breathing vapors or dust, keep upwind. Wash away any material which may have come into contact with the body with copious amounts of soap and water. Consider appropriate evacuation.

TITANIUM TETRACHLORIDE

DOT Number: UN 1838 *DOT Hazard Class:* Corrosive *DOT Guide Number:* 39
Synonyms: none given
STCC Number: 4932385 *Reportable Qty:* n/a
Mfg Name: Stauffer Chemical *Phone No:* 1-203-222-3000

Physical Description:

Physical Form: Liquid *Color:* Colorless to light yellow *Odor:* Pungent
Other Information: n/a

Chemical Properties:

Specific Gravity: 1.72 *Vapor Density:* n/a *Boiling Point: 277° F(136.1° C)*
Melting Point: n/a *Vapor Pressure:* 10 mm Hg at 70° F(21.1° C) *Solubility in water:* No
Other Information: Violently reacts with water and produces dense fumes in air.

Health Hazards:

Inhalation Hazard: Coughing, headache.
Ingestion Hazard: Nausea and vomiting.
Absorption Hazard: Will burn the skin and the eyes.
Hazards to Wildlife: n/a
Decontamination Procedures: Wash away any material with copious amounts of soap and water.
First Aid Procedures: Remove victim to fresh air, call emergency medical care. If not breathing give CPR. If breathing is difficult administer oxygen. Treat for shock.

Fire Hazards:

Flashpoint: n/a *Ignition temperature:* n/a
Flammable Explosive High Range: n/a *Low Range:* n/a
Toxic Products of Combustion: n/a
Other Hazards: Avoid contact with liquid and vapor.
Possible *Possible extinguishing agents:* Do not use water. Use dry chemical, dry sand, or carbon dioxide.

Reactivity Hazards:

Reactive With: Water to form a dense white fume and hydrochloric acid. *Other Reactions:* Moisture

Corrosivity Hazards:

Corrosive With: Metal and tissue. *Neutralizing Agent:* Crushed limestone, soda ash, or lime.

Radioactivity Hazards:

Radiation Emitted: n/a *Other Hazards:* n/a

Recommended Protection for Response Personnel:

Avoid breathing vapors, keep upwind. Wear a sealed chemical suit (PVC, viton). Wash away any material which may have come into contact with the body with copious amounts of soap and water. Consider appropriate evacuation.

TOLUENE

DOT Number: UN 1294 *DOT Hazard Class:* Flammable liquid *DOT Guide Number:* 27
Synonyms: methylbenzene, toluol
STCC Number: 4909305 *Reportable Qty:* 1000/454
Mfg Name: Shell Chemical *Phone No:* 1-713-241-6161

Physical Description:

Physical Form: Liquid *Color:* Colorless *Odor:* Pleasant
Other Information: It is used in aviation and automotive fuels, in solvents, and to make other chemicals.

Chemical Properties:

Specific Gravity: .9 *Vapor Density:* 3.1 *Boiling Point: 231° F(110.5° C)*
Melting Point: −139° F(−95° C) *Vapor Pressure:* 22 mm Hg at 68° F(20° C)
Solubility in water: No *Degree of Solubility: .05%*
Other Information: Lighter than and insoluble in water. Vapors are heavier than air. weighs 7.2 lbs/3.2 kg per gallon/3.8 l.

Health Hazards:

Inhalation Hazard: Nausea, vomiting, difficulty in breathing, loss consciousness.
Ingestion Hazard: Nausea and vomiting.
Absorption Hazard: Irritating to the skin and eyes.
Hazards to Wildlife: Dangerous to aquatic life.
Decontamination Procedures: Wash away any material with copious amounts of soap and water.
First Aid Procedures: Remove victim to fresh air, call emergency medical care. If not breathing give CPR. If breathing is difficult administer oxygen. Treat for shock.

Fire Hazards:

Flashpoint: 40° F(4.4° C) *Ignition temperature:* 896° F(480° C)
Flammable Explosive High Range: 7.1 *Low Range:* 1.2
Toxic Products of Combustion: n/a
Other Hazards: Flashback along vapor trail may occur. Vapors may explode if ignited in an enclosed area.
Possible extinguishing agents: Apply water from as far a distance as possible. Use foam, dry chemical, or carbon dioxide.

Reactivity Hazards:

Reactive With: n/a *Other Reactions:* n/a

Corrosivity Hazards:

Corrosive With: n/a *Neutralizing Agent:* n/a

Radioactivity Hazards:

Radiation Emitted: n/a *Other Hazards:* n/a

Recommended Protection for Response Personnel:

Avoid breathing vapors, keep upwind. Structural protective clothing provides limited protection. Wash away any material which may have come into contact with the body with copious amounts of soap and water. Consider appropriate evacuation.

TOLUENE-2,4-DIISOCYANATE

DOT Number: UN 2078 *DOT Hazard Class:* Poison B *DOT Guide Number:* 57
Synonyms: Hylene T, Mondur TDS, TDI
STCC Number: 4921575 *Reportable Qty:* 100/45.4
Mfg Name: Eastman Organic *Phone No:* 1-800-455-6325

Physical Description:

Physical Form: Liquid *Color:* Colorless to light yellow *Odor:* Sharp sweet fruity
Other Information: It is used to make polyurethane foam and paints.

Chemical Properties:

Specific Gravity: 1.2 *Vapor Density:* 6 *Boiling Point: 484° F(251.1° C)*
Melting Point: 71° F(21.6° C) *Vapor Pressure:* .04 mm Hg at 68° F(20° C) *Solubility in water:* No
Other Information: Reacts with water. Vapors are heavier than air.

Health Hazards:

Inhalation Hazard: n/a
Ingestion Hazard: Poisonous if swallowed.
Absorption Hazard: Will burn the skin and eyes.
Hazards to Wildlife: n/a
Decontamination Procedures: Wash away any material with copious amounts of soap and water.
First Aid Procedures: Remove victim to fresh air, call emergency medical care. If not breathing give CPR. If breathing is difficult administer oxygen. Treat for shock.

Fire Hazards:

Flashpoint: 260° F(126.6° C) *Ignition temperature:* 300° F(148.8° C)
Flammable Explosive High Range: 9.5 *Low Range:* .9
Toxic Products of Combustion: Toxic oxides of nitrogen are produced in a fire.
Other Hazards: Avoid contact with liquid and vapor.
Possible extinguishing agents: Apply water from as far a distance as possible. Use alcohol foam, dry chemical, or carbon dioxide.

Reactivity Hazards:

Reactive With: Water to form carbon dioxide gas. *Other Reactions:* n/a

Corrosivity Hazards:

Corrosive With: n/a *Neutralizing Agent:* n/a

Radioactivity Hazards:

Radiation Emitted: n/a *Other Hazards:* n/a

Recommended Protection for Response Personnel:

Avoid breathing vapors, keep upwind. Wear a sealed chemical suit (polycarbonate, butyl rubber, viton). Wash away any material which may have come into contact with the body with copious amounts of soap and water. Consider appropriate evacuation.

TOLUENE SULFONIC ACID

DOT Number: UN 2583 *DOT Hazard Class:* Corrosive *DOT Guide Number:* 60
Synonyms: methylbenzenesulfonic acid, p-toluene sulfonate, tosic acid, p-TSA
STCC Number: n/a *Reportable Qty:* n/a
Mfg Name: Witco Chemical Corp. *Phone No:* 1-212-605-3800

Physical Description:

Physical Form: Solid *Color:* Colorless to black *Odor:* Odorless or slight
Other Information: n/a

Chemical Properties:

Specific Gravity: 1.45 *Vapor Density:* 9.5 *Boiling Point: Decomposes*
Melting Point: 219-221° F(103-105° C) *Vapor Pressure:* n/a
Solubility in water: Yes
Other Information: Mixes with water.

Health Hazards:

Inhalation Hazard: n/a
Ingestion Hazard: Harmful if swallowed.
Absorption Hazard: Irritating to the skin and eyes.
Hazards to Wildlife: n/a
Decontamination Procedures: Wash away any material with copious amounts of soap and water.
First Aid Procedures: Remove victim to fresh air, call emergency medical care. If not breathing give CPR. If breathing is difficult administer oxygen. Treat for shock.

Fire Hazards:

Flashpoint: Solid with low flammability *Ignition temperature:* n/a
Flammable Explosive High Range: n/a *Low Range:* n/a
Toxic Products of Combustion: Irritating oxides of sulfur may be formed.
Other Hazards: n/a
Possible *Possible extinguishing agents:* Flood with water.

Reactivity Hazards:

Reactive With: n/a *Other Reactions:* n/a

Corrosivity Hazards:

Corrosive With: Metals. *Neutralizing Agent:* Crushed limestone, soda ash, or lime.

Radioactivity Hazards:

Radiation Emitted: n/a *Other Hazards:* n/a

Recommended Protection for Response Personnel:

Avoid breathing vapors, keep upwind. Wear appropriate sealed chemical clothing. Wash away any material which may have come into contact with the body with copious amounts of soap and water. If the fire becomes uncontrollable, consider appropriate evacuation.

TOLUIDINE-m

DOT Number: UN 1708 *DOT Hazard Class:* n/a *DOT Guide Number:* 55
Synonyms: none given
STCC Number: n/a *Reportable Qty:* n/a
Mfg Name: E.I. Du Pont *Phone No:* 1-800-441-3637

Physical Description:

Physical Form: Liquid *Color:* Colorless to yellow brown *Odor:* Chemical
Other Information: n/a

Chemical Properties:

Specific Gravity: 1 *Vapor Density:* 3.7 *Boiling Point: 392° F(200° C)*
Melting Point: −3° F(−19.4° C) *Vapor Pressure:* 1 mm Hg at 68° F(20° C)
Solubility in water: No *Degree of Solubility:* 1.5%
Other Information: May float or sink in water.

Health Hazards:

Inhalation Hazard: n/a
Ingestion Hazard: Nausea, vomiting, loss of consciousness.
Absorption Hazard: Irritating to the skin and eyes.
Hazards to Wildlife: Dangerous to aquatic life.
Decontamination Procedures: Wash away any material with copious amounts of soap and water.
First Aid Procedures: Remove victim to fresh air, call emergency medical care. If not breathing give CPR. If breathing is difficult administer oxygen. Treat for shock.

Fire Hazards:

Flashpoint: 185° F(85° C) *Ignition temperature:* 900° F(482.2° C)
Flammable Explosive High Range: n/a *Low Range:* 1.5
Toxic Products of Combustion: Toxic oxides of nitrogen are produced in a fire.
Other Hazards: Avoid contact with liquid and vapor.
Possible extinguishing agents: Apply water from as far a distance as possible. Use foam, dry chemical, or carbon dioxide. Note: Water may be ineffective on fire.

Reactivity Hazards:

Reactive With: n/a *Other Reactions:* n/a

Corrosivity Hazards:

Corrosive With: n/a *Neutralizing Agent:* n/a

Radioactivity Hazards:

Radiation Emitted: n/a *Other Hazards:* n/a

Recommended Protection for Response Personnel:

Avoid breathing vapors, keep upwind. Wear appropriate chemical clothing. Wash away any material which may have come into contact with the body with copious amounts of soap and water. Consider appropriate evacuation.

TOLUIDINE-o

DOT Number: UN 1708 | *DOT Hazard Class:* n/a | *DOT Guide Number:* 55
Synonyms: none given
STCC Number: n/a | *Reportable Qty:* n/a
Mfg Name: E.I. Du Pont | *Phone No:* 1-800-441-3637

Physical Description:

Physical Form: Liquid | *Color:* Colorless to yellow brown | *Odor:* Chemical
Other Information: n/a

Chemical Properties:

Specific Gravity: 1 | *Vapor Density:* 3.7 | *Boiling Point: 392° F(200° C)*
Melting Point: –3° F(–19.4° C) | *Vapor Pressure:* 1 mm Hg at 68° F(20° C)
Solubility in water: No | *Degree of Solubility:* 1.5%
Other Information: May float or sink in water.

Health Hazards:

Inhalation Hazard: n/a
Ingestion Hazard: Nausea, vomiting, loss of consciousness.
Absorption Hazard: Irritating to the skin and eyes.
Hazards to Wildlife: Dangerous to aquatic life.
Decontamination Procedures: Wash away any material with copious amounts of soap and water.
First Aid Procedures: Remove victim to fresh air, call emergency medical care. If not breathing give CPR. If breathing is difficult administer oxygen. Treat for shock.

Fire Hazards:

Flashpoint: 185° F(85° C) | *Ignition temperature:* 900° F(482.2° C)
Flammable Explosive High Range: n/a | *Low Range:* 1.5
Toxic Products of Combustion: Toxic oxides of nitrogen are produced in a fire.
Other Hazards: Avoid contact with liquid and vapor.
Possible extinguishing agents: Apply water from as far a distance as possible. Use foam, dry chemical, or carbon dioxide. Note: Water may be ineffective on fire.

Reactivity Hazards:

Reactive With: n/a | *Other Reactions:* n/a

Corrosivity Hazards:

Corrosive With: n/a | *Neutralizing Agent:* n/a

Radioactivity Hazards:

Radiation Emitted: n/a | *Other Hazards:* n/a

Recommended Protection for Response Personnel:

Avoid breathing vapors, keep upwind. Wear appropriate chemical clothing. Wash away any material which may come into contact with the body with copious amounts of soap and water. Consider appropriate evacuation.

TOLUIDINE-p

DOT Number: UN 1708 *DOT Hazard Class:* n/a *DOT Guide Number:* 55
Synonyms: none given
STCC Number: n/a *Reportable Qty:* n/a
Mfg Name: E.I. Du Pont *Phone No:* 1-800-441-3637

Physical Description:

Physical Form: Liquid *Color:* Colorless to yellow brown *Odor:* Chemical
Other Information: n/a

Chemical Properties:

Specific Gravity: 1 *Vapor Density:* 3.7 *Boiling Point: 392° F(200° C)*
Melting Point: −3° F(−19.4° C) *Vapor Pressure:* 1 mm Hg at 68° F(20° C)
Solubility in water: No *Degree of Solubility:* 1.5%
Other Information: May float or sink in water.

Health Hazards:

Inhalation Hazard: n/a
Ingestion Hazard: Nausea, vomiting, loss of consciousness.
Absorption Hazard: Irritating to the skin and eyes.
Hazards to Wildlife: Dangerous to aquatic life.
Decontamination Procedures: Wash away any material with copious amounts of soap and water.
First Aid Procedures: Remove victim to fresh air, call emergency medical care. If not breathing give CPR. If breathing is difficult administer oxygen. Treat for shock.

Fire Hazards:

Flashpoint: 185° F(85° C) *Ignition temperature:* 900° F(482.2° C)
Flammable Explosive High Range: n/a *Low Range:* 1.5
Toxic Products of Combustion: Toxic oxides of nitrogen are produced in a fire.
Other Hazards: Avoid contact with liquid and vapor.
Possible extinguishing agents: Apply water from as far a distance as possible. Use foam, dry chemical, or carbon dioxide. Note: Water may be ineffective on fire.

Reactivity Hazards:

Reactive With: n/a *Other Reactions:* n/a

Corrosivity Hazards:

Corrosive With: n/a *Neutralizing Agent:* n/a

Radioactivity Hazards:

Radiation Emitted: n/a *Other Hazards:* n/a

Recommended Protection for Response Personnel:

Avoid breathing vapors, keep upwind. Wear appropriate chemical clothing. Wash away any material which may come into contact with the body with copious amounts of soap and water. Consider appropriate evacuation.

TOXAPHENE (liquid)

DOT Number: NA 2761 *DOT Hazard Class:* ORM-A *DOT Guide Number:* 55
Synonyms: octachlorocamphene
STCC Number: 4941188 *Reportable Qty:* 1/.454
Mfg Name: Hercules Inc. *Phone No:* 1-302-594-5000

Physical Description:

Physical Form: Liquid *Color:* Amber *Odor:* n/a
Other Information: In case of damage to, or leaking from these containers, contact the Pesticide Safety Team Network at 1-800-424-9300

Chemical Properties:

Specific Gravity: 1.6 *Vapor Density:* n/a *Boiling Point: Decomposes*
Melting Point: n/a *Vapor Pressure:* n/a *Solubility in water:* No
Other Information: Floats on water.

Health Hazards:

Inhalation Hazard: Poisonous if inhaled.
Ingestion Hazard: Poisonous if swallowed.
Absorption Hazard: Irritating to the skin and eyes.
Hazards to Wildlife: Dangerous to both aquatic life and waterfowl.
Decontamination Procedures: Wash away any material with copious amounts of soap and water.
First Aid Procedures: Remove victim to fresh air, call emergency medical care. If not breathing give CPR. If breathing is difficult administer oxygen. Treat for shock.

Fire Hazards:

Flashpoint: 84° F(28.8° C) *Ignition temperature:* 986° F(530° C)
Flammable Explosive High Range: 6.4 *Low Range:* 1.1
Toxic Products of Combustion: Toxic vapors are generated when heated.
Other Hazards: n/a
Possible extinguishing agents: Extinguish fire using suitable agent for the type of surrounding fire.

Reactivity Hazards:

Reactive With: n/a *Other Reactions:* n/a

Corrosivity Hazards:

Corrosive With: n/a *Neutralizing Agent:* n/a

Radioactivity Hazards:

Radiation Emitted: n/a *Other Hazards:* n/a

Recommended Protection for Response Personnel:

Avoid breathing vapors, keep upwind. Wear a sealed chemical suit (polycarbonate, butyl rubber). Wash away any material which may have come into contact with the body with copious amounts of soap and water. Consider appropriate evacuation.

TOXAPHENE (solid)

DOT Number: NA 2761 *DOT Hazard Class:* ORM-A *DOT Guide Number:* 55
Synonyms: octachlorocamphene
STCC Number: 4941189 *Reportable Qty:* 1/.454
Mfg Name: Hercules Inc. *Phone No:* 1-302-594-5000

Physical Description:

Physical Form: Solid *Color:* Amber *Odor:* n/a
Other Information: In case of damage to, or leaking from these containers, contact the Pesticide Safety Team Network at 1-800-424-9300.

Chemical Properties:

Specific Gravity: 1.6 *Vapor Density:* 14 *Boiling Point: Decomposes*
Melting Point: n/a *Vapor Pressure:* n/a *Solubility in water:* n/a
Other Information: Sinks in water.

Health Hazards:

Inhalation Hazard: n/a
Ingestion Hazard: Poisonous if swallowed.
Absorption Hazard: Irritating to the skin and eyes.
Hazards to Wildlife: Dangerous to both aquatic life and waterfowl.
Decontamination Procedures: Wash away any material with copious amounts of soap and water.
First Aid Procedures: Remove victim to fresh air, call emergency medical care. If not breathing give CPR. If breathing is difficult administer oxygen. Treat for shock.

Fire Hazards:

Flashpoint: n/a *Ignition temperature:* n/a
Flammable Explosive High Range: n/a *Low Range:* n/a
Toxic Products of Combustion: Toxic vapors are generated when heated.
Other Hazards: n/a
Possible extinguishing agents: Extinguish fire using suitable agent for the type of surrounding fire.

Reactivity Hazards:

Reactive With: n/a *Other Reactions:* n/a

Corrosivity Hazards:

Corrosive With: n/a *Neutralizing Agent:* n/a

Radioactivity Hazards:

Radiation Emitted: n/a *Other Hazards:* n/a

Recommended Protection for Response Personnel:

Avoid breathing vapors, keep upwind. Wear a sealed chemical suit (polycarbonate, butyl rubber). Wash away any material which may have come into contact with the body with copious amounts of soap and water. Consider appropriate evacuation.

TOXAPHENE (liquid)

DOT Number: NA 2761 *DOT Hazard Class:* ORM-A *DOT Guide Number:* 55
Synonyms: octachlorocamphene
STCC Number: 4941188 *Reportable Qty:* 1/.454
Mfg Name: Hercules Inc. *Phone No:* 1-302-594-5000

Physical Description:

Physical Form: Liquid *Color:* Amber *Odor:* n/a
Other Information: In case of damage to, or leaking from these containers, contact the Pesticide Safety Team Network at 1-800-424-9300

Chemical Properties:

Specific Gravity: 1.6 *Vapor Density:* n/a *Boiling Point: Decomposes*
Melting Point: n/a *Vapor Pressure:* n/a *Solubility in water:* No
Other Information: Floats on water.

Health Hazards:

Inhalation Hazard: Poisonous if inhaled.
Ingestion Hazard: Poisonous if swallowed.
Absorption Hazard: Irritating to the skin and eyes.
Hazards to Wildlife: Dangerous to both aquatic life and waterfowl.
Decontamination Procedures: Wash away any material with copious amounts of soap and water.
First Aid Procedures: Remove victim to fresh air, call emergency medical care. If not breathing give CPR. If breathing is difficult administer oxygen. Treat for shock.

Fire Hazards:

Flashpoint: 84° F(28.8° C) *Ignition temperature:* 986° F(530° C)
Flammable Explosive High Range: 6.4 *Low Range:* 1.1
Toxic Products of Combustion: Toxic vapors are generated when heated.
Other Hazards: n/a
Possible extinguishing agents: Extinguish fire using suitable agent for the type of surrounding fire.

Reactivity Hazards:

Reactive With: n/a *Other Reactions:* n/a

Corrosivity Hazards:

Corrosive With: n/a *Neutralizing Agent:* n/a

Radioactivity Hazards:

Radiation Emitted: n/a *Other Hazards:* n/a

Recommended Protection for Response Personnel:

Avoid breathing vapors, keep upwind. Wear a sealed chemical suit (polycarbonate, butyl rubber). Wash away any material which may have come into contact with the body with copious amounts of soap and water. Consider appropriate evacuation.

TOXAPHENE (solid)

DOT Number: NA 2761 *DOT Hazard Class:* ORM-A *DOT Guide Number:* 55
Synonyms: octachlorocamphene
STCC Number: 4941189 *Reportable Qty:* 1/.454
Mfg Name: Hercules Inc. *Phone No:* 1-302-594-5000

Physical Description:

Physical Form: Solid *Color:* Amber *Odor:* n/a
Other Information: In case of damage to, or leaking from these containers, contact the Pesticide Safety Team Network at 1-800-424-9300.

Chemical Properties:

Specific Gravity: 1.6 *Vapor Density:* 14 *Boiling Point: Decomposes*
Melting Point: n/a *Vapor Pressure:* n/a *Solubility in water:* n/a
Other Information: Sinks in water.

Health Hazards:

Inhalation Hazard: n/a
Ingestion Hazard: Poisonous if swallowed.
Absorption Hazard: Irritating to the skin and eyes.
Hazards to Wildlife: Dangerous to both aquatic life and waterfowl.
Decontamination Procedures: Wash away any material with copious amounts of soap and water.
First Aid Procedures: Remove victim to fresh air, call emergency medical care. If not breathing give CPR. If breathing is difficult administer oxygen. Treat for shock.

Fire Hazards:

Flashpoint: n/a *Ignition temperature:* n/a
Flammable Explosive High Range: n/a *Low Range:* n/a
Toxic Products of Combustion: Toxic vapors are generated when heated.
Other Hazards: n/a
Possible extinguishing agents: Extinguish fire using suitable agent for the type of surrounding fire.

Reactivity Hazards:

Reactive With: n/a *Other Reactions:* n/a

Corrosivity Hazards:

Corrosive With: n/a *Neutralizing Agent:* n/a

Radioactivity Hazards:

Radiation Emitted: n/a *Other Hazards:* n/a

Recommended Protection for Response Personnel:

Avoid breathing vapors, keep upwind. Wear a sealed chemical suit (polycarbonate, butyl rubber). Wash away any material which may have come into contact with the body with copious amounts of soap and water. Consider appropriate evacuation.

TRICHLORFON

DOT Number: NA 2783 *DOT Hazard Class:* ORM-A *DOT Guide Number:* 55
Synonyms: Bayer 13/59: O,O-dimethyl-chlorophos, Dipterex, Dylox,
STCC Number: 4940376 *Reportable Qty:* 100/45.4 *CHEMTREC Phone No:* 1-800-424-9300

Physical Description:

Physical Form: Solid *Color:* White *Odor:* Ethyl ether
Other Information: In case of damage to, or leaking from these containers, contact the Pesticide Safety Team Network at 1-800-424-9300.

Chemical Properties:

Specific Gravity: 1.73 *Vapor Density:* 8.9 *Boiling Point: 212° F(100° C)*
Melting Point: n/a *Vapor Pressure:* n/a *Solubility in water:* Yes
Other Information: Sinks and mixes with water.

Health Hazards:

Inhalation Hazard: Poisonous if inhaled.
Ingestion Hazard: Poisonous if swallowed.
Absorption Hazard: Poisonous if the skin is exposed.
Hazards to Wildlife: Dangerous to both aquatic life and waterfowl.
Decontamination Procedures: Wash away any material with copious amounts of soap and water.
First Aid Procedures: Remove victim to fresh air, call emergency medical care. If not breathing give CPR. If breathing is difficult administer oxygen. Treat for shock.

Fire Hazards:

Flashpoint: n/a *Ignition temperature:* n/a
Flammable Explosive High Range: n/a *Low Range:* n/a
Toxic Products of Combustion: n/a
Other Hazards: n/a
Possible extinguishing agents: Extinguish fire using suitable agent for the type of surrounding fire.

Reactivity Hazards:

Reactive With: n/a *Other Reactions:* n/a

Corrosivity Hazards:

Corrosive With: n/a *Neutralizing Agent:* n/a

Radioactivity Hazards:

Radiation Emitted: n/a *Other Hazards:* n/a

Recommended Protection for Response Personnel:

Avoid breathing vapors, keep upwind. Wear appropriate chemical clothing. Wash away any material which may come into contact with the body with copious amounts of soap and water. Consider appropriate evacuation.

TRICHLORO-s-TRIAZINETRIONE

DOT Number: UN 2468 *DOT Hazard Class:* Oxidizer *DOT Guide Number:* 42
Synonyms: trichloro-s-triazine-2,4,6-(1H,3H,5H)-trione, trichlorotriazinetrione, trichloroisocyanuric acid
STCC Number: n/a *Reportable Qty:* n/a
Mfg Name: Monsanto Chemical *Phone No:* 1-314-694-1000

Physical Description:

Physical Form: Solid *Color:* White *Odor:* Bleachlike
Other Information: n/a

Chemical Properties:

Specific Gravity: 1 *Vapor Density:* 8 *Boiling Point: Decomposes*
Melting Point: n/a *Vapor Pressure:* 2 mm Hg at 158° F(70° C)
Solubility in water: Yes
Other Information: Sinks and mixes slowly with water.

Health Hazards:

Inhalation Hazard: Coughing or difficulty in breathing.
Ingestion Hazard: Harmful if swallowed.
Absorption Hazard: Irritating to the skin and eyes.
Hazards to Wildlife: n/a
Decontamination Procedures: Wash away any material with copious amounts of soap and water.
First Aid Procedures: Remove victim to fresh air, call emergency medical care. If not breathing give CPR. If breathing is difficult administer oxygen. Treat for shock.

Fire Hazards:

Flashpoint: n/a *Ignition temperature:* n/a
Flammable Explosive High Range: n/a *Low Range:* n/a
Toxic Products of Combustion: Toxic chlorine or nitrogen trichloride may be formed in fires.
Other Hazards: Containers may explode in fire. May cause fire upon contact with combustibles.
Possible extinguishing agents: Flood with water.

Reactivity Hazards:

Reactive With: Water to form a bleach solution.
Other Reactions: Foreign material, organic matter or easily chlorinated or oxidized materials may result in fire.

Corrosivity Hazards:

Corrosive With: n/a *Neutralizing Agent:* n/a

Radioactivity Hazards:

Radiation Emitted: n/a *Other Hazards:* n/a

Recommended Protection for Response Personnel:

Avoid breathing vapors, keep upwind. Structural protective clothing provides limited protection. Wash away any material which may have come into contact with the body with copious amounts of soap and water. Consider appropriate evacuation.

TRICHLOROETHYLENE

DOT Number: UN 1710 *DOT Hazard Class:* ORM-A *DOT Guide Number:* 74
Synonyms: gemalgene, chlorylen, trethylene, trichloroethylene
STCC Number: 4941171 *Reportable Qty:* 1000/454
Mfg Name: Dow Chemical *Phone No:* 1-517-636-4400

Physical Description:

Physical Form: Liquid *Color:* Colorless *Odor:* Sweet
Other Information: It is used as a solvent, fumigant, in the manufacture of other chemicals, and for many other uses.

Chemical Properties:

Specific Gravity: 1.5 *Vapor Density:* 4.5 *Boiling Point: 188° F(86.6° C)*
Melting Point: −123° F(−86.1° C) Vapor Pressure: 58 mm Hg at 68° F(20° C)
Solubility in water: Yes *Degree of Solubility:* .1%
Other Information: Sinks in water. Irritating vapor is produced.

Health Hazards:

Inhalation Hazard: Nausea, vomiting, difficulty in breathing, loss of consciousness.
Ingestion Hazard: Nausea, vomiting, difficulty in breathing, loss of consciousness.
Absorption Hazard: Irritating to the skin and eyes.
Hazards to Wildlife: Dangerous to aquatic life.
Decontamination Procedures: Wash away any material with copious amounts of soap and water.
First Aid Procedures: Remove victim to fresh air, call emergency medical care. If not breathing give CPR. If breathing is difficult administer oxygen. Treat for shock.

Fire Hazards:

Flashpoint: 90° F(32.2° C) *Ignition temperature:* 788° F(420° C)
Flammable Explosive High Range: 10.5 *Low Range:* 8
Toxic Products of Combustion: Toxic and irritating gases are produced in a fire.
Other Hazards: n/a
Possible extinguishing agents: Extinguish fire using suitable agent for the type of surrounding fire.

Reactivity Hazards:

Reactive With: n/a *Other Reactions:* n/a

Corrosivity Hazards:

Corrosive With: n/a *Neutralizing Agent:* n/a

Radioactivity Hazards:

Radiation Emitted: n/a *Other Hazards:* n/a

Recommended Protection for Response Personnel:

Avoid breathing vapors, keep upwind. Wear a sealed chemical suit (viton). Wash away any material which may come into contact with the body with copious amounts of soap and water. Consider appropriate evacuation.

TRICHLOROFLUOROMETHANE

DOT Number: n/a *DOT Hazard Class:* n/a *DOT Guide Number:* n/a
Synonyms: Arcton 9, Eskimon 11, F-11, Freon 11, Frigen 11, Genetron 11, Isotron 11, Isreon 11, Ucon 11
STCC Number: n/a *Reportable Qty:* n/a
Mfg Name: Allied Corp. *Phone No:* 1-201-455-2000

Physical Description:

Physical Form: Liquid *Color:* Colorless *Odor:* Odorless
Other Information: n/a

Chemical Properties:

Specific Gravity: n/a *Vapor Density:* 4.7 *Boiling Point: 75° F(23.8° C)*
Melting Point: n/a *Vapor Pressure:* n/a *Solubility in water:* n/a
Other Information: Sinks in water. Harmful vapor is produced.

Health Hazards:

Inhalation Hazard: Dizziness or difficulty in breathing.
Ingestion Hazard: n/a
Absorption Hazard: n/a
Hazards to Wildlife: Not harmful.
Decontamination Procedures: Wash away any material with copious amounts of soap and water.
First Aid Procedures: Remove victim to fresh air, call emergency medical care. If not breathing give CPR. If breathing is difficult administer oxygen. Treat for shock.

Fire Hazards:

Flashpoint: n/a *Ignition temperature:* n/a
Flammable Explosive High Range: n/a *Low Range:* n/a
Toxic Products of Combustion: Produces irritating and toxic products when heated to decomposition temperatures.
Other Hazards: n/a
Possible extinguishing agents: Water, alcohol foam, dry chemical and carbon dioxide.

Reactivity Hazards:

Reactive With: n/a *Other Reactions:* n/a

Corrosivity Hazards:

Corrosive With: n/a *Neutralizing Agent:* n/a

Radioactivity Hazards:

Radiation Emitted: n/a *Other Hazards:* n/a

Recommended Protection for Response Personnel:

Avoid breathing vapors, keep upwind. Structural protective clothing provides limited protection. Wash away any material which may have come into contact with the body with copious amounts of soap and water. Consider appropriate evacuation.

TRICHLOROPHENOL

DOT Number: NA 2020 *DOT Hazard Class:* ORM-A *DOT Guide Number:* 53
Synonyms: Dowicide 2, omalphenachlor, 2,4,5-trichlorophenol
STCC Number: 4940325 *Reportable Qty:* 10/4.54
Mfg Name: Dow Chemical *Phone No:* 1-517-636-4400

Physical Description:

Physical Form: Solid *Color:* Yellow *Odor:* Strong disinfectant
Other Information: It is used as a fungicide and a herbicide.

Chemical Properties:

Specific Gravity: 1.7 *Vapor Density:* 6.8 *Boiling Point: 485° F(251.6° C)*
Melting Point: n/a *Vapor Pressure:* 1 mm Hg at 161° F(71.6° C) *Solubility in water:* Yes
Other Information: Sinks and mixes with water.

Health Hazards:

Inhalation Hazard: Irritating to the nose and throat.
Ingestion Hazard: n/a
Absorption Hazard: Irritating to the skin and eyes.
Hazards to Wildlife: Dangerous to aquatic life.
Decontamination Procedures: Wash away any material with copious amounts of soap and water.
First Aid Procedures: Remove victim to fresh air, call emergency medical care. If not breathing give CPR. If breathing is difficult administer oxygen. Treat for shock.

Fire Hazards:

Flashpoint: n/a *Ignition temperature:* n/a
Flammable Explosive High Range: n/a *Low Range:* n/a
Toxic Products of Combustion: n/a
Other Hazards: n/a
Possible extinguishing agents: Extinguish fire using suitable agent for the type of surrounding fire.

Reactivity Hazards:

Reactive With: n/a *Other Reactions:* n/a

Corrosivity Hazards:

Corrosive With: n/a *Neutralizing Agent:* n/a

Radioactivity Hazards:

Radiation Emitted: n/a *Other Hazards:* n/a

Recommended Protection for Response Personnel:

Avoid breathing vapors, keep upwind. Wear a sealed chemical suit (polycarbonate, butyl rubber). Wash away any material which may have come into contact with the body with copious amounts of soap and water. Consider appropriate evacuation.

TRICHLOROPHENOXYACETIC ACID AMINE

DOT Number: NA 2765 *DOT Hazard Class:* ORM-A *DOT Guide Number:* 55
Synonyms: 2,4,5-T sodium salt
STCC Number: 4941185 *Reportable Qty:* 1000/454
Mfg Name: Dow Chemical *Phone No:* 1-517-636-4400

Physical Description:

Physical Form: Solid *Color:* Light tan *Odor:* Odorless
Other Information: It is used as a herbicide, defoliant, and a plant growth regulator.

Chemical Properties:

Specific Gravity: n/a *Vapor Density:* n/a *Boiling Point: n/a*
Melting Point: n/a *Vapor Pressure:* n/a *Solubility in water:* No
Other Information: n/a

Health Hazards:

Inhalation Hazard: n/a
Ingestion Hazard: Harmful if swallowed.
Absorption Hazard: Irritating to the skin and eyes.
Hazards to Wildlife: Dangerous to aquatic life.
Decontamination Procedures: Wash away any material with copious amounts of soap and water.
First Aid Procedures: Remove victim to fresh air, call emergency medical care. If not breathing give CPR. If breathing is difficult administer oxygen. Treat for shock.

Fire Hazards:

Flashpoint: n/a *Ignition temperature:* n/a
Flammable Explosive High Range: n/a *Low Range:* n/a
Toxic Products of Combustion: Emits noxious fumes, including chloride.
Other Hazards: n/a
Possible extinguishing agents: Extinguish fire using suitable agent for the type of surrounding fire.

Reactivity Hazards:

Reactive With: n/a *Other Reactions:* n/a

Corrosivity Hazards:

Corrosive With: n/a *Neutralizing Agent:* n/a

Radioactivity Hazards:

Radiation Emitted: n/a *Other Hazards:* n/a

Recommended Protection for Response Personnel:

Avoid breathing vapors, keep upwind. Wear appropriate chemical clothing. Wash away any material which may come into contact with the body with copious amounts of soap and water. Consider appropriate evacuation.

TRICHLOROPHENOXYACETIC ACID ESTER

DOT Number: NA 2765 *DOT Hazard Class:* ORM-E *DOT Guide Number:* 55
Synonyms: Silvex; 2,4,5-TP acid esters, isooctyl ester
STCC Number: 4926180 *Reportable Qty:* 100/45.4
Mfg Name: Dow Chemical *Phone No:* 1-517-636-4400

Physical Description:

Physical Form: Solid *Color:* Amber to dark brown *Odor:* Odorless
Other Information: It is used as a herbicide, defoliant, and a plant growth regulator.

Chemical Properties:

Specific Gravity: 1.18 *Vapor Density:* 8.8 *Boiling Point: 320° F(160° C)*
Melting Point: n/a *Vapor Pressure:* n/a *Solubility in water:* No
Other Information: Sinks in water.

Health Hazards:

Inhalation Hazard: n/a
Ingestion Hazard: Harmful if swallowed.
Absorption Hazard: Irritating to the skin and eyes.
Hazards to Wildlife: Dangerous to aquatic life.
Decontamination Procedures: Wash away any material with copious amounts of soap and water.
First Aid Procedures: Remove victim to fresh air, call emergency medical care. If not breathing give CPR. If breathing is difficult administer oxygen. Treat for shock.

Fire Hazards:

Flashpoint: 405° F(207.2° C) *Ignition temperature:* n/a
Flammable Explosive High Range: n/a *Low Range:* n/a
Toxic Products of Combustion: Poisonous gases may be produced in a fire.
Other Hazards: n/a
Possible extinguishing agents: Apply water from as far a distance as possible. Use foam, dry chemical or carbon dioxide.

Reactivity Hazards:

Reactive With: n/a *Other Reactions:* n/a

Corrosivity Hazards:

Corrosive With: n/a *Neutralizing Agent:* n/a

Radioactivity Hazards:

Radiation Emitted: n/a *Other Hazards:* n/a

Recommended Protection for Response Personnel:

Avoid breathing vapors, keep upwind. Wear appropriate chemical clothing. Wash away any material which may have come into contact with the body with copious amounts of soap and water. Consider appropriate evacuation.

TRICHLOROPHENOXYPROPIONIC ACID ESTER

DOT Number: NA 2765 *DOT Hazard Class:* ORM-A *DOT Guide Number:* 55
Synonyms: Fenoprop, Kurosaig, Silvex, 2,4,5-TP
STCC Number: 4941179 *Reportable Qty:* 100/45.4
Mfg Name: Dow Chemical *Phone No:* 1-517-636-4400

Physical Description:

Physical Form: Solid *Color:* White *Odor:* Odorless
Other Information: It is used as a herbicide, and a plant growth regulator.

Chemical Properties:

Specific Gravity: 1.20 *Vapor Density:* n/a *Boiling Point: 300° F(148.8° C)*
Melting Point: n/a *Vapor Pressure:* n/a *Solubility in water:* Yes
Other Information: Sinks and mixes slowly with water.

Health Hazards:

Inhalation Hazard: Harmful if inhaled.
Ingestion Hazard: Harmful if swallowed.
Absorption Hazard: Irritating to the skin and eyes.
Hazards to Wildlife: Dangerous to aquatic life.
Decontamination Procedures: Wash away any material with copious amounts of soap and water.
First Aid Procedures: Remove victim to fresh air, call emergency medical care. If not breathing give CPR. If breathing is difficult administer oxygen. Treat for shock.

Fire Hazards:

Flashpoint: n/a *Ignition temperature:* n/a
Flammable Explosive High Range: n/a *Low Range:* n/a
Toxic Products of Combustion: Hydrogen chloride may be liberated.
Other Hazards: n/a
Possible extinguishing agents: Extinguish fire using suitable agent for the type of surrounding fire.

Reactivity Hazards:

Reactive With: n/a *Other Reactions:* n/a

Corrosivity Hazards:

Corrosive With: n/a *Neutralizing Agent:* n/a

Radioactivity Hazards:

Radiation Emitted: n/a *Other Hazards:* n/a

Recommended Protection for Response Personnel:

Avoid breathing vapors, keep upwind. Wear appropriate chemical clothing. Wash away any material which may have come into contact with the body with copious amounts of soap and water. Consider appropriate evacuation.

TRICHLOROSILANE

DOT Number: UN 1295 *DOT Hazard Class:* Flammable liquid *DOT Guide Number:* 38
Synonyms: trichloromonosilane
STCC Number: 4907675 *Reportable Qty:* n/a
Mfg Name: Union Carbide *Phone No:* 1-203-794-2000

Physical Description:

Physical Form: Liquid *Color:* Colorless *Odor:* Sharp choking
Other Information: n/a

Chemical Properties:

Specific Gravity: 1.3 *Vapor Density:* 4.7 *Boiling Point: 89° F(31.6° C)*
Melting Point: n/a *Vapor Pressure:* 400 mm Hg at 58° F(14.4° C) *Solubility in water:* Decomposes
Other Information: Vapors are heavier than air.

Health Hazards:

Inhalation Hazard: Harmful if inhaled.
Ingestion Hazard: Harmful if swallowed.
Absorption Hazard: Will burn the skin and the eyes.
Hazards to Wildlife: n/a
Decontamination Procedures: Wash away any material with copious amounts of soap and water.
First Aid Procedures: Remove victim to fresh air, call emergency medical care. If not breathing give CPR. If breathing is difficult administer oxygen. Treat for shock.

Fire Hazards:

Flashpoint: 7° F(−13.8° C) *Ignition temperature:* 220° F(104.4° C)
Flammable Explosive High Range: 90.5 *Low Range:* 1.2
Toxic Products of Combustion: Toxic nitrogen chloride and phosgene gases.
Other Hazards: Containers may explode in fire. Flashback along vapor trail may occur.
Possible extinguishing agents: Use alcohol foam, dry chemical, or carbon dioxide.

Reactivity Hazards:

Reactive With: Water to form hydrochloric acid. *Other Reactions:* n/a

Corrosivity Hazards:

Corrosive With: Surface moisture to form hydrochloric acid which corrodes most metals and gives off flammable hydrogen gas. *Neutralizing Agent:* Crushed limestone, soda ash, or lime.

Radioactivity Hazards:

Radiation Emitted: n/a *Other Hazards:* n/a

Recommended Protection for Response Personnel:

Avoid breathing vapors, keep upwind. Structural protective clothing provides limited protection. Wash away any material which may have come into contact with the body with copious amounts of soap and water. If the fire becomes uncontrollable, or if the containers are exposed to direct flame, consider appropriate evacuation.

TRICRESYLPHOSPHATE

DOT Number: UN 2574 *DOT Hazard Class:* n/a *DOT Guide Number:* 55
Synonyms: TCP, tri-p-cresyl phosphate, tri-p-tolyl phosphate
STCC Number: n/a *Reportable Qty:* n/a
Mfg Name: Monsanto Chemical *Phone No:* 1-314-694-1000

Physical Description:

Physical Form: Liquid *Color:* Colorless *Odor:* Odorless
Other Information: n/a

Chemical Properties:

Specific Gravity: 1.16 *Vapor Density:* 12 *Boiling Point: 770° F(410° C)*
Melting Point: −27° F(−32.7° C) *Vapor Pressure:* n/a *Solubility in water:* n/a
Other Information: Sinks in water.

Health Hazards:

Inhalation Hazard: n/a
Ingestion Hazard: n/a
Absorption Hazard: Harmful to the skin and eyes.
Hazards to Wildlife: n/a
Decontamination Procedures: Wash away any material with copious amounts of soap and water.
First Aid Procedures: Remove victim to fresh air, call emergency medical care. If not breathing give CPR. If breathing is difficult administer oxygen. Treat for shock.

Fire Hazards:

Flashpoint: 410° F(210° C) *Ignition temperature:* n/a
Flammable Explosive High Range: n/a *Low Range:* n/a
Toxic Products of Combustion: n/a
Other Hazards: n/a
Possible extinguishing agents: Foam, dry chemical, carbon dioxide. Water and foam may cause frothing. Water may be ineffective on fire. Cool all exposed containers with water.

Reactivity Hazards:

Reactive With: n/a *Other Reactions:* n/a

Corrosivity Hazards:

Corrosive With: n/a *Neutralizing Agent:* n/a

Radioactivity Hazards:

Radiation Emitted: n/a *Other Hazards:* n/a

Recommended Protection for Response Personnel:

Avoid breathing vapors, keep upwind. Structural protective clothing provides limited protection. Wash away any material which may have come into contact with the body with copious amounts of soap and water. If the fire becomes uncontrollable, consider appropriate evacuation.

TRIDECANOL

DOT Number: n/a *DOT Hazard Class:* n/a *DOT Guide Number:* n/a
Synonyms: isotridecanol, isotridecyl alcohol, oxotridecyl alcohol, 1-tridecanol
STCC Number: n/a *Reportable Qty:* n/a
Mfg Name: Exxon Chemical America *Phone No:* 1-713-870-6000

Physical Description:

Physical Form: Liquid *Color:* Colorless *Odor:* Mild pleasant
Other Information: n/a

Chemical Properties:

Specific Gravity: .85 *Vapor Density:* 9.5 *Boiling Point: 525° F(273.8° C)*
Melting Point: n/a *Vapor Pressure:* n/a *Solubility in water:* n/a
Other Information: Floats on water.

Health Hazards:

Inhalation Hazard: Not harmful.
Ingestion Hazard: Not harmful.
Absorption Hazard: Not harmful.
Hazards to Wildlife: n/a
Decontamination Procedures: Wash away any material with copious amounts of soap and water.
First Aid Procedures: Remove victim to fresh air, call emergency medical care. If not breathing give CPR. If breathing is difficult administer oxygen. Treat for shock.

Fire Hazards:

Flashpoint: 250° F(121.1° C) *Ignition temperature:* n/a
Flammable Explosive High Range: n/a *Low Range:* n/a
Toxic Products of Combustion: n/a
Other Hazards: n/a
Possible extinguishing agents: Alcohol foam, dry chemical, carbon dioxide. Water and foam may cause frothing. Water may be ineffective on fire. Cool all exposed containers with water.

Reactivity Hazards:

Reactive With: n/a *Other Reactions:* n/a

Corrosivity Hazards:

Corrosive With: n/a *Neutralizing Agent:* n/a

Radioactivity Hazards:

Radiation Emitted: n/a *Other Hazards:* n/a

Recommended Protection for Response Personnel:

Avoid breathing vapors, keep upwind. Structural protective clothing provides limited protection. Wash away any material which may have come into contact with the body with copious amounts of soap and water. If the fire becomes uncontrollable, consider appropriate evacuation.

TRIDECENE

DOT Number: n/a *DOT Hazard Class:* n/a *DOT Guide Number:* n/a
Synonyms: undecylethylene
STCC Number: n/a *Reportable Qty:* n/a
Mfg Name: Phillips Petroleum *Phone No:* 1-918-661-6600

Physical Description:

Physical Form: Liquid *Color:* Colorless *Odor:* Mild pleasant
Other Information: n/a

Chemical Properties:

Specific Gravity: .76 *Vapor Density:* n/a *Boiling Point: 451° F(232.7° C)*
Melting Point: −11° F(−23.8° C) *Vapor Pressure:* n/a *Solubility in water:* n/a
Other Information: Floats on water.

Health Hazards:

Inhalation Hazard: n/a
Ingestion Hazard: n/a
Absorption Hazard: Irritating to the eyes.
Hazards to Wildlife: n/a
Decontamination Procedures: Wash away any material with copious amounts of soap and water.
First Aid Procedures: Remove victim to fresh air, call emergency medical care. If not breathing give CPR. If breathing is difficult administer oxygen. Treat for shock.

Fire Hazards:

Flashpoint: 175° F(79.4° C) *Ignition temperature:* n/a
Flammable Explosive High Range: n/a *Low Range:* n/a
Toxic Products of Combustion: n/a
Other Hazards: n/a
Possible extinguishing agents: Foam, dry chemical, carbon dioxide. Water may be ineffective on fire. Cool all exposed containers with water.

Reactivity Hazards:

Reactive With: n/a *Other Reactions:* n/a

Corrosivity Hazards:

Corrosive With: n/a *Neutralizing Agent:* n/a

Radioactivity Hazards:

Radiation Emitted: n/a *Other Hazards:* n/a

Recommended Protection for Response Personnel:

Avoid breathing vapors, keep upwind. Structural protective clothing provides limited protection. Wash away any material which may have come into contact with the body with copious amounts of soap and water. If the fire becomes uncontrollable, consider appropriate evacuation.

TRIETHANOLAMINE

DOT Number: NA 9151 *DOT Hazard Class:* ORM-E *DOT Guide Number:* 31
Synonyms: triethylolamine, trihydroxytriethylamine, tris(hydroxyethyl)amine
STCC Number: 4963379 *Reportable Qty:* 1000/454
Mfg Name: Dow Chemical *Phone No:* 1-513-636-4400

Physical Description:

Physical Form: Liquid *Color:* Colorless *Odor:* Mild ammonia
Other Information: It is used as a detergent.

Chemical Properties:

Specific Gravity: 1.13 *Vapor Density:* 5.1 *Boiling Point: Decomposes*
Melting Point: 70.9° F(21.6° C) *Vapor Pressure:* 10 mm Hg at 401° F(205° C) *Solubility in water:* Yes
Other Information: Sinks and mixes with water.

Health Hazards:

Inhalation Hazard: n/a
Ingestion Hazard: Harmful if swallowed.
Absorption Hazard: Irritating to the skin and eyes.
Hazards to Wildlife: n/a
Decontamination Procedures: Wash away any material with copious amounts of soap and water.
First Aid Procedures: Remove victim to fresh air, call emergency medical care. If not breathing give CPR. If breathing is difficult administer oxygen. Treat for shock.

Fire Hazards:

Flashpoint: 355° F(179.4° C) *Ignition temperature:* n/a
Flammable Explosive High Range: n/a *Low Range:* n/a
Toxic Products of Combustion: Poisonous gases may be formed in a fire.
Other Hazards: n/a
Possible extinguishing agents: Use alcohol foam, dry chemical, or carbon dioxide. extinguish fire using suitable agent for the type of surrounding fire.

Reactivity Hazards:

Reactive With: n/a *Other Reactions:* n/a

Corrosivity Hazards:

Corrosive With: n/a *Neutralizing Agent:* n/a

Radioactivity Hazards:

Radiation Emitted: n/a *Other Hazards:* n/a

Recommended Protection for Response Personnel:

Avoid breathing vapors, keep upwind. Structural protective clothing provides limited protection. Wash away any material which may have come into contact with the body with copious amounts of soap and water. Consider appropriate evacuation.

TRIETHYLALUMINUM

DOT Number: UN 1103 *DOT Hazard Class:* Flammable liquid *DOT Guide Number:* 40
Synonyms: aluminum triethyl, ATE, TEA
STCC Number: n/a *Reportable Qty:* n/a
Mfg Name: Stauffer Chemical *Phone No:* 1-203-222-3000

Physical Description:

Physical Form: Liquid *Color:* Colorless *Odor:* n/a
Other Information: Ignites when exposed to air.

Chemical Properties:

Specific Gravity: .8 *Vapor Density:* 3.9 *Boiling Point: 367.9° F(186.6° C)*
Melting Point: n/a *Vapor Pressure:* 4 mm Hg at 181° F(82.7° C) *Solubility in water:* n/a
Other Information: n/a

Health Hazards:

Inhalation Hazard: n/a
Ingestion Hazard: Harmful if swallowed.
Absorption Hazard: Will burn the skin and the eyes.
Hazards to Wildlife: n/a
Decontamination Procedures: Wash away any material with copious amounts of soap and water.
First Aid Procedures: Remove victim to fresh air, call emergency medical care. If not breathing give cpr, If breathing is difficult administer oxygen. Treat for shock.

Fire Hazards:

Flashpoint: n/a *Ignition temperature:* n/a
Flammable Explosive High Range: n/a *Low Range:* n/a
Toxic Products of Combustion: Poisonous gases may be produced in a fire.
Other Hazards: Ignites when exposed to air.
Possible extinguishing agents: Use inert powders, sand, limestone and dry chemical.

Reactivity Hazards:

Reactive With: Air to form fire. Water to form flammable ethane gas. *Other Reactions:* n/a

Corrosivity Hazards:

Corrosive With: n/a *Neutralizing Agent:* n/a

Radioactivity Hazards:

Radiation Emitted: n/a *Other Hazards:* n/a

Recommended Protection for Response Personnel:

Avoid breathing vapors, keep upwind. Wear a sealed chemical suit (viton, chlorinated polyethylene, PVC). Wash away any material which may have come into contact with the body with copious amounts of soap and water. If the fire becomes uncontrollable, or if the containers are exposed to direct flame, consider appropriate evacuation.
Do not apply water !

TRIETHYLAMINE

DOT Number: UN 1296 *DOT Hazard Class:* Flammable liquid *DOT Guide Number:* 29
Synonyms: TEN
STCC Number: 4907880 *Reportable Qty:* 100/45.4
Mfg Name: Union Carbide *Phone No:* 1-203-794-2000

Physical Description:

Physical Form: Liquid *Color:* Colorless *Odor:* Fishy
Other Information: n/a

Chemical Properties:

Specific Gravity: .7 *Vapor Density:* 3.5 *Boiling Point: 193° F(89.4° C)*
Melting Point: −175° F(−115° C) *Vapor Pressure:* 54 mm Hg at 68° F(20° C)
Solubility in water: Yes *Degree of Solubility: 5.5%*
Other Information: Solution is lighter, soluble in water. Vapors are heavier than air. Weighs 7.4 lbs/3.3 kg per gallon/3.8 l.

Health Hazards:

Inhalation Hazard: Coughing, difficulty in breathing, loss of consciousness.
Ingestion Hazard: Harmful if swallowed.
Absorption Hazard: Will burn the skin and eyes.
Hazards to Wildlife: Dangerous to aquatic life.
Decontamination Procedures: Wash away any material with copious amounts of soap and water.
First Aid Procedures: Remove victim to fresh air, call emergency medical care. If not breathing give CPR. If breathing is difficult administer oxygen. Treat for shock.

Fire Hazards:

Flashpoint: 16° F(−8.8° C) *Ignition temperature:* 480° F(248.8° C)
Flammable Explosive High Range: 8 *Low Range:* 1.2
Toxic Products of Combustion: Toxic oxides of nitrogen are produced.
Other Hazards: Flashback along vapor trail may occur. Vapors may ignite in an enclosed area.
Possible extinguishing agents: Apply water from as far a distance as possible. Use alcohol foam, dry chemical, or carbon dioxide.

Reactivity Hazards:

Reactive With: n/a *Other Reactions:* n/a

Corrosivity Hazards:

Corrosive With: n/a *Neutralizing Agent:* n/a

Radioactivity Hazards:

Radiation Emitted: n/a *Other Hazards:* n/a

Recommended Protection for Response Personnel:

Avoid breathing vapors, keep upwind. Wear a sealed chemical suit (butyl rubber). Wash away any material which may have come into contact with the body with copious amounts of soap and water. If fire becomes uncontrollable, or if containers are exposed to direct flame, consider appropriate evacuation.

TRIETHYLBENZENE

DOT Number: n/a *DOT Hazard Class:* n/a *DOT Guide Number:* n/a
Synonyms: 1,3,5-triethylbenzene, sym-triethylbenzene
STCC Number: n/a *Reportable Qty:* n/a *CHEMTREC Phone No:* 1-800-424-9300

Physical Description:

Physical Form: Liquid *Color:* Colorless *Odor:* Weak chemical
Other Information: n/a

Chemical Properties:

Specific Gravity: .86 *Vapor Density:* 5.1 *Boiling Point: 421° F(216.1° C)*
Melting Point: n/a *Vapor Pressure:* n/a *Solubility in water:* n/a
Other Information: Floats on water.

Health Hazards:

Inhalation Hazard: n/a
Ingestion Hazard: n/a
Absorption Hazard: Irritating to the skin and eyes.
Hazards to Wildlife: n/a
Decontamination Procedures: Wash away any material with copious amounts of soap and water.
First Aid Procedures: Remove victim to fresh air, call emergency medical care. If not breathing give CPR. If breathing is difficult administer oxygen. Treat for shock.

Fire Hazards:

Flashpoint: 181° F(82.7° C) *Ignition temperature:* n/a
Flammable Explosive High Range: n/a *Low Range:* n/a
Toxic Products of Combustion: n/a
Other Hazards: n/a
Possible extinguishing agents: Foam, dry chemical, carbon dioxide. Water may be ineffective on fire. Cool all exposed containers with water.

Reactivity Hazards:

Reactive With: n/a *Other Reactions:* n/a

Corrosivity Hazards:

Corrosive With: n/a *Neutralizing Agent:* n/a

Radioactivity Hazards:

Radiation Emitted: n/a *Other Hazards:* n/a

Recommended Protection for Response Personnel:

Avoid breathing vapors, keep upwind. Wear a sealed chemical suit (polycarbonate, butyl rubber). Wash away any material which may have come into contact with the body with copious amounts of soap and water. If the fire becomes uncontrollable, consider appropriate evacuation.

TRIETHYLENE GLYCOL

DOT Number: n/a *DOT Hazard Class:* n/a *DOT Guide Number:* n/a
Synonyms: glycol-bis(hydroxyethyl)ether, ethylene glycol hydroxydiethyl ether, TEG, trigylcol
STCC Number: n/a *Reportable Qty:* n/a
Mfg Name: Ashland Chemical *Phone No:* 1-614-889-3333

Physical Description:

Physical Form: Liquid *Color:* Colorless *Odor:* Mild
Other Information: n/a

Chemical Properties:

Specific Gravity: 1.12 *Vapor Density:* 5.7 *Boiling Point: 550° F(287.7° C)*
Melting Point: 24° F(−4.4° C) *Vapor Pressure:* 1 mm Hg at 237° F(113.8° C) *Solubility in water:* Yes
Other Information: Sinks and mixes with water.

Health Hazards:

Inhalation Hazard: Not harmful.
Ingestion Hazard: Not harmful.
Absorption Hazard: Not harmful.
Hazards to Wildlife: n/a
Decontamination Procedures: Wash away any material with copious amounts of soap and water.
First Aid Procedures: Remove victim to fresh air, call emergency medical care. If not breathing give CPR. If breathing is difficult administer oxygen. Treat for shock.

Fire Hazards:

Flashpoint: 330° F(165.5° C) *Ignition temperature:* 700° F(371.1° C)
Flammable Explosive High Range: 9.2 *Low Range:* .9
Toxic Products of Combustion: n/a
Other Hazards: n/a
Possible extinguishing agents: Alcohol foam, dry chemical, carbon dioxide. Water and foam may cause frothing. Water may be ineffective on fire. Cool all exposed containers with water.

Reactivity Hazards:

Reactive With: n/a *Other Reactions:* n/a

Corrosivity Hazards:

Corrosive With: n/a *Neutralizing Agent:* n/a

Radioactivity Hazards:

Radiation Emitted: n/a *Other Hazards:* n/a

Recommended Protection for Response Personnel:

Avoid breathing vapors, keep upwind. Structural protective clothing provides limited protection. Wash away any material which may have come into contact with the body with copious amounts of soap and water. If the fire becomes uncontrollable, consider appropriate evacuation.

TRIETHYLENE TETRAMINE

DOT Number: UN 2259 *DOT Hazard Class:* n/a *DOT Guide Number:* 60
Synonyms: N,N′-bis(2-aminoethyl)ethylenediamine, trien, TETA
STCC Number: n/a *Reportable Qty:* n/a
Mfg Name: Allied Corp. *Phone No:* 1-201-445-2000

Physical Description:

Physical Form: Liquid *Color:* Light straw to amber *Odor:* Ammonia
Other Information: n/a

Chemical Properties:

Specific Gravity: .98 *Vapor Density:* 5 *Boiling Point: 531° F(277.2° C)*
Melting Point: −31° F(−35° C) *Vapor Pressure:* .01 mm Hg at 68° F(20° C) *Solubility in water:* Yes
Other Information: Floats and mixes with water.

Health Hazards:

Inhalation Hazard: n/a
Ingestion Hazard: Harmful if swallowed.
Absorption Hazard: Will burn the skin and eyes.
Hazards to Wildlife: n/a
Decontamination Procedures: Wash away any material with copious amounts of soap and water.
First Aid Procedures: Remove victim to fresh air, call emergency medical care. If not breathing give CPR. If breathing is difficult administer oxygen. Treat for shock.

Fire Hazards:

Flashpoint: 275° F(135° C) *Ignition temperature:* 640° F(337.7° C)
Flammable Explosive High Range: n/a *Low Range:* n/a
Toxic Products of Combustion: n/a
Other Hazards: n/a
Possible extinguishing agents: Alcohol foam, dry chemical, carbon dioxide. Water and foam may cause frothing. Water may be ineffective on fire.

Reactivity Hazards:

Reactive With: n/a *Other Reactions:* n/a

Corrosivity Hazards:

Corrosive With: n/a *Neutralizing Agent:* n/a

Radioactivity Hazards:

Radiation Emitted: n/a *Other Hazards:* n/a

Recommended Protection for Response Personnel:

Avoid breathing vapors, keep upwind. Wear a sealed chemical suit (butyl rubber). Wash away any material which may have come into contact with the body with copious amounts of soap and water. if the fire becomes uncontrollable, consider appropriate evacuation.

TRIFLUOROCHLOROETHYLENE

DOT Number: UN 1082 *DOT Hazard Class:* Flammable gas *DOT Guide Number:* 17
Synonyms: CTFE, Genetron 1113, Kel F monomer
STCC Number: 4905785 *Reportable Qty:* n/a
Mfg Name: Allied Corp. *Phone No:* 1-201-455-2000

Physical Description:

Physical Form: Gas *Color:* Colorless *Odor:* Ethereal
Other Information: It is shipped as a liquefied gas under its own vapor pressure.

Chemical Properties:

Specific Gravity: 1.3 *Vapor Density:* 4.02 *Boiling Point: – 18° F(–27.7° C)*
Melting Point: n/a *Vapor Pressure:* n/a *Solubility in water:* No
Other Information: Vapors are heavier than air.

Health Hazards:

Inhalation Hazard: Dizziness, nausea or vomiting.
Ingestion Hazard: Will cause frostbite.
Absorption Hazard: Will cause frostbite.
Hazards to Wildlife: No
Decontamination Procedures: Wash away any material with copious amounts of soap and water.
First Aid Procedures: Remove victim to fresh air, call emergency medical care. If not breathing give CPR. If breathing is difficult administer oxygen. Treat for shock.

Fire Hazards:

Flashpoint: n/a *Ignition temperature:* n/a
Flammable Explosive High Range: 16 *Low Range:* 8.4
Toxic Products of Combustion: Poisonous gases are produced in a fire.
Other Hazards: Under fire conditions, cylinders may violently rupture and or rocket, (BLEVE) flashback along vapor trail may occur.
Possible extinguishing agents: Apply water from as far a distance as possible.

Reactivity Hazards:

Reactive With: n/a *Other Reactions:* n/a

Corrosivity Hazards:

Corrosive With: n/a *Neutralizing Agent:* n/a

Radioactivity Hazards:

Radiation Emitted: n/a *Other Hazards:* n/a

Recommended Protection for Response Personnel:

Avoid breathing vapors, keep upwind. Structural protective clothing provides limited protection. Wash away any material which may have come into contact with the body with copious amounts of soap and water. If the fire becomes uncontrollable, or if the containers are exposed to direct flame, consider appropriate evacuation. (BLEVE)

TRIFLURALIN

DOT Number: UN 1609 *DOT Hazard Class:* Poison *DOT Guide Number:* n/a
Synonyms: Treflan
STCC Number: n/a *Reportable Qty:* n/a
Mfg Name: Elanco Products Co. *Phone No:* 1-617-261-3784

Physical Description:

Physical Form: Solid *Color:* Orange yellow *Odor:* n/a
Other Information: n/a

Chemical Properties:

Specific Gravity: 1.29 *Vapor Density:* n/a *Boiling Point: Decomposes*
Melting Point: 108° F(42.2° C) *Vapor Pressure:* n/a *Solubility in water:* n/a
Other Information: Sinks in water.

Health Hazards:

Inhalation Hazard: Poisonous if inhaled.
Ingestion Hazard: Poisonous if swallowed.
Absorption Hazard: Irritating to the skin and eyes.
Hazards to Wildlife: Dangerous to aquatic life.
Decontamination Procedures: Wash away any material with copious amounts of soap and water.
First Aid Procedures: Remove victim to fresh air, call emergency medical care. If not breathing give CPR. If breathing is difficult administer oxygen. Treat for shock.

Fire Hazards:

Flashpoint: 185° F(85° C) *Ignition temperature:* n/a
Flammable Explosive High Range: n/a *Low Range:* n/a
Toxic Products of Combustion: n/a
Other Hazards: n/a
Possible extinguishing agents: Extinguish fire using suitable agent for the type of surrounding fire.

Reactivity Hazards:

Reactive With: n/a *Other Reactions:* n/a

Corrosivity Hazards:

Corrosive With: n/a *Neutralizing Agent:* n/a

Radioactivity Hazards:

Radiation Emitted: n/a *Other Hazards:* n/a

Recommended Protection for Response Personnel:

Avoid breathing vapors, keep upwind. Structural protective clothing provides limited protection. Wash away any material which may have come into contact with the body with copious amounts of soap and water. If the fire becomes uncontrollable, consider appropriate evacuation.

TRIISOBUTYLALUMINUM

DOT Number: UN 1930 *DOT Hazard Class:* Flammable liquid *DOT Guide Number:* 40
Synonyms: aluminum triisobutyl, TIBA, TIBAL
STCC Number: n/a *Reportable Qty:* n/a
Mfg Name: Texas Alkyls Inc. *Phone No:* 1-713-479-8411

Physical Description:

Physical Form: Liquid *Color:* Colorless *Odor:* n/a
Other Information: Ignites when exposed to air.

Chemical Properties:

Specific Gravity: .8 *Vapor Density:* 6.8 *Boiling Point: 414° F(212.2° C)*
Melting Point: n/a *Vapor Pressure:* 1 mm Hg at 117° F(47.2° C)
Solubility in water: n/a
Other Information: n/a

Health Hazards:

Inhalation Hazard: n/a
Ingestion Hazard: Harmful if swallowed.
Absorption Hazard: Will burn the skin and the eyes.
Hazards to Wildlife: n/a
Decontamination Procedures: Wash away any material with copious amounts of soap and water.
First Aid Procedures: Remove victim to fresh air, call emergency medical care. If not breathing give CPR. If breathing is difficult administer oxygen. Treat for shock.

Fire Hazards:

Flashpoint: n/a *Ignition temperature:* n/a
Flammable Explosive High Range: n/a *Low Range:* n/a
Toxic Products of Combustion: Poisonous gases may be produced in a fire.
Other Hazards: Ignites when exposed to air.
Possible extinguishing agents: Use inert powders, sand, limestone and dry chemical.

Reactivity Hazards:

Reactive With: Air to form fire. *Other Reactions:* n/a

Corrosivity Hazards:

Corrosive With: Silicon rubber or urethane rubber. *Neutralizing Agent:* n/a

Radioactivity Hazards:

Radiation Emitted: n/a *Other Hazards:* n/a

Recommended Protection for Response Personnel:

Avoid breathing vapors, keep upwind. Wear a sealed chemical suit (polycarbonate, butyl rubber). Wash away any material which may have come into contact with the body with copious amounts of soap and water. If the fire becomes uncontrollable, or if the containers are exposed to direct flame, consider appropriate evacuation.
Do not apply water !

TRIMETHYLCHLOROSILANE

DOT Number: UN 1298 *DOT Hazard Class:* Flammable liquid *DOT Guide Number:* 29
Synonyms: chlorotrimethylsilane
STCC Number: 4907680 *Reportable Qty:* n/a
Mfg Name: Dow Chemical *Phone No:* 1-517-636-4400

Physical Description:

Physical Form: Liquid *Color:* Colorless *Odor:* Sharp irritating
Other Information: n/a

Chemical Properties:

Specific Gravity: .9 *Vapor Density:* 3.75 *Boiling Point: 135° F(57.2° C)*
Melting Point: n/a *Vapor Pressure:* n/a *Solubility in water:* Yes
Other Information: Heavier than and decomposed by water. Vapors are heavier than air.

Health Hazards:

Inhalation Hazard: Harmful if inhaled.
Ingestion Hazard: Harmful if swallowed.
Absorption Hazard: Will burn the skin and the eyes.
Hazards to Wildlife: n/a
Decontamination Procedures: Wash away any material with copious amounts of soap and water.
First Aid Procedures: Remove victim to fresh air, call emergency medical care. If not breathing give CPR. If breathing is difficult administer oxygen. Treat for shock.

Fire Hazards:

Flashpoint: –18° F(–27.7° C) *Ignition temperature:* 743° F(395° C)
Flammable Explosive High Range: n/a *Low Range:* 1.8
Toxic Products of Combustion: Poisonous gases may be produced in a fire.
Other Hazards: Containers may explode in a fire. Flashback along vapor trail may occur.
Possible extinguishing agents: Apply water from as far a distance as possible. Use alcohol foam, dry chemical, or carbon dioxide.

Reactivity Hazards:

Reactive With: Water to form hydrochloric acid.
Other Reactions: Surface moisture to form hydrogen chloride.

Corrosivity Hazards:

Corrosive With: Surface moisture to form hydrogen chloride which is corrosive to common metals to form flammable hydrogen gas. *Neutralizing Agent:* Crushed limestone, soda ash, or lime.

Radioactivity Hazards:

Radiation Emitted: n/a *Other Hazards:* n/a

Recommended Protection for Response Personnel:

Avoid breathing vapors, keep upwind. Wear appropriate chemical clothing, Wash away any material which may come into contact with the body with copious amounts of soap and water. If fire becomes uncontrollable, or if containers are exposed to direct flame, consider appropriate evacuation. Do not apply water!

TRIPROPYLENE GLYCOL

DOT Number: n/a *DOT Hazard Class:* n/a *DOT Guide Number:* n/a
Synonyms: none given
STCC Number: n/a *Reportable Qty:* n/a
Mfg Name: Union Carbide *Phone No:* 1-203-794-2000

Physical Description:

Physical Form: Liquid *Color:* Colorless *Odor:* Characteristic
Other Information: n/a

Chemical Properties:

Specific Gravity: 1.02 *Vapor Density:* 6.6 *Boiling Point: 523° F(272.7° C)*
Melting Point: −49° F(−45° C) *Vapor Pressure:* 1 mm Hg at 205° F(96.1° C) *Solubility in water:* Yes
Other Information: May float or sink with water.

Health Hazards:

Inhalation Hazard: Not harmful.
Ingestion Hazard: Not harmful.
Absorption Hazard: Not harmful.
Hazards to Wildlife: n/a
Decontamination Procedures: Wash away any material with copious amounts of soap and water.
First Aid Procedures: Remove victim to fresh air, call emergency medical care. If not breathing give CPR. If breathing is difficult administer oxygen. Treat for shock.

Fire Hazards:

Flashpoint: 285° F(140.5° C) *Ignition temperature:* n/a
Flammable Explosive High Range: 5 *Low Range:* .8
Toxic Products of Combustion: Acrid fumes of acids and aldehydes may form in fires.
Other Hazards: n/a
Possible extinguishing agents: Water may be ineffective on fires. Use alcohol foam, dry chemical, carbon dioxide. Cool all affected containers with water.

Reactivity Hazards:

Reactive With: n/a *Other Reactions:* n/a

Corrosivity Hazards:

Corrosive With: May attack some forms of plastic. *Neutralizing Agent:* n/a

Radioactivity Hazards:

Radiation Emitted: n/a *Other Hazards:* n/a

Recommended Protection for Response Personnel:

Avoid breathing vapors, keep upwind. Structural protective clothing provides limited protection. Wash away any material which may have come into contact with the body with copious amounts of soap and water. If the fire becomes uncontrollable, consider appropriate evacuation.

TURPENTINE

DOT Number: UN 1299 *DOT Hazard Class:* Flammable liquid *DOT Guide Number:* 27
Synonyms: spirit of gum, wood, D.D., turps
STCC Number: 4910313 *Reportable Qty:* n/a
Mfg Name: Georgia Pacific *Phone No:* 1-404-521-4000

Physical Description:

Physical Form: Liquid *Color:* Colorless *Odor:* Penetrating unpleasant
Other Information: It is used as a paint remover.

Chemical Properties:

Specific Gravity: 1 *Vapor Density:* 4.8 *Boiling Point: 300° F(148.8° C)*
Melting Point: −58/−76° F(−50/−60° C) *Vapor Pressure:* 5 mm Hg at 68° F(20° C)
Solubility in water: No
Other Information: Lighter than and insoluble in water. Vapors are heavier than air.

Health Hazards:

Inhalation Hazard: Nausea, vomiting, difficulty in breathing, loss of consciousness.
Ingestion Hazard: Poisonous if swallowed.
Absorption Hazard: Irritating to the skin and eyes.
Hazards to Wildlife: Dangerous to aquatic life.
Decontamination Procedures: Wash away any material with copious amounts of soap and water.
First Aid Procedures: Remove victim to fresh air, call emergency medical care. If not breathing give CPR. If breathing is difficult administer oxygen. Treat for shock.

Fire Hazards:

Flashpoint: 95° F(35° C) *Ignition temperature:* 488° F(253.3° C)
Flammable Explosive High Range: n/a *Low Range:* .8
Toxic Products of Combustion: n/a
Other Hazards: Flashback along vapor trail may occur. Vapors may ignite if confined in an enclosed area.
Possible extinguishing agents: Apply water from as far a distance as possible. Use foam, dry chemical, or carbon dioxide.

Reactivity Hazards:

Reactive With: n/a *Other Reactions:* n/a

Corrosivity Hazards:

Corrosive With: n/a *Neutralizing Agent:* n/a

Radioactivity Hazards:

Radiation Emitted: n/a *Other Hazards:* n/a

Recommended Protection for Response Personnel:

Avoid breathing vapors, keep upwind. Structural protective clothing provides limited protection. Wash away any material which may have come into contact with the body with copious amounts of soap and water. Consider appropriate evacuation.

UNDECANOL

DOT Number: n/a *DOT Hazard Class:* n/a *DOT Guide Number:* n/a
Synonyms: Alcohol C-11, 1-hendecanol, undecyl alcohol, 1-undecanol, undecanoic alcohol, undecylic alcohol
STCC Number: n/a *Reportable Qty:* n/a
Mfg Name: Union Carbide *Phone No:* 1-203-794-2000

Physical Description:

Physical Form: Solid *Color:* Colorless *Odor:* Mild
Other Information: n/a

Chemical Properties:

Specific Gravity: .83 *Vapor Density:* 5.9 *Boiling Point: 473° F(245° C)*
Melting Point: 60° F(15.5° C) *Vapor Pressure:* n/a *Solubility in water:* n/a
Other Information: Floats on water.

Health Hazards:

Inhalation Hazard: n/a
Ingestion Hazard: n/a
Absorption Hazard: Irritating to the eyes.
Hazards to Wildlife: n/a
Decontamination Procedures: Wash away any material with copious amounts of soap and water.
First Aid Procedures: Remove victim to fresh air, call emergency medical care. If not breathing give CPR. If breathing is difficult administer oxygen. Treat for shock.

Fire Hazards:

Flashpoint: 200° F(93.3° C) *Ignition temperature:* n/a
Flammable Explosive High Range: n/a *Low Range:* n/a
Toxic Products of Combustion: n/a
Other Hazards: n/a
Possible extinguishing agents: Foam, dry chemical, carbon dioxide. Water or foam may cause frothing.

Reactivity Hazards:

Reactive With: n/a *Other Reactions:* n/a

Corrosivity Hazards:

Corrosive With: n/a *Neutralizing Agent:* n/a

Radioactivity Hazards:

Radiation Emitted: n/a *Other Hazards:* n/a

Recommended Protection for Response Personnel:

Avoid breathing vapors, keep upwind. Structural protective clothing provides limited protection. Wash away any material which may have come into contact with the body with copious amounts of soap and water. if the fire becomes uncontrollable, consider appropriate evacuation.

UNDECENE

DOT Number: n/a *DOT Hazard Class:* n/a *DOT Guide Number:* n/a
Synonyms: n-nonylethylene
STCC Number: n/a *Reportable Qty:* n/a
Mfg Name: Humphrey Chemical Co. *Phone No:* 1-203-281-0012

Physical Description:

Physical Form: Liquid *Color:* Colorless *Odor:* Mild
Other Information: n/a

Chemical Properties:

Specific Gravity: .75 *Vapor Density:* n/a *Boiling Point: 379° F(192.7° C)*
Melting Point: −56° F(−48.8° C) *Vapor Pressure:* n/a *Solubility in water:* n/a
Other Information: Floats on water.

Health Hazards:

Inhalation Hazard: n/a
Ingestion Hazard: Harmful if swallowed.
Absorption Hazard: Irritating to the skin and eyes.
Hazards to Wildlife: n/a
Decontamination Procedures: Wash away any material with copious amounts of soap and water.
First Aid Procedures: Remove victim to fresh air, call emergency medical care. If not breathing give CPR. If breathing is difficult administer oxygen. Treat for shock.

Fire Hazards:

Flashpoint: 160° F(71.1° C) *Ignition temperature:* n/a
Flammable Explosive High Range: n/a *Low Range:* n/a
Toxic Products of Combustion: n/a
Other Hazards: n/a
Possible extinguishing agents: Foam, dry chemical, carbon dioxide. Water may be ineffective on fire. Cool all exposed containers with water.

Reactivity Hazards:

Reactive With: n/a *Other Reactions:* n/a

Corrosivity Hazards:

Corrosive With: n/a *Neutralizing Agent:* n/a

Radioactivity Hazards:

Radiation Emitted: n/a *Other Hazards:* n/a

Recommended Protection for Response Personnel:

Avoid breathing vapors, keep upwind. Structural protective clothing provides limited protection. Wash away any material which may have come into contact with the body with copious amounts of soap and water. If the fire becomes uncontrollable, consider appropriate evacuation.

UNDECYLBENZENE

DOT Number: n/a *DOT Hazard Class:* n/a *DOT Guide Number:* n/a
Synonyms: 1-phenylundecane
STCC Number: n/a *Reportable Qty:* n/a
Mfg Name: Humphrey Chemical Co. *Phone No:* 1-203-281-0012

Physical Description:

Physical Form: Liquid *Color:* Colorless *Odor:* Mild
Other Information: n/a

Chemical Properties:

Specific Gravity: .85 *Vapor Density:* n/a *Boiling Point: 601° F(316.1° C)*
Melting Point: 23° F(−5° C) *Vapor Pressure:* n/a *Solubility in water:* n/a
Other Information: Floats on water.

Health Hazards:

Inhalation Hazard: n/a
Ingestion Hazard: Nausea and vomiting.
Absorption Hazard: Irritating to the skin and eyes.
Hazards to Wildlife: n/a
Decontamination Procedures: Wash away any material with copious amounts of soap and water.
First Aid Procedures: Remove victim to fresh air, call emergency medical care. If not breathing give CPR. If breathing is difficult administer oxygen. Treat for shock.

Fire Hazards:

Flashpoint: 285° F(140.5° C) *Ignition temperature:* n/a
Flammable Explosive High Range: n/a *Low Range:* n/a
Toxic Products of Combustion: n/a
Other Hazards: n/a
Possible extinguishing agents: Foam, dry chemical, carbon dioxide. Water may be ineffective on fire. Cool all exposed containers with water.

Reactivity Hazards:

Reactive With: n/a *Other Reactions:* n/a

Corrosivity Hazards:

Corrosive With: May attack some forms of plastic. *Neutralizing Agent:* n/a

Radioactivity Hazards:

Radiation Emitted: n/a *Other Hazards:* n/a

Recommended Protection for Response Personnel:

Avoid breathing vapors, keep upwind. Structural protective clothing provides limited protection. Wash away any material which may have come into contact with the body with copious amounts of soap and water. if the fire becomes uncontrollable, consider appropriate evacuation.

URANIUM HEXAFLUORIDE

DOT Number: UN 2977 *DOT Hazard Class:* Radioactive Material *DOT Guide Number:* 66
Synonyms: uranium fluoride, uranium(VI) fluoride
STCC Number: 4927490 *Reportable Qty:* n/a *CHEMTREC Phone No:* 1-800-424-9300

Physical Description:

Physical Form: Solid *Color:* Colorless, white, or pale yellow *Odor:* n/a
Other Information: The primary hazard associated with uranium hexafluoride is its corrosivity.

Chemical Properties:

Specific Gravity: 4.7 to 5.1 *Vapor Density:* n/a *Boiling Point: 132° F(55.5° C)*
Melting Point: 148° F(64.4° C) *Vapor Pressure:* 109 mm Hg at 68° F(20° C) *Solubility in water:* Reacts.
Other Information: Contains more than 1% U-235.

Health Hazards:

Inhalation Hazard: Radioactive poisoning.
Ingestion Hazard: Radioactive poisoning.
Absorption Hazard: Radioactive poisoning.
Hazards to Wildlife: Radioactive poisoning.
Decontamination Procedures: No procedures available!!
First Aid Procedures: Remove victim to fresh air, call emergency medical care. If not breathing give CPR. If breathing is difficult administer oxygen. Treat for shock.

Fire Hazards:

Flashpoint: n/a *Ignition temperature:* n/a
Flammable Explosive High Range: n/a *Low Range:* n/a
Toxic Products of Combustion: Radiation fallout
Other Hazards: High emitance of radioactive energy.
Possible extinguishing agents: Apply water from as far a distance as possible. extinguish fire using suitable agent for the type of surrounding fire.

Reactivity Hazards:

Reactive With: Itself!! *Other Reactions:* n/a

Corrosivity Hazards:

Corrosive With: Metal and tissue. *Neutralizing Agent:* Crushed limestone, soda ash, or lime.

Radioactivity Hazards:

Radiation Emitted: Alpha, beta, and gamma!
Other Hazards: It is shipped in special steel containers. All unauthorized personnel should be kept as far a distance as possible until qualified personnel and equipment can be obtained.

Recommended Protection for Response Personnel:

Avoid breathing vapors or dust, keep upwind. Wash away any material which may have come into contact with the body with copious amounts of soap and water. Consider appropriate evacuation.

URANIUM PEROXIDE

DOT Number: n/a *DOT Hazard Class:* Radioactive Material *DOT Guide Number:* n/a
Synonyms: uranium oxide, uranium oxide peroxide
STCC Number: n/a *Reportable Qty:* n/a
Mfg Name: Atomergic Chemetals Corp. *Phone No:* 1-516-349-8800

Physical Description:

Physical Form: Solid *Color:* Yellow *Odor:* n/a
Other Information: n/a

Chemical Properties:

Specific Gravity: 11.66 *Vapor Density:* n/a *Boiling Point: Decomposes*
Melting Point: 239° F(115° C) *Vapor Pressure:* n/a *Solubility in water:* No
Other Information: Sinks in water.

Health Hazards:

Inhalation Hazard: Radioactive poisoning.
Ingestion Hazard: Radioactive poisoning.
Absorption Hazard: Radioactive poisoning.
Hazards to Wildlife: Radioactive poisoning.
Decontamination Procedures: No procedures available!!
First Aid Procedures: Remove victim to fresh air, call emergency medical care. If not breathing give CPR. If breathing is difficult administer oxygen. Treat for shock.

Fire Hazards:

Flashpoint: n/a *Ignition temperature:* n/a
Flammable Explosive High Range: n/a *Low Range:* n/a
Toxic Products of Combustion: Radiation fallout, decomposes to form U_2O_7 then to UO_3 and oxygen.
Other Hazards: High emitance of radioactive energy.
Possible extinguishing agents: Apply water from as far a distance as possible. extinguish fire using suitable agent for the type of surrounding fire.

Reactivity Hazards:

Reactive With: Itself!! *Other Reactions:* n/a

Corrosivity Hazards:

Corrosive With: Metal and tissue *Neutralizing Agent:* Crushed limestone, soda ash, or lime.

Radioactivity Hazards:

Radiation Emitted: Alpha, beta, and gamma!
Other Hazards: It is shipped in special steel containers. All unauthorized personnel should be kept as far a distance as possible until qualified personnel and equipment can be obtained.

Recommended Protection for Response Personnel:

Avoid breathing vapors or dust, keep upwind. Wash away any material which may have come into contact with the body with copious amounts of soap and water. Consider appropriate evacuation.

URANYL ACETATE

DOT Number: NA 9180 *DOT Hazard Class:* Radioactive Material *DOT Guide Number:* 62
Synonyms: uranium acetate, uranyl acetate dihydrate
STCC Number: 4927455 *Reportable Qty:* 100/45.4
Mfg Name: J.T. Baker Chemical *Phone No:* 1-201-859-2151

Physical Description:

Physical Form: Solid *Color:* Yellow *Odor:* Vinegar
Other Information: Radioactive material emits certain rays that are hazardous. These can only be detected with special equipment.

Chemical Properties:

Specific Gravity: 2.89 *Vapor Density:* n/a *Boiling Point: Decomposes*
Melting Point: n/a *Vapor Pressure:* n/a *Solubility in water:* Yes
Other Information: Chemical name uranyl acetate is used in dry copying ink and as a activator in bacterial oxidation process.

Health Hazards:

Inhalation Hazard: Radioactive poisoning.
Ingestion Hazard: Radioactive poisoning.
Absorption Hazard: Radioactive poisoning.
Hazards to Wildlife: Radioactive poisoning.
Decontamination Procedures: None available!!
First Aid Procedures: Remove victim to fresh air, call emergency medical care. If not breathing give CPR. If breathing is difficult administer oxygen. Treat for shock.

Fire Hazards:

Flashpoint: n/a *Ignition temperature:* n/a
Flammable Explosive High Range: None *Low Range:* None
Toxic Products of Combustion: Radiation fallout!!
Other Hazards: High emittance of radioactive energy!!
Possible extinguishing agents: Apply water from as far a distance as possible. extinguish fire using suitable agent for type of surrounding fire.

Reactivity Hazards:

Reactive With: Water *Other Reactions:* None

Corrosivity Hazards:

Corrosive With: n/a *Neutralizing Agent:* n/a

Radioactivity Hazards:

Radiation Emitted: n/a *Other Hazards:* In case of trouble with shipment of these containers, all unauthorized personnel should be kept as far away as possible until qualified personnel and equipment can be obtained.

Recommended Protection for Response Personnel:

Avoid breathing vapors, keep upwind. Structural protective clothing provides limited protection. Wash away any material which may have come into contact with the body with copious amounts of soap and water. Consider appropriate evacuation.

URANYL NITRATE

DOT Number: UN 2981 *DOT Hazard Class:* Radioactive Material *DOT Guide Number:* 64
Synonyms: uranium nitrate
STCC Number: 4926320 *Reportable Qty:* 100/45.4
Mfg Name: J.T. Baker Chemical *Phone No:* 1-201-859-2151

Physical Description:

Physical Form: Solid *Color:* Light yellow *Odor:* Odorless
Other Information: Radioactive material emits certain rays that are hazardous. these can only be detected with special equipment.

Chemical Properties:

Specific Gravity: 2.81 *Vapor Density:* n/a *Boiling Point: Decomposes*
Melting Point: n/a *Vapor Pressure:* n/a *Solubility in water:* Yes
Other Information: Used to produce ceramic glazes.

Health Hazards:

Inhalation Hazard: Radioactive poisoning.
Ingestion Hazard: Radioactive poisoning.
Absorption Hazard: Radioactive poisoning.
Hazards to Wildlife: Radioactive poisoning.
Decontamination Procedures: None available!!
First Aid Procedures: Remove victim to fresh air, call emergency medical care. If not breathing give CPR. If breathing is difficult administer oxygen. Treat for shock.

Fire Hazards:

Flashpoint: n/a *Ignition temperature:* n/a
Flammable Explosive High Range: None *Low Range:* None
Toxic Products of Combustion: Toxic oxides of nitrogen are produced.
Other Hazards: Unirradiated uranium is only mildly radioactive, and radiation hazard in transportation is minimal.
Possible extinguishing agents: Apply water from as far a distance as possible. extinguish fire using suitable agent for type of surrounding fire.

Reactivity Hazards:

Reactive With: Water forming a weak solution of nitric acid. *Other Reactions:* Easily oxidizable substances.

Corrosivity Hazards:

Corrosive With: Water solution is corrosive with metals. *Neutralizing Agent:* n/a

Radioactivity Hazards:

Radiation Emitted: n/a *Other Hazards:* In case of trouble with shipment of these containers, all unauthorized personnel should be kept as far away as possible until qualified personnel and equipment can be obtained.

Recommended Protection for Response Personnel:

Avoid breathing vapors, keep upwind. Wear a sealed chemical suit (polycarbonate, butyl rubber). Wash away any material which may have come into contact with the body with copious amounts of soap and water. Consider appropriate evacuation.

URANYL SULFATE

DOT Number: n/a *DOT Hazard Class:* Radioactive Material *DOT Guide Number:* n/a
Synonyms: uranium sulfate, uranium sulfate trihydrate, uranyl sulfate trihydrate
STCC Number: n/a *Reportable Qty:* n/a
Mfg Name: Gallard-Schlesinger *Phone No:* 1-516-333-5600

Physical Description:

Physical Form: Solid *Color:* Yellow *Odor:* Odorless
Other Information: n/a

Chemical Properties:

Specific Gravity: 3.28 *Vapor Density:* n/a *Boiling Point: Decomposes*
Melting Point: n/a *Vapor Pressure:* n/a *Solubility in water:* Yes
Other Information: Sinks and mixes with water.

Health Hazards:

Inhalation Hazard: Coughing or difficulty in breathing.
Ingestion Hazard: Harmful if swallowed.
Absorption Hazard: Irritating to the skin and eyes.
Hazards to Wildlife: n/a
Decontamination Procedures: No procedures available!!
First Aid Procedures: Remove victim to fresh air, call emergency medical care. If not breathing give CPR. If breathing is difficult administer oxygen. Treat for shock.

Fire Hazards:

Flashpoint: n/a *Ignition temperature:* n/a
Flammable Explosive High Range: n/a *Low Range:* n/a
Toxic Products of Combustion: n/a
Other Hazards: n/a
Possible extinguishing agents: Apply water from as far a distance as possible. extinguish fire using suitable agent for the type of surrounding fire.

Reactivity Hazards:

Reactive With: Itself!! *Other Reactions:* n/a

Corrosivity Hazards:

Corrosive With: Metal and tissue *Neutralizing Agent:* Crushed limestone, soda ash, or lime.

Radioactivity Hazards:

Radiation Emitted: Alpha, beta, and gamma!
Other Hazards: It is shipped in special steel containers. All unauthorized personnel should be kept as far away as possible until qualified personnel and equipment can be obtained.

Recommended Protection for Response Personnel:

Avoid breathing vapors or dust, keep upwind. Wear a sealed chemical suit (polycarbonate, butyl rubber). Wash away any material which may have come into contact with the body with copious amounts of soap and water. If the fire becomes uncontrollable, consider appropriate evacuation.

URANYL NITRATE

DOT Number: UN 2981 *DOT Hazard Class:* Radioactive Material *DOT Guide Number:* 64
Synonyms: uranium nitrate
STCC Number: 4926320 *Reportable Qty:* 100/45.4
Mfg Name: J.T. Baker Chemical *Phone No:* 1-201-859-2151

Physical Description:

Physical Form: Solid *Color:* Light yellow *Odor:* Odorless
Other Information: Radioactive material emits certain rays that are hazardous. these can only be detected with special equipment.

Chemical Properties:

Specific Gravity: 2.81 *Vapor Density:* n/a *Boiling Point: Decomposes*
Melting Point: n/a *Vapor Pressure:* n/a *Solubility in water:* Yes
Other Information: Used to produce ceramic glazes.

Health Hazards:

Inhalation Hazard: Radioactive poisoning.
Ingestion Hazard: Radioactive poisoning.
Absorption Hazard: Radioactive poisoning.
Hazards to Wildlife: Radioactive poisoning.
Decontamination Procedures: None available!!
First Aid Procedures: Remove victim to fresh air, call emergency medical care. If not breathing give CPR. If breathing is difficult administer oxygen. Treat for shock.

Fire Hazards:

Flashpoint: n/a *Ignition temperature:* n/a
Flammable Explosive High Range: None *Low Range:* None
Toxic Products of Combustion: Toxic oxides of nitrogen are produced.
Other Hazards: Unirradiated uranium is only mildly radioactive, and radiation hazard in transportation is minimal.
Possible extinguishing agents: Apply water from as far a distance as possible. extinguish fire using suitable agent for type of surrounding fire.

Reactivity Hazards:

Reactive With: Water forming a weak solution of nitric acid. *Other Reactions:* Easily oxidizable substances.

Corrosivity Hazards:

Corrosive With: Water solution is corrosive with metals. *Neutralizing Agent:* n/a

Radioactivity Hazards:

Radiation Emitted: n/a *Other Hazards:* In case of trouble with shipment of these containers, all unauthorized personnel should be kept as far away as possible until qualified personnel and equipment can be obtained.

Recommended Protection for Response Personnel:

Avoid breathing vapors, keep upwind. Wear a sealed chemical suit (polycarbonate, butyl rubber). Wash away any material which may have come into contact with the body with copious amounts of soap and water. Consider appropriate evacuation.

URANYL SULFATE

DOT Number: n/a *DOT Hazard Class:* Radioactive Material *DOT Guide Number:* n/a
Synonyms: uranium sulfate, uranium sulfate trihydrate, uranyl sulfate trihydrate
STCC Number: n/a *Reportable Qty:* n/a
Mfg Name: Gallard-Schlesinger *Phone No:* 1-516-333-5600

Physical Description:

Physical Form: Solid *Color:* Yellow *Odor:* Odorless
Other Information: n/a

Chemical Properties:

Specific Gravity: 3.28 *Vapor Density:* n/a *Boiling Point: Decomposes*
Melting Point: n/a *Vapor Pressure:* n/a *Solubility in water:* Yes
Other Information: Sinks and mixes with water.

Health Hazards:

Inhalation Hazard: Coughing or difficulty in breathing.
Ingestion Hazard: Harmful if swallowed.
Absorption Hazard: Irritating to the skin and eyes.
Hazards to Wildlife: n/a
Decontamination Procedures: No procedures available!!
First Aid Procedures: Remove victim to fresh air, call emergency medical care. If not breathing give CPR. If breathing is difficult administer oxygen. Treat for shock.

Fire Hazards:

Flashpoint: n/a *Ignition temperature:* n/a
Flammable Explosive High Range: n/a *Low Range:* n/a
Toxic Products of Combustion: n/a
Other Hazards: n/a
Possible extinguishing agents: Apply water from as far a distance as possible. extinguish fire using suitable agent for the type of surrounding fire.

Reactivity Hazards:

Reactive With: Itself!! *Other Reactions:* n/a

Corrosivity Hazards:

Corrosive With: Metal and tissue *Neutralizing Agent:* Crushed limestone, soda ash, or lime.

Radioactivity Hazards:

Radiation Emitted: Alpha, beta, and gamma!
Other Hazards: It is shipped in special steel containers. All unauthorized personnel should be kept as far away as possible until qualified personnel and equipment can be obtained.

Recommended Protection for Response Personnel:

Avoid breathing vapors or dust, keep upwind. Wear a sealed chemical suit (polycarbonate, butyl rubber). Wash away any material which may have come into contact with the body with copious amounts of soap and water. If the fire becomes uncontrollable, consider appropriate evacuation.

UREA

DOT Number: n/a *DOT Hazard Class:* n/a *DOT Guide Number:* n/a
Synonyms: carbamide, carbonyldiamide
STCC Number: n/a *Reportable Qty:* n/a
Mfg Name: Allied Corp. *Phone No:* 1-201-455-2000

Physical Description:

Physical Form: Solid *Color:* White *Odor:* Odorless
Other Information: n/a

Chemical Properties:

Specific Gravity: 1.34 *Vapor Density:* 1 *Boiling Point: Decomposes*
Melting Point: 271° F(132.7° C) *Vapor Pressure:* n/a *Solubility in water:* Yes
Other Information: Sinks and mixes with water.

Health Hazards:

Inhalation Hazard: Not harmful.
Ingestion Hazard: Not harmful.
Absorption Hazard: Not harmful.
Hazards to Wildlife: n/a
Decontamination Procedures: Wash away any material with copious amounts of soap and water.
First Aid Procedures: Remove victim to fresh air, call emergency medical care. If not breathing give CPR. If breathing is difficult administer oxygen. Treat for shock.

Fire Hazards:

Flashpoint: n/a *Ignition temperature:* n/a
Flammable Explosive High Range: n/a *Low Range:* n/a
Toxic Products of Combustion: n/a
Other Hazards: Melts and decomposes generating ammonia.
Possible *Possible extinguishing agents:* Water.

Reactivity Hazards:

Reactive With: n/a *Other Reactions:* n/a

Corrosivity Hazards:

Corrosive With: n/a *Neutralizing Agent:* n/a

Radioactivity Hazards:

Radiation Emitted: n/a *Other Hazards:* n/a

Recommended Protection for Response Personnel:

Avoid breathing vapors, keep upwind. Structural protective clothing provides limited protection. Wash away any material which may have come into contact with the body with copious amounts of soap and water. Consider appropriate evacuation.

UREA PEROXIDE

DOT Number: NA 1511 *DOT Hazard Class:* Organic peroxide *DOT Guide Number:* 35
Synonyms: percarbamide, urea; urea hydrogen peroxide
STCC Number: 4919595 *Reportable Qty:* n/a
Mfg Name: Gallard-Schlesinger *Phone No:* 1-516-333-5600

Physical Description:

Physical Form: Solid *Color:* White *Odor:* Odorless
Other Information: n/a

Chemical Properties:

Specific Gravity: .8 *Vapor Density:* 3.2 *Boiling Point: Decomposes*
Melting Point: n/a *Vapor Pressure:* n/a *Solubility in water:* Yes
Other Information: Certain solvents, e.g. esters, acetone, can leach hydrogen peroxide from urea peroxide and form exceedingly flammable solutions.

Health Hazards:

Inhalation Hazard: n/a
Ingestion Hazard: n/a
Absorption Hazard: Irritating to the skin and eyes.
Hazards to Wildlife: n/a
Decontamination Procedures: Wash away any material with copious amounts of soap and water.
First Aid Procedures: Remove victim to fresh air, call emergency medical care. If not breathing give CPR. If breathing is difficult administer oxygen. Treat for shock.

Fire Hazards:

Flashpoint: n/a *Ignition temperature:* 680° F(360° C)
Flammable Explosive High Range: n/a *Low Range:* n/a
Toxic Products of Combustion: Irritating ammonia gas may be formed in a fire.
Other Hazards: Melts and decomposes giving off oxygen and ammonia increasing severity of fire. Containers may explode.
Possible extinguishing agents: Inert powders.

Reactivity Hazards:

Reactive With: Water forming hydrogen peroxide. *Other Reactions:* Dust and rubbish at 122° F(50° C)

Corrosivity Hazards:

Corrosive With: n/a *Neutralizing Agent:* n/a

Radioactivity Hazards:

Radiation Emitted: n/a *Other Hazards:* n/a

Recommended Protection for Response Personnel:

Avoid breathing vapors, keep upwind. Structural protective clothing provides limited protection. Wash away any material which may have come into contact with the body with copious amounts of soap and water. If the material is involved in a fire, consider appropriate evacuation.

VALERALDEHYDE

DOT Number: UN 2058 *DOT Hazard Class:* Flammable liquid *DOT Guide Number:* 26
Synonyms: amyl aldehyde, pentanal, valeral, valeric aldehyde,
STCC Number: n/a *Reportable Qty:* n/a
Mfg Name: Union Carbide *Phone No:* 1-203-794-2000

Physical Description:

Physical Form: Liquid *Color:* Colorless *Odor:* Fruity
Other Information: n/a

Chemical Properties:

Specific Gravity: .81 *Vapor Density:* 3 *Boiling Point: 217° F(102.7° C)*
Melting Point: 132° F(55.5° C) *Vapor Pressure:* n/a *Solubility in water:* n/a
Other Information: Floats on water. Flammable, irritating vapor is produced.

Health Hazards:

Inhalation Hazard: n/a
Ingestion Hazard: n/a
Absorption Hazard: Irritating to the skin and eyes.
Hazards to Wildlife: n/a
Decontamination Procedures: Wash away any material with copious amounts of soap and water.
First Aid Procedures: Remove victim to fresh air, call emergency medical care. If not breathing give CPR. If breathing is difficult administer oxygen. Treat for shock.

Fire Hazards:

Flashpoint: 54° F(12.2° C) *Ignition temperature:* n/a
Flammable Explosive High Range: n/a *Low Range:* n/a
Toxic Products of Combustion: n/a
Other Hazards: Flashback along vapor trail may occur. Vapors may explode if ignited in an enclosed area.
Possible extinguishing agents: Foam, dry chemical, or carbon dioxide. Water may be ineffective on fires. Cool all exposed containers with water. Apply water from as far a distance as possible.

Reactivity Hazards:

Reactive With: n/a *Other Reactions:* n/a

Corrosivity Hazards:

Corrosive With: n/a *Neutralizing Agent:* n/a

Radioactivity Hazards:

Radiation Emitted: n/a *Other Hazards:* n/a

Recommended Protection for Response Personnel:

Avoid breathing vapors, keep upwind. Structural protective clothing provides limited protection. Wash away any material which may have come into contact with the body with copious amounts of soap and water. Consider appropriate evacuation.

VANADIUM OXYTRICHLORIDE

DOT Number: UN 2443 *DOT Hazard Class:* Corrosive *DOT Guide Number:* 39
Synonyms: vanadyl chloride, vanadyl trichloride
STCC Number: 4932388 *Reportable Qty:* n/a
Mfg Name: Stauffer Chemical *Phone No:* 1-203-222-3000

Physical Description:

Physical Form: Liquid *Color:* Lemon yellow *Odor:* Sharp, unpleasant
Other Information: n/a

Chemical Properties:

Specific Gravity: 1.83 *Vapor Density:* 6.5 *Boiling Point: 259° F(126.1° C)*
Melting Point: n/a *Vapor Pressure:* n/a *Solubility in water:* Yes
Other Information: Sinks and mixes violently in water.

Health Hazards:

Inhalation Hazard: Coughing, difficulty in breathing.
Ingestion Hazard: Poisonous if swallowed.
Absorption Hazard: Poisonous if the skin is exposed.
Hazards to Wildlife: n/a
Decontamination Procedures: Wash away any material with copious amounts of soap and water.
First Aid Procedures: Remove victim to fresh air, call emergency medical care. If not breathing give CPR. If breathing is difficult administer oxygen. Treat for shock.

Fire Hazards:

Flashpoint: n/a *Ignition temperature:* n/a
Flammable Explosive High Range: n/a *Low Range:* n/a
Toxic Products of Combustion: Irritating fumes of hydrogen chloride may be formed in a fire.
Other Hazards: n/a
Possible extinguishing agents: Use dry chemical, dry sand, carbon dioxide. Do not use water.

Reactivity Hazards:

Reactive With: Water forming hydrochloric acid. *Other Reactions:* n/a

Corrosivity Hazards:

Corrosive With: Metals and tissue. *Neutralizing Agent:* Crushed limestone, soda ash, or lime.

Radioactivity Hazards:

Radiation Emitted: n/a *Other Hazards:* n/a

Recommended Protection for Response Personnel:

Avoid breathing vapors, keep upwind. Wear a sealed chemical suit (butyl rubber, viton, PVC). Wash away any material which may have come into contact with the body with copious amounts of soap and water. Consider appropriate evacuation.

VANADIUM PENTOXIDE

DOT Number: NA 2862 *DOT Hazard Class:* ORM-E *DOT Guide Number:* 55
Synonyms: vanadic anhydride, vanadium pentaoxide
STCC Number: 4963385 *Reportable Qty:* 1000/454
Mfg Name: Foote Mineral Co. *Phone No:* 1-215-363-6500

Physical Description:

Physical Form: Solid *Color:* Yellowish brown *Odor:* Odorless
Other Information: It is used as a catalyst in chemical processes, a pigment in ceramics, in the manufacture of other vanadium compounds, and in many other uses.

Chemical Properties:

Specific Gravity: 3.36 *Vapor Density:* 3.6 *Boiling Point: 3182° F(1750° C)*
Melting Point: 1274° F(690° C) *Vapor Pressure:* 0 mm Hg at 68° F(20° C)
Solubility in water: Slightly *Degree of Solubility:* .1%
Other Information: Sinks and mixes with water.

Health Hazards:

Inhalation Hazard: Coughing, difficulty in breathing.
Ingestion Hazard: Causes nausea.
Absorption Hazard: Irritating to the skin and eyes.
Hazards to Wildlife: Dangerous to aquatic life.
Decontamination Procedures: Wash away any material with copious amounts of soap and water.
First Aid Procedures: Remove victim to fresh air, call emergency medical care. If not breathing give CPR. If breathing is difficult administer oxygen. Treat for shock.

Fire Hazards:

Flashpoint: n/a *Ignition temperature:* n/a
Flammable Explosive High Range: n/a *Low Range:* n/a
Toxic Products of Combustion: n/a
Other Hazards: Will increase the intensity of a fire.
Possible extinguishing agents: Extinguish fire using suitable agent for the type of surrounding fire.

Reactivity Hazards:

Reactive With: n/a *Other Reactions:* n/a

Corrosivity Hazards:

Corrosive With: n/a *Neutralizing Agent:* n/a

Radioactivity Hazards:

Radiation Emitted: n/a *Other Hazards:* n/a

Recommended Protection for Response Personnel:

Avoid breathing vapors, keep upwind. Wear a sealed chemical suit (polycarbonate, butyl rubber). Wash away any material which may have come into contact with the body with copious amounts of soap and water. Consider appropriate evacuation.

VANADYL SULFATE

DOT Number: NA 2931 *DOT Hazard Class:* ORM-E *DOT Guide Number:* 55
Synonyms: vanadium oxysulfate, vanadyl sulfate dihydrate
STCC Number: 4963384 *Reportable Qty:* 1000/454
Mfg Name: Gallard-Schlesinger *Phone No:* 1-516-333-5600

Physical Description:

Physical Form: Solid *Color:* Pale blue *Odor:* Odorless
Other Information: It is used as a pigment in glasses or ceramics and in chemical processing.

Chemical Properties:

Specific Gravity: 2.56 *Vapor Density:* 5.6 *Boiling Point: Decomposes*
Melting Point: n/a *Vapor Pressure:* n/a *Solubility in water:* Yes
Other Information: Sinks and mixes with water.

Health Hazards:

Inhalation Hazard: Coughing, difficulty in breathing.
Ingestion Hazard: Causes nausea.
Absorption Hazard: Irritating to the skin and eyes.
Hazards to Wildlife: Dangerous to aquatic life.
Decontamination Procedures: Wash away any material with copious amounts of soap and water.
First Aid Procedures: Remove victim to fresh air, call emergency medical care. If not breathing give CPR. If breathing is difficult administer oxygen. Treat for shock.

Fire Hazards:

Flashpoint: n/a *Ignition temperature:* n/a
Flammable Explosive High Range: n/a *Low Range:* n/a
Toxic Products of Combustion: n/a
Other Hazards: n/a
Possible extinguishing agents: Extinguish fire using suitable agent for the type of surrounding fire.

Reactivity Hazards:

Reactive With: n/a *Other Reactions:* n/a

Corrosivity Hazards:

Corrosive With: n/a *Neutralizing Agent:* n/a

Radioactivity Hazards:

Radiation Emitted: n/a *Other Hazards:* n/a

Recommended Protection for Response Personnel:

Avoid breathing vapors, keep upwind. Wear a sealed chemical suit (polycarbonate, butyl rubber). Wash away any material which may have come into contact with the body with copious amounts of soap and water. Consider appropriate evacuation.

VINYL ACETATE

DOT Number: UN 1301 *DOT Hazard Class:* Flammable liquid *DOT Guide Number:* 26
Synonyms: VAM, vinyl A monomer, VyAc
STCC Number: 4907270 *Reportable Qty:* 5000/2270
Mfg Name: Celanese Chemical *Phone No:* 1-214-689-4818

Physical Description:

Physical Form: Liquid *Color:* Colorless *Odor:* Fruity
Other Information: It is used to make adhesives, paints, and plastics.

Chemical Properties:

Specific Gravity: .9 *Vapor Density:* 3 *Boiling Point: 161° F(71.6° C)*
Melting Point: n/a *Vapor Pressure:* 100 mm Hg at 71° F(21.6° C) *Solubility in water:* Slight
Other Information: Lighter than and slightly soluble in water. Vapors are heavier than air.

Health Hazards:

Inhalation Hazard: Dizziness, difficulty in breathing.
Ingestion Hazard: Harmful if swallowed.
Absorption Hazard: Irritating to the skin and eyes.
Hazards to Wildlife: Dangerous to aquatic life.
Decontamination Procedures: Wash away any material with copious amounts of soap and water.
First Aid Procedures: Remove victim to fresh air, call emergency medical care. If not breathing give CPR. If breathing is difficult administer oxygen. Treat for shock.

Fire Hazards:

Flashpoint: 18° F(−7.7° C) *Ignition temperature:* 756° F(402.2° C)
Flammable Explosive High Range: 13.4 *Low Range:* 2.6
Toxic Products of Combustion: n/a
Other Hazards: Containers may explode in fire. Flashback along vapor trail may occur. Vapors may explode if ignited in an enclosed area.
Possible extinguishing agents: Use alcohol foam, dry chemical or carbon dioxide.

Reactivity Hazards:

Reactive With: Heat by polymerization. *Other Reactions:* n/a

Corrosivity Hazards:

Corrosive With: n/a *Neutralizing Agent:* n/a

Radioactivity Hazards:

Radiation Emitted: n/a *Other Hazards:* n/a

Recommended Protection for Response Personnel:

Avoid breathing vapors, keep upwind. Structural protective clothing provides limited protection. Wash away any material which may have come into contact with the body with copious amounts of soap and water. If the fire becomes uncontrollable, consider appropriate evacuation. If polymerization takes place, the container is subject to violent rupture.

VINYL CHLORIDE

DOT Number: UN 1086 *DOT Hazard Class:* Flammable gas *DOT Guide Number:* 17
Synonyms: VCL, VCM, vinyl C monomer
STCC Number: 4905792 *Reportable Qty:* 1/.454
Mfg Name: Dow Chemical *Phone No:* 1-517-636-4400

Physical Description:

Physical Form: Gas *Color:* Colorless *Odor:* Sweet
Other Information: It is used to make adhesives, plastics, and other chemicals.

Chemical Properties:

Specific Gravity: .91 *Vapor Density:* 2.2 *Boiling Point: 7° F(−13.8° C)*
Melting Point: −245° F(−153.8° C) *Vapor Pressure:* 2580 mm Hg at 68° F(20° C)
Solubility in water: No
Other Information: It is shipped as a liquefied gas under its own vapor pressure.

Health Hazards:

Inhalation Hazard: Dizziness, difficulty in breathing.
Ingestion Hazard: n/a
Absorption Hazard: Will cause frostbite.
Hazards to Wildlife: n/a
Decontamination Procedures: Wash away any material with copious amounts of soap and water.
First Aid Procedures: Remove victim to fresh air, call emergency medical care. If not breathing give CPR. If breathing is difficult administer oxygen. Treat for shock.

Fire Hazards:

Flashpoint: −110° F(−78.8° C) *Ignition temperature:* 882° F(472.2° C)
Flammable Explosive High Range: 33 *Low Range:* 3.6
Toxic Products of Combustion: Hydrogen chloride, phosgene, and carbon dioxide are formed in a fire.
Other Hazards: Flashback along vapor trail may occur. Vapors may explode if ignited in an enclosed area. Under fire conditions, cylinders or tank cars may rupture, rocket. (BLEVE.)
Possible extinguishing agents: Use alcohol foam, dry chemical, or carbon dioxide.

Reactivity Hazards:

Reactive With: n/a *Other Reactions:* n/a

Corrosivity Hazards:

Corrosive With: n/a *Neutralizing Agent:* n/a

Radioactivity Hazards:

Radiation Emitted: n/a *Other Hazards:* n/a

Recommended Protection for Response Personnel:

Avoid breathing vapors, keep upwind. Wear a sealed chemical suit (viton). Wash away any material which may have come into contact with the body with copious amounts of soap and water. If the fire becomes uncontrollable, Consider appropriate evacuation. (BLEVE.)

VINYL ETHYL ETHER

DOT Number: UN 1302 *DOT Hazard Class:* Flammable liquid *DOT Guide Number:* 26
Synonyms: ethyl vinyl ester, Vinamar
STCC Number: 4907275 *Reportable Qty:* n/a
Mfg Name: Union Carbide *Phone No:* 1-203-794-2000

Physical Description:

Physical Form: Liquid *Color:* Colorless *Odor:* Disagreeable
Other Information: n/a

Chemical Properties:

Specific Gravity: .8 *Vapor Density:* 2.5 *Boiling Point: 96° F(35.5° C)*
Melting Point: n/a *Vapor Pressure:* 428 mm Hg at 68° F(20° C)
Solubility in water: No *Degree of Solubility:* Slight
Other Information: Lighter than and slightly soluble in water. Vapors are heavier than air.

Health Hazards:

Inhalation Hazard: Harmful if inhaled.
Ingestion Hazard: Harmful if swallowed.
Absorption Hazard: Irritating to the skin and eyes.
Hazards to Wildlife: n/a
Decontamination Procedures: Wash away any material with copious amounts of soap and water.
First Aid Procedures: Remove victim to fresh air, call emergency medical care. If not breathing give CPR. If breathing is difficult administer oxygen. Treat for shock.

Fire Hazards:

Flashpoint: 50° F(10° C) *Ignition temperature:* 395° F(201.6° C)
Flammable Explosive High Range: 28 *Low Range:* 1.7
Toxic Products of Combustion: n/a
Other Hazards: Flashback along vapor trail may occur. Vapors may explode if ignited in an enclosed area. It is subject to polymerization. If polymerization occurs, containers may violently rupture.
Possible extinguishing agents: Use alcohol foam, dry chemical, or carbon dioxide.

Reactivity Hazards:

Reactive With: n/a *Other Reactions:* n/a

Corrosivity Hazards:

Corrosive With: n/a *Neutralizing Agent:* n/a

Radioactivity Hazards:

Radiation Emitted: n/a *Other Hazards:* n/a

Recommended Protection for Response Personnel:

Avoid breathing vapors, keep upwind. Wear appropriate chemical clothing, Wash away any material which may have come into contact with the body with copious amounts of soap and water. If fire becomes uncontrollable, consider appropriate evacuation. (BLEVE.)

VINYL FLUORIDE

DOT Number: UN 1860 *DOT Hazard Class:* Flammable gas *DOT Guide Number:* 17
Synonyms: fluoroethylene
STCC Number: 4905793 *Reportable Qty:* n/a
Mfg Name: E.I. Du Pont *Phone No:* 1-800-441-3637

Physical Description:

Physical Form: Gas *Color:* Colorless *Odor:* Faint
Other Information: It is shipped as a liquefied gas under its own vapor pressure.

Chemical Properties:

Specific Gravity: .7 *Vapor Density:* 1.6 *Boiling Point: −97.5° F(−71.9° C)*
Melting Point: n/a *Vapor Pressure:* n/a *Solubility in water:* Slight
Other Information: Vapors are heavier than air.

Health Hazards:

Inhalation Hazard: Headache, dizziness.
Ingestion Hazard: n/a
Absorption Hazard: Will cause frostbite.
Hazards to Wildlife: n/a
Decontamination Procedures: Wash away any material with copious amounts of soap and water.
First Aid Procedures: Remove victim to fresh air, call emergency medical care. If not breathing give CPR. If breathing is difficult administer oxygen. Treat for shock.

Fire Hazards:

Flashpoint: n/a *Ignition temperature:* 725° F(385° C)
Flammable Explosive High Range: 21.7 *Low Range:* 2.6
Toxic Products of Combustion: Toxic hydrogen fluoride gases are produced in a fire.
Other Hazards: Flashback along vapor trail may occur. Vapors may explode if ignited in an enclosed area. Containers may explode in a fire. (BLEVE.)
Possible extinguishing agents: Apply water from as far a distance as possible.

Reactivity Hazards:

Reactive With: Heat to cause polymerization. *Other Reactions:* n/a

Corrosivity Hazards:

Corrosive With: n/a *Neutralizing Agent:* n/a

Radioactivity Hazards:

Radiation Emitted: n/a *Other Hazards:* n/a

Recommended Protection for Response Personnel:

Avoid breathing vapors, keep upwind. Wear appropriate chemical clothing. Wash away any material which may come into contact with the body with copious amounts of soap and water. If the fire becomes uncontrollable, or if container is exposed to direct flame, consider appropriate evacuation. (BLEVE.)

VINYLIDENE CHLORIDE

DOT Number: UN 1303 *DOT Hazard Class:* Flammable liquid *DOT Guide Number:* 26
Synonyms: 1,1-dichloroethylene
STCC Number: 4907280 *Reportable Qty:* 5000/2270
Mfg Name: Dow Chemical *Phone No:* 1-517-636-4400

Physical Description:

Physical Form: Liquid *Color:* Colorless *Odor:* Sweet
Other Information: n/a

Chemical Properties:

Specific Gravity: 1.2 *Vapor Density:* 3.4 *Boiling Point: 89° F(31.6° C)*
Melting Point: n/a *Vapor Pressure:* n/a *Solubility in water:* No
Other Information: Heavier than and insoluble in water. Vapors are heavier than air. Weighs 10.1 lbs/4.5 kg per gallon/3.8 l.

Health Hazards:

Inhalation Hazard: Dizziness, difficulty in breathing.
Ingestion Hazard: Harmful if swallowed.
Absorption Hazard: Will burn the skin and the eyes.
Hazards to Wildlife: n/a
Decontamination Procedures: Wash away any material with copious amounts of soap and water.
First Aid Procedures: Remove victim to fresh air, call emergency medical care. If not breathing give CPR. If breathing is difficult administer oxygen. Treat for shock.

Fire Hazards:

Flashpoint: −19° F(−28.3° C) *Ignition temperature:* 1058° F(570° C)
Flammable Explosive High Range: 15.5 *Low Range:* 6.5
Toxic Products of Combustion: Toxic nitrogen chloride and phosgene gases are produced in a fire.
Other Hazards: Flashback along vapor trail may occur. Vapors may explode if ignited in an enclosed area. Containers may explode in a fire. (BLEVE.)
Possible extinguishing agents: Apply water from as far a distance as possible. Use foam, dry chemical, or carbon dioxide.

Reactivity Hazards:

Reactive With: Copper and aluminum to cause polymerization. *Other Reactions:* Sunlight, air, and heat.

Corrosivity Hazards:

Corrosive With: n/a *Neutralizing Agent:* n/a

Radioactivity Hazards:

Radiation Emitted: n/a *Other Hazards:* n/a

Recommended Protection for Response Personnel:

Avoid breathing vapors, keep upwind. Wear a sealed chemical suit (butyl rubber). Wash away any material which may have come into contact with the body with copious amounts of soap and water. If fire becomes uncontrollable, or if container is exposed to direct flame, consider appropriate evacuation. (BLEVE.)

VINYL METHYL ETHER

DOT Number: UN 1087 *DOT Hazard Class:* Flammable gas *DOT Guide Number:* 17
Synonyms: methyl vinyl ether
STCC Number: 4905795 *Reportable Qty:* n/a
Mfg Name: Union Carbide *Phone No:* 1-203-794-2000

Physical Description:

Physical Form: Gas *Color:* Colorless *Odor:* Sweet
Other Information: It is shipped as a liquefied gas under its own pressure.

Chemical Properties:

Specific Gravity: .77 *Vapor Density:* 2 *Boiling Point: 43° F(6.1° C)*
Melting Point: n/a *Vapor Pressure:* 1052 mm Hg at 68° F(20° C) *Solubility in water:* Slight
Other Information: Floats and may boil on water.

Health Hazards:

Inhalation Hazard: Dizziness, headache, loss of consciousness.
Ingestion Hazard: Harmful if swallowed.
Absorption Hazard: Will cause frostbite.
Hazards to Wildlife: n/a
Decontamination Procedures: Wash away any material with copious amounts of soap and water.
First Aid Procedures: Remove victim to fresh air, call emergency medical care. If not breathing give CPR. If breathing is difficult administer oxygen. Treat for shock.

Fire Hazards:

Flashpoint: −69° F(−56.1° C) *Ignition temperature:* 549° F(287.2° C)
Flammable Explosive High Range: 39 *Low Range:* 2.6
Toxic Products of Combustion: n/a
Other Hazards: Flashback along vapor trail may occur. Vapors may explode if ignited in an enclosed area. Containers may explode in a fire. (BLEVE.)
Possible extinguishing agents: Apply water from as far a distance as possible. Use foam, dry chemical, or carbon dioxide.

Reactivity Hazards:

Reactive With: Water to form acetaldehyde. *Other Reactions:* Acid to cause polymerization.

Corrosivity Hazards:

Corrosive With: n/a *Neutralizing Agent:* n/a

Radioactivity Hazards:

Radiation Emitted: n/a *Other Hazards:* n/a

Recommended Protection for Response Personnel:

Avoid breathing vapors, keep upwind. Wear appropriate chemical clothing. Wash away any material which may have come into contact with the body with copious amounts of soap and water. If the fire becomes uncontrollable, or if container is exposed to direct flame, consider appropriate evacuation. (BLEVE.)

VINYL TOLUENE

DOT Number: NA 2618 *DOT Hazard Class:* Combustible liquid *DOT Guide Number:* 27
Synonyms: p-methylstyrene
STCC Number: 4912275 *Reportable Qty:* n/a
Mfg Name: Dow Chemical *Phone No:* 1-517-636-4400

Physical Description:

Physical Form: Liquid *Color:* Colorless *Odor:* Unpleasant
Other Information: It is used as a solvent, and to make other chemicals.

Chemical Properties:

Specific Gravity: .9 *Vapor Density:* 4.08 *Boiling Point: 334° F(167.7° C)*
Melting Point: −106° F(−76.6° C) *Vapor Pressure:* 1.1 mm Hg at 68° F(20° C)
Solubility in water: No *Degree of Solubility:* .009%
Other Information: Lighter than and insoluble in water. Vapors are heavier than air. Weighs 7.5 lbs/3.4 kg per gallon/3.8 l.

Health Hazards:

Inhalation Hazard: n/a
Ingestion Hazard: Harmful if swallowed.
Absorption Hazard: Irritating to the skin and eyes.
Hazards to Wildlife: n/a
Decontamination Procedures: Wash away any material with copious amounts of soap and water.
First Aid Procedures: Remove victim to fresh air, call emergency medical care. If not breathing give CPR. If breathing is difficult administer oxygen. Treat for shock.

Fire Hazards:

Flashpoint: 120° F(48.8° C) *Ignition temperature:* 921° F(493.8° C)
Flammable Explosive High Range: 11 *Low Range:* .8
Toxic Products of Combustion: n/a
Other Hazards: Containers may explode in a fire.
Possible extinguishing agents: Apply water from as far a distance as possible. Use foam, dry chemical, or carbon dioxide.

Reactivity Hazards:

Reactive With: n/a *Other Reactions:* n/a

Corrosivity Hazards:

Corrosive With: n/a *Neutralizing Agent:* n/a

Radioactivity Hazards:

Radiation Emitted: n/a *Other Hazards:* n/a

Recommended Protection for Response Personnel:

Avoid breathing vapors, keep upwind. Structural protective clothing provides limited protection. Wash away any material which may have come into contact with the body with copious amounts of soap and water. If the fire becomes uncontrollable, or if the container is exposed to direct flame, consider appropriate evacuation.

VINYL TRICHLOROSILANE

DOT Number: UN 1305 *DOT Hazard Class:* Flammable liquid *DOT Guide Number:* 29
Synonyms: trichlorovinylsilane
STCC Number: 4907685 *Reportable Qty:* n/a
Mfg Name: Dow Chemical *Phone No:* 1-517-636-4400

Physical Description:

Physical Form: Liquid *Color:* Colorless to light yellow *Odor:* Sharp, choking
Other Information: n/a

Chemical Properties:

Specific Gravity: 1.3 *Vapor Density:* 5.61 *Boiling Point: 195° F(90.5° C)*
Melting Point: n/a *Vapor Pressure:* n/a *Solubility in water:* Reacts
Other Information: Heavier, decomposed by water. Vapors are heavier than air.

Health Hazards:

Inhalation Hazard: Harmful if inhaled.
Ingestion Hazard: Harmful if swallowed.
Absorption Hazard: Will burn the skin and the eyes.
Hazards to Wildlife: n/a
Decontamination Procedures: Wash away any material with copious amounts of soap and water.
First Aid Procedures: Remove victim to fresh air, call emergency medical care. If not breathing give CPR. If breathing is difficult administer oxygen. Treat for shock.

Fire Hazards:

Flashpoint: 60° F(15.5° C) *Ignition temperature:* 550° F(287.7° C)
Flammable Explosive High Range: n/a *Low Range:* 3
Toxic Products of Combustion: Toxic chlorine and phosgene gases are produced in a fire.
Other Hazards: Containers may explode in a fire. Flashback along vapor trail may occur. Vapors may explode if ignited in an enclosed area.
Possible extinguishing agents: Apply water from as far a distance as possible. Use alcohol foam, dry chemical, or carbon dioxide.

Reactivity Hazards:

Reactive With: Water to form hydrochloric acid. *Other Reactions:* n/a

Corrosivity Hazards:

Corrosive With: Metals with surface moisture to form flammable hydrogen gas. *Neutralizing Agent:* Crushed limestone, soda ash, or lime.

Radioactivity Hazards:

Radiation Emitted: n/a *Other Hazards:* n/a

Recommended Protection for Response Personnel:

Avoid breathing vapors, keep upwind. Wear a sealed chemical suit (butyl rubber). Wash away any material which may have come into contact with the body with copious amounts of soap and water. If fire becomes uncontrollable, or if container is exposed to direct flame, consider appropriate evacuation.

WAX

DOT Number: UN 1993 *DOT Hazard Class:* Combustible liquid *DOT Guide Number:* 27
Synonyms: petroleum wax
STCC Number: 4915387 *Reportable Qty:* n/a
Mfg Name: Witco Chemical Corp. *Phone No:* 1-212-605-3800

Physical Description:

Physical Form: Liquid *Color:* Yellow to white *Odor:* Waxy
Other Information: n/a

Chemical Properties:

Specific Gravity: .9 *Vapor Density:* n/a *Boiling Point: 700° F(371.1° C)*
Melting Point: n/a *Vapor Pressure:* n/a *Solubility in water:* n/a
Other Information: It is generally lighter than and partly soluble in water. Vapors are heavier than air.

Health Hazards:

Inhalation Hazard: n/a
Ingestion Hazard: n/a
Absorption Hazard: Will burn the skin and eyes.
Hazards to Wildlife: n/a
Decontamination Procedures: Wash away any material with copious amounts of soap and water.
First Aid Procedures: Remove victim to fresh air, call emergency medical care. If not breathing give CPR. If breathing is difficult administer oxygen. Treat for shock.

Fire Hazards:

Flashpoint: 390° F(198.8° C) *Ignition temperature:* 473° F(245° C)
Flammable Explosive High Range: n/a *Low Range:* n/a
Toxic Products of Combustion: n/a
Other Hazards: n/a
Possible extinguishing agents: Apply water from as far a distance as possible. Use alcohol foam, dry chemical, or carbon dioxide.

Reactivity Hazards:

Reactive With: n/a *Other Reactions:* n/a

Corrosivity Hazards:

Corrosive With: n/a *Neutralizing Agent:* n/a

Radioactivity Hazards:

Radiation Emitted: n/a *Other Hazards:* n/a

Recommended Protection for Response Personnel:

Avoid breathing vapors, keep upwind. Structural protective clothing provides limited protection. Wash away any material which may have come into contact with the body with copious amounts of soap and water. Consider appropriate evacuation.

XYLENE-m

DOT Number: UN 1307 *DOT Hazard Class:* Flammable liquid *DOT Guide Number:* 27
Synonyms: 1,3-dimethylbenzene, xylol
STCC Number: 4909350 *Reportable Qty:* 1000/454
Mfg Name: Shell Chemical *Phone No:* 1-713-241-6161

Physical Description:

Physical Form: Liquid *Color:* Colorless *Odor:* Sweet
Other Information: It is used as a solvent for paints and adhesives, and to make other chemicals.

Chemical Properties:

Specific Gravity: .9 *Vapor Density:* 3.7 *Boiling Point: 282° F(138.8° C)*
Melting Point: −54° F(−47.7° C) *Vapor Pressure:* 9 mm Hg at 68° F(20° C)
Solubility in water: No *Degree of Solubility:* .00003%
Other Information: Lighter than and insoluble in water. Vapors are heavier than air. Weighs 7.2 lbs/3.2 kg per gallon/3.8 l.

Health Hazards:

Inhalation Hazard: Headache, difficulty in breathing, loss of consciousness.
Ingestion Hazard: Nausea, vomiting, loss of consciousness.
Absorption Hazard: Irritating to the skin and eyes.
Hazards to Wildlife: Dangerous to aquatic life.
Decontamination Procedures: Wash away any material with copious amounts of soap and water.
First Aid Procedures: Remove victim to fresh air, call emergency medical care. If not breathing give CPR. If breathing is difficult administer oxygen. Treat for shock.

Fire Hazards:

Flashpoint: 81° F(27.2° C) *Ignition temperature:* 982° F(527.7° C)
Flammable Explosive High Range: 7 *Low Range:* 1.1
Toxic Products of Combustion: n/a
Other Hazards: Flashback along vapor trail may occur. Vapors may explode if ignited in an enclosed area.
Possible extinguishing agents: Apply water from as far a distance as possible. Use foam, dry chemical, or carbon dioxide.

Reactivity Hazards:

Reactive With: n/a *Other Reactions:* n/a

Corrosivity Hazards:

Corrosive With: n/a *Neutralizing Agent:* n/a

Radioactivity Hazards:

Radiation Emitted: n/a *Other Hazards:* n/a

Recommended Protection for Response Personnel:

Avoid breathing vapors, keep upwind. Structural protective clothing provides limited protection. Wash away any material which may have come into contact with the body with copious amounts of soap and water. If the fire becomes uncontrollable, or if container is exposed to direct flame, consider appropriate evacuation.

XYLENOL (liquid)

DOT Number: UN 2261 *DOT Hazard Class:* ORM-A *DOT Guide Number:* 55
Synonyms: cresylic acid, 2,6-xylenol
STCC Number: 4941193 *Reportable Qty:* 1000/454
Mfg Name: Crowley Tar Products Co. *Phone No:* 1-212-682-1200

Physical Description:

Physical Form: Liquid *Color:* Light yellowish to brown *Odor:* Sweet, tarry
Other Information: It is used in disinfectants, insecticides, and to make other chemicals.

Chemical Properties:

Specific Gravity: 1.1 *Vapor Density:* n/a *Boiling Point: 413° F(211.6° C)*
Melting Point: n/a *Vapor Pressure:* n/a *Solubility in water:* n/a
Other Information: May float or sink in water.

Health Hazards:

Inhalation Hazard: Harmful if inhaled.
Ingestion Hazard: Cause nausea and vomiting.
Absorption Hazard: Cause nausea and vomiting.
Hazards to Wildlife: Dangerous to aquatic life.
Decontamination Procedures: Wash away any material with copious amounts of soap and water.
First Aid Procedures: Remove victim to fresh air, call emergency medical care. If not breathing give CPR. If breathing is difficult administer oxygen. Treat for shock.

Fire Hazards:

Flashpoint: 186° F(85.5° C) *Ignition temperature:* 1110° F(598.8° C)
Flammable Explosive High Range: n/a *Low Range:* 1.4
Toxic Products of Combustion: Toxic vapors of unburned material may be formed in fires.
Other Hazards: n/a
Possible extinguishing agents: Apply water from as far a distance as possible. Use alcohol foam, dry chemical, or carbon dioxide.

Reactivity Hazards:

Reactive With: n/a *Other Reactions:* n/a

Corrosivity Hazards:

Corrosive With: n/a *Neutralizing Agent:* n/a

Radioactivity Hazards:

Radiation Emitted: n/a *Other Hazards:* n/a

Recommended Protection for Response Personnel:

Avoid breathing vapors, keep upwind. Wear a sealed chemical suit. (polycarbonate, butyl rubber) Wash away any material which may have come into contact with the body with copious amounts of soap and water. Consider appropriate evacuation.

XYLENOL (solid)

DOT Number: UN 2261 *DOT Hazard Class:* ORM-A *DOT Guide Number:* 55
Synonyms: cresylic acid, 2,6-xylenol
STCC Number: 4941193 *Reportable Qty:* 1000/454
Mfg Name: Crowley Tar Products Co. *Phone No:* 1-212-682-1200

Physical Description:

Physical Form: Solid *Color:* Light yellowish to brown *Odor:* Sweet, tarry
Other Information: It is used in disinfectants, insecticides, and to make other chemicals.

Chemical Properties:

Specific Gravity: 1.1 *Vapor Density:* n/a *Boiling Point: 413° F(211.6° C)*
Melting Point: n/a *Vapor Pressure:* n/a *Solubility in water:* n/a
Other Information: May float or sink in water.

Health Hazards:

Inhalation Hazard: Harmful if inhaled.
Ingestion Hazard: Cause nausea and vomiting.
Absorption Hazard: Cause nausea and vomiting.
Hazards to Wildlife: Dangerous to aquatic life.
Decontamination Procedures: Wash away any material with copious amounts of soap and water.
First Aid Procedures: Remove victim to fresh air, call emergency medical care. If not breathing give CPR. If breathing is difficult administer oxygen. Treat for shock.

Fire Hazards:

Flashpoint: 186° F(85.5° C) *Ignition temperature:* 1110° F(598.8° C)
Flammable Explosive High Range: n/a *Low Range:* 1.4
Toxic Products of Combustion: Toxic vapors of unburned material may be formed in fires.
Other Hazards: n/a
Possible extinguishing agents: Apply water from as far a distance as possible. Use alcohol foam, dry chemical, or carb-on dioxide.

Reactivity Hazards:

Reactive With: n/a *Other Reactions:* n/a

Corrosivity Hazards:

Corrosive With: n/a *Neutralizing Agent:* n/a

Radioactivity Hazards:

Radiation Emitted: n/a *Other Hazards:* n/a

Recommended Protection for Response Personnel:

Avoid breathing vapors, keep upwind. Wear a sealed chemical suit. (polycarbonate, butyl rubber) Wash away any material which may have come into contact with the body with copious amounts of soap and water. Consider ap-propriate evacuation.

ZECTRAN

DOT Number: UN 1615 *DOT Hazard Class:* Poison B *DOT Guide Number:* n/a
Synonyms: mexacarbate, zactran, zectane, zextran
STCC Number: n/a *Reportable Qty:* n/a
Mfg Name: Dow Chemical *Phone No:* 1-517-636-4400

Physical Description:

Physical Form: Solid *Color:* White to tan *Odor:* Odorless
Other Information: n/a

Chemical Properties:

Specific Gravity: n/a *Vapor Density:* n/a *Boiling Point: Decomposes*
Melting Point: 185° F(85° C) *Vapor Pressure:* n/a *Solubility in water:* n/a
Other Information: Avoid contact with solid, keep people away!!

Health Hazards:

Inhalation Hazard: Poisonous if inhaled.
Ingestion Hazard: Poisonous if swallowed.
Absorption Hazard: Poisonous to the skin and eyes.
Hazards to Wildlife: Dangerous to aquatic life.
Decontamination Procedures: Wash away any material with copious amounts of soap and water.
First Aid Procedures: Remove victim to fresh air, call emergency medical care. If not breathing give CPR. If breathing is difficult administer oxygen. Treat for shock.

Fire Hazards: Data is not available!!

Flashpoint: n/a *Ignition temperature:* n/a
Flammable Explosive High Range: n/a *Low Range:* n/a
Toxic Products of Combustion: n/a
Other Hazards: n/a
Possible extinguishing agents: Alcohol foam, dry chemical, carbon dioxide, or water.

Reactivity Hazards:

Reactive With: n/a *Other Reactions:* n/a

Corrosivity Hazards:

Corrosive With: n/a *Neutralizing Agent:* n/a

Radioactivity Hazards:

Radiation Emitted: n/a *Other Hazards:* n/a

Recommended Protection for Response Personnel:

Avoid breathing vapors, keep upwind. Wear appropriate sealed chemical suit. Wash away any material which may have come into contact with the body with copious amounts of soap and water. Consider appropriate evacuation.

ZINC ACETATE

DOT Number: NA 9153 *DOT Hazard Class:* ORM-E *DOT Guide Number:* 31
Synonyms: acetic acid, zinc salt; zinc acetate dihydrate
STCC Number: 4963387 *Reportable Qty:* 1000/454
Mfg Name: Allied Corp. *Phone No:* 1-201-455-2000

Physical Description:

Physical Form: Solid *Color:* White *Odor:* Faint vinegar
Other Information: It is used to preserve wood, to make other zinc compounds, as a food and feed additive, and for many other uses.

Chemical Properties:

Specific Gravity: 1.74 *Vapor Density:* 6.3 *Boiling Point: Decomposes*
Melting Point: n/a *Vapor Pressure:* n/a *Solubility in water:* Yes
Other Information: Sinks and mixes with water.

Health Hazards:

Inhalation Hazard: Cause coughing, difficulty in breathing.
Ingestion Hazard: Cause nausea and vomiting.
Absorption Hazard: Irritating to the skin and eyes.
Hazards to Wildlife: Dangerous to aquatic life.
Decontamination Procedures: Wash away any material with copious amounts of soap and water.
First Aid Procedures: Remove victim to fresh air, call emergency medical care. If not breathing give CPR. If breathing is difficult administer oxygen. Treat for shock.

Fire Hazards:

Flashpoint: n/a *Ignition temperature:* n/a
Flammable Explosive High Range: n/a *Low Range:* n/a
Toxic Products of Combustion: n/a
Other Hazards: n/a
Possible extinguishing agents: Extinguish fire using suitable agent for the type of surrounding fire.

Reactivity Hazards:

Reactive With: n/a *Other Reactions:* n/a

Corrosivity Hazards:

Corrosive With: n/a *Neutralizing Agent:* n/a

Radioactivity Hazards:

Radiation Emitted: n/a *Other Hazards:* n/a

Recommended Protection for Response Personnel:

Avoid breathing vapors, keep upwind. Structural protective clothing provides limited protection. Wash away any material which may have come into contact with the body with copious amounts of soap and water. Consider appropriate evacuation.

ZINC AMMONIUM CHLORIDE

DOT Number: NA 9154 *DOT Hazard Class:* ORM-E *DOT Guide Number:* 31
Synonyms: ammonium pentachlorozincate, ammonium zinc chloride
STCC Number: 4966386 *Reportable Qty:* 1000/454
Mfg Name: Chemical & Pigment Co. *Phone No:* 1-415-689-2030

Physical Description:

Physical Form: Solid *Color:* White *Odor:* Odorless
Other Information: It is used in welding, in soldering fluxes, and in galvanizing steel.

Chemical Properties:

Specific Gravity: 1.81 *Vapor Density:* 1 *Boiling Point: 644° F(340° C)*
Melting Point: n/a *Vapor Pressure:* n/a *Solubility in water:* Yes
Other Information: Sinks and mixes with water.

Health Hazards:

Inhalation Hazard: Cause coughing, difficulty in breathing.
Ingestion Hazard: Cause nausea and vomiting.
Absorption Hazard: n/a
Hazards to Wildlife: Dangerous to aquatic life.
Decontamination Procedures: Wash away any material with copious amounts of soap and water.
First Aid Procedures: Remove victim to fresh air, call emergency medical care. If not breathing give CPR. If breathing is difficult administer oxygen. Treat for shock.

Fire Hazards:

Flashpoint: n/a *Ignition temperature:* n/a
Flammable Explosive High Range: n/a *Low Range:* n/a
Toxic Products of Combustion: n/a
Other Hazards: n/a
Possible extinguishing agents: Extinguish fire using suitable agent for the type of surrounding fire.

Reactivity Hazards:

Reactive With: n/a *Other Reactions:* n/a

Corrosivity Hazards:

Corrosive With: n/a *Neutralizing Agent:* n/a

Radioactivity Hazards:

Radiation Emitted: n/a *Other Hazards:* n/a

Recommended Protection for Response Personnel:

Avoid breathing vapors, keep upwind. Structural protective clothing provides limited protection. Wash away any material which may have come into contact with the body with copious amounts of soap and water. Consider appropriate evacuation.

ZINC ARSENATE

DOT Number: UN 1712 *DOT Hazard Class:* Poison B *DOT Guide Number:* 53
Synonyms: None given
STCC Number: 4923490 *Reportable Qty:* n/a *CHEMTREC Phone No:* 1-800-424-9300

Physical Description:

Physical Form: Solid *Color:* Colorless *Odor:* Odorless
Other Information: n/a

Chemical Properties:

Specific Gravity: 3.31 *Vapor Density:* 21 *Boiling Point: Decomposes*
Melting Point: n/a *Vapor Pressure:* n/a *Solubility in water:* No
Other Information: Sinks in water

Health Hazards:

Inhalation Hazard: Poisonous if inhaled.
Ingestion Hazard: Poisonous if swallowed.
Absorption Hazard: Irritating to the skin and eyes.
Hazards to Wildlife: n/a
Decontamination Procedures: Wash away any material with copious amounts of soap and water.
First Aid Procedures: Remove victim to fresh air, call emergency medical care. If not breathing give CPR. If breathing is difficult administer oxygen. Treat for shock.

Fire Hazards:

Flashpoint: n/a *Ignition temperature:* n/a
Flammable Explosive High Range: n/a *Low Range:* n/a
Toxic Products of Combustion: n/a
Other Hazards: n/a
Possible extinguishing agents: Extinguish fire using suitable agent for the type of surrounding fire.

Reactivity Hazards:

Reactive With: n/a *Other Reactions:* n/a

Corrosivity Hazards:

Corrosive With: n/a *Neutralizing Agent:* n/a

Radioactivity Hazards:

Radiation Emitted: n/a *Other Hazards:* n/a

Recommended Protection for Response Personnel:

Avoid breathing vapors, keep upwind. Wear a sealed chemical suit. (polycarbonate, butyl rubber) Wash away any material which may have come into contact with the body with copious amounts of soap and water. Consider appropriate evacuation.

ZINC ARSENITE

DOT Number: UN 1712 *DOT Hazard Class:* Poison B *DOT Guide Number:* 53
Synonyms: None given
STCC Number: 4923492 *Reportable Qty:* n/a *CHEMTREC Phone No:* 1-800-424-9300

Physical Description:

Physical Form: Solid *Color:* Colorless *Odor:* Odorless
Other Information: n/a

Chemical Properties:

Specific Gravity: 3.31 *Vapor Density:* n/a *Boiling Point: Decomposes*
Melting Point: n/a *Vapor Pressure:* n/a *Solubility in water:* No
Other Information: Sinks in water

Health Hazards:

Inhalation Hazard: Poisonous if inhaled.
Ingestion Hazard: Poisonous if swallowed.
Absorption Hazard: Irritating to the skin and eyes.
Hazards to Wildlife: n/a
Decontamination Procedures: Wash away any material with copious amounts of soap and water.
First Aid Procedures: Remove victim to fresh air, call emergency medical care. If not breathing give CPR. If breathing is difficult administer oxygen. Treat for shock.

Fire Hazards:

Flashpoint: n/a *Ignition temperature:* n/a
Flammable Explosive High Range: n/a *Low Range:* n/a
Toxic Products of Combustion: n/a
Other Hazards: n/a
Possible extinguishing agents: Extinguish fire using suitable agent for type of surrounding fire. Use water, foam, dry chemical, or carbon dioxide.

Reactivity Hazards:

Reactive With: n/a *Other Reactions:* n/a

Corrosivity Hazards:

Corrosive With: n/a *Neutralizing Agent:* n/a

Radioactivity Hazards:

Radiation Emitted: n/a *Other Hazards:* n/a

Recommended Protection for Response Personnel:

Avoid breathing vapors, keep upwind. Wear appropriate chemical clothing, Wash away any material which may come into contact with the body with copious amounts of soap and water. Consider appropriate evacuation.

ZINC BICHROMATE

DOT Number: n/a *DOT Hazard Class:* Oxidizer *DOT Guide Number:* n/a
Synonyms: zinc dichromate
STCC Number: n/a *Reportable Qty:* n/a
Mfg Name: Atomergic Chemetals Corp. *Phone No:* 1-516-349-8800

Physical Description:

Physical Form: Solid *Color:* Reddish brown, orange-yellow *Odor:* None given
Other Information: n/a

Chemical Properties:

Specific Gravity: n/a *Vapor Density:* n/a *Boiling Point: n/a*
Melting Point: n/a *Vapor Pressure:* n/a *Solubility in water:* Yes
Other Information: Mixes with water

Health Hazards:

Inhalation Hazard: Harmful if inhaled.
Ingestion Hazard: Dizziness, nausea, convulsions, coma.
Absorption Hazard: Will burn the skin and eyes.
Hazards to Wildlife: Dangerous to aquatic life.
Decontamination Procedures: Wash away any material with copious amounts of soap and water.
First Aid Procedures: Remove victim to fresh air, call emergency medical care. If not breathing give CPR. If breathing is difficult administer oxygen. Treat for shock.

Fire Hazards:

Flashpoint: n/a *Ignition temperature:* n/a
Flammable Explosive High Range: n/a *Low Range:* n/a
Toxic Products of Combustion: n/a
Other Hazards: Will cause fire upon contact with combustibles.
Possible extinguishing agents: Flood discharge area with water. Cool exposed containers with water.

Reactivity Hazards:

Reactive With: Combustibles which may cause fire. *Other Reactions:* n/a

Corrosivity Hazards:

Corrosive With: n/a *Neutralizing Agent:* n/a

Radioactivity Hazards:

Radiation Emitted: n/a *Other Hazards:* n/a

Recommended Protection for Response Personnel:

Avoid breathing vapors, keep upwind. Wear appropriate sealed chemical suit. Wash away any material which may have come into contact with the body with copious amounts of soap and water. Consider appropriate evacuation.

ZINC BORATE

DOT Number: NA 1988 *DOT Hazard Class:* ORM-E *DOT Guide Number:* n/a
Synonyms: None given
STCC Number: 4963389 *Reportable Qty:* 1000/454
Mfg Name: A.D. MacKay Inc. *Phone No:* 1-203-655-3000

Physical Description:

Physical Form: Solid *Color:* White *Odor:* Odorless
Other Information: It is used as a fungus and mildew inhibitor, to fireproof textiles, and for many other uses.

Chemical Properties:

Specific Gravity: 2.7 *Vapor Density:* n/a *Boiling Point: Decomposes*
Melting Point: n/a *Vapor Pressure:* n/a *Solubility in water:* Slight
Other Information: Sinks in water

Health Hazards:

Inhalation Hazard: Cause coughing or difficulty in breathing.
Ingestion Hazard: Cause nausea and vomiting.
Absorption Hazard: Irritating to the skin and eyes.
Hazards to Wildlife: n/a
Decontamination Procedures: Wash away any material with copious amounts of soap and water.
First Aid Procedures: Remove victim to fresh air, call emergency medical care. If not breathing give CPR. If breathing is difficult administer oxygen. Treat for shock.

Fire Hazards:

Flashpoint: n/a *Ignition temperature:* n/a
Flammable Explosive High Range: n/a *Low Range:* n/a
Toxic Products of Combustion: n/a
Other Hazards: n/a
Possible extinguishing agents: Extinguish fire using suitable agent for the type of surrounding fire.

Reactivity Hazards:

Reactive With: n/a *Other Reactions:* n/a

Corrosivity Hazards:

Corrosive With: n/a *Neutralizing Agent:* n/a

Radioactivity Hazards:

Radiation Emitted: n/a *Other Hazards:* n/a

Recommended Protection for Response Personnel:

Avoid breathing vapors, keep upwind. Structural protective clothing provides limited protection. Wash away any material which may have come into contact with the body with copious amounts of soap and water. Consider appropriate evacuation.

ZINC BROMIDE

DOT Number: NA 9156 *DOT Hazard Class:* ORM-E *DOT Guide Number:* 60
Synonyms: None given
STCC Number: 4966780 *Reportable Qty:* 1000/454
Mfg Name: A.D. MacKay Inc *Phone No:* 1-203-655-3000

Physical Description:

Physical Form: Solid *Color:* White *Odor:* Odorless
Other Information: It is used in medicine, in photography, and for many other uses.

Chemical Properties:

Specific Gravity: 4.22 *Vapor Density:* 7.8 *Boiling Point: Decomposes*
Melting Point: n/a *Vapor Pressure:* n/a *Solubility in water:* Yes
Other Information: Sinks and mixes with water.

Health Hazards:

Inhalation Hazard: Cause coughing or difficulty in breathing.
Ingestion Hazard: Cause nausea and vomiting.
Absorption Hazard: Irritating to the skin and eyes.
Hazards to Wildlife: n/a
Decontamination Procedures: Wash away any material with copious amounts of soap and water.
First Aid Procedures: Remove victim to fresh air, call emergency medical care. If not breathing give CPR. If breathing is difficult administer oxygen. Treat for shock.

Fire Hazards:

Flashpoint: n/a *Ignition temperature:* n/a
Flammable Explosive High Range: n/a *Low Range:* n/a
Toxic Products of Combustion: n/a
Other Hazards: n/a
Possible extinguishing agents: Extinguish fire using suitable agent for the type of surrounding fire.

Reactivity Hazards:

Reactive With: n/a *Other Reactions:* n/a

Corrosivity Hazards:

Corrosive With: n/a *Neutralizing Agent:* n/a

Radioactivity Hazards:

Radiation Emitted: n/a *Other Hazards:* n/a

Recommended Protection for Response Personnel:

Avoid breathing vapors, keep upwind. Structural protective clothing provides limited protection. Wash away any material which may have come into contact with the body with copious amounts of soap and water. Consider appropriate evacuation.

ZINC CARBONATE

DOT Number: NA 9157 *DOT Hazard Class:* ORM-E *DOT Guide Number:* 31
Synonyms: calamine, smithsonite
STCC Number: 4963890 *Reportable Qty:* 1000/454
Mfg Name: J.T. Baker Chemical *Phone No:* 1-201-857-2151

Physical Description:

Physical Form: Solid *Color:* White *Odor:* Odorless
Other Information: It is used in pharmaceuticals, to make other zinc compounds, as a feed additive, and for many other uses.

Chemical Properties:

Specific Gravity: 4.4 *Vapor Density:* 4.3 *Boiling Point: n/a*
Melting Point: n/a *Vapor Pressure:* n/a *Solubility in water:* No
Other Information: Sinks in water

Health Hazards:

Inhalation Hazard: n/a
Ingestion Hazard: Cause nausea and vomiting.
Absorption Hazard: n/a
Hazards to Wildlife: Dangerous to aquatic life.
Decontamination Procedures: Wash away any material with copious amounts of soap and water.
First Aid Procedures: Remove victim to fresh air, call emergency medical care. If not breathing give CPR. If breathing is difficult administer oxygen. Treat for shock.

Fire Hazards:

Flashpoint: n/a *Ignition temperature:* n/a
Flammable Explosive High Range: n/a *Low Range:* n/a
Toxic Products of Combustion: Could decompose to liberate carbon dioxide.
Other Hazards: n/a
Possible extinguishing agents: Extinguish fire using suitable agent for the type of surrounding fire.

Reactivity Hazards:

Reactive With: n/a *Other Reactions:* n/a

Corrosivity Hazards:

Corrosive With: n/a *Neutralizing Agent:* n/a

Radioactivity Hazards:

Radiation Emitted: n/a *Other Hazards:* n/a

Recommended Protection for Response Personnel:

Avoid breathing vapors, keep upwind. Structural protective clothing provides limited protection. Wash away any material which may have come into contact with the body with copious amounts of soap and water. Consider appropriate evacuation.

ZINC CHLORIDE

DOT Number: NA 1840 *DOT Hazard Class:* ORM-E *DOT Guide Number:* 60
Synonyms: None given
STCC Number: 4966790 *Reportable Qty:* 1000/454
Mfg Name: E.I. Du Pont *Phone No:* 1-800-441-3637

Physical Description:

Physical Form: Solid *Color:* White *Odor:* Odorless
Other Information: It is used for preserving wood, in soldering fluxes, as a catalyst in chemical manufacturing, and for many other uses.

Chemical Properties:

Specific Gravity: 2.9 *Vapor Density:* 4.7 *Boiling Point: Very high*
Melting Point: n/a *Vapor Pressure:* 1 mm Hg at 802° F(427.7° C) *Solubility in water:* Yes
Other Information: Sinks and mixes with water.

Health Hazards:

Inhalation Hazard: n/a
Ingestion Hazard: Cause nausea and vomiting.
Absorption Hazard: Irritating to the skin and eyes.
Hazards to Wildlife: Dangerous to aquatic life.
Decontamination Procedures: Wash away any material with copious amounts of soap and water.
First Aid Procedures: Remove victim to fresh air, call emergency medical care. If not breathing give CPR. If breathing is difficult administer oxygen. Treat for shock.

Fire Hazards:

Flashpoint: n/a *Ignition temperature:* n/a
Flammable Explosive High Range: n/a *Low Range:* n/a
Toxic Products of Combustion: n/a
Other Hazards: n/a
Possible extinguishing agents: Extinguish fire using suitable agent for the type of surrounding fire.

Reactivity Hazards:

Reactive With: n/a *Other Reactions:* n/a

Corrosivity Hazards:

Corrosive With: n/a *Neutralizing Agent:* n/a

Radioactivity Hazards:

Radiation Emitted: n/a *Other Hazards:* n/a

Recommended Protection for Response Personnel:

Avoid breathing vapors, keep upwind. Wear a sealed chemical suit. (polycarbonate, butyl rubber, viton, pvc, chlorinated polyethylene, nitrile, neoprene) Wash away any material which may have come into contact with the body with copious amounts of soap and water. Consider appropriate evacuation.

ZINC CHROMATE

DOT Number: n/a *DOT Hazard Class:* n/a *DOT Guide Number:* n/a
Synonyms: buttercup yellow, zinc chromate(VI) hydroxide, zinc yellow
STCC Number: n/a *Reportable Qty:* n/a
Mfg Name: A.D. MacKay Inc. *Phone No:* 1-203-655-3000

Physical Description:

Physical Form: Solid *Color:* Yellow *Odor:* Odorless
Other Information: n/a

Chemical Properties:

Specific Gravity: 3.43 *Vapor Density:* n/a *Boiling Point: Decomposes*
Melting Point: n/a *Vapor Pressure:* n/a *Solubility in water:* n/a
Other Information: Sinks in water

Health Hazards:

Inhalation Hazard: Coughing or difficulty in breathing.
Ingestion Hazard: Nausea and vomiting.
Absorption Hazard: Irritating to the skin and eyes.
Hazards to Wildlife: n/a
Decontamination Procedures: Wash away any material with copious amounts of soap and water.
First Aid Procedures: Remove victim to fresh air, call emergency medical care. If not breathing give CPR. If breathing is difficult administer oxygen. Treat for shock.

Fire Hazards:

Flashpoint: n/a *Ignition temperature:* n/a
Flammable Explosive High Range: n/a *Low Range:* n/a
Toxic Products of Combustion: n/a
Other Hazards: n/a
Possible extinguishing agents: Extinguish fire using suitable agent for the type of surrounding fire.

Reactivity Hazards:

Reactive With: n/a *Other Reactions:* n/a

Corrosivity Hazards:

Corrosive With: n/a *Neutralizing Agent:* n/a

Radioactivity Hazards:

Radiation Emitted: n/a *Other Hazards:* n/a

Recommended Protection for Response Personnel:

Avoid breathing vapors, keep upwind. Wear a sealed chemical suit. (polycarbonate, butyl rubber) Wash away any material which may have come into contact with the body with copious amounts of soap and water. Consider appropriate evacuation.

ZINC CYANIDE

DOT Number: UN 1713 *DOT Hazard Class:* Poison B *DOT Guide Number:* 53
Synonyms: cyanide of zinc, zinc dicyanide
STCC Number: 4923495 *Reportable Qty:* 10/4.54
Mfg Name: Ashland Chemical *Phone No:* 1-614-889-3333

Physical Description:

Physical Form: Solid *Color:* Greyish white to white *Odor:* Odorless
Other Information: It is used in medicine, metal plating, and in chemical analysis.

Chemical Properties:

Specific Gravity: 1.85 *Vapor Density:* 4 *Boiling Point: Decomposes*
Melting Point: n/a *Vapor Pressure:* n/a *Solubility in water:* No
Other Information: Insoluble in water.

Health Hazards:

Inhalation Hazard: Poisonous if inhaled.
Ingestion Hazard: Poisonous if swallowed.
Absorption Hazard: Poisonous if absorbed.
Hazards to Wildlife: Dangerous to aquatic life.
Decontamination Procedures: Wash away any material with copious amounts of soap and water.
First Aid Procedures: Remove victim to fresh air, call emergency medical care. If not breathing give CPR. If breathing is difficult administer oxygen. Treat for shock.

Fire Hazards:

Flashpoint: n/a *Ignition temperature:* n/a
Flammable Explosive High Range: n/a *Low Range:* n/a
Toxic Products of Combustion: n/a
Other Hazards: n/a
Possible extinguishing agents: Extinguish fire using suitable agent for type of surrounding fire. Use foam, dry chemical, or carbon dioxide. Do not use water on the material itself.

Reactivity Hazards:

Reactive With: Acids or acid salts which liberates highly flammable and toxic hydrogen cyanide gas.
Other Reactions: n/a

Corrosivity Hazards:

Corrosive With: n/a *Neutralizing Agent:* n/a

Radioactivity Hazards:

Radiation Emitted: n/a *Other Hazards:* n/a

Recommended Protection for Response Personnel:

Avoid breathing vapors, keep upwind. Wear appropriate chemical clothing, Wash away any material which may come into contact with the body with copious amounts of soap and water. Consider appropriate evacuation.

ZINC DIALKYLDITHIOPHOSPHATE

DOT Number: UN 1893 *DOT Hazard Class:* Poison *DOT Guide Number:* n/a
Synonyms: zinc dihexyldithiophosphate, zinc dihexylphosprodithioate, zinc O,O-di-n-butylphosphorodithioate
STCC Number: n/a *Reportable Qty:* n/a
Mfg Name: Monsanto Chemical *Phone No:* 1-314-694-1000

Physical Description:

Physical Form: Solid or liquid *Color:* Straw yellow to green *Odor:* Sweet alcohol like
Other Information: n/a

Chemical Properties:

Specific Gravity: 1.10 *Vapor Density:* n/a *Boiling Point: Decomposes*
Melting Point: n/a *Vapor Pressure:* n/a *Solubility in water:* n/a
Other Information: Sinks in water

Health Hazards:

Inhalation Hazard: Coughing or difficulty in breathing.
Ingestion Hazard: Nausea and vomiting.
Absorption Hazard: Irritating to the skin and eyes.
Hazards to Wildlife: n/a
Decontamination Procedures: Wash away any material with copious amounts of soap and water.
First Aid Procedures: Remove victim to fresh air, call emergency medical care. If not breathing give CPR. If breathing is difficult administer oxygen. Treat for shock.

Fire Hazards:

Flashpoint: 360° F(182.2° C) *Ignition temperature:* n/a
Flammable Explosive High Range: n/a *Low Range:* n/a
Toxic Products of Combustion: Irritating oxides of sulfur and phosphorus may be formed in fires.
Other Hazards: n/a
Possible extinguishing agents: Extinguish fire using suitable agent for the type of surrounding fire.

Reactivity Hazards:

Reactive With: n/a *Other Reactions:* n/a

Corrosivity Hazards:

Corrosive With: n/a *Neutralizing Agent:* n/a

Radioactivity Hazards:

Radiation Emitted: n/a *Other Hazards:* n/a

Recommended Protection for Response Personnel:

Avoid breathing vapors, keep upwind. Structural protective clothing provides limited protection. Wash away any material which may have come into contact with the body with copious amounts of soap and water. If the fire becomes uncontrollable, consider appropriate evacuation.

ZINC FLUORIDE

DOT Number: NA 9158 *DOT Hazard Class:* ORM-E *DOT Guide Number:* 31
Synonyms: zinc difluoride
STCC Number: 4963195 *Reportable Qty:* 1000/454
Mfg Name: Shepherd Chemical Co. *Phone No:* 1-513-731-1110

Physical Description:

Physical Form: Solid *Color:* Colorless to white *Odor:* n/a
Other Information: It is used for galvanizing steel, and in the making of ceramics.

Chemical Properties:

Specific Gravity: 4.95 *Vapor Density:* 3.5 *Boiling Point: 2732° F(1500° C)*
Melting Point: n/a *Vapor Pressure:* 1 mm Hg at 1778° F(970° C) *Solubility in water:* Yes
Other Information: Sinks and mixes with water.

Health Hazards:

Inhalation Hazard: n/a
Ingestion Hazard: Harmful if swallowed.
Absorption Hazard: Irritating to the skin and eyes.
Hazards to Wildlife: Dangerous to aquatic life.
Decontamination Procedures: Wash away any material with copious amounts of soap and water.
First Aid Procedures: Remove victim to fresh air, call emergency medical care. If not breathing give CPR. If breathing is difficult administer oxygen. Treat for shock.

Fire Hazards:

Flashpoint: n/a *Ignition temperature:* n/a
Flammable Explosive High Range: n/a *Low Range:* n/a
Toxic Products of Combustion: n/a
Other Hazards: n/a
Possible extinguishing agents: Extinguish fire using suitable agent for the type of surrounding fire.

Reactivity Hazards:

Reactive With: n/a *Other Reactions:* n/a

Corrosivity Hazards:

Corrosive With: n/a *Neutralizing Agent:* n/a

Radioactivity Hazards:

Radiation Emitted: n/a *Other Hazards:* n/a

Recommended Protection for Response Personnel:

Avoid breathing vapors, keep upwind. Wear appropriate sealed chemical suit. Wash away any material which may have come into contact with the body with copious amounts of soap and water. Consider appropriate evacuation.

ZINC FLUOROBORATE

DOT Number: n/a *DOT Hazard Class:* n/a *DOT Guide Number:* n/a
Synonyms: zinc fluoroborate solution
STCC Number: n/a *Reportable Qty:* n/a
Mfg Name: Allied Corp. *Phone No:* 1-201-455-2000

Physical Description:

Physical Form: Liquid *Color:* Colorless *Odor:* Odorless
Other Information: n/a

Chemical Properties:

Specific Gravity: 1.45 *Vapor Density:* n/a *Boiling Point: 212° F(100° C)*
Melting Point: n/a *Vapor Pressure:* n/a *Solubility in water:* Yes
Other Information: Sinks and mixes with water.

Health Hazards:

Inhalation Hazard: n/a
Ingestion Hazard: Nausea and vomiting.
Absorption Hazard: Irritating to the skin and eyes.
Hazards to Wildlife: n/a
Decontamination Procedures: Wash away any material with copious amounts of soap and water.
First Aid Procedures: Remove victim to fresh air, call emergency medical care. If not breathing give CPR. If breathing is difficult administer oxygen. Treat for shock.

Fire Hazards:

Flashpoint: n/a *Ignition temperature:* n/a
Flammable Explosive High Range: n/a *Low Range:* n/a
Toxic Products of Combustion: n/a
Other Hazards: n/a
Possible extinguishing agents: Extinguish fire using suitable agent for the type of surrounding fire.

Reactivity Hazards:

Reactive With: n/a *Other Reactions:* n/a

Corrosivity Hazards:

Corrosive With: n/a *Neutralizing Agent:* n/a

Radioactivity Hazards:

Radiation Emitted: n/a *Other Hazards:* n/a

Recommended Protection for Response Personnel:

Avoid breathing vapors, keep upwind. Wear appropriate sealed chemical clothing. Wash away any material which may have come into contact with the body with copious amounts of soap and water. Consider appropriate evacuation.

ZINC FORMATE

DOT Number: NA 9159 *DOT Hazard Class:* ORM-E *DOT Guide Number:* 31
Synonyms: zinc salt, formatic acid
STCC Number: 4963895 *Reportable Qty:* 1000/454
Mfg Name: Shepherd Chemical Co. *Phone No:* 1-513-731-1110

Physical Description:

Physical Form: Solid *Color:* Colorless to white *Odor:* Odorless
Other Information: It is used for galvanizing steel and making ceramics.

Chemical Properties:

Specific Gravity: 4.95 *Vapor Density:* 5.3 *Boiling Point: 2732° F(1500° C)*
Melting Point: n/a *Vapor Pressure:* n/a *Solubility in water:* Yes
Other Information: Sinks and mixes with water.

Health Hazards:

Inhalation Hazard: n/a
Ingestion Hazard: Harmful if swallowed.
Absorption Hazard: Irritating to the skin and eyes.
Hazards to Wildlife: Dangerous to aquatic life.
Decontamination Procedures: Wash away any material with copious amounts of soap and water.
First Aid Procedures: Remove victim to fresh air, call emergency medical care. If not breathing give CPR. If breathing is difficult administer oxygen. Treat for shock.

Fire Hazards:

Flashpoint: n/a *Ignition temperature:* n/a
Flammable Explosive High Range: n/a *Low Range:* n/a
Toxic Products of Combustion: n/a
Other Hazards: n/a
Possible extinguishing agents: Extinguish fire using suitable agent for the type of surrounding fire.

Reactivity Hazards:

Reactive With: n/a *Other Reactions:* n/a

Corrosivity Hazards:

Corrosive With: n/a *Neutralizing Agent:* n/a

Radioactivity Hazards:

Radiation Emitted: n/a *Other Hazards:* n/a

Recommended Protection for Response Personnel:

Avoid breathing vapors, keep upwind. Structural protective clothing provides limited protection. Wash away any material which may have come into contact with the body with copious amounts of soap and water. Consider appropriate evacuation.

ZINC HYDROSULFITE

DOT Number: UN 1931 *DOT Hazard Class:* ORM-A *DOT Guide Number:* 32
Synonyms: zinc dithionite
STCC Number: 4941195 *Reportable Qty:* 1000/454
Mfg Name: Virginia Chemical Co. *Phone No:* 1-804-483-7000

Physical Description:

Physical Form: Solid *Color:* White *Odor:* Slight SO_2
Other Information: It is used to bleach wood pulp, textiles and other naturally occurring materials.

Chemical Properties:

Specific Gravity: n/a *Vapor Density:* 6.6 *Boiling Point: n/a*
Melting Point: n/a *Vapor Pressure:* n/a *Solubility in water:* Yes
Other Information: Mixes with water.

Health Hazards:

Inhalation Hazard: n/a
Ingestion Hazard: Cause nausea and vomiting.
Absorption Hazard: Irritating to the skin and eyes.
Hazards to Wildlife: Dangerous to aquatic life.
Decontamination Procedures: Wash away any material with copious amounts of soap and water.
First Aid Procedures: Remove victim to fresh air, call emergency medical care. If not breathing give CPR. If breathing is difficult administer oxygen. Treat for shock.

Fire Hazards:

Flashpoint: n/a *Ignition temperature:* n/a
Flammable Explosive High Range: n/a *Low Range:* n/a
Toxic Products of Combustion: Decomposes giving off SO_2.
Other Hazards: n/a
Possible extinguishing agents: Use dry chemical, dry sand, or carbon dioxide.

Reactivity Hazards:

Reactive With: Water to form irritating SO_2 gas. *Other Reactions:* Oxidizing agents and acids.

Corrosivity Hazards:

Corrosive With: n/a *Neutralizing Agent:* n/a

Radioactivity Hazards:

Radiation Emitted: n/a *Other Hazards:* n/a

Recommended Protection for Response Personnel:

Avoid breathing vapors, keep upwind. Structural protective clothing provides limited protection. Wash away any material which may have come into contact with the body with copious amounts of soap and water. Consider appropriate evacuation.

ZINC NITRATE

DOT Number: UN 1514 *DOT Hazard Class:* Oxidizer *DOT Guide Number:* 35
Synonyms: cyanide of zinc, zinc dicyanide
STCC Number: 4918790 *Reportable Qty:* 1000/454
Mfg Name: J.T. Baker Chemical *Phone No:* 1-201-859-2151

Physical Description:

Physical Form: Solid *Color:* White *Odor:* Odorless
Other Information: It is used as a catalyst, in the manufacture of other chemicals, in medicines, and in dyeing.

Chemical Properties:

Specific Gravity: 2.07 *Vapor Density:* 8.4 *Boiling Point: Decomposes*
Melting Point: n/a *Vapor Pressure:* n/a *Solubility in water:* Yes
Other Information: Sinks and mixes with water.

Health Hazards:

Inhalation Hazard: Coughing, difficulty in breathing.
Ingestion Hazard: Nausea and vomiting.
Absorption Hazard: Irritating to the skin and eyes.
Hazards to Wildlife: Dangerous to aquatic life.
Decontamination Procedures: Wash away any material with copious amounts of soap and water.
First Aid Procedures: Remove victim to fresh air, call emergency medical care. If not breathing give CPR. If breathing is difficult administer oxygen. Treat for shock.

Fire Hazards:

Flashpoint: n/a *Ignition temperature:* n/a
Flammable Explosive High Range: n/a *Low Range:* n/a
Toxic Products of Combustion: Toxic oxides of nitrogen are produced.
Other Hazards: Will accelerate the combustion of other materials. Prolonged exposure of fire or heat may result in an explosion.
Possible extinguishing agents: Apply water from as far a distance as possible. Flood with water.

Reactivity Hazards:

Reactive With: n/a *Other Reactions:* None

Corrosivity Hazards:

Corrosive With: n/a *Neutralizing Agent:* n/a

Radioactivity Hazards:

Radiation Emitted: n/a *Other Hazards:* n/a

Recommended Protection for Response Personnel:

Avoid breathing vapors, keep upwind. Structural protective clothing provides limited protection. Wash away any material which may have come into contact with the body with copious amounts of soap and water. If the fire becomes uncontrollable, consider appropriate evacuation.

ZINC PHENOLSULFONATE

DOT Number: NA 9160 *DOT Hazard Class:* ORM-E *DOT Guide Number:* 31
Synonyms: zinc p-phenolsulfonate, zinc sulfocarbolate
STCC Number: 4966389 *Reportable Qty:* 5000/2270
Mfg Name: Allied Corp. *Phone No:* 1-201-445-2000

Physical Description:

Physical Form: Solid *Color:* White *Odor:* Odorless
Other Information: Turns pink on exposure to air. Used as an insecticide.

Chemical Properties:

Specific Gravity: 1 *Vapor Density:* 1 *Boiling Point: Decomposes*
Melting Point: n/a *Vapor Pressure:* n/a *Solubility in water:* Yes
Other Information: Mixes with water

Health Hazards:

Inhalation Hazard: Cause coughing or difficulty in breathing.
Ingestion Hazard: Cause nausea and vomiting.
Absorption Hazard: Irritating to the skin and eyes.
Hazards to Wildlife: n/a
Decontamination Procedures: Wash away any material with copious amounts of soap and water.
First Aid Procedures: Remove victim to fresh air, call emergency medical care. If not breathing give CPR. If breathing is difficult administer oxygen. Treat for shock.

Fire Hazards:

Flashpoint: n/a *Ignition temperature:* n/a
Flammable Explosive High Range: n/a *Low Range:* n/a
Toxic Products of Combustion: Irritating oxides of sulfur may be formed in a fire.
Other Hazards: n/a
Possible extinguishing agents: Extinguish fire using suitable agent for the type of surrounding fire.

Reactivity Hazards:

Reactive With: n/a *Other Reactions:* n/a

Corrosivity Hazards:

Corrosive With: n/a *Neutralizing Agent:* n/a

Radioactivity Hazards:

Radiation Emitted: n/a *Other Hazards:* n/a

Recommended Protection for Response Personnel:

Avoid breathing vapors, keep upwind. Structural protective clothing provides limited protection. Wash away any material which may have come into contact with the body with copious amounts of soap and water. Consider appropriate evacuation.

ZINC PHOSPHIDE

DOT Number: UN 1714 *DOT Hazard Class:* Poison B *DOT Guide Number:* 41
Synonyms: None given
STCC Number: 4923496 *Reportable Qty:* 100/45.4
Mfg Name: A.D. MacKay Inc. *Phone No:* 1-203-655-3000

Physical Description:

Physical Form: Solid *Color:* Gray to black *Odor:* Faint
Other Information: It is used in medicine and in rat poison.

Chemical Properties:

Specific Gravity: 4.55 *Vapor Density:* 8.9 *Boiling Point: 2012° F(1100° C)*
Melting Point: n/a *Vapor Pressure:* n/a *Solubility in water:* Slowly decomposes
Other Information: n/a

Health Hazards:

Inhalation Hazard: Dizziness, difficulty in breathing, loss of consciousness.
Ingestion Hazard: Poisonous if swallowed.
Absorption Hazard: Irritating to the skin and eyes.
Hazards to Wildlife: Dangerous to waterfowl.
Decontamination Procedures: Wash away any material with copious amounts of soap and water.
First Aid Procedures: Remove victim to fresh air, call emergency medical care. If not breathing give CPR. If breathing is difficult administer oxygen. Treat for shock.

Fire Hazards:

Flashpoint: n/a *Ignition temperature:* n/a
Flammable Explosive High Range: n/a *Low Range:* n/a
Toxic Products of Combustion: Irritating gases may be produced when heated.
Other Hazards: Avoid contact with solid or dust.
Possible extinguishing agents: Use water, foam, dry chemical or carbon dioxide.

Reactivity Hazards:

Reactive With: Water to form phosgene gas which is toxic and spontaneously flammable.
Other Reactions: None

Corrosivity Hazards:

Corrosive With: n/a *Neutralizing Agent:* n/a

Radioactivity Hazards:

Radiation Emitted: n/a *Other Hazards:* n/a

Recommended Protection for Response Personnel:

Avoid breathing vapors, keep upwind. Structural protective clothing provides Limited protection. Wash away any material which may have come into contact with the body with copious amounts of soap and water. If the fire becomes uncontrollable, consider appropriate evacuation.

ZINC POTASSIUM CHROMATE

DOT Number: n/a *DOT Hazard Class:* n/a *DOT Guide Number:* n/a
Synonyms: potassium zinc chromate, zinc yellow Y-539-D
STCC Number: n/a *Reportable Qty:* n/a *CHEMTREC Phone No:* 1-800-424-9300

Physical Description:

Physical Form: Solid *Color:* Yellow *Odor:* Odorless
Other Information: n/a

Chemical Properties:

Specific Gravity: 3.5 *Vapor Density:* n/a *Boiling Point: n/a*
Melting Point: n/a *Vapor Pressure:* n/a *Solubility in water:* Yes
Other Information: Sinks and mixes with water.

Health Hazards:

Inhalation Hazard: Harmful if inhaled.
Ingestion Hazard: Harmful if swallowed.
Absorption Hazard: Irritating to the skin, eyes, nose and throat.
Hazards to Wildlife: Dangerous to aquatic life.
Decontamination Procedures: Wash away any material with copious amounts of soap and water.
First Aid Procedures: Remove victim to fresh air, call emergency medical care. If not breathing give CPR. If breathing is difficult administer oxygen. Treat for shock.

Fire Hazards: Data is not available!!

Flashpoint: n/a *Ignition temperature:* n/a
Flammable Explosive High Range: n/a *Low Range:* n/a
Toxic Products of Combustion: n/a
Other Hazards: n/a
Possible extinguishing agents: Extinguish fire using suitable agent for the type of surrounding fire.

Reactivity Hazards:

Reactive With: n/a *Other Reactions:* n/a

Corrosivity Hazards:

Corrosive With: n/a *Neutralizing Agent:* n/a

Radioactivity Hazards:

Radiation Emitted: n/a *Other Hazards:* n/a

Recommended Protection for Response Personnel:

Avoid breathing vapors, keep upwind. Structural protective clothing provides Limited protection. Wash away any material which may have come into contact with the body with copious amounts of soap and water. Consider appropriate evacuation.

ZINC SILICOFLUORIDE

DOT Number: NA 2855 *DOT Hazard Class:* ORM-E *DOT Guide Number:* 53
Synonyms: zinc fluosilicate, zinc hexafluorosilicate
STCC Number: 4966392 *Reportable Qty:* 5000/2270
Mfg Name: W.R. Grace & Co. *Phone No:* 1-212-819-5500

Physical Description:

Physical Form: Solid *Color:* White *Odor:* Odorless
Other Information: It is used as a concrete hardener, as a preservative, and for many other uses.

Chemical Properties:

Specific Gravity: 2.10 *Vapor Density:* 11 *Boiling Point: Decomposes*
Melting Point: n/a *Vapor Pressure:* n/a *Solubility in water:* Yes
Other Information: Sinks and mixes with water.

Health Hazards:

Inhalation Hazard: Cause coughing or difficulty in breathing.
Ingestion Hazard: Poisonous if swallowed.
Absorption Hazard: Irritating to the skin and eyes.
Hazards to Wildlife: n/a
Decontamination Procedures: Wash away any material with copious amounts of soap and water.
First Aid Procedures: Remove victim to fresh air, call emergency medical care. If not breathing give CPR. If breathing is difficult administer oxygen. Treat for shock.

Fire Hazards:

Flashpoint: n/a *Ignition temperature:* n/a
Flammable Explosive High Range: n/a *Low Range:* n/a
Toxic Products of Combustion: Toxic and irritating hydrogen fluoride and silicon tetrafluoride are formed in fires.
Other Hazards: n/a
Possible extinguishing agents: Extinguish fire using suitable agent for the type of surrounding fire.

Reactivity Hazards:

Reactive With: n/a *Other Reactions:* n/a

Corrosivity Hazards:

Corrosive With: n/a *Neutralizing Agent:* n/a

Radioactivity Hazards:

Radiation Emitted: n/a *Other Hazards:* n/a

Recommended Protection for Response Personnel:

Avoid breathing vapors, keep upwind. Wear a sealed chemical suit. (polycarbonate, butyl rubber) Wash away any material which may have come into contact with the body with copious amounts of soap and water. Consider appropriate evacuation.

ZINC SULFATE

DOT Number: NA 9161 *DOT Hazard Class:* ORM-E *DOT Guide Number:* 31
Synonyms: white vitriol, zinc sulfate heptahydrate, zinc vitriol
STCC Number: 4963786 *Reportable Qty:* 1000/454
Mfg Name: J.T. Baker Chemical *Phone No:* 1-201-859-2151

Physical Description:

Physical Form: Solid *Color:* White *Odor:* Odorless
Other Information: It is used in the production of rayon, as a feed preservative, and as a fertilizer ingredient.

Chemical Properties:

Specific Gravity: 1.96 *Vapor Density:* 5.5 *Boiling Point: Decomposes*
Melting Point: n/a *Vapor Pressure:* n/a *Solubility in water:* Yes
Other Information: Sinks and mixes with water.

Health Hazards:

Inhalation Hazard: Cause coughing or difficulty in breathing.
Ingestion Hazard: Cause nausea and vomiting.
Absorption Hazard: Irritating to the skin and eyes.
Hazards to Wildlife: Dangerous to aquatic life.
Decontamination Procedures: Wash away any material with copious amounts of soap and water.
First Aid Procedures: Remove victim to fresh air, call emergency medical care. If not breathing give CPR. If breathing is difficult administer oxygen. Treat for shock.

Fire Hazards:

Flashpoint: n/a *Ignition temperature:* n/a
Flammable Explosive High Range: n/a *Low Range:* n/a
Toxic Products of Combustion: n/a
Other Hazards: n/a
Possible extinguishing agents: Extinguish fire using suitable agent for the type of surrounding fire.

Reactivity Hazards:

Reactive With: n/a *Other Reactions:* n/a

Corrosivity Hazards:

Corrosive With: n/a *Neutralizing Agent:* n/a

Radioactivity Hazards:

Radiation Emitted: n/a *Other Hazards:* n/a

Recommended Protection for Response Personnel:

Avoid breathing vapors, keep upwind. Structural protective clothing provides limited protection. Wash away any material which may have come into contact with the body with copious amounts of soap and water. Consider appropriate evacuation.

ZIRCONIUM ACETATE

DOT Number: n/a *DOT Hazard Class:* n/a *DOT Guide Number:* n/a
Synonyms: zirconium acetate solution
STCC Number: n/a *Reportable Qty:* n/a
Mfg Name: Magnesium Elektron Inc. *Phone No:* 1-201-782-5800

Physical Description:

Physical Form: Liquid *Color:* White *Odor:* Weak vinegar
Other Information: n/a

Chemical Properties:

Specific Gravity: 1.37 *Vapor Density:* n/a *Boiling Point: n/a*
Melting Point: n/a *Vapor Pressure:* n/a *Solubility in water:* Yes
Other Information: Sinks and mixes with water.

Health Hazards:

Inhalation Hazard: n/a
Ingestion Hazard: Harmful if swallowed.
Absorption Hazard: Irritating to the skin and eyes.
Hazards to Wildlife: n/a
Decontamination Procedures: Wash away any material with copious amounts of soap and water.
First Aid Procedures: Remove victim to fresh air, call emergency medical care. If not breathing give CPR. If breathing is difficult administer oxygen. Treat for shock.

Fire Hazards:

Flashpoint: n/a *Ignition temperature:* n/a
Flammable Explosive High Range: n/a *Low Range:* n/a
Toxic Products of Combustion: n/a
Other Hazards: n/a
Possible extinguishing agents: Extinguish fire using suitable agent for the type of surrounding fire.

Reactivity Hazards:

Reactive With: n/a *Other Reactions:* n/a

Corrosivity Hazards:

Corrosive With: n/a *Neutralizing Agent:* n/a

Radioactivity Hazards:

Radiation Emitted: n/a *Other Hazards:* n/a

Recommended Protection for Response Personnel:

Avoid breathing vapors, keep upwind. Structural protective clothing provides limited protection. Wash away any material which may have come into contact with the body with copious amounts of soap and water. Consider appropriate evacuation.

ZIRCONIUM NITRATE

DOT Number: UN 2728 *DOT Hazard Class:* Oxidizer *DOT Guide Number:* 35
Synonyms: zirconium nitrate, pentahydrate
STCC Number: 4918791 *Reportable Qty:* 5000/2270
Mfg Name: Magnesium Elektron Inc. *Phone No:* 1-201-782-5800

Physical Description:

Physical Form: Solid *Color:* White *Odor:* Odorless
Other Information: It is used as a preservative.

Chemical Properties:

Specific Gravity: 1 *Vapor Density:* 15 *Boiling Point: Decomposes*
Melting Point: n/a *Vapor Pressure:* n/a *Solubility in water:* Yes
Other Information: Sinks and mixes with water.

Health Hazards:

Inhalation Hazard: Coughing or difficulty in breathing.
Ingestion Hazard: Cause nausea and vomiting.
Absorption Hazard: Irritating to the skin and eyes.
Hazards to Wildlife: n/a
Decontamination Procedures: Wash away any material with copious amounts of soap and water.
First Aid Procedures: Remove victim to fresh air, call emergency medical care. If not breathing give CPR. If breathing is difficult administer oxygen. Treat for shock.

Fire Hazards:

Flashpoint: n/a *Ignition temperature:* n/a
Flammable Explosive High Range: n/a *Low Range:* n/a
Toxic Products of Combustion: Toxic oxides of nitrogen are produced in a fire.
Other Hazards: Will accelerate the burning of combustible materials. Prolonged exposure to heat or fire may result in an explosion.
Possible extinguishing agents: Flood with water. Apply water from as far a distance as possible.

Reactivity Hazards:

Reactive With: Water to form a weak acid solution. *Other Reactions:* None

Corrosivity Hazards:

Corrosive With: Most metals *Neutralizing Agent:* Crushed limestone, soda ash, or lime.

Radioactivity Hazards:

Radiation Emitted: n/a *Other Hazards:* n/a

Recommended Protection for Response Personnel:

Avoid breathing vapors, keep upwind. Structural protective clothing provides limited protection. Wash away any material which may have come into contact with the body with copious amounts of soap and water. If the fire becomes uncontrollable, consider appropriate evacuation.

ZIRCONIUM OXYCHLORIDE

DOT Number: n/a *DOT Hazard Class:* n/a *DOT Guide Number:* n/a
Synonyms: basic zirconium chloride, zirconium oxide chloride, zirconium oxychloride hydrate, zirconyl chloride
STCC Number: n/a *Reportable Qty:* n/a
Mfg Name: Magnesium Elektron Inc. *Phone No:* 1-201-782-5800

Physical Description:

Physical Form: Solid *Color:* White to yellow *Odor:* Odorless
Other Information: n/a

Chemical Properties:

Specific Gravity: 1 *Vapor Density:* 15 *Boiling Point: Decomposes*
Melting Point: n/a *Vapor Pressure:* n/a *Solubility in water:* Yes
Other Information: Sinks and mixes with water.

Health Hazards:

Inhalation Hazard: Coughing or difficulty in breathing.
Ingestion Hazard: Nausea and vomiting.
Absorption Hazard: Irritating to the skin and eyes.
Hazards to Wildlife: n/a
Decontamination Procedures: Wash away any material with copious amounts of soap and water.
First Aid Procedures: Remove victim to fresh air, call emergency medical care. If not breathing give CPR. If breathing is difficult administer oxygen. Treat for shock.

Fire Hazards:

Flashpoint: n/a *Ignition temperature:* n/a
Flammable Explosive High Range: n/a *Low Range:* n/a
Toxic Products of Combustion: n/a
Other Hazards: n/a
Possible extinguishing agents: Extinguish fire using suitable agent for the type of surrounding fire.

Reactivity Hazards:

Reactive With: n/a *Other Reactions:* n/a

Corrosivity Hazards:

Corrosive With: n/a *Neutralizing Agent:* n/a

Radioactivity Hazards:

Radiation Emitted: n/a *Other Hazards:* n/a

Recommended Protection for Response Personnel:

Avoid breathing vapors, keep upwind. Structural protective clothing provides limited protection. Wash away any material which may have come into contact with the body with copious amounts of soap and water. Consider appropriate evacuation.

ZIRCONIUM POTASSIUM FLUORIDE

DOT Number: NA 9162 *DOT Hazard Class:* ORM-E *DOT Guide Number:* 31
Synonyms: potassium fluozirconate
STCC Number: 4966395 *Reportable Qty:* 1000/454
CHEMTREC Phone No: 1-800-424-9300

Physical Description:

Physical Form: Solid *Color:* Colorless *Odor:* Odorless
Other Information: It is used in metal processing, as a catalyst in chemical manufacturing, and for many other uses.

Chemical Properties:

Specific Gravity: 3.48 *Vapor Density:* 12 *Boiling Point: n/a*
Melting Point: n/a *Vapor Pressure:* n/a *Solubility in water:* Yes
Other Information: Sinks and mixes with water.

Health Hazards:

Inhalation Hazard: Harmful if inhaled.
Ingestion Hazard: Harmful if swallowed.
Absorption Hazard: Irritating to the skin.
Hazards to Wildlife: Dangerous to aquatic life.
Decontamination Procedures: Wash away any material with copious amounts of soap and water.
First Aid Procedures: Remove victim to fresh air, call emergency medical care. If not breathing give CPR. If breathing is difficult administer oxygen. Treat for shock.

Fire Hazards:

Flashpoint: n/a *Ignition temperature:* n/a
Flammable Explosive High Range: n/a *Low Range:* n/a
Toxic Products of Combustion: n/a
Other Hazards: n/a
Possible extinguishing agents: Extinguish fire using suitable agent for the type of surrounding fire.

Reactivity Hazards:

Reactive With: n/a *Other Reactions:* n/a

Corrosivity Hazards:

Corrosive With: n/a *Neutralizing Agent:* n/a

Radioactivity Hazards:

Radiation Emitted: n/a *Other Hazards:* n/a

Recommended Protection for Response Personnel:

Avoid breathing vapors, keep upwind. Wear appropriate chemical clothing, wash away any material which may come into contact with the body with copious amounts of soap and water. Consider appropriate evacuation.

ZIRCONIUM SULFATE

DOT Number: NA 9163 *DOT Hazard Class:* ORM-B *DOT Guide Number:* 31
Synonyms: disulfatozirconic acid, zirconium sulfate tetrahydrate
STCC Number: 4944185 *Reportable Qty:* 5000/2270
Mfg Name: Magnesium Elektron Inc. *Phone No:* 1-201-782-5800

Physical Description:

Physical Form: Solid *Color:* White *Odor:* Odorless
Other Information: It is used in chemical analysis, as an additive for lubricants, and for many other uses.

Chemical Properties:

Specific Gravity: 3 *Vapor Density:* 12.0 *Boiling Point: Decomposes*
Melting Point: n/a *Vapor Pressure:* n/a *Solubility in water:* Yes
Other Information: Sinks and mixes with water.

Health Hazards:

Inhalation Hazard: Cause coughing or difficulty in breathing.
Ingestion Hazard: Nausea and vomiting.
Absorption Hazard: Irritating to the skin and eyes.
Hazards to Wildlife: Dangerous to aquatic life.
Decontamination Procedures: Wash away any material with copious amounts of soap and water.
First Aid Procedures: Remove victim to fresh air, call emergency medical care. If not breathing give CPR. If breathing is difficult administer oxygen. Treat for shock.

Fire Hazards:

Flashpoint: n/a *Ignition temperature:* n/a
Flammable Explosive High Range: n/a *Low Range:* n/a
Toxic Products of Combustion: n/a
Other Hazards: n/a
Possible extinguishing agents: Extinguish fire using suitable agent for the type of surrounding fire.

Reactivity Hazards:

Reactive With: n/a *Other Reactions:* n/a

Corrosivity Hazards:

Corrosive With: Aluminum *Neutralizing Agent:* Crushed limestone, soda ash, or lime.

Radioactivity Hazards:

Radiation Emitted: n/a *Other Hazards:* n/a

Recommended Protection for Response Personnel:

Avoid breathing vapors, keep upwind. Structural protective clothing provides limited protection. Wash away any material which may come into contact with the body with copious amounts of soap and water. Consider appropriate evacuation.

ZIRCONIUM POTASSIUM FLUORIDE

DOT Number: NA 9162 *DOT Hazard Class:* ORM-E *DOT Guide Number:* 31
Synonyms: potassium fluozirconate
STCC Number: 4966395 *Reportable Qty:* 1000/454
CHEMTREC Phone No: 1-800-424-9300

Physical Description:

Physical Form: Solid *Color:* Colorless *Odor:* Odorless
Other Information: It is used in metal processing, as a catalyst in chemical manufacturing, and for many other uses.

Chemical Properties:

Specific Gravity: 3.48 *Vapor Density:* 12 *Boiling Point: n/a*
Melting Point: n/a *Vapor Pressure:* n/a *Solubility in water:* Yes
Other Information: Sinks and mixes with water.

Health Hazards:

Inhalation Hazard: Harmful if inhaled.
Ingestion Hazard: Harmful if swallowed.
Absorption Hazard: Irritating to the skin.
Hazards to Wildlife: Dangerous to aquatic life.
Decontamination Procedures: Wash away any material with copious amounts of soap and water.
First Aid Procedures: Remove victim to fresh air, call emergency medical care. If not breathing give CPR. If breathing is difficult administer oxygen. Treat for shock.

Fire Hazards:

Flashpoint: n/a *Ignition temperature:* n/a
Flammable Explosive High Range: n/a *Low Range:* n/a
Toxic Products of Combustion: n/a
Other Hazards: n/a
Possible extinguishing agents: Extinguish fire using suitable agent for the type of surrounding fire.

Reactivity Hazards:

Reactive With: n/a *Other Reactions:* n/a

Corrosivity Hazards:

Corrosive With: n/a *Neutralizing Agent:* n/a

Radioactivity Hazards:

Radiation Emitted: n/a *Other Hazards:* n/a

Recommended Protection for Response Personnel:

Avoid breathing vapors, keep upwind. Wear appropriate chemical clothing, wash away any material which may come into contact with the body with copious amounts of soap and water. Consider appropriate evacuation.

ZIRCONIUM SULFATE

DOT Number: NA 9163 *DOT Hazard Class:* ORM-B *DOT Guide Number:* 31
Synonyms: disulfatozirconic acid, zirconium sulfate tetrahydrate
STCC Number: 4944185 *Reportable Qty:* 5000/2270
Mfg Name: Magnesium Elektron Inc. *Phone No:* 1-201-782-5800

Physical Description:

Physical Form: Solid *Color:* White *Odor:* Odorless
Other Information: It is used in chemical analysis, as an additive for lubricants, and for many other uses.

Chemical Properties:

Specific Gravity: 3 *Vapor Density:* 12.0 *Boiling Point: Decomposes*
Melting Point: n/a *Vapor Pressure:* n/a *Solubility in water:* Yes
Other Information: Sinks and mixes with water.

Health Hazards:

Inhalation Hazard: Cause coughing or difficulty in breathing.
Ingestion Hazard: Nausea and vomiting.
Absorption Hazard: Irritating to the skin and eyes.
Hazards to Wildlife: Dangerous to aquatic life.
Decontamination Procedures: Wash away any material with copious amounts of soap and water.
First Aid Procedures: Remove victim to fresh air, call emergency medical care. If not breathing give CPR. If breathing is difficult administer oxygen. Treat for shock.

Fire Hazards:

Flashpoint: n/a *Ignition temperature:* n/a
Flammable Explosive High Range: n/a *Low Range:* n/a
Toxic Products of Combustion: n/a
Other Hazards: n/a
Possible extinguishing agents: Extinguish fire using suitable agent for the type of surrounding fire.

Reactivity Hazards:

Reactive With: n/a *Other Reactions:* n/a

Corrosivity Hazards:

Corrosive With: Aluminum *Neutralizing Agent:* Crushed limestone, soda ash, or lime.

Radioactivity Hazards:

Radiation Emitted: n/a *Other Hazards:* n/a

Recommended Protection for Response Personnel:

Avoid breathing vapors, keep upwind. Structural protective clothing provides limited protection. Wash away any material which may come into contact with the body with copious amounts of soap and water. Consider appropriate evacuation.

WAX

DOT Number: UN 1993 *DOT Hazard Class:* Combustible liquid *DOT Guide Number:* 27
Synonyms: petroleum wax
STCC Number: 4915387 *Reportable Qty:* n/a
Mfg Name: Witco Chemical Corp. *Phone No:* 1-212-605-3800

Physical Description:

Physical Form: Liquid *Color:* Yellow to white *Odor:* Waxy
Other Information: n/a

Chemical Properties:

Specific Gravity: .9 *Vapor Density:* n/a *Boiling Point: 700° F(371.1° C)*
Melting Point: n/a *Vapor Pressure:* n/a *Solubility in water:* n/a
Other Information: It is generally lighter than and partly soluble in water. Vapors are heavier than air.

Health Hazards:

Inhalation Hazard: n/a
Ingestion Hazard: n/a
Absorption Hazard: Will burn the skin and eyes.
Hazards to Wildlife: n/a
Decontamination Procedures: Wash away any material with copious amounts of soap and water.
First Aid Procedures: Remove victim to fresh air, call emergency medical care. If not breathing give CPR. If breathing is difficult administer oxygen. Treat for shock.

Fire Hazards:

Flashpoint: 390° F(198.8° C) *Ignition temperature:* 473° F(245° C)
Flammable Explosive High Range: n/a *Low Range:* n/a
Toxic Products of Combustion: n/a
Other Hazards: n/a
Possible extinguishing agents: Apply water from as far a distance as possible. Use alcohol foam, dry chemical, or carbon dioxide.

Reactivity Hazards:

Reactive With: n/a *Other Reactions:* n/a

Corrosivity Hazards:

Corrosive With: n/a *Neutralizing Agent:* n/a

Radioactivity Hazards:

Radiation Emitted: n/a *Other Hazards:* n/a

Recommended Protection for Response Personnel:

Avoid breathing vapors, keep upwind. Structural protective clothing provides limited protection. Wash away any material which may have come into contact with the body with copious amounts of soap and water. Consider appropriate evacuation.

XYLENE-m

DOT Number: UN 1307 *DOT Hazard Class:* Flammable liquid *DOT Guide Number:* 27
Synonyms: 1,3-dimethylbenzene, xylol
STCC Number: 4909350 *Reportable Qty:* 1000/454
Mfg Name: Shell Chemical *Phone No:* 1-713-241-6161

Physical Description:

Physical Form: Liquid *Color:* Colorless *Odor:* Sweet
Other Information: It is used as a solvent for paints and adhesives, and to make other chemicals.

Chemical Properties:

Specific Gravity: .9 *Vapor Density:* 3.7 *Boiling Point: 282° F(138.8° C)*
Melting Point: −54° F(−47.7° C) *Vapor Pressure:* 9 mm Hg at 68° F(20° C)
Solubility in water: No *Degree of Solubility:* .00003%
Other Information: Lighter than and insoluble in water. Vapors are heavier than air. Weighs 7.2 lbs/3.2 kg per gallon/3.8 l.

Health Hazards:

Inhalation Hazard: Headache, difficulty in breathing, loss of consciousness.
Ingestion Hazard: Nausea, vomiting, loss of consciousness.
Absorption Hazard: Irritating to the skin and eyes.
Hazards to Wildlife: Dangerous to aquatic life.
Decontamination Procedures: Wash away any material with copious amounts of soap and water.
First Aid Procedures: Remove victim to fresh air, call emergency medical care. If not breathing give CPR. If breathing is difficult administer oxygen. Treat for shock.

Fire Hazards:

Flashpoint: 81° F(27.2° C) *Ignition temperature:* 982° F(527.7° C)
Flammable Explosive High Range: 7 *Low Range:* 1.1
Toxic Products of Combustion: n/a
Other Hazards: Flashback along vapor trail may occur. Vapors may explode if ignited in an enclosed area.
Possible extinguishing agents: Apply water from as far a distance as possible. Use foam, dry chemical, or carbon dioxide.

Reactivity Hazards:

Reactive With: n/a *Other Reactions:* n/a

Corrosivity Hazards:

Corrosive With: n/a *Neutralizing Agent:* n/a

Radioactivity Hazards:

Radiation Emitted: n/a *Other Hazards:* n/a

Recommended Protection for Response Personnel:

Avoid breathing vapors, keep upwind. Structural protective clothing provides limited protection. Wash away any material which may have come into contact with the body with copious amounts of soap and water. If the fire becomes uncontrollable, or if container is exposed to direct flame, consider appropriate evacuation.

XYLENE-o

DOT Number: UN 1307 *DOT Hazard Class:* Flammable liquid *DOT Guide Number:* 27
Synonyms: 1,2-dimethylbenzene, xylol
STCC Number: 4909350 *Reportable Qty:* 1000/454
Mfg Name: Shell Chemical *Phone No:* 1-713-241-6161

Physical Description:

Physical Form: Liquid *Color:* Colorless *Odor:* Sweet
Other Information: It is used as a solvent for paints and adhesives, and to make other chemicals.

Chemical Properties:

Specific Gravity: .9 *Vapor Density:* 3.7 *Boiling Point: 292° F(144.4° C)*
Melting Point: −12° F(−24.4° C) *Vapor Pressure:* 7 mm Hg at 68° F(20° C)
Solubility in water: No *Degree of Solubility:* .00003%
Other Information: Lighter than and insoluble in water. Vapors are heavier than air. Weighs 7.2 lbs/3.2 kg per gallon/3.8 l.

Health Hazards:

Inhalation Hazard: Headache, difficulty in breathing, loss of consciousness.
Ingestion Hazard: Nausea, vomiting, loss of consciousness.
Absorption Hazard: Irritating to the skin and eyes.
Hazards to Wildlife: Dangerous to aquatic life.
Decontamination Procedures: Wash away any material with copious amounts of soap and water.
First Aid Procedures: Remove victim to fresh air, call emergency medical care. If not breathing give CPR. If breathing is difficult administer oxygen. Treat for shock.

Fire Hazards:

Flashpoint: 90° F(32.2° C) *Ignition temperature:* 867° F(463.8° C)
Flammable Explosive High Range: 7 *Low Range:* 1
Toxic Products of Combustion: n/a
Other Hazards: Flashback along vapor trail may occur. Vapors may explode if ignited in an enclosed area.
Possible extinguishing agents: Apply water from as far a distance as possible. Use foam, dry chemical, or carbon dioxide.

Reactivity Hazards:

Reactive With: n/a *Other Reactions:* n/a

Corrosivity Hazards:

Corrosive With: n/a *Neutralizing Agent:* n/a

Radioactivity Hazards:

Radiation Emitted: n/a *Other Hazards:* n/a

Recommended Protection for Response Personnel:

Avoid breathing vapors, keep upwind. Structural protective clothing provides limited protection. Wash away any material which may have come into contact with the body with copious amounts of soap and water. If the fire becomes uncontrollable, or if container is exposed to direct flame, consider appropriate evacuation.

XYLENE-p

DOT Number: UN 1307 *DOT Hazard Class:* Flammable liquid *DOT Guide Number:* 27
Synonyms: 1,4-dimethylbenzene, xylol
STCC Number: 4909350 *Reportable Qty:* 1000/454
Mfg Name: Shell Chemical *Phone No:* 1-713-241-6161

Physical Description:

Physical Form: Liquid *Color:* Colorless *Odor:* Sweet
Other Information: It is used as a solvent for paints and adhesives, and to make other chemicals.

Chemical Properties:

Specific Gravity: .9 *Vapor Density:* 3.7 *Boiling Point: 281° F(138.3° C)*
Melting Point: 55° F(12.7° C) *Vapor Pressure:* 9 mm Hg at 68° F(20° C)
Solubility in water: No *Degree of Solubility:* .00003%
Other Information: Lighter than and insoluble in water. Vapors are heavier than air. Weighs 7.2 lbs/3.2 kg per gallon/3.8 l.

Health Hazards:

Inhalation Hazard: Headache, difficulty in breathing, loss of consciousness.
Ingestion Hazard: Nausea, vomiting, loss of consciousness.
Absorption Hazard: Irritating to the skin and eyes.
Hazards to Wildlife: Dangerous to aquatic life.
Decontamination Procedures: Wash away any material with copious amounts of soap and water.
First Aid Procedures: Remove victim to fresh air, call emergency medical care. If not breathing give CPR. If breathing is difficult administer oxygen. Treat for shock.

Fire Hazards:

Flashpoint: 81° F(27.2° C) *Ignition temperature:* 984° F(528.8° C)
Flammable Explosive High Range: 7 *Low Range:* 1.1
Toxic Products of Combustion: n/a
Other Hazards: Flashback along vapor trail may occur. Vapors may explode if ignited in an enclosed area.
Possible extinguishing agents: Apply water from as far a distance as possible. Use foam, dry chemical, or carbon dioxide.

Reactivity Hazards:

Reactive With: n/a *Other Reactions:* n/a

Corrosivity Hazards:

Corrosive With: n/a *Neutralizing Agent:* n/a

Radioactivity Hazards:

Radiation Emitted: n/a *Other Hazards:* n/a

Recommended Protection for Response Personnel:

Avoid breathing vapors, keep upwind. Structural protective clothing provides limited protection. Wash away any material which may have come into contact with the body with copious amounts of soap and water. If the fire becomes uncontrollable, or if container is exposed to direct flame, consider appropriate evacuation.

ZIRCONIUM TETRACHLORIDE

DOT Number: UN 2503 *DOT Hazard Class:* Corrosive *DOT Guide Number:* 39
Synonyms: zirconium chloride
STCC Number: 4923395 *Reportable Qty:* 5000/2270
Mfg Name: Atomergic Chemetals Corp. *Phone No:* 1-516-349-8800

Physical Description:

Physical Form: Solid *Color:* White *Odor:* n/a
Other Information: n/a

Chemical Properties:

Specific Gravity: 2.08 *Vapor Density:* 8 *Boiling Point: 627.8° F(331° C)*
Melting Point: n/a *Vapor Pressure:* 1 mm Hg at 662° F(350° C) *Solubility in water:* Decomposes
Other Information: Sinks and decomposes in water. An irritating vapor is produced.

Health Hazards:

Inhalation Hazard: Harmful if inhaled.
Ingestion Hazard: Harmful if swallowed.
Absorption Hazard: Irritating to the skin and eyes.
Hazards to Wildlife: Dangerous to aquatic life.
Decontamination Procedures: Wash away any material with copious amounts of soap and water.
First Aid Procedures: Remove victim to fresh air, call emergency medical care. If not breathing give CPR. If breathing is difficult administer oxygen. Treat for shock.

Fire Hazards:

Flashpoint: n/a *Ignition temperature:* n/a
Flammable Explosive High Range: n/a *Low Range:* n/a
Toxic Products of Combustion: n/a
Other Hazards: n/a
Possible extinguishing agents: Use dry chemical, dry sand, or carbon dioxide. Apply water from as far a distance as possible.

Reactivity Hazards:

Reactive With: Water to vigorously form hydrochloric acid. *Other Reactions:* None

Corrosivity Hazards:

Corrosive With: Metals in the presence of moisture, and tissue.
Neutralizing Agent: Crushed limestone, lime, or sodium bicarbonate.

Radioactivity Hazards:

Radiation Emitted: n/a *Other Hazards:* n/a

Recommended Protection for Response Personnel:

Avoid breathing vapors, keep upwind. Wear appropriate chemical clothing, wash away any material which may come into contact with the body with copious amounts of soap and water. If the fire becomes uncontrollable, consider appropriate evacuation.

CHEMICAL NAME INDEX

CHEMICAL NAME	DOT#	PAGE
ACETAL	1088	1
ACETALDEHYDE	1089	2
ACETIC ACID (aqueous solution)	2790	3
ACETIC ACID (glacial)	2789	4
ACETIC ANHYDRIDE	1715	5
ACETONE	1090	6
ACETONE CYANOHYDRIN	1541	7
ACETONITRILE	1648	8
ACETOPHENONE	9207	9
ACETYL BROMIDE	1716	10
ACETYL CHLORIDE	1717	11
ACETYL PEROXIDE	2084	12
ACETYL PEROXIDE SOLUTION	2084	13
ACETYLACETONE	2080	14
ACETYLENE	1001	15
ACRIDINE	2713	16
ACROLEIN	1092	17
ACRYLAMIDE	2074	18
ACRYLIC ACID	2218	19
ACRYLONITRILE	1093	20
ADIPIC ACID	9077	21
ADIPONITRILE	2205	22
ALDRIN	2761	23
ALLYL ALCOHOL	1098	24
ALLYL BROMIDE	1099	25
ALLYL CHLORIDE	1100	26
ALLYL CHLOROFORMATE	1722	27
ALLYL TRICHLOROSILANE	1724	28
ALUMINUM SULFATE-(liquid)	1760	29
ALUMINUM SULFATE-(solid)	9078	30
ALUMINUM CHLORIDE	1726	31
ALUMINUM NITRATE	1438	32
2(2)AMINOETHOXY(ETHANOL)	1760	33
AMINOETHYLETHANOLAMINE	1760	34
AMMONIA, ANHYDROUS	1005	35
AMMONIUM ACETATE	9079	36
AMMONIUM BENZOATE	9080	37
AMMONIUM BICARBONATE	9081	38
AMMONIUM BIFLUORIDE	1727	39
AMMONIUM BISULFITE	2693	40
AMMONIUM CARBAMATE	9083	41
AMMONIUM CARBONATE	9084	42
AMMONIUM CHLORIDE	9085	43
AMMONIUM CHROMATE	9086	44
AMMONIUM CITRATE	9087	45
AMMONIUM DICHROMATE	1439	46
AMMONIUM FLUOBORATE	9088	47
AMMONIUM FLUORIDE	2505	48
AMMONIUM HYDROXIDE	2672	49
AMMONIUM NITRATE-PHOSPHATE	2070	50
AMMONIUM NITRATE	1942	51
AMMONIUM OXALATE	2449	52
AMMONIUM PERCHLORATE	1442	53
AMMONIUM PERSULFATE	1444	54
AMMONIUM SILICOFLUORIDE	2854	55
AMMONIUM SULFAMATE	9188	56
AMMONIUM SULFIDE	2683	57
AMMONIUM SULFITE	9090	58
AMMONIUM TARTRATE	9091	59
AMMONIUM THIOCYANATE	9188	60

CHEMICAL NAME	DOT#	PAGE
AMMONIUM THIOCYANATE LIQUOR	9092	61
AMMONIUM THIOSULFATE-(liquid)	9188	62
AMMONIUM THIOSULFATE-(solid)	9093	63
AMYL ACETATE	1104	64
AMYL ALCOHOL	1105	65
AMYL CHLORIDE	1107	66
AMYL MERCAPTAN	1111	67
AMYL METHYL KETONE	1110	68
AMYL NITRATE	1112	69
AMYL NITRITE	1113	70
AMYLTRICHLORSILANE	1728	71
ANILINE	1547	72
ANISOYL CHLORIDE	1729	73
ANTIMONY PENTACHLORIDE	1730	74
ANTIMONY PENTAFLUORIDE	1732	75
ANTIMONY POTASSIUM TARTRATE	1551	76
ANTIMONY TRIBROMIDE	1549	77
ANTIMONY TRICHLORIDE	1733	78
ANTIMONY TRIFLUORIDE	1549	79
ANTIMONY TRIOXIDE	9201	80
ARSENIC ACID-(liquid)	1553	81
ARSENIC ACID-(solid)	1554	82
ARSENIC DISULFIDE	1557	83
ARSENIC PENTOXIDE	1559	84
ARSENIC TRICHLORIDE	1560	85
ARSENIC TRIOXIDE	1561	86
ARSENIC TRISULFIDE	1557	87
ARSINE	2188	88
ASBESTOS, BLUE	2212	89
ASPHALT	1999	90
AZINPHOS METHYL	2783	91
BARIUM CARBONATE	1564	92
BARIUM CHLORATE	1445	93
BARIUM CYANIDE	1565	94
BARIUM NITRATE	1446	95
BARIUM PERCHLORATE	1447	96
BARIUM PERMANGANATE	1448	97
BARIUM PEROXIDE	1449	98
BENZALDEHYDE	1989	99
BENZENE	1114	100
BENZENE HEXACHLORIDE	2761	101
BENZENE PHOSPHORUS DICHLORIDE	2798	102
BENZENE PHOSPHORUS THIODICHLORIDE	2799	103
BENZOIC ACID	9094	104
BENZONITRILE	2224	105
BENZOYL CHLORIDE	1736	106
BENZOYL PEROXIDE	2085	107
BENZYL BROMIDE	1737	108
BENZYL CHLORIDE	1738	109
BENZYL CHLORFORMATE	1739	110
BENZYLAMINE	na	111
BERYLLIUM CHLORIDE	1566	112
BERYLLIUM FLUORIDE	1556	113
BERYLLIUM NITRATE	2464	114
BERYLLIUM OXIDE	1566	115
BERYLLIUM SULFATE	1566	116
BORON TRIBROMIDE	2692	117
BORON TRICHLORIDE	1741	118
BROMINE	1744	119

CHEMICAL NAME	DOT#	PAGE
BROMINE PENTAFLUORIDE	1745	120
BROMINE TRIFLUORIDE	1746	121
BROMOBENZENE	2514	122
BRUCINE	1570	123
BUTADIENE, INHIBITED	1010	124
BUTANE	1011	125
BUTANEDIOL	1987	126
BUTENE	1012	127
BUTENEDIOL	1987	128
BUTYL ACETATE	1123	129
sec-BUTYL ACETATE	1124	130
BUTYL ACRYLATE	2348	131
iso-BUTYL ACRYLATE	2527	132
n-BUTYL ACRYLATE	2348	133
BUTYLENE	1012	134
BUTYLENE OXIDE	3022	135
BUTYL PHENOL	2229	136
BUTYL TRICHLOROSILANE	1747	137
BUTYRALDEHYDE	1129	138
iso-BUTYRALDEHYDE	1129	139
BUTYRIC ACID	2820	140
CACODYLIC ACID	1572	141
CADMIUM ACETATE	2570	142
CADMIUM BROMIDE	2570	143
CADMIUM CHLORIDE	2570	144
CADMIUM FLUOROBORATE	2570	145
CADMIUM NITRATE	2570	146
CADMIUM OXIDE	2570	147
CADMIUM SULFATE	2570	148
CALCIUM ARSENATE	1573	149
CALCIUM ARSENITE	1574	150
CALCIUM CARBIDE	1402	151
CALCIUM CHLORATE	1452	152
CALCIUM CHROMATE	9096	153
CALCIUM CYANIDE	1575	154
CALCIUM HYPOCHLORITE	1748	155
CALCIUM NITRATE	1454	156
CALCIUM OXIDE	1910	157
CALCIUM PEROXIDE	1457	158
CALCIUM PHOSPHIDE	1360	159
CALCIUM RESINATE	1313	160
CAMPHENE	9011	161
CAMPHOR OIL	1130	162
CAPTAN-(liquid)	9099	163
CAPTAN-(solid)	9099	164
CARBARYL-(liquid)	2757	165
CARBARYL-(solid)	2757	166
CARBOFURAN	2757	167
CARBOLIC ACID	1671	168
CARBOLIC OIL	2821	169
CARBON BISULFIDE	1131	170
CARBON DIOXIDE	1013	171
CARBON DISULFIDE	1131	172
CARBON MONOXIDE	1016	173
CARBON TETRACHLORIDE	1846	174
CAUSTIC POTASH SOLUTION	1824	175
CAUSTIC SODA SOLUTION	1824	176
CHARCOAL	1361	177
CHLORDANE	2762	178
CHLORINE	1017	179
CHLORINE TRIFLUORIDE	1749	180
CHLOROACETIC ACID	1751	181
CHLOROACETOPHENONE	1697	182
CHLOROACETYL CHLORIDE	1752	183
CHLOROANILINE	2018	184
CHLOROBENZENE	1134	185
CHLOROFORM	1888	186
CHLOROHYDRINS	2023	187
CHLOROMETHYL METHYL ETHER	1239	188
CHLORONITROBENZENE	1578	189
CHLOROPHENOL	2020	190
CHLOROPICRIN	1580	191
CHLOROPRENE	1991	192
CHLOROSULPHONIC ACID	1754	193
CHLOROTOLUENE-(o-m-p)	2239	194
CHROMIC ACETATE	9101	195
CHROMIC ACID	1463	196
CHROMIC ANHYDRIDE	1463	197
CHROMIC SULFATE	9100	198
CHROMOUS CHLORIDE	9102	199
CHROMYL CHLORIDE	1758	200
COLLODION	2059	201
COPPER ACETATE	9106	202
COPPER ACETOARSENITE	1585	203
COPPER ARSENITE	1586	204
COPPER CHLORIDE	2802	205
COPPER CYANIDE	1587	206
COPPER NAPHTHENATE	1168	207
COPPER NITRATE	1479	208
COPPER OXALATE	2449	209
COPPER SULFATE	9109	210
COUMAPHOS	2783	211
CREOSOTE, COAL TAR	1993	212
CRESOL-(o-m-p)	2067	213
CROTONALDEHYDE	1143	214
CUMENE	1918	215
CUMENE HYDROPEROXIDE	2116	216
CUPRIC NITRATE	1479	217
CUPRIC SULFATE	9109	218
CUPRIETHYLENEDIAMINE SOLUTION	1761	219
CYANOACETIC ACID	1935	220
CYANOGEN	1026	221
CYANOGEN BROMIDE	1889	222
CYANOGEN CHLORIDE	1589	223
CYCLOHEXANE	1145	224
CYCLOHEXANOL	1993	225
CYCLOHEXANONE	1915	226
CYCLOHEXANONE PEROXIDE	2118	227
CYCLOHEXENYL TRICHLOROSILANE	1762	228
CYCLOHEXENYLAMINE	2357	229
CYCLOPENTANE	1146	230
CYCLOPROPANE	1027	231
CYMENE	2046	232
2,4-D ESTERS	2765	233
DALAPON	1760	234
DDD	2761	235
DDT	2761	236
DECABORANE	1868	237
DECAHYDRONAPHTHALENE	1147	238
DECYL ALCOHOL	1987	239
DI-(2-ETHYLHEXYL) PHOSPHORIC ACID	1902	240
DIBUTYL ETHER	1149	241
DIACETONE ALCOHOL	1148	242
DIAZINON	2783	243
DIBENZOYL PEROXIDE	2085	244
DIBORANE	1911	245
DIBUTYL PHTHALATE	9095	246
DICAMBA-(liquid)	2769	247
DICAMBA-(solid)	2769	248
DICHLOBENIL-(liquid)	2769	249

CHEMICAL NAME	DOT#	PAGE
DICHLOBENIL-(solid)	2769	250
DICHLONE-(liquid)	2761	251
DICHLONE-(solid)	2761	252
1,2-DICHLOROETHYLENE	1150	253
o-DICHLOROBENZENE	1591	254
p-DICHLOROBENZENE	1592	255
DICHLOROBENZOYL PEROXIDE-2,4	2138	256
DICHLOROBUTENE	2924	257
DICHLORODIFLUOROMETHANE	1028	258
DICHLOROETHANE	2362	259
DICHLOROETHYL ETHER	1916	260
DICHLOROMETHANE	1593	261
DICHLOROPHENOL	2020	262
DICHLOROPHENOXYACETIC ACID	2765	263
1,2-DICHLOROPROPANE	1279	264
1,3-DICHLOROPROPENE	2047	265
DICHLORVOS-(liquid)	2783	266
DICHLORVOS-(solid)	2783	267
DICYCLOPENTADIENE	2048	268
DIELDRIN-(liquid)	2761	269
DIELDRIN-(solid)	2761	270
DIETHYL CARBONATE	2366	271
DIETHYLAMINE	1154	272
DIETHYLBENZENE	2049	273
DIETHYLENETRIAMINE	2079	274
DIETHYLZINC	1366	275
1,1-DIFLUOROETHANE	1030	276
DIFLUOROPHOSPHORIC ACID	1768	277
DIISOBUTYL KETONE	1157	278
DIISOBUTYLCARBINOL	1993	279
DIISOBUTYLENE	2050	280
DIISOPROPYLAMINE	1158	281
DIISOPROPYLBENZENE HYDROPEROXIDE	2171	282
DIMETHYL ETHER	1033	283
DIMETHYL SULFATE	1595	284
DIMETHYL SULFIDE	1164	285
DIMETHYLAMINE	1032	286
DIMETHYLDICHLOROSILANE	1162	287
DIMETHYLFORMAMIDE	2265	288
DIMETHYLHEXANE DIHYDROPEROXIDE	2174	289
DIMETHYLHYDRAZINE	1163	290
DIMETHYLZINC	1370	291
DINITROANILINE	1596	292
DINITROBENZENE-(liquid)	1597	293
DINITROBENZENE-(solid)	1597	294
DINITROCRESOL	1598	295
DINITROPHENOL	1320	296
DINITROTOULENE-(liquid)	1600	297
DINITROTOULENE-(solid)	2038	298
DIOXANE	1165	299
DIPENTENE	2052	300
DIPHENYLAMINE	na	301
DIPHENYLDICHLOROSILANE	1769	302
DIPHENYLMETHYL DIOSOCYANATE	2489	303
DIPROPYLAMINE	2383	304
DIQUAT-(liquid)	2781	305
DIQUAT-(solid)	2781	306
DISULFOTON-(liquid)	2783	307
DISULFOTON-(solid)	2783	308
DIURON	2767	309
DODECYLBENZENESULFONIC ACID	2584	310
DODECYL TRICHLOROSILANE	1771	311
DURSBAN	1615	312
ENDOSULFAN-(liquid)	2761	313
ENDOSULFAN-(solid)	2761	314
ENDRIN-(liquid)	2761	315
ENDRIN-(solid)	2761	316
EPICHLOROHYDRIN	2023	317
ETHANE	1035	318
ETHION	2783	319
ETHYL ACETATE	1173	320
ETHYL ACETOACETATE	1993	321
ETHYL ACRYLATE, INHIBITED	1917	322
ETHYL ACRYLATE-(liquid)	1173	323
ETHYL ALCOHOL	1170	324
ETHYL ALUMINUM SESQUICHLORIDE	1925	325
ETHYLAMINE	1036	326
ETHYLBENZENE	1175	327
ETHYLBUTANOL	2275	328
ETHYL BUTYRATE	1180	329
ETHYL CHLORIDE	1037	330
ETHYL CHLOROACETATE	1181	331
ETHYL CHLOROFORMATE	1182	332
ETHYL CYANOHYDRIN	na	333
ETHYLDICHLOROSILANE	1183	334
ETHYLENE	1962	335
ETHYLENE CHLOROHYDRIN	1135	336
ETHYLENEDIAMINE	1604	337
ETHYLENEDIAMINE TETRACETIC ACID	9117	338
ETHYLENE DIBROMIDE	1605	339
ETHYLENE DICHLORIDE	1184	340
ETHYLENE GLYCOL DIETHYL ETHER	1153	341
ETHYLENE GLYCOL MONOETHYL ETHER	1171	342
ETHYLENE GLYCOL MONOETHYL ETHER ACETATE	1172	343
ETHYLENE IMINE	1185	344
ETHYLENE OXIDE	1040	345
ETHYLENETRICHLOROSILANE	1196	346
ETHYL ETHER	1155	347
ETHYL FORMATE	1190	348
ETHYL HEXALDEHYDE	1191	349
ETHYL LACTATE	1192	350
ETHYL MERCAPTAN	2363	351
ETHYL METHACRYLATE	2277	352
ETHYL NITRATE	1993	353
ETHYL PHENYL DICHLOROSILANE	2435	354
ETHYL PHOSPHONOTHIOIC DICHLORIDE	1760	355
ETHYL PHOSPHORODICHLORIDATE	1760	356
ETHYL PROPIONATE	1195	357
ETHYL PROPYLACROLENE	1191	358
ETHYL SILICATE	1292	359
ETIOLOGIC AGENT	2814	360
EXPLOSIVES A	na	361
FERRIC AMMONIUM CITRATE	9118	362
FERRIC AMMONIUM OXILATE	9119	363
FERRIC CHLORIDE-(liquid)	2582	364
FERRIC CHLORIDE-(solid)	1773	365
FERRIC FLUORIDE	9120	366
FERRIC GYLCEROPHOSPHATE	na	367
FERRIC NITRATE	1466	368
FERRIC SULFATE	9121	369
FERROUS AMMONIUM SULFATE	9122	370
FERROUS CHLORIDE-(liquid)	1760	371
FERROUS CHLORIDE-(solid)	1759	372
FERROUS FLUOROBORATE	na	373
FERROUS OXALATE	na	374
FERROUS SULFATE	9125	375
FLUORINE	1045	376
FLUOROSILICIC ACID	1778	377
FLUROSULPHONIC ACID	1777	378

CHEMICAL NAME	DOT#	PAGE
FORMALDEHYDE	1198	379
FORMIC ACID	1779	380
FUEL OIL NUMBER 1	1993	381
FUEL OIL NUMBER 2	1993	382
FUEL OIL NUMBER 4	1993	383
FUEL OIL NUMBER 5	1993	384
FUMARIC ACID	9126	385
FURFURAL	1199	386
FURFURYL ALCOHOL	2874	387
GALLIC ACID	na	388
GAS OIL	1202	389
GASOLINE	1203	390
GASOLINE AVIATION	1203	391
GASOLINE CASINGHEAD	1203	392
GELATINE DYNAMITE	na	393
GLYCERINE	na	394
GLYCIDALDEHYDE	2622	395
GLYCIDYL METHACRYLATE	na	396
GLYOXAL	na	397
GRENADE(WITH POISON B CHARGE)	2016	398
HAZARDOUS WASTE	9189	399
HEPTACHLOR	2761	400
HEPTANE	1206	401
HEPTANOL	na	402
HEPTENE	2278	403
HEXACHLOROCYCLOPENTADIENE	2646	404
HEXACHLOROETHANE	9037	405
HEXADECYLTRIMETHYLAMMONIUM CHLORIDE	na	406
HEXALDEHYDE	1207	407
HEXAMETHYLENE DIAMINE	2280	408
HEXAMETHYLENIMINE	2493	409
HEXAMETHYLENETETRAMINE	1328	410
HEXANE	1208	411
HEXANOL	2282	412
HEXENE	2370	413
HEXYLENE GLYCOL	2030	414
HYDROCHLORIC ACID	1789	415
HYDRAZINE	2029	416
HYDROCYANIC ACID	1051	417
HYDROFLUORIC ACID	1790	418
HYDROGEN	1049	419
HYDROGEN BROMIDE	1048	420
HYDROGEN CHLORIDE	1050	421
HYDROGEN CYANIDE	1051	422
HYDROGEN FLUORIDE	1052	423
HYDROGEN PEROXIDE (52–100%)	2015	424
HYDROGEN SULFIDE	1053	425
HYDROQUINONE	2662	426
HYDROSILICOFLUORIC ACID	1778	427
HYDROXYETHYL ACRYLATE	na	428
HYDROXYLAMINE	na	429
HYDROXYLAMINE SULFATE	2865	430
HYDROXYPROPYL ACRYLATE	1760	431
HYDROXYPROPYL METHACRYLATE	na	432
ISOAMYL ALCOHOL	1105	433
ISOBUTANE	1969	434
ISOBUTYL ACETATE	1213	435
ISOBUTYL ALCOHOL	1212	436
ISOBUTYLAMINE	1214	437
ISOBUTYLENE	1055	438
ISOBUTYLNITRILE	2284	439
ISOBUTYRALDEHYDE	2045	440
ISOBUTYRIC ACID	2529	441
ISODECALDEHYDE	na	442

CHEMICAL NAME	DOT#	PAGE
ISODECYL ACRYLATE	na	443
ISODECYL ALCOHOL	na	444
ISOHEXANE	1208	445
ISOOCTALDEHYDE	na	446
ISOOCTYL ALCOHOL	1987	447
ISOPENTANE	1265	448
ISOPHRONE	na	449
ISOPHTHALIC ACID	na	450
ISOPRENE	1218	451
ISOPROPANOL	1219	452
ISOPROPYL ACETATE	1220	453
ISOPROPYL ALCOHOL	1219	454
ISOPROPYLAMINE	1221	455
ISOPROPYL ETHER	1159	456
ISOPROPYL MERCAPTAN	2703	457
ISOPROPYL PERCARBONATE (STABILIZED)	2134	458
ISOPROPYL PERCARBONATE (UNSTABILIZED)	2133	459
ISOVALERALDEHYDE	1989	460
JET FUELS, JP-1	2761	461
JET FUELS, JP-3	1863	462
JET FUELS, JP-4	1223	463
JET FUELS, JP-5	1223	464
KEPONE	2761	465
KEROSENE	1223	466
LACTIC ACID	1760	467
LATEX LIQUID SYNTHETIC	na	468
LAUROYL PEROXIDE	2124	469
LAUROYL PEROXIDE (LESS THAN 42%)	2893	470
LAURYL MERCAPTAN	na	471
LEAD ACETATE	1616	472
LEAD ARSENATE	1617	473
LEAD CHLORIDE	2291	474
LEAD FLUORIDE	2811	475
LEAD FLUOROBORATE	2291	476
LEAD IODINE	2811	477
LEAD NITRATE	1469	478
LEAD STEARATE	2811	479
LEAD SULFIDE	2811	480
LEAD TETRAACETATE	na	481
LEAD THIOCYANATE	2291	482
LEAD THIOSULFATE	na	483
LEAD TUNGSTATE	na	484
LINDANE	2761	485
LINEAR ALCOHOL	na	486
LIQUIFIED NATURAL GAS (LNG)	1972	487
LIQUIFIED PETROLEUM	1075	488
LITHARGE	na	489
LITHIUM	1415	490
LITHIUM ALUMINUM HYDRIDE	1410	491
LITHIUM BICHROMATE	na	492
LITHIUM CHROMATE	9134	493
LITHIUM HYDRIDE	1414	494
MAGNESIUM	1869	495
MAGNESIUM PERCHLORATE	1475	496
MALATHION	2783	497
MALEIC ACID	2215	498
MALEIC ANHYDRIDE	2215	499
MALEIC HYDRAZIDE	na	500
MERCAPTODIMETHUR	2757	501
MERCURIC ACETATE	1629	502
MERCURIC CHLORIDE	1624	503
MERCURIC CYANIDE	1636	504
MERCURIC IODIDE-(liquid)	1638	505

CHEMICAL NAME	DOT#	PAGE
MERCURIC IODIDE-(solid)	1638	506
MERCURIC NITRATE	1625	507
MERCURIC OXIDE	1641	508
MERCURIC SULFATE	1645	509
MERCURIC SULFIDE	2025	510
MERCURIC THIOCYANATE	1646	511
MERCUROUS CHLORIDE	2025	512
MERCUROUS NITRATE	1627	513
MERCURY	2809	514
MERCURY AMMONIUM CHLORIDE	1630	515
MESITYL OXIDE	1229	516
METHALLYL ALCOHOL	2614	517
METHANE	1971	518
METHANEARSONIC ACID, SODIUM SALT	1557	519
METHANOL	1230	520
METHOXYCHLOR	2761	521
METHYL ACETATE	1231	522
METHYL ACETYLENE	1060	523
METHYL ACRYLATE	1919	524
METHYL ALCOHOL	1230	525
METHYL ALLYL CHLORIDE	2554	526
METHYLAMINE	1061	527
METHYL AMYL ACETATE	1233	528
METHYL AMYL ALCOHOL	2053	529
METHYL AMYL KETONE	1110	530
METHYLANILINE	2294	531
METHYL BROMIDE	1062	532
METHYL BUTENE	2460	533
METHYL CHLORIDE	1063	534
METHYL CHLOROFORMATE	1238	535
METHYL CYANIDE	1648	536
METHYL CYCLOPENTANE	2298	537
METHYL DICHLOROSILANE	1242	538
METHYLENE CHLORIDE	1593	539
METHYL ETHYL KETONE	1193	540
METHYL ETHYL PYRIDINE	2300	541
METHYL FORMAL	1234	542
METHYL FORMATE	1243	543
METHYLHYDRAZINE	1244	544
METHYL ISOBUTYL CARBINOL	2053	545
METHYL ISOBUTYL KETONE	1245	546
METHYL ISOCYANATE	2480	547
METHYL ISOPROPENYL KETONE	1246	548
METHYL MERCAPTAN	1064	549
METHYL METHACRYLATE	1247	550
METHYL NBUTYL KETONE	1993	551
METHYL PARATHION	2783	552
METHYL PHOSPHONOTHIOIC DICHLORIDE	1760	553
METHYLPYRROLIDONE	na	554
METHYLSTRYENE	na	555
METHYL TRICHLOROSILANE	1250	556
METHYL VINYL KETONE	1251	557
MINERAL SPIRITS	1300	558
MIREX	1615	559
MOLBDIC TRIOXIDE	na	560
MONOCHLOROACETIC ACID-(liquid)	1750	561
MONOCHLOROACETIC ACID-(solid)	1750	562
MONOCHLORODIFLUOROMETHANE	1018	563
MONOETHANOLAMINE	2491	564
MONOISOPROPANOLIMINE	na	565
MONOMETHYLAMINE	1061	566
MORPHOLINE	2054	567
MOTOR FUEL ANTIKNOCK COMPOUND	1649	568
NABAM	1609	569
NALED-(liquid)	2783	570
NALED-(solid)	2783	571
NAPHTHA	2553	572
NAPHTHA, SOLVENT	1256	573
NAPHTHA (STODDARD SOLVENT)	1268	574
NAPHTHA (WMP)	1255	575
NAPHTHALENE	2304	576
NAPHTHENIC ACID	9137	577
NAPHTHYLAMINE	2077	578
NATURAL GAS	1972	579
NEOHEXANE	1208	580
NICKEL ACETATE	na	581
NICKEL AMMONIUM SULFATE	9138	582
NICKEL BROMIDE	na	583
NICKEL CARBONYL	1259	584
NICKEL CHLORIDE	9139	585
NICKEL CYANIDE	1653	586
NICKEL FLUOROBORATE	na	587
NICKEL FORMATE	na	588
NICKEL HYDROXIDE	9140	589
NICKEL NITRATE	2725	590
NICKEL SULFATE	9141	591
NICOTINE	1654	592
NICOTINE SULFATE-(liquid)	1658	593
NICOTINE SULFATE-(solid)	1658	594
NITRALIN	1609	595
NITRATING ACID	1796	596
NITRIC ACID	2031	597
NITRIC OXIDE	1660	598
NITRILOTRIACETIC ACID AND SALTS	na	599
NITROANILINE	1661	600
NITROBENZENE	1662	601
NITROCHLOROBENZENE-META	1578	602
NITROCHLOROBENZENE-ORTHO	1578	603
NITROCHLOROBENZENE-PARA	1578	604
NITROETHANE	2842	605
NITROGEN	1977	606
NITROGEN TETROXIDE	1067	607
NITROMETHANE	1261	608
NITROPHENOL	1663	609
NITROPROPANE	2608	610
NITROSYL CHLORIDE	1069	611
NITROTOLUENE	1664	612
NITROUS OXIDE	1070	613
NONANE	1920	614
NONANOL	na	615
NONYLPHENOL	na	616
NUCLEAR REACTOR FUEL	2918	617
OCTANE	1262	618
OCTANOL	na	619
OCTENE	na	620
OCTYL EPOXY TALLATE	na	621
OIL	1270	622
OLEIC ACID	na	623
OLEIC ACID, POTASSIUM SALT	na	624
OLEIC ACID, SODIUM SALT	na	625
OLEUM	1831	626
OXALIC ACID	na	627
OXYGEN	1073	628
PARAFORMALDEHYDE	2213	629
PARALDEHYDE	1264	630
PARATHION	2783	631
PENTABORANE	1380	632
PENTACHLOROPHENOL	2020	633
PENTADECANOL	na	634

CHEMICAL NAME	DOT#	PAGE
PENTAERYTHRITOL	na	635
PENTANE	1265	636
PENTENE	1108	637
PERACETIC ACID	2131	638
PERCHLORETHYLENE	1897	639
PERCHLORIC ACID LESS THAN 50%	1802	640
PERCHLORIC ACID OVER 50%	1873	641
PERCHLOROMETHYL MERCAPTAN	1670	642
PETROLATUM	na	643
PETROLEUM NAPHTHA	1255	644
PHENOL	1671	645
PHENYLDICHLOROARSINE	1556	646
PHENYLHYDRAZINE HYDROCHLORIDE	2572	647
PHOSDRIN	2783	648
PHOSGENE	1076	649
PHOSPHORIC ACID	1805	650
PHOSPHORIC ACID TRIETHYLENEIMINE	2105	651
PHOSPHORUS BLACK	1381	652
PHOSPHORUS OXYCHLORIDE	1810	653
PHOSPHORUS PENTASULFIDE	1340	654
PHOSPHORUS RED	1338	655
PHOSPHORUS TRIBROMIDE	1808	656
PHOSPHORUS TRICHLORIDE	1809	657
PHOSPHORUS WHITE	1381	658
PHTHALIC ANHYDRIDE	2214	659
PINE OIL	1272	660
PIPERAZINE	2579	661
POLYBUTENE	na	662
POLYCHLORINATED BIPHENYL	2315	663
POLYMETHYLENE POLYPHENYL ISOCYANATE	na	664
POLYPHOSPHORIC ACID	na	665
POLYPROPYLENE	na	666
POTASSIUM	2257	667
POTASSIUM ARSENATE	1677	668
POTASSIUM ARSENITE	1678	669
POTASSIUM CHLORATE	1485	670
POTASSIUM CHROMATE	9142	671
POTASSIUM CYANIDE-(liquid)	1680	672
POTASSIUM CYANIDE-(solid)	1680	673
POTASSIUM DICHLOROISOCYANURATE	2465	674
POTASSIUM DICHROMATE	1479	675
POTASSIUM HYDROXIDE-(liquid)	1814	676
POTASSIUM HYDROXIDE-(solid)	1813	677
POTASSIUM NITRATE	1486	678
POTASSIUM OXALATE	na	679
POTASSIUM PERMANGANATE	1490	680
POTASSIUM PEROXIDE	1491	681
PROPANE	1078	682
PROPARGITE	2765	683
PROPIOLACTONEBETA	na	684
PROPIONALDEHYDE	1275	685
PROPIONIC ACID	1848	686
PROPIONIC ANHYDRIDE	2496	687
PROPYL ACETATE	1276	688
PROPYL ALCOHOL	1274	689
PROPYLAMINE	1277	690
PROPYLENE	1077	691
PROPYLENE GLYCOL	na	692
PROPYLENE OXIDE	1280	693
PROPYLENE TETRAMER	2850	694
PROPYLENEIMINE	1921	695
PROPYL MERCAPTAN	2402	696
PYRETHRINS ACID-(liquid)	9184	697
PYRETHRINS ACID-(solid)	9184	698
PYRIDINE	1282	699
PYROGALLIC ACID	na	700
QUINOLINE	2656	701
RADIOACTIVE MATERIAL (special form) N.O.S.	2974	702
RESORCINOL	2876	703
ROSIN SOLUTION	1993	704
SALICYLIC ACID	na	705
SELENIUM DIOXIDE	na	706
SELENIUM TRIOXIDE	na	707
SILICON CHLORIDE	1818	708
SILVER ACETATE	na	709
SILVER CARBONATE	na	710
SILVER FLUORIDE	na	711
SILVER IODATE	na	712
SILVER NITRATE	1493	713
SILVER OXIDE	na	714
SILVER SULFATE	na	715
SODIUM	1428	716
SODIUM ALKYLBENZENESULFONATES	na	717
SODIUM ALKYL SULFATES	na	718
SODIUM AMIDE	1425	719
SODIUM ARSENATE	1685	720
SODIUM ARSENITE-(liquid)	1686	721
SODIUM ARSENITE-(solid)	2027	722
SODIUM AZIDE	1687	723
SODIUM BIFLUORIDE	2439	724
SODIUM BISULFIDE-(liquid)	2693	725
SODIUM BISULFIDE-(solid)	1821	726
SODIUM BORATE	na	727
SODIUM BOROHYDRIDE	1426	728
SODIUM CACODYLATE	1688	729
SODIUM CHLORATE	1495	730
SODIUM CHROMATE	9145	731
SODIUM CYANIDE-(liquid)	1689	732
SODIUM CYANIDE-(solid)	1689	733
SODIUM DICHLORO-s-TRIAZINETRIONE	2465	734
SODIUM DICHROMATE	1479	735
SODIUM FERROCYANIDE	na	736
SODIUM FLUORIDE	1690	737
SODIUM HYDRIDE	1427	738
SODIUM HYDROSULFIDE SOLUTION	2922	739
SODIUM HYDROSULFITE	1384	740
SODIUM HYDROXIDE	1823	741
SODIUM HYPOCHLORITE	1791	742
SODIUM METHYLATE	1431	743
SODIUM NITRATE	1498	744
SODIUM NITRITE	1500	745
SODIUM OXALATE	na	746
SODIUM PHOSPHATE	9148	747
SODIUM SELENITE	2630	748
SODIUM SILICATE	na	749
SODIUM SILICOFLUORIDE	2674	750
SODIUM SULFIDE	1385	751
SODIUM THIOCYANATE	na	752
SORBITOL	na	753
STANNOUS FLUORIDE	na	754
STEARIC ACID	na	755
STRONTIUM CHROMATE	9149	756
STRYCHINE	1692	757
STYRENE	2055	758
SUCROSE	na	759
SULFOLANE	na	760
SULFUR	1350	761
SULFUR CHLORIDE	1828	762

CHEMICAL NAME	DOT#	PAGE
SULFUR DIOXIDE	1079	763
SULFUR MONOCHLORIDE	1828	764
SULFURIC ACID	1831	765
SULFURIC ACID, SPENT	1832	766
SULFURYL CHLORIDE	1834	767
TANNIC ACID	na	768
TETRABUTYL TITANATE	na	769
TETRACHLOROETHANE	1702	770
TETRACHLOROETHYLENE	1897	771
TETRADECANOL	na	772
TETRADECENE	na	773
TETRAETHYL DITHIOPYROPHOSPHATE-(liquid)	1704	774
TETRAETHYL DITHIOPYROPHOSPHATE-(solid)	1704	775
TETRAETHYL LEAD	1649	776
TETRAETHYL PYROPHOSPHATE	1705	777
TETRAETHYLENE GLYCOL	na	778
TETRAETHYLENEPENTAMINE	2320	779
TETRAFLUOROETHYLENE	1081	780
TETRAHYDROFURAN	2056	781
TETRAHYDRONAPHTHALENE	na	782
THALLIUM SULFATE	1707	783
THIOPHOSGENE	2474	784
THIRAM	2771	785
THORIUM NITRATE	2976	786
THORIUM ORE	2912	787
TITANIUM TETRACHLORIDE	1838	788
TOLUENE	1294	789
TOLUENE-2,4-DIISOCYANATE	2078	790
TOLUENE SULFONIC ACID	2583	791
TOLUIDINE-M	1708	792
TOLUIDINE-O	1708	793
TOLUIDINE-P	1708	794
TOXAPHENE-(liquid)	2761	795
TOXAPHENE-(solid)	2761	796
TRICHLORFON	2783	797
TRICHLOROPHENOXY ACETIC ACID AMINE	2765	798
TRICHLORO-S-TRIAZINETRIONE	2468	799
TRICHLOROETHYLENE	1710	800
TRICHLOROFLUOROMETHANE	na	801
TRICHLOROPHENOL	2020	802
TRICHLOROPHENOXY ACETIC ACID ESTER	2765	803
TRICHLOROPHENOXYPROPIONIC ACID ESTER	2765	804
TRICHLOROSILANE	1295	805
TRICRESYLPHOSPHATE	2574	806
TRIDECANOL	na	807
TRIDECENE	na	808
TRIETHANOLAMINE	9151	809
TRIETHYLALUMINUM	1103	810
TRIETHYLAMINE	1296	811
TRIETHYLBENZENE	na	812
TRIETHYLENE GLYCOL	na	813
TRIETHYLENE TETRAMINE	2259	814
TRIFLUOROCHLOROETHYLENE	1082	815
TRIFLURALIN	1609	816
TRIISOBUTYLALUMINUM	1930	817
TRIMETHYLCHLOROSILANE	1298	818
TRIPROPYLENE GLYCOL	na	819
TURPENTINE	1299	820
UNDECANOL	na	821
UNDECENE	na	822
UNDECYLBENZENE	na	823
URANIUM HEXAFLUORIDE	2977	824
URANIUM PEROXIDE	na	825
URANYL ACETATE	9180	826
URANYL NITRATE	2981	827
URANYL SULFATE	na	828
UREA	na	829
UREA PEROXIDE	1511	830
VALERALDEHYDE	2058	831
VANADIUM OXYTRICHLORIDE	2443	832
VANADIUM PENTOXIDE	2862	833
VANADYL SULFATE	2931	834
VINYL ACETATE	1301	835
VINYL CHLORIDE	1086	836
VINYL ETHYL ETHER	1302	837
VINYL FLUORIDE	1860	838
VINYLIDENE CHLORIDE	1303	839
VINYL METHYL ETHER	1087	840
VINYL TOLUENE	2618	841
VINYL TRICHLOROSILANE	1305	842
WAX	1993	843
XYLENE-M	1307	844
XYLENE-O	1307	845
XYLENE-P	1307	846
XYLENOL-(liquid)	2261	847
XYLENOL-(solid)	2261	848
ZECTRAN	1615	849
ZINC ACETATE	9153	850
ZINC AMMONIUM CHLORIDE	9154	851
ZINC ARSENATE	1712	852
ZINC ARSENITE	1712	853
ZINC BICHROMATE	na	854
ZINC BORATE	1988	855
ZINC BROMIDE	9156	856
ZINC CARBONATE	9157	857
ZINC CHLORIDE	1840	858
ZINC CHROMATE	na	859
ZINC CYANIDE	1713	860
ZINC DIALKYLDITHIOPHOSPHATE	1893	861
ZINC FLUORIDE	9158	862
ZINC FLUOROBORATE	na	863
ZINC FORMATE	9159	864
ZINC HYDROSULFITE	1931	865
ZINC NITRATE	1514	866
ZINC PHENOLSULFONATE	9160	867
ZINC PHOSPHIDE	1714	868
ZINC POTASSIUM CHROMATE	na	869
ZINC SILICOFLUORIDE	2855	870
ZINC SULFATE	9161	871
ZIRCONIUM ACETATE	na	872
ZIRCONIUM NITRATE	2728	873
ZIRCONIUM OXYCHLORIDE	na	874
ZIRCONIUM POTASSIUM FLUORIDE	9162	875
ZIRCONIUM SULFATE	9163	876
ZIRCONIUM TETRACHLORIDE	2503	877

DOT NUMBER INDEX

DOT#	PAGE	CHEMICAL NAME
1001	15	ACETYLENE
1005	35	AMMONIA, ANHYDROUS
1010	124	BUTADIENE, INHIBITED
1011	125	BUTANE
1012	127	BUTENE
1012	134	BUTYLENE
1013	172	CARBON DIOXIDE
1016	173	CARBON MONOXIDE
1017	179	CHLORINE
1018	563	MONOCHLORODIFLUOROMETHANE
1026	221	CYANOGEN
1027	231	CYCLOPROPANE
1028	258	DICHLORODIFLUOROMETHANE
1030	276	1,1-DIFLUOROETHANE
1032	286	DIMETHYLAMINE
1033	283	DIMETHYL ETHER
1035	318	ETHANE
1036	326	ETHYLAMINE
1037	330	ETHYL CHLORIDE
1040	345	ETHYLENE OXIDE
1045	376	FLUORINE
1048	420	HYDROGEN BROMIDE
1049	419	HYDROGEN
1050	421	HYDROGEN CHLORIDE
1051	417	HYDROCYANIC ACID
1051	422	HYDROGEN CYANIDE
1052	423	HYDROGEN FLUORIDE
1053	425	HYDROGEN SULFIDE
1055	438	ISOBUTYLENE
1060	523	METHYL ACETYLENE
1061	527	METHYLAMINE
1061	566	MONOMETHYLAMINE
1062	532	METHYL BROMIDE
1063	534	METHYL CHLORIDE
1064	549	METHYL MERCAPTAN
1067	607	NITROGEN TETROXIDE
1069	611	NITROSYL CHLORIDE
1070	613	NITROUS OXIDE
1073	628	OXYGEN
1075	488	LIQUIFIED PETROLEUM
1076	649	PHOSGENE
1077	691	PROPYLENE
1078	682	PROPANE
1079	763	SULFUR DIOXIDE
1081	780	TETRAFLUOROETHYLENE
1082	815	TRIFLUOROCHLOROETHYLENE
1086	836	VINYL CHLORIDE
1087	840	VINYL METHYL ETHER
1088	1	ACETAL
1089	2	ACETALDEHYDE
1090	6	ACETONE
1092	17	ACROLEIN
1093	20	ACRYLONITRILE
1098	24	ALLYL ALCOHOL
1099	25	ALLYL BROMIDE
1100	26	ALLYL CHLORIDE
1103	810	TRIETHYLALUMINUM
1104	64	AMYL ACETATE
1105	65	AMYL ALCOHOL
1105	433	ISOAMYL ALCOHOL

DOT#	PAGE	CHEMICAL NAME
1107	66	AMYL CHLORIDE
1108	637	PENTENE
1110	68	AMYL METHYL KETONE
1110	530	METHYL AMYL KETONE
1111	67	AMYL MERCAPTAN
1112	69	AMYL NITRATE
1113	70	AMYL NITRITE
1114	100	BENZENE
1120	134	BUTYL ALCOHOL
1123	129	BUTYL ACETATE
1124	130	sec-BUTYL ACETATE
1129	139	iso-BUTYRALDEHYDE
1129	146	BUTYRALDEHYDE
1130	162	CAMPHOR OIL
1131	170	CARBON BISULFIDE
1131	172	CARBON DISULFIDE
1134	185	CHLOROBENZENE
1135	336	ETHYLENE CHLOROHYDRIN
1143	214	CROTONALDEHYDE
1145	224	CYCLOHEXANE
1146	230	CYCLOPENTANE
1147	238	DECAHYDRONAPHTHALENE
1148	242	DIACETONE ALCOHOL
1149	241	DIBUTYL ETHER
1150	253	1,2-DICHLOROETHYLENE
1153	341	ETHYLENE GLYCOL DIETHYL ETHER
1154	272	DIETHYLAMINE
1155	347	ETHYL ETHER
1157	278	DIISOBUTYL KETONE
1158	281	DIISOPROPYLAMINE
1159	456	ISOPROPYL ETHER
1162	287	DIMETHYLDICHLOROSILANE
1163	290	DIMETHYLHYDRAZINE
1164	285	DIMETHYL SULFIDE
1165	299	DIOXANE
1168	207	COPPER NAPHTHENATE
1170	324	ETHYL ALCOHOL
1171	342	ETHYLENE GLYCOL MONOETHYL ETHER
1172	343	ETHYLENE GLYCOL MONOETHYL ETHER ACETATE
1173	320	ETHYL ACETATE
1173	323	ETHYL ACRYLATE (liquid)
1175	327	ETHYLBENZENE
1180	329	ETHYL BUTYRATE
1181	331	ETHYL CHLOROACETATE
1182	332	ETHYL CHLOROFORMATE
1183	334	ETHYLDICHLOROSILANE
1184	340	ETHYLENE DICHLORIDE
1185	344	ETHYLENE IMINE
1190	348	ETHYL FORMATE
1191	349	ETHYL HEXALDEHYDE
1191	358	ETHYL PROPYLACROLENE
1192	350	ETHYL LACTATE
1193	540	METHYL ETHYL KETONE
1195	357	ETHYL PROPIONATE
1196	346	ETHYLENETRICHLOROSILANE
1198	379	FORMALDEHYDE
1199	386	FURFURAL
1202	389	GAS OIL
1203	390	GASOLINE

DOT#	PAGE	CHEMICAL NAME
1203	391	GASOLINE AVIATION
1203	392	GASOLINE CASINGHEAD
1206	401	HEPTANE
1207	407	HEXALDEHYDE
1208	411	HEXANE
1208	445	ISOHEXANE
1208	580	NEOHEXANE
1212	436	ISOBUTYL ALCOHOL
1213	435	ISOBUTYL ACETATE
1214	437	ISOBUTYLAMINE
1218	451	ISOPRENE
1219	452	ISOPROPANOL
1219	454	ISOPROPYL ALCOHOL
1220	453	ISOPROPYL ACETATE
1221	455	ISOPROPYLAMINE
1223	461	JET FUELS, JP-1
1223	462	JET FUELS, JP-3
1223	466	KEROSENE
1229	516	MESITYL OXIDE
1230	520	METHANOL
1230	525	METHYL ALCOHOL
1231	522	METHYL ACETATE
1233	528	METHYL AMYL ACETATE
1234	542	METHYL FORMAL
1238	535	METHYL CHLOROFORMATE
1239	188	CHLOROMETHYL METHYL ETHER
1242	538	METHYL DICHLOROSILANE
1243	543	METHYL FORMATE
1244	544	METHYLHYDRAZINE
1245	546	METHYL ISOBUTYL KETONE
1246	548	METHYL ISOPROPENYL KETONE
1247	550	METHYL METHACRYLATE
1250	556	METHYL TRICHLOROSILANE
1251	557	METHYL VINYL KETONE
1255	575	NAPHTHA (WMP)
1255	644	PETROLEUM NAPHTHA
1256	573	NAPHTHA, SOLVENT
1259	584	NICKEL CARBONYL
1261	608	NITROMETHANE
1262	618	OCTANE
1264	630	PARALDEHYDE
1265	448	ISOPENTANE
1265	636	PENTANE
1268	574	NAPHTHA (STODDARD SOLVENT)
1270	622	OIL
1272	660	PINE OIL
1274	689	PROPYL ALCOHOL
1275	685	PROPIONALDEHYDE
1276	688	PROPYL ACETATE
1277	690	PROPYLAMINE
1279	264	1,2-DICHLOROPROPANE
1280	693	PROPYLENE OXIDE
1282	699	PYRIDINE
1292	359	ETHYL SILICATE
1294	789	TOLUENE
1295	805	TRICHLOROSILANE
1296	811	TRIETHYLAMINE
1298	818	TRIMETHYLCHLOROSILANE
1299	820	TURPENTINE
1300	558	MINERAL SPIRITS
1301	835	VINYL ACETATE
1302	837	VINYL ETHYL ETHER
1303	839	VINYLIDENE CHLORIDE
1305	842	VINYL TRICHLOROSILANE
1307	844	XYLENE-M
1307	845	XYLENE-O
1307	846	XYLENE-P
1313	160	CALCIUM RESINATE
1320	296	DINITROPHENOL
1328	410	HEXAMETHYLENETETRAMINE
1338	655	PHOSPHORUS RED
1340	654	PHOSPHORUS PENTASULFIDE
1350	761	SULFUR
1360	159	CALCIUM PHOSPHIDE
1361	177	CHARCOAL
1366	275	DIETHYLZINC
1370	291	DIMETHYLZINC
1380	632	PENTABORANE
1381	652	PHOSPHORUS BLACK
1381	658	PHOSPHORUS WHITE
1384	740	SODIUM HYDROSULFITE
1385	751	SODIUM SULFIDE
1402	151	CALCIUM CARBIDE
1410	491	LITHIUM ALUMINUM HYDRIDE
1414	494	LITHIUM HYDRIDE
1415	490	LITHIUM
1425	719	SODIUM AMIDE
1426	728	SODIUM BOROHYDRIDE
1427	738	SODIUM HYDRIDE
1428	716	SODIUM
1431	743	SODIUM METHYLATE
1438	32	ALUMINUM NITRATE
1439	46	AMMONIUM DICHROMATE
1442	53	AMMONIUM PERCHLORATE
1444	54	AMMONIUM PERSULFATE
1445	93	BARIUM CHLORATE
1446	95	BARIUM NITRATE
1447	96	BARIUM PERCHLORATE
1448	97	BARIUM PERMANGANATE
1449	98	BARIUM PEROXIDE
1452	152	CALCIUM CHLORATE
1454	156	CALCIUM NITRATE
1457	158	CALCIUM PEROXIDE
1463	196	CHROMIC ACID
1463	197	CHROMIC ANHYDRIDE
1466	368	FERRIC NITRATE
1469	478	LEAD NITRATE
1475	496	MAGNESIUM PERCHLORATE
1479	208	COPPER NITRATE
1479	217	CUPRIC NITRATE
1479	675	POTASSIUM DICHROMATE
1479	735	SODIUM DICHROMATE
1485	670	POTASSIUM CHLORATE
1486	678	POTASSIUM NITRATE
1490	680	POTASSIUM PERMANGANATE
1491	681	POTASSIUM PEROXIDE
1493	713	SILVER NITRATE
1495	730	SODIUM CHLORATE
1498	744	SODIUM NITRATE
1500	745	SODIUM NITRITE
1511	830	UREA PEROXIDE
1514	866	ZINC NITRATE
1541	7	ACETONE CYANOHYDRIN
1547	72	ANILINE
1549	77	ANTIMONY TRIBROMIDE
1549	79	ANTIMONY TRIFLUORIDE
1551	76	ANTIMONY POTASSIUM TARTRATE
1553	81	ARSENIC ACID (liquid)
1554	82	ARSENIC ACID (solid)
1556	113	BERYLLIUM FLUORIDE

DOT#	PAGE	CHEMICAL NAME
1556	646	PHENYLDICHLOROARSINE
1557	83	ARSENIC DISULFIDE
1557	87	ARSENIC TRISULFIDE
1557	519	METHANEARSONIC ACID, SODIUM SALT
1559	84	ARSENIC PENTOXIDE
1560	85	ARSENIC TRICHLORIDE
1561	86	ARSENIC TRIOXIDE
1564	92	BARIUM CARBONATE
1565	94	BARIUM CYANIDE
1566	112	BERYLLIUM CHLORIDE
1566	115	BERYLLIUM OXIDE
1566	116	BERYLLIUM SULFATE
1570	123	BRUCINE
1572	141	CACODYLIC ACID
1573	149	CALCIUM ARSENATE
1574	150	CALCIUM ARSENITE
1575	154	CALCIUM CYANIDE
1578	189	CHLORONITROBENZENE
1578	602	NITROCHLOROBENZENEMETA
1578	603	NITROCHLOROBENZENEORTHO
1578	604	NITROCHLOROBENZENEPARA
1580	191	CHLOROPICRIN
1585	203	COPPER ACETOARSENITE
1586	204	COPPER ARSENITE
1587	206	COPPER CYANIDE
1589	223	CYANOGEN CHLORIDE
1591	254	o-DICHLOROBENZENE
1592	255	p-DICHLOROBENZENE
1593	261	DICHLOROMETHANE
1593	539	METHYLENE CHLORIDE
1595	284	DIMETHYL SULFATE
1596	292	DINITROANILINE
1597	293	DINITROBENZENE (liquid)
1597	294	DINITROBENZENE (solid)
1598	295	DINITROCRESOL
1600	297	DINITROTOULENE (liquid)
1604	337	ETHYLENEDIAMINE
1605	339	ETHYLENE DIBROMIDE
1609	569	NABAM
1609	595	NITRALIN
1609	816	TRIFLURALIN
1615	319	DURSBAN
1615	566	MIREX
1615	849	ZECTRAN
1616	472	LEAD ACETATE
1617	473	LEAD ARSENATE
1624	503	MERCURIC CHLORIDE
1625	507	MECURIC NITRATE
1627	513	MERCUROUS NITRATE
1629	502	MERCURIC ACETATE
1630	515	MERCURY AMMONIUM CHLORIDE
1636	504	MERCURIC CYANIDE
1638	505	MERCURIC IODIDE (liquid)
1638	506	MERCURIC IODIDE (solid)
1641	508	MERCURIC OXIDE
1645	509	MERCURIC SULFATE
1646	511	MERCURIC THIOCYANATE
1648	8	ACETONITRILE
1648	536	METHYL CYANIDE
1649	568	MOTOR FUEL ANTIKNOCK COMPOUND
1649	776	TETRAETHYL LEAD
1653	586	NICKEL CYANIDE
1654	592	NICOTINE
1658	593	NICOTINE SULFATE (liquid)
1658	594	NICOTINE SULFATE (solid)
1660	599	NITRIC OXIDE
1661	600	NITROANILINE
1662	601	NITROBENZENE
1663	609	NITROPHENOL
1664	612	NITROTOLUENE
1670	642	PERCHLOROMETHYL MERCAPTAN
1671	168	CARBOLIC ACID
1671	645	PHENOL
1677	668	POTASSIUM ARSENATE
1678	669	POTASSIUM ARSENITE
1680	672	POTASSIUM CYANIDE (liquid)
1680	673	POTASSIUM CYANIDE (solid)
1685	720	SODIUM ARSENATE
1686	721	SODIUM ARSENITE (liquid)
1687	723	SODIUM AZIDE
1688	729	SODIUM CACODYLATE
1689	732	SODIUM CYANIDE (liquid)
1689	733	SODIUM CYANIDE (solid)
1690	737	SODIUM FLUORIDE
1692	757	STRYCHINE
1697	182	CHLOROACETOPHENONE
1702	770	TETRACHLOROETHANE
1704	774	TETRAETHYL DITHIOPYROPHOSPHATE (liquid)
1704	785	TETRAETHYL DITHIOPYROPHOSPHATE (solid)
1705	777	TETRAETHYL PYROPHOSPHATE
1707	783	THALLIUM SULFATE
1708	792	TOLUIDINE-M
1708	793	TOLUIDINE-O
1708	794	TOLUIDINE-P
1710	800	TRICHLOROETHYLENE
1712	852	ZINC ARSENATE
1712	853	ZINC ARSENITE
1713	860	ZINC CYANIDE
1714	868	ZINC PHOSPHIDE
1715	5	ACETIC ANHYDRIDE
1716	10	ACETYL BROMIDE
1717	11	ACETYL CHLORIDE
1722	27	ALLYL CHLOROFORMATE
1724	28	ALLYL TRICHLOROSILANE
1726	31	ALUMINUM CHLORIDE
1727	39	AMMONIUM BIFLUORIDE
1728	71	AMYLTRICHLORSILANE
1729	73	ANISOYL CHLORIDE
1730	74	ANTIMONY PENTACHLORIDE
1732	75	ANTIMONY PENTAFLUORIDE
1733	78	ANTIMONY TRICHLORIDE
1736	106	BENZOYL CHLORIDE
1737	108	BENZYL BROMIDE
1738	109	BENZYL CHLORIDE
1739	110	BENZYL CHLORFORMATE
1741	118	BORON TRICHLORIDE
1744	119	BROMINE
1745	120	BROMINE PENTAFLUORIDE
1746	121	BROMINE TRIFLUORIDE
1747	145	BUTYL TRICHLOROSILANE
1748	155	CALCIUM HYPOCHLORITE
1749	180	CHLORINE TRIFLUORIDE
1750	561	MONOCHLOROACETIC ACID (liquid)
1750	562	MONOCHLOROACETIC ACID (solid)
1751	181	CHLOROACETIC ACID
1752	183	CHLOROACETYL CHLORIDE
1754	193	CHLOROSULPHONIC ACID
1758	200	CHROMYL CHLORIDE

DOT#	PAGE	CHEMICAL NAME
1759	372	FERROUS CHLORIDE (solid)
1760	29	ALUMINUM SULFATE (liquid)
1760	33	2-(2)-AMINOETHOXY(ETHANOL)
1760	34	AMINOETHYLETHANOLAMINE
1760	234	DALAPON
1760	355	ETHYL PHOSPHONOTHIOIC DICHLORIDE
1760	356	ETHYL PHOSPHORODICHLORIDATE
1760	364	FERROUS CHLORIDE (liquid)
1760	431	HYDROXYPROPYL ACRYLATE
1760	467	LACTIC ACID
1760	553	METHYL PHOSPHONOTHIOIC DICHLORIDE
1761	219	CUPRIETHYLENEDIAMINE SOLUTION
1762	228	CYCLOHEXENYL TRICHLOROSILANE
1768	277	DIFLUOROPHOSPHORIC ACID
1769	302	DIPHENYLDICHLOROSILANE
1771	311	DODECYL TRICHLOROSILANE
1773	365	FERRIC CHLORIDE (solid)
1777	378	FLUROSULPHONIC ACID
1778	377	FLUOROSILICIC ACID
1778	427	HYDROSILICOFLUORIC ACID
1779	380	FORMIC ACID
1789	415	HYDROCHLORIC ACID
1790	418	HYDROFLUORIC ACID
1791	741	SODIUM HYPOCHLORITE
1796	596	NITRATING ACID
1802	640	PERCHLORIC ACID LESS THAN 50%
1805	650	PHOSPHORIC ACID
1808	656	PHOSPHORUS TRIBROMIDE
1809	657	PHOSPHORUS TRICHLORIDE
1810	653	PHOSPHORUS OXYCHLORIDE
1813	676	POTASSIUM HYDROXIDE (liquid)
1814	677	POTASSIUM HYDROXIDE (solid)
1818	708	SILICON CHLORIDE
1821	726	SODIUM BISULFIDE (solid)
1823	741	SODIUM HYDROXIDE
1824	175	CAUSTIC POTASH SOLUTION
1824	176	CAUSTIC SODA SOLUTION
1828	762	SULFUR CHLORIDE
1828	764	SULFUR MONOCHLORIDE
1831	626	OLEUM
1831	765	SULFURIC ACID
1832	766	SULFURIC ACID, SPENT
1834	767	SULFURYL CHLORIDE
1838	788	TITANIUM TETRACHLORIDE
1840	858	ZINC CHLORIDE
1846	174	CARBON TETRACHLORIDE
1848	686	PROPIONIC ACID
1860	838	VINYL FLUORIDE
1863	463	JET FUELS, JP-4
1868	237	DECABORANE
1869	495	MAGNESIUM
1873	641	PERCHLORIC ACID OVER 50%
1888	186	CHLOROFORM
1889	222	CYANOGEN BROMIDE
1893	861	ZINC DIALKYLDITHIOPHOSPHATE
1897	639	PERCHLORETHYLENE
1897	771	TETRACHLOROETHYLENE
1902	240	DI(2ETHYLHEXYL) PHOSPHORIC ACID
1910	157	CALCIUM OXIDE
1911	245	DIBORANE
1915	226	CYCLOHEXANONE
1916	261	DICHLOROETHYL ETHER
1917	322	ETHYL ACRYLATE, INHIBITED
1918	215	CUMENE
1919	524	METHYL ACRYLATE
1920	614	NONANE
1921	695	PROPYLENEIMINE
1925	325	ETHYL ALUMINUM SESQUICHLORIDE
1930	817	TRIISOBUTYLALUMINUM
1931	865	ZINC HYDROSULFITE
1935	220	CYANOACETIC ACID
1942	51	AMMONIUM NITRATE
1962	335	ETHYLENE
1969	434	ISOBUTANE
1971	518	METHANE
1972	487	LIQUIFIED NATURAL GAS (LNG)
1972	579	NATURAL GAS
1977	606	NITROGEN
1987	126	BUTANEDIOL
1987	128	BUTENEDIOL
1987	239	DECYL ALCOHOL
1987	447	ISOOCTYL ALCOHOL
1988	855	ZINC BORATE
1989	99	BENZALDEHYDE
1989	460	ISOVALERALDEHYDE
1991	192	CHLOROPRENE
1993	212	CREOSOTE, COAL TAR
1993	231	CYCLOHEXANOL
1993	279	DIISOBUTYLCARBINOL
1993	321	ETHYL ACETOACTEATE
1993	353	ETHYL NITRATE
1993	381	FUEL OIL NUMBER 1
1993	382	FUEL OIL NUMBER 2
1993	383	FUEL OIL NUMBER 4
1993	384	FUEL OIL NUMBER 5
1993	551	METHYL-n-BUTYL KETONE
1993	704	ROSIN SOLUTION
1993	843	WAX
1999	90	ASPHALT
2015	424	HYDROGEN PEROXIDE 52–100%
2016	398	GRENADE(WITH POISON B CHARGE)
2018	184	CHLOROANILINE
2020	190	CHLOROPHENOL
2020	262	DICHLOROPHENOL
2020	633	PENTACHLOROPHENOL
2020	802	TRICHLOROPHENOL
2023	187	CHLOROHYDRINS
2023	317	EPICHLOROHYDRIN
2025	510	MERCURIC SULFIDE
2025	512	MERCUROUS CHLORIDE
2027	722	SODIUM ARSENITE (solid)
2029	416	HYDRAZINE
2030	414	HEXYLENE GLYCOL
2031	597	NITRIC ACID
2038	298	DINITROTOULENE (solid)
2045	440	ISOBUTYRALDEHYDE
2046	215	CYMENE
2047	265	1,3-DICHLOROPROPENE
2048	268	DICYCLOPENTADIENE
2049	273	DIETHYLBENZENE
2050	280	DIISOBUTYLENE
2052	300	DIPENTENE
2053	529	METHYL AMYL ALCOHOL
2053	545	METHYL ISOBUTYL CARBINOL
2054	567	MORPHOLINE
2055	758	STYRENE
2056	781	TETRAHYDROFURAN
2058	831	VALERALDEHYDE

DOT#	PAGE	CHEMICAL NAME
2059	201	COLLODION
2067	213	CRESOL (o-m-p)
2070	50	AMMONIUM NITRATEPHOSPHATE
2074	18	ACRYLAMIDE
2077	578	NAPHTHYLAMINE
2078	790	TOLUENE-2,4-DIISOCYANATE
2079	274	DIETHYLENETRIAMINE
2080	14	ACETYLACETONE
2084	12	ACETYL PEROXIDE
2084	13	ACETYL PEROXIDE SOLUTION
2085	107	BENZOYL PEROXIDE
2085	244	DIBENZOYL PEROXIDE
2105	651	PHOSPHORIC ACID TRIETHYLENEIMINE
2116	216	CUMENE HYDROPEROXIDE
2118	227	CYCLOHEXANONE PEROXIDE
2124	469	LAUROYL PEROXIDE
2131	638	PERACETIC ACID
2133	459	ISOPROPYL PERCARBONATE (UNSTABILIZED)
2134	458	ISOPROPYL PERCARBONATE (STABILIZED)
2138	256	DICHLOROBENZOYL PEROXIDE-2,4
2171	282	DIISOPROPYLBENZENE HYDROPEROXIDE
2174	289	DIMETHYLHEXANE DIHYDROPEROXIDE
2188	88	ARSINE
2205	22	ADIPONITRILE
2212	89	ASBESTOS, BLUE
2213	629	PARAFORMALDEHYDE
2214	659	PHTHALIC ANHYDRIDE
2215	498	MALEIC ACID
2215	499	MALEIC ANYDRIDE
2218	19	ACRYLIC ACID
2224	105	BENZONITRILE
2229	136	BUTYL PHENOL
2239	194	CHLOROTOULENE (o-m-p)
2257	667	POTASSIUM
2259	814	TRIETHYLENE TETRAMINE
2261	847	XYLENOL (liquid)
2261	848	XYLENOL (solid)
2265	288	DIMETHYLFORMAMIDE
2275	328	ETHYLBUTANOL
2277	352	ETHYL METHACRYLATE
2278	403	HEPTENE
2280	408	HEXAMETHYLENE DIAMINE
2282	402	HEXANOL
2284	439	ISOBUTYLNITRILE
2291	474	LEAD CHLORIDE
2291	476	LEAD FLUOROBORATE
2291	482	LEAD THIOCYANATE
2294	531	METHYLANILINE
2298	537	METHYL CYCLOPENTANE
2300	541	METHYL ETHYL PYRIDINE
2304	576	NAPHTHALENE
2315	663	POLYCHLORINATED BIPHENYL
2320	779	TETRAETHYLENEPENTAMINE
2348	131	BUTYL ACRYLATE
2348	133	n-BUTYL ACRYLATE
2357	229	CYCLOHEXENYLAMINE
2362	259	DICHLOROETHANE
2363	351	ETHYL MERCAPTAN
2366	271	DIETHYL CARBONATE
2370	413	HEXENE
2383	304	DIPROPYLAMINE
2402	696	PROPYL MERCAPTAN
2435	354	ETHYL PHENYL DICHLOROSILANE
2439	724	SODIUM BIFLUORIDE
2443	832	VANADIUM OXYTRICHLORIDE
2449	52	AMMONIUM OXALATE
2449	209	COPPER OXALATE
2460	533	METHYL BUTENE
2464	114	BERYLLIUM NITRATE
2465	674	POTASSIUM DICHLOROISOCYANURATE
2465	734	SODIUM DICHLORO-s-TRIAZINETRIONE
2474	784	THIOPHOSGENE
2480	547	METHYL ISOCYANATE
2489	303	DIPHENYLMETHYL DIOSOCYANATE
2491	564	MONOETHANOLAMINE
2493	409	HEXAMETHYLENIMINE
2496	687	PROPIONIC ANHYDRIDE
2503	877	ZIRCONIUM TETRACHLORIDE
2505	48	AMMONIUM FLUORIDE
2514	122	BROMOBENZENE
2527	132	iso-BUTYL ACRYLATE
2529	441	ISOBUTYRIC ACID
2553	572	NAPHTHA
2554	526	METHYL ALLYL CHLORIDE
2570	142	CADMIUM ACETATE
2570	143	CADMIUM BROMIDE
2570	144	CADMIUM CHLORIDE
2570	145	CADMIUM FLUOROBORATE
2570	146	CADMIUM NITRATE
2570	147	CADMIUM OXIDE
2570	148	CADMIUM SULFATE
2572	647	PHENYLHYDRAZINE HYDROCHLORIDE
2574	806	TRICRESYLPHOSPHATE
2579	661	PIPERAZINE
2582	364	FERRIC CHLORIDE (liquid)
2583	791	TOLUENE SULFONIC ACID
2584	310	DODECYLBENZENESULFONIC ACID
2608	610	NITROPROPANE
2614	525	METHALLYL ALCOHOL
2618	841	VINYL TOLUENE
2622	395	GLYCIDALDEHYDE
2630	748	SODIUM SELENITE
2646	404	HEXACHLOROCYCLOPENTADIENE
2656	701	QUINOLINE
2662	426	HYDROQUINONE
2672	49	AMMONIUM HYDROXIDE
2674	750	SODIUM SILICOFLUORIDE
2683	57	AMMONIUM SULFIDE
2692	117	BORON TRIBROMIDE
2693	40	AMMONIUM BISULFITE
2693	725	SODIUM BISULFIDE (liquid)
2703	457	ISOPROPYL MERCAPTAN
2713	16	ACRIDINE
2725	590	NICKEL NITRATE
2728	873	ZIRCONIUM NITRATE
2757	165	CARBURYL (liquid)
2757	166	CARBURYL (solid)
2757	167	CARBOFURAN
2757	501	MERCAPTODIMETHUR
2761	23	ALDRIN
2761	101	BENZENE HEXACHLORIDE
2761	235	DDD
2761	236	DDT
2761	251	DICHLONE (liquid)
2761	252	DICHLONE (solid)
2761	269	DIELDRIN (liquid)
2761	270	DIELDRIN (solid)

DOT#	PAGE	CHEMICAL NAME
2761	313	ENDOSULFAN (liquid)
2761	314	ENDOSULFAN (solid)
2761	315	ENDRIN (liquid)
2761	316	ENDRIN (solid)
2761	400	HEPTACHLOR
2761	464	JET FUELS, JP-5
2761	465	KEPONE
2761	485	LINDANE
2761	521	METHOXYCHLOR
2761	795	TOXAPHENE (liquid)
2761	796	TOXAPHENE (solid)
2762	178	CHLORDANE
2765	233	2,4-D ESTERS
2765	263	DICHLOROPHENOXYACETIC ACID
2765	683	PROPARGITE
2765	798	TRICHLOROPHENOXY ACETIC ACID AMINE
2765	803	TRICHLOROPHENOXY ACETIC ACID ESTER
2765	814	TRICHLOROPHENOXYPROPIONIC ACID ESTER
2767	309	DIURON
2769	247	DICAMBA (liquid)
2769	248	DICAMBA (solid)
2769	249	DICHLOBENIL (liquid)
2769	250	DICHLOBENIL (solid)
2771	785	THIRAM
2781	305	DIQUAT (liquid)
2781	306	DIQUAT (solid)
2783	91	AZINPHOS METHYL
2783	211	COUMAPHOS
2783	243	DIAZINON
2783	266	DICHLORVOS (liquid)
2783	267	DICHLORVOS (solid)
2783	307	DISULFOTON (liquid)
2783	308	DISULFOTON (solid)
2783	319	ETHION
2783	497	MALATHION
2783	552	METHYL PARATHION
2783	570	NALED (liquid)
2783	571	NALED (solid)
2783	631	PARATHION
2783	648	PHOSDRIN
2783	797	TRICHLORFON
2789	4	ACETIC ACID (glacial)
2790	3	ACETIC ACID (aqueous solution)
2798	102	BENZENE PHOSPHORUS DICHLORIDE
2799	103	BENZENE PHOSPHORUS THIODICHLORIDE
2802	205	COPPER CHLORIDE
2809	514	MERCURY
2811	475	LEAD FLUORIDE
2811	477	LEAD IODINE
2811	479	LEAD STEARATE
2811	480	LEAD SULFIDE
2814	360	ETIOLOGIC AGENT
2820	140	BUTYRIC ACID
2821	169	CARBOLIC OIL
2842	605	NITROETHANE
2850	694	PROPYLENE TETRAMER
2854	55	AMMONIUM SILICOFLUORIDE
2855	870	ZINC SILICOFLUORIDE
2862	833	VANADIUM PENTOXIDE
2865	430	HYDROXYLAMINE SULFATE
2874	387	FURFURYL ALCOHOL
2876	703	RESORCINOL
2893	470	LAUROYL PEROXIDE (LESS THAN 42%)
2912	787	THORIUM ORE
2918	617	NUCLEAR REACTOR FUEL
2922	738	SODIUM HYDROSULFIDE SOLUTION
2924	257	DICHLOROBUTENE
2931	834	VANADYL SULFATE
2974	702	RADIOACTIVE MATERIAL (special form) N.O.S.
2976	786	THORIUM NITRATE
2977	824	URANIUM HEXAFLUORIDE
2981	827	URANYL NITRATE
3022	135	BUTYLENE OXIDE
9011	161	CAMPHENE
9037	405	HEXACHLOROETHANE
9077	21	ADIPIC ACID
9078	30	ALUMINUM SULFATE (solid)
9079	36	AMMONIUM ACETATE
9080	37	AMMONIUM BENZOATE
9081	38	AMMONIUM BICARBONATE
9083	41	AMMONIUM CARBAMATE
9084	42	AMMONIUM CARBONATE
9085	43	AMMONIUM CHLORIDE
9086	44	AMMONIUM CHROMATE
9087	45	AMMONIUM CITRATE
9088	47	AMMONIUM FLUOBORATE
9090	58	AMMONIUM SULFITE
9091	59	AMMONIUM TARTRATE
9092	61	AMMONIUM THIOCYANATE LIQUOR
9093	63	AMMONIUM THIOSULFATE (solid)
9094	104	BENZOIC ACID
9095	246	DIBUTYL PHTHALATE
9096	153	CALCIUM CHROMATE
9099	163	CAPTAN (liquid)
9099	164	CAPTAN (solid)
9100	198	CHROMIC SULFATE
9101	195	CHROMIC ACETATE
9102	199	CHROMOUS CHLORIDE
9106	202	COPPER ACETATE
9109	218	CUPRIC SULFATE
9109	210	COPPER SULFATE
9117	338	ETHYLENEDIAMINE TETRACETIC ACID
9118	362	FERRIC AMMONIUM CITRATE
9119	363	FERRIC AMMONIUM OXILATE
9120	366	FERRIC FLUORIDE
9121	369	FERRIC SULFATE
9122	370	FERROUS AMMONIUM SULFATE
9125	375	FERROUS SULFATE
9126	385	FUMARIC ACID
9134	493	LITHIUM CHROMATE
9137	577	NAPHTHENIC ACID
9138	582	NICKEL AMMONIUM SULFATE
9139	585	NICKEL CHLORIDE
9140	589	NICKEL HYDROXIDE
9141	591	NICKEL SULFATE
9142	671	POTASSIUM CHROMATE
9145	731	SODIUM CHROMATE
9148	747	SODIUM PHOSPHATE
9149	756	STRONTIUM CHROMATE
9151	809	TRIETHANOLAMINE
9153	850	ZINC ACETATE
9154	851	ZINC AMMONIUM CHLORIDE
9156	856	ZINC BROMIDE
9157	857	ZINC CARBONATE
9158	862	ZINC FLUORIDE
9159	864	ZINC FORMATE
9160	867	ZINC PHENOLSULFONATE
9161	871	ZINC SULFATE

DOT#	PAGE	CHEMICAL NAME
9162	875	ZIRCONIUM POTASSIUM FLUORIDE
9163	876	ZIRCONIUM SULFATE
9180	826	URANYL ACETATE
9184	697	PYRETHRINS ACID (liquid)
9184	698	PYRETHRINS ACID (solid)
9188	56	AMMONIUM SULFAMATE
9188	60	AMMONIUM THIOCYANATE
9188	62	AMMONIUM THIOSULFATE (liquid)
9189	399	HAZARDOUS WASTE
9201	80	ANTIMONY TRIOXIDE
9207	9	ACETOPHENONE

SYNONYM INDEX

SYNONYM	PAGE	CHEMICAL NAME
3A-4-7-7A-TETRAHYDRO-4-7-METHANOINDENE	268	DICYCLOPENTADIENE
ACETIC ACID	830	ZINC ACETATE
ACETIC ACID	688	PROPYL ACETATE
ACETIC ACID	581	NICKEL ACETATE
ACETIC ACID	522	METHYL ACETATE
ACETIC ACID	453	ISOPROPYL ACETATE
ACETIC ACID	435	ISOBUTYL ACETATE
ACETIC ACID	202	COPPER ACETATE
ACETIC ACID	195	CHROMIC ACETATE
ACETIC ACID	130	sec-BUTYL ACETATE
ACETIC ACID	36	AMMONIUM ACETATE
ACETIC ESTER	320	ETHYL ACETATE
ACETIC ETHER	320	ETHYL ACETATE
ACETOACETIC ACID	321	ETHYL ACETOACETATE
ACETOACETIC ESTER	321	ETHYL ACETOACETATE
ACETONITRILE	536	METHYL CYANIDE
ACETYLBENZENE	9	ACETOPHENONE
ACETYLENE TETRA-CHLORIDE	770	TETRACHLOROETHANE
ACETYLENOGEN	151	CALCIUM CARBIDE
ACID AMMONIA CARBONATE	38	AMMONIUM BICARBONATE
ACID AMMONIUM FLUORIDE	39	AMMONIUM BIFLUORIDE
ACID MIXTURES	596	NITRATING ACID
ACID PLUMBOUS ARSENATE	473	LEAD ARSENATE
ACRALDEHYDE	17	ACROLEIN
ACRYALDEHYDE	17	ACROLEIN
ACRYLIC ACID	524	METHYL ACRYLATE
ACRYLIC ACID	323	ETHYL ACRYLATE (liquid)
ACRYLIC ACID	329	ETHYL ACRYLATE, INHIBITED
ACRYLIC ACID AMINE 50	18	ACRYLAMIDE
ACRYLIC ALDEHYDE	17	ACROLEIN
ACRYLIC AMIDE 50	18	ACRYLAMIDE
ACRYLIC ESTER	524	METHYL ACRYLATE
ADIPINIC ACID	21	ADIPIC ACID
ADRONAL	225	CYCLOHEXANOL
ALBONE	424	HYDROGEN PEROXIDE 52–100%
ALBUS	575	MERCURY AMMONIUM CHLORIDE
ALCOHOL C-11	821	UNDECANOL
ALCOHOL C-8	619	OCTANOL
ALCOHOL C-10	239	DECYL ALCOHOL
ALDIFEN	296	DINITROPHENOL
ALFA-TOX	243	DIAZINON
ALGEON 22	563	MONOCHLORODIFLUORO-METHANE
ALKYLBENZENESULFONIC ACID SODIUM SALT	717	SODIUM ALKYLBENZENESULFONATES
ALLOMALEIC ACID	385	FUMARIC ACID
ALLY CHLOROCARBONATE	27	ALLYL CHLOROFORMATE
ALLYL ALDEHYDE	17	ACROLEIN
ALLYLSILICONE TRICHLORIDE	28	ALLYL TRICHLOROSILANE
ALPHA	216	CUMENE HYDROPEROXIDE
ALPHA-AMINOTOLUENE	111	BENZYLAMINE
ALPHA-BROMOTOLUENE	108	BENZYL BROMIDE
ALPHA-BUTENE	127	BUTENE
ALPHA-BUTYLENE OXIDE	135	BUTYLENE OXIDE
ALPHA-CHLOROTOLUENE	109	BENZYL CHLORIDE
ALPHA-DINITROPHENOL	296	DINITROPHENOL
ALPHA-HEXENE	413	HEXENE
ALPHA-HYDROXYPROPIONIC ACID	467	LACTIC ACID
ALPHA-N-AMYLENE	637	PENTENE
ALPHA-NAPHTHYLAMINE	578	NAPHTHYLAMINE
ALPHA-OCTYLENE	620	OCTENE
ALUMINUM SALT,NITRIC ACID	32	ALUMINUM NITRATE
ALUMINUM SULFATE	29	ALUMINUM SULFATE (solid)
ALUMINUM SULFATE	30	ALUMINUM SULFATE (liquid)
ALUMINUM TRIETHYL	810	TRIETHYLALUMINUM
ALUMINUM TRIISOBUTYL	817	TRIISOBUTYLALUMINUM
AMCHLOR	43	AMMONIUM CHLORIDE
AMCHLORIDE	43	AMMONIUM CHLORIDE
1-AMINO-4-CHLOROBENZENE	184	CHLOROANILINE
AMINOCYCLOHEXANE	229	CYCLOHEXENYLAMINE
AMINOETHANE	326	ETHYLAMINE
1-AMINOPROPANE	690	PROPYLAMINE
1-AMINO-2-PROPANOL	565	MONOISOPROPANOLIMINE
AMMATE	56	AMMONIUM SULFAMATE
AMMINOFORM	410	HEXAMETHYLENE-TETRAMINE
AMMMONIATED MERCURY	515	MERCURY AMMONIUM CHLORIDE
AMMONERIC	43	AMMONIUM CHLORIDE
AMMONIA WATER	49	AMMONIUM HYDROXIDE
AMMONIOFORMALDEHYDE	410	HEXAMETHYLENE-TETRAMINE
AMMONIUM ACID FLUORIDE	39	AMMONIUM BIFLUORIDE
AMMONIUM AMINOFORMATE	41	AMMONIUM CARBAMATE
AMMONIUM BOROFLUORIDE	47	AMMONIUM FLUOBORATE
AMMONIUM DICHROMATE	46	AMMONIUM DICHROMATE
AMMONIUM DISULFATONICKELATE	582	NICKEL AMMONIUM SULFATE
AMMONIUM FERRIC CITRATE	362	FERRIC AMMONIUM CITRATE
AMMONIUM FERRIC OXILATE TRIHYDRIDE	363	FERRIC AMMONIUM OXILATE
AMMONIUM HYDROGEN	40	AMMONIUM BISULFITE
AMMONIUM HYDROGEN CARBONATE	38	AMMONIUM BICARBONATE
AMMONIUM HYDROGEN FLUORIDE	39	AMMONIUM BIFLUORIDE
AMMONIUM HYDROGEN SULFIDE SOLUTION	57	AMMONIUM SULFIDE
AMMONIUM HYPOSULFITE	62	AMMONIUM THIOSULFATE (solid)
AMMONIUM HYPOSULFITE	63	AMMONIUM THIOSULFATE (liquid)
AMMONIUM NICKEL SULFATE	589	NICKEL AMMONIUM SULFATE
AMMONIUM OXILATE HYDRATE	52	AMMONIUM OXALATE
AMMONIUM PENTACHLOROZINCATE	851	ZINC AMMONIUM CHLORIDE
AMMONIUM PEROXYDISULFATE	54	AMMONIUM PERSULFATE
AMMONIUM RHODANATE	61	AMMONIUM THIOCYANATE LIQUOR
AMMONIUM RHODANATE	60	AMMONIUM THIOCYANATE
AMMONIUM RHODANIDE	61	AMMONIUM THIOCYANATE LIQUOR

SYNONYM	PAGE	CHEMICAL NAME
AMMONIUM RHODANIDE	60	AMMONIUM THIOCYANATE
AMMONIUM SALT	61	AMMONIUM THIOCYANATE LIQUOR
AMMONIUM SALT	60	AMMONIUM THIOCYANATE
AMMONIUM SALT	59	AMMONIUM TARTRATE
AMMONIUM SALT	41	AMMONIUM CARBAMATE
AMMONIUM SALT	37	AMMONIUM BENZOATE
AMMONIUM SALT	36	AMMONIUM ACETATE
AMMONIUM SULFO-CYANATE	61	AMMONIUM THIOCYANATE LIQUOR
AMMONIUM SULFO-CYANATE	60	AMMONIUM THIOCYANATE
AMMONIUM TETRAFLURBORATE	47	AMMONIUM FLUOBORATE
AMMONIUM THIOCYANATE LIQUOR	61	AMMONIUM THIOCYANATE LIQUOR
AMMONIUM THIOCYANATE LIQUOR	60	AMMONIUM THIOCYANATE
AMMONIUM ZINC CHLORIDE	851	ZINC AMMONIUM CHLORIDE
AMORPHOUS PHOSPHORUS	655	PHOSPHORUS RED
AMS	56	AMMONIUM SULFAMATE
AMYL ALDEHYDE	831	VALERALDEHYDE
AMYL CARBINOL	412	HEXANOL
AND XYLENES	572	NAPHTHA
ANHYDRONE	496	MAGNESIUM PERCHLORATE
ANHYDROUS	188	CHLOROMETHYL METHYL ETHER
ANHYDROUS AMMONIA CHLORIDE	31	ALUMINUM CHLORIDE
ANHYDROUS METHYL-AMINE	527	METHYLAMINE
ANILINE OIL	72	ANILINE
ANILINOBENZENE	301	DIPHENYLAMINE
ANILINOMETHANE	531	METHYLANILINE
ANONEHYTROL O	226	CYCLOHEXANONE
ANSAR	141	CACODYLIC ACID
ANTHON	797	TRICHLORFON
ANTIMONOUS BROMIDE	77	ANTIMONY TRIBROMIDE
ANTIMONY 111 CHLORIDE	78	ANTIMONY TRICHLORIDE
ANTIMONY PERCHLORIDE	74	ANTIMONY PENTACHLORIDE
APO	659	PHOSPHORIC ACID TRIETHYLENEIMINE
AQUACIDE	305	DIQUAT (liquid)
AQUACIDE	306	DIQUAT (solid)
AQUEOUS WATER	49	AMMONIUM HYDROXIDE
ARCTON 9	801	TRICHLOROFLUORO-METHANE
ARGENTOUS FLUORIDE	711	SILVER FLUORIDE
ARGENTOUS OXIDE	714	SILVER OXIDE
AROCHLOR	663	POLYCHLORINATED BIPHENYL
AROTON 4	563	MONOCHLORODIFLUORO-METHANE
ARSENIC ACID ANHYDRIDE	84	ARSENIC PENTOXIDE
ARSENIC OXIDE	84	ARSENIC PENTOXIDE
ARSENIC PENTAOXIDE	84	ARSENIC PENTOXIDE
ARSENIC PENTOXIDE	81	ARSENIC ACID (liquid)
ARSENIC PENTOXIDE	82	ARSENIC ACID (solid)
ARSENIC YELLOW	87	ARSENIC TRISULFIDE
ARSENIOUS ACID	669	POTASSIUM ARSENITE
ARSENOUS ACID	150	CALCIUM ARSENITE
ARSENOUS ACID	86	ARSENIC TRIOXIDE
ARSICODILE	729	SODIUM CACODYLATE
ARTHODIBROM	570	NALED (liquid)

SYNONYM	PAGE	CHEMICAL NAME
ARTHODIBROM	571	NALED (solid)
ARTIC	534	METHYL CHLORIDE
ARTIFICIAL CINNABAR	510	MERCURIC SULFIDE
ASBESTOS	89	ASBESTOS, BLUE
ASPHALT CEMENT	90	ASPHALT
ATE	810	TRIETHYLALUMINUM
ATHANECARBOXYLIC ACID	686	PROPIONIC ACID
AV GAS	391	GASOLINE AVIATION
10-AZAANTHRACENE	55	AMMONIUM SILICO-FLUORIDE
10-AZAANTHRACENE	16	ACRIDINE
AZIRANE	344	ETHYLENE IMINE
AZIRIDINE	344	ETHYLENE IMINE
BANVIL D	247	DICAMBA (liquid)
BANVIL D	248	DICAMBA (solid)
BARIUM BINOXIDE	98	BARIUM PEROXIDE
BARIUM CHLORATE MONOHYDRATE	93	BARIUM CHLORATE
BARIUM CYANIDE SOLID	94	BARIUM CYANIDE
BARIUM DIOXIDE	98	BARIUM PEROXIDE
BARIUM PERCHLORATE TRIHYDRATE	96	BARIUM PERCHLORATE
BARIUM SUPER OXIDE	98	BARIUM PEROXIDE
BASIC ZIRCONIUM CHLORIDE	874	ZIRCONIUM OXYCHLORIDE
BATTERY ACID	765	SULFURIC ACID
BAY 37344	501	MERCAPTODIMETHUR
BAYER 13 59	797	TRICHLORFON
BEET SUGAR	769	SUCROSE
1-BENZAZINE	701	QUINOLINE
BENZENE	602	NITROCHLOROBENZENE-META
BENZENE	603	NITROCHLOROBENZENE-ORTHO
BENZENE	604	NITROCHLOROBENZENE-PARA
BENZENECARBONYL CHLORIDE	106	BENZOYL CHLORIDE
BENZENECARBOXYLIC ACID	104	BENZOIC ACID
BENZENE-1,3-DIECAR-BOXYLAC ACID	450	ISOPHTHALIC ACID
1,3 BENZENEDIOL	703	RESORCINOL
1,4-BENZENEDIOL	426	HYDROQUINONE
1-2-3-BENZENETRIOL	700	PYROGALLIC ACID
BENZINOFORM	174	CARBON TETRACHLORIDE
BENZO B QUINOLINE	55	AMMONIUM SILICOFLUORIDE
BENZO B QUINOLINE	16	ACRIDINE
BENZOIC ACID	107	BENZOYL PEROXIDE
BENZOIC ACID	37	AMMONIUM BENZOATE
BENZOIC ACID NITRILE	105	BENZONITRILE
BENZOIC ALDEHYDE	99	BENZALDEHYDE
BENZOL	100	BENZENE
BENZOLE	100	BENZENE
BENZOYL PEROXIDE	244	DIBENZOYL PEROXIDE
BENZOYL SUPEROXIDE	244	DIBENZOYL PEROXIDE
BENZOYL SUPEROXIDE	107	BENZOYL PEROXIDE
BENZYLCARBONYL CHLORIDE	110	BENZYL CHLORFORMATE
BENZYL CHLOROCARBONATE	110	BENZYL CHLORFORMATE
BENZYL ESTER	110	BENZYL CHLORFORMATE
BERYLLIA	115	BERYLLIUM OXIDE
BERYLLIUM SULFATE TETRAHYDRATE	116	BERYLLIUM SULFATE
BETA-AMINEOETHYL ALCOHOL	564	MONOETHANOLAMINE

SYNONYM	PAGE	CHEMICAL NAME
BETA-HYDROXYETHYL ACRYLATE	428	HYDROXYETHYL ACRYLATE
BETA-METHALLYL CHLORIDE	526	METHYL ALLYL CHLORIDE
BETA PROPIOLACTONE	684	PROPIOLACTONE-BETA
BETA PROPIONOLACTONE	684	PROPIOLACTONE-BETA
BHC	101	BENZENE HEXACHLORIDE
BICHROME	675	POTASSIUM DICHROMATE
BICYCLO 4.4.0 DECANE	238	DECAHYDRONAPHTHALENE
BIFORMAL	397	GLYOXAL
BIFORMYL	397	GLYOXAL
BIS 2-AMINOETHYL AMINE	274	DIETHYLENETRIAMINE
BIS 2-CHLOROETHYL ETHER	260	DICHLOROETHYL ETHER
BIS 2-ETHYLHEXYL HYDROGEN PHOSPHATE	247	DI-(2-ETHYLHEXYL) PHOSPHORIC ACID
BIS-2-2-HYDROXYETHOXY ETHYL ETHER	778	TETRAETHYLENE GLYCOL
BIVINYL,DIVINAL	124	BUTADIENE,INHIBITED
B-KETOPROPANE	6	ACETONE
BLACK LEAF 40 40 WATER SOLUTION	593	NICOTINE SULFATE (liquid)
BLACK LEAF 40 40 WATER SOLUTION	594	NICOTINE SULFATE (solid)
BLASTING	361	EXPLOSIVES A
BLUE OIL	72	ANILINE
BLUE VITOL	218	CUPRIC SULFATE
BLUE VITOL	210	COPPER SULFATE
BOLETIC ACID	385	FUMARIC ACID
BOMB	361	EXPLOSIVES A
BORAX ANHYDROUS	726	SODIUM BORATE
BOROETHANE	245	DIBORANE
BOROHYDRIDE	728	SODIUM BOROHYDRIDE
BORON CHLORIDE	118	BORON TRICHLORIDE
BOTTLED GAS	488	LIQUIFIED PETROLEUM
BOX TOE GUM	201	COLLODION
BP	244	DIBENZOYL PEROXIDE
BPO	244	DIBENZOYL PEROXIDE
BRIMSTONE	761	SULFUR
BROALLYLENE	25	ALLYL BROMIDE
BROCIDE	340	ETHYLENE DICHLORIDE
BROMELITE	115	BERYLLIUM OXIDE
BROMOBENZOL	122	BROMOBENZENE
BROMOFUME	339	ETHYLENE DIBROMIDE
2-BROMOPROPANE	25	ALLYL BROMIDE
BRONEX	570	NALED (liquid)
BRONEX	571	NALED (solid)
BRUCINE DIHYDRATE	123	BRUCINE
BUNKER C	384	FUEL OIL NUMBER 5
3-BUTADIENE	451	ISOPRENE
BUTANIC ACID	140	BUTYRIC ACID
2-BUTANONE	540	METHYL ETHYL KETONE
1-BUTEEN OXIDE	135	BUTYLENE OXIDE
1-BUTENE	134	BUTYLENE
1-BUTENE	127	BUTENE
2-BUTENE-1,4-IDOL	128	BUTENEDIOL
1-BUTENE-3-ONE ISOPROPENYL METHYLKETONE	546	METHYL ISOPROPENYL KETONE
BUTENE RESINS	662	POLYBUTENE
3-BUTEN-2-ONE	557	METHYL VINYL KETONE
1-BUTOXY BUTANE	241	DI-BUTYL ETHER
BUTOXYETHYL ESTER	233	2,4-D ESTERS
BUTTERCUP YELLOW	859	ZINC CHROMATE
BUTTER OF ANTIMONY	78	ANTIMONY TRICHLORIDE
BUTTER OF ARSENIC	85	ARSENIC TRICHLORIDE
BUTYL ACRYLATE	133	n-BUTYL ACRYLATE
BUTYL ALCOHOL	436	ISOBUTYL ALCOHOL

SYNONYM	PAGE	CHEMICAL NAME
BUTYL 2,4-DICHLOROPHENOXYACETATE	233	2,4-D ESTERS
BUTYLENE	127	BUTENE
2-BUTYLENE DICHLORIDE	257	DICHLOROBUTENE
BUTYL ETHER	241	DI-BUTYL ETHER
BUTYLETHYLACETALDEHYDE	349	ETHYL HEXALDEHYDE
BUTYL ETHYLENE	413	HEXENE
BUTYL PHTHALATE	246	DIBUTYL PHTHALATE
BUTYL 2 PROPENATE	133	n-BUTYL ACRYLATE
BUTYL-2-PROPENATE	131	BUTYL ACRYLATE
BUTYL TITANATE	769	TETRABUTYL TITANATE
BUTYLTITANATE MONOMER	769	TETRABUTYL TITANATE
BUTYRIC ACID	329	ETHYL BUTYRATE
n-BUTYRIC ACID	140	BUTYRIC ACID
C-56	404	HEXACHLOROCYCLOPENTADIENE
C-4	393	GELATINE DYNAMITE
CADMIUM ACETATE DIHYDRATE	142	CADMIUM ACETATE
CADMIUM BROMIDE TETRAHYDRATE	143	CADMIUM BROMIDE
CADMIUM FLUOBORATE	145	CADMIUM FLUOROBORATE
CADMIUM FLUOBORATE SOLUTION	145	CADMIUM FLUOROBORATE
CADMIUM FUME	147	CADMIUM OXIDE
CADMIUM NITRATE TETRAHYDRATE	146	CADMIUM NITRATE
CADOX HDP	227	CYCLOHEXANONE PEROXIDE
CADOX-PS	256	DICHLOROBENZOYL PEROXIDE-2,4
CAKE ALUMINUM	30	ALUMINUM SULFATE (solid)
CALAMINE	857	ZINC CARBONATE
CALCIUM CHROMATE DIHYDRATE	153	CALCIUM CHROMATE
CALCIUM CHROMATE VI	153	CALCIUM CHROMATE
CALCIUM DIOXIDE	158	CALCIUM PEROXIDE
CALCIUM NITRATE TETRAHYDRATE	156	CALCIUM NITRATE
CALCIUM ROSIN	160	CALCIUM RESINATE
CALCIUM SALT	150	CALCIUM ARSENITE
CALOCHLOR	503	MERCURIC CHLORIDE
CALOMEL	512	MERCUROUS CHLORIDE
CANE SUGAR	759	SUCROSE
CAPRIC ALCOHOL	239	DECYL ALCOHOL
CAPRYLENE	620	OCTENE
CARBAMIC ACID	41	AMMONIUM CARBAMATE
CARBAMIDE	829	UREA
CARBIDE	151	CALCIUM CARBIDE
CARBINAMINE	566	MONOMETHYLAMINE
CARBOBENZOXY CHLORIDE	110	BENZYL CHLORFORMATE
CARBOLIC ACID	645	PHENOL
CARBOLIC ACID	169	CARBOLIC OIL
CARBON BISULFIDE	172	CARBON DISULFIDE
CARBON DISULFATE	170	CARBON BISULFIDE
CARBON HEXACHLORIDE	405	HEXACHLOROETHANE
CARBONIC ACID	271	DIETHYL CARBONATE
CARBONIC ACID	38	AMMONIUM BICARBONATE
CARBONIC ACID GAS	171	CARBON DIOXIDE
CARBON TET	174	CARBON TETRACHLORIDE
CARBONYL CHLORIDE	649	PHOSGENE
CARBONYLDIMAIDE	829	UREA
CARBOXYLBENZENE	104	BENZOIC ACID
CARWINATE 125 M	303	DIPHENYLMETHYL DIOSOCYANATE

SYNONYM	PAGE	CHEMICAL NAME
CASORON	249	DICHLOBENIL (liquid)
CASORON	250	DICHLOBENIL (solid)
CAT CRACKER FEED STOCK	383	FUEL OIL NUMBER 4
CAUSTIC POTASH	676	POTASSIUM HYDROXIDE (solid)
CAUSTIC POTASH	677	POTASSIUM HYDROXIDE (liquid)
CAUSTIC SODA	741	SODIUM HYDROXIDE
CD 68	178	CHLORDANE
CELLOSOLVE ACETATE	343	ETHYLENE GLYCOL MONOETHYL ETHER ACETATE
CETYLTRIMETHYL-AMMONIUM CHLORIDE SOLUTION	413	HEXADECYLTRIMETHYL-AMMONIUM CHLORIDE
CHEM BAM	569	NABAM
CHILE SALT PETER	744	SODIUM NITRATE
CHINESE RED	510	MERCURIC SULFIDE
CHINESE TANNIN	768	TANNIC ACID
CHLORACETIC ACID	561	MONOCHLOROACETIC ACID (liquid)
CHLORACETIC ACID	562	MONOCHLOROACETIC ACID (solid)
CHLORACETIC ACID	181	CHLOROACETIC ACID
CHLORACETYL CHLORIDE	183	CHLOROACETYL CHLORIDE
CHLORATE OF POTASH	670	POTASSIUM CHLORATE
CHLORATE OF POTASSIUM	670	POTASSIUM CHLORATE
CHLORATE OF SODA	730	SODIUM CHLORATE
CHLORDAN	178	CHLORDANE
CHLORDECONE	465	KEPONE
CHLOREX	260	DICHLOROETHYL ETHER
CHLORFORMIC ACID	535	METHYL CHLOROFORMATE
CHLORIDE OF AMYL	66	AMYL CHLORIDE
CHLORINATED HYDRO-CHLORIC ETHER	259	DICHLOROETHANE
CHLOROACETIC ACID	561	MONOCHLOROACETIC ACID (liquid)
CHLOROACETIC ACID	562	MONOCHLOROACETIC ACID (solid)
CHLOROACETIC ACID	331	ETHYL CHLOROACETATE
4-CHLOROANILINE	184	CHLOROANILINE
2-CHLORO-1,3-BUTADIENE	192	CHLOROPRENE
1-CHLORO-2-3-EPOXYPROPANE	317	EPICHLOROHYDRIN
CHLOROETHANOIC ACID	561	MONOCHLOROACETIC ACID (liquid)
CHLOROETHANOIC ACID	562	MONOCHLOROACETIC ACID (solid)
2-CHLOROETHANOL	336	ETHYLENE CHLOROHYDRIN
CHLOROFORMIC ACID	332	ETHYL CHLOROFORMATE
CHLOROFORMIC ACID	110	BENZYL CHLORFORMATE
CHLOROMETHANE	534	METHYL CHLORIDE
1-CHLORO-4-METHYLBENZENE	194	CHLOROTOLUENE o-m-p
4-CHLORO-1-METHYLBENZENE	194	CHLOROTOLUENE o-m-p
CHLOROMETHYLOXIRANE	317	EPICHLOROHYDRIN
CHLORO-M-NITROBENZENE	602	NITROCHLOROBENZENE-META
1-CHLORO-2-NITRILE	603	NITROCHLOROBENZENE-ORTHO
1-CHLORO-3-NITRILE	602	NITROCHLOROBENZENE-META
4-CHLORO-1-NITRO-BENZENE	604	NITROCHLOROBENZENE-PARA
CHLORO-1-NITROBENZENE	603	NITROCHLOROBENZENE-ORTHO
1-CHLORO-2-NITROBENZENE	189	CHLORONITROBENZENE
4-CHLOROPHENOL	190	CHLOROPHENOL
4-CHLOROPHENYLAMINE	184	CHLOROANILINE
CHLOROPHOS	797	TRICHLORFON
CHLORO-2-PROPENE	26	ALLYL CHLORIDE
3-CHLORO-1-20PROPYLENE OXIDE	317	EPICHLOROHYDRIN
4-CHLOROTOLUENE	194	CHLOROTOLUENE o-m-p
CHLOROTOULENE M	194	CHLOROTOLUENE o-m-p
CHLOROTOULENE O	194	CHLOROTOLUENE o-m-p
CHLOROTOULENE P	194	CHLOROTOLUENE o-m-p
CHLOROTRIMETHYLSILANE	818	TRIMETHYLCHLOROSILANE
CHLOROX	741	SODIUM HYPOCHLORITE
CHLORPYRIFOS	312	DURSBAN
CHLORSULFONIC ACID	193	CHLOROSULPHONIC ACID
CHLORSULFURIC ACID	193	CHLOROSULPHONIC ACID
CHLORYLEN	800	TRICHLOROETHYLENE
CHOROETHANE	330	ETHYL CHLORIDE
CHP	216	CUMENE HYDROPEROXIDE
CHROMIC ACID	756	STRONTIUM CHROMATE
CHROMIC ACID	493	LITHIUM CHROMATE
CHROMIC ACID	197	CHROMIC ANHYDRIDE
CHROMIC ANHYDRIDE	196	CHROMIC ACID
CHROMIC III ACETATE	195	CHROMIC ACETATE
CHROMIC OXIDE	197	CHROMIC ANHYDRIDE
CHROMIC OXIDE	196	CHROMIC ACID
CHROMIUM ACETATE	195	CHROMIC ACETATE
CHROMIUM CHLORIDE	199	CHROMOUS CHLORIDE
CHROMIUM DICHLORIDE	199	CHROMOUS CHLORIDE
CHROMIUM LITHIUM OXIDE	493	LITHIUM CHROMATE
CHROMIUM OXYCHLORIDE	200	CHROMYL CHLORIDE
CHROMIUM SALT	195	CHROMIC ACETATE
CHROMIUM SULFATE	198	CHROMIC SULFATE
CHROMIUM TRIACETATE	195	CHROMIC ACETATE
CHROMIUM VI DIOXYCHLORIDE	200	CHROMYL CHLORIDE
CIANURINA	504	MERCURIC CYANIDE
CIS-ACETYLENE	253	1,2-DICHLOROETHYLENE
CIS-2-BUTENE-1,4-DIOL	126	BUTENEDIOL
CIS-8-HEPTADECYLENE CARBOXYLIC ACID	623	OLEIC ACID
CIS-9-OCTADECENOIC ACID	623	OLEIC ACID
CITRIC ACID	45	AMMONIUM CITRATE
COAL OIL	381	FUEL OIL NUMBER 1
COAL TAR	212	CREOSOTE, COAL TAR
COLD TAR	572	NAPHTHA
COLOPHONY SOLUTION	704	ROSIN SOLUTION
CONDENSED PHOSPHORIC ACID	665	POLYPHOSPHORIC ACID
CONOCO SA597	310	DODECYLBENZENE-SULFONIC ACID
COPPERAS	369	FERROUS SULFATE
COPPER NITRATE TRIHYDRATE	217	CUPRIC NITRATE
COPPER SULFATE	218	CUPRIC SULFATE
COPPER SULFATE PENTAHYDRATE	218	CUPRIC SULFATE
COPPER SULFATE PENTAHYDRATE	210	COPPER SULFATE
CO-RAL	211	COUMAPHOS
CREOSOTE	212	CREOSOTE, COAL TAR
CREOSOTE OIL	212	CREOSOTE, COAL TAR
3-CRESOL	213	CRESOL (o-m-p)

SYNONYM	PAGE	CHEMICAL NAME
CRESOL M	213	CRESOL (o-m-p)
CRESOL O	213	CRESOL (o-m-p)
CRESOL P	213	CRESOL (o-m-p)
CRESYLIC ACID	847	XYLENOL (liquid)
CRESYLIC ACID	848	XYLENOL (solid)
CROTENALDEHYDE	214	CROTONALDEHYDE
CROTONIC ALDEHYDE	214	CROTONALDEHYDE
CRUDE EPICHLOROHYDRIN	187	CHLOROHYDRINS
CRYSTALLIZED VERDIGRIS	202	COPPER ACETATE
CTF	180	CHLORINE TRIFLUORIDE
CTFE	815	TRIFLUOROCHLORO-ETHYLENE
CUCUMBER DUST	149	CALCIUM ARSENATE
CUMOL	215	CUMENE
CUMYL HYDROPEROXIDE	216	CUMENE HYDROPEROXIDE
CUPRIC ACETATE MONOHYDRATE	202	COPPER ACETATE
CUPRIC ARSENITE	204	COPPER ARSENITE
CUPRIC CHLORIDE DIHYDRATE	205	COPPER CHLORIDE
CUPRIC NITRATE TRIHYDRATE	208	COPPER NITRATE
CUPRIC OXILATE	209	COPPER OXALATE
CUPRIC SALT	202	COPPER ACETATE
CUPRIC SULFATE	210	COPPER SULFATE
CUPROUS CYANIDE, CUPRICIN CYANIDE	206	COPPER CYANIDE
CURATERR	167	CARBOFURAN
CYANACETIC ACID	220	CYANOACETIC ACID
CYANIDE	672	POTASSIUM CYANIDE (liquid)
CYANIDE	673	POTASSIUM CYANIDE (solid)
CYANIDE OF CALCIUM	154	CALCIUM CYANIDE
CYANIDE OF ZINC	866	ZINC NITRATE
CYANIDE OF ZINC	860	ZINC CYANIDE
CYANOBENZENE	105	BENZONITRILE
CYANOGAS A-DUST	154	CALCIUM CYANIDE
CYANO-METHANE	8	ACETONITRILE
CYCLODAN	307	ENDOSULFAN (liquid)
CYCLODAN	308	ENDOSULFAN (solid)
CYCLOHEXY ALCOHOL	225	CYCLOHEXANOL
CYCLOPENTANE METHYL	537	METHYL CYCLOPENTANE
CYMOL	232	CYMENE
CYPRIETHYLENDIAMINE HYDOXIDE SOLUTION	219	CUPRIETHYLENEDIAMINE SOLUTION
CYTHION INSECTICIDE	497	MALATHION
2,4-D	263	DICHLOROPHENOXYACETIC ACID
DALMATION-INSECT POWDER	697	PYRETHRINS ACID (liquid)
DALMATION-INSECT POWDER	698	PYRETHRINS ACID (solid)
2,6-DBN	249	DICHLOBENIL (liquid)
2,6-DBN	250	DICHLOBENIL (solid)
DBP	246	DIBUTYL PHTHALATE
DCEE	260	DICHLOROETHYL ETHER
DD	820	TURPENTINE
DDVP	266	DICHLORVOS (liquid)
DDVP	267	DICHLORVOS (solid)
DEAD OIL	212	CREOSOTE, COAL TAR
DECHLORANE	559	MIREX
DEEP LEMON YELLOW	756	STRONTIUM CHROMATE
DEHPA	240	DI-(2-ETHYLHEXYL) PHOSPHORIC ACID
DE KALIN	238	DECAHYDRONAPHTHALENE
DELSICOL	400	HEPTACHLOR
DEN	272	DIETHYLAMINE
DEXTRONE	305	DIQUAT (liquid)
DEXTRONE	306	DIQUAT (solid)
D-GLUCITOL	753	SORBITOL
DIACETIC ETHER	321	ETHYL ACETOACETATE
DIACETYLMETHANE	14	ACETYLACETONE
DIACETYL PEROXIDE SOLUTION	13	ACETYL PEROXIDE SOLUTION
DIALUMINUM SALT	52	AMMONIUM OXALATE
2,2-DIAMINODIETHYL-AMINE	281	DIETHYLENETRIAMINE
1,2-DIAMINOETHANE	274	ETHYLENEDIAMINE
1,11-DIAMINO-3,6,9-TRIAZAUNDECANE	779	TETRAETHYLENEPENT-AMINE
DIAMMONIUM CHROMATE	44	AMMONIUM CHROMATE
DIAMMONIUM CITRATE	45	AMMONIUM CITRATE
DIAMMONIUM OXILATE	52	AMMONIUM OXALATE
DIAMMONIUM SALT	54	AMMONIUM PERSULFATE
DIAMMONIUM SALT	45	AMMONIUM CITRATE
DIANTIMONY TRIOXIDE	80	ANTIMONY TRIOXIDE
DIBASIC	45	AMMONIUM CITRATE
DIBENZO B,E PYRIDINE	55	AMMONIUM SILICO-FLUORIDE
DIBENZO B,E PYRIDINE	16	ACRIDINE
DIBK	278	DIISOBUTYL KETONE
DIBROM	570	NALED (liquid)
DIBROM	571	NALED (solid)
1,2-DIBROMOETHANE	339	ETHYLENE DIBROMIDE
DIBUTYL ESTER	246	DIBUTYL PHTHALATE
DIBUTYL ETHER	241	DI-BUTYL ETHER
DIBUTYL OXIDE	241	DI-BUTYL ETHER
DICAMBA	247	DICAMBA (liquid)
DICAMBA	248	DICAMBA (solid)
DICARBOXYLIC ACID	627	OXALIC ACID
DICHLONE	251	DICHLONE (liquid)
DICHLONE	252	DICHLONE (solid)
DICHLORIDE	103	BENZENE PHOSPHORUS THIODICHLORIDE
DICHLOROBENIL	249	DICHLOBENIL (liquid)
DICHLOROBENIL	250	DICHLOBENIL (solid)
2,6-DICHLOROBENZONITRILE	249	DICHLOBENIL (liquid)
2,6-DICHLOROBENZONITRILE	250	DICHLOBENIL (solid)
1,1-DICHLORO-2,2-BIS P-CHLOROPHENYL ETHER	235	DDD
1,4-DICHLORO-2-BUTENE	257	DICHLOROBUTENE
DICHLORODIETHYL ETHER	260	DICHLOROETHYL ETHER
DICHLORODIPHENYL-DICHLOROETHANE	235	DDD
DICHLORODIPHENYLSILANE	302	DIPHENYLDICHLORO-SILANE
DICHLORODIPHENYLTRI-CHLOROETHANE	236	DDT
DICHLOROETHER	260	DICHLOROETHYL ETHER
1,1-DICHLOROETHYLENE	839	VINYLIDENE CHLORIDE
DI-2-CHLOROETHYL ETHER	260	DICHLOROETHYL ETHER
DICHLOROMETHANE	539	METHYLENE CHLORIDE
2,3-DICHLORO-1,4-NAPHTHO-QUINONE	251	DICHLONE (liquid)
2,3-DICHLORO-1,4-NAPHTHO-QUINONE	252	DICHLONE (solid)
3,6-DICHLORO-O-ANISIC ACID	247	DICAMBA (liquid)
3,6-DICHLORO-O-ANISIC ACID	248	DICAMBA (solid)
2,4-DICHLOROPHENOXY ACETIC ACID	233	2,4-D ESTERS
DICHLOROPHENYL-PHOSPHINE	102	BENZENE PHOSPHORUS DICHLORIDE
DICHLOROPROPANE	264	1,2-DICHLOROPROPANE

SYNONYM	PAGE	CHEMICAL NAME
2,2-DICHLOROPROPANIC ACID	234	DALAPON
2,2-DICHLOROPROPANOIC ACID	234	DALAPON
DICHLOROPROPENE	265	1,3-DICHLOROPROPENE
DICHROMIUM SULFATE	198	CHROMIC SULFATE
DICY	268	DICYCLOPENTADIENE
DICYAN	221	CYANOGEN
1,4-DICYANOBUTANE	22	ADIPONITRILE
DICYCLOHEXANONE DIPEROXIDE	227	CYCLOHEXANONE PEROXIDE
DIESEL IGNITION IMPROVER	69	AMYL NITRATE
DIESEL OIL	382	FUEL OIL NUMBER 2
DIETHION	319	ETHION
1,2-DIETHOXYETHANE	341	ETHYLENE GLYCOL DIETHYL ETHER
DIETHYL CELLOSOLVE	341	ETHYLENE GLYCOL DIETHYL ETHER
DIETHYLENE OXIDE	781	TETRAHYDROFURAN
DIETHYL ESTER	271	DIETHYL CARBONATE
DIETHYL ETHER,ETHYL OXIDE,ETHER	347	ETHYL ETHER
DI-2-ETHYLHEXYL PHOSPHATE	240	DI-(2-ETHYLHEXYL) PHOSPHORIC ACID
DIFLUOROPHOSPHORUS ACID	277	DIFLUOROPHOSPHORIC ACID
DIFORMYL	397	GLYOXAL
DIGLYCOLAMINE	33	2-(2)-AMINOETHOXY (ETHANOL)
2,5-DIHYDROPEROXY-2,5-DIMETHYLHEXANE	289	DIMETHYLHEXANE DIHYDROPEROXIDE
1-2-DIHYDRO-3-6-PYRIDAZINEDIONE	500	MALEIC HYDRAZIDE
1,2-DIHYDROXPROPANE	692	PROPYLENE GLYCOL
1,4-DIHYDROXYBUTANE	126	BUTANEDIOL
1,4-DIHYDROXY-2-BUTENE	128	BUTENEDIOL
DIISOPROPYL ETHER	456	ISOPROPYL ETHER
DIISOPROPYL PERCARBONATE	459	ISOPROPYL PERCARBONATE (UNSTABILIZED)
DIISOPROPYL PERCARBONATE	458	ISOPROPYL PERCARBONATE (STABILIZED)
DIISOPROPYL PEROXYDICARBONATE	459	ISOPROPYL PERCARBONATE (UNSTABILIZED)
DIISOPROPYL PEROXYDICARBONATE	458	ISOPROPYL PERCARBONATE (STABILIZED)
DILAUROYL PEROXIDE	470	LAUROYL PEROXIDE (LESS THAN 42%)
DILAUROYL PEROXIDE	469	LAUROYL PEROXIDE
DILITHIUM SALT	493	LITHIUM CHROMATE
DILUTED SULFURIC ACID	766	SULFURIC ACID, SPENT
DIMAZINE	290	DIMETHYLHYDRAZINE
DIMETHOXYMETHANE	542	METHYL FORMAL
10,11-DIMETHOXYSTRYCHNINE	123	BRUCINE
2,5-DIMETHYHEXANE-2,5-DIHYDROPEROXIDE	289	DIMETHYLHEXANE DIHYDROPEROXIDE
DIMETHYL ACETATIC ACID	441	ISOBUTYRIC ACID
DIMETHYLAMINE ANHYDROUS	286	DIMETHYLAMINE
DIMETHYLARSINIC ACID	141	CACODYLIC ACID
1,4-DIMETHYLBENZENE	846	XYLENE-P
1,2-DIMETHYLBENZENE	845	XYLENE-O
1,3-DIMETHYLBENZENE	844	XYLENE-M
2,2-DIMETHYLBUTANE	580	NEOHEXANE
DIMETHYLFORMAL	542	METHYL FORMAL
2,6-DIMETHYL-4-HEPTANOL	279	DIISOBUTYLCARBINOL
DIMETHYLHEXAMALS	446	ISOOCTALDEHYDE
DIMETHYL-1-HEXANOLS	447	ISOOCTYL ALCOHOL
DIMETHYL KETONE	6	ACETONE
DIMETHYL METHANE	682	PROPANE
2,2-DIMETHYL-3-METHYLENE-NORBORANE	161	CAMPHENE
3,3-DIMETHYL-2,METHYLENE NORCAMPHANE	161	CAMPHENE
2,4-DINITRANILINE	292	DINITROANILINE
2,6-DINITRO-O-CRESOL	295	DINITROCRESOL
3,5-DINITRO-O-CRESOL	295	DINITROCRESOL
4,6-DINITRO-O-CRESOL	295	DINITROCRESOL
2,4-DINITROTOLUOL	297	DINITROTOULENE (liquid)
2,4-DINITROTOLUOL	298	DINITROTOULENE (solid)
DIOFORM	253	1,2-DICHLOROETHYLENE
11-DIOL	778	TETRAETHYLENE GLYCOL
DI-ON	309	DIURON
1,4,DIOXANE	299	DIOXANE
DIOXONIUM PERCHLORATE SOLUTION	641	PERCHLORIC ACID OVER 50%
DIOXONIUM PERCHLORATE SOLUTION	640	PERCHLORIC ACID LESS THAN 50%
DIPROPYLAMINE	304	DIPROPYLAMINE
DIPTEREX	797	TRICHLORFON
DISODIUM ARSENATE HEPTAHYDRATE	720	SODIUM ARSENATE
DISODIUM ETHYLENEBIS	569	NABAM
DISODIUM METHANE-ARSONATE	519	METHANEARSONIC ACID, SODIUM SALT
DISODIUM METHYL ARSENATE	519	METHANEARSONIC ACID, SODIUM SALT
DISODIUM NITRILOTRI-ACETATE	599	NITRILOTRIACETIC ACID AND SALTS
DISODIUM SALT	748	SODIUM SELENITE
DISODIUM SALT	746	SODIUM OXALATE
DISODIUM SALT	740	SODIUM HYDROSULFITE
DISODIUMSELENITE	748	SODIUM SELENITE
DISULFATOZIRCONIC ACID	876	ZIRCONIUM SULFATE
DISULFUR DICHLORIDE	772	SULFUR CHLORIDE
DI-SYSTON	307	DISULFOTON (liquid)
DI-SYSTON	308	DISULFOTON (solid)
DITHANE	569	NABAM
DITHIOCARBONIC ANHYDRIDE	170	CARBON BISULFIDE
DITHIOPYROPHOSPHORIC ACID	774	TETRAETHYL DITHIOPYROPHOSPHATE (liquid)
DITHIOPYROPHOSPHORIC ACID	784	TETRAETHYL DITHIOPYROPHOSPHATE (solid)
DIUREX	309	DIURON
DMDT	521	METHOXYCHLOR
DMF	288	DIMETHYLFORMAMIDE
DNPA	304	DIPROPYLAMINE
DNT	297	DINITROTOULENE (liquid)
DNT	298	DINITROTOULENE (solid)
DO 14	683	PROPARGITE
1-DODECANETHIOL	471	LAURYL MERCAPTAN
DODECANOL	486	LINEAR ALCOHOL
DODECEEN NONLINEAR	694	PROPYLENE TETRAMER
DODECYLETHYLENE	773	TETRADECENE
DODECYL MERCAPTAN	471	LAURYL MERCAPTAN
DOT	877	ZIRCONIUM TETRA-CHLORIDE

SYNONYM	PAGE	CHEMICAL NAME
DOWANOL EM	342	ETHYLENE GLYCOL MONOETHYL ETHER
DOWCO 179	312	DURSBAN
DOW-FUME 40	339	ETHYLENE DIBROMIDE
DOWICIDE 2	802	TRICHLOROPHENOL
DOWICIDE 7	633	PENTACHLOROPHENOL
DRACYCLIC ACID	104	BENZOIC ACID
DRY CLEANERS NAPTHA	574	NAPHTHA (STODDARD SOLVENT)
DRY ICE	171	CARBON DIOXIDE
DSMA	519	METHANEARSONIC ACID, SODIUM SALT
DU-SPREX	249	DICHLOBENIL (liquid)
DU-SPREX	250	DICHLOBENIL (solid)
DUTCH LIQUID	340	ETHYLENE DICHLORIDE
DYLOX	797	TRICHLORFON
DYTOL S91	239	DECYL ALCOHOL
E3314	400	HEPTACHLOR
EAA	321	ETHYL ACETOACETATE
EASE	325	ETHYL ALUMINUM SESQUICHLORIDE
EBDC	569	NABAM
ECRINITRIT	745	SODIUM NITRITE
EDC	340	ETHYLENE DICHLORIDE
EDTA	338	ETHYLENEDIAMINE TETRACETIC ACID
EMBAFUME	532	METHYL BROMIDE
ENANTHIC ALCOHOL	402	HEPTANOL
ENDO,	269	DIELDRIN (liquid)
ENDO,	270	DIELDRIN (solid)
ENDRATE	338	ETHYLENEDIAMINE TETRACETIC ACID
ENT 25,719	559	MIREX
ENT-16391	465	KEPONE
ENT 27,311	312	DURSBAN
ENT 4225	235	DDD
EPICLEAR	107	BENZOYL PEROXIDE
EPOXIDIZED TALL OIL	621	OCTYL EPOXY TALLATE
1,2-EPOXYBUTANE	135	BUTYLENE OXIDE
1,2-EPOXYETHANE	345	ETHYLENE OXIDE
1,2-EPOXYPROPANE	698	PROPYLENE OXIDE
ERIOCHOLCITE ANHYDROUS	205	COPPER CHLORIDE
ESKIMON 11	801	TRICHLOROFLUORO-METHANE
ETHANAL,ACETIC ALDEHYDE	2	ACETALDEHYDE
ETHANE-DI- ACID	627	OXALIC ACID
ETHANEDIAL	397	GLYOXAL
ETHANE DINITRILE	221	CYANOGEN
ETHANEDIOIC ACID	746	SODIUM OXALATE
ETHANEDIOIC ACID	627	OXALIC ACID
ETHANENITRILE	8	ACETONITRILE
ETHANETHIOL	351	ETHYL MERCAPTAN
ETHANOIC ANHYDRIA	5	ACETIC ANHYDRIDE
ETHANOL-INTERNATIONAL NAME	324	ETHYL ALCOHOL
ETHANOYL CHLORIDE	11	ACETYL CHLORIDE
ETHENE	335	ETHYLENE
ETHINE	15	ACETYLENE
ETHIOPS MINERAL	510	MERCURIC SULFIDE
2-ETHOXYETHYL ACETATE	343	ETHYLENE GLYCOL MONOETHYL ETHER ACETATE
ETHYL ALDEHYDE	2	ACETALDEHYDE
ETHYLBENZOL	327	ETHYLBENZENE
ETHYL BUTANOATE	329	ETHYL BUTYRATE
2-ETHYL-1-BUTANOL	328	ETHYLBUTANOL
2-ETHYLBUTYL ALCOHOL	328	ETHYLBUTANOL
ETHYL CARBONATE	271	DIETHYL CARBONATE
ETHYL CHLOROCARBONATE	332	ETHYL CHLOROFORMATE
ETHYL DICHLOROPHOS-PHATE	356	ETHYL PHOSPHORO-DICHLORIDATE
ETHYLENE ALDEHYDE	17	ACROLEIN
1,2-ETHYLENEDIAMINE	337	ETHYLENEDIAMINE
ETHYLENE GLYCOL HIDYDROXYDIETHYL ETHER	813	TRIETHYLENE GLYCOL
ETHYL ESTER	321	ETHYL ACETOACETATE
ETHYL ESTER	322	ETHYL ACRYLATE, INHIBITED
ETHYL ESTER	323	ETHYL ACRYLATE (liquid)
ETHYL ESTER	332	ETHYL CHLOROFORMATE
ETHYL ESTER	348	ETHYL FORMATE
ETHYL ESTER	350	ETHYL LACTATE
ETHYL ESTER	352	ETHYL METHACRYLATE
ETHYLETHYLENE	127	BUTENE
2-ETHYL HEXALDEHYDE	349	ETHYL HEXALDEHYDE
2-ETHYLHEXANAL	349	ETHYL HEXALDEHYDE
2-ETHYL-1- HEXANOL HYDROGEN PHOSPHATE	240	DI-(2-ETHYLHEXYL) PHOSPHORIC ACID
2-ETHYL-2-HEXENAL	358	ETHYL PROPYLACROLENE
ETHYL 2-HYDROXY-PROPANOATE	350	ETHYL LACTATE
ETHYL 2-HYDROXY-PROPIONATE	350	ETHYL LACTATE
ETHYLIDENE CHLORIDE	259	DICHLOROETHANE
ETHYLIDENE DICHLORIDE	259	DICHLOROETHANE
ETHYL 2-METHACRYLATE	352	ETHYL METHACRYLATE
ETHYL METHACRYLATE-INHIBITED	352	ETHYL METHACRYLATE
5-ETHYL-2-METHYL PYRIDINE 5-ETHYL 2-2 P	541	METHYL ETHYL PYRIDINE
ETHYL NITRILE	8	ACETONITRILE
ETHYL 3-OXOBUTANOATE	321	ETHYL ACETOACETATE
ETHYL PARATHION	631	PARATHION
ETHYL-2-PROPENOATE	322	ETHYL ACRYLATE, INHIBITED
2-ETHYL-3-PROPYL-ACRYLALDEHYDE	358	ETHYL PROPYLACROLENE
ETHYLSILICON TRICHLORIDE	346	ETHYLENETRICHLORO-SILANE
ETHYL SULFHYDRATE	351	ETHYL MERCAPTAN
ETHYL THIONOPHOSPHORYL DICHLORIDE	355	ETHYL PHOSPHONOTHIOIC DICHLORIDE
ETHYL VINYL ESTHER	837	VINYL ETHYL ETHER
ETHYLZINC	275	DIETHYLZINC
ETHYNE	15	ACETYLENE
EUFIN	271	DIETHYL CARBONATE
EUNATROL	625	OLEIC ACID, SODIUM SALT
EXITELITE	80	ANTIMONY TRIOXIDE
F-11	801	TRICHLOROFLUORO-METHANE
F22	563	MONOCHLORODI-FLUOROMETHANE
FENOPROP	803	TRICHLOROPHENOXY-PROPIONIC ACID ESTER
FERMENTATION AMYL ALCOHOL	433	ISOAMYL ALCOHOL
FERRIC AMMONIUM CITRATE GREEN, BROWN	362	FERRIC AMMONIUM CITRATE
FERRIC CHLORIDE ANHYDROUS	364	FERRIC CHLORIDE (liquid)

SYNONYM	PAGE	CHEMICAL NAME
FERRIC CHLORIDE HEXAHYDRATE	364	FERRIC CHLORIDE (liquid)
FERRIC NITRATE NONAHYDRATE	368	FERRIC NITRATE
FERROUS BOROFLUORIDE	373	FERROUS FLUOROBORATE
FERROUS OXILATE DIHYDRATE	374	FERROUS OXALATE
FERROUS SALT	374	FERROUS OXALATE
FERROX	374	FERROUS OXALATE
FERTILIZER ACID	765	SULFURIC ACID
FILMERINE	745	SODIUM NITRITE
FLOWERS OF ANTIMONY	80	ANTIMONY TRIOXIDE
FLOWERS OF SULFUR	761	SULFUR
FLUORISTAN	754	STANNOUS FLUORIDE
FLUOROETHYLENE	838	VINYL FLUORIDE
FLUOROHYDRIC ACID	423	HYDROGEN FLUORIDE
FLUOROSULFONIC ACID	378	FLUROSULPHONIC ACID
FLUOROSULFURIC ACID	378	FLUROSULPHONIC ACID
FLUROSILIC ACID	377	FLUOROSILICIC ACID
FORMALDEHYDE DIMETHYLACETAL	542	METHYL FORMAL
FORMALIN	379	FORMALDEHYDE
FORMALITH	379	FORMALDEHYDE
FORMATIC ACID	864	ZINC FORMATE
FORMAYLIC ACID	380	FORMIC ACID
FORMIC ACID	543	METHYL FORMATE
FORMIC ACID	348	ETHYL FORMATE
FOWLERS SOLUTION	669	POTASSIUM ARSENITE
FREON 11	801	TRICHLOROFLUORO-METHANE
FREON 22	563	MONOCHLORODI-FLUOROMETHANE
FREON 12	258	DICHLORODIFLUORO-METHANE
FREON 20	186	CHLOROFORM
FRIGEN 11	801	TRICHLOROFLUORO-METHANE
FUEL	617	NUCLEAR REACTOR FUEL
FUEL OIL 1	461	JET FUELS, JP-1
FUEL OIL NUMBER ONE	466	KEROSENE
FUMING LIQUID ARSENIC	85	ARSENIC TRICHLORIDE
FUMING SULFURIC ACID	626	OLEUM
FURADAN	167	CARBOFURAN
FURAL	386	FURFURAL
FURFURALCOHOL	387	FURFURYL ALCOHOL
FURFUROLE	386	FURFURAL
FUSEL OIL	433	ISOAMYL ALCOHOL
GALENA	480	LEAD SULFIDE
GALLIC ACID MONOHYDRATE	388	GALLIC ACID
GALLOTANNIC ACID	768	TANNIC ACID
GALLTANNIN	768	TANNIC ACID
GAMMA-CHLOROPRO-PYLENE OXIDE	317	EPICHLOROHYDRIN
GAMMEXANE	101	BENZENE HEXACHLORIDE
GC-1189	465	KEPONE
GELBIN YELLOW ULTRA MARINE	153	CALCIUM CHROMATE
GEMALGENE	800	TRICHLOROETHYLENE
GENETRON 1113	815	TRIFLUOROCHLORO-ETHYLENE
GENETRON 11	801	TRICHLOROFLUORO-METHANE
GERHARDITE	217	CUPRIC NITRATE
GERHARDITE	208	COPPER NITRATE
GLACIAL ACETIC ACID	12	ACETYL PEROXIDE
GLACIAL ACETIC ACID	4	ACETIC ACID (GLACIAL)
GLYCIDYL ALPHA-METHYL-ACRYLATE	396	GLYCIDYL METHACRYLATE
GLYCOL CHLOROHYDRIN	336	ETHYLENE CHLOROHYDRIN
GLYCOL CYANOHYDRIN	333	ETHYL CYANOHYDRIN
GLYCOL DICHLORIDE	340	ETHYLENE DICHLORIDE
GLYCOL MONOETHYL ETHER ACETATE	343	ETHYLENE GLYCOL MONOETHYL ETHER ACETATE
GREEN NICKEL OXIDE	589	NICKEL HYDROXIDE
GREEN VITRIOL	375	FERROUS SULFATE
GUTHION	91	AZINPHOS METHYL
GYCOL MONOMETHYLETHER	342	ETHYLENE GLYCOL MONOETHYL ETHER
GYLCERITE	768	TANNIC ACID
GYLCEROL	394	GLYCERINE
GYLCOL BES HYDDROXY-ETHYL ETHER	813	TRIETHYLENE GLYCOL
HAH	491	LITHIUM ALUMINUM HYDRIDE
HALOGENATED WAXES	663	POLYCHLORINATED BIPHENYL
HALON 122	258	DICHLORODIFLUORO-METHANE
HARTSHORN	42	AMMONIUM CARBONATE
HCCPD	404	HEXACHLORO-CYCLOPENTADIENE
HCE	405	HEXACHLOROETHANE
HEHYDRITE	496	MAGNESIUM PERCHLORATE
HEMIHYDRATE	209	COPPER OXALATE
HENDECANOIC ALCOHOL	821	UNDECANOL
1-HENDECANOL	821	UNDECANOL
HEOD	269	DIELDRIN (liquid)
HEOD	270	DIELDRIN (solid)
HEPATIC GAS	425	HYDROGEN SULFIDE
1-HEPTADECANECAR-BOXYLIC ACID	755	STEARIC ACID
1-HEPTANOL	402	HEPTANOL
2-HEPTANONE	530	METHYL AMYL KETONE
2-HEPTANONE	68	AMYL METHYL KETONE
HEPTYL ALCOHOL	402	HEPTANOL
HEPTYLCARBINOL	619	OCTANOL
HEPTYLENE	403	HEPTENE
HETRA CARBONYL	584	NICKEL CARBONYL
HEX	404	HEXACHLOROCYCLO-PENTADIENE
HEXA	410	HEXAMETHYLENETETRAMINE
1,2,3,4,5,6,-HEXACHLORO-HEXACHLOROCYCL	101	BENZENE HEXACHLORIDE
HEXADRIN	315	ENDRIN (liquid)
HEXADRIN	316	ENDRIN (solid)
HEXAFLUOSILICIC ACID	377	FLUOROSILICIC ACID
HEXAHYDRIC ALCOHOL	753	SORBITOL
HEXAHYDROANILINE	229	CYCLOHEXENYLAMINE
HEXAHYDROAZEPINE	409	HEXAMETHYLENIMINE
HEXAHYDROBENZENE	224	CYCLOHEXANE
HEXAHYDROPHENOR	225	CYCLOHEXANOL
HEXALINANOL	225	CYCLOHEXANOL
HEXAMINE	410	HEXAMETHYLENE-TETRAMINE
HEXANAL	407	HEXALDEHYDE
1,6-HEXANEDIAMINE	408	HEXAMETHYLENE DIAMINE
1,2,3,4,5,6-HEXANNEHEXOL	753	SORBITOL
1-HEXANOL	412	HEXANOL
2-HEXANONE	551	METHYL n-BUTYL KETONE

SYNONYM	PAGE	CHEMICAL NAME
1-HEXENE	413	HEXENE
HEXONE	546	METHYL ISOBUTYL KETONE
HEXYL ACETATE	528	METHYL AMYL ACETATE
HEXYLDYRIDE	411	HEXANE
HEXYLENE	413	HEXENE
HHDN	23	ALDRIN
HI-DRY	778	TETRAETHYLENE GLYCOL
HMDA	408	HEXAMETHYLENE DIAMINE
HOME HEATING	382	FUEL OIL NUMBER 2
HOUSEHOLD AMMONIA	49	AMMONIUM HYDROXIDE
HTH	155	CALCIUM HYPOCHLORITE
HTH DRY CHLORINE	155	CALCIUM HYPOCHLORITE
HYDE	379	FORMALDEHYDE
HYDHRACRYLONITRILE, 2-CYANOETHANOL	333	ETHYL CYANOHYDRIN
HYDRAZINE ANHYDROUS	416	HYDRAZINE
HYDRAZOIC ACID, SODIUM SALT	723	SODIUM AZIDE
HYDROBROMIC ACID ANHYDROUS	420	HYDROGEN BROMIDE
HYDROCHLORIC ACID	421	HYDROGEN CHLORIDE
HYDROCYANIC ACID	732	SODIUM CYANIDE (liquid)
HYDROCYANIC ACID	733	SODIUM CYANIDE (solid)
HYDROCYANIC ACID	422	HYDROGEN CYANIDE
HYDROFLUORIC ACID ANHYDROUS	423	HYDROGEN FLUORIDE
HYDROGEN BROMIDE ANHYDROUS	420	HYDROGEN BROMIDE
HYDROGEN DIOXIDE	424	HYDROGEN PEROXIDE 52–100%
HYDROGEN FLUORIDE ANHYDROUS	423	HYDROGEN FLUORIDE
HYDROGEN HEXAFLUOROSILICATE	377	FLUOROSILICIC ACID
HYDROQUINOL	426	HYDROQUINONE
HYDROSILICOFLUORIC ACID	427	HYDROSILICOFLUORIC ACID
HYDROSULFITE OF SODA	740	SODIUM HYDROSULFITE
HYDROSULFURIC ACID	425	HYDROGEN SULFIDE
HYDROXYCYCLOHEXANE	225	CYCLOHEXANOL
HYDROXYDIMETHYLARSINE OXIDE	141	CACODYLIC ACID
2-HYDROXYETHYL-2-PROPENOATE	428	HYDROXYETHYL ACRYLATE
1-HYDROXYHEPTANE	402	HEPTANOL
1-HYDROXYHEXANE	412	HEXANOL
6-HYDROXY-3-2H-PYRIDAZINONE	500	MALEIC HYDRAZIDE
HYDROXYL AMMONIUM SULFATE	430	HYDROXYLAMINE SULFATE
4-HYDROXY-4-METHYL-2-PENTANONE	242	DIACETONE ALCOHOL
2-HYDROXY NITROBENZENE	609	NITROPHENOL
2-HYDROXY-2-PETHYL PROPIONITRILE	7	ACETONE CYANOHYDRIN
3-HYDROXYPROPANENITRILE	333	ETHYL CYANOHYDRIN
2-HYDROXYPROPANOIC ACID	467	LACTIC ACID
2-HYDROXYPROPYLAMINE	565	MONOISOPROPANOLIMINE
HYLENE-M50	303	DIPHENYLMETHYL DIOSOCYANATE
HYLENE T	790	TOLUENE-2,4-DIISOCYANATE
HYPONE	9	ACETOPHENONE
IBN	439	ISOBUTYLNITRILE
I-DECANOL	239	DECYL ALCOHOL
II	582	NICKEL AMMONIUM SULFATE
II,CHLORIDE	474	LEAD CHLORIDE
ILLUMINATING OIL	466	KEROSENE
IMPERIAL GREEN	203	COPPER ACETOARSENITE
INFECTIOUS SUBSTANCE	360	ETIOLOGIC AGENT
IPA	452	ISOPROPANOL
IPA	450	ISOPHTHALIC ACID
IRON AMMONIUM SULFATE	370	FERROUS AMMONIUM SULFATE
IRON DICHLORIDE	371	FERROUS CHLORIDE (liquid)
IRON DICHLORIDE	372	FERROUS CHLORIDE (solid)
IRON FLUORIDE	366	FERRIC FLUORIDE
IRON III CHLORIDE	365	FERRIC CHLORIDE (solid)
IRON III, SULFATE	369	FERRIC SULFATE
IRON OUS SULFATE	375	FERROUS SULFATE
IRON PERCHLORIDE	365	FERRIC CHLORIDE (solid)
IRON PROTOCHLORIDE	371	FERROUS CHLORIDE (liquid)
IRON PROTOCHLORIDE	372	FERROUS CHLORIDE (solid)
IRON PROTOXALATE	374	FERROUS OXALATE
IRON SESQUISULFATE	369	FERRIC SULFATE
IRON TRICHLORIDE	365	FERRIC CHLORIDE (solid)
IRON TRISULFATE	369	FERRIC SULFATE
IRON VITRIOL	375	FERROUS SULFATE
ISOBUTANAL	440	ISOBUTYRALDEHYDE
ISOBUTANOL	436	ISOBUTYL ALCOHOL
ISOBUTENE	438	ISOBUTYLENE
ISOBUTENYL KETONE	516	MESITYL OXIDE
ISOBUTYLALDEHYDE	440	ISOBUTYRALDEHYDE
ISOBUTYLALDEHYDE	139	iso-BUTYRALDEHYDE
ISOBUTYLALDEHYDE	138	BUTYRALDEHYDE
ISO-BUTYL AMINE	437	ISOBUTYLAMINE
ISOBUTYLCARBINOL	433	ISOAMYL ALCOHOL
ISOBUTYL ESTER	435	ISOBUTYL ACETATE
ISOBUTYL 2-PROPENOATE	132	iso-BUTYL ACRYLATE
ISOBUTYRALADEHYDE	138	BUTYRALDEHYDE
ISOBUTYRALDEHYDE	139	iso-BUTYRALDEHYDE
ISOBUTYRIC ALDEHYDE	440	ISOBUTYRALDEHYDE
ISOBUTYRIC ALDEHYDE	139	iso-BUTYRALDEHYDE
ISOCYANATOMETHANE	547	METHYL ISOCYANATE
ISOCYANIC ACID	547	METHYL ISOCYANATE
ISODECALDEYDE MIXED ISOMERS	442	ISODECALDEHYDE
ISOOCTYLALDEHYDE	446	ISOOCTALDEHYDE
ISOOCTYL ESTER	804	TRICHLOROPHENOXY ACETIC ACID ESTER
ISOPENTYL ALCHOHOL	433	ISOAMYL ALCOHOL
ISOPENTYL NITRATE	70	AMYL NITRITE
ISOPROPANAL	454	ISOPROPYL ALCOHOL
ISOPROPANOLAMINE	565	MONOISOPROPANOLIMINE
ISOPROPENYL BENZENE	555	METHYLSTRYENE
ISOPROPYLACETONE	546	METHYL ISOBUTYL KETONE
ISOPROPYL ALCOHOL	452	ISOPROPANOL
ISOPROPYLBENZENE	215	CUMENE
ISOPROPYLCUMLY HYDROPEROXIDE	282	DIISOPROPYLBENZENE HYDROPEROXIDE
ISOPROPYL CYANIDE	439	ISOBUTYLRONITRILE
ISOPROPYL 2,4-DICHLOROPHENOXY ACETATE	233	2,4-D ESTERS
ISOPROPYL ESTER	453	ISOPROPYL ACETATE
ISOPROPYL PERCARBONATE	459	ISOPROPYL PERCARBON-ATE (UNSTABILIZED)
ISOPROPYL PERCARBONATE	458	ISOPROPYL PERCARBON-ATE (STABILIZED)
ISOPROPYL PEROXYDICARBONATE	459	ISOPROPYL PERCARBON-ATE (UNSTABILIZED)

SYNONYM	PAGE	CHEMICAL NAME
ISOPROPYL PEROXYDICARBONATE	458	ISOPROPYL PERCARBONATE (STABILIZED)
ISOPROPYLTOLUOL	232	CYMENE
ISOTRIDECANOL	807	TRIDECANOL
ISOTRIDECYL ALCOHOL	807	TRIDECANOL
ISOTRON 11	801	TRICHLOROFLUOROMETHANE
ISOVALERAL	460	ISOVALERALDEHYDE
ISOVALERIC ALDEHYDE	460	ISOVALERALDEHYDE
ISOVALERONE	278	DIISOBUTYL KETONE
ISREON 11	801	TRICHLOROFLUOROMETHANE
JET FUEL	466	KEROSENE
JET FUEL	461	JET FUELS, JP-1
JET FUEL	462	JET FUELS, JP-3
JET FUEL	463	JET FUELS, JP-4
JET FUEL	464	JET FUELS, JP-5
JP-1	466	KEROSENE
JP-1	381	FUEL OIL NUMBER 1
KARMEX	309	DIURON
KEL F MONMOER	815	TRIFLUOROCHLOROETHYLENE
KEROSENE	461	JET FUELS, JP-1
KEROSENE	381	FUEL OIL NUMBER 1
KEROSENE HEAVY	464	JET FUELS, JP-5
KEROSINE	461	JET FUELS, JP-1
2-KETOHEPTANE	68	AMYL METHYL KETONE
KILLMASTER	312	DURSBAN
KING S GOLD	87	ARSENIC TRISULFIDE
KING S YELLOW	87	ARSENIC TRISULFIDE
KUROSAIG	804	TRICHLOROPHENOXYPROPIONIC ACID ESTER
KWIK-KILL	757	STRYCHINE
LACCONOL 988 A	310	DODECYLBENZENESULFONIC ACID
LACTIC ACID	350	ETHYL LACTATE
LAUGHING GAS	613	NITROUS OXIDE
LAUROYL PEROXIDE	470	LAUROYL PEROXIDE (LESS THAN 42%)
LAUROYL PEROXIDE	469	LAUROYL PEROXIDE
LEAD	474	LEAD CHLORIDE
LEAD DICHLORIDE	474	LEAD CHLORIDE
LEAD DIFLUORIDE	475	LEAD FLUORIDE
LEAD FLUOBORATE	476	LEAD FLUOROBORATE
LEAD FLUOROBORATE SOLUTION	476	LEAD FLUOROBORATE
LEAD HYPOSULFITE	483	LEAD THIOSULFATE
LEAD IV ACETATE	481	LEAD TETRAACETATE
LEAD LL SALT	478	LEAD NITRATE
LEAD MONOXIDE	489	LITHARGE
LEAD OXIDE	489	LITHARGE
LEAD PROTOXIDE	489	LITHARGE
LEAD SALT	483	LEAD THIOSULFATE
LEAD SALT	479	LEAD STEARATE
LEAD SULFOCYANATE	482	LEAD THIOCYANATE
LEAD TETRATHYL	776	TETRAETHYL LEAD
LEAD WOLFRAMATE	484	LEAD TUNGSTATE
LEUCOL	701	QUINOLINE
LICHENIC ACID	385	FUMARIC ACID
LIGHT NAPTHA	573	NAPHTHA, SOLVENT
LIGHT NAPTHA	575	NAPHTHA (WMP)
LIMONENE	300	DIPENTENE
LINDANE	101	BENZENE HEXACHLORIDE
LIQUID AMMONIA	35	AMMONIA, ANHYDROUS
LIQUID BLEACH	742	SODIUM HYPOCHLORITE
LIQUID CAMPHOR	162	CAMPHOR OIL
LIQUID GUM CAMPHOR	162	CAMPHOR OIL
LIQUID HYDROGEN	419	HYDROGEN
LIQUID IMPURE CAMPHOR	162	CAMPHOR OIL
LIQUID NITROGEN	606	NITROGEN
LIQUID OXYGEN	628	OXYGEN
LIQUID PITCH	212	CREOSOTE, COAL TAR
LIQUIFIED PHENOL	169	CARBOLIC OIL
LITHIUM BICHROMATE DIHYDRATE	492	LITHIUM BICHROMATE
LITHIUM DICHROMATE	492	LITHIUM BICHROMATE
LNG	487	LIQUIFIED NATURAL GAS (LNG)
LOROL-22	239	DECYL ALCOHOL
LORSBAN	312	DURSBAN
LOX	628	OXYGEN
LPG	488	LIQUIFIED PETROLEUM
LP GAS	488	LIQUIFIED PETROLEUM
LUCIDOL OXILITE	244	DIBENZOYL PEROXIDE
LUMERICL	661	PIPERAZINE
LUNAR CAUSTIC	713	SILVER NITRATE
LUPERCO JDB-50-T	227	CYCLOHEXANONE PEROXIDE
LYE	175	CAUSTIC POTASH SOLUTION
LYE	176	CAUSTIC SODA SOLUTION
LYE	676	POTASSIUM HYDROXIDE (liquid)
LYE	677	POTASSIUM HYDROXIDE (solid)
LYE	741	SODIUM HYDROXIDE
MAAC	528	METHYL AMYL ACETATE
MACQUER S SALT	668	POTASSIUM ARSENATE
MALEIC ACID HYDRAZIDE	500	MALEIC HYDRAZIDE
MALEINIC ACID	498	MALEIC ACID
MALENIC ACID	498	MALEIC ACID
MALIX	313	ENDOSULFAN (liquid)
MALIX	314	ENDOSULFAN (solid)
MALONIC MONONITRILE	220	CYANOACETIC ACID
MAOH	545	METHYL ISOBUTYL CARBINOL
MAOH	529	METHYL AMYL ALCOHOL
MAPP GAS	523	METHYL ACETYLENE
MARLATE 50	521	METHOXYCHLOR
MARMER	309	DIURON
MARSH GAS	579	NATURAL GAS
MARSH GAS	518	METHANE
MASSICOT	489	LITHARGE
M-B-C FUMIGANT	532	METHYL BROMIDE
MCB PHENYL CHLORIDE	185	CHLOROBENZENE
M-CRESYLIC ACID	213	CRESOL (o-m-p)
MDI	303	DIPHENYLMETHYL DIOSOCYANATE
MDMS	285	DIMETHYL SULFIDE
MEK	540	METHYL ETHYL KETONE
MENDRIN	315	ENDRIN (liquid)
MENDRIN	316	ENDRIN (solid)
MENITE	648	PHOSDRIN
MENITE	652	PHOSPHORUS BLACK
MEP	541	METHYL ETHYL PYRIDINE
MERCAPTOMETHANE	549	METHYL MERCAPTAN
MERCURIC CHLORIDE AMMONIATED	515	MERCURY AMMONIUM CHLORIDE
MERCURIC IODIDE RED	505	MERCURIC IODIDE (liquid)
MERCURIC IODIDE RED	506	MERCURIC IODIDE (solid)
MERCURIC OXIDE RED	508	MERCURIC OXIDE
MERCURIC OXIDE YELLOW	508	MERCURIC OXIDE
MERCURIC SOLFOCYANATE	511	MERCURIC THIOCYANATE
MERCURIC SULFIDE BLACK	510	MERCURIC SULFIDE

SYNONYM	PAGE	CHEMICAL NAME
MERCURIC SULFIDE RED	510	MERCURIC SULFIDE
MERCURIC SULFOCYANIDE	511	MERCURIC THIOCYANATE
MERCUROUS NITRATE MONOHYDRATE	513	MERCUROUS NITRATE
MERCURY AMIDE CHLORIDE	515	MERCURY AMMONIUM CHLORIDE
MERCURY AMMONIUM CHLORIDE	515	MERCURY AMMONIUM CHLORIDE
MERCURY BINIODIDE	505	MERCURIC IODIDE (liquid)
MERCURY BINIODIDE	506	MERCURIC IODIDE (solid)
MERCURY BISULFIDE	509	MERCURIC SULFATE
MERCURY, LL CYANIDE	504	MERCURIC CYANIDE
MERCURY MONOCHLORIDE	512	MERCUROUS CHLORIDE
MERCURY MONONITRATE	507	MECURIC NITRATE
MERCURY OXIDE	508	MERCURIC OXIDE
MERCURY PERNITRATE	507	MECURIC NITRATE
MERCURY PERSULFIDE	509	MERCURIC SULFATE
MERCURY PROTOCHLORIDE	512	MERCUROUS CHLORIDE
MERCURY PROTONITRATE	513	MERCUROUS NITRATE
MERCURY RHODANIDE	511	MERCURIC THIOCYANATE
MERCURY SUBCHLORIDE	512	MERCUROUS CHLORIDE
MERCURY VICHLORIDE	503	MERCURIC CHLORIDE
MERX	465	KEPONE
MESUROL	501	MERCAPTODIMETHUR
METALLIC RESINATE	160	CALCIUM RESINATE
METALLIC SODIUM	716	SODIUM
META-NITROTOLUENE	612	NITROTOLUENE
METHACRYLIC ACID	550	METHYL METHACRYLATE
METHACRYLIC ACID	332	ETHYL METHACRYLATE
METHACRYLIC ACID-2-3-EPOXY PROPYL ESTER	396	GLYCIDYL METHACRYLATE
METHANE	579	NATURAL GAS
METHANEAMINE	410	HEXAMETHYLENETETRAMINE
METHANETHIOL	549	METHYL MERCAPTAN
METHANOIC ACID	380	FORMIC ACID
METHANOL	525	METHYL ALCOHOL
METHIOCARB	501	MERCAPTODIMETHUR
METHOXY DDT	521	METHOXYCHLOR
2-METHYL	548	METHYL ISOPROPENYL KETONE
METHYL	516	MESITYL OXIDE
2-METHYL-1	451	ISOPRENE
METHYLACETIC ANHYDRIDE	687	PROPIONIC ANHYDRIDE
METHYLAL	542	METHYL FORMAL
METHYL ALCOHOL	520	METHANOL
METHYL AMYL KETONE	68	AMYL METHYL KETONE
METHYLANILINE MONO	531	METHYLANILINE
2-METHYLAZARIDINE	695	PROPYLENEIMINE
METHYLBENZENESULFONIC ACID	791	TOLUENE SULFONIC ACID
METHYLBENZINE	789	TOLUENE
3-METHYLBUTANAL	460	ISOVALERALDEHYDE
2-METHYLBUTANE	533	METHYLBUTENE
2-METHYL BUTANE	448	ISOPENTANE
3-METHYL-1-BUTANOL	433	ISOAMYL ALCOHOL
2-METHYL BUTENE-2	637	PENTENE
3-METHYLBUTYL NITRITE	70	AMYL NITRITE
3-METHYLBUTYRALDEHYDE	460	ISOVALERALDEHYDE
METHYL CELLOSOLVE	342	ETHYLENE GLYCOL MONOETHYL ETHER
METHYL CHLORO CARBONATE	535	METHYL CHLOROFORMATE
METHYL CHLOROMETHYL ETHER	188	CHLOROMETHYL METHYL ETHER
METHYL CYANIDE	8	ACETONITRILE
METHYLENE CHLORIDE	261	DICHLOROMETHANE
METHYLENE DICHLORIDE	261	DICHLOROMETHANE
METHYLENE DIMETHYL ETHER	542	METHYL FORMAL
METHYLENE GLYCOL	692	PROPYLENE GLYCOL
METHYL ESTER	550	METHYL METHACRYLATE
METHYL ESTER	543	METHYL FORMATE
METHYL ESTER	535	METHYL CHLOROFORMATE
METHYL ESTER	522	METHYL ACETATE
METHYL ETHER	283	DIMETHYL ETHER
6-METHYL-1-HEPTANAL	446	ISOOCTALDEHYDE
6-METHYL-1-HEPTANOL	447	ISOOCTYL ALCOHOL
METHYL HYDROXIDE	520	METHANOL
1-METHYL-4-ISOPROPYLBENZENE	232	CYMENE
METHYLMETHANE	318	ETHANE
2-METHYLPENTANE	445	ISOHEXANE
2-METHYL PENTANEDIOL-2-4	414	HEXYLENE GLYCOL
4-METHYL-2-PENTANOL ACETATE	528	METHYL AMYL ACETATE
4-METHYL-2-PENTYL ACETATE	528	METHYL AMYL ACETATE
METHYL PENTYL KETONE	68	AMYL METHYL KETONE
METHYL PHENYLAMINE	531	METHYLANILINE
1-METHYL-1-PHENYL-ETHYLENE	555	METHYLSTRYENE
2-METHYLPROPANAL	139	iso-BUTYRALDEHYDE
2-METHYL PROPANE	434	ISOBUTANE
METHYL PROPYL BENZENE	232	CYMENE
1-METHYL-2-3-PYRIDYL PYRROLIDINE	592	NICOTINE
1-METHYL-2-PYRROLIDINONE	554	METHYLPYRROLIDONE
METHYL SULFATE	284	DIMETHYL SULFATE
METHYL SULFHYDRATE	549	METHYL MERCAPTAN
METHYL SULFIDE 2-THIAPROPANE	285	DIMETHYL SULFIDE
4-METHYLSULFONYL-2-6-DINITRO-N-N-DIPR	595	NITRALIN
METHYL THIRAM	785	THIRAM
METHYL TUADS	785	THIRAM
METHYL VINYL ETHER	840	VINYL METHYL ETHER
METHYLZINC	291	DIMETHYLZINC
METRON	552	METHYL PARATHION
MEXACARBATE	849	ZECTRAN
MH	500	MALEIC HYDRAZIDE
MIBC	545	METHYL ISOBUTYL CARBINOL
MIBC	529	METHYL AMYL ALCOHOL
MIBK	546	METHYL ISOBUTYL KETONE
MIC	529	METHYL AMYL ALCOHOL
MIC	545	METHYL ISOBUTYL CARBINOL
MIC	547	METHYL ISOCYANATE
MIDDLE OIL	169	CARBOLIC OIL
MIDIBEN	247	DICAMBA (liquid)
MIDIBEN	248	DICAMBA (solid)
MIK	546	METHYL ISOBUTYL KETONE
MILD MERCURY CHLORIDE	512	MERCUROUS CHLORIDE
MILK ACID	467	LACTIC ACID
MINE	361	EXPLOSIVES A
MIPA	565	MONOISOPROPANOLIMINE
MIXED ACID	596	NITRATING ACID
MIXED PRIMARY AMYL NITRATES	69	AMYL NITRATE

SYNONYM	PAGE	CHEMICAL NAME
MIXTURE OF BENZENE	572	NAPHTHA
MMH	544	METHYLHYDRAZINE
MOHR S SALT	370	FERROUS AMMONIUM SULFATE
MOLYBDENUM TRIOXIDE	560	MOLBDIC TRIOXIDE
MOLYBDIC ANHYDRIDE	560	MOLBDIC TRIOXIDE
MONAZITE SAND	787	THORIUM ORE
MONDUR TDS	790	TOLUENE-2,4-DIISOCYANATE
MONOALUMINUM SALT	38	AMMONIUM BICARBONATE
MONOBROMOBENZENE	122	BROMOBENZENE
MONOCHLOROMETHYL ETHER	188	CHLOROMETHYL METHYL ETHER
MONOETHYLAMINE	326	ETHYLAMINE
MONOISOPROPYLAMINE	455	ISOPROPYLAMINE
MONOMER	780	TETRAFLUOROETHYLENE
MONOMETHYL ANHYDROUS	566	MONOMETHYLAMINE
MONOMETHYLHYDRAZINE	544	METHYLHYDRAZINE
MONOPE	635	PENTAERYTHRITOL
MONOSODIUM METHANEARSONATE	519	METHANEARSONIC ACID, SODIUM SALT
MONOSODIUM METHYL ARSENATE	519	METHANEARSONIC ACID, SODIUM SALT
MONOXIDE	173	CARBON MONOXIDE
MOTOR SPIRIT	390	GASOLINE
MOTOR SPIRIT	392	GASOLINE CASINGHEAD
MOUSE-TOX	757	STRYCHINE
M-PHTHALIC ACID	450	ISOPHTHALIC ACID
MPT	552	METHYL PARATHION
MPTD	553	METHYL PHOSPHONOTHIOIC DICHLORIDE
MSMA	519	METHANEARSONIC ACID, SODIUM SALT
MULTRATHANE M	303	DIPHENYLMETHYL DIOSOCYANATE
MURIATIC ACID	415	HYDROCHLORIC ACID
MURIATIC ETHER	330	ETHYL CHLORIDE
MYRISTIC ALCOHOL	772	TETRADECANOL
MYRISTYL ALCOHOL	772	TETRADECANOL
NACCONATE 300	303	DIPHENYLMETHYL DIOSOCYANATE
NADONE	226	CYCLOHEXANONE
N-AMYL CARBINOL	412	HEXANOL
NAPHTHANE	238	DECAHYDRONAPHTHALENE
NAPTHA	558	MINERAL SPIRITS
NAPTHALANE	238	DECAHYDRONAPHTHALENE
1-NAPTHYL N-METHYL-CARBAMATE	165	CARBARYL (liquid)
1-NAPTHYL N-METHYL-CARBAMATE	166	CARBARYL (solid)
NATRIUM	716	SODIUM
NATURAL AMMONIUM FLUORIDE	48	AMMONIUM FLUORIDE
NATURAL GAS	518	METHANE
NATURAL LEAD ACETATE	472	LEAD ACETATE
NATURAL LEAD STEARATE	479	LEAD STEARATE
NATURAL NICOTINE SULFATE	593	NICOTINE SULFATE (liquid)
NATURAL NICOTINE SULFATE	594	NICOTINE SULFATE (solid)
NATURAL SODIUM CHROMATE ANNHYDROUS	731	SODIUM CHROMATE
NAUGATUCK-DO 14 PROPARGIL LIQUID	683	PROPARGITE
NAVY SPECIAL FUEL OIL	384	FUEL OIL NUMBER 5
N-BUTYL ACETATE	129	BUTYL ACETATE
N-BUTYL ACRYLATE	131	BUTYL ACRYLATE
N-BUTYL ETHER	241	DI-BUTYL ETHER
N-BUTYL METHYL KETONE	551	METHYL N-BUTYL KETONE
N-BUTYL TRICHLORORSILANE	137	BUTYL TRICHLOROSILANE
N-DIBUTYL ETHER	241	DI-BUTYL ETHER
N-DIPROPYLAMINE	304	DIPROPYLAMINE
NECATORINA	174	CARBON TETRACHLORIDE
NERKOL	266	DICHLORVOS (liquid)
NERKOL	267	DICHLORVOS (solid)
NEUTRAL AMMONIUM CHROMATE	44	AMMONIUM CHROMATE
NEUTRAL VERDIGRIS	202	COPPER ACETATE
N-HEPTANE	401	HEPTANE
N-HEXANE	411	HEXANE
N-HEXYL ALCOHOL	412	HEXANOL
NIA 12 40	319	ETHION
NIA 5996	249	DICHLOBENIL (liquid)
NIA 5996	250	DICHLOBENIL (solid)
NIAGARA 10242	167	CARBOFURAN
NIALATE	319	ETHION
NICKEL	584	NICKEL CARBONYL
NICKEL ACETATE TETRAHYDRATE	581	NICKEL ACETATE
NICKEL BROMIDE TRIHYDRATE	583	NICKEL BROMIDE
NICKEL CHLORIDE HEXAHYDRATE	585	NICKEL CHLORIDE
NICKEL 2 FLUOROBORATE	587	NICKEL FLUOROBORATE
NICKEL FLUOROBORATE SOLUTION	587	NICKEL FLUOROBORATE
NICKEL FORMATE DIHYDRATE	588	NICKEL FORMATE
NICKEL HYDROXIDE	589	NICKEL HYDROXIDE
NICKEL NITRATE HEXAHYDRIDE	590	NICKEL NITRATE
NICKELOUS ACETATE	581	NICKEL ACETATE
NICKELOUS HYDROXIDE	589	NICKEL HYDROXIDE
NICKELOUS SULFATE	591	NICKEL SULFATE
NICKEL 11 SALT	581	NICKEL ACETATE
NITER	678	POTASSIUM NITRATE
NITOS,KILLAX	777	TETRAETHYL PYROPHOSPHATE
NITRAM	51	AMMONIUM NITRATE
NITRE	678	POTASSIUM NITRATE
NITRIC ACID	478	LEAD NITRATE
NITRIC ACID IRON II SALT	368	FERRIC NITRATE
NITRIC ACID IRON +3 SALT	368	FERRIC NITRATE
NITRIC ACID, POTASSIUM SALT	677	POTASSIUM HYDROXIDE (solid)
NITRIC SULFURIC ACID MIXTURES	596	NITRATING ACID
3-NITROBENZENE	602	NITROCHLOROBENZENE-META
1,2 NITROBENZENE	294	DINITROBENZENE (solid)
NITROBENZOL	601	NITROBENZENE
NITROCARBOL	608	NITROMETHANE
NITROCELLULOSE GUM	201	COLLODION
NITROCHLOROBENZENE	189	CHLORONITROBENZENE
NITROCHLOROFORM	191	CHLOROPICRIN
NITROGEN DIOXIDE	607	NITROGEN TETROXIDE
NITROGEN MONOXIDE	598	NITRIC OXIDE
NITROGEN PEROXIDE	607	NITROGEN TETROXIDE
0-NITROPHENOL	609	NITROPHENOL
NITROUS ETHER	353	ETHYL NITRATE
N,M-DIMETHYLFORMAIDE	288	DIMETHYLFORMAMIDE

SYNONYM	PAGE	CHEMICAL NAME
N-METHYL-ALPHA-PYRROLIDONE	554	METHYLPYRROLIDONE
N-METHYLAMINOBENZENE	531	METHYLANILINE
N-METHYLPYRROLIDINONE	554	METHYLPYRROLIDONE
N-METHYLPYRROLIDONE	554	METHYLPYRROLIDONE
N,N -BIS 2-AMINOETHYL ETHYLENEDIAMINE	814	TRIETHYLENE TETRAMINE
N-NONANE	614	NONANE
N-NONYLETHYLENE	822	UNDECENE
N-OCTADECYLIC ACID	755	STEARIC ACID
1-NONANOL	615	NONANOL
NONYL ALCOHOL	615	NONANOL
NONYLCARBINOL	239	DECYL ALCOHOL
NORMAL OCTANE	618	OCTANE
NORMAL PENTANE	636	PENTANE
NOS	702	RADIOACTIVE MATERIAL
NOS	622	OIL
NOS	360	ETIOLOGIC AGENT
2-NP	610	NITROPROPANE
N-PHENYLANILINE	301	DIPHENYLAMINE
N-PROPYL ALCOHOL	689	PROPYL ALCOHOL
N-PROPYL ESTER	688	PROPYL ACETATE
N-PROPY 1-1-PROPANAMINE	304	DIPROPYLAMINE
NSFO	384	FUEL OIL NUMBER 5
NTA	599	NITRILOTRIACETIC ACID AND SALTS
N-TETRADECYL ALCOHOL	772	TETRADECANOL
NUX-VOMICA	757	STRYCHINE
OCTACHLOROCAMPHENE	795	TOXAPHENE (liquid)
OCTACHLOROCAMPHENE	796	TOXAPHENE (solid)
OCTADECANOIC ACID	755	STEARIC ACID
OCTAKLOR	178	CHLORDANE
OCTALENE	23	ALDRIN
1-OCTANOL	619	OCTANOL
OCTYCARBINOL	615	NONANOL
OCTYL ALCOHOL	619	OCTANOL
OCTYL ALDEHYDE	349	ETHYL HEXALDEHYDE
OCTYL ESTER	621	OCTYL EPOXY TALLATE
O-HYDROXYBENZOIC ACID	705	SALICYLIC ACID
OIL OF BITTER ALMOND	99	BENZALDEHYDE
OIL OF MIRBANE	601	NITROBENZENE
OIL OF PINE	660	PINE OIL
OIL OF 5-SIBERIAN	660	PINE OIL
OLE FAINT GAS	335	ETHYLENE
OMALPHENACHLOR	802	TRICHLOROPHENOL
OMEGA-BROMOTOLUENE	108	BENZYL BROMIDE
OMITE	683	PROPARGITE
ONCB	603	NITROCHLOROBENZENE-ORTHO
O-NITROBENZOYL	294	DINITROBENZENE (solid)
ONP	609	NITROPHENOL
O,O-DIETHYL	211	COUMAPHOS
O,O-DIMETHYL	797	TRICHLORFON
OR PCNB	604	NITROCHLOROBENZENE-PARA
ORPIMENT	87	ARSENIC TRISULFIDE
2- OR P- NITROANILINE	600	NITROANILINE
ORTHOARSENIC ACID	81	ARSENIC ACID (liquid)
ORTHOARSENIC ACID	82	ARSENIC ACID (solid)
ORTHOCIDE	163	CAPTAN (liquid)
ORTHOCIDE	164	CAPTAN (solid)
ORTHODICHLORBENZENE	254	O-DICHLOROBENZENE
ORTHO PHOSPHORIC	650	PHOSPHORIC ACID
ORTHOTICANIC ACID	769	TETRABUTYL TITANATE
OXAL	397	GLYOXAL
OXALADEHYDE	397	GLYOXAL
OXALIC ACID	374	FERROUS OXALATE
OXALIC ACID	52	AMMONIUM OXALATE
OXALIC ACID DIHYDRATE	627	OXALIC ACID
OXAMMONIUM	429	HYDROXYLAMINE
OXAMMONIUM SULFATE	430	HYDROXYLAMINE SULFATE
4-OXAZINE	567	MORPHOLINE
OXIRANE	345	ETHYLENE OXIDE
OXO OCTALDEHYDE	446	ISOOCTALDEHYDE
OXO OCTYL ALCOHOL	447	ISOOCTYL ALCOHOL
OXOTRIDECYL ALCOHOL	807	TRIDECANOL
PAINT DRYER	207	COPPER NAPHTHENATE
PAINTERS NAPTHA	575	NAPHTHA (WMP)
PAN	659	PHTHALIC ANHYDRIDE
P-ANISOYL CHLORIDE	73	ANISOYL CHLORIDE
PAPI	664	POLYMETHYLENE POLY-PHENYL ISOCYANATE
PARACETALDEHYDE	630	PARALDEHYDE
PARADI	255	P-DICHLOROBENZENE
PARADOW	255	P-DICHLOROBENZENE
PARA HYDROGEN	419	HYDROGEN
PARIDOL	552	METHYL PARATHION
PARIMOTH	255	P-DICHLOROBENZENE
PARIS GREEN	203	COPPER ACETOARSENITE
PATENT ALUMINUM	30	ALUMINUM SULFATE (solid)
PCB	663	POLYCHLORINATED BIPHENYL
P-DIHYDROXYBENZENE	426	HYDROQUINONE
P-DIOXANE	299	DIOXANE
PE	635	PENTAERYTHRITOL
PELARGONIC ALCOHOL	615	NONANOL
PENETEK	635	PENTAERYTHRITOL
PENTA	633	PENTACHLOROPHENOL
9-PENTABORON NONAHYDRIDE	632	PENTABORANE
1-PENTADECANOL	634	PENTADECANOL
PENTADECANOL	486	LINEAR ALCOHOL
PENTADECYL ALCOHOL	634	PENTADECANOL
PENTAERYTHRITE	635	PENTAERYTHRITOL
PENTAMETHYLENE	230	CYCLOPENTANE
PENTANAL	831	VALERALDEHYDE
1,5-PENTANEDIAL	395	GLYCIDALDEHYDE
2,4-PENTANEDIONE	14	ACETYLACETONE
1-PENTANETHIOL	67	AMYL MERCAPTAN
PENTANOL	65	AMYL ALCOHOL
PENTANOL ACETATE	64	AMYL ACETATE
PENTEK	635	PENTAERYTHRITOL
PENTYL CHLORIDE	66	AMYL CHLORIDE
PENTYL METHYL KETONE	68	AMYL METHYL KETONE
PENTYLSILICON TRICHLORIDE	71	AMYLTRICHLORSILANE
PENYLETHANE	327	ETHYLBENZENE
PERCARBAMIDE	830	UREA PEROXIDE
PERCHLORATE ACID SOLUTION	640	PERCHLORIC ACID LESS THAN 50%
PERCHLORATE ACID SOLUTION	641	PERCHLORIC ACID OVER 50%
PERCHLOROCYCLO-PENTADIENE	404	HEXACHLOROCYCLO-PENTADIENE
PERCHLORODIHOMOCUBANE	559	MIREX
PERCHLOROETHANE	405	HEXACHLOROETHANE
PERCHLOROETHYLENE	771	TETRACHLOROETHYLENE
PERCLENE	771	TETRACHLOROETHYLENE
PERCLENE	639	PERCHLORETHYLENE
PERHYDRONAPTHALENE	238	DECAHYDRONAPHTHALENE
PERK	771	TETRACHLOROETHYLENE

SYNONYM	PAGE	CHEMICAL NAME
PERK	639	PERCHLORETHYLENE
PEROXIDE	424	HYDROGEN PEROXIDE (52–100%)
PEROXYACETIC ACID	638	PERACETIC ACID
PEROXYDISULFURIC ACID	54	AMMONIUM PERSULFATE
PERSIAN-INSECT POWDER	697	PYRETHRINS ACID (liquid)
PERSIAN-INSECT POWDER	698	PYRETHRINS ACID (solid)
PETROHOL	454	ISOPROPYL ALCOHOL
PETROL	390	GASOLINE
PETROL	392	GASOLINE CASINGHEAD
PETROLATUM JELLY	643	PETROLATUM
PETROLEUM ASPHALT	90	ASPHALT
PETROLEUM BENZINE	558	MINERAL SPIRITS
PETROLEUM ETHER	558	MINERAL SPIRITS
PETROLEUM JELLY	643	PETROLATUM
PETROLEUM NAPTHA	558	MINERAL SPIRITS
PETROLEUM OIL	622	OIL
PETROLEUM SOLVENT	573	NAPTHA, SOLVENT
PETROLEUM SOLVENT	574	NAPHTHA (STODDARD SOLVENT)
PETROLEUM SOLVENT	575	NAPHTHA (WMP)
PETROLEUM SOLVENT	644	PETROLEUM NAPTHA
PETROLEUM SPIRITS	558	MINERAL SPIRITS
PETROLEUM WAX	843	WAX
PEVINPHOS	648	PHOSDRIN
PEVINPHOS	652	PHOSPHORUS BLACK
PHENIC ACID	645	PHENOL
PHENOL	168	CARBOLIC ACID
PHENYLARSENIC DICHLORIDE	646	PHENYLDICHLOROARSINE
PHENYL BROMIDE	122	BROMOBENZENE
PHENYL CHLOROMETHYL-KETONE	182	CHLOROACETOPHENONE
PHENYLCYANIDE	105	BENZONITRILE
PHENYLHYDRAZINIUM CHLORIDE	647	PHENYLHYDRAZINE HYDROCHLORIDE
PHENYLIC ACID	168	CARBOLIC ACID
PHENYLMETHYL AMINE	111	BENZYLAMINE
PHENYL METHYL KETONE	9	ACETOPHENONE
PHENYLPHOSPHONOTHIOIC	103	BENZENE PHOSPHORUS
THIODICHLORIDE	555	METHYLSTRYENE
PHENYLPROPYLENE		
1-PHENYLUNDECANE	823	UNDECYLBENZENE
PHENYPHOSPHORUS DICHLORIDE	102	BENZENE PHOSPHORUS DICHLORIDE
PHOSFENE	648	PHOSDRIN
PHOSFENE	652	PHOSPHORUS BLACK
PHOSPHORIC SULFIDE	654	PHOSPHORUS PENTA-SULFIDE
PHOSPHORUS BROMIDE	656	PHOSPHORUS TRIBROMIDE
PHOSPHORUS CHLORIDE	657	PHOSPHORUS TRICHLORIDE
PHOSPHORYL CHLORIDE	653	PHOSPHORUS OXYCHLORIDE
PHOTOPHOR	159	CALCIUM PHOSPHIDE
PHTHALIC ACID	246	DIBUTYL PHTHALATE
PHTHALIC ACID ANHYDRIDE	659	PHTHALIC ANHYDRIDE
PHYGON	251	DICHLONE (liquid)
PHYGON	252	DICHLONE (solid)
PHYGON-XL	251	DICHLONE (liquid)
PHYGON-XL	252	DICHLONE (solid)
PHYTAR	729	SODIUM CACODYLATE
PICFUME	191	CHLOROPICRIN
PIMELIC KETONE	226	CYCLOHEXANONE
PIPERAZIDINE	661	PIPERAZINE
P-ISOPROPYLTOLUENE	232	CYMENE
PLANAVIN	595	NITRALIN
PLASTIC LATEX	468	LATEX LIQUID SYNTHETIC
PLUMBOUS FLUORIDE	475	LEAD FLUORIDE
PLUMBOUS SULFIDE	480	LEAD SULFIDE
P-METHYLSTYRENE	841	VINYL TOLUENE
PNCB	604	NITROCHLOROBENZENE-PARA
POLY DIMETHYL SILOXANE	468	LATEX LIQUID SYNTHETIC
POLY FORMALDEHYDE	629	PARAFORMALDEHYDE
POLYISOBUTYLENE PLASTICS	662	POLYBUTENE
POLYISOBUTYLENE RESINS	662	POLYBUTENE
POLYISOBUTYLENE WAXES	662	POLYBUTENE
POLY OXYMETHYLENE	629	PARAFORMALDEHYDE
POLY-SOLV EE ACETATE	343	ETHYLENE GLYCOL MONOETHYL ETHER ACETATE
POLY-SOLV EM	342	ETHYLENE GLYCOL MONOETHYL ETHER
POTASSIUM ACID ARSENATE	668	POTASSIUM ARSENATE
POTASSIUM ANTIMONYL TARTRATE	76	ANTIMONY POTASSIUM TARTRATE
POTASSIUM BICHROMATE	675	POTASSIUM DICHROMATE
POTASSIUM CHROMATE, VI NATURAL POTASSIUM	671	POTASSIUM CHROMATE
POTASSIUM DICHLORO-S-TRIAZINETRIONE	674	POTASSIUM DICHLOROISOCYANURATE
POTASSIUM FLUOZIR-CONATE	875	ZIRCONIUM POTASSIUM FLUORIDE
POTASSIUM HYDROXIDE SOLUTION	175	CAUSTIC POTASH SOLUTION
POTASSIUM OLEATE	624	OLEIC ACID, POTASSIUM SALT
POTASSIUM OXILATE MONOHYDRATE	679	POTASSIUM OXILATE
POTASSIUM SALT	669	POTASSIUM ARSENITE
POTASSIUM SUPEROXIDE	681	POTASSIUM PEROXIDE
POTASSIUM ZINC CHROMATE	869	ZINC POTASSIUM CHROMATE
POTATO SPIRIT	433	ISOAMYL ALCOHOL
POTCRATE	670	POTASSIUM CHLORATE
PROJECTILE	361	EXPLOSIVES A
PROPANAL	685	PROPIONALDEHYDE
1,2-PROPANEDIOL	692	PROPYLENE GLYCOL
1,2-PROPANEDIOL-1-ACRYLATE	431	HYDROXYPROPYL ACRYLATE
1,2-PROPANEDIOL 1-METH-ACRYLATE	432	HYDROXYPROPYL METHACRYLATE
PROPANE-1-THIOL	696	PROPYL MERCAPTAN
1-PROPANETHIOL	696	PROPYL MERCAPTAN
PROPANE-2-THIOL	457	ISOPROPYL MERCAPTAN
2-PROPANETHIOL	457	ISOPROPYL MERCAPTAN
1,2,3-PROPANETRIOL	394	GLYCERINE
PROPANOIC ACID	686	PROPIONIC ACID
PROPANOIC ANHYDRIDE	687	PROPIONIC ANHYDRIDE
1-PROPANOL	689	PROPYL ALCOHOL
2-PROPANOL	452	ISOPROPANOL
2-PROPANONE	6	ACETONE
PROPEAE	691	PROPYLENE
2-PROPENAL	17	ACROLEIN
PROPENAMIDE 50	18	ACRYLAMIDE
PROPIONIC ALDEHYDE	685	PROPIONALDEHYDE
PROPYL ALDEHYDE	685	PROPIONALDEHYDE
PROPYLENE GLYCOL MONOACRYLATE	431	HYDROXYPROPYL ACRYLATE

SYNONYM	PAGE	CHEMICAL NAME
PROPYLENE GLYCOL MONOMETHACRYLATE	432	HYDROXYPROPYL METHACRYLATE
PROPYLENEIMINE	695	PROPYLENEIMINE
PROPYLENE POLYMER	666	POLYPROPYLENE
PROPYLTHYLENE	637	PENTENE
PRUSSIC ACID	422	HYDROGEN CYANIDE
PSEUDOHEXYL ALCOHOL	328	ETHYLBUTANOL
P-TOLUENE SULFONATE	791	TOLUENE SULFONIC ACID
P-TOLYL CHLORIDE	194	CHLOROTOLUENE
P-TSA	791	TOLUENE SULFONIC ACID
PYRETHRUM FLOWERS	697	PYRETHRINS ACID (liquid)
PYRETHRUM FLOWERS	698	PYRETHRINS ACID (solid)
PYROACETIC ETHER	6	ACETONE
PYROGALLOL	700	PYROGALLIC ACID
PYROGENTISIC ACID	426	HYDROQUINONE
PYROXYLIN SOLUTION	201	COLLODION
QUAKERAL	386	FURFURAL
QUICKLIME	157	CALCIUM OXIDE
QUICK SILVER	514	MERCURY
QUINOL	426	HYDROQUINONE
R22	563	MONOCHLORODIFLUORO-METHANE
R 20	186	CHLOROFORM
RACEMIC LACTIC ACID	467	LACTIC ACID
RANGE OIL	466	KEROSENE
RANGE OIL	461	JET FUELS, JP-1
RANGE OIL	381	FUEL OIL NUMBER 1
RASPITE	484	LEAD TUNGSTATE
RC PLASTICIZER DBP	246	DIBUTYL PHTHALATE
RED ARSENIC GLASS	83	ARSENIC DISULFIDE
RED NITROGEN	607	NITROGEN TETROXIDE
RED OIL	623	OLEIC ACID
RED ORPIMENT	83	ARSENIC DISULFIDE
REFRIGERANT 52A	276	1,1-DIFLUOROETHANE
REGLGAR	83	ARSENIC DISULFIDE
REGLONE	305	DIQUAT (liquid)
REGLONE	306	DIQUAT (solid)
REGULOX	500	MALEIC HYDRAZIDE
RESIDENTIAL FUEL OIL 5	384	FUEL OIL NUMBER 5
RESIDENTIAL FUEL OIL	383	FUEL OIL NUMBER 4
RESORCIN	703	RESORCINOL
RETARDER W	705	SALICYLIC ACID
RHODANATE	752	SODIUM THIOCYANATE
RHOTHANE	235	DDD
RUBBING ALCOHOL	452	ISOPROPANOL
RUBY ARSENIC	83	ARSENIC DISULFIDE
SACCHAROSE	759	SUCROSE
SACCHARUM	759	SUCROSE
SAL AMMONIAC	43	AMMONIUM CHLORIDE
SALMAIC	43	AMMONIUM CHLORIDE
SALT OF SATURN	472	LEAD ACETATE
SALT PETER	678	POTASSIUM NITRATE
SALTPETER	678	POTASSIUM NITRATE
SALUFER	750	SODIUM SILICOFLUORIDE
SAL VOLATILE	42	AMMONIUM CARBONATE
SAND ACID	427	HYDROSILICOFLUORIC ACID
SAND ACID	377	FLUOROSILICIC ACID
SANTOPHEN 20	633	PENTACHLOROPHENOL
SARALEX	243	DIAZINON
SCHEELITE	484	LEAD TUNGSTATE
SEC-BUTYL ESTER	130	SEC-BUTYL ACETATE
SEC-HEXYL ALCOHOL	328	ETHYLBUTANOL
SEC-NITROPROPANE	610	NITROPROPANE
SEC- PENTYLCARBINOL	328	ETHYLBUTANOL
SELENIC ANHYDRIDE	707	SELENIUM TRIOXIDE
SELENIOUS ANHYDRIDE	706	SELENIUM DIOXIDE
SELENIUM OXIDE	706	SELENIUM DIOXIDE
SENTRY	155	CALCIUM HYPOCHLORITE
SEVIN	165	CARBARYL (liquid)
SEVIN	166	CARBARYL (solid)
SEXTONE	226	CYCLOHEXANONE
SHELL CHARCOAL	177	CHARCOAL
SILICOFLUORIC ACID	427	HYDROSILICOFLUORIC ACID
SILICOFLUORIC ACID	377	FLUOROSILICIC ACID
SILICON TETRACHLORIDE	708	SILICON CHLORIDE
SILVER MONOFLUORIDE	711	SILVER FLUORIDE
SILVEX	804	TRICHLOROPHENOXY-PROPIONIC ACID ESTER
SILVEX	803	TRICHLOROPHENOXY ACETIC ACID ESTER
SILVISAR-510	141	CACODYLIC ACID
SMITHSONITE	857	ZINC CARBONATE
SODA LYE	741	SODIUM HYDROXIDE
SODAMIDE	719	SODIUM AMIDE
SODA NITER	744	SODIUM NITRATE
SODIUM ACID SULFITE	725	SODIUM BISULFIDE (liquid)
SODIUM ACID SULFITE	726	SODIUM BISULFIDE (solid)
SODIUM ATOM	716	SODIUM
SODIUM BIBORATE	727	SODIUM BORATE
SODIUM BICHROMATE	735	SODIUM DICHROMATE
SODIUM BISUIFIDE	739	SODIUM HYDROSULFIDE SOLUTION
SODIUM CHROMATE VI	731	SODIUM CHROMATE
SODIUM DICHLOROISOCYANURATE	734	SODIUM DICHLORO-S-TRIAZINETRIONE
SODIUM DIFLUORIDE	724	SODIUM BIFLUORIDE
SODIUM FLUOSILICATE	750	SODIUM SILICOFLUORIDE
SODIUM HEXAFLUORO-SILICATE	750	SODIUM SILICOFLUORIDE
SODIUM HYDROGEN ALKYL SULFATE	718	SODIUM ALKYL SULFATES
SODIUM HYDROSULPHITE	740	SODIUM HYDROSULFITE
SODIUM HYDROXIDE SOLUTION	176	CAUSTIC SODA SOLUTION
SODIUM META ARSENITE	721	SODIUM ARSENITE (liquid)
SODIUM META ARSENITE	722	SODIUM ARSENITE (solid)
SODIUM METHOXIDE	743	SODIUM METHYLATE
SODIUM OLEATE	625	OLEIC ACID, SODIUM SALT
SODIUM PHOSPHATE DIBASIC	747	SODIUM PHOSPHATE
SODIUM PHOSPHATE MONOBASIC	747	SODIUM PHOSPHATE
SODIUM PHOSPHATE TRIBASIC	747	SODIUM PHOSPHATE
SODIUM PYROBORATE	727	SODIUM BORATE
SODIUM PYROSULFIDE	725	SODIUM BISULFIDE (liquid)
SODIUM PYROSULFIDE	726	SODIUM BISULFIDE (solid)
SODIUM RHODANIDE	752	SODIUM THIOCYANATE
SODIUM SALT	732	SODIUM CYANIDE (liquid)
SODIUM SALT	733	SODIUM CYANIDE (solid)
SODIUM SALT	569	NABAM
SODIUM SULFHYDRATE	739	SODIUM HYDROSULFIDE SOLUTION
SODIUM SULFOCYANATE	752	SODIUM THIOCYANATE
SODIUM SULFOXYLATE	740	SODIUM HYDROSULFITE
SODIUM TETRABORATE ANHYDROUS	727	SODIUM BORATE
SOLFONATED ALKY-BENZENE SODIUM SALT	717	SODIUM ALKYLBENZENE-SULFONATES
SOLUABLE GLASS	749	SODIUM SILICATE

SYNONYM	PAGE	CHEMICAL NAME
SORBIT	753	SORBITOL
SORBOL	753	SORBITOL
SPECTRACIDE	243	DIAZINON
SPENT FUEL	617	NUCLEAR REACTOR FUEL
SPIRIT OF ETHER NITRATE	346	ETHYL NITRATE
SPIRIT OF GUM	820	TURPENTINE
SPOTTING NAPTHA	574	NAPHTHA (STODDARD SOLVENT)
STEAROPHANIC ACID	755	STEARIC ACID
STEINBULN YELLOW	153	CALCIUM CHROMATE
STERIC ACID	479	LEAD STEARATE
STOLZITE	484	LEAD TUNGSTATE
STRONTIUM SALT 11	756	STRONTIUM CHROMATE
STRONTIUM YELLOW	756	STRONTIUM CHROMATE
STYROL	758	STYRENE
STYROLENE	758	STYRENE
SUGAR	759	SUCROSE
SUGAR OF LEAD	472	LEAD ACETATE
SULFAMIC ACID	56	AMMONIUM SULFAMATE
SULFATE OF COPPER	218	CUPRIC SULFATE
SULFATE OF COPPER	210	COPPER SULFATE
SULFITE	40	AMMONIUM BISULFITE
SULFOLANE-W	760	SULFOLANE
SULFUR CHLORIDE	764	SULFUR MONOCHLORIDE
SULFURETTED HYDROGEN	425	HYDROGEN SULFIDE
SULFURIC ACID	783	THALLIUM SULFATE
SULFURIC ACID	198	CHROMIC SULFATE
SULFUR MONOCHLORIDE	762	SULFUR CHLORIDE
SULFUR OF FLOWER	761	SULFUR
SULFUR SUBCHLORIDE	762	SULFUR CHLORIDE
SULPHUR	761	SULFUR
SUPEROXYL	424	HYDROGEN PEROXIDE 52–100%
SWEDISH GREEN	204	COPPER ARSENITE
SYM-TRIETHYLBENZENE	812	TRIETHYLBENZENE
SYNTHETIC PINE OIL	660	PINE OIL
SYNTHETIC RUBBER LATEX	468	LATEX LIQUID SYNTHETIC
TANNIN	768	TANNIC ACID
TAR CAMPHOR	576	NAPHTHALENE
TAR OIL	212	CREOSOTE, COAL TAR
TARTAR EMETIC	76	ANTIMONY POTASSIUM TARTRATE
1-TARTARIC ACID	59	AMMONIUM TARTRATE
TARTARIZED ANTIMONY	76	ANTIMONY POTASSIUM TARTRATE
TARTRATED ANTIMONY	76	ANTIMONY POTASSIUM TARTRATE
TCP	139	TRICRESYLPHOSPHATE
TDE	235	DDD
TDI	790	TOLUENE-2,4-DIISOCYANATE
TEA	810	TRIETHYLALUMINUM
TEAR GAS	182	CHLOROACETOPHENONE
TEDP	774	TETRAETHYL DITHIOPYROPHOSPHATE (liquid)
TEDP	775	TETRAETHYL DITHIOPYROPHOSPHATE (solid)
TEFLON	780	TETRAFLUOROETHYLENE
TEG	813	TRIETHYLENE GLYCOL
TEL	776	TETRAETHYL LEAD
TEN	811	TRIETHYLAMINE
TEP TEPP	777	TETRAETHYL PYROPHOS-PHATE
TERPENENE	300	DIPENTENE
TETA	814	TRIETHYLENE TETRAMINE
TETRABUTYL	769	TETRABUTYL TITANATE
TETRACAP	771	TETRACHLOROETHYLENE
TETRACAP	639	PERCHLORETHYLENE
1,1,2,2-TETRACHLORO-ETHANE	770	TETRACHLOROETHANE
TETRACHLOROETHYLENE	639	PERCHLORETHYLENE
TETRACHLOROMETHANE	174	CARBON TETRACHLORIDE
TETRACHLOROSILANE	708	SILICON CHLORIDE
1-TETRADECANOL	772	TETRADECANOL
TETRADECANOL	486	LINEAR ALCOHOL
TETRADECYL ALCOHOL	772	TETRADECANOL
TETRADECYLENE	773	TETRADECENE
TETRAETHYL SILICATE	359	ETHYL SILICATE
TETRAHYDRO-2H-1	567	MORPHOLINE
1,2,3,4,TETRAHYDRON-APHTHALENE	782	TETRAHYDRONAPHTHALENE
TETRAHYDROTHIOPHENE-1-1-DIOXIDE	760	SULFOLANE
TETRALIN	782	TETRAHYDRONAPHTHALENE
TETRAMETHYLENECYANIDE	22	ADIPONITRILE
TETRAMETHYLENE GLYCOL	126	BUTANEDIOL
TETRAMETHYLENE SULFONE	760	SULFOLANE
TETRAMETHYLENE-SULFONE	760	SULFOLANE
TETRAMP	782	TETRAHYDRONAPHTHALENE
TETRANAP	782	TETRAHYDRONAPHTHALENE
TETRAPROPYLENE	694	PROPYLENE TETRAMER
TETRINE ACID	338	ETHYLENEDIAMINE TETRACETIC ACID
THALLIUM SALT	783	THALLIUM SULFATE
THF	781	TETRAHYDROFURAN
THIOCARBONYL CHLORIDE	794	THIOPHOSGENE
THIOCYANIC ACID	62	AMMONIUM THIOCYANATE LIQUOR (liquid)
THIOCYANIC ACID	61	AMMONIUM THIOCYANATE LIQUOR
THIODAN	313	ENDOSULFAN (liquid)
THIODAN	314	ENDOSULFAN (solid)
THIODEMETON	307	DISULFOTON (liquid)
THIODEMETON	308	DISULFOTON (solid)
THIOMETHYL ALCOHOL	549	METHYL MERCAPTAN
THIOSULFURIC ACID	483	LEAD THIOSULFATE
THIURAM	785	THIRAM
THORIUM NITRATE TETRAHYDRIDE	786	THORIUM NITRATE
TIBA	817	TRIISOBUTYLALUMINUM
TIBAL	817	TRIISOBUTYLALUMINUM
TIN BIFLUORIDE	754	STANNOUS FLUORIDE
TIN DIFLUORIDE	754	STANNOUS FLUORIDE
TIRCHLORMETHYL SULFUR CHLORIDE	642	PERCHLOROMETHYL MERCAPTAN
TITANIUM BUTOXIDE	769	TETRABUTYL TITANATE
TITANIUM TETRABUTOXIDE	769	TETRABUTYL TITANATE
TOLUOL	789	TOLUENE
TORPEDO	361	EXPLOSIVES A
TOSIC ACID	791	TOLUENE SULFONIC ACID
TOULENE	572	NAPHTHA
TOXICHLOR	178	CHLORDANE
TOXILIC ACID	498	MALEIC ACID
TOXILIC ANHYDRIDE	499	MALEIC ANHYDRIDE
2,4,5-TP	804	TRICHLOROPHENOXY-PROPIONIC ACID ESTER
2,4,5-TP ACID ESTERS	803	TRICHLOROPHENOXY ACETIC ACID ESTER
TREFLAN	816	TRIFLURALIN

SYNONYM	PAGE	CHEMICAL NAME
TRETHYLENE	800	TRICHLOROETHYLENE
TRI-6	101	BENZENE HEXACHLORIDE
TRICHLOROAMYLSILANE	71	AMYLTRICHLORSILANE
TRICHLOROETHYL SILICONE	346	ETHYLENETRICHLORO-SILANE
TRICHLOROETYL SILANE	346	ETHYLENETRICHLORO-SILANE
TRICHLOROISOCYANURIC ACID	799	TRICHLORO-S-TRIAZINE-TRIONE
TRICHLOROMETHANE	186	CHLOROFORM
TRICHLOROMETHYLSILANE	556	METHYL TRICHLORO-SILANE
TRICHLOROMONOSILANE	805	TRICHLOROSILANE
TRICHLORONITROMETHANE	191	CHLOROPICRIN
TRICHLOROPENTYLSILANE	71	AMYLTRICHLORSILANE
2,4,5-TRICHLOROPHENOL	802	TRICHLOROPHENOL
TRICHLORO-S-TRIAZINE-2,4,6-1H,3H,5H-T	799	TRICHLORO-S-TRIAZINE-TRIONE
TRICHLOROTHYLENE	800	TRICHLOROETHYLENE
TRICHLOROTRIAZINETRIONE	799	TRICHLORO-S-TRIAZINE-TRIONE
TRICHLOROVINYLSILANE	842	VINYL TRICHLOROSILANE
1-TRIDECANOL	807	TRIDECANOL
TRIDECONOL	486	LINEAR ALCOHOL
TRIEN	814	TRIETHYLENE TETRAMINE
1,3,5-TRIETHYLBENZENE	812	TRIETHYLBENZENE
TRIETHYLOLAMINE	809	TRIETHANOLAMINE
TRIEUHYLENEPHOSPHOR-AMIDE	651	PHOSPHORIC ACID TRIETHYLENEIMINE
TRIGLYCINE	599	NITRILOTRIACETIC ACID AND SALTS
TRIGYLCOL	813	TRIETHYLENE GLYCOL
1-2-3-TRIHYDROXYBENZENE	700	PYROGALLIC ACID
3,4,5-TRIHYDROXYBENZOIC ACID	388	GALLIC ACID
1,2,3-TRIHYDROXYPROPANE	394	GLYCERINE
TRIHYDROXYTRIETHYLAMINE	809	TRIETHANOLAMINE
3,5,5-TRIMETHYL-2-CYCLOHEXANE-1-100	449	ISOPHRONE
TRIMETHYLENE	231	CYCLOPROPANE
TRIMETHYLHEPTANALS	442	ISODECALDEHYDE
2,4,4-TRIMETHYL-1-PENTENE	280	DIISOBUTYLENE
TRIOSULFUROUS DICHLORIDE	762	SULFUR CHLORIDE
1,3,5-TRIOXANE	630	PARALDEHYDE
3,6,9-TRIOXAUNDECAN-1	778	TETRAETHYLENE GLYCOL
TRI-P-CRESYL PHOSPHATE	806	TRICRESYLPHOSPHATE
TRI-P-TOLYL PHOSPHATE	806	TRICRESYLPHOSPHATE
TRIS HYDROXYETHYL AMINE	809	TRIETHANOLAMINE
TRISODIUM NITRILOTRIACETATE	599	NITRILOTRIACETIC ACID AND SALTS
2,4,5-T SODIUM SALT	798	TRICHLOROPHENOXY ACETIC ACID
AMINE T-STUFF	424	HYDROGEN PEROXIDE 52–100%
TURPS	820	TURPENTINE
UCON 11	801	TRICHLOROFLUOROMETHANE
UDMH	290	DIMETHYLHYDRAZINE
UNDDECYLIC ALCOHOL	821	UNDECANOL
1-UNDECANOL	821	UNDECANOL
UNDECYL ALCOHOL	821	UNDECANOL
UNDECYLETHYLENE	808	TRIDECENE
UNSLAKED LIME	157	CALCIUM OXIDE
URANIUM ACETATE	826	URANYL ACETATE
URANIUM FLUORIDE	824	URANIUM HEXAFLUORIDE
URANIUM NITRATE	827	URANYL NITRATE
URANIUM OXIDE	825	URANIUM PEROXIDE
URANIUM OXIDE PEROXIDE	825	URANIUM PEROXIDE
URANIUM SULFATE	828	URANYL SULFATE
URANIUM SULFATE TRIHYDRATE	828	URANYL SULFATE
URANIUM VI FLUORIDE	824	URANIUM HEXAFLUORIDE
URANYL ACETATE DIHYDRATE	826	URANYL ACETATE
URANYL SULFATE TRIHYDRATE	828	URANYL SULFATE
UREA	830	UREA PEROXIDE
UREA HYDROGEN PEROXIDE	830	UREA PEROXIDE
UROTROPIN	410	HEXAMETHYLENE-TETRAMINE
VALENORE	278	DIISOBUTYL KETONE
VALENTINITE	80	ANTIMONY TRIOXIDE
VALERAL	831	VALERALDEHYDE
VALERIC ALDEHYDE	831	VALERALDEHYDE
VAM	835	VINYL ACETATE
VANADIC ANHYDRIDE	833	VANADIUM PENTOXIDE
VANADIUM OXYSULFATE	834	VANADYL SULFATE
VANADIUM PENTAOXIDE	833	VANADIUM PENTOXIDE
VANADYL CHLORIDE	832	VANADIUM OXYTRI-CHLORIDE
VANADYL SULFATE DIHYDRATE	834	VANADYL SULFATE
VANADYL TRICHLORIDE	832	VANADIUM OXYTRI-CHLORIDE
VANCIDE	163	CAPTAN (liquid)
VANCIDE	164	CAPTAN (solid)
VAPONA	266	DICHLORVOS (liquid)
VAPONA	267	DICHLORVOS (solid)
VASOLINE	643	PETROLATUM
VCL	836	VINYL CHLORIDE
VCM	836	VINYL CHLORIDE
VEGETABLE, ANIMAL, MINERAL, CARBON ACTIVAT	177	CHARCOAL
VELSICOL 1068	178	CHLORDANE
VERMILION	510	MERCURIC SULFIDE
VERSENE ACID	338	ETHYLENEDIAMINE TETRACETIC ACID
VILRATHANE 4300	303	DIPHENYLMETHYL DIOSOCYANATE
VINAMAR	837	VINYL ETHYL ETHER
VINEGAR ACID	13	ACETYL PEROXIDE
VINEGAR ACID	3	ACETIC ACID (aqueous solution)
VINEGAR ACID	4	ACETIC ACID (glacial)
VINYL A MONOMER	835	VINYL ACETATE
VINYLBENZINE	758	STYRENE
VINYL CARBINOL	24	ALLYL ALCOHOL
VINYL C MONOMER	836	VINYL CHLORIDE
VINYL CYANIDE	20	ACRYLONITRILE
VYAC	835	VINYL ACETATE
W-40	339	ETHYLENE DIBROMIDE
W-15	339	ETHYLENE DIBROMIDE
W-10	339	ETHYLENE DIBROMIDE
WATER GLASS	749	SODIUM SILICATE
WHITE ARSENIC	86	ARSENIC TRIOXIDE
WHITE TAR	576	NAPHTHALENE
WHITE VITRIOL	871	ZINC SULFATE
WILLIE P	658	PHOSPHORUS WHITE
WITICIZER 300	246	DIBUTYL PHTHALATE
WOOD	820	TURPENTINE

SYNONYM	PAGE	CHEMICAL NAME
WOOD	177	CHARCOAL
WOOD ALCOHOL	525	METHYL ALCOHOL
WOOD ALCOHOL	520	METHANOL
WOOD ETHER	283	DIMETHYL ETHER
WOOD SPIRIT	520	METHANOL
2,6-XYLENOL	847	XYLENOL (liquid)
2,6-XYLENOL	848	XYLENOL (solid)
XYLOL	844	XYLENE-M
XYLOL	845	XYLENE-O
XYLOL	846	XYLENE-P
YARMAR	660	PINE OIL
YELLOW PETROLATUM	643	PETROLATUM
YELLOW PHOSPHORUS	658	PHOSPHORUS WHITE
YELLOW PLUMBOUS OXIDE	489	LITHARGE
YUKON 22	563	MONOCHLORODI-FLUOROMETHANE
YUKON 12	258	DICHLORODIFLUORO-METHANE
ZACTRAN	849	ZECTRAN
ZECTANE	849	ZECTRAN
ZEXTRAN	849	ZECTRAN
ZINC ACETATE DIHYDRATE	850	ZINC ACETATE
ZINC CHROMATE VI HYDROXIDE	859	ZINC CHROMATE
ZINC DICHROMATE	854	ZINC BICHROMATE
ZINC DICYANIDE	866	ZINC NITRATE
ZINC DICYANIDE	860	ZINC CYANIDE
ZINC DIETHYL	275	DIETHYLZINC
ZINC DIFLUORIDE	862	ZINC FLUORIDE
ZINC DIHEXYLDITHIO-PHOSPHATE	861	ZINC DIALKYLDITHIOPHOSPHATE
ZINC DIHEXYLPHOS-PRODITHIOATE	861	ZINC DIALKYLDITHIOPHOSPHATE
ZINC DIMETHYL	291	DIMETHYLZINC
ZINC DITHIONITE	865	ZINC HYDROSULFITE
ZINC ETHYL	275	DIETHYLZINC
ZINC FLUOROBORATE SOLUTION	863	ZINC FLUOROBORATE
ZINC FLUOSILICATE	870	ZINC SILICOFLUORIDE
ZINC HEXAFLUOROSILICATE	870	ZINC SILICOFLUORIDE
ZINC METHYL	291	DIMETHYLZINC
ZINC O-O-DI-N-BUTYLPHOSPHORO-DITHIOATE	861	ZINC DIALKYLDITHIOPHOSPHATE
ZINC P-PHENOLSULFONATE	867	ZINC PHENOLSULFONATE
ZINC SALT	864	ZINC FORMATE
ZINC SALT	850	ZINC ACETATE
ZINC SULFATE HEPTA-HYDRATE	871	ZINC SULFATE
ZINC SULFOCARBOLATE	867	ZINC PHENOLSULFONATE
ZINC VITRIOL	871	ZINC SULFATE
ZINC YELLOW	859	ZINC CHROMATE
ZINC YELLOW Y-539-D	869	ZINC POTASSIUM CHROMATE
ZIRCONIUM ACETATE SOLUTION	872	ZIRCONIUM ACETATE
ZIRCONIUM CHLORIDE	877	ZIRCONIUM TETRA-CHLORIDE
ZIRCONIUM NITRATE PENTAHYDRATE	873	ZIRCONIUM NITRATE
ZIRCONIUM OXIDE CHLORIDE	874	ZIRCONIUM OXYCHLORIDE
ZIRCONIUM OXYCHLORIDE HYDRATE	874	ZIRCONIUM OXYCHLORIDE
ZIRCONIUM SULFATE TETRAHYDRATE	876	ZIRCONIUM SULFATE
ZIRCONYL CHLORIDE	874	ZIRCONIUM OXYCHLORIDE